TOXIC METALS
IN THE ATMOSPHERE

Volume

17

in the Wiley Series in

Advances in Environmental Science and Technology

JEROME O. NRIAGU, Series Editor

TOXIC METALS IN THE ATMOSPHERE

Edited by

Jerome O. Nriagu

National Water Research Institute
Burlington, Ontario, Canada

Cliff I. Davidson

Department of Civil Engineering
Carnegie-Mellon University
Schenley Park
Pittsburgh, Pennsylvania

A Wiley-Interscience Publication

JOHN WILEY & SONS

New York • Chichester • Brisbane • Toronto • Singapore

Library of Congress Cataloging in Publication Data:

Main entry under title:

Toxic metals in the atmosphere.

(Advances in environmental science and technology, ISSN 0065-2563 ; v. 17)
"A Wiley-Interscience publication."
1. Metals—Environmental aspects. 2. Metals—Toxicity. 3. Trace elements—Environmental aspects. I. Nriagu, Jerome O. II. Davidson, Cliff I. III. Series.

TD180.A38 vol. 17 628 s [628.5′3] 85-17783
[TD887.M4]
ISBN 0-471-82654-5

Printed in the United States of America

10 9 8 7 6 5 4 3 2 1

CONTRIBUTORS

ADAMS, F. C. Department of Chemistry, University of Antwerp, Universiteitsplein 1, B-2610 Wilrijk, Belgium.

BOUTRON, CLAUDE F., Laboratoire de Glaciologie et, Géophysique de l'Environnement, Centre National de la Recherche Scientifique, 2 rue Très-Cloîtres 38031 Grenoble-Cedex France

CASS, GLEN R., Environmental Engineering Science Department and Environmental Quality Laboratory, 206-40, California Institute of Technology, Pasadena, California 91125

CHAN, WALTER H., Air Resources Branch, Ontario Ministry of the Environment, 880 Bay Street, Toronto, Ontario M5S 1Z8

CHU, LIH-CHING, Illinois Department of Energy and Natural Resources, State Water Survey Division, P.O. Box 5050, Station A, Champaign, Illinois 61820

DAVIDSON, CLIFF I., Departments of Civil Engineering, and Engineering & Public Policy, Schenley Park, Carnegie-Mellon University, Pittsburgh, Pennsylvania 15213

ENSOR, D. S., Aerosol Technology Program, Chemical Engineering Unit, Research Triangle Institute, Research Triangle Park, North Carolina 27709

GATZ, DONALD F., Illinois Department of Energy and Natural Resources, State Water Survey Division, P.O. Box 5050, Station A, Champaign, Illinois 61820

GLOOSCHENKO, WALTER A., Aquatic Ecology Division, National Water Research Institute, P.O. Box 5050, Burlington, Ontario L7R 4A6, Canada

HARRISON, ROY M., Department of Chemistry, University of Essex, Wivenhoe Park, Colchester, C04 3SQ, England

HEIDAM, NEILS Z., National Agency of Environmental Protection, Air Pollution Laboratory, Risø National Laboratory, DK-4000 Roskilde, Denmark

HOPKE, PHILIP K., Institute for Environmental Studies, Department of Civil Engineering and Nuclear Engineering Program, University of Illinois, Urbana, Illinois 61801

HUSAIN, LIAQUAT, Center for Laboratories and Research, New York State Department of Health, Albany, New York 12201

De JONGHE, W. R. A., Department of Chemistry, University of Antwerp, Universiteitsplein 1, B-2610 Wilrijk, Belgium

LINDBERG, S. E., Environmental Sciences Division, Oak Ridge National Laboratory, Oak Ridge, Tennessee 37830

LUSIS, MARIS A., Air Resources Branch, Ontario Ministry of the Environment, 880 Bay Street, Toronto, Ontario M5S 1Z8

MCRAE, GREGORY R., Departments of Chemical Engineering, Carnegie-Mellon University, Pittsburgh, Pennsylvania 15213

OSBORN, J. F., Departments of Civil Engineering, and Engineering & Public Policy, Carnegie-Mellon University, Pittsburgh, Pennsylvania 15213

PACYNA, JOSEF M., Norwegian Institute for Air Research, Elvegaten 52, 2000 Lillestrøm, Norway

SALOMONS, WIM, Delft Hydraulics Laboratory, Haren Branch, c/o Institute for Soil Fertility, P.O. Box 30003, 9750 RA Haren (Gr), The Netherlands

SERVANT, JEAN, Laboratoire d'Aérologie, Université Paul Sabatier, 118 Route de Narbonne, 31062 Toulouse Cedex, France

SHENDRIKAR, A. D., Mead CompuChem, Research Triangle Park, North Carolina 27709

SUGIMAE, AKIYOSHI, Environmental Pollution Control Center, Osaka Prefecture, 3-62, 1-chome, Nakamichi, Higashinari-ku, Osaka 537, Japan

WIERSMA, G. B., Earth and Life Sciences, EG&G Idaho, P.O. Box 1625, Idaho Falls, Idaho 83415

INTRODUCTION TO THE SERIES

The deterioration of environmental quality, which began when mankind first congregated into villages, has existed as a serious problem since the industrial revolution. In the second half of the twentieth century, under the ever-increasing impacts of exponentially growing population and of industrializing society, environmental contamination of the air, water, soil, and food has become a threat to the continued existence of many plant and animal communities of various ecosystems and may ultimately threaten the very survival of the human race. Understandably, many scientific, industrial, and governmental communities have recently committed large resources of money and human power to the problems of environmental pollution and pollution abatement by effective control measures.

Advances in Environmental Sciences and Technology deals with creative reviews and critical assessments of all studies pertaining to the quality of the environment and to the technology of its conservation. The volumes published in the series are expected to service several objectives: (1) stimulate interdisciplinary cooperation and understanding among the environmental scientists; (2) provide the scientists with a periodic overview of environmental developments that are of general concern or that are of relevance to their own work or interests; (3) provide the graduate student with a critical assessment of past accomplishment which may help stimulate him or her toward the career opportunities in this vital area; and (4) provide the research manager and the legislative or administrative official with an assured awareness of newly developing research work on the critical pollutants, and with the background information important to their responsibility.

As the skills and techniques of many scientific disciplines are brought to bear on the fundamental and applied aspects of the environmental issues, there is a heightened need to draw together the numerous threads and to present a coherent picture of the various research endeavors. This need and the recent tremendous growth in the field of environmental studies have clearly made some editorial adjustments necessary. Apart from the changes in style and format, each future volume in the series will focus on one particular theme or timely topic, starting with Volume 12. The author(s) of each pertinent section will be expected to critically review the literature and the most important recent developments in the particular field; to critically evaluate new concepts, methods, and data; and to focus attention on important unresolved or controversial questions and on probable future trends. Monographs embodying the results of unusually extensive and well-rounded investigations will also be published in the series. The net result of the new editorial policy should be more integrative and comprehensive volumes on key environmental issues and pollutants. Indeed, the development of realistic standards of environmental quality for many pollutants often entails such a holistic treatment.

JEROME O. NRIAGU, Series Editor

PREFACE

Keeping the air clean has become a major problem for industrialized societies. The growing apprehension about air quality has heightened the concern for the long-term ecological and health effects of the less ubiquitous but highly toxic metals which are now being released in large quantities into the atmosphere. This volume presents some of the most recent findings on the types of metals and their distribution in the atmosphere. It deals comprehensively with the origins, behavior, and fate of such airborne metals, their physical and chemical characteristics, and their interactions with, transport in, and removal from the ambient air. Trends as well as routine and special monitoring of atmospheric metal levels are also covered in some detail. The atmosphere serves as an important medium for the transfer of toxic metals from their sources to receptors at distant ecosystems. Special emphasis has therefore been given to the contrasts in rates of atmospheric inputs of metals into polluted versus unpolluted environments.

Papers dealing with the atmospheric cycle of metals are widely scattered in the journals of ecology, agriculture, biology, chemistry, geology, geochemistry, atmospheric chemistry, oceanography, physics, geophysics, and meteorology. This volume represents a systematic endeavor to interface these various studies and should thus serve as an introduction and guide to the efficient utilization of the large volume of literature on atmospheric metal pollution. It is addressed to engineers, chemists, public health physicians, meteorologists, economists, sociologists, agronomists, toxicologists, ecologists, and geochemists. Indeed, the material in the volume should be of interest to anyone concerned about the quality of our air.

Any success of this volume belongs to our distinguished group of contributors. Our appreciation goes to the staff at Wiley for invaluable editorial assistance.

JEROME O. NRIAGU
CLIFF I. DAVIDSON

Burlington, Ontario
Pittsburgh, Pennsylvania
January 1986

CONTENTS

TOXIC METALS
IN THE ATMOSPHERE

1

EMISSION FACTORS OF ATMOSPHERIC ELEMENTS

Jozef M. Pacyna

Norwegian Institute for Air Research
Lillestrøm, Norway

1. INTRODUCTION

In the past few decades, there has been a growing concern about environmental contamination by trace elements. Industrial production, energy generation, and vehicular traffic have increased significantly the levels of these pollutants in the environment. In order to assess perturbations of biogeochemical cycles of trace elements in the human environment, accurate data on the quantity and characteristics of emissions from several sources are required.

There are several methods for estimating emissions, namely, material balances, emission factors, and source measurements. All of these are partially linked with each other. Emission factors are often based on either stack measurements or material balances, including information about raw material characteristics, collector removal efficiencies, and industrial technology profile. Until recently, a majority of air pollutant emission factors have been compiled for particulate matter without any chemical characterization. The EPA particulate emission factors cover most of the common emission categories, but only information on lead emission rates from certain sources is available (EPA, 1979a). The EPA emission factors have been used to calculate a nationwide inventory of air pollutant emissions in Canada (Environment Canada, 1978). Where available, emission factors developed for Canadian conditions by the Air Pollution Control Directorate were applied. In West Germany, particulate emission factors were developed using emission declarations, according to the Federal Emission Protection Act (Umwelt Bundesamt, 1980). Nowadays, the use of emission factors and emission inventories has changed. As it was concluded during the APCA 1979 Conference on Emission Factors and Inventories (APCA, 1979), "now, emissions are critical in obtaining emission offsets from specific existing facilities" and "are also used for determining permit fees." To meet these new requirements, the accuracy of emission factors is being improved. As a result, more information on emission rates of various trace elements are available in the literature. However, the main efforts of several authors are concentrated on either a single metal or a certain emission source.

In this chapter, trace element emission factors are presented for several sources based on an extensive literature survey and, in a few cases, stack sampling data. From among a number of trace elements contained in fuels, ores, and rocks, only a few are considered here. These elements either are the most toxic or have particularly high enrichment factors in ambient aerosols relative to the earth's crust. The degree of public health concern associated with the element was also considered. The industrial categories and emission sources discussed in this chapter have been chosen from a ranking of sources based on the total yearly tonnage and chemical forms of trace elements emitted.

The main source of information for the data presented here is an earlier report by the author on emission factors of trace elements (Pacyna, 1982a).

2. EMISSION FACTORS OF TRACE ELEMENTS FROM ANTHROPOGENIC SOURCES

For a majority of trace elements, emissions from manmade sources significantly exceed emissions from natural sources. All the anthropogenic sources can be divided into six groups, namely: (1) stationary fuel combustion, (2) internal combustion engines, (3) nonferrous metal manufacturing, (4) iron, steel, and ferroalloy plants and foundries, (5) refuse incineration, and (6) cement production. For certain trace elements, other sources may also be important, for example, fertilizer production or industrial application of metals.

2.1. Stationary Fuel Combustion

Fuel combustion is divided into two groups: (1) external combustion of coal, oil, and wood in electric utility plants, industrial boilers, and process heaters, and space heating units; and (2) internal combustion units. Gas combustion is not considered due to almost zero emission of particles.

2.1.1. Power Generation from Coal

Trace element behavior during coal combustion depends on such parameters as (1) affinity of elements for pure coal and mineral matter, (2) physicochemical properties of elements and their concentrations in coal, (3) combustion conditions, and (4) emission control devices in power plant.

Almost 50 years ago, Goldschmidt (1935) suggested that source elements in coal have either a high organic or inorganic affinity. Today, elements are generally described as (1) associated with the organic fraction, (2) mainly associated with the inorganic fraction, and (3) elements that could be associated with either or both fractions (Nicholls, 1968). Organic sulfur, Br, Ge, Be, Sb, and B consistently fall in the organic phase (Kuhn et al., 1980). The sulfide-forming elements Zn, As, Cd, Fe, Zr, Hg, Pb, Hf, Mn, and pyritic sulfur are found mostly in the inorganic fraction. A number of other elements, Al, Si, Ti, V, Mo, K, P, Ga, Ca, Cr, Co, Ni, Cu, Mg, and Se, are either intermediate in their association or highly variable. From among these, P, Ga, Ti, and V tend to be allied with the other elements having organic affinities, and Co, Ni, Cr, Se, and Cu are more closely associated with the inorganically combined elements (Zubovic, 1966). The affinity of trace elements for pure coal and mineral matter affects the trace element emissions during coal combustion by being responsible for a chemical form of pollutants released and various ranges of element concentrations in other types of coals, such as bituminous and subbituminous coals and lignites.

The fate of trace elements during coal combustion depends not only on the affinity, concentration, and distribution of each element within the coal matrix, but also on process conditions such as temperature, heating rate, exposure time at elevated temperatures, and the surrounding environment of either oxidizing or reducing conditions. The volatile species in the coal are evaporated in the boiler and recondensed as submicron aerosol particles or on the surface of ash particles as the flue gas cools in the convective sections (Flagan and Friedlander, 1976; Kaakinen, 1979; Davison et al., 1974; Pacyna, 1980). The concentrations of chalcophiled Pb, Sb, Cd, Se, As, Zn, and Mo increase markedly with decreasing particle size, while lithophiled Mn and Zr show little or no enrichment with decreasing particle size. The other elements such as Be, Co, Cr, Cu, Ni, and V display an intermediate behavior. Apparently, more than 90% of mercury in coal is released as vapor (Billings et al., 1973; Kaakinen et al., 1975; Pacyna, 1980). Thus, the higher temperatures are in a boiler, the larger the emission factors of volatile elements. Combustion temperature is closely related to the boiler design. It should also be noted with respect to the fine particle emissions that 80% of the ash from pulverized coal firing becomes entrained in the flue gases. This value is relatively high compared to stoker and cyclone furnaces. This is due to the fact that pulverized coal is burned in suspension. In cyclone furnaces, only 20 to 30% of the total ash is entrained in the fly ash (Babcock and Wilcox Corp., 1972). Hence, the percentage distribution of various size particles in stack dust differs from one type of boiler to another.

The type of fly ash control system and its efficiency also influence the trace element emissions. From among several types of control devices, electrostatic precipitators (ESP) and wet scrubbers are mainly installed in coal-fired power plants. The fly ash escaping from electrostatic precipitators and venturi wet scrubbers (generally 1% or less of the total) is smaller than 2 μm in size. Generally, a venturi wet scrubber system is more efficient in removing As, Cd, Mn, Ni, Pb, and Zn from a flue gas stream than electrostatic precipitators. The latter is, however, more useful for removing Se from a flue gas stream (Ondov et al., 1979).

Considering all the parameters listed above, the trace element emission factors for coal-fired power plants have been calculated and presented in Table 1.1. The details on the methodology of calculations are published by the author in an earlier report (Pacyna, 1981a). Data presented in Table 1.1 are derived for coal with 10% ash burned in a power plant equipped with 99%-efficient ESP. For other efficiencies of dust removal and other ash contents, similar results can be easily obtained (by simple multiplications). It must be admitted that average values of trace element concentrations in coal and stack dust were employed for all three types of coals when calculating the emission

Table 1.1. Emission Factors of Trace Elements for Coal-Fired Power Plants (in $\mu g/MJ$)

	Bituminous			Subbituminous			Lignite		
Element	Cyclone	Stoker	Pulverized	Cyclone	Stoker	Pulverized	Cyclone	Stoker	Pulverized
As	24	28	16	28	34	19.2	40	47	27
Be	2.5	3.7	1.6	3	4.5	2.0	4.1	6.2	2.7
Cd	7.3	8.7	5.1	8.8	10.5	6.1	12.3	14.7	8.5
Co	44	51	25	54	62	31	75	86	43
Cr	120	200	85	145	242	103	201	337	143
Cu	94	164	63	114	198	76	158	276	106
Hg[a]	0.5	1.0	0.4	0.6	1.3	0.4	0.9	1.8	0.6
Mn	102	186	70	123	226	85	172	314	118
Mo	31	42	19	37	51	23	52	71	32
Ni	150	243	96	182	294	117	253	409	163
Pb	85	128	55	103	156	66	144	217	92
Sb	15	23	9.3	17	28	11.3	24	39	15.7
Se[a]	11	18.7	7.3	13	23	8.9	18.5	31	12.4
V	84	162	58	101	197	71	141	274	98
Zn	119	191	79	144	231	96	200	321	133
Zr	82	179	60	100	217	73	139	301	99

[a]Only emission with particles is considered.

factors in Table 1.1. These average values have been selected from the reviewed literature (Bertine and Goldberg, 1971; Lee and Lehmden, 1973; Capes et al., 1974; Davison et al., 1974; Natusch et al., 1974; Bolton et al., 1975; Kaakinen et al., 1975; Klein et al., 1975; Holland et al., 1975; Block and Dams, 1976; Gluskoter et al., 1977; Rottman et al., 1977; Campbell et al., 1978; Dvorak and Lewis, 1978; Coles et al., 1979; Murthy et al., 1979; Smith et al., 1979; Ensor et al., 1981; KHM Prosjekt, 1981; and Roy et al., 1981). However, as the element concentrations in coal change from one coal field to

Table 1.2. Penetration (in %) of Elements Contained in Particles[a]

	Penetration Range			Penetration Range	
Element	ESP	Venturi	Element	ESP	Venturi
As	4.3–11.5	2.5–7.5	Sb	3.1–7.7	3.0–6.6
Be	0.5–0.9		Se	3.8–8.1	10–21
Cd	3.3–8.8		V	1.6–3.7	0.5–1.1
Co	1.2–3.2	0.06–2.1	W	3.1–7.2	1.7–3.5
Cr	1.2–12.1	0.6–36	Zn	2.3–6.3	0.3–8.6
Mo	1.8–6.8	0.9–2.2	Zr	0.5–1.6	0.05–0.14
Mn	0.3–1.6	0.07–4.6			
Pb	2.2–5.5				

[a]Particles emitted from a coal-fired generation unit equipped with a Venturi wet scrubber and an ESP. From Ondov et al. (1979).

another, and even from one location in the field to another, average values of these concentrations should be regarded with caution. When a Venturi wet scrubber system is installed, the trace element emission factors can be estimated using the data from Table 1.1 and information on penetration of elements through control devices. The data on penetration of elements through both control devices are shown in Table 1.2 (Ondov et al., 1979).

2.1.2. Power Generation from Oil

The trace element emission during oil combustion depends primarily upon the element concentrations in crude oils, physicochemical properties of trace elements, and combustion conditions.

The trace element contents of crude oils from several fields in the world are presented in Table 1.3. The data have been collected from the literature (Baker, 1964; Jacquin et al., 1973; de Pereyra, 1973; Filby, 1975; Yen, 1975; Smith et al., 1975; Brown, 1981; Speight, 1981; Torgaard, 1983).

As can be seen from Table 1.3, there are large differences between concentrations of certain metals measured in several fields of the world. Even in the same field (or country) the concentrations vary from one place of the field (or country) to another. As an example, the vanadium concentration in the Venezuelan crude oil in Boscan (which probably has the highest vanadium concentration in crude oil in the world (1400 ppm)) is 13 times as high as that in the crude from another Venezuelan field in Pilon. If one is going to use trace element concentrations in crude petroleum in certain areas for further calculations (e.g., emission factor estimations), one must have accurate data for this region. As a general rule, the heavier the crude, the more metal present.

During refining, metals concentrate in the heavy distillate residuals, such as residual fuel oils, asphalts, and in the liquid and solid waste streams. The limited information available indicates that probably 90% of the metals in crude oil are retained in residual fuel oil and asphalt (Smith et al. 1975). Since asphalt is used primarily in road construction and for roofing, a majority of potentially hazardous trace metals is discharged by combustion of residual oil by both electric utilities and industrial boilers.

Table 1.3. Trace Metal Contents of Crude Oils (in ppm)

Metal	Concentration Range	Metal	Concentration Range
As	0.0024–1.63	Mn	0.021–3.1
Cd	0.03[a]	Mo	0.008–7.85
Co	0.0027–14.5	Ni	0.2–345.0
Cr	0.0016–0.729	Pb	0.001–0.31
Cu	0.03–12.0	Se	0.009–1.40
Sb	0.051–0.3	V	0.07–1400.0
Hg	0.014–30.0	Zn	0.1–86.0

[a] Value estimated from zinc content.

The second parameter influencing the trace element emissions from oil-firing power plants deals with physicochemical properties of trace elements. Of the more abundant trace metal compounds formed, only the amounts of silicon dioxide and vanadium pentoxide emitted remain constant for combustion temperatures in the range from 500 to 1800 K. Nickel emerges predominantly as the oxide at 1800 K, but mostly as sulfate at lower temperatures. Other metals form sulfates. When the SO_2 concentration increases at higher temperatures, there are less metal sulfates; at lower temperatures there are more metal sulfates and less SO_2. The trace element loading of stack gases also depends upon the efficiency of combustion and the buildup of boiler deposits. Poor mixing, low flame temperatures, and short residence time in the combustion zone result in larger particles with a higher content of combustible matter and higher particulate loadings (Smith, 1962). Low-pressure atomization in the burner produces larger fly ash particles and a higher particulate loading. High-pressure atomization produces smaller particles, fewer cenospheres, and lower particulate loadings. Considering other boiler designs, the trace element emission rates from two commonly used types of boilers, tangential and horizontal, working under similar conditions are comparable. Oil-fired boilers do not require ash hoppers and ash pits (Dvorak and Lewis, 1978). Fly ash removal equipment is not generally required (Witkowski, 1977), although such equipment is used at some oil-burning plants (Dvorak and Lewis, 1978). Dust collectors are, however, used during soot blowing. This equipment serves principally to collect particulate matter larger than 10 μm. The emission of particulate matter from an oil-fired unit without stack gas cleaning is comparable to a coal-fired unit of better than 99% collection efficiency (PSEG Comp., 1967).

Taking into account all the above discussed parameters of the trace element emissions from oil combustion and information on movement of crude petroleum in the world in 1981 (OECD, 1982), the trace element emission factors have been calculated for the European oil-fired power plants and are listed in Table 1.4.

The emission factors in Table 1.4 were calculated for oil containing 1% sulfur. Sulfur content of crude oil highly affects the trace element emissions during oil combustion. It is due to SO_3 adsorption, causing increased

Table 1.4. Emission Factors of Trace Elements for the European Oil-Firced Power Plants (in μg/MJ)

Element	Emission Factor	Element	Emission Factor
As	24	Mo	28
Cd	12	Ni	1020
Co	130	Pb	126
Cr	43	Se	18
Cu	174	V	3700
Mn	41	Zn	89

particulate formation for oils of higher sulfur content. This relationship is expressed for particle emissions from residual fuel oil combustion in electric utility furnaces and industrial and commercial boilers by Eq. (1) (U.S.EPA, 1977):

$$e_p = 1.25S + 0.38 \tag{1}$$

where e_p = particulate emission factor (kilograms of dust per 10^3 liters of oil) and
S = percentage by weight of sulfur in crude oil

Thus, the trace element emission factors for oils of sulfur content other than 1% can be recalculated using Eq. (1). The methodology of calculations is presented elsewhere (Pacyna, 1982b).

2.1.3. Combustion of Coal and Oil in Industry

Large industrial plants often generate their own electric power or process steam. Trace element emissions from industrial units burning coal or fuel oil parallel those discussed under electric utilities. The main differences are brought about by the type of boilers employed in industry and the type and collection efficiency of control devices. The stoker-type boiler is the dominant unit used in industrial plants firing coal. The pulverized and cyclone boiler units are generally associated with larger industrial complexes, and are similar in design to those discussed under electric utilities. Mechanical collectors are mostly employed as control devices, but electrostatic precipitators and wet scrubbers are also used. Applications of control devices to industrial boilers have been reviewed recently by Roeck and Dennis (1979). For stoker-fired boilers, collection efficiencies of even 90 to 95% may be achieved with a

Table 1.5. Emission Factors of Trace Elements from Coal- and Oil-Fired Industrial Boilers

	Coal-fired units[a]					Coal-fired units[a]			
Element	Cyclone	Stoker	Pulverized	Oil furnace[b]	Element	Cyclone	Stoker	Pulverized-	Oil furnace[b]
As	0.3	1.7	1.3	1.2	Mo	0.4	2.5	0.6	1.4
Be	0.04	0.2	0.13		Ni	2.2	14.5	8	52
Cd	0.1	0.5	0.4	0.6	Pb	1.2	7.7	4.5	6.4
Co	0.6	3.1	2.1	6.6	Sb	0.2	1.4	0.8	
Cr	1.7	12	7	2.2	Se[c]	0.2	1.1	0.6	0.9
Cu	1.4	9.8	5.2	8.8	V	1.2	9.7	4.8	187
Hg[c]	0.01	0.06	0.03		Zn	1.7	11.4	6.5	4.5
Mn	1.5	11	5.8	2.1	Zr	1.2	11	5	

[a] Assuming coal with 10% of ash in plants equipped with 85% efficient control devices. Unit used is grams of trace element per 1 tonne of coal.

[b] Assuming oil with the sulphur content of 1%. Unit used is grams of trace element per 1000 liters of oil.

[c] Only emissions of Hg and Se with particles are considered.

mechanical collector. Higher efficiencies and finer particle sizes (as found in pulverized coal units) require use of a scrubber, ESP, or fabric filter (Rubin, 1981).

Large industrial complexes use essentially the same design of oil-fired furnaces as electric utility plants. Smaller industrial operations utilize lower capacity units with a lower flame temperature. The stack dust loadings for industrial sources of oil combustion are higher than those for electricity generating plants. According to the literature, the average level of particulate emission from industrial boilers is 2.3 times as high as the level of emissions from power plants (MRJ, 1971).

The trace element emission factors for coal and oil combustion in industrial units are presented in Table 1.5. The same chemical composition of fuel as that of coal and oil burned in electricity generating plants is assumed.

2.1.4. Combustion of Coal and Oil in Commercial and Residential Furnaces

Commercial and residential furnaces are mainly used for space heating. Small stoker-type boilers and hand-fired units are the main furnaces used in commercial and residential operations of coal burning. Boiler types for fuel-oil firing are similar to the small units discussed in previous sections. Control equipment is not generally used on these small furnaces. The trace element emission factors have been calculated assuming the same chemical composition of fuels and stack dusts as used for emission factors in Sections 2.1.1. and 2.1.2., and the particulate emission factors of 10 kg dust/coal and 1.2 kg dust/10^3 L oil for coal and oil combustion, respectively (U.S. EPA, 1973a). The results are presented in Table 1.6.

Table 1.6. Emission Factors of Trace Elements from Coal and Oil Combustion in Commercial and Residential Units

Element	Coal (g/t)	Oil (g/10^3 L)	Element	Coal (g/t)	Oil (g/10^3 L)
As	0.6	0.6	Mo	0.9	0.7
Be	0.1		Ni	5.1	27
Cd	0.2	0.3	Pb	2.7	3.3
Co	1.1	3.4	Sb	0.5	0.5
Cr	4.2	1.1	V	3.4	98
Cu	3.5	4.6	Zn	4.0	2.3
Mn	4.0	1.1	Zr	3.8	

2.1.5. Wood Combustion

In the developed countries wood is no longer regarded as a primary source of heat and energy; however, wood-fuel is still the largest biomass energy source used in many parts of the world today (Reckard, 1979). Currently available information suggests a substantial environmental impact from residential

Table 1.7. Emission Factors of Trace Elements from Wood Combustion (in g/t of wood)

	Source			Source	
Element	Wood Stove	Fireplace	Element	Wood Stove	Fireplace
As	0.5	0.3	Ni	4.7	3.1
Cd	0.3	0.2	Pb	7	4.7
Cu	19	12	Zn	58	39

wood combustion emissions (Cooper, 1980; De Angelis et al., 1980; Rudling and Løfroth, 1981).

The main sources of emissions from residential wood combustion are wood-burning stoves and fireplaces. Very limited information exists on the emissions of trace elements from these sources. Generally, the trace element emissions from residential wood combustion can be estimated using data on particulate emissions from these sources and the mean concentrations of trace elements in vegetation. Particulate emissions from wood burning are related to inefficient combustion and vary mainly due to the type and amount of fuel burned, draft setting, fuel moisture, and type of stove. In this work, 3.6 and 2.4 g/kg of wood were considered as the particulate emission factors for wood-burning stoves and fireplaces, respectively (Butcher and Sorensen, 1979). The values do not include condensable organic emissions. The trace element emission factors for wood combustion can be calculated from the above-mentioned particulate emission rates and the trace element concentrations in ash of trees and vegetation. Using the concentrations presented by Nriagu (1979a) and Walsh et al. (1979), the emission factors for As, Cd, Cu, Pb, Ni, and Zn from wood combustion have been estimated (see Table 1.7).

2.2. Internal Combustion Engines

The largest source of trace elements entering the environment from petroleum products are tetraethyl lead and other gasoline additives, diesel and jet fuel combustion, metal compound additives for lubricants, and worn metals that accumulate in spent lubricants.

The fate of lead in gasoline has been studied extensively. Studies by Huntzicker et al. (1975) and Kowalczyk et al. (1978) show that lead emissions can serve as a tracer for all highway-derived emissions, including both gasoline and diesel-powered vehicles. A very interesting source profile for fine aerosol emissions from highway vehicles was constructed recently by Cass and McRae (1983). This profile used the following assumptions: (1) 1.2 g Pb/gal gasoline; (2) an average mileage of 13.6 miles/gal, giving Pb emission rate equivalent of 0.147 per mile; (3) 70.5% Pb emitted as aerosol and only 43% of that material in particle sizes less than 9 μm in aerodynamic diameter (after

Huntzicker et al., 1975); and (4) fine-particle lead emitted as lead salts (after Habibi, 1973). Cass and McRae (1983) have found that lead constitutes 21.1% of the mass of fine-particulated matter emitted from cars burning leaded gasoline. Their profile showing the chemical composition of fine aerosol emissions from highway vehicles is presented in Table 1.8.

The aerosol components shown in the profile were taken from Watson's leaded automobile exhaust profile fine particle fraction (Watson, 1979), except for Pb, Br, Cl, C, and sulfates. The limitation of the presented profile is that much of the automotive lead is emitted in large particles. The mass fraction of Pb in the total vehicle-derived aerosol can range from 4 to 50% (Kowalczyk et al., 1978; Watson, 1979; Pierson and Brachaczek, 1976), depending on the lead content of the fuel, the ratio of leaded to unleaded gasoline used, and the mix of vehicles present. In many countries, the maximum permitted lead content of gasoline is 0.4 g/L (Harrison and Laxen, 1981); however, in some countries, e.g., West Germany and Sweden, a far lower limit of 0.15 g/L has recently been adopted.

Methylocyclopentadienyl manganese tricarbonyl (MMT) is sometimes used as a substitute for tetraethyl lead in some antiknock preparations. The maximum level of MMT in gasoline is 16 mg Mn used for 1 L of gasoline. Thus, the Mn emission factor ranges from 0.3 to 1.5 μg/km or 1.74 to 8.7 μg/L.

Table 1.8. Chemical Composition of Fine Aerosol Emissions from Highway Vehicles (in %)[a]

Compound	Gasoline Autos and Trucks (Leaded Fuel)	Automobile (Unleaded Fuel)	Diesel Engine	Tire Tread	Brake Lining	Highway Composite
Al	0.043	0.12	0.34			0.074
Br	8.2		0.031			4.98
Ca		0.17	0.84		5.5	0.86
Cl	5.4		1.69			3.51
Cu	0.004	0.024	0.73			0.10
Fe	0.25	0.11	1.32			0.333
Pb	21.1		0.095			12.8
Mg					8.25	1.12
Mn		0.015	0.027			0.0039
Ni		0.015				0.0002
K		0.044				0.001
Si	0.075	0.51	0.17		15.4	2.17
Na			0.37			0.05
V			0.01			0.00136
Zn	0.021	0.08	0.23	1.0		0.151
sulfates	0.213	50.0	4.2			1.40
nitrates			0.72			0.1
carbon	54.5	39.0	70.0	87.0	28.3	56.2

[a]From Cass and McRae (1983).

Table 1.9. Lubricant Additives in Use[a]

Metal	Representative Compounds	Purpose
Antimony	Antimony dialkyl dithiocarbamates	Antiwear, extreme pressure, and antioxidant additives in conventional and low-ash-type automatic crankcase oils, industrial and automotive gear oils, greases (amounts $\leq$1–3%)
Barium	Barium diorgano dithiophosphates, Barium petroleum sulfonates, Barium phenolates, Barium phosphonates or thiophosphonates	Corrosion inhibitors, detergents, rust inhibitor; Automatic transmission fluids, greases
Boron	Borax, boric acid esters	Antiwear agents, antioxidant, deodorant cutting oils, greases, brake fluid
Cadmium	Cadmium dithiophosphates	Steam turbine oils
Chromium	Chromium salts	Grease additive
Lead	Lead naphthenate	Extreme pressure additive, greases, gear oils
Mercury	Organic mercury compounds	Bactericide, e.g., cutting oil emulsions
Molybdenum	MoS_2–Mo dibutyl dithiocarbamate and phosphate	Greases, extreme pressure additives
Nickel	Cyclopentadienylnickel complexes	Antiwear agents, carbon deposits, minimizers, lubrication and combustion improvers
Selenium	Selenides	Oxidation and bearing corrosion inhibitors
Tin	Organotin compounds	Antiscuffing additive, metal deactivators
Zinc	Zinc diorgano dithiophosphates, Zinc dithiocarbamates, Zinc phenolates	Antioxidant, corrosion inhibitors, antiwear additives, detergent, extreme pressure additives, in crankcase oils, hypoid gear lubricants, greases, aircraft piston-engine oils, turbine oils, automatic transmission fluids, railroad diesel-engine oils, differential and wet-brake lubricants

[a]From Smith et al. (1975).

Other sources of trace element emissions from petroleum product uses, such as the metal compounds added to lubricants and worn metals that accumulate in used lubricants, have been discussed by Smith et al. (1975). They concluded that there are simply not enough reliable data on the total amounts of trace elements lost to the environment each year from processing, use, and disposal of petroleum products. However, they list some of the types of metal-containing additives now in use. These lubricant additives are given in Table 1.9.

2.3. Production of Nonferrous Metals

In this section the trace element emission factors are presented for the smelting and refining of nonferrous metals. Both primary processing from mineral concentrates and secondary processing from scrap and residual materials are covered, as well as mining of nonferrous metals and brass and bronze ingot production.

In terms of production volume, the major nonferrous metals are Al, Cu, Zn, Pb and Ni (U.S. Department of the Interior, 1970, and other editions). The main chemical pollutants of concern in aluminum production are fluorides. As fluorides are not discussed in this work, aluminum production is not further considered.

2.3.1. Mining of Nonferrous Metals

When nonferrous metal mines are considered as a significant source of atmospheric trace elements, attention is particularly paid to the exploitation of lead–zinc ores. These ores contain high concentrations of Cd and As, varying widely from one field to another. As an example, Wedepohl (1970) gives the arsenic concentration range from 10 to 10,000 μg/g. Thus, it is a very hard task to calculate an average concentration that may be used to estimate an emission factor. Additionally, there is very insufficient information on the trace element emissions from nonferrous metal mines. However, on the basis of data given in the literature, the emission factors of As, Cd, Cu, Mn, Ni, Pb, Se, and Zn have been obtained. The results are shown in Table 1.10.

Table 1.10. Emission Factors of Trace Elements from Nonferrous Metal Mines

Element	Emission Factor	Unit	Reference
As	100	g As/ t As mined	US EPA (1973b)
Cd	0.5	g Cd/ t Zn mined	Nriagu (1980)
Cu	100	g Cu/ t Cu mined	Nriagu (1976b)
Mn	90	g Mn/ t Mn mined	WHO (1981)
Ni	9000[a]	g Ni/ t Ni mined	Sittig (1975); NRCC (1981)
Pb	910	g Pb/ t Pb mined	EC (1973)
Zn	100	g Zn/ t (Zn+Cu+Pb) mined	NAS (1978)
Se	8	mg Se/ t Cu mined	Hawley and Nichol, (1959); Jonasson and Sangster (1975); NAS (1976)
	8	mg Se/ t (Cu–Ni) ore mined	
	25	mg Se/ t (Cu–Zn) ore mined	
	25	mg Se/ t (Pb–Zn) ore mined	

[a]The value includes emission from mining plus refining.

2.3.2. *Primary Nonferrous Metal Production*

Copper, Zn, Pb, and Ni occur as sulfide ores and the feed to the metal winning plant is typically a sulfide concentrate. The treatment involves oxidation, which yields SO_2, the main chemical pollutant to be controlled. However, significant amounts of dust and fumes are also released, causing a serious hazard of heavy metals.

There are three main factors which contribute to large emissions of many elements and their enrichment in the plumes of a metal smelter: (1) type of technology employed in a nonferrous metal production, (2) concentrations of trace elements in ores, and (3) type of air pollution control devices installed and their efficiencies.

Roasting of ore concentrates and low-grade ore enrichment, the initial steps in smelters, are common in Cu, Ni, Pb, Zn, and Cd production. The next steps in the copper–nickel metallurgy are smelting, converting, and refining. Thus, the three main sources of trace elements are roasters, smelting furnaces, and converters. The trace elements of environmental concern released from these sources are Cu, As, Cd, Mo (almost completely discharged in the furnace slag), Pb, Sb, Se, Bi, Ni, and Zn.

Further steps in primary production of zinc and cadmium depend on the process used. Generally, two distinct processes are employed: thermal smelting and electrolytic extraction. In the thermal process, ore concentrates from roasters are sintered and subsequently smelted to reduce zinc and cadmium oxides. The last step is purification. In the electrolytic process, the roasted ore is loaded with sulfuric acid to produce zinc sulfate, and cadmium is precipitated from the solution. The cadmium-rich precipitate and the zinc sulfate solution are then processed separately in electrolytic cells. The lead ore concentrate from roasting is extracted in a sintering plant. Sometimes the sintering machine or electric furnace receives the lead sulfide ore directly. Lead sulfide is then oxidized in a blast furnace to lead oxide and reduced to produce lead metal. In the last step, the lead is refined either by electrolysis or by chemical precipitation.

The main sources of atmospheric emission from the zinc, cadmium, and lead metallurgy are the roasting, sintering, and reduction processes.

Recently, a major study has been carried out in Poland to assess the contributions of several emission sources within the copper smelter complex and the lead smelter to the total emissions of trace elements from the copper and lead metallurgy (Pacyna et al., 1981b). The copper smelter complex under study is one of the biggest and most modern in Europe, equipped with high efficiency electrostatic precipitators. The emission rates of Cu, Pb, Zn, As, Bi, and Sb have been measured for the sources in this copper smelter complex. The highest rates were obtained for smelting furnaces and roasters. The emission factors of the above-mentioned trace elements are presented in Table 1.11, together with the factors based on literature studies. A similar study has also been performed around a lead smelter in Poland (Pacyna et al., 1981b). The emission factors obtained are shown in the same table and compared with the data available in the literature.

Table 1.11. Emission Factors of Trace Elements from Primary Nonferrous Metal Production (in g/t of metal produced)

	Copper–Nickel Smelters		Zinc–Cadmium Smelters	Lead Smelters	
Element	Pacyna et al. (1981b)	Literature Study	Literature Study	Pacyna et al. (1981b)	Literature Study
As	1000	1000–3000 (Walsh et al., 1979) (NAS, 1977)	590 (NAS, 1977) (Walsh et al., 1979)	180	364 (NAS, 1977)
Cd		320–1320 (U.S. EPA 1979b)	0.2–100 (Hutton, 1982)		5 (Nriagu, 1979a)
Cu	1700–3600	2500 (Schwitzgebel et al., 1978) (Nriagu, 1979b)	140 (Nriagu, 1979b)	65	72 (Nriagu, 1979b)
Hg			8.4^a–45^a (MacLaren, 1973) (NRCC, 1979)		1.95^a–3.9^a (MacLaren, 1973) (NRCC, 1977)
Ni		9000 (Schmidt and Andren, 1980)			85 (Nriagu, 1979a)
Pb	2300–3600	3090 (Nriagu, 1978)	1200–25000 (U.S. EPA, 1977)	3000	6360–7750 (Nriagu, 1979a) (Pacyna, 1982a)
Sb	100				
Se			4.1		
Zn	970	845 (Nriagu and Davidson, 1980)	15720–17600 (Pacyna, 1982a) (Jacko and Neuendorf, 1977)	50	110 (Nriagu, 1979a)
Bi	170				

[a] Uncontrolled factor.

As can be seen from Table 1.11, there are substantial differences between the various factors found in the literature and the factors based on measurements by the author. In the former case the differences are mainly due to various processes in the technology, namely, sintering, smelting, or roasting. The emission rates are of course highly sensitive to differences in the efficiency of control devices. High efficiency control devices are often employed in smelters. Roaster facilities, as well as sintering processes, are mainly controlled with fabric filters and electrostatic precipitators. The efficiencies of fabric filters at the copper smelter in Poland varied from 80 to 99.6% and of electrostatic precipitators from 93 to 94%. The range of efficiencies of fabric filters installed in the lead smelter was from 86 to 99.3%.

2.3.3. *Secondary Nonferrous Metal Production*

The share of nonferrous metals produced by secondary processing in the total metal production varies from 20% for zinc and nickel, to 37% for lead and 38% for copper (Barbour et al., 1978).

The major sources of trace element emissions during secondary copper production are ascribed to three processes: (1) the melting of the scrap in blast furnaces (mainly cupola furnaces); (2) the oxidation of impurities in the scrap in a convertor; and (3) the refining of the copper. The largest emissions are released from the blast furnaces. The basic operations in secondary zinc plants, releasing trace elements, are (1) distillation (muffle furnaces or retorts), (2) sweat furnace (kettle, reverberatory, rotary kilns), and (3) pot furnace operations.

In the case of secondary lead productions, the most severe emissions of several pollutants are from (1) rotary furnaces or sweating tubes for materials which contain a small percentage of the metal, (2) reverberatory furnaces for materials of high lead content, and (3) the blast furnaces in secondary smelting of lead storage batteries.

The chemical composition of input scrap and the type and efficiency of control devices are two factors that affect the trace element emissions most from this source. The chemical composition of input scrap varies considerably. Its influence on the releases of trace elements can be compared to the influence of trace element concentrations in fossil fuel on the emission from coal or oil combustion. Several types of control devices are used in secondary nonferrous metal production. The most commonly installed are fabric filters and electrostatic precipitators. The trace element emission factors for the secondary nonferrous metal production have been calculated using the EPA

Table 1.12. Emission Factors of Trace Elements from Secondary Nonferrous Metal Production (in g/t of metal produced)

	Secondary Nonferrous Metal		
Element	Copper	Zinc	Lead
Cd	4		2.5
Cu	150		
Pb	54–214		770
Sb	3		
Zn	500–1630	9000	300

particulate emission factors (1980) and the metal concentrations in dust from factories, cited in the literature (Pacyna, 1982a). The results are presented in Table 1.12.

2.3.4. *Brass and Bronze Ingot Production*

Brass and bronze ingots are produced by melting, smelting, refining, and alloying domestic and industrial copper-bearing scrap. Air contaminants released from brass and bronze furnaces consist of combustion products from the fuel, particulate matter in the form of dust, and metallic fumes. According to Hash (1973), the emission rates of particles depend on the following parameters: (1) alloy composition, (2) furnace type, (3) foundry practices, (4) pouring temperature, and (5) type and efficiency of control devices used. One of the biggest air pollution problems is created by zinc-rich alloys when poured at the temperatures, which are just slightly below the zinc alloy boiling points. Therefore, zinc oxide often escapes into the atmosphere. Another volatile element which follows zinc is lead. Among the various furnace types, reverberatory, rotary, and crucible furnaces are the ones most widely used. The lead and zinc emission factors for several types of furnaces have been calculated using data on particulate emission factors for brass and bronze melting furnaces (U.S. EPA, 1973a) and lead and zinc concentrations in outlet fumes (MRJ, 1971; Duncan et al., 1973). The results are presented in Table 1.13.

The factors in Table 1.13 were calculated assuming that the furnace and the pouring station were equipped with 97.3%-efficient baghouse collectors. This type of control devices is apparently the only air pollution control equipment to receive general acceptance in brass and bronze ingot production. Other types of control devices have not proven entirely satisfactory (Barbour et al., 1978).

The data for lead in Table 1.13 are very close to the lead emission factor of 90 g Pb/t of ingot produced, which has been calculated on the basis of estimates made by Harrison and Laxen (1981) and brass and bronze production figures (Metallgesellschaft Aktiengesellschaft, 1980).

Table 1.13. Zinc and Lead Emission Factors for the Production of Brass and Bronze Ingots (in g/t of ingot produced)

	Zinc Emission Factor		Lead Emission Factor	
Type of furnace	Range	Average	Range	Average
Blast	22–240	130	3–29	16
Crucible	13–160	80	2–19	10
Cupola	79–985	530	10–117	65
Reverberatory	76–945	510	9–112	60
Rotary	65–810	440	8–96	60

2.4 Iron, Steel, and Ferroalloy Manufacturing

The manufacture of iron and steel from iron ore in an integrated operation involves many processes with the potential of creating atmospheric pollution. A large amount of dust (50 kg/t of steel produced) is produced at integrated steelworks, and 65% of it comes from the raw materials and pig iron department, especially the sintering plant (Tsubaki, 1982). As these emissions contain many toxic metallic oxides, the metal emission factors are considered here separately for iron making, iron foundries, and steel making. Finally, the emission rates of trace elements are presented for ferroalloy manufacturing.

2.4.1. Iron Works

There are five major processes in iron works that are responsible for dust emissions: (1) ore crushing, screening, and dumping; (2) sintering and pellitizing; (3) blast furnace; (4) blast furnace gas; and (5) direct reduction of iron ore (Speight, 1978). As far as emission of trace elements is concerned, however, only dust from blast furnaces will be considered. This dust contains certain amounts of zinc, lead, and manganese. The emission rates of zinc, lead, and manganese from blast furnaces vary chiefly according to the type of ore operated and the type and efficiency of control devices. Using information on the chemical analysis of top gas dust burden (Campbell, 1982) and particulate emission factor of 580 g/t of pig iron for blast furnace (Pacyna, 1982a), it was found that zinc emission factors are from 0.3 to 33 g/t of pig iron. The range of lead emission factors is from 0.1 to 8 g/t of pig iron. The lower values are for processes processing haematite and ironstone; the higher ones for siderite or magnetite. From the chemical composition of the blast furnace emissions it was also found that the concentration of manganese in dusts ranges from 0.5 to 1.0% (MRJ, 1971). Thus, the manganese emission factors for iron production are within the range of 2.9 to 5.8 g/t of pig iron. All the above-mentioned emission factors were calculated for iron works equipped with 99%-efficient control devices.

2.4.2. Iron Foundries

The cupola furnace is the most important source of air pollutants in the production of iron castings. In many ways the cupola is similar to a small blast furnace. There are two main types of cupolas, namely, the cold-blast and the hot-blast cupolas. The emissions from the latter are generally higher than from the former, due to the larger amounts of small steel scrap being charged into the hot-blast cupolas.

Cupola dust is a very heterogeneous mixture. The particulate emission factor for a hot-blast cupola is 4.9 kg/t of hot metal and 7.5 kg/t of hot metal for a cold-blast cupola (MRJ, 1971), assuming a control efficiency of 80%. From among the trace elements contained in the dust emitted, only manganese appears to be important. The MnO concentrations in this dust vary from 1 to 2% (Engels and Weber, 1969). Using these values, manganese emission factors for iron foundries vary from 37 to 75 g/t of hot metal for a hot-blast cupola and from 58 to 115 g/t of hot metal for a cold-blast cupola.

2.4.3. *Steel Works*

Steel today is produced in open-hearth, electric-arc, and basic oxygen furnaces, and the type of technology used is the major parameter affecting the trace element emissions. Because of the decline in use of open-hearth furnaces in favor of electric-arc and basic oxygen ones, it is probably safe to say that no new open-hearth furnaces will be constructed (Steiner, 1973). However, open-hearth furnaces are still in operation in several European countries (Prater, 1982). The amount of dust generated during the open-hearth process varies at different stages of the process and depends on the operating practices. Information on trace element emissions from these operations is available in literature (Jacko et al., 1976; Rauhut, 1978). A comparison of atmospheric emissions from electric arc steelmaking with those from basic oxygen steelmaking clearly indicates higher emission factors for the former technology. It reflects the differences in the charge materials used in the two processes. Electric-arc furnaces receive a charge consisting mainly of iron (Hutton, 1982). Thus, the impurities in steel scrap create a potential of atmospheric pollution by trace elements. The trace element emission factors from electric-arc furnaces are affected by the refining procedure used (e.g., with or without oxygen lancing to decarburize the iron), the refining time, and the type and efficiency of particulate control device. Using data for particulate emission factors (U.S. EPA, 1979a) and the chemical composition of the dust from the steelworking furnaces (Midwest Research Institute, 1971; Lee et al., 1975; Jacko et al., 1976), trace element emission factors for steel mills have been calculated. The results are shown in Table 1.14.

In the case of electric-arc furnaces, the following efficiencies of control devices were considered: 98% for venturi scrubbers, and 99% for electrostatic precipitators and bag filters. The efficiencies of electrostatic precipitators

Table 1.14. Trace Element Emission Factors for Steel Works (in g/t of steel)

	Electric-Arc Furnace			Open-Hearth Furnace
Element	Venturi Scrubber	Electrostatic Precipitator	Baghouse	Electrostatic Precipitator
Cd		0.4		0.04–0.05
Cu	1.0	1.3–3.6	0.5	0.2–0.3
Cr	9.0	13.5–36.1	4.0	
Mn	2.6–10.2	15.3–40-9	1.3–5.1	
Ni	2.6	3.9–10.4	1.3	0.05–0.1
Pb	4.1	6.1–16.3	2.0	1.7–3.6
Zn	39	58–155	20	8.6–24

employed in the open-hearth furnace were varying from 97.2 to 99.7% with respect to physicochemical properties of the individual trace metals. Data from Table 1.14 for electric-arc furnace were estimated for oxygen lancing. The emission factors of trace elements from the non-oxygen lance electric-arc furnaces are approximately 18% less than those in Table 1.14 (Pacyna, 1981a).

2.4.4. Ferroalloy Manufacture

Several metals are used for deoxidation, alloying, and graphitization of steel, mainly Mn, Si, Cr, and P, followed by Mo, W, Ti, Zr, V, B, and Cb. During ferroalloy manufacturing, these elements are released into the atmosphere in varying amounts. The emitted amounts depend on (1) type of alloy produced, (2) type of process (i.e., continuous or batch), (3) choice of raw materials, (4) operating techniques, and (5) physicochemical properties of the metals. From among the elements mentioned above, the highest emission rates are for Mn and Pb. For ferromanganese alloys, produced by the two most commonly used techniques (i.e., blast furnace and electric-arc furnace), the Mn emission factors were calculated as 0.4 and 2.45 kg/t of alloy produced, respectively (Pacyna, 1982a). A similar factor for silicomanganese produced in electric-arc furnace was estimated to be 7.1 kg/t of alloy produced. The factors have been obtained assuming a control device efficiency of 99% for ferromanganese production in blast furnaces, and only 40% for electric-arc furnaces. The Pb emission factor for ferroalloys was found to be 0.41 kg/t of ferroalloy produced (Nriagu, 1978).

2.5. Refuse Incineration

Very limited information is available on the trace element emissions from refuse incineration. The process, where large amounts of solid wastes may be combusted, will, like all combustion processes, cause air pollution unless carefully controlled. This section summarizes available data on metal emissions from municipal and sewage sludge incinerations.

2.5.1. Municipal Incineration

Trace element emissions for municipal incinerators depend on the proportion of combustible and noncombustible material in the refuse input, the chemical composition of the refuse input, the incinerator chamber design (combustion temperature), and the efficiency of control devices (if any). From field experiments, it was found that Ag, As, Cd, Cr, Mn, Pb, Sn, and Zn in incinerator effluents came from noncombustible materials. The emissions of Al, Ba, Co, Fe, Li, Na, Ni, and Sb were attributed to the combustible part of the refuse input. Other elements, such as Ca, Cu, Hg, K, and Mg, probably result from the combustible components (Greenberg et al., 1978a; Law and Gordon, 1979).

Municipal incinerators are usually equipped with some type of particulate

Table 1.15. Trace Element Concentrations in Dust and Trace Element Emission Factors for Municipal Incinerators

Metal	Concentration ($\mu g/g$[a]		Emission Factor
	Range	Average	(g/t of refuse)
As	200–310	240	0.52
Cd	1100–1900	1500	2.25
Co	2.3–12.0	6.6	0.01
Cr	105–870	490	1.10
Cu	1500–2000	1700	3.70
Mn	270–1500	730	1.60
Ni	79–200	150	0.33
Pb, %	6.9–8.9	8.1	17.6
Sb	1600–2400	2100	4.55
Se	23–49	37	0.08
Sn, %	1.1–1.3	1.15	25
Zn, %	11–13	12	260

[a] Unless % is indicated.

control device, such as a spray chamber or an electrostatic precipitator. Investigations carried out by Scott (1979) show the importance of using a proper form of dust collector in refuse-incineration plants. Generally, the control devices in these plants are the same as those employed in electric power plants. However, incinerators differ from power stations by the combustion at higher temperatures (Johnson and Burnet, 1978), which present particular problems in collecting volatile trace elements. Taking into consideration the particulate emission factor of 2.2 kg/t of refuse burned, chosen from the range of 0.45 to 4.00 kg/t of refuse and average values of the trace element concentrations in dust emitted from municipal incinerators (Greenberg et al., 1978b; Law and Gordon, 1979; Scott, 1979), the calculated trace element emission factors are shown in Table 1.15.

The emission factors in Table 15 were estimated assuming an 85% efficient control device. The actual efficiency will be very dependent upon a feature of a certain trace element and the size range of particles released during the incineration.

2.5.2. Sewage Sludge Incineration

A few of the recent studies on the trace element behavior during sewage sludge incineration show that atmospheric metal emissions are related to (1) amount of metal in the sludge, (2) combustion temperature, (3) volatility of the metal, and (4) type of air pollution controls used (Copeland, 1975; Takeda and Hiraoka, 1976; Wall and Farrell, 1979; Greenberg et al., 1981; Gerstle and Albrinck, 1982). It was found that the content of volatile metals increases in

Table 1.16. Trace Element Concentration in Dry Sludge and Trace Element Emission Factors from Sewage Sludge Incinerators

	Concentration, (μg/g of dry sludge)		
Metal	Range	Average	Emission Factor from Sludge Incinerator (g/t)
Ag	ND[a]–960	225	
As	10–50	9	
Cd	ND–1,100	87	12
Co	ND–800	350	1.2
Cr	22–30,000	1,800	10
Cu	45–16,000	1,250	58
Hg	0.1–90	7	3.5[b]
Mn	100–8,800	1,190	26
Ni	ND–2,800	410	
Pb	80–26,000	1,940	140
Se	10–180	26	10
V	ND–2,100	510	6.2
Zn	51–28,400	3,500	104

[a] ND = values at the detection limit.
[b] From Trout (1975).

the flue gas with increasing temperature. Conventional incineration temperatures are from 1033 to 1088 K. The venturi scrubber-controlled sewage sludge incinerators release only a small fraction of the metals found in the incoming sludge, except for mercury. Collection efficiencies of these installations are from 90 to greater than 99%. Other control devices (electrostatic precipitators and baghouses) are not generally found on sewage sludge incinerators (Gerstle and Albrinck, 1982).

Table 1.16 shows typical metal concentrations in dry sludge, together with emissions factors. The data were calculated from informations by the U.S. Environmental Protection Agency (1976), and Greenberg et al., (1981).

2.6. Cement Production

Air pollutants from the cement industry ore consist mainly of rather nontoxic dust, emitted in very fine particles. However, there are certain units in the cement plants emitting dust with several trace elements: the feed system, the fuel-firing kiln system, and the clinker-cooling and handling system. In some of these, a mixture of shale and limestone is exposed to high temperatures. Information in the literature is poor on the quantity of trace element emissions from the cement factories. Any available data show that the type of

Table 1.17. Emission Factors of Lead and Cadmium from Cement Manufacturing (in g/t of cement)

	Lead Emission Factor			Cadmium Emission Factor		
Process	Multi-cyclones	ESP	Baghouse	Multi-cyclones	ESP	Baghouse
Dry process (total)	16.0	4.0	0.16			
Kiln/cooler	12.0	3.0	0.12	0.60	0.15	0.01
Dryer/grinder	4.0	1.0	0.04			
Wet process (total)	12.0	3.0	0.12	0.05	0.02	0.02
Kiln/cooler	10.00	2.5	0.10	0.04	0.01	0.01
Dryer/grinder	2.0	0.5	0.02	0.01	0.01	0.01

production process (wet or dry), the type of fuel used in the grinding mill, the type of fuel-firing system employed, and the type of control equipment all affect the emission rates of trace elements during cement production. The degree of this variability is evident from Table 1.17, where lead and cadmium emission factors are listed for cement manufacturing plants (U.S. EPA, 1977; U.S. EPA, 1979b; Pacyna, 1982a).

The type of control device is especially important when comparing the emission factors in Table 1.17. Fabric filters are apparently the most effective control devices in the cement industry (Gilliland, 1973; Gates, 1978).

Little is known about emission of other trace elements from cement manufacturing. Because coal is the most commonly used fuel for powering the machinery in grinding mills, certain portions of Mn, Zn, Cu, As, and Cd are also expected in dust from these sources. In addition, trace elements are also released as a result of application of several additives in cement production. As much as 2% of thallium has been measured in flue dusts from the West German factories (Nriagu, 1984). This high concentration resulted in thallium emission of 16 t in the FRG in 1982. Emission factors for trace elements from the use of additives are hard to obtain because of lack of data on their chemical composition in the open literature.

3. NATURAL SOURCES OF TRACE ELEMENTS

Natural sources appear to dominate the atmospheric emissons of elements such as Se, As, Hg, and Cd. Unfortunately, there is a lack of hard, reliable information about the quantities and qualities of those emissions. In many cases the estimates of emission rates or emission fluxes are based on educated guesses and are not confirmed by measurements. Even in those cases where some measured data are available for a certain natural source (e.g., volcanic eruptions), vast disagreements between measurements by different authors make this information hard to utilize. Zoller (1983), in his review paper on anthropogenic perturbation of metal fluxes into the atmosphere, concludes that the emission rates of individual volcanoes can vary by more than two orders of magnitude. Consequently, fluxes of elements, such as Cd and Zn from volcanoes, can vary by a factor of one hundred.

In this section, any available data on the trace element emission from volcanoes, windblown dust, forest wildfires, vegetation, and airborne seasalt are used to assess emission factors of these pollutants from several natural sources. The results are presented in Table 1.18.

The most extensively studied source among those in Table 1.18 appears to be volcanic eruptions. However, there is the high variability of the chemical composition of emissions from both different volcanoes and the same volcano during activity eruptions and quiescent periods, which results in the differences in the trace element emission factors in Table 1.18. For some elements, such as As, Cd, and Se, a part of the emissions is in the gaseous state, volcanoes being high-temperature emission sources. During eruption the most volatile elements evaporate and then condense on the very small particles. The gaseous part of the emissions of As, Cd, Se, and other metals is missing in Table 1.18.

Table 1.18. Emission Factors of Trace Elements to the Atmosphere from Natural Sources (in ng/kg of dust)

Metal	Volcanic Particles[a]	Windblown Dust[b]	Forest Wildfires[c]	Vegetation[d]	Seasalt Sprays[e]
As	300–800	0.5–2.0	0.5–4.4	3.5	0.1–0.6
Cd	30–800	0.002–1.7	0.03–2	2.7–36	0.001–0.003
Co	140	8			
Cu	200–5,400	0.3–52	0.7–47	33–440	0.02–0.2
Cr	390	100			
Mn		850			
Ni	200–5,600	0.5–88	1.4–92	21–280	0.01–0.08
Pb	100–9,600	0.4–70	1.1–78	21–280	0.001–0.09
Se	10–1,700	0.6			
Sb	30				
V	690	100			
Zn	500–10,500	0.6–110	5–330	125–1,700	0.004–0.03

[a] From Lantzy and Mackenzie (1979), Nriagu (1979a), Jaworowski et al. (1981), Gordon et al. (1983), Walsh et al. (1979), Arnold et al. (1981), Phelan et al. (1982), Servant (1982), and Pacyna (1982a).

[b] From Lantzy and Mackenzie (1979), Nriagu (1979a), Jaworowski et al. (1981), Gordon et al. (1983), Pacyna (1982a), and Walsh et al. (1979).

[c] From Lantzy and MacKenzie (1979), Nriagu (1979a), Pacyna (1982a), and Davidson et al. (1981).

[d] From Nriagu (1979a) and Pacyna (1982a).

[e] From Nriagu (1979a), Jaworowski et al. (1981), Gordon et al. (1983), Walsh et al. (1979), and Pacyna (1982a).

The emission factors for windblown dust are calculated from the concentration of each metal in different soils and the amount of continental dust annually wafted into the atmosphere. The former parameter is mainly responsible for the differences in the trace element emission-factor estimations. The latter is often referred to be 5×10^{14} g, after Goldberg (1971).

Emission factors for forest wildfires are estimated from the average acreage that is burned by forest wildfires and the concentrations of trace elements in forest stock. Because the concentrations vary from one forest stock to another, a wide range of factors is shown in Table 1.18. The potential emissions of As, Cd, Cu, Ni, Pb, and Zn from plant leaf sources can be estimated on the basis of their concentrations in plant matter and the metabolism in plants and soil. This source is coded as vegetation. The biogenic emissions of trace elements from vegetation depend on the presence of microbes in the soil, the availability of trace elements, the creation of mobile forms of metals in soil, and physicochemical transformation of the trace elements in soil and vegetation. The trace element emission factors for seasalt are very small, as there are differences in the estimates by several authors. However, this is due to the fact that all the estimates are based on similar data on the flux of seasalt to the atmosphere.

4. FINAL REMARKS

The intention of this review was to present a range of trace element emissions from several sources, as given here indirectly in terms of emission factors. Such presentation gives an opportunity to analyze important parameters that affect the magnitudes of emissions, while direct presentation of amount of trace elements released shows chiefly absolute numbers. Application of the trace element emission factors depends upon the scale of emission inventory uses. In general, emission factors are not precise indicators of emissions from a single source, but are most valid when used as an average for large number of sources. Thus, the trace element emission factors presented here would be mostly useful for emissions inventories and estimates of the trace element perturbations on global and regional scales. As an example, the factors from the sections of this chapter can be utilized to calculate the trace element emissions in Europe for studying the long-range transport of air pollutants. These data are presented in the next chapter of this book (Pacyna, 1983). There are limitations to the applicability of emission factors when analyzing behavior of trace elements on the local scale, as for instance, the effects of trace element emissions from certain power plant on the surrounding environment. In this case, emission values are used in several models to assess a migration of pollutants through the particular environmental medium. Emission factors can be considered as a way to obtain the emission values; however, it is necessary to develop these factors on the basis of direct measurements. Another problem with trace element emission factors is their applicability from one case to another, which may be limited by differences in the extent of trace element contamination of raw materials, technologies utilized, and efficiencies of control devices.

There is an urgent need to improve the accuracy of emission factor calculations. This accuracy is most directly related to the number of studies performed on a particular source. Thus, more emission measurements are required by both the extractive method and nonextractive on-site systems, which will make emission factors more broadly and readily available.

REFERENCES

Air Pollution Control Association (1979). "Emission Factors and Inventories". *Proceedings of a Specialty Conference on Emission Factors and Inventories.* Frederick, E.R., Ed., Anaheim. Calif.

Arnold, M., Buat-Menard, P., and Chesselet, R. (1981). "An estimate of the input of trace metals to the global atmosphere by volcanic activity." International Association of Meteorology and Atmospheric Physics, Third Scientific Assembly. Hamburg, FRG.

Babcock and Wilcox Corp. (1972). *Steam, its generation and use.* 38th Ed. Babcock and Wilcox, New York.

Baker, E.W. (1964). "Vanadium and nickel in crude petroleum of South American and Middle East origin." *J. Chem. Eng. Data* **9,** 307–308.

Barbour, A.K., Castle, J.F., and Woods, S.E. (1978). "Production of non-ferrous metal." In Parker, A., Ed., *Industrial air pollution handbook.* McGraw-Hill Book Company, (U.K.) Limited, London.

Billings, C.E., Sacco, A.M., Matson, W.R., Griffin, R.M., Coniglio, W.R., and Harley, R.A. (1973). "Mercury balance on a large pulverized coal-fired furnace." *J. Air Pollut. Control Assoc.* **23,** 773–777.

Block, Ch., and Dams, R. (1976) "Study of fly ash emission during combustion of coal." *Environ. Sci. Technol.* **10,** 1011–1017.

Bolton, N.E., Carter, J.A., Emery, J.E., Feldman, C., Fulkerson, W., Hulett, U.D., and Lyon, W.S. (1975). "Trace element mass balance around a coal-fired steam plant." In *Trace elements in fuel.* S.B. Babu, Ed., Advances in Chemistry Series No. 141, American Chemical Society, Washington, D.C.

Brown, R.D. (1981). "Health and environmental effects of oil and gas technologies: Research needs." A report to the Federal Interagency Committee on the Health and Environmental Effects of Energy Technologies. The MITRE Corp., McLean, Va.

Butcher, S.S., and Sorensen, E.M. (1979). "A study of wood stove particulate emissions." *J. Air Pollut. Control Assoc.* **29,** 724.

Campbell, J.A., Laul, J.C., Nielson, K.K., and Smith, R.D. (1978). "Separation and chemical characterization of finely-sized fly ash particles." *Anal. Chem.* **50,** 1032–1040.

Campbell, J.M. (1982) "Advances in water pollution control and waste recovery in blast furnace operations." *UNEP Ind. Environ.* **5,** (4), 9–12.

Capes, C.E., Mc Ilhinney, A.E., Russell, D.S., and Sirianni, A.F. (1974). "Rejection of trace metals from coal during beneficiation by agglomeration." *Environ. Sci. Technol.* **8,** 35–38.

Coles, D.G., Ragaini, R.C., Ondov, J.M., Fisher, G.L. Silberman, D., and Prentice, B.A. (1979). "Chemical studies of stack fly ash from coal-fired power plant." *Environ. Sci. Technol.* **13,** 455–459.

Cooper, J.A. (1980). "Environmental impact of residential wood combustion emissions and its implications." *Environ. Sci. Technol.* **8,** 855–861.

Copeland, B.J. (1975) "A study of heavy metal emissions from fluidized-bed incinerators." *The Proceedings of Purdue Industrial Waste Conference.* West Lafayette, In.

Davidson, C.I., Grimm, T.C., and Nasta, M.A. (1981) "Airborne lead and other elements derived from local fires in the Himalayas." *Science* **214,** 1344–1346.

Davison, R.L., Natusch, D.F.S., Wallace, J.R., and Evans, Ch.A., Jr. (1974). "Trace elements in fly ash. Dependence of concentration on particle size." *Environ. Sci. Technol.* **8,** 1100–1113.

DeAngelis, D.G., Ruffin, D.S., and Reznik, R.B. (1980). *Preliminary characterization of emissions from wood-fired residential combustion equipment.* U.S. EPA Report 600/7-80-040. U.S. Environmental Protection Agency, Washington, DC.

De Pereyra, B. (1973). "Difficulties of desulfurization of Venezuelan petroleum." In *Proceedings of International Symposium on Vanadium and Other Metals in Petroleum,* Maracaibo, Venezuela.

Duncan, L.J., Keitz, E.L., and Krajeski, E.P. (1973). *Selected characteristics of hazardous pollutant emissions.* Report MTR-6401, Vol. II. The MITRE Corporation, Washington, D.C.

Dvorak, A.J., and Lewis, B.G. (1978). *Impacts of coal-fired power plants on fish wildlife and their habitats.* Report FWS/OBS-78/29. U.S. Dept. of Interior, Washington, D.C.

Engels, G., and Weber, E. (1969). *Cupola emission control.* Gray and Ductile Iron Foundries Society, Inc., Des Plaines, Illinois, p. 60.

Ensor, D.S., Cowen, S., Shendrikar A., Markowski, G., and Waffinden, G. (1981). *Kramer station fabric filter evaluation.* EPRJ Report CS-1669. Electric Power Research Institute, Palo Alto, Calif.

Environment Canada (1973). *National inventory of sources and emissions of lead.* Air Pollution Control Directorate, Report No. APCD 73-7. Ottawa, Can.

Environment Canada (1978). *A nationwide inventory of emissions of air contaminants.* Economic and Technical Review Report EPS 3-AP-78-2. Environmental Protection Service, Ottawa, Can.

Flagan, R.G., and Friedlander, S.K. (1976). "Particle formation in pulverized coal combustion—a review". In *Proceedings of the symposium on aerosol science and technology of the 82nd National Meeting of the American Institute of Chemical Engineers,* Atlantic City, N.J.

Filby, R.H (1975). "The nature of metals in petroleum". In *The role of trace metals in petroleum.* Yen, T.F. Ed. Ann Arbor Press, Ann Arbor, Mich., pp. 31–58.

Gates, R.J. (1978). "Manufacture of portland cement." In Parker, A., Ed., *Industrial air pollution handbook.* McGraw-Hill Book Company, (U.K.) Limited, London.

Gerstle, R.W., and Albrinck, D.N. (1982). "Atmosphereic emissions of metals from sewage sludge incineration." *J. Air Pollut. Control Assoc.* **32,** 1119–1123.

Gilliland, J.L. (1973). "Particulate emission controls in portland cement manufacturing." In Noll, K. and Duncan, J., Eds., *Industrial Air Pollution Control.* Ann Arbor Science Publishers Inc., Ann Arbor, Mich., pp. 195–202.

Gluskoter, H.J., Ruch, R.R., Miller, W.G., Cahill, R.A., Dreher, G.B., and Kuhn, J.K. (1977). *Trace elements in coal; occurrence and distribution.* Illinois State Geological Survey. Circular 499. Urbana, Il.

Goldberg, E.D. (1971). "Atmospheric dust, the sedimentary cycle and man." *Geophysics* **1,** 117–132.

Goldschmidt, W.M. (1935) "Rare elements in coal ashes." *Ind. Eng. Chem.* **27,** 1100–1102.

Gordon, G.E., Moyers, J.L., Rahn, K.A., Gatz, D.F., Dzubay, T.G., Zoller, W.H., and Corrin, M.H. (1983). "Atmospheric trace elements: cycles and measurements." *Rev. Geophys. Space Phys.* (in press).

Greenberg, R.R., Gordon, G.E., Zoller, W.H., Jacko, R.B., Neuendorf, D.W., and Yost, K.J. (1978a). "Composition of particles emitted from the Nicosia Municipal Incinerator." *Environ. Sci. Technol.* **12,** 1329–1332.

Greenberg, R. R., Zoller, W.H., and Gordon, G.E. (1978b) "Composition and size distribution of particles released in refuse incineration." *Environ. Sci. Technol.* **12,** 566–573.

Greenberg, R.R., Zoller, W.H., and Gordon, G.E. (1981). "Atmospheric emissons of elements on particles from the Parkway Sewage-Sludge Incinerator." *Environ. Sci. Technol.* **15,** 64.

Habibi, K. (1973). *Characterization of particulate matter in vehicle exhaust.* Chapman and Hall Ltd., London.

Hash, R.T. (1973). "Brass and bronze emission control." In Noll, K. and Duncan, J., Eds., *Industrial Air Pollution Control.* Ann Arbor Science Publishers Inc., Ann Arbor, Mich., pp. 171–184.

Hawley, J.E., and Nichol, J. (1959). "Selenium in some Canadian sulfides." *Economic Geology* **54,** 608–628.

Holland, W.F., Wilde, K.A., Parr, J.L., Lowell, P.S., and Pohler, R.F. (1975). *The environmental effects of trace elements in the pond disposal of ash and flue gas desulfurization sludge.* EPRI Report. Electric Power Research Institute, Palo Alto, Calif.

Huntzicker, J.J., Friedlander, S.K., and Davidson, C.J. (1975). "Material balance for automobile-emitted lead in Los Angeles Basin." *Environ. Sci. Technol.* **5,** 448–457.

Hutton, M. (1982). *Cadmium in the European Community: a prospective assessment of sources, human exposure and environmental impact.* MARC Report 26. Monitoring and Assessment Research Centre, Chelsea College, University of London, London.

Isubaki, H. (1982). "Resource recycling in the Japanese steel industry." *UNEP Ind. Environ.* **5,** (4), 18–23.

Jacko, R.B., and Neuendorf, D.W. (1977). "Trace metal particulate emission test results from a number of industrial and municipal point sources." *J. Air Pollut. Control. Assoc.* **27,** 989–994.

Jacquin, Y., Deschamps, A., Le Page, J.F., and Billon, A. (1973). "Problems of pollution by sulphur from crude oil in European countries." In *Proceedings of International Symposium on Vanadium and Other Metals in Petroleum,* Maracaibo, Venezuela.

Jaworowski, Z., Bysiek, M., and Kownacka, L. (1981). "Flow of metals into the global atmosphere." *Geochim. Cosmochim. Acta* **45,** 2185–2199.

Johnson, H.B., and Burnett, J.M. (1978). "Incineration of refuse." In Parker, A., Ed., *Industrial Air Pollution Handbook.* McGraw-Hill Book Company, (U.K.) Limited, London.

Jonasson, J.R., and Langster, D.F. (1975). *Selenium in sulphides from some Canadian base metal deposits.* Geological Survey Canada, Paper 75-1c, Ottawa, Can.

Kaakinen, J.W., Jorden, R.M., Lawasani, M.H., and West, R.E. (1975). "Trace element behavior in coal-fired power plant." *Environ. Sci. Technol.* **9,** 862–863.

Kaakinen, J.W. (1979). "Trace element study on a pulverized coal-fired power plant." PhD dissertation, University of Colorado, Boulder, Colo.

KHM Prosjekt (1981). *Kolets hälso- och miljöeffekter.* Lägesrapport, Stockholm.

Klein, D.H., Andren, A.W., Carter, J.A., Emery, J.F., Feldman, C., Fulkerson, W., Lyon, W.S., Ogle, J.C., Talmi, Y., Van Hook, R.J., and Bolton, N. (1975). "Pathways of thirty-seven trace elements through coal-fired power plant." *Environ. Sci. Technol.* **9,** 973–979.

Kowalczyk, G.S., Choquette, C.E., and Gordon, G.E. (1978). "Chemical element balances and identification of air pollution sources in Washington, DC." *Atmospher. Environ.* **12,** 1143–1154.

Kuhn, J.K., Fiene, F.L., Cahill, R.A., Gluskoter, H.J. and Shimp, N.F. (1980). "Abundance of trace and minor elements in organic and mineral fractions of coal." Illinois Institute of Natural Resources, *Environ. Geol. Notes* No. **88.**

Lantzy, R.J., and Mackenzie, F.T. (1979). "Atmospheric trace metals: global cycles and assessment of man's impact." *Geochimi. Cosmochim. Acta* **43,** 511–523.

Law, S.L., and Gordon, G.E. (1979). "Sources of metals in municipal incinerator emissions." *Environ. Sci. Technol.* **13,** 432–438.

Lee, R.E., and Lehmden, D.J. (1973). "Trace metal pollution in the environment." *J. Air Pollut. Control Assoc.* **23,** 853–857.

Lee, R.E. Jr., Crist, H.L., Riley, A.E., and Mac Leod, K.E. (1975). "Concentration and size of trace metal emissions from a power plant, a steel plant, and a cotton gin." *Environ. Sci. Technol.* **9,** 643–647.

MacLaren, J.F., Ltd. (1973). *National inventory of sources and emissions of mercury.* Air Pollution Control Directorate, Report No. APCD 73-6. Environmental Protection Service, Ottawa, Can.

Metallgesellschaft Aktiengesellschaft (1980). *Metal statistics 1969–1979.* 67th Ed. Frankfurt am Main, FRG.

Midwest Research Institute (1971). *Particulate pollutant system study,* Vol. III of *Handbook of Emission Property.* MRJ Project 3326–C. Midwest Research Institute, Durham, N.C.

Murthy, K.S., Howes, J.E., and Nack, H. (1979). "Emissions from pressurized fludized-bed combustion processes." *Environ. Sci. Technol.* **13,** 197–204.

National Academy of Sciences, National Research Council (1976). *Selenium.* Subcommittee on Selenium, Committee on Medical and Biological Effects of Environmental Pollutants. PB-251-318. Washington, D.C.

National Academy of Sciences, National Research Council (1977). *Arsenic.* Committee on Medical and Biological Effects of Environmental Pollutants. Washington, D.C.

National Academy of Sciences, National Research Council (1978). *Zinc.* Committee on Medical and Biological Effects of Environmental Pollutants. Washington, D.C.

National Research Council of Canada (1979). *Effects of mercury in the Canadian environment.* NRCC Report No. 16739. Ottawa, Can.

National Research Council of Canada (1981). *Effects of nickel in the Canadian Environment.* NRCC Report No. 18568. Ottawa, Can.

Natusch, D.F.S., Wallace, J.R., and Evans, C.A. Jr. (1974). "Toxic trace elements: preferential concentration in respirable particles." *Science* **183,** 202–204.

Nicholls, G. C. (1968). "The geochemistry of coal-bearing strata." In *Coal and coal-bearing strata.* American Elsevier, New York, pp. 269–307.

Nriagu, J.O. (1978). "Lead in the atmosphere." In Nriagu, J.O., Ed., *The biogeochemistry of lead in the environment,* Part A of *Ecological cycles,* Elsevier/North-Holland Biomedical Press. Amsterdam, New York, Oxford.

Nriagu, J.O. (1979a). "Global inventory of natural and anthropogenic emissions of trace metals to the atmosphere." *Nature* **279,** 409–411.

Nriagu, J.O. (1979b) "Copper in the atmosphere and precipitation". In Nriagu, J.O., Ed., *Copper in the Environment. Environmental Science and Technology Series.* John Wiley & Sons, Inc., New York.

Nriagu, J.O. (1980) "Cadmium in the atmosphere and in precipitation." In Nriagu, J.O., Ed., *Cadmium in the Environment, Environmental Science and Technology Series.* John Wiley & Sons, Inc., New York.

Nriagu, J.O., and Davidson, C.J. (1980). "Zinc in the atmosphere." In Nriagu, J.O., Ed., *Zinc in the Environment, Environmental Science and Technology Series.* John Wiley & Sons, Inc., New York.

Nriagu, J.O., Ed. (1984). "Changing metal cycles and human health." Dahlem Konferenzen. Springer Verlag, Berlin.

Organization for Economic Cooperation and Development (1982). *Quarterly oil statistics.* International Energy Agency, Paris.

Ondov, J.M., Ragaini, R.C., and Bierman, A.H. (1979). "Elemental emissions from a coal-fired power plant. Comparison of a venturi wet scrubber system with a cold-side electrostatic precipitator." *Environ. Sci. Technol.* **13,** 588–601.

Pacyna, J.M., (1980). "Coal-fired power plants as a source of environmental contamination by trace metals and radionuclides." Habilitation thesis. Technical University of Wroclaw, Wroclaw, Poland.

Pacyna, J.M. (1981a). *Emission factors for trace metals from coal-fired power plants.* NILU Report 14/81. Norwegian Institute for Air Research, Lillestrφm, Norway.

Pacyna, J.M., Zwozdziak, A., Zwozdziak, J., Matyniak, Z., Kuklinski, A., and Kmiec, G. (1981b). *Zagadnienia ochrony powietrza atmosferycznego dla stanu istniejacego i perspektywicznego na obszarze pilotowym wojewodztwa legnickiego* (Air pollution problems caused by the LGOM copper smelter complex—present situation and perspectives). Report SPR 14-81. Technical University of Wroclaw, Wroclaw, Poland.

Pacyna, J.M. (1982a). *Emission Factors of Trace Elements.* MARC Report. Monitoring and Assessment Research Centre, Chelsea College, University of London, London.

Pacyna, J.M. (1982b) *Estimation of emission factors of trace metals form oil-fired power plants.* NILU Report 2/82. Norwegian Institute for Air Research, Lillestrφm, Norway.

Pacyna, J.M. (1983) "Atmospheric trace elements from natural and antropogenic sources." In Nriagu, J.O., Ed., *Metals in the Air, Environmental Science and Technology Series,* John Wiley & Sons, Inc., New York.

Phelan, J.M., and Brachaczek, W.W. (1976). *Particulate matter associated with vehicles on the road.* Paper 760039. Society of Automotive Engineers, Warrendale, Pa.

Prater, B.E. (1982). "Cadmium in U.K. steelmaking operations." In Hutton, M., *Cadmium in the European Community,* MARC Report 26. Monitoring and Assessment Research Centre, Chelsea College, University of London, London.

PSEG Co. (1967). *Report on sulfur dioxide and fly ash. Emissions from electric utility boilers.* Public Service Electric and Gas Company, N.J.

Rauhut, A. (1978). "Survey of industrial emission of cadmium in the European Economic Community." ENV/2223/74/E. Commission of the European Communities, Brussels, Belgium.

Reckard, M.K. (1979) *Decentralized energy: technology assessment and systems description.* Report BNL 50987. Brookhaven National Laboratory, Upton, NY.

Rottman, H.K., Kary, R.E., and Hudgins, T. (1977). "Ecological distribution of trace elements emitted by coal-burning power generating units employing scrubbers and electrostatic precipitators." In *Fourth Symposium on Coal Utilization.* National Coal Association/Bituminous Coal Research Inc., Washington, D.C.

Roeck, D.R., and Dennis, R. (1979) *Technology assessment report for industrial boiler applications: particulate collection.* Report No. EPA-600/7-79-178h. U.S. Environmental Protection Agency, Research Triangle Park, N.C.

Roy, W.R., Thiery, R.G., Schuller, R.M., and Suloway, J.J. (1981) "Coal fly ash, a review of the literature and proposed classification system with emphasis on environmental impacts." Illinois Institute of Natural Resources, *Environ. Geol. Notes,* No. 96.

Rubin, E.S. (1981). "Air pollution constraints on increased coal use by industry." *J. Air Pollut. Control Assoc.* **31,** 349–360.

Rudling, L., and Löfroth, G. (1981). *Chemical and biochemical characterization of emissions from combustion of peat and wood-chips.* Naturvårdsverket Rapport PM 1449. Stockholm, Sweden.

Schmidt, J.A., and Andren, A.W. (1980). "The atmospheric chemistry of nickel." In Nriagu, J.O., Ed., *Nickel in the environment, Environmental Science and Technology Series,* John Wiley & Sons, Inc., New York.

Schwitzgebel, K., Coleman, R.T., Collins, R.V., Mann, R.M., and Thompson, C.M. (1978). *Trace element study at a primary copper smelter.* Radian Report, 68-01-4136. Austin, Texas.

Scott, D.W. (1979). *The fate of cadmium in municipal refuse incinerators.* Report CR 1710 (AP). Warren Spring Laboratory, Stevenage, Herts.

Servant, J. (1982). *Atmospheric trace elements from natural and industrial sources.* MARC Report No. 27. Monitoring and Assessment Research Centre, Chelsea College, University of London, London.

Sittig, M. (1975). *Environmental sciences handbook.* Environ. Technol. Handbook No. 2. Noyes Data Corp., Park Ridge, N.J. and London.

Smith, W.C. (1962). *Atmospheric emissions from fuel oil combustion—an inventory guide.* Public Health Service, Publication No. 999-AP-2. Washington, D.C.

Smith, J.C., Ferguson, T.L., and Carson, B.L. (1975). "Metals in new and used petroleum products and by-products. Quantities and consequences." In *The Role of Trace Metals in Petroleum.* T.F. Yen, Ed. Ann Arbor Science Publishers, Ann Arbor, Mich., pp. 123–148.

Smith, R.D., Campbell, J.A., and Nielson, K.K. (1979). "Characterization and formation of submicron particles in coal-fired plants." *Atmospher. Environ.* **13,** 607–617.

Speight, G.E. (1978). "Iron and steel works." In Parker, A., Ed., *Industrial Air Pollution Handbook,* McGraw-Hill Book Company, (U.K.) Limited, London.

Speight, J.G. (1981) "The chemistry and technology of petroleum." *Chem. Ind.* **3.**

Steiner, B.A. (1973). "Particulate emission control in the steel industry." In Noll, K. and Duncan, J., Eds., *Industrial Air Pollution Control.* Ann Arbor Science Publishers Inc., Ann Arbor, Mich. pp. 163–170.

Takeda, N., and Hiraoka, M. (1976). "Combined process of pyrolysis and combustion for sludge disposal." *Environ. Sci. Technol.* **10,** 1147.

Torgaard, H. (1983). "North Sea residue tested as FFC feed-stock", *Oil Gas J.* January 10.

Trout, D.A. (1975). *Analysis of ambient air quality in the vicinity of the parkway incinerator.* Battelle Memorial Institute Report of the Washington Suburban Sanitary Commission. Columbus, Ohio.

Umwelt Bundes Amt (1980). *Emissionsfaktoren für Luftverunreinigungen.* Materialien 2/80. Erich Schmidt Verlag, Berlin.

U.S. Department of the Interior (1970). *Mineral facts and problems.* U.S. Department of the Interior, Bureau of Mines, Washington, D.C., Metalgesellschaft A.G. Metalstrahshes, 1965–1975.

U.S. Environmental Protection Agency (1973a). *Complication of air pollutant emission factors.* 2nd Ed. AP-42, Research Triangle Park, N.C.

U.S. Environmental Protection Agency (1973b). *Emission factors for trace substances.* EPA-450/2-73-001. Washington, D.C.

U.S. Environmental Protection Agency (1976). *Municipal sludge management, an overview of the sludge management situation.* EPA Construction Grants Program, EPA 430/9-76-009. Washington, D.C.

U.S. Environmental Protection Agency (1977). *Control technologies for lead air emissions.* EPA-450/2-77-012. Research Triangle Park, N.C.

U.S. Environmental Protection Agency (1979a). *Compilation of air pollutant emission factors.* 3rd Ed. AP-42, Supplement No. 9. Research Triangle Park, N.C.

Coleman R. et al. (1979b) *Sources of atmospheric cadmium.* EPA-450/5-79-006. U.S. Environmental Protection Agency, Research Triangle Park, N.C.

U.S. Environmental Protection Agency (1980). *Source category survey: secondary copper smelting and refining industry.* EPA-450/3-80-011, Research Triangle Park, N.C.

Wall, H., and Farrell, J. (1979). *Particulate emissions from municipal wastewater sludge incinerators.* U.S. Environmental Protection Agency, Cincinnati, Ohio.

Walsh, P.R., Duce, R.A., and Fashing, J.L. (1979). "Considerations of the Enrichment, Sources, and Flux of Arsenic in the Troposphere." *J. Geophys. Res.* **84,** 1719–1726.

Watson, J.G. (1979). Ph.D. Dissertation. Oregon Graduate Center, Beaverton, Ore.

Wedepohl, K.H. (1979). "Arsenic." *Handbook of Geochemistry,* Vol. 2. New York.

Witkowski, S.J. (1977). "Conversion of oil- and gas-fired units to coal firing." In *Fourth Symposium on Coal Utilization.* National Coal Association/Bituminous Coal Research Inc., Washington, D.C.

World Health Organization (1981). "Manganese." *Environmental Health Criteria* **17,** Geneva.

Yen, T.F. (1975). "Chemical aspects of metals in native petroleum." In *The Role of Trace Metals in Petroleum,* T.F. Yen, Ed. Ann Arbor Science Publishers, Ann Arbor, Mich., pp. 1–30.

Zoller, W.H. (1983). "Anthropogenic perturbation of metal fluxes into the atmosphere." In Nriagu, J.O. Ed., *Changing metal cycles and human health.* Dahlem Konferenzen, Springer Verlag, Berlin.

Zubovic, P. (1966). "Physiochemical properties of certain minor elements as controlling factors of their distribution in coal." *Adv. Chem. Ser.* **66,** 221–246.

2

ATMOSPHERIC TRACE ELEMENTS FROM NATURAL AND ANTHROPOGENIC SOURCES

Jozef M. Pacyna

Norwegian Institute for Air Research
Lillestrøm, Norway

1. INTRODUCTION

Emission inventories are the key to effective air quality management programs, serving as a basis for the planning of control strategies to achieve ambient air quality goals. So far, air pollution authorities have mainly focused on gaseous pollutants, such as SO_2 and NO_x, and total particulate matter. Increasing energy demands being met by a greater use of fossil fuels, particularly coal (World Bank, 1979; Wilson, 1980; UN, 1981; Häfele, 1981), have resulted in more environmental hazards from trace elements.Improved analytical methodology in the 1970s provided a higher quality and quantity of data on trace metals, making it possible to assess the impact of these pollutants on our health. Recently, a number of papers has appeared in print dealing with many aspects of the atmospheric cycles of trace metals.

A broad literature review on the fluxes of the metals into the atmosphere is presented in this chapter.

2. NATURAL SOURCES OF TRACE ELEMENTS

An accurate presentation of natural source strengths is both important and difficult to assess. The importance stems from the fact that for many elements, natural emissions exceed those from anthropogenic sources. Accurate assessment of the amounts of trace elements emitted from natural sources is hampered by the lack of reliable information and measurements. This is especially true when considering natural emission sources, such as vegetative exudates or forest wildfires. Very few worldwide emission surveys of trace elements from natural sources are found in the literature. The results are summarized in Table 2.1.

Numbers in brackets show the range of emissons found in the literature. The numbers over the ranges are regarded by the author as the most acceptable. Large discrepancies exist between estimates made by several authors. These differences are mainly due to the variety of parameters influencing the trace element releases, such as the trace element concentrations in rocks, soils and dust, and global production of dust. A comprehensive review of trace element concentrations in several environmental media has been made by Bowen (1966). From among natural sources of trace elements in Table 2.1, windblown dusts and volcanic eruption deserve special attention. These sources release the largest amounts of trace elements compared to other sources in Table 2.1. Furthermore, extensive studies on contamination of the environment by windblown dust and volcanoes make these sources relatively well known. During recent years, several cases of a substantial long-range transport of the Saharan dust have been documented by direct measurements (SCOPE, 1979). Rough estimates show that the Saharan area may contribute annually (60 to 200) $\times 10^6$ t of soil dust to the troposphere. Rahn et al. (1979) found that the total concentrations of the Sahara aerosol were higher than expected and surprisingly constant. The composition of the Saharan aerosol was even more constant than its total airborne concentration. This constancy of the Saharan aerosol makes the trace elements amounts transported with the dust easier and more reliable to assess. Rahn et al. conclude that the Saharan aerosol has a composition close to that of the average crust. Long-range transport of Asian dust was also studied by Rahn et al. (1979). The mass of Asian desert dust transported into the Arctic seemed to be very great and

Table 2.1. Worldwide Emissions of Trace Elements from Natural Sources

Source	Global Production[a] ($\times 10^{-9}$) (kg/y)	Annual Emission ($\times 10^{-6}$) (kg) As[b]	Cd[c]	Co[d]	Cu[e]	Cr[f]	Mn[g]	Ni[h]	Pb[i]	Se[j]	V[k]	Zn[l]	Hg[m]
Windblown	6–1100	0.24	0.25	4	12	50	425	20	10	0.3	50	25	0.03
dust			(0.05–0.85)		(0.14–26)		(200–425)	(0.2–44)	(3–35)		(30–50)	(25–80)	(0.005–0.035)
Volcanogenic	6.5–150	7	0.50	1.4	4	3.9	82	3.8	6.4	0.1	6.9	10	0.03
particles		(0.3–39)	(0.04–7.8)		(2.3–54)		(9–82)	(2.4–56)	(0.4–96)	(0.1–0.4)	(0.7–6.9)	(4.6–105)	(0.01–0.24)
Forest wild-	2–200	0.16	0.01	—	0.3	—	—	0.6	0.5	—	—	0.5	0.1
fires			(0.01–1.5)		(0.025–1.7)			(0.05–3.3)	(0.04–6.8)			(0.2–12)	(0.04–0.16)
Vegetation	75–1000	0.26	0.2	—	2.5	—	5	1.6	1.6	—	0.2	10	—
			(0.05–2.7)		(2.5–33)			(1.6–21)	(0.2–21)			(9.4–125)	
Seasalt	300–2000	0.14	0.002	—	0.1	—	4	0.04	0.1	—	9	0.02	0.003
			(0.001–0.4)		(0.02–0.2)			(0.01–0.05)	(0.001–5)			(0.01–0.09)	(0.001–0.006)
Total		7.8	0.96	5.4	18.9	8.9	516	26	18.6	0.4	66.1	4	0.16

[a] From Nriagu (1979), Peterson and Junge (1971), Study of Man's Impact On Climate (1971), and Jaenicke (1980).

[b] From Walsh et al. (1979), Lantzy and Mackenzie (1979), Gordon et al. (1983), and Phelan et al. (1982).

[c] From Nriagu (1979), Lantzy and Mackenzie (1979), Jaworowski et al. (1981), Gordon et al. (1983), and Weisel (1981).

[d] From Pacyna (1982a).

[e] From Nriagu (1979).

[f] From Pacyna (1982a).

[g] From Weisel (1981) and Pacyna (1982a).

[h] From Nriagu (1979).

[i] From Nriagu (1979), Lantzy and Mackenzie (1979), Jaworowski et al. (1981), Gordon et al. (1983), and Weisel (1981).

[j] From Phelan et al. (1982), Pacyna (1982a), and Arnold et al. (1981).

[k] From Weisel (1981) and Pacyna (1981a).

[l] From Nriagu (1979), Lantzy and Mackenzie (1979), Gordon et al. (1983), Weisel (1981), and Phelan et al. (1982).

[m] From Lantzy and Mackenzie (1979), Jaworowski et al. (1981), Gordon et al. (1983), and Phelan et al. (1982).

Table 2.2. Comparison of Volcanic Emission Estimates from Individual Volcanoes[a]

	Cd		Zn		Se		As		Hg	
	($\times 10^{-3}$) (g/day)	($\times 10^{-9}$) (g/year)	($\times 10^{-3}$) (g/day)	($\times 10^{-9}$) (g/year)	($\times 10^{-3}$) (g/day)	($\times 10^{-9}$) (g/year)	($\times 10^{-3}$) (g/day)	($\times 10^{-9}$) (g/year)	($\times 10^{-3}$) (g/day)	($\times 10^{-9}$) (g/year)
Mt. St. Helens (USA)	5.0	0.06	1.000	10	100	1.11	800	8.9	80	0.9
Poas (Costa Rica)	2.4	0.03	40	0.54	7.6	0.10	3.0	0.04	76	1.0
Arenal (Costa Rica)	50	2.7	50	2.4	14	0.76	7.0	0.38	55	3.0
Colima (Mexico)	6.7	0.23	10	0.34	6.2	0.21	3.1	0.11		
El Chicon (Mexico)			100	1.11	17	0.19	845	9.4	88	2.2

[a] After Zoller (1983).

could be considered as a major contributor to the Arctic aerosol. Taking into account the composition of the Asian dust, the authors suggest that it completely alters the aerosol chemistry of that region when it enters. Thus, information on the trace element composition of the Saharan and Asian dusts enables one to estimate the emissions of these pollutants ascribed to windblown dusts. The information has been used for calculating data in Table 2.1.

Data on trace element emissions during volcanic eruptions are being collected by researchers around the world. A comparison of volcanic emission estimates found in the literature (after Zoller, 1983) is given in Table 2.2

In Europe, few measurements have been carried out around Mount Etna in Sicily. One of the investigations by Arnold et al. (1978) suggests total emission of 2.8×10^4 g/day, or 10×10^9 g/yr cadmium from this volcano. Releases of other elements from Etna are also high, relative to data in Table 2.2. Generally, the emission rates of individual volcanoes can vary to a very great degree, which makes detailed calculations rather difficult. The numbers for volcanic emissions in Table 2.1 referred as the most acceptable ones were achieved using the ratio of trace element fluxes to the volcanic sulfur flux for all volcanoes considered.

In certain parts of the world forest fires are the major emission sources, and an intense turbulence of the atmosphere in desert regions can result in very high concentrations of soil dust in air. Thus, it is important to know how large these emissions are. The values in Table 2.1 have been based on emission factors presented in Chapter 1 (Pacyna, 1983a) and total yearly ash production of 36×10^{12} g (Nriagu, 1979). The investigations on airborne lead and other elements from local fires in the Himalayas (Davidson et al., 1981) were also considered when calculating emissions of lead and copper.

Other possible natural processes which could be important for some of the metals include biological mobilization (including methylation) and airborne seasalt. Two seasalt production processes have been considered as natural sources of trace elements: (1) "bubble bursting" (Cd, Cu, Ni, Pb, and Zn) and (2) gas exchange (As). The calculations for bubble bursting have been done using the trace element concentrations in surface ocean waters and the enrichment of trace elements in the atmospheric seasalt particles. The potential significance of arsenic vapor emissions is usually assessed from ambient measurements of arsenic in marine aerosols.

3. ANTHROPOGENIC SOURCES OF TRACE ELEMENTS

Energy generation, industrial metal production, and vehicular traffic have brought about a serious increase in trace element emissions. To compare these emissions with trace element releases from natural sources, one must take into account the scale of pollutant perturbations. For some trace elements, such as Se, Hg, and Mn, global natural emissions exceed total releases from anthropogenic sources. However, Mn emission from man-made sources in Europe is several times larger than that from natural sources in this part of the world. In this paper, three geographical scales of trace element emissons are considered, namely, global (worldwide), regional, and local.

3.1. Global Emissons of Trace Elements from Anthropogenic Sources

Global emissions of trace elements from anthropogenic sources have been calculated using emission factors presented in Chapter 1 (Pacyna, 1983a) and statistical information on the world consumption of ores, rocks, and fuels and on the world production of various types of industrial goods. The results for several trace elements are presented in Table 2.3. Data on emissions of Cd, Cu, Ni, Pb, and Zn are from Nriagu (1979), As from Walsh et al. (1979), and Se from NAS (1976).

The main objectives of global emission calculations are (1) to assess the current emission situation in the world, such as determining the contribution from the various emitter categories, (2) to predict changes in the emission

Table 2.3. Worldwide Anthropogenic Emissions of Trace Elements during 1975[a] (10^{-9}) (g/yr)

Source	As[b]	Cd	Cu	Ni	Pb	Se[c]	Zn
Mining, nonferrous metals	0.013	0.002	0.8	—	8.2	0.005	1.6
Primary nonferrous metal production	15.2	4.71	20.8	9.4	76.5	0.28	106.7
Secondary nonferrous metal production	—	0.60	0.33	0.2	0.8	—	9.5
Iron and steel production	4.2	0.07	5.9	1.2	50	0.01	35
Industrial applications	0.02	0.05	4.9	1.9	7.4	0.06	26
Coal combustion	0.55	0.06	4.7	0.7	14	0.68	15
Oil combustion (including gasoline)	0.004	0.003	0.74	27	273	0.06	0.1
Wood combustion	0.60	0.20	12	3.0	4.5	—	75
Waste incineration	0.43	1.40	5.3	3.4	8.9	—	37
Manufacture, phosphate fertilizer	2.66	0.21	0.6	0.6	0.05	—	1.8
Miscellaneous	—	—	—	—	5.9	—	6.7
Total	23.6	7.3	56	47	449	1.1	314

[a] After Nriagu (1979).

[b] After Walsh et al. (1979).

[c] After NAS (1976).

patterns in view of the addition of new air pollution sources, and (3) to design and manage air quality monitoring networks. Thus, the global emission inventories show that anthropogenic Be, Co, Mo, Sb, and Se are chiefly emitted from coal combustion.

On the other hand, Ni and V are released mainly during oil firing. Smelters and secondary nonferrous metal plants emit the largest amounts of As, Cd, Cu, and Zn when compared with other sources. Cr and Mn are released mainly from factories producing iron, steel, and ferroalloys. Finally, Pb enters the environment primarily as a result of gasoline combustion.

The prediction of changes in the emission patterns are closely related to changes in energy generation and industrial activities (McCarroll, 1980; Lincoln and Rubin, 1980). The projected increased use of coal in many countries has raised the risk of increased metal emissions into the atmosphere. Replacement of older boilers and the introduction of new, more efficient control devices can be expected to reduce these emissions. New coal technologies, such as *in situ* gasification, coal pyrolysis, and chemical precleaning, may also decrease environmental contamination to some degree.

Future industrial growth will enhance the risk of health hazard due to increases of trace element emissions. Thus, this growth may only be feasible if (1) it occurs in areas not traditionally industrial or (2) sufficient emission reductions are achieved from other sources to permit new additional industrial emissions at existing locations.

Several international air quality monitoring networks are under operation (e.g., GEMS, OECD network, FAO network), and emission inventories prepared on a global scale were used to the network designs.

3.2. Inventories of Trace Element Emissions on a Regional Scale

Usually, "regional" effects refer to a scale of 10 1000 km (Nriagu, ed., 1983). Thus, emission inventories calculated for large industrial areas or particular countries can be regarded as inventories on a regional scale. Recently, emissions of the trace elements As, Be, Cd, Co, Cr, Cu, Mn, Mo, Ni, Pb, Se, Sb, V, Zn, and Zr to the atmosphere were estimated for all European countries (Pacyna, 1983b). The European emission of trace elements from various sources in 1979 are presented in Table 2.4, while emissions from all sources in the various countries are given in Table 2.5.

The data in Tables 2.4 and 2.5 were based on trace element emission factors and information on the consumption of ores, rocks, and fuels in Europe and on the production of various types of industrial goods. What is different from the application of trace element emisson factors in global emission inventories is that in the European survey emission factors were estimated for each of the European countries separately. The emission factors from Chapter 1 (Pacyna, 1983a) were used as a basis for further estimations. Then, in the case of fossil-fuel combustion, information was collected on (1) the trace element concen-

Table 2.4. European Emissions of Trace Elements from Various Sources in 1979 (t/yr)

Source	As	Be	Cd	Co	Cr	Cu	Mn	Mo	Ni	Pb	Sb	Se[a]	V	Zn	Zr
Conventional thermal power plants	284	21	101	787	1,196	1,377	1,011	352	4,580	1,138	122	155	12,600	1,316	732
Industrial commercial and residential combustion of fuels	394	29	155	1,214	1,580	2,038	1,378	493	7,467	1,652	154	218	21,800	1,824	939
Wood combustion	40		25			1,500			375	562				4,590	
Gasoline combustion			31				92		1,330	74,300					
Mining			1			192	275		1,640	1,090		0.2		460	
Primary nonferrous metal production															
Copper–nickel	4,490		595			7,850				9,250				2,500	
Zinc–cadmium	910		1,550			440				7,880		13		48,800	
Lead	300		8			120			140	10,450				180	
Second nonferrous metal production															
Copper			2			61				55	1			660	
Zinc														2,630	
Lead			1							387				150	
Iron, steel ferroalloys, manufacturing			58		15,400	1,710	10,770		340	14,660				10,250	
Refuse incineration	11		84	4	53	260	114		10	804	100	32	19	5,880	
Phosphate fertilizers			27			77			77	6		[b]		230	
Cement production			15		663					746					
Total	6,500	50	2,700	2,000	18,900	15,500	17,600	850	16,000	123,000	380	420	34,500	80,000	1,700

[a] Se emission with particles.

[b] Very small.

Table 2.5. Emissions of Trace Elements for All Sources in Europe in 1979 (t/yr)

Country	As	Be	Cd	Co	Cr	Cu	Mn	Mo	Ni	Pb	Sb	Se	V	Zn	Zr
Albania	31	0.1	1	3	5	71	1	1	92	134	0.4	0.5	43	72	2
Austria	103	0.2	137	22	200	134	182	6	184	1,933	1.1	4.5	552	4,370	7
Belgium	360	0.5	171	55	642	613	613	25	381	3,986	10.9	11.4	908	4,736	52
Bulgaria	152	1.4	67	47	161	208	218	22	291	2,234	7.5	9.7	701	1,722	45
Czechoslovakia	86	3.1	23	86	791	323	712	44	472	1,726	16.8	18.0	943	635	99
Denmark	7	0.1	9	23	50	38	37	6	185	753	12.2	3.8	596	706	5
Finland	127	0.2	84	23	115	246	109	7	237	1,621	1.2	4.1	565	2,460	8
France	228	1.4	170	103	1,095	450	1,192	34	903	10,545	30.3	18.0	2,338	6,127	51
German Democratic Republic	133	4.7	37	108	528	376	432	61	549	2,084	25.1	24.1	965	746	148
German Federal Republic	782	3.9	328	136	2,153	1,552	2,054	60	1,103	9,308	49.5	46.6	2,222	11,689	133
Greece	10	0.2	4	17	77	55	45	6	273	1,303	1.6	3.1	372	121	10
Hungary	34	0.6	8	24	198	509	160	10	162	888	3.4	4.6	389	280	20
Iceland[a]	73	—	81	378	336	514	—	84	4,130	36,300	—	53	10,900	264	—
Ireland	2	0.1	1	8	11	13	8	2	65	456	0.4	1	199	33	3
Italy	93	0.8	124	150	1,055	385	925	38	1,300	9,365	16	24	3,952	4,420	25
Netherlands	58	0.3	88	38	255	105	253	10	321	2,427	9.3	7.9	979	3,067	8
Norway	36	[b]	39	6	40	56	45	2	66	803	0.2	1.2	160	1,188	1
Poland	656	8.2	207	151	1,161	1,313	1,009	97	653	4,568	43.0	37.0	672	4,725	254
Portugal	7	[b]	3	10	27	29	20	2	97	525	0.1	1.4	268	39	1
Romania	35	2.4	13	61	619	228	554	33	338	1,827	12.7	13.1	660	614	76
Spain	302	0.9	126	61	571	565	427	20	510	5,534	4.8	10.9	1,373	3,255	30
Sweden	147	0.1	16	36	195	237	172	9	323	2,270	0.4	5.4	1,003	346	3
Switzerland	1	[b]	1	5	40	18	25	1	51	1,083	0.03	0.6	130	50	0
Turkey	62	1.1	17	30	147	427	126	15	277	1,180	6.0	5.1	419	994	39
USSR	2,812	15.0	816	631	7,147	6,535	6,874	257	6,014	43,842	80.0	120.0	11,262	21,281	471
United Kingdom	164	4	99	130	1,134	580	1,032	60	899	10,098	40.0	36.0	2,074	3,488	150
Yugoslavia	134	0.8	65	40	205	287	177	16	284	2,423	5.3	7.6	718	2,013	29
Luxemburg	0.4	[b]	1.1	1.3	196	24	192	0.4	15	301	0.6	0.2	30	158	0.5
Total	6,500	50	2,700	2,000	18,900	15,500	17,700	850	16,000	123,000	380	420	34,500	80,000	1700

[a] Data in kg/yr.

[b] Very small.

trations in coal and oil burned in several countries; (2) the ash contents of coal and sulfur contents of oil (the sulfur content of oil is closely related to the trace element emissions during oil combustion (Pacyna, 1982b)); (3) physico-chemical properties of metals, emphasizing the volatile nature of some trace elements; (4) combustion conditions with special attention paid to type of boiler; and (5) type and efficiency of control devices employed in power plants. This information was applied to recalculate "basic" factors for obtaining the values for the particular countries. Similar calculations of emission factors have been made for other sources, considering differences in industrial technologies employed in particular European countries, differences in the chemical composition of input materials in industry, types and efficiencies of fly ash control systems, etc.

It must be admitted that for certain sources information on emission factors was difficult to obtain. This applies especially to emissions from factories utilizing metals in the production process. As an example, cadmium is emitted from plants producing pigments, stabilizers, nickel–cadmium batteries, and alloys. The amounts of these releases are not included in Table 2.4 because the author believes that the trace element emissions from these sources are relatively small and can be neglected. Purposes of the trace element emission surveys, calculated on a regional scale, are similar to those for the global emissions. However, an additional aim of the former estimates is to use trace elements as tracers for aerosols transported at long distances, such as to the Arctic. Then, the spatial distributions of trace element emissions are necessary. An example of two different methods of spatial distribution presentations is shown in Figures 2.1–2.3. The spatial distributions of vanadium and beryllium emissions in Europe are given in Figures 2.1 and 2.2, respectively (Pacyna, 1983b). The EMEP grid system (150 km by 150 km) was used (Dovland and Saltbones, 1979). The shaded areas in Figures 2.1 and 2.2 represent locations with emissions higher than the average for Europe. The second method of the spatial distribution presentation shows maps of places with only extremely high emissions. The spatial distributions of arsenic, cadmium, copper, chromium, zinc, and manganese by the second method are presented in Figure 2.3. All distributions were estimated from information on location and capacity of power plants and industrial factories (Pacyna, 1983b). In addition, any available data from the literature were used, such as distribution of European SO_2 and NO_x emissions from fossil-fuel combustion, population density, and location of industrial areas. As can be seen from these figures, there are several areas with concentrated emission sources in Europe. The three major emission areas are (1) the Soviet Union (quite far from western Europe), (2) Poland and Czechoslovakia, and (3) the Benelux countries and the western part of the FRG. In the case of zinc and vanadium, significant emission areas are also located in the United Kingdom, Spain, and Italy. Generally, zinc and vanadium emissions are distributed very evenly in Europe, whereas emissions of arsenic, cadmium, copper, chromium, manganese, and beryllium are rather concentrated in Central and Eastern Europe.

The spatial distributions of trace element emissions in Europe, supplemented by wind trajectories, are now used to analyze long-range transport of pollutants from Europe to the Arctic.

The trace element emission surveys are gradually becoming available in the literature for particular countries or groups of countries. Special attention has been paid to emissions of lead and cadmium (U.S. EPA, 1977; Van Enk, 1980; U.S. EPA, 1981; SNV, 1982; Hutton, 1982). The emission levels and toxicity of both elements are high. Lead and cadmium are very volatile elements, having particularly high enrichment factors in ambient aerosols relative to the earth's crust.

3.3. Inventories of Trace Element Emissions on a Local Scale

Local scale of emission surveys covers the effluents of pollutants from single-point sources, such as power plants and individual factories. Trace element emisson data, collected for particular-point sources, are mainly based on source measurements. Two currently available systems can be utilized to measure emissions from stationary sources: (1) the extractive system and (2) the in-site system. Other procedures of data collection include questionnaire forms and emission factors which are broadly used to calculate emission on global and regional scales play less important role.

Inventories of trace elements on the local scale are being used to quantify the consequences of pollutant emissions in the environment around the emission source. Several models, having emission values as an input data, can then be employed to assess the environmental pathways of trace elements in the vicinity of a certain plant or factory. These migration models are often applied to calculate human exposure to trace elements, and then to analyze dose-response relationships. There are several emission inventories for single-point sources described in the literature. As an example, recently, a comprehensive study of a coal-fired power plant as a source of environmental contamination by trace elements was conducted in Poland (Pacyna, 1980). Field measurements were conducted from 1976 to 1980 in a 2000-MWe lignite-fired power plant. During the first part of investigations inside the power station, the flow rates of flue gas and ash were determined at sampling sites located along three streams, namely, bottom ash, fly ash, and stack ash. The trace element concentrations were measured in samples taken at the same locations. The chemical composition of coal burned was also analyzed. The rates of stream flows and concentrations of trace elements in coal and ashes were used to establish the mass balance of the pollutants in the power plant. This balance is presented in Table 2.6. For the majority of elements from Table 2.6, total amounts in bottom ash, fly ash, and stack ash did not equal the amount of the pollutants introduced with coal. This imbalance is caused either by the behavior of trace elements during coal combustion or simply by inaccuracy of measurements. The largest imbalances were found for mercury

EMISSIONS OF V

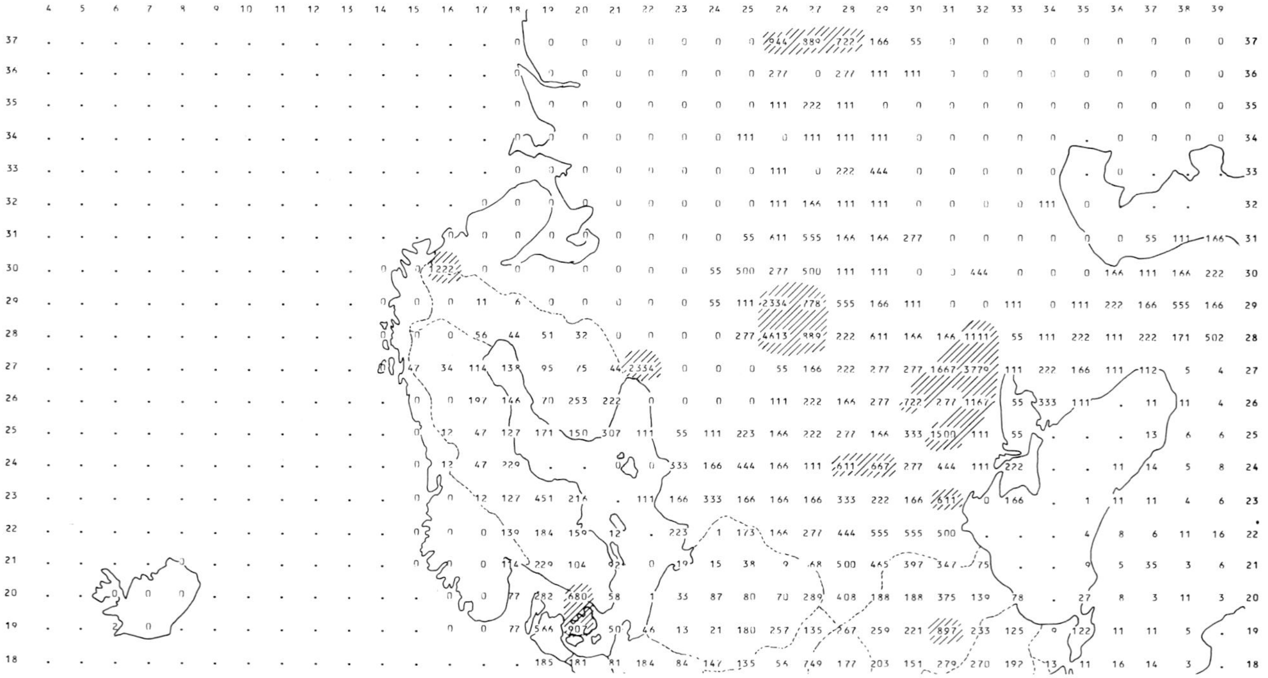

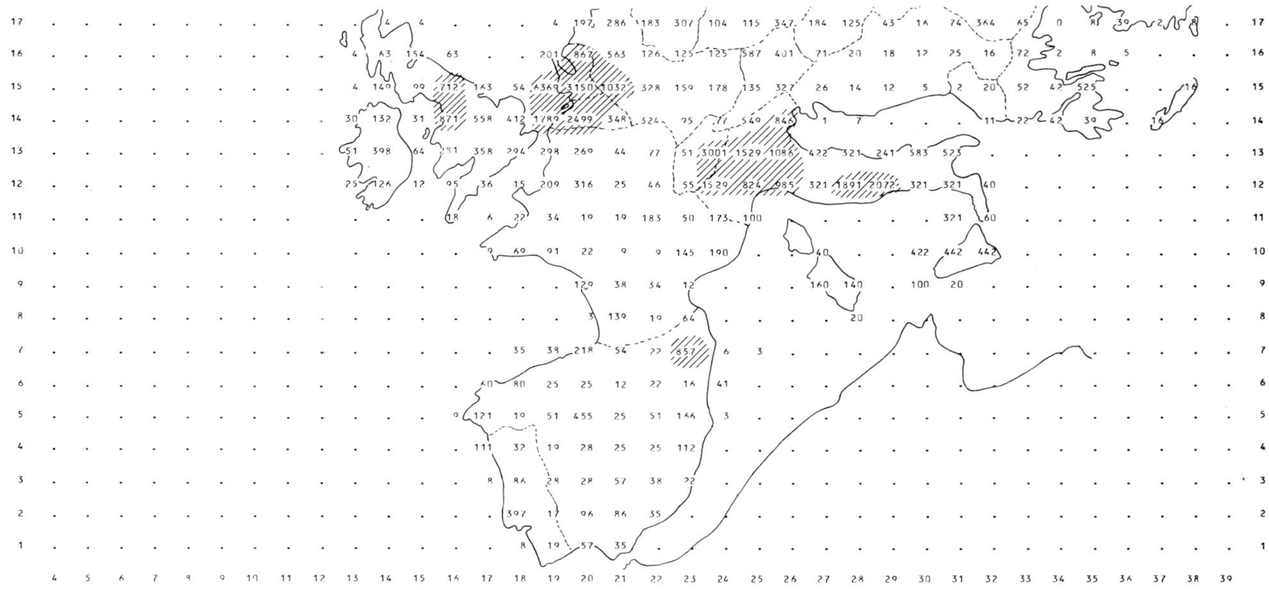

Figure 2.1. Spatial distribution of vanadium emission in Europe in 1979 in 10^2 kg/yr. The shaded areas represent locations with emissions higher than the average for Europe.

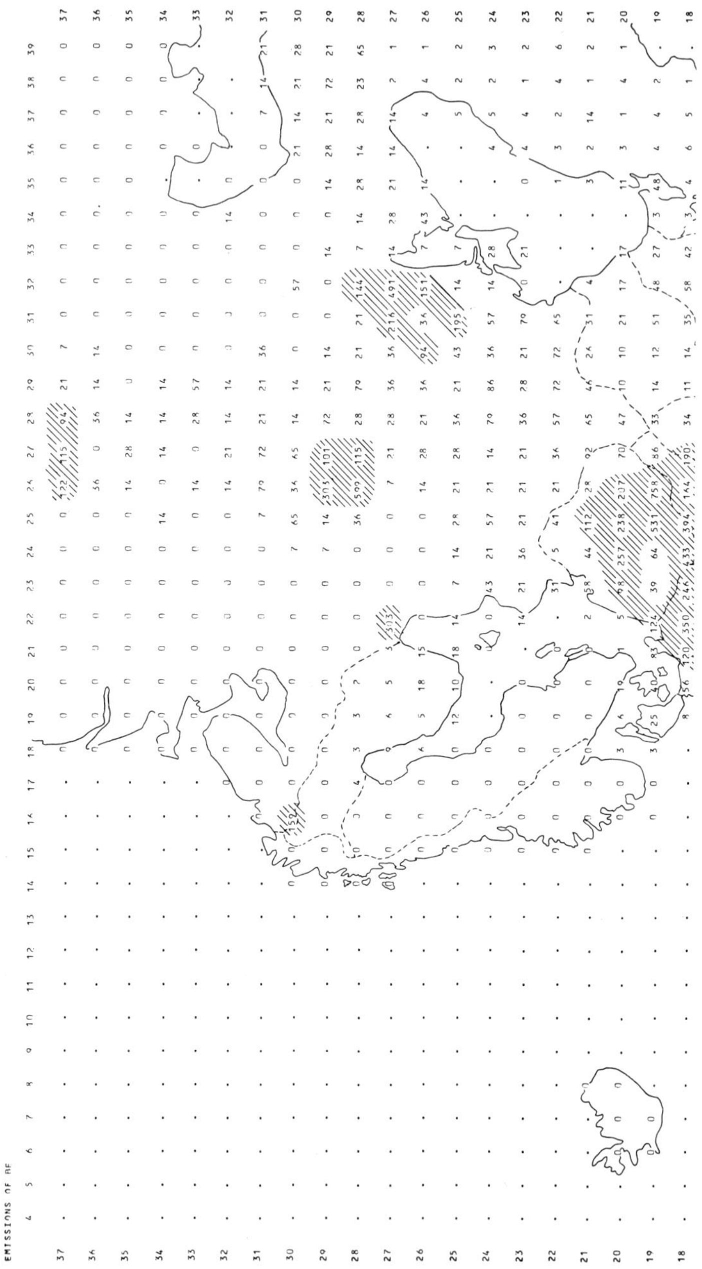
EMISSIONS OF HF

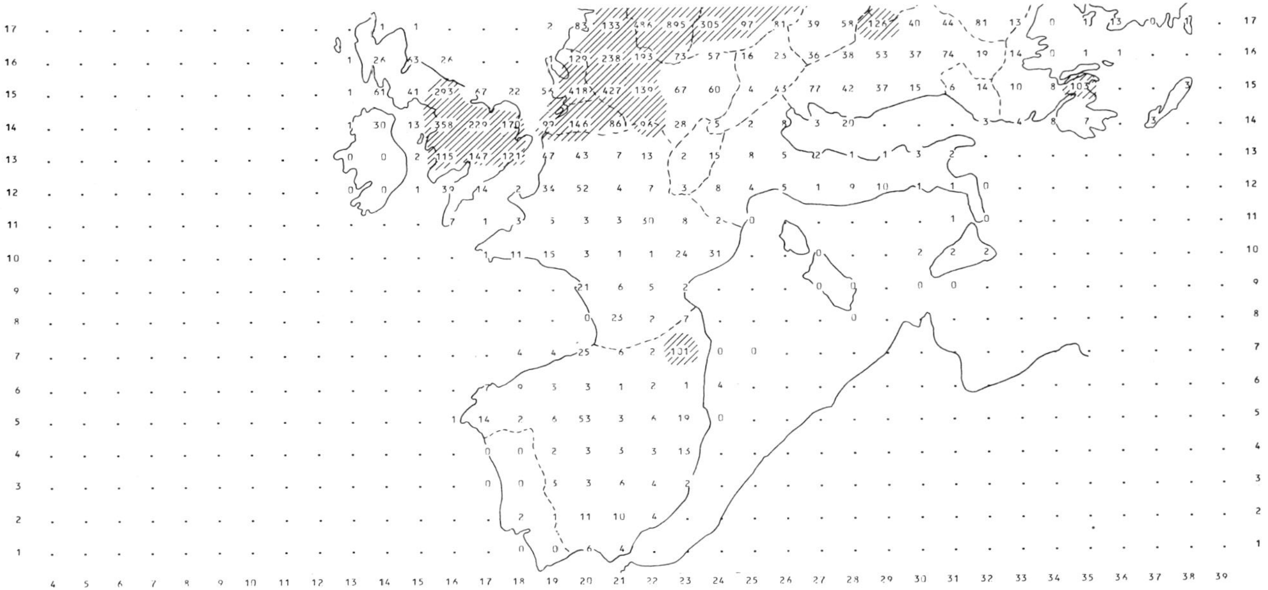

Figure 2.2. Spatial distribution of beryllium emission in Europe in 1979 in kg/yr. The shaded areas represent locations with emissions higher than the average for Europe.

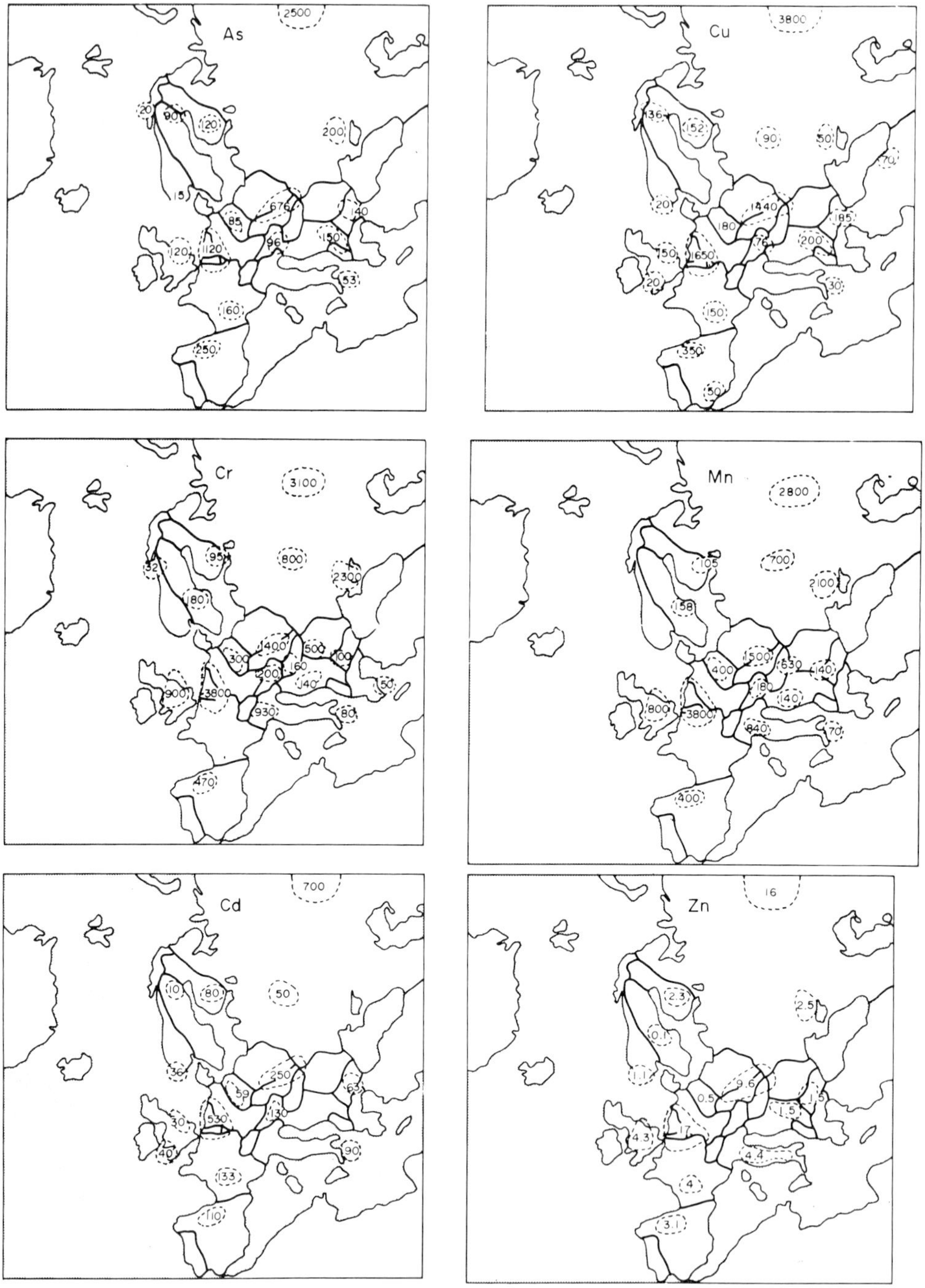

Figure 2.3. Spatial distributions of the European emissions of arsenic, copper, chromium, manganese, cadmium, and zinc. The As, Cu, Cr, Mn, and Cd emissions are in t/yr and the Zn emission in 10^3 t/yr.

Table 2.6. Trace Element Emission Rates by a 2000-MWe Lignite-Fired Power Plant in Poland[a]

Elements	Coal	Bottom Ash	Fly Ash	Stack Ash	Imbalance	Emission (kg/day)
Al (kg/min)	495.0	136.5	367.2	3.8	−12.5	5500
As (g/min)	2948.0	432.2	2335.8	53.8	+126.2	77.5
Ca (kg/min)	484.0	205.0	266.2	3.2	+9.6	4600
Cu (g/min)	2585.0	491.4	2205.2	32.8	−142.4	47.2
Fe (kg/min)	385.0	91.0	265.2	2.7	+26.1	3900
Hg (g/min)	4.4	0.0	0.2	0.0	+4.2	
Mg (kg/min)	99.0	31.8	67.3	0.0	−0.2	
Mo (g/min)	1078.0	134.2	938.4	29.1	−23.7	41.9
Na (kg/min)	99.0	18.2	81.6	1.1	−1.9	1600
Nb (g/min)	374.0	91.9	263.1	2.6	+16.4	3.7
Pb (g/min)	814.0	112.8	637.2	37.0	+26.5	53.3
Rb (g/min)	1364.0	455.0	979.2	9.6	−79.8	13.8
Sb (g/min)	385.0	69.2	298.9	6.8	+10.1	9.8
Se (g/min)	418.0	40.5	127.5	2.9	+247.1	4.2
Sr (kg/min)	18.1	5.4	11.4	0.1	+1.2	138
Y (g/min)	1171.5	389.5	887.4	8.5	+113.9	12.2
Zn (g/min)	2546.5	369.0	2305.2	53.1	−202.7	76.5
Zr (g/min)	4840.0	1638.0	3366.0	30.0	−194.0	43.2

[a] From Pacyna (1980).

and selenium. The results in Table 6 suggest that 95% of mercury and 59% of selenium enter the atmosphere either as vapors or as very fine particles. The last column in Table 2.6 presents daily emission of trace elements from the 2000-MWe lignite-fired power plant in Poland.

Very high emissions of arsenic in Table 2.6 are due to extremely high concentrations of this element in the lignites mined in southern Poland and

Table 2.7. Annual Human Exposure to As, Cd, and Pb through Ingestion in the Area of the 2000-MWe Power Plant in Poland[a]

Element	Unit	Estimated Exposure Around the Power Plant		Literature Values[b]	Effects Level[b]
		Adults	Children		
As	μg/kg of body	14.0	11.4	~10.0	15–340
Cd	μg/g of kidneys	60.0	52.8	14.0–16.0	~200
Pb	μg/dL of blood	16.7	14.5	10.0–13.0	10.0–60.0

[a] From Pacyna and Silvertsen (1981).

[b] From Bennett (1981).

northern Czechoslovakia. These concentrations are the highest noted anywhere in the world. The emission data from Table 2.6 were then used as inputs in dispersion models to calculate distributions of trace element concentrations in the air, and in migration models to assess contamination of particular environmental media. Finally, a simple, quasi-stationary compartment model was applied to estimate the level of human exposure to some of the measured elements (Bennett, 1981). The results for As, Cd, and Pb are shown in Table 2.7. In this area, inhalation contributes to human exposure about 23% Pb, 11% Cd, and 2% As.

4. SUMMARY

This broad literature review shows that in the field of trace element emission inventories much more research should be done. In the case of natural sources, measurements are rather scarce. The majority of details are available from more or less accurate estimates. The accuracy of calculations differs depending on the type of source and the elements in question.

Emission surveys of trace elements from anthropogenic sources are frequently described in the literature, with measurements and calculations done either for a certain country, for a given element, or for an individual source. So far, special attention has been paid to the elements lead and cadmium. Estimates for arsenic, mercury, and vanadium have also become available. Fuel combustion and metal smelting are the most investigated types of emission sources.

Future investigations should improve the accuracy of existing inventories and add to the list other important pollutants, such as Se, Sb, Sn, Ag, Ge, and Tl. These elements are volatilized during industrial processes and thus represent a significant health concern. Sources that require more attention include waste incinerators, biomass burning, and cement manufacturing.

Other problems for future research concern trace element emission surveys for fine particles and chemical speciation of trace elements, emphasizing their organometallic forms, physicochemical associations, and the interactions between individual pollutants during the emission process.

ACKNOWLEDGMENTS

The author thanks Dr. Val Vitols for helpful discussions.

REFERENCES

Arnold, M., Buat-Menard, P., and Chesselet, R. (1981). "An estimate of the input of trace metals to the global atmosphere by volcanic activity." International Association of Meteorology and Atmospheric Physics, Third Scientific Assembly, Hamburg, FRG.

Bennett, B.G. (1981). *Exposure commitment assessments of environmental pollutants,* Vol. 1, No. 1. MARC Report 23. Monitoring and Assessment Research Centre, Chelsea College, University of London, London.

Bowen, H.J.M. (1966). *Trace elements in biochemistry.* Academic Press, London, New York, San Francisco.

Buat-Menard, P., and Arnold, M. (1978). "The heavy met;al chemistry of atmospheric particulate matter emitted by Mount Etna volcano", *Geophys. Res. Lett.* **5,** 245–248.

Davidson, C.I., Grimm, T.C., and Nasta, M.A. (1981). "Airborne lead and other elements derived from local fires in the Himalayas", *Science* **214,** 1344–1346.

Dovland, H., and Saltbones, J. (1979). *Emissions of sulphur dioxide in Europe in 1978.* NILU/EMEP/CCC-Report 2/79. The Norwegian Institute for Air Research, Lillestrøm, Norway.

van Enk (1980). "The pathway of cadmium in the European community." *European Appl. Res. Rept. Environ. Nat. Res.* Sect. **1.**

Gordon, G.E., Moyers, J.L., Rahn, K.A., Gatz, D.F., Dzubay, T.G., Zoller, W.H., and Corrin, M.H. (1983). "Atmospheric trace elements: cycles and measurements." *Rev. Geophys. Space Phys.* (in press).

Häfele, W. (1981). *Energy in a finite world: a global systems analysis.* Ballinger, Cambridge, Mass.

Hutton, M. (1982). *Cadmium in the European community.* MARC Report 26. Monitoring and Assessment Research Centre, Chelsea College, University of London, London.

Jaenicke, R. (1980). "Atmospheric aerosols and global climate." *J.Aerosol Sci.* **11,** 577–588.

Jaworowski, Z., Bysiek, M., and Kownacka, L. (1981). "Flow of metals into the global atmosphere." *Geochim. Cosmochim. Acta* **45,** 2185–2199.

Lantzy, R.J., and Mackenzie, F.T. (1979). "Atmospheric trace metals: global cycles and assessment of man's impact." *Geochim. Cosmochim. Acta* **43,** 511–523.

Lincoln, D.R., and Rubin, E.S. (1980). "Air pollution emissions from increased industrial coal use in the Northeastern United States." *J.Air Pollut. Control Assoc.* **30,** 1310–1315.

McCarroll, J. (1980). "Health effects associated with increased use of coal." *J.Air Pollut. Control Assoc.* **30,** 652–656.

National Academy of Sciences, National Research Council (1976). *Selenium.* PB-251. Subcommittee on Selenium, Committee on Medical and Biological Effects of Environmental Pollutants, Washington, D.C,

Nriagu, J.O. (1979). "Global inventory of natural and anthropogenic emissions of trace metals to the atmosphere." *Nature* **279,** 409–411.

Nriagu, J.O., ed. (1983). *Changing metal cycles and human health.* Dahlem Konferenzen, Springer Verlag, Berlin.

Pacyna, J.M. (1980). "Coal-fired power plant as a source of environmental contamination by trace metals and radionuclides." *Habilitation Monography No. 17–47.* Technical University of Wroclaw, Wroclaw, Poland.

Pacyna, J.M. and Sivertsen, B. (1981). *Determination of human exposure using measured data of Cd, As, and Pb.* NILU Report 15/81. The Norwegian Institute for Air Research, Lillestrøm, Norway.

Pacyna, J.M. (1982a). *Emission factors of trace elements.* MARC Report. Monitoring and Assessment Research Centre, Chelsea College, University of London, London.

Pacyna, J.M. (1982b). *Estimation of emission factors of trace metals from oil-fired power plants.* NILU Report 2/82. The Norwegian Institute for Air Research, Lillestrøm, Norway.

Pacyna, J.M. (1983a). "Emission factors of atmospheric elements." In Nriagu, J.O., ed., *Toxic metals in the air, Environmental Science and Technology Series,* John Wiley & Sons, Inc., New York.

Pacyna, J.M. (1983b). *Trace element emission from anthropogenic sources in Europe.* NILU Report 10/82. The Norwegian Institute for Air Research, Lillestrøm, Norway.

Peterson, J.T., and Junge, C.E. (1971). *Man's impact on the climate.* Matthew et al., ed. MIT Press, Cambridge, Mass.

Phelan, J.M., Finnegan, D.L., Ballantine, D.S., Zoller, W.H., Hart, M.A., and Moyers, J.L. (1982). "Airborne aerosol measurements in the quiescent plume of Mount St. Helens: September, 1980." *J.Geophys.Res.* **9,** 1093–1096.

Rahn, K.A., Borys, R.D., Shaw, G.E., Schütz, L., and Jaenicke, R. (1979). "Long range impact of desert aerosol on atmospheric chemistry: two examples." In *Saharan dust, SCOPE 14.* John Wiley & Sons, Inc., New York.

SCOPE (1979). *Saharan Dust. Scientific Committee on Problems of the Environment.* John Wiley & Sons, Inc., New York.

SMIC (1971). *Inadvertent climate modification. Report of the Study of Man's Impact on Climate.* MIT Press, Cambridge, Mass.

SNV (1982). "Tungmetaller och organiske miljögifter i svensk natur" (Heavy metals and organic compounds in the Swedish environment). *Monitor* **1982.** Statens naturvårdsverk, Solna, Sweden.

United Nations (1981). *The coal situation in the ECE region in 1979 and its prospects.* Report ECE/COAL/57. United Nations, New York.

U.S. Environmental Protection Agency (1977). *Air quality criteria for lead.* U.S. EPA Report 600/8-77-017, Washington, D.C.

U.S. Environmental Protection Agency (1981). *Health assessment document for cadmium.* U.S. EPA Report 600/8-81-023. Research Triangle Park, N.C.

Walsh, P.R., Duce, R.A., and Fasching, J.L. (1979). "Considerations of the enrichment, sources, and flux of arsenic in the troposphere." *J.Geophys. Res.* **84,** 1719–1726.

Weisel, C.P. (1981). "The atmospheric flux of elements from the ocean." Ph.D. Thesis, University of Rhode Island, 174 pp.

Wilson, C. (1980). *Coal bridge to the future: report of the world coal study.* Ballinger, Cambridge, Mass.

World Bank (1979). *Coal development potential and prospects in the developing countries.* World Bank, Washington, D.C.

Zoller, W.H. (1983). "Anthropogenic perturbation of metal fluxes into the atmosphere." In Nriagu, J.O., ed., *Changing metal cycles and human health,* Dahlem Konferenzen, Springer Verlag, Berlin.

3

SAMPLING AND MEASUREMENT OF TRACE ELEMENT EMISSIONS FROM PARTICULATE CONTROL DEVICES

A.D. Shendrikar

CompuChem Laboratories
Research Triangle Park, North Carolina

D.S. Ensor

Center for Aerosol Technology
Research Triangle Institute
Research Triangle Park, North Carolina

1. INTRODUCTION

1.1. Statement of Problem

Recent studies of air pollutant problems indicate that fossil fuel combustion in utility boilers is a major source of trace elements to our environment (Goldberg, 1972; Magee, 1976; Malte, 1977; Smith, 1980). The environmental impact of trace element emissions is of concern because they are emitted from stacks as fine particles and vapors which intensify the potential adverse health effects for the following reasons:

1. Fine particles containing trace elements penetrate the natural filters of the respiratory tract and the air spaces of the lung (Task Group of Lung Dynamics, 1966).

2. Fine particle emissions, which are believed to be largely formed by condensation (Flagen and Friedlander, 1976; Natusch et al., 1974; Linton et al., 1976) of volatile minerals within the gas passages of the boiler, are enriched in toxic elements such as As, Se, Ag, Cd, and Cr in comparison to the average composition of the earth's crust (a measure of the acceptable metabolic tolerance levels in humans).
3. Trace elements in the form of fine particles are also capable of serving as nuclei which may initiate gas-to-particle conversion reactions, thereby increasing the concentration of fine particles in the atmosphere.

Airborne trace elements are essentially due to (1) dust storms, (2) volcanic eruptions, (3) evaporation of ocean spray, and (4) forest fires and human activity on this planet. Anthropogenic trace element sources are incineration, chemical processes, smelting of ores, and primarily oil and coal combustion to generate steam and electricity. In this chapter, the emissions of trace elements due to coal combustion are of concern; however, the technical discussion included may be applied with slight modifications to other fuel combustion used for energy production and hazardous waste incineration impact evaluations. In addition, the chemical and physical properties of the emission source are believed to be important in source/receptor modeling. Accurate elemental data as a function of particle size is often the limiting information in receptor modeling (Watson, 1981).

Coal is inherently a dirty fuel. It contains mostly fossilized organic matter and a wide range of inorganic minerals derived from the earth's crust in quantities and proportions relative to its nature and sources (Swain, 1976; Ruch et al., 1963). Although to date the exact nature of coal is not fully known, there is a large body of data on its chemical composition (Francis, 1961; Ode, 1963; Walt, 1968). Comparatively less is known about the chemical forms in which inorganic minerals are found in coal (Filby et al., 1978). Such information is considered important for understanding the behavior of inorganic minerals during coal combustion (Kaakinen, 1974). To generate steam or electricity, coal is combusted at a high temperature (2000–3000° F) in the fire box of utility boilers. This process releases inorganic minerals as fly ash particles, most of which are collected by the control equipment; however, a small fraction, about 2%, is emitted into the atmosphere in the form of fine particles and as vapors.

The usual classification of particulate control devices include electrostatic precipitators with a wide range of design temperatures, fabric filters, and wet scrubbers. For further description of these technologies and their operating principles, the reader should consult standard references such as the ones by Billings and Wilder (1971) for fabric filtration, Yung et al. (1977) for wet scrubbers, and McDonald (1978) for electrostatic precipitators. Each of these control devices has a unique particle-size-dependent removal efficiency. This removal efficiency coupled with the enrichment of the fine particles with respect to certain toxic elements offers intriguing possibilities as far as the enrichments by the control devices are concerned (Klein et al., 1975; Natusch et al., 1974) because the minimum control efficiency coincides with the maximum enrichment. In addition, wet scrubbing often uses recycled water; thus, water-soluble elements are enriched in the water phase and may be introduced into the flue gas (Maddalone et al., 1981; Ondov et al., 1979).

The importance of fine particle emissions must be quantified through some methodology for assessment of potential risk similar to that developed for total emissions, as reported by Cleland and Kingsbury (1977) and Duke et al. (1977). A complication recognized with the chemistry of fine particles is that the particle composition may depend on the size and the combustion conditions. In addition, some trace elements may be in oxidation states that exhibit high vapor pressures under the process conditions and may exist either as a vapor or as particles.

The potential impact of coal combustion as a source of environmental pollution is significant. At an estimated rate of coal consumption of 700×10^9 kg/yr in utility boilers in the United States (Smith, 1980) and an assumed 10% average ash content of coal, assuming an 80% carryover as fly ash, the yield of uncontrolled fly ash emissions would be approximately 56×10^9 kg/yr. If such emissions are controlled by the use of a particulate control device operating at 98% efficiency, even then the particulate emissions would amount to 1.0×10^9 kg. For a trace element present in coal at 1 ppm level and assuming 98% efficiency of the control device, stack emissions of the trace element would be aproximately 1000 kg/yr. If this element is mercury, where the efficiency of the particulate control device is poor (Diehl et al., 1972; Billings and Matson, 1972) and 98% of the coal-bound mercury is emitted into the atmosphere, then the 1 ppm component would result in stack emissions of 55,000 kg/yr.

1.2. Review Focus and Chapter Organization

This chapter includes the discussion of sampling and measurement of inorganic species in utility boiler flue gas. The total and size-dependent trace element emissions due to coal combustion are considered. The processes occurring during coal combustion through fly ash particle formation, volatilization, and the fate and distribution of trace elements in the process streams, as relevant to sampling and analysis, are also considered. The physical and chemical properties of major and trace elements in coal and fly ash are also covered in this chapter. Often, it is desirable to compute mass balances to determine relative flows of the elements through various combustion processes, which requires source sampling and sample analysis protocols.

Most of the discussion relates to experiences in sampling the processes associated with coal combustion; however, it is believed that this experience and technical discussion may offer some guidance to other applications and combustion processes where trace element emissions occur (e.g., waste incineration, retorting of oil shale, and tar sands).

This chapter also includes analytical protocol requirements for the characterization of various kinds of samples with simple to complex matrices. Popularly used analytical methods are discussed, and some data on their applicability, limitations, and validations are also given. A few emerging analytical methods that may be used for fly ash particle surface characterization in terms of trace element chemical speciation are pointed out.

The chapter organization includes discussion of trace element emission problems, examination of the coal combustion processes leading to fly ash, trace elements and vaporous trace element formation, sampling point sources in a typical power plant, sample analysis, data interpretation, and associated quality-assurance requirements to meet program objectives.

2. BACKGROUND

2.1. Pathways of Trace Elements through the Process

A typical coal-fired steam-electric generating plant comprises a boiler, economizer, fuel-handling equipment, dust collection and disposal equipment, water handling and treatment facilities, heat recovery systems, particulate control devices, stack, etc. A flow diagram of a full-scale power plant with major process components is shown in Figure 3.1.

Inorganic minerals in coal vary widely in amount, composition, and source (see Table 3.1; Gluskoter, 1977). Some of the inorganic matter is assimilated during plant growth and is in its elemental form, but the major portion occurs as physically identifiable inclusions in the coal matrix (O'Gorman et al., 1972;

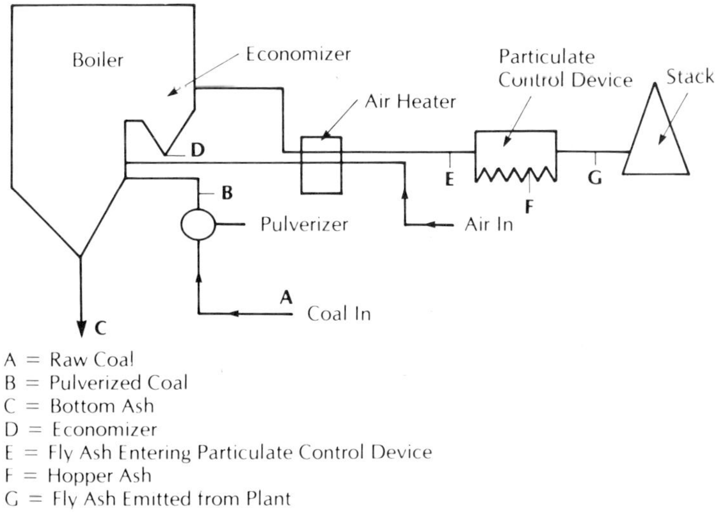

Figure 3.1. A flow diagram of a full-scale power plant with major process components (Shendrikar and Ensor, 1981).

Table 3.1. Selected Coal Sources and Composition[a]

	Western (28 samples)		Illinois (144 samples)		Appalachian (23 samples)	
	Arithmetic mean	Min:Max	Arithmetic mean	Min:Max	Arithmetic mean	Min:Max
Aluminum, %	1.0	0.3:2.2	1.2	0.43:3.0	1.7	1.1:3.1
Calcium, %	1.7	0.44:3.8	0.67	0.01:2.7	0.47	0.09:2.6
Chlorine, %	0.03	0.01:0.13	0.14	0.01:0.54	0.17	0.01:0.80
Iron, %	0.53	0.03:1.2	2.0	0.45:4.1	1.5	0.50:2.6
Potassium, %	0.05	0.01:0.32	0.17	0.04:0.56	0.25	0.06:0.68
Magnesium, %	0.14	0.03:0.39	0.05	0.01:0.17	0.06	0.02:0.15
Sodium, %	0.14	0.01:1.60	0.05	-:0.2	0.04	0.01:0.08
Silicon, %	1.7	0.38:4.7	2.4	0.58:4.7	2.8	1.0:6.3
Titanium, %	0.05	0.02:0.13	0.06	0.02:0.15	0.09	0.05:0.15
Moisture, %	18	4.1:137	9.4	0.5:18	2.7	1.0:6.8
Volatiles, %	44	33:53	40	27:46	33	17:42
Fixed carbon, %	46	35.55	49	41:61	55	45:72
Ash, %	9.6	4.1:20	11	4.6:20	12	6.1:25
Sulfur, %	0.76	0.34:1.9	3.6	0.56:6.4	2.3	0.55:5.0
Heat value, Btu/lb	11,409	10,084:12,901	12,712	11,562:14,362	13,111	11,374:13,816

[a] From Gluskoter et al. (1977).

Headlee and Hunter, 1955; Mezey et al., 1976). In addition, when the coal seam is mined, overburden material may be mixed with the coal. Finkelman (1970) found that minerals are dispersed in the coal with diameters ranging down into the submicrometer region. Flagen and Friedlander (1978) also reported that inorganic minerals are widely distributed in size with a mean diameter of 1 μm. In addition, clays with submicrometer particle sizes may also be present in the coal.

In general, very limited qualitative data on coal-mineral size distribution exist in the literature. Major forms of minerals appear to be aluminosilicates with pyrites, calcites, and magnesites in various proportions (Gluskoter et al., 1981). The coal structure and composition have a definite influence on combustion emissions, and this may change or depend on coal mining, handling, cleaning, etc.

The mechanical firing methods for coal-fired boilers in present day usage include pulverized, cyclone, and stoker types with pulverized firing comprising nearly 90% of the total (Babcock and Wilcox, 1978). The crushed coal from the mine is pulverized into fine powder, usually a 60-μm mean particle diameter, and introduced into the furnace with carrier air. In the furnace, coal particles are heated both radiatively and by mixing with hot gases. Combustion temperatures up to 3500° F are usual, and these generally depend upon percent of excess air, coal quality, and effectiveness of mixing. Different coal particles may be subjected to varying temperatures due to differences in the particle size and nonuniformity of mixing.

As the coal particle is being heated, a number of processes can occur. A general overall description will be presented. It is believed that the fine particles and coarse particles are formed by distinctively different mechanisms. The schematic of coal particle combustion and fine particle formation through a number of proposed mechanisms is shown in Figure 3.2 (Damle et al., 1982). Fine particle formation mechanisms are homogeneous nucleation of volatile species and condensation on existing particles to produce surface enrichments with respect to volatile inorganic species (Flagan and Friedlander, 1978; Biermann and Ondov, 1980). Bubbles of ash bursting at high temperatures (Ramsden, 1969; Smith et al., 1979) is another fine particle formation mechanism that has been advanced.

The large fly ash particles are believed to be formed by the semiempirical coalescence-breakup model presented by Flagen and Friedlander (1978). This model considers melting of coal mineral inclusions followed by coalescence as the combustion front recedes on the coal particle. Sarofim et al. (1977) conducted laboratory studies supporting this model.

The vaporization condensation model is considered to be the best candidate to predict fine submicrometer particle formation. In this type of mechanism, a fraction of ash is assumed to be vaporized and homogeneously condensed to form primary particles on the order of 10 Å diameter, followed by coagulation to yield a self-preserving particle size distribution within a few seconds (Flagen and Friedlander, 1978). This model predicts a sharp submicrometer

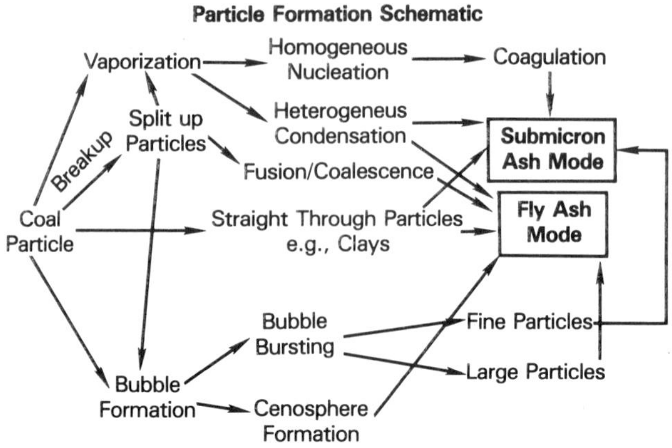

Figure 3.2. Schematic of coal-combustion aerosol formation (Damle et al., 1982). Reprinted by permission of the publisher from "Coal combustion aerosol formation mechanisms: a review" by A.S. Damle D.S. Ensor, and M.B. Ranade *Aerosol Sci. Technol* **1,** 124. Copyright 1982 by Elsevier Science Publishing Co., Inc.

peak in mass concentration at about 0.1 μm. Vaporization of inorganic minerals depends upon their relative volatilities. Therefore, such a submicrometer ash mode that forms is expected to be composed primarily of volatile elements and their compounds. The chemical composition of such a submicrometer mode is expected to be significantly different from large particles formed due to breakup. The vaporization-condensation model also has been supported by laboratory combustion studies (Sarofim et al., 1977; Mims et al., 1979; Flagen and Taylor, 1980; Neville et al. 1980).

Field tests of particle size distribution and elemental analysis of ash emitted from six boilers tend to support the laboratory data (McElroy et al., 1982). The size distributions are typically bimodal in nature. The larger particles are predominantly oxides of elements, such as Al, Si, Fe, Na, Mg, and K; smaller particles, representing only a small fraction of the total mass emitted are enriched in condensed volatile trace elements such as As, Sb, Hg,, Se, Cd, Ni, and Zn.

2.2. Process Dynamics

A power plant or any other process is a dynamic operation. The composition of coal and its thermal and mineral content can often be highly variable. The tracing of a small sample of coal with its associated ash through the process can be instructive:

1. The coal as received is usually stored on a pile for a few days to several months.
2. After conveying into the plant, the coal may be stored in a bunker or storage bin for several hours.

3. The residence time of the gas-phase-contained material is generally a few seconds after pulverization and combustion.
4. The bottom ash may be accumulated 24 h before being discharged as a slurry.
5. The fly ash that is collected in the gas-cleaning equipment may reside in the hoppers for a period of a few minutes to several hours as the ash accumulates between removal cycles.

The objective of obtaining elemental balances by tracing a specific sample of coal through the process is complicated by the mismatch of material flow rates. An additional problem in evaluating gas-cleaning equipment, as pointed out by Ensor et al. (1979), is the characteristic times to reach steady state. For example, baghouses often have long "break-in" times of a month or more.

There are two approaches to overcome source compositional variation problems: (1) obtain composite samples over times consistent with some of the long residence times, or (2) obtain a number of samples and evaluate the process variations. A difficulty with the former approach is that often, compositing some streams for consistency with others is impractical. For example, if the bottom ash sample can be obtained from a 24-h accumulation in the fire box, it would be impractical to obtain a 24-h flue-gas sample to maintain the same time frequency. A second problem is that a transient may occur and affect one stream more than another, thus confusing the results because of the loss of information. The major difficulty with the frequent sampling method is the high cost of analysis.

Another approach is the "staged" sampling method. The process streams are sampled when it is believed (based on calculation) that the parcel of coal of interest is passing a certain point in the process. For example, if the residence time in the coal bunkers is 6 h, then the flue gas would be sampled 6 h after the sample of coal was obtained during filling of the boilers.

3. SAMPLING FOR TRACE ELEMENTS IN PROCESS STREAMS

3.1. Introduction

Coal combustion in a utility boiler results in the distribution of trace elements in various process streams of the power plant. Therefore, the sampling approach that is employed should be consistent with the objectives of the study. In sampling for trace elements, the major problem area, both in the field and in the analytical laboratory, is contamination of the samples. Every step of the sampling task can be a source of contamination; thus, care should be exercised. For example, metal contaminants may be introduced through abrasion of sampling train components, sample storage containers can absorb certain elements, volatile trace elements can be lost at different stages, unclean glassware or apparatus can introduce contaminants, and so on.

In general, the nature of samples collected for evaluation purposes falls into categories of liquids, solids, and gaseous samples.

3.2. Liquid and Slurry Sampling

The composition of liquid streams that might be sampled for their trace element content in a power plant is wastewater from an ash removal system or a scrubber system or steam condensate and cooling water. The factors that must be considered in accurately sampling a fluid stream include

stream homogeneity,
stream flow rate variations,
prevention of sample loss,
sources of contamination, and
sample size required for trace element quantification.

Of these, stream homogeneity should be of great concern, since liquid streams are more likely to be stratified. A flow-proportional composite–sampling technique may be used for sampling liquid and slurry streams. A composite sample can be also taken by using several different methods such as positioned probes, a single multiported probe, or a combination of these. In the case of slurry stream sampling, it is important to avoid segregation of liquid and solid phases. Presently, several commercially available sampling units perform well in the field (Maddalone and Quinlivan, 1976).

Several studies (Starik, 1959; Gould, 1968; Robertson, 1968; Shendrikar and West, 1975) have shown that trace elements are lost from liquid solutions on the walls of the storage container. Shendrikar and West (1974) have studied adsorption characteristics of Pyrex glass, borosilicate glass, and polyethylene surfaces and found the last two kinds of materials to be particularly effective in removing traces of Cr(III) and Cr(VI) from dilute solutions. Benes and Rajam (1969) investigated adsorption losses of bivalent mercury on polyethylene surfaces and concluded that a 5-day-old solution loses about 85% of its initial mercury ions. However, mercury in solution can be stabilized by acidification and a 3-% addition of NaCl (Matsunoga et al., 1979). Shendrikar et al. (1976) studied adsorption patterns of Ba, Be, Cd, Mn, Pb, and Zn at different hydrogen ion concentrations in Pyrex glass, flint glass, and polyethylene. A recent study by Shendrikar et al. (1984) indicates that freeze-drying of liquid samples from the flue-gas environment eliminates trace element adsorption losses up to a period of 100 days after sampling.

Just as trace elements can be lost from the liquid samples due to adsorption, so may a sample be contaminated with them from storage container surfaces. This may result from chemical extraction of the wall material by chemical components in the sample or by physical abrasion of the storage container walls. Such problems can be minimized by cleaning the containers for sample collection and storage before use, as shown in Figure 3.3 (Shendrikar and West, 1974).

Another important factor that must be considered in sampling liquid streams is sample size. This depends essentially on the objectives of the

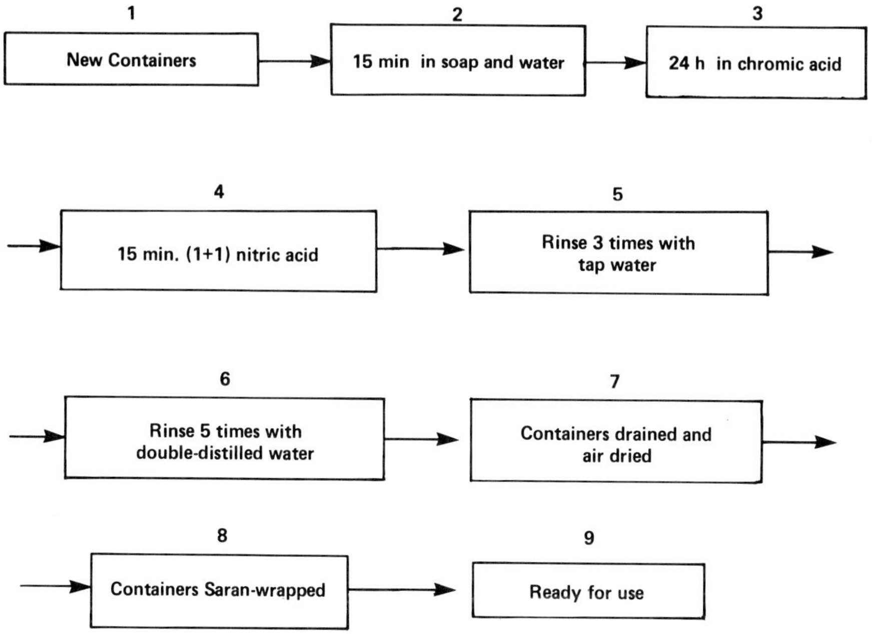

Figure 3.3. Treatment of containers for sample collection/storage (Shendrikar and West, 1974).

sampling and analysis program, that is, number of metals to be quantified in the samples, analytical sensitivity, and determining other parameters requiring measurement.

3.3. Solid Sampling

The range of material and sample sites that can be encountered in sampling for solid materials in a power plant are so diverse that a unified sampling approach is generally impossible. Solid samples incorporate a broad spectrum of materials from large lumps of coal, down to powder and dusts of pulverized coal and ashes from hoppers, to submicrometer fly ash suspended in the flue gas. There is also a diverse assortment of potential sample sites that need to sampled, such as railroad cars, large piles, plant hoppers, conveyer belts, and process stream pipes. There are several locations requiring sampling such as coal piles or coal on a conveyer belt, fly ash from hoppers, or pulverized coal. The main consideration of solid sampling, where no sample size limitations exist, is the problem of collecting samples that are representative of the source and the site. There are several sampling methods that provide a representative sample, some of which are grab sampling, coning and quartering, and fractional shoveling. The equipment and sample collection details can be found in books by Peele (1966), Taggart (1945), Perry et al. (1966), and in a report by Flegal et al. (1975).

Solid samples collected by any of the above-mentioned methods should be stored in prewashed and dried Nalgene bottles (see Fig. 3.3 for procedural details) or polyethylene-lined drums.

3.4. Gas Sampling

Sampling to determine trace element levels in a gas stream is similar to determining particulate loading of the flue gas in a utility stack. The problems encountered and the sampling methods used for obtaining representative samples of such point sources are common to both particulate and trace metal sampling. The differences and special considerations related to trace metal source sampling in the gaseous streams are sample contamination, sample alteration, equipment selection, and properties of trace element with respect to sampling systems. Most of the commonly used sampling methods for trace elements are developed and used for particulate mass sampling in gaseous streams. In fact, in practice, particulate mass samples are collected using well-established sampling methods, and samples collected are subsequently analyzed for trace elements of interest. The sampling methods discussed below are for both particulate mass as well as trace elements in the flue gas.

3.4.1. Sampling Approaches for Gas Streams

Field sampling for trace elements is based on the mass emissions procedures given in the Federal Register (1971). Methods 1 and 2 are used to determine the number of traverse test points and to measure the gas velocity. Method 3 is the Orsat method to determine O_2, CO_2, and CO to compute the pseudomolecular weight of the stack gas. The mass sampling procedure is described in the Method 5 documentation. The sampling rates are continually adjusted to maintain isokinetic nozzle velocities.

The out-of-duct sampling for trace elements that are present in the flue gas as particulate matter and vapors can be carried out by using an EPA Method 5 train or a high sampling volume version of EPA Method 5 called Source Assessment Sampling System (SASS) (Duke et al., 1977; Flegal et al., 1975; Hamersma et al., 1977). Both types of sampling trains are essentially the same in operating principles except the sampling rate for the EPA Method 5 is generally less than 27 L/min (1 cfm), and in the SASS train the rate is in the range of 112 to 162 L/min (4 to 6 cfm). The SASS train assembly also includes a module containing an ion-exchange resin to capture trace organics and a few vaporous trace elements such as Hg and Se. Also, the stainless steel construction of the SASS train makes it more useful for trace organic sampling than for trace metals. Although the internal passages of the train are coated with Teflon, a possibility exists for contamination from breaching of this protection and exposure of the sample to the metal wall.

Ensor et al. (1981) have slightly modified the commercially available

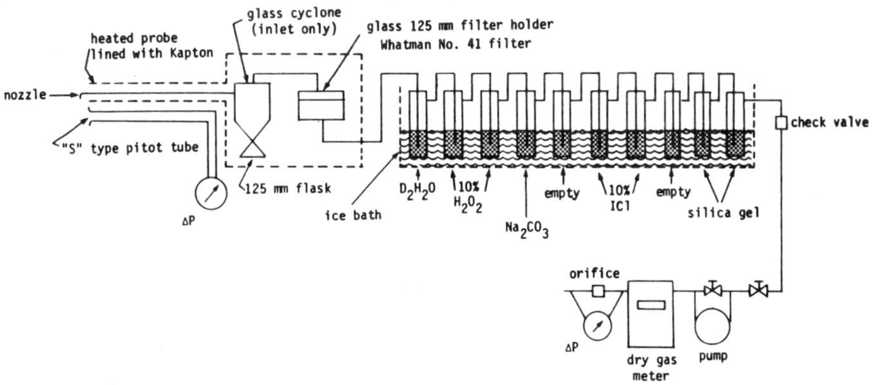

Figure 3.4. Modified EPA Method 5 trace element sampling train (Ensor et al., 1981). Copyright © 1981, CS-1669, Kramer station fabric filter evaluation. Reprinted with permission.

Method 5 train and sampled utility stacks for trace elements. The sampling assembly used is shown in Figure 3.4. Essentially, the modification includes incorporation of 10 Pyrex glass impingers to increase collection efficiencies of various volatile trace elements contained in the flue gases. Also, it was believed that the conventional impingers operated at rated flow rates would be very effective absorbers because of the favorable height-to-width ratio. The train assembly consists of (1) a stainless steel probe, (2) a temperature-controlled oven containing a cyclone (optional) and a filter holder for a Whatman No. 41 filter with low trace element background (see Table 3.2), and (3) ten impingers connected in series. The sample contamination and sample recovery is facilitated by lining the probe with Kapton (Flegal et al., 1975). Kapton, a polyimide film (DuPont Bulletin H-1D), is capable of withstanding high flue-gas temperatures, and it is relatively inert in terms of trace element contamination (see Table 3.2) (Ensor et al., 1981). Note that the Kapton film-insertion technique described by Flegal et al. (1975) of spiraling the film on a rod which is pushed through the probe and then released of the coiled film is generally hard to use in the field. A better technique is to select a strip of film slightly wider than the inside circumference of the stainless steel probe and simply to pull the strip of film through the probe with a string. The film is folded to form a tube with overlapping edges running the length of the probe. Weights of the probe catch can be obtained by preweighing the strip and string, and then weighing the recovered film together with the sample and the material that was trimmed from the original strip of film.

To avoid sample contamination, the entire train, except for the probe, is made of Pyrex glass. The flue gas is drawn through the nozzle and passed through the probe, cyclone, and filter, all maintained at 121° C. Any solid particles passing the cyclone are collected by the filter. The flue gas is then

Table 3.2. Trace Element Content of Blank Kapton and Whatman No. 41 Filter[a]

Element	Kapton Blank[b]	Whatman No. 41 Filter Blank
Al	24	12
Sb	<0.017	<0.008
As	<0.04	0.015
Ba	<6.2	<8.7
Br	0.081	0.15
Ca	42	<54
Ce	<0.013	<0.076
Cs	0.004	0.004
Cl	19	20
Cr	0.468	0.168
Co	<0.017	<0.013
Cu	0.001	0.54
Eu	0.003	0.002
Ga	<0.1	<0.0004
Au	0.0006	0.0007
Hf	<0.003	<0.004
Fe	5.95	3.78
La	0.011	0.033
Mg	<24	<30
Mn	0.10	0.09
Hg	<0.013	<0.015
Ni	<1.7	<0.26
K	<8	<5
Rb	<0.22	<0.31
Sm	<0.002	<0.001
Sc	0.001	0.001
Ag	<0.02	<0.003
Se	0.012	0.022
Na	4.92	4.06
Sr	<1	<1.3
Ta	<0.002	<0.004
Tb	<0.002	<0.002
Th	<0.003	<0.006
Ti	<6	<6
U	<0.03	<0.02
V	<0.06	<0.06
Zn	<0.60	0.32
Zr	NA	0.63

[a] All reported values are given as total micrograms determined on the entire analyzed sample in question. Neutron activation analysis method was used. NA, analysis not available. (Ensor et al., 1981).

[b] 0.5-mil thick Kapton coated with Apiezon-L grease and "Blair" adhesive spray.

pulled through the impingers where volatile species of trace element are collected and retained. The impingers contain chemical solutions, and their configuration is as follows (Ensor et al., 1981):

1. The first impinger contains 100 mL of double-distilled water.
2. The second and third impingers each are filled with 250mL of 10% H_2O_2 solution.
3. The fourth impinger contains 250 mL of 10% Na_2CO_3 solution.
4. The fifth impinger is empty.
5. A 250-mL solution of ICl (10%) is placed in each sixth and seventh impinger.
6. The eighth impinger is empty.
7. The ninth and tenth impingers each contain 250 g of silica gel.

Between the fifth and sixth impingers, a side stream to a silver wool collector is used to obtain an elemental mercury sample from the flue gas (Wroblewski et al., 1974).

The rationale behind this impinger configuration is that the water-soluble volatile trace elements such as Se are collected in the first impinger which is also expected to collect stack moisture. The H_2O_2 impingers are to collect SO_2 that is contained in the stack gas and also to provide an acidic media for the collection of transition-group elements that are not trapped by the filter. A recent paper by McQuaker and Sandberg (1982) also recommends that H_2O_2 be used in the first impinger to scrub SO_2. The Na_2CO_3 impinger is used to collect the excess SO_2 and additionally to stabilize some of the volatile trace elements. The empty impinger is incorporated to avoid back-flushing problems. Silver wool and ICl impingers are used to trap elemental mercury and mercury compounds. Silica gel impingers are included to absorb moisture and protect the pump assembly.

Ensor et al. (1981) reported that during field sampling at a full-scale utility boiler, the Na_2CO_3 solution from the fourth impinger of the sampling train was found to be precipitating, and therefore, it needed to be replaced. Although a few studies (Bolton et al., 1973; Cato and Venezia, 1976) report the use of the Na_2CO_3 solution, none mention precipitation problems. The chemical basis for replacement of the Na_2CO_3 impinger is that two impingers, each containing 250 mL of 10% H_2CO_2, are more than sufficient to oxidize all the SO_2 contained in the flue gas for sampling times up to 24 h.

The other sampling problem with the modified Method 5 train indicated by Ensor et al. (1981) is with the silver wool collector. It was found to be the source of leaks, and the mercury data obtained were erratic and difficult to interpret. As a result, the silver wool Hg collector that was implemented as a side stream was replaced. The new design of mercury impinger (see Fig. 3.5) includes placing iodized activated charcoal with multiple layers of silver wool over it. With this arrangement, stack gases are first passed through charcoal where most of the mercury and its compounds are collected (Murthy, 1975; Moffitt and Kupel, 1970) and the remaining, if any, are trapped by the silver

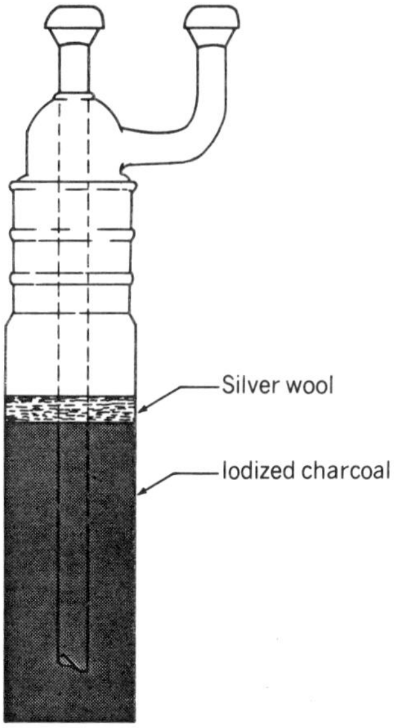

Figure 3.5. Mercury collection impinger.

wool. Based on the above discussion, the following changes are suggested in the impinger configuration (Ensor et al., 1981):

1. The first impinger has 100 mL 6 N HNO_3.
2. The fourth impinger has 200 mL of double-distilled water.
3. The seventh impinger is empty.
4. The eighth impinger is filled with 125 g of activated charcoal, the top layer of which is covered with multilayered silver wool.

Ideally, Method 5 sampling is simultaneously carried out at the inlet and outlet of the particulate control device. Such an effort not only permits the evaluation of effectiveness of the particulate control device with respect to trace element removal but also helps to understand behavior of trace elements, particularly volatile ones, within the control device. In the Method 5 tests, the sample is generally distributed into nozzle, probe liner, probe wash, cyclone, filter, and front half of the train which includes glass connections from probe to cyclone, and cyclone to the filter. The back half of the train includes the connection from the filter to the first impinger and the impinger connecting glasswares. The particulate mass needs to be recovered from each of these sources and analyzed individually or as a whole.

3.4.2. Size-Dependent Trace Element Sampling

Present particulate matter emissions standards are based on a total mass per unit of thermal output and do not differentiate with regard to chemical composition or particle size. However, fine particles (less than 10-μm aerodynamic diameter) have been considered potentially hazardous to health (Amdur and Corn, 1963), particularly because they contain some toxic elements (U.S. EPA, 1977). Furthermore, fine particles can contribute to visibility degradation. The U.S. Environmental Protection Agency recently announced that it is considering promulgation of a size-dependent inhalable particle standard in the 1980s (Cowherd, 1980). Therefore, data on size-dependent particulate matter and trace element emissions from stationary sources are desirable.

Various sampling approaches that are used to obtain size-segregated particle samples include cascade impactors and cyclones. Cascade impactors use a jet directed at a substrate (transects of jets and substrates) to collect particles by impaction, whereas a cyclone causes the gas to spin, forcing the particles to the wall for collection. These two devices separate the particles by their inertial properties and are useful for particles greater than 0.3 μm.

A third sampling device called the diffusion battery consists of arrays of screens which offer a large surface area for the deposition of particles by Brownian diffusion. However, the diffusion battery is still in a developmental stage for sampling particles and Knapp (1982) recently reported a new diffusion battery for source testing to obtain samples for chemical analysis.

Most of the size-segregated trace element data reported to date have been taken with cascade impactors. Cascade impactors specifically developed for source testing have been commercially available since the early 1970s. The instrument offer the advantages of a well-developed theoretical basis, a compact design allowing the use of a number of stages, and minimal chance of sample contamination from the walls. A major limitation of cascade impactors is the small sample of less than 10 mg that is collected per stage.

Cascade cyclones have also been used to obtain samples of particulate matter is broad size bands. The SASS train contains three cyclones, and a cascade system of five cyclones based on the design reported by Smith and Wilson (1978) has become commercially available (Sierra Instruments, Inc.). The major limitation of cyclones is the lack of a theoretical basis for operation; thus, each cyclone must be calibrated for the planned sampling conditions (Smith and Wilson, 1978). Also, the cyclone systems are inadequate for sampling very dilute streams, and the need to remove physically the sample from the walls of the collector makes recovery of samples uncontaminated with wall material difficult.

3.4.2.1. Cascade Impactor Sampling. Cascade impaction is widely used, and the theory of operation is well understood. An example of some of the early work in developing and applying impactors to source testing was reported by Pilat et al. (1970). The cascade impactor consists of a series of jets, each directed toward a flat surface covered with a substrate. An example impactor (Meteorology Research, Inc.) is shown in Figure 3.6. The design is based on a simple annular arrangement of jets and disk collectors as reported by Cohen and Montan (1967).

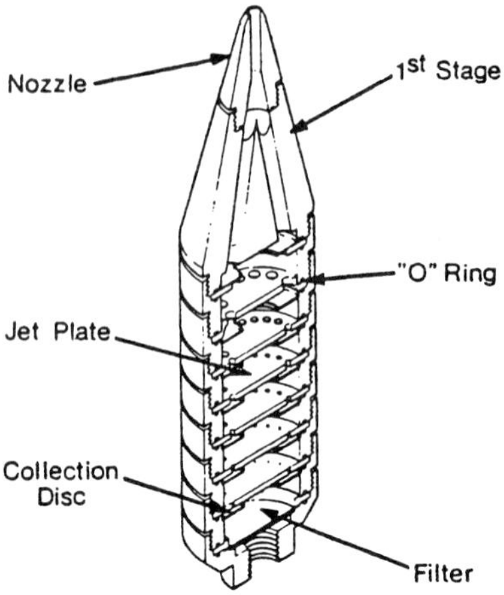

Figure 3.6.. MRI cascade impactor (Ensor et al., 1981).

The simple jet-collection plate geometry has been thoroughly studied by Marple and Liu (1974). The equation describing the impactor collection is given by

$$D_p^2 = \frac{9 \text{ Stk } \mu W}{\rho C V_o}$$

where D_p^2 = particle diameter,
Stk = Stokes number,
W = diameter of jet
μ = viscosity
ρ = particle density
V_o = jet velocity
C = correction for slip.

The calibration curve for an impactor is the collection efficiency as a function of the square root of the Stokes number. The slope of the curve depends on the sample collection surface. A sticky surface will have a very sharp cutoff, whereas a filter surface will exhibit a flatter curve. The selection of cascade impactor substrates is an important part of impactor use. Felix et al. (1977) reported the evaluation of a number of substrate materials including various greases and filter papers. They reported that Apiezon grease-coated stainless steel foil or fiberglass filters treated to prevent sulfation were suitable for mass collection. For trace element sampling, fiberglass filter papers may exhibit excessive and variable trace element backgrounds and therefore are not suitable. However, Apiezon greases are limited to applications with temperature less than 177° C. At the present time, no suitable trace element substrate with low backgrounds exists that also has weight stability for high-temperature sampling.

An additional consideration is particle bounce from the collection substrates in the impactor. Atmospheric aerosol samples have been demonstrated by Dzubay et al. (1976) to have significant contamination of the fine particle on the final filter by large particles. Ondov et al. (1978) also discussed the probable result of particle bounce on fly ash results. The mass of sample on the backup filter of the impactor is often less than 1% of the total mass collected in the impactor. Therefore, large particle contamination of the lower stages and backup filter of the impactor will obscure enrichment results. Ondov et al. (1978) estimated that enrichment may be understated by a factor of 10 or more. Particle bounce effects can be generally reduced by avoiding collection of more than 10 mg per stage, using grease whenever possible, and keeping the jet velocities to less than about 50 m/s. Some guidelines for impactor use have been offered by Lundgren and Balfour (1980).

Another consideration is artifact formation on the stages by chemical reactions on the surface resulting in weight gains and for weighing problems or degradation of the substrates evidenced by weight losses. In a quality-control program reported by Ensor and Hooper (1976), the accuracy of the weighing was determined by control substrates and blank tests. Control substrates are preweighed and transported to the field but not used. Blank tests are conducted by passing filtered flue gas through the impactor to detect chemical reactions on the substrate during the tests. The control disks are analyzed to detect contamination in the laboratory during the preparation and recovery steps. The blank disks are analyzed to detect sampling artifacts caused by chemical reactions of the substrate.

Modifications of the cascade impactor to operate the last stages at low pressure allow collection of particles less than 0.1 μm because of reduced air resistance (Pilat et al., 1978). The University of Washington (UW) Mark 10 and Mark 20 low-pressure cascade impactors developed by Pilat (Pollution Control Systems Corp.) are ideally used to aerodynamically size-segregate the fine particle samples in the range of 0.02 to 10 μm. The Mark 20 is used at the outlet of the control device and the Mark 10 at the inlet.

The low-pressure impaction is still a new method, and some differences exist in the literature with respect to the particle diameters actually collected by the low-pressure stages. For example, Hering et al. (1978, 1979) reported a somewhat larger particle diameter than Pilat et al. (1978) under similar

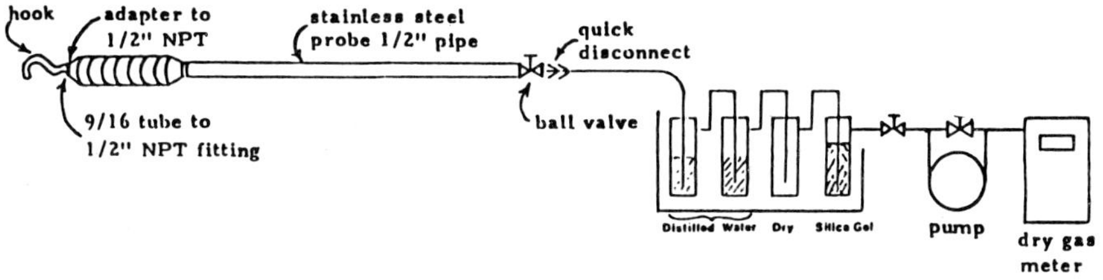

Figure 3.7. Impactor sampling train.

temperature, pressure, and flow. More calibration studies of low-pressure impactors are still required.

Figure 3.7 illustrates the impactor train similar to that described by Harris (1977). Following the impactor is a series of stainless steel impingers immersed in an ice bath. One or two impingers contain distilled water, and the rest are each filled with 400 g of silica gel. A high-vacuum pump draws the sample through the sampling assembly. Next is a dry gas meter. A control unit houses the electronics for the temperature read-out and controls.

3.4.2.2. Preparation for Impactor Sampling. The impactor collection substrates for Meteorology Research, Inc. (MRI) and the UW impactors are generally prepared in the laboratory. The collection disks are covered with peelable Kapton film, 0.5 mil thick. The Kapton is coated (Ensor et al., 1975) with Apiezon-L grease which provides a base for sample collection and its retention on the impactor stage. Kapton is a low trace element background (see Table 3.2). It is stable up to 400° C and is generally resistive to stack gases. However, the film may absorb moisture under some conditions, and hot H_2SO_4 mist may react with the substrate.

3.4.2.3. Sample Collection. The impactor trace element sampling procedure is identical to that used in sampling for particle mass size distribution. The sampling procedures are described in the MRI Inertial Cascade Impactor Instruction Manual (MRI, 1974) and by Pilat et al. (1978) which follow the guidelines described by Harris (1977). The impactors are operated at an isokinetic flow rate at a single point near the center of the duct. Traversing is not done with impactors as with the mass sampling because adjusting the flow through the impactor to match the velocity distribution in the duct would invalidate the particle size data. Because of the large differences in the concentrations between inlet and outlet of the control device, the outlet tests are conducted for a period of one to several hours to obtain sufficient sample, and the inlet tests generally run for 2 to 3 min. This short sampling period at the inlet is necessary to prevent overloading of the impactor stages. The optimum stage load is about 10 mg, depending on the stage and adhesive properties of the particulate matter. Ususally, the optimum loading cannot be achieved for all stages, and sampling times are changed on a trial basis.

3.4.2.4. Sample Recovery. All samples are recovered in the field. After the test, the impactor is removed from the duct, detached from the probe, and allowed to cool. The impactor is carefully disassembled in a dust-free area, and the collection disks are placed into 60-× 15-mm petri dishes for protection and storage. Samples collected on the inside wall of the impactor are brushed onto the appropriate collection stages. The samples are generally desiccated for at least 24 h before determining weight gains.

3.4.2.5. Sample Preparation for Analysis. The sample preparation involves carefully aliquoting the impactor stages, filter, and blanks for trace metal analysis. A quadrant is cut from each impactor stage and folded from outside and placed between two glass slides. The glass slides are taped using masking tape and placed in a slide tray to hold the samples in place during transportation to the analysis laboratory.

3.5. Quality Assurance and Quality Control

The role of quality control in any program that involves sample collection and analysis must be considered an important one. Here, the program requirements dictate sample collection of various kinds (solid, liquid, or gas), matrices, and sizes; therefore, an overall quality-control effort must include details and valid considerations for reliability of the sampling procedures in addition to reliability of trace element analyses of the samples. This is important because sampling is an integral part of chemical analysis, and the results of any analytical test procedure are no better than the sample on which it is performed.

For sampling, valid and well-tested procedures must be used, and care must be exercised so that samples are not contaminated during sampling, recovery, and storage steps. The analysis procedures that are used should be validated for precision and accuracy using actual field samples.

4. SAMPLE STORAGE

After completion of the test series, samples are commonly recovered in the field. For sample collection and storage, Nalgene bottles that are cleaned following the procedure in Figure 3.3 are used. The scope of a field-testing program is such that a long time lapse of 1 to 4 months may result between the actual test and the completion of the sample analysis. A survey of the literature indicates that very little is known about the storage stability (or instability) of such solid, liquid, and flue-gas samples.

In related areas, Shendrikar et al. (1984) recently have investigated adsorption losses of EPA Method 5 impinger samples from a full-scale power plant test. The losses of As, Hg, Sb, Se, Cd, and Zn onto the polyethylene containers from media of double-distilled water, and 10% solutions of H_2O_2, Na_2CO_3, and ICl are investigated. The conclusions of the study are

1. Trace element losses due to storage can be significant for flue-gas samples.
2. Adsorption losses varied greatly with element and chemical nature of the storage medium.
3. Cadmium and zinc adsorption losses amounted to about 64 and 50%, respectively, during the 100-day storage period.
4. Mercury showed significant losses from all media except ICl.
5. Up to 10% losses of mercury occurred from double-distilled water and Na_2CO_3 due to reduction to elemental mercury and its subsequent

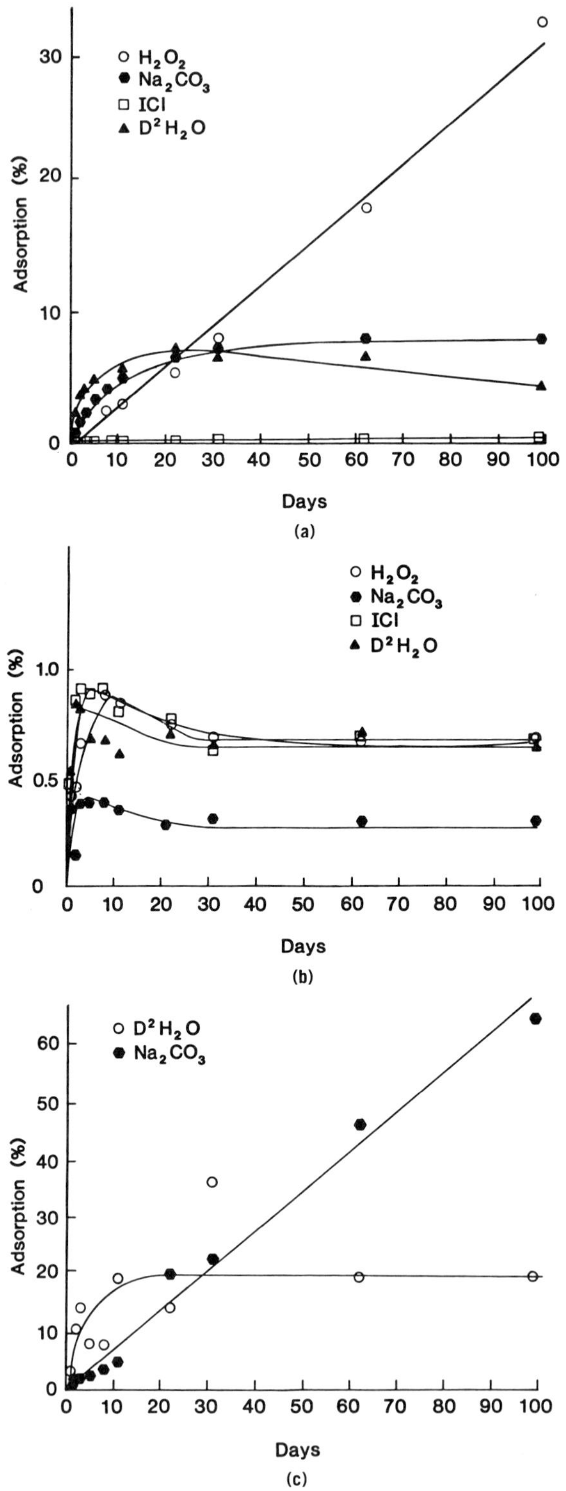

Figure 3.8. Summary of absorption patterns for (**a**) Hg, (**b**) Se, and (**c**) Cd from four storage media (Shendrikar et al., 1984).

vaporization through Saran wrap. This material was used for covering the polyethylene storage containers to minimize solution evaporization loss during the 100-day storage period.

6. Adsorption losses of Se, Sb, and As were found to be negligible ($<1\%$) from all four media studied.
7. Preliminary experiments indicated that freeze-drying of impinger samples eliminated adsorption losses of all the elements surveyed in the study from all four media.

The adsorption patterns of Hg, Se, and Cd in the four media are summarized in Figure 3.8 (Shendrikar et al., 1984).

Literature that was reviewed relevant to sample integrity changes during storage of coal, pulverized coal, bulk fly ash, and fly ash collected from stack tests indicated that workers seemed to have paid very little attention to this problem. Fisher et al. (1976), while examining microcrystalline structures in fly ash samples, observed extensive growths on the surfaces of fly ash particles after a 4-month storage period in closed containers at ambient temperature. This growth must be considered in terms of physical and chemical changes in fly ash particles with respect to storage time. This report appears to be the first observation that indicates instability of fly ash particles during storage. However, this evidence does not appear to have been followed up by the original investigators nor any other workers in the field; therefore, the magnitude of this phenomenon (Fisher et al., 1976) or its significance apparently remains unknown. In related work, Huelett and Weinberger (1980) have demonstrated the formation of crystal phases in some fly ash particles. The formation of these crystal phases clearly indicates again a mechanism for migration of species in the fly ash and, in turn, an instability of ash particles with respect to time while in storage.

Even to date, very little, if any, data exist on sample storage; therefore, systematic and detailed studies of this nature are urgently needed (Shendrikar, 1982). This kind of information particularly becomes important, because presently an *in situ* sampling and chemical analysis approach does not exist, and samples are routinely stored before analysis. Maintenance of sample integrity in terms of chemical composition during exposure studies of humans and animals to determine effects of air pollution is of prime importance, because it is only then that the true effects of pollutants can be assessed.

5. TRACE ELEMENT SAMPLE ANALYSIS

5.1. Introduction

The trace element characterization of samples from the power plant involves analysis of samples of different chemical composition and matrices. The samples collected may be liquids, solids, or gases. In terms of chemical quantification, the samples may also vary in size from kilogram quantities to fractions of a milligram. Solids include belt coal, pulverized coal, and ash samples from the hoppers. Samples from such process streams generally have no size limitations. Liquid samples result when bottom ash from the boiler and fly ash from the economizers are slurried for piping to a pond for disposal. The liquid samples are also collected from various process streams in and around the power plant. Like solids, liquid samples generally have no size limitations. Gas samples, the third kind of sample, are commonly collected from the flue gas and represent that portion of solid fly ash resulting from the coal combustion. Fly ash particles in such samples range in diameter from submicrometer to 100 μm. These samples are also considered to represent that portion of the coal combustion material that may exist in the vaporous and/or liquid state under firebox conditions but converts into solid particles during condensation or similar processes because of cooling in the convective heat transfer sections and in the air pollution control equipment. Samples from these sources impose severe sample size limitations in terms of analytical quantification requirements.

5.2. Sample Storage and Preparation for Chemical Analysis

Sample storage effects of solid and liquid samples are discussed earlier in this chapter. Preparation of samples for chemical analysis is another area that needs consideration. Preferably, sample preparation should be minimal; and when required, care should be exercised to see that:

the sample preparation approach is compatible with the analytical method being used;

the preparation method does not introduce contaminants to the sample nor alter the sample chemistry;

losses of trace elements of interest do not occur; and

the sample preparation approach is simple, quick, and efficient.

In summary, sample preparation is an important step in chemical analysis. Unless samples are prepared properly, analytical data may not be valid.

5.3. Chemical Analysis and Analysis Techniques

A large portion of the trace element data included in this chapter are obtained by analyzing numerous samples collected at a full-scale utility power plant controlled by a baghouse (see Ensor et al., 1981).

The art of analyzing samples from fossil-fuel power plants is still in the developmental stages. The applications of both classical and modern trace element analysis methods for coal and fly ash samples, although they have grown rapidly, still show the problems that are generally associated with samples from any other field (i.e., precision and accuracy of analysis).

Additionally, because of the diversity of sample nature and available sizes, a unified analytical approach is difficult to formulate. Generally, field sampling provides two types of samples: bulk samples and flue-gas samples. The latter by virtue of its nature and source implies sample-size limitation.The scope and extent of the sample characterization program often dictates the selection of analytical methods. Ideally, the chemical analysis protocol formulated should include the following considerations:

1. The analytical method should require minimal or no sample preparation; this will avoid sample contamination problems.
2. The analytical method should not be limited by sample size and should accept samples from a fraction of a milligram to gram quantities.
3. The analytical method should be free from interferences that are likely to give erroneous data.
4. The analytical method should be fast, simple, cost-effective, and capable of producing chemical data with the highest degree of precision and accuracy.
5. The analytical method should be sensitive, since samples may often contain nanogram or even picogram quantities of the trace elements of interest.
6. The analytical method should be nondestructive and capable of providing accurate multielement analysis on a time and cost-effectiveness basis.

Considerations of the above discussions and evaluation of the available methods indicate that in spite of recent rapid developments in the field of analytical chemistry, no one method meets the most desired qualities of the analysis protocols as well as program requirements. Often, the scope of the sample analysis program, the analytical limitations, and the sample matrices are such that it is advantageous to use more than one analytical method. An important point to be stressed here is that such an approach often provides more than one value for a trace element, and this can help to uncover experimental and analytical errors in the same analysis step.

Table 3.3 lists some of the commonly used analytical methods with their assets and limitations for the trace element characterization of coal and ash samples. In this chapter, a few of these methods are only briefly described and discussed, since comprehensive surveys of the techniques can be found in many standard books and articles. In addition to methods included in Table 3.3, there are a number of semiquantitative optical methods for trace elements and their compounds (speciation), and a few are mentioned so that readers can assess their capabilities in trace element and compound speciation.

5.3.1. Atomic Absorption Spectrometer (AAS)

The atomic absorption analysis technique is accepted as one of the best trace element analysis methods (U.S. EPA, 1979). It is relatively interference-free and can produce data with good precision and accuracy (Mann et al., 1978). Typically, the instrument consists of a hollow cathode discharge tube to

Table 3.3. A Few of the Commonly Used Analytical Methods for Trace Element Analysis of Field Samples[a]

METHOD	REMARKS
Atomic Absorption Spectrometry	o Sample solutions are aspirated into the flame.
	o Sample size 25 mg to 1 g.
	o Trace elements are determined individually.
	o Can give accurate data but determinations are done individually.
Inductively Coupled Plasma	o Sample solutions are aspirated into a high-temperature (10,000°K) argon plasma.
	o Multielement capability from 23 to 32 elements.
	o Sensitive to sample matrices.
Proton-Induced X-ray Emission	o Nondestructive, sensitive to several elements including S.
	o Beam irradiation of sample.
	o Sample spot should match beam diameter.
	o Sample should be monolayer for accurate data.
Neutron Activation Analysis	o Nondestructive.
	o Sample preparation minimal.
	o Induced activity is counted for quantification.
	o No sample size limitations.
	o Can produce accurate data.
	o Multielement capability.
	o Expensive and time-consuming.
X-ray Fluorescence	o Sample irradiation with beam.
	o No sample preparation.
	o Fast multielement method.
	o Can accept only small sample size.
	o Sample should be monolayer for accuracy and precision.
	o Sensitive to many elements.
	o Relatively inexpensive.
Spark Source Spectroscopy	o Multielement.
	o Can handle only small sample size.
	o Fast and inexpensive.
	o Accuracy of the data is not that high.

[a] From Shendrikar and Ensor (1981).

provide a desired wavelength of light, a flame in which the sample solution is vaporized and atomized, and a detector to read out or record signals. Comparatively recent innovations in AAS include the use of a graphite furnace, aerosol sample depositor, and autosampler to increase sensitivities and to decrease sample analysis times (Conley et al., 1981).

The disadvantages of AAS are as follows: It is time consuming, requires sample destruction, and needs relatively large sample sizes for analysis. AAS multielemental capabilities still needs to be fully tested using samples of complex matrices.

5.3.2. *Inductively Coupled Plasma (ICP)*

Among emerging analytical methods, inductively coupled plasma and associated instrumentation has shown impressive progress in the past few years for the analysis of major and minor trace elements in samples of all kinds and matrices. The technique can simultaneously and sequentially determine up to 48 elements with detection limits of 1 ppb or less (Barnes, 1979; Robinson, 1978). Generally, liquid samples are aspirated into a high-temperature (10,000 K) argon plasma which causes molecular breakdown atomization and/or ionization and excitation of metals in the solution. The resultant radiation that is produced as the excited atoms relax is passed through the entrance slit of a dispersive device where it is separated into discreet wavelengths. The intensity of each of the characteristic wavelengths is detected by a photomultiplier tube. The photocurrent is transformed into concentration values by reference to a standard. The high operating temperature produces a large number of atoms and thus generally enables ICP to achieve greater sensitivities than flame (but not flameless) AAS.

At present, ICP appears to be a thoroughly established multielemental technique (Watters and Norris, 1978; Moore, 1981); if properly used, this analysis method can be free from most physical and chemical interferences (Barnes, 1978). Recently, Tzavaras and Shendrikar (1984) developed a quick and simple sample preparation method and analyzed environmental samples for 26 trace elements using the ICP's semiquantitative approach.

The method is simple, and fast, and can provide multielemental data, but sample destruction is required, and sample size requirements ranges from 100 mg to 1 g; thus, it is unsuited for impactor sample analysis.

5.3.3. *Proton-Induced X-Ray Emission (PIXE)*

This technique is ideally suited for fly ash sample analysis and can provide multielemental data for up to 20 elements. An average analysis normally requires about 100 seconds per sample of the accelerator time and about 25 s of computer time, run concurrently for data reduction. Sodium and heavier elements can be routinely detected with detection limits ranging from a few nanograms to several hundred nanograms, depending upon the interferences present. This analysis method essentially requires no sample preparation and is nondestructive in nature. However, it cannot analyze large masses of material, and the sample needs to be placed on a Mylar or similar thin film, ideally as a monolayer. Also, the emitted X rays are absorbed to some extent by the sample deposit (self-attenuation) which requires a size-dependent correction to the data.

In this analysis method (Cahill, 1975), a collimated 18-MeV beam of alpha particles from a cyclotron impinges on the thin sample (target) which is mounted at an angle of 45° to the incoming beam. The energy on the sample is made uniform by the use of an aluminum diffusion foil. The beam is collected by the Faraday cup and integrated to a precision of about 2% to give the total charge that passes through the sample. X rays are passed into a detector and a 25-μm beryllium window and then converted into electrical pulses by a liquid nitrogen-cooled Si(Li) detector and associated pulsed-optical feedback circuitry. Data are accumulated in a computer, giving a spectrum of characteristic X rays and a smooth background.

5.3.4. Neutron Activation Analysis (NAA)

Instrumental neutron activation analysis provides the best accuracy and precision in trace element analysis of samples of all kinds. One of the NAA advantages is that it requires minimum sample preparation prior to irradiation; thus, contamination of the sample and loss of volatile elements are minimized. The technique is sensitive, nondestructive, and multielemental in nature, and the precision and accuracy of the method has been well established (Ondov et al., 1974).

The same preparation involves inserting an aliquot of the sample into a small polyethylene vial which is heat-sealed and irradiated to induce radioactivity. Generally, the NAA analysis involves two sample irradiation periods (5–10 min and 8 hr) and four counting intervals. Gamma-ray spectroscopy is used for all gamma-ray-emitting radionuclide activity measurements. The peak areas and net peak areas are calculated and converted with proper decay time corrections to elemental weight (ppm) in a sample by a comparison to known standards.

The apparent disadvantages of NAA are that the technique is time-consuming, expensive, and unable to provide data on certain elements, such as Pb (the product of radionuclide decay) and light elements (S, Si, etc.), with acceptable accuracy and precision.

5.3.5. X-Ray Fluorescence Analysis (XRF)

This analysis method is based on the X-ray excitation of the solid sample, causing inner shell electrons to be ejected. The radiation emitted is then analyzed. aliquots of samples and standards are typically analyzed for 60 min.

X-Ray fluorescence is rapidly gaining popularity for elemental analysis of various samples, including impactor samples, because of relatively low analysis cost and time and minimal sample preparation (Wolfe and Flocchini, 1977; Van Espen et al., 1981; Adams et al., 1983).

5.3.6. *Spark Source Mass Spectroscopy (SSMS)*

The spark source mass spectroscopy has successfully served the purpose of providing semiquantitative trace element data for many years on a variety of samples. It is a multielemental technique that has been used extensively in the analysis of coal and other samples including soil and the environmental and metallurgical samples. For a long time in the field of analytical chemistry, there has been no method to challenge SSMS in terms of cost per sample analysis, comprehensive elemental analysis capability, and data quality (within the stated accuracy, precision, and detection limits).

In SSMS, ions are produced by 20-kV, 500-kHz sparks that are generated between sample electrodes. These sparks produce essentially atomic ions of the elements present in the samples and are extracted into the mass analyzer section of the mass spectrometer. Here, the ions are detected on a photoplate or by electrical detection methods.

5.3.7. *Other Analysis Methods*

1. Proximate, ultimate, and mineral analyses of coal samples provide data on the nature and composition of coal used for combustion. For such analyses, ASTM-271 methods are routinely used. Bulk fly ash samples are subjected to loss on ignition determination by heating to 725° C (Method ASTM-271).

2. Fly ash density measurements are performed using a commercially available helium–air pycnometer.

3. The behavior of coal and fly ash samples on heating can provide information related to the volatility and the bulk thermochemical properties. For such determination, thermogravimetric analysis (TGA) and differential thermal analysis (DTA) techniques are commonly used. In these analyses, samples of coal and fly ash are placed in an alumina crucible and heated preferably at the rate of 10° C/min from room temperature to 1400° C in dry air or argon using Al_2O_3 as a reference element. Simultaneous DTA/TGA curves can be obtained using commercially available instrumentation such as the Mettler Vacuum Thermal Balance.

5.3.8. *Particle Characterization Techniques*

High-temperature coal combustion produces a variety of particulate pollutants, some of which have the potential for health, ecological, and material effects. To truly assess the effect of pollutant particles from power plant emissions, more information on the particle composition is obviously needed. This particularly becomes important in the light of the knowledge that toxicity and carcinogenicity potentials of trace elements depend on their chemical nature.

In the past decade or so, several analytical techniques have been designed to provide the chemical and compositional surface characterization of

Table 3.4. Methods for Particle Surface Characterization

o Low-energy electron diffraction (LEED)	Makes use of electron energy (20 to 1000 ev) that can penetrate only the outermost layers of particles. The technique provides information on the arrangement of and spacing between surface atoms and the degree of surface symmetry. Also measured are distances between atom layers near surface, and the distances between adsorbed species.
o Atom beam scattering	A monoenergetic light beam of low energy strikes the particles and is diffracted by surface atoms. Angle of diffraction measured provides information on the structure and periodicity of ordered surface layers.
o Low-energy ion scattering (LEIS)	A low energy beam of inert gas ions is directed at particle surfaces. Measurements of energy loss and scattering angles can be used to determine the positions and masses of atoms in the first few layers.
o Rutherford back scattering (RBS)	High energy ions are used to bombard the surface. By measuring the "shadows" cast by the surface atoms, the position of atoms can be pinpointed.
o Field ion microscopy (FIM)	The tip of a tiny crystal (500 A° in diameter) having a number of crystallographic planes exposed is subjected to strong electrical field in the presence of an inert imaging gas at low pressure. Atomic sites on the crystal planes produce their own local fields which promote ionization of the gas atoms. The ions produced form an emission pattern on a fluorescent screen that contains information on the surface arrangement of atoms.
o Angular resolved photoelectron spectroscopy (ARPES)	Here, monoenergetic photons (10 to 1500 ev) strike a particle surface and eject photoelectrons from chemisorbed molecular species. By measuring the angles and intensities of photoelectrons, one can determine the molecular orientation of the adsorbed species.

Table 3.4. ***(Continued)***

o Surface extended x-ray absorption fine structure spectroscopy (SEXAFSS)	A tunable x-ray source aimed at the surface scans through a range of x-ray energies. At characteristic energies associated with certain atomic transitions, the radiation is absorbed, causing atoms in the sample to emit electrons. The only electrons measured are the Auger electrons, which characteristically are detected as coming from surface atoms. As scanning continues small fluctuations in x-ray absorptivity are observed. These variations are related to quantum machanical interference effects between neighboring atoms and are used to determine the position and bond length of an atom or molecule adsorbed on the particle surface.
o Electron-stimulated desorption ion angular distribution (ESDIAD)	Relies on the fact that adsorbed molecules may be ionized by electron impact. The ions escape from the particle surface in directions corresponding closely to the orientations of the original bonds in the adsorbed species. By observing the ion's trajectories, one can determine how the parent molecule or atom was oriented relative to crystal planes on the surface.
o Surface-enhanced Raman spectroscopy (SERS)	This technique is based on the finding that the intensity of Raman scaterring by molecules adsorbed onto microscopically rough metal surfaces. This method is useful in identifying molecular structures.
o Electron energy loss spectroscopy (EELS)	Uses monoenergetic low-energy electrons (2 ev) to probe the vibrational energy spectrum of molecular species adsorbed on a surface.
o Reflection-absorption infrared spectroscopy (RAIS)	Infrared radiation reflected from a surface is used to measure the infrared spectrum of the absorbed species.

particles (Henry and Knapp, 1980). Many workers in the field have used such techniques for the characterization of trace element inclusions in coals, bulk fly ash particles, and chemical composition of surface, and as a function of depth in fly ash particles. The intent here is only to mention briefly a few of these promising techniques without providing operational details or applications to combustion-originated particle samples (see Table 3.4). Apart from the methods included in this table, techniques such as scanning electron microscope (SEM), energy-dispersive X-ray spectrometer (EDX), SEM-EDX, electron spectroscopy for chemical analysis (ESCA), and secondary ion mass spectrometry (SIMS) are finding applications in fly ash particle characterization. Modulated molecular beam mass spectrometric techniques offer the potential for direct identification of actual species volatilized during heating. In this analysis, samples of fly ash or coal are placed in alumina and tungsten Knudsen cells and heated to temperatures up to 1500° C under high-vacuum conditions. The vapors can be analyzed directly using modulated beam mass spectrometry (Smith and Street, 1978).

5.4. Comparison of Trace Element Analytical Techniques

Recently, Shendrikar and Ensor (1981) and Ensor et al. (1981) compared four methods by analyzing field samples from tests of utility plants. The methods are

Neutron Activation Analysis (NAA),
Proton-Induced X-Ray Emission (PIXE),
Atomic Absorption Spectroscopy (AAS), and
Inductively Coupled Plasma (ICP).

Both NAA and PIXE are nondestructive techniques. The former method exhibits sensitivity problems for S, Si, Pb, and so on, while the latter method shows good sensitivity for these elements. Additionally, NAA is sensitive to a wide range of trace elements (Ondov et al., 1974) that are not practical to determine by PIXE. These two methods together form a complementary analysis approach. Since precision and accuracy of AAS in trace element analysis are well documented (EPA, 1979), the applicability of this technique to coal combustion research programs was evaluated. Among emerging analytical methods, ICP shows potential of a good multielement technique; therefore, this technique was also incorporated into the study. Comparison of analytical methods requires that parameters such as precision, accuracy, and equivalency be established.

5.4.1. Precision and Accuracy

Precision expresses the reproducibility of the analytical measurement; that is, the agreement of several measurements of the same sample. The precision of analytical methods is evaluated by analyzing duplicate field samples. The data obtained are included in Table 3.5. The laboratories that were contracted for analysis were not aware of sample duplication. One composite of hopper ash sample that was collected at a power plant was duplicated and analyzed using NAA, PIXE, and AAS. Similarly, one composite fly ash sample from another power plant was duplicated and analyzed by ICP to evaluate its precision.

Table 3.5. Precision of Four Trace Element Analytical Methods[a]

Element (ppm)	NAA		PIXE		AAS		ICP	
	1	1-D	1	1-D	1	1-D	2	2-D
Al	96300	94000	1691	2634	81400	67000	91144	126904
Sb	0.80	0.73	-	-	<1.0	<1.0		
As	4.95	4.61	-	-	23.0	9.0		
Ba	1907	1943	<441	<271	1800	1500	148	397
Be	-	-	-	-	<0.5	<0.5	4.81	6.9
Bi	-	-	-	-	44	6.6		
B	-	-	-	-	1000	900		
Br	11.6	11.2	<222	<137	-	-		
Cd	-	-	-	-	1.50	1.60	9.46	<17
Ca	43400	40500	3418	3428	11100	15000	1555	4482
Ce	65	61	-	-	35.7	41.0		
Cs	0.80	0.80	-	-	-	-		
Cl	<40	<40	<348	<212	-	-		
Cr	26.5	30.2	<121	<73	-	-		
Co	9.1	8.7			6.9	7.6	42	42
Cu	19.2	18.81	<901	<551	41.3	47.6	56	92
Eu	0.80	0.80	-	-	-	-		
Ga	11.2	10.7	-	-	-	-		
Au	<7.5	5.15	-	-	<0.51	<0.5		
Hf	8.1	7.8	-	-	-	-		
I	2.1	<1.0	-	-	-	-		
Fe	24688	24658	2615	2186	22600	20800	71724	86777
La	37.6	35.4	-	-	-	-		
Pb	-	-	<478	<294	41	38	41	2.7
Mg	17100	16600	<402	<245	12000	11700	1504	1417
Mn	250	266	<114	44	257	287	66	98
Hg	<0.08	<0.03	-	-	0.07	0.08		
Mo	-	-	-	-	2.50	2.70	64	68
Ni	20.0	16.3	<83	<58	106	27	69	91
Pt	-	-	-	-	<1	<1		
K	2374	2312	<225	111	460	371	12303	16985
Rb	13.2	13.8	-	-	-	-		
Sm	4.93	4.65	-	-	-	-		
Sc	6.0	5.8	-	-	-	-		
Se	9.1	8.7	<181	86	3.90	4.40		
Si	-	-	2564	5273	68692	67757		
Ag	<0.2	<0.1	-	-	<0.2	<0.2	1.93	0.87
Na	770	732	<2238	<1367	582	512	1559	1871
Sr	591	636	-	-	16.3	33.6	131	163
S	<100000	<100000	672	177	-	-		
Ta	0.90	0.84	-	-	-	-		
Th	12.0	10.0	-	-	-	-		
Sn	-	-	-	-	1.7	2.3	18.6	17.4
Ti	3330	2940	308	202	4500	3800	9293	9902
V	47	39	<135	<83	37	27	166	249
U	3.02	5.03	-	-	-	-		
Zn	71.0	65.0	<124	76	66.0	72.0	134	170
Zr	204	197	-	-	-	-		

[a] Note: Samples 1 and 2 are from two different sites. Pixe = Proton-induced X-ray Emission; ICP = Inductively Coupled Plasma; NAA = Neutron Activation Analysis; AAS = Atomic Absorption Spectrometry. (Shendrikar and Ensor, 1981)

The data in Table 3.5 indicate that NAA shows better precision than does AAS, ICP, and PIXE for most of the elements determined. The precision of AAS for elements such as As, Bi, Ni, and Sr is poor. Similarly, ICP shows poor precision for Ba, Ca, Cr, Cu, Pb, Ag, and V. PIXE has good precision for only Ca and Fe, but its precision for other trace elements appeared to be poor.

The sample preparation for PIXE analysis involved placing a sample of known weight (3–15 mg) on a Kapton substrate, which was placed on a 35-mm slide holder. The sample was affixed to the substrate by the use of a trace-element-free spray (Blair Spray Clean). The sample size, its weight, and the way that the sample was prepared for PIXE analysis might be the reason for PIXE's poor precision. Self-attenuation corrections for the X rays that are absorbed by the sample requires a known particle size distribution. Similarly, samples must be reasonably homogeneous, and all of it should be within the beam area for reliable PIXE data.

Accuracy deals with the correctness of a measurement or the agreement of the measurement with the "true" value of the quantity measured. The accuracy of the four methods is evaluated by analyzing replicate samples of NBS-SRM coal. The results that are included in Figure 3.9 indicate that AAS and NAA (and to a certain extent, ICP) agree with NBS-reported values within an acceptable error of ±20%. Exceptions exhibited by AAS include As, ±22%; Cu, +25%; Hg, −38%; and Ni, +57%. Exceptions for NAA include Cu, −35%; Hg, −91%; Eu, −22%; and U, +47%. Only V shows good agreement with the NBS-reported values for PIXE. Trace elements by ICP showing poor agreement are Ca, −32%; Cr, −32%; Pb, −21%; and Zn, +26%.

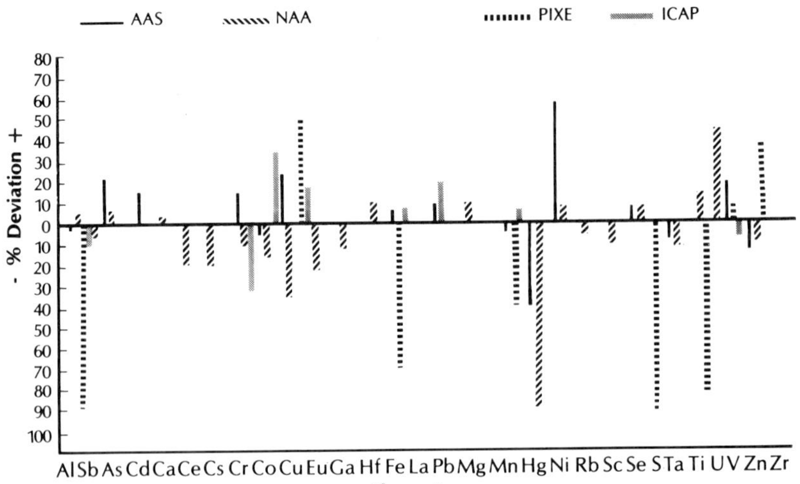

Figure 3.9. Evaluation of accuracy of four analytical methods (Shendrikar and Ensor, 1981).

Table 3.6. Replicate Analysis of NBS-SRM Coal No. 1632[a] by Neutron Activation Analysis (NAA)[a]

ement ppm)	1	1-A	1-B	1-C	1-D	1-E	Avg. Found	Reported	% Deviation
Al	33200±653	32200±684	32870±372	<39380	32056±496	31486±606	32362±562	30700	+5
Sb	0.46±0.01	0.56±0.02	0.52±0.02	0.52±0.01	0.7±0.03	0.48±0.04	0.54±0.02	0.58	7
As	8.13±0.11	8.64±0.12	11.4±0.4	10.4±0.3	11.8±0.3	9.09±0.32	9.91±0.3	9.3±1	+6
Ba	89±2.2	96.0±3.7	117±4	109±3	177±6	117±9	118±5	VN	--
Br	45.6±0.4	43.4±0.4	52.9±1.1	53.8±1.0	62.9±1.1	49.4±1.0	51±0.8	VN	--
Ca	2570±261	2530±283	2422±496	<5490	2442±369	3086±410	2610±290		
Ce	22.6±0.7	24.4±0.06	27.6±0.08	22.8±0.05	32.5±0.1	26±0.12	26±0.2	30	-14
Cs	1.75±0.01	1.90±0.05	1.81±0.08	1.92±0.07	2.4±0.8	1.91±0.09	1.95±0.06	2.4	-19
Cl	710±12	748±8	634±20	683±31	781±13	822±15	722±17		
Cr	28.8±9.1	31.4±1.0	33.14±0.2	30±0.12	41±0.2	32.10±0.3	32.7±2	34.4±1.5	-5
Co	5.1±0.13	5.2±0.2	5.73±0.04	5.4±0.3	7.4±0.04	5.76±0.06	5.77±0.1	6.8	-15
Cu	10.9±1	10.7±1	<1.02	0.57	<0.05	<1.12	10.8±1	16.5±1	-35
u	0.4±0.04	0.42±0.03	0.44±0.008	0.42±0.006	0.58±0.01	0.48±0.01	0.47±0.02	0.54	-22
a	6.14±0.12	6.19±0.15	7.89±0.45	8.02±0.36	10.1±0.4	6.30±0.43	7.44±0.3	8.49	-12
u	<7.73	<8.15	<16.5	<15.2	<18.3	<14.9	<13	VN	--
f	1.2±0.01	1.34±0.01	1.44±0.02	1.29±0.01	1.90±0.02	1.45± 0.03	1.44±0.02	1.6	+10
	<2	0.96±0.2	0.8±0.2	<2	1.17±0.3	1.58±0.5	1.13±0.3		
e	8990±55	9511±285	10792±98	9377±107	13043±118	11236±112	10492±129	11100±200	-5
a	13.6±0.2	13.0±0.2	17.7±0.5	18.8±0.4	20.9±0.5	17.3±0.5	17±4	VN	--
g	1140±83	1780±110	627±88	<4860	245±33	1469±148	1052±92		
n	31±0.4	31±0.3	29±0.6	33±0.7	30±2	29±2	31±1	28±2	+10
g	0.02±0.0007	0.02±0.001	0.0008±0.0004	0.011±.004	<.002	0.035±0.002	0.012±0.0009	0.13±0.03	-91
i	10.6±0.9	10±0.8	24.6±3	15.1±2	35.2±4	29.4±5	20.8±3	19.4±1	+7
	3762±41	3774±44	5240±139	5248±119	6425±140	4311±118	4793±100	VN	--
b	26.0±0.3	30.88±0.5	28.60±0.6	24.5±0.4	35.5±0.8	31.1±1.14	29.3±0.6	31	-5
n	2.18±0.01	2.05±0.1	2.95±0.04	2.77±0.03	3.75±0.04	3.14±0.04	2.81±0.03	VN	--
c	5.12±0.01	5.13±0.007	5.6±0.01	4.82±0.008	7.25±0.01	6.0±0.01	5.65±0.009	6.3	-10
g	<0.2	<0.07	NA	NA	NA	NA	<0.14	VN	--
e	1.96±0.03	2.00±0.03	2.10±0.05	2.0±0.04	2.86±0.06	2.29±0.09	2.20±0.05	2.6±0.7	+15
	767±13	774±13	1118±12	1134±12	1376±14	986±11	1026±13	VN	--
	88.0±3.4	75.0±3.6	124±7	82±4	117±7	115±12	100±6	VN	--
	<40000	<50000	<85000	<122000	<37000	<64000	<66333		
	0.31±0.02	0.31±0.02	0.34±0.03	0.33±0.02	0.44±0.03	0.36±0.03	0.35±0.03	VN	
	3.6±0.03	3.05±0.01	4.0±0.01	4±0.01	5.28±0.02	4.25±0.03	4.16±0.02	4.5±0.1	-8
	1750±98	1580±105	2263±233	<2600	2271±160	2354±174	2104±154	1750	+17
	2.6±0.06	1.66±0.06	3.22±0.13	2.62±0.07	2.54±0.14	1.83±0.02	2.42±0.1	1.28±0.02	+47
	41±1.5	44±1.8	54±3	<43	43±2	46±2	46±2	44±3	+2
	27.1±2	<1.1	<1.37	24±2	<1.7	<2.5	26±2	28±2	-7
	38±3	36.2±3.4	70±8	46±5	61±7	47±9	50±6	VN	--

om Shendrikar and Ensor (1981). NA, Analysis not available. VN, Value not reported by NBS.

An interesting observation can be made with NAA data from the replicate analysis on NBS-SRM coal samples (see Table 3.6). For example, the As data indicate that replicate analysis of the NBS coal sample is essential to obtain accurate analytical results. If the As results from columns 3, 4, and 5 (of table 3.6) are averaged, then the average is 11 ± 2 ppm, and the deviation from the NBS-reported value becomes 20% instead of 6% as reported. Therefore, it appears that in order to improve the chances of getting the most accurate data, a sample needs to be analyzed repeatedly. This is an old rule in analytical chemistry, and eighteenth century chemists probably practiced it on a routine basis. However, because the modern multielemental trace methods are expensive and time-consuming, sample replication is often limited, thus often reducing the value of the study.

5.4.2. Equivalency of Methods

Because of the analytical program requirements, two analytical methods are frequently used for trace metal analysis. In a situation like this, establishing equivalency methods becomes imperative so that data from the methods can be favorably compared. The data presented in Table 3.6 and Figure 3.9 indicate that NAA and AAS are equivalent methods for most of the trace elements except Ba, Ca, Mg, Ag, Sr, Ti, and Zn.

Table 3.7 includes comparison results for NBS fly ash by NAA and XRF (Ondov et al., 1975). In general, there is good agreement between the results of each analytical method and the reported values.

In summary, methods evaluation indicated that NAA is capable of producing precise and accurate data for most of the trace elements; some exceptions are Hg, Cu, and U. The method is found to perform well for all kinds of samples including impactor samples. NAA is a nondestructive and sensitive technique for most of the trace elements of interest, but it is expensive, time-consuming, and unable to provide analytical data for Si, Pb, S, Cd, Be, and Bi, among others. XRF and NAA show good comparisons to one another. The former method also has been found to provide fairly accurate trace element data, and this method appears to have all the advantages of NAA, plus it is simple, less time-consuming, and cost-effective. AAS and ICP are also found to perform well with some exceptions. Both are sample-destructive techniques, and sample-size requirements make them nonapplicable for size-segregated sample analysis. PIXE appears to be inadequate for quantitative work, but sample preparation may have significantly contributed to this failure.

6. DATA PRESENTATION AND INTERPRETATION

6.1. Introduction

In a typical power plant sampling program, trace element data due to the following sample analyses are obtained:

1. Belt coal and pulverized coal.
2. Hopper and bottom fly ash.
3. Liquid samples from ash slurries.
4. Fly ash samples collected in the stack using EPA Method 5 or any other equivalent method. These samples are generally collected concurrently at the inlet and outlet of the particulate control device.

Table 3.7. Comparison of NAA and XRF for NBS Fly Ash Sample[a]

Element	NAA	XRF	Reported
Al (%)	12.6	NA	VN
Sb (PPM)	7.2	NA	VN
As (PPM)	61	62	61
Ba (PPM)	3400	2500	VN
Br (PPM)	12	9.5	VN
Ca (%)	4.7	5.04	VN
Ce (PPM)	146	154	VN
Cl (PPM)	42	NA	VN
Cr (PPM)	131	112	131
Co (PPM)	40	<52	VN
Cu (PPM)	NA	124	128
Fe (%)	6.5	5.94	VN
La (PPM)	82	78	VN
Mn (PPM)	489	508	493
Hg (PPM)	NA	11	VN
Ni (PPM)	98	94	98
K (%)	1.71	1.63	VN
Rb (PPM)	124	114	VN
Se (PPM)	8.8	8.8	9.4
Si (%)	NA	21.9	VN
Sr (PPM)	1900	1390	VN
S (PPM)	NA	7800	VN
Ti (%)	0.76	0.73	VN
V (PPM)	220	182	214
Zn (PPM)	216	202	210
Zr (PPM)	301	305	VN
Y (PPM)	62	66	VN

[a]From Ondov et al. (1975). NA, Analysis not available. VN, Value not reported by NBS.

5. Size-segregated fly ash samples collected in the stack using impactors. Generally, these samples are also collected simultaneously from the inlet and outlet of the particular control device.

Analysis of the collected samples for trace elements generally results in masses of data that need to be reduced, tabulated, and interpreted to achieve the desired objective to describe the coal combustion process. In this section, an example of trace element data from a power plant controlled with a baghouse are presented which provides information on the partitioning of trace elements during coal combustion and effectiveness of control device in controlling trace element emissions into the atmosphere.

6.2. Trace Element Removal Performance of the Control Device

The ability of a flue-gas cleaning device to control trace element emissions is often expressed as the collection efficiency. In other words, trace element collection efficiency of a control device is the percentage removal from flue gas of the total trace element mass entering the control device. Another measure of trace element emission control is percent penetration, which is the percentage of the total trace element mass entering a control device that remains uncollected in the flue gas. Thus, percent penetration is 100 percent minus percent collection efficiency. Often, trace element percent penetration calculations reveal more of the process conditions than percent efficiency. For example, differences between trace element collection efficiency of 99 and 95% may not appear as great as the difference between the respective penetrations of 1 and 5%. The latter number indicates a fivefold difference in the trace element emissions that would leave the utility stack for the same amount entering the control device.

If the mass flow rates of trace elements in the particulate control device inlet M_i, outlet M_o, and ash collected M_a are known, then the collection efficiency E can be calculated by any of the following formulas:

$$E = 100\,(M_i - M_o)/M_i \qquad (1)$$
$$E = 100\,M_a/M_i \qquad (2)$$
$$E = 100\,M_a/(M_o + M_a) \qquad (3)$$

The percent penetration P is calculated by

$$P = 100\,M_o/M_i \qquad (4)$$
$$P = 100\,(M_i - M_a)/M_i \qquad (5)$$
$$P = 100\,M_o/(M_o + M_a) \qquad (6)$$

The overall trace element collection efficiency is determined by sampling concurrently at the inlet and outlet of the control device using EPA Method 5 or a similar sampling method. Collection efficiency calculations are performed by summing up individual elemental concentrations in samples from various sections of the sampling train such as nozzle, probe, probe wash, front-half wash, cyclone, filter, back-half wash, and impingers. This step is repeated to obtain individual trace element concentrations due to Method 5 sampling at the outlet of the control device.

Included in Table 3.8 is trace element data due to simultaneous inlet and outlet sampling at a power plant controlled with a baghouse (Ensor et al., 1981). The individual elemental concentrations that are obtained by analyzing samples from various sections (e.g.,nozzle, probe liner, cyclone) are summed up to obtain the elemental concentrations contained in the flue gas. The Method 5 impinger catch representing volatile trace elements is not included in the table. Examination of data in Table 3.8 indicates that most of the elements are collected in the particulate control device with about the same efficiency or a slightly better efficiency than total particulate mass. Exceptions are the volatile elements.

The flue-gas concentrations of volatile elements such as Sb, As, Br, Cr, Hg, Ni, Se, and Zn that are collected in the impinger train and analyzed using the neutron activation analysis technique are given in Table 3.9 (Ensor et al., 1981). Although impinger samples from inlet and outlet trains are analyzed individually, the data presented in this table represents the sum concentration of individual elements from all impingers of the trains. Only As and Cr could not be detected in the impinger samples, but these results confirm at least partial volatility of these elements. The particulate (or nonvolatile) concentrations of these elements are also included in the table, and this indicates that these elements are present both as vapors and particulates in the flue gas. Clearly, more work is yet needed in this area to determine why and under what conditions a trace element is converted into vapor, fully or partially, during coal combustion.

Sampling for volatile trace elements in stack flue gases is generally performed by permitting these elements to react and be retained in the impinger train. The impingers and the chemical solutions that are used are designed for maximum collection efficiency of volatile trace elements through laboratory studies (Flegal et al., 1975). However, evaluation of data reported in the literature (Cato and Venezia, 1976; Andren and Klein, 1975) indicates that elements such as Hg and Se are not effectively collected in the impinger train. Ensor et al. (1981) recently carried out a systematic study to investigate the utility and effectiveness of the impingers and chemical solutions contained in them. They have analyzed both inlet and outlet impinger samples collected after passing separately through double-distilled water; 10% H_2O_2, 10% Na_2CO_3, and 10% ICl for volatile elements and the data obtained are given in Table 3.10. The data indicates that As, Sb, Se, and Hg are not efficiently collected in the impinger trains. For example, As is not collected by the impingers containing water but appears to be partially collected in the H_2O_2 and Na_2CO_3 impingers. Selenium appears to be distributed in all impingers both in inlet and outlet trains. Relatively high concentrations of Se in No. 1

Table 3.8. Individual Trace Element Collection Efficiency by the Baghouse[a]

Element	METHOD 5 (without impinger catch) % Efficiency = 99.63 Total Inlet Mass = 1.09 g/m^3 Total Outlet Mass = 0.0041 g/m^3			IMPACTOR (LPI) % Efficiency = 99.68 Total Inlet Mass = 1.57 g/m^3 Total Outlet Mass = 0.0051 g/m^3		
	Inlet (ug/m^3)	Outlet (ug/m^3)	Efficiency (%)	Inlet (ug/m3)	Outlet (ug/m^3)	Efficiency (%)
Al	120,188	55	99.95	86,000	202	99.77
*Sb	2.24	0.0063	97.72	33.9	0.0125	99.96
*As	14.02	0.009	99.94	112	0.0639	99.94
Ba	2,570	1	99.96	2,440	6.26	99.74
*Br	3.22	0.04	98.76	(428)	0.0864	(99.98)
Ca	70,914	249	99.65	57,700	85.4	99.85
Ce	74	0.03	99.92	97.6	0.152	99.84
Cs	1.44	0.0007	99.95	3.89	0.00489	99.87
Cl	<87	2	NO [b]	1,780	3.44	99.81
*Cr	63.37	0.19	99.70	183	0.103	99.94
Co	17.02	0.008	99.95	23.2	0.0555	99.76
Cu	<15.6	0.21	NO [b]	(787)	2.68	(99.66)
Eu	0.783	0.003	99.62	0.118	0.00268	97.73
Ga	28.7	0.056	99.80	(408)	0.105	(99.97)
Au	<14.36	0.25	NO [b]	0.0825	0.000139	99.83
Hf	6.69	0.0026	99.96	--	0.00936	NO [c]
I	2.1	0.05	97.62			
Fe	20,516	13.43	99.93	29,900	56.2	99.81
La	49.4	0.016	99.97	(1,730)	0.0904	(99.99)
Mg	146,063	9.4	99.99	27,800	35.6	99.87
Mn	551	0.32	99.94	491	1.13	99.78
*Hg	0.0007	0.00023	67.14	0.307	<0.0007	>99.77
*Ni	53.56	0.09	99.83	173	<0.51	>99.71
K	4,971	2.25	99.95	9,700	11.4	99.88
Rb	37.23	1.63	95.63	50.4	0.0368	99.93
Sm	6.1	0.003	99.95	202	0.0109	99.99
Sc	6.70	0.004	99.94	14.3	0.021	99.85
*Se	13.80	0.204	98.52	13.1	0.0781	99.40
Ag	2.21	0.027	98.78	13.5	<0.0005	>99.99
Na	1,808	12.9	99.29	3,600	6.84	99.81
Sr	615	0.33	99.95	1,200	3.11	99.74
Ta	1.48	0.0008	99.95	2.14	0.0035	99.63
Th	12.09	0.004	99.97	13.9	0.00333	99.98
Tb				0.961	0.0035	99.63
Ti	4,893	2.08	99.96	2,530	7.82	99.69
U	5.7	0.004	99.93	6.14	<0.006	>99.90
V	94	0.051	99.95	91.7	0.212	99.77
*Zn	1,321	0.96	99.93	394	<1.10	99.72
Zr	218	0.33	99.85	245	0.408	99.83

[a]From Ensor et al. (1981). Neutron activation analysis used.

[b]The inlet concentration is less than indicated levels and thus efficiency cannot be calculated.

[c]The outlet concentration is higher than the inlet.

* Volatile trace elements.

Table 3.9. Distribution of Volatile Trace Elements in the Stack Flue Gas[a]

	Inlet (ug/m^3)			Outlet (ug/m^3)			
Element	Volatile [b]	Nonvolatile [c]	Total	Volatile [b]	Nonvolatile [c]	Total	Percent Efficiency
Sb	0.016	2.24	2.256	0.005	0.0063	0.0113	99.50
As	<0.13	14.02	14.02	<0.091 [d]	0.009	0.009	99.94
Br	2.13	3.22	5.35	9.26	0.04	9.30	NO [e]
Cr	<1.0	63.37	63.37	<0.85 [d]	0.19	0.19	99.70
Hg	0.0007	<0.0007 [d]	0.0007	<0.0066 [d]	0.00023	0.00023	67.14
Ni	0.34	53.56	53.9	0.16	0.09	0.26	99.51
Se	1.70	13.80	15.50	0.817	0.204	1.021	93.41
Zn	1.10	1,319.9	1,321	0.11	0.96	1.07	99.92

[a]From Ensor et al. (1981). Neutron activation analysis was used for analysis.
Copyright © 1981, CS-1669, Kramer station fabric filter evaluation. Reprinted with permission.

[b]Fraction collected in the impinger train. Data from inlet impingers 3 and 4, and outlet impinger 4 are not included because of sodium contamination problems. High sodium activity prevented any measurements.

[c]Fraction present in the Method 5 particulate catch.

[d]Variable sample matrices affected detection limits of the analytical method.

[e]The outlet concentration is higher than the inlet.

Table 3.10. Evaluation of Collection Efficiency of the EPA Method 5 Impinger System[a]

Element	D_2H_2O IMPINGER NO. 1 INLET	OUTLET	10% H_2O_2 IMPINGER NO. 2 INLET	OUTLET	10% H_2O_2 IMPINGER NO. 3 INLET	OUTLET	10% Na_2CO_3 IMPINGER NO. 4 INLET	OUTLET	10% ICl IMPINGER NO. 6 INLET	OUTLET	10% ICl IMPINGER NO. 7 INLET	OUTLET
	MICROGRAM PER CUBIC METER											
Cadmium	<0.25	<0.22	<0.09	<0.06	<0.06	<0.05	<0.06	<0.03	<0.06	<0.03	<0.07	<0.03
Zinc	2.44	1.14	<0.09	<0.06	<0.06	<0.05	<0.06	0.04	<0.05	<0.03	<0.07	<0.03
Mercury	0.10	0.44	0.26	<0.01	<0.01	0.02	0.10	<0.03	0.18	0.07	0.20	0.07
Arsenic	<0.25	<0.22	0.42	0.27	0.63	0.07	0.29	0.07	<0.06	<0.03	<0.07	<0.03
Antimony	<0.25	<0.22	<0.09	0.80	0.12	<0.05	<0.06	<0.03	<0.06	<0.03	<0.07	<0.03
Selenium	5.69	3.16	0.69	0.27	2.45	0.47	7.24	2.63	<0.06	0.19	<0.07	0.13

[a]Individual impingers were analyzed using atomic absorption spectrometry (Ensor, et al; 1981).

impinger indicates that Se may be present in stack gases as an oxide that shows greater solubility in water. Mercury, again, shows its distribution throughout the impinger train, and the two 10% ICl impingers connected in series are not efficient in collecting the Hg contained in the flue gas. Mercury vapors are also collected partially in the H_2O_2 impingers designed for SO_2 collection. Recently, Shendrikar et al. (1983) have generated known and stable Hg atmospheres and found H_2O_2 impingers in a SASS train assembly to collect up to 20% Hg contained in the test atmosphere. This work and the data in Table 3.10 indicate that all impingers in the EPA Method 5 train must be analyzed for all volatile trace elements; otherwise, lower concentration bias of volatile elements may exist in the reported data.

Some possible causes of the impinger collection inefficiencies may be

high sampling rate through the collecting substrate which does not provide sufficient time for reaction to occur,

flue gas containing chemical compounds that render the collection media ineffective, and

flue gas containing volatile trace element compounds other than those for which optimum substrate collection efficiency was evaluated in the laboratory.

In general, volatile trace element sample-collection efficiency, chemical solutions used in the impingers, and the impinger utility and usefulness are areas where more research is needed.

The trace element collection efficiency of the control device can also be determined by impactor sampling conducted concurrently both at the inlet and outlet of the control device. Impactor trace element concentrations are obtained by adding up the concentration of each element from each impactor stage. Typical overall collection efficiencies of individual trace elements obtained at a utility baghouse due to impactor (UW Mark 10 at the inlet, and Mark III–IV at the outlet) sampling are also included in Table 3.8 (Ensor et al., 1981). The neutron activation analysis method is used for sample analysis. Here, the EPA Method 5 impinger catch representing volatile trace elements has been omitted to permit a direct comparison between Method 5 and impactor results. The data in Table 3.8 indicate that almost all of the trace elements are collected in the baghouse with about the same or slightly better efficiency than total particulate mass (data for particulate mass are also included in this table). The exceptions in the impactor include Br, Eu, Se, and Ta; in Method 5, the exceptions include Eu, I, Hg, Se, and Na. Some of these (Se, Hg, I, etc.) might be present in the flue gas as vapors and thus could not be collected on the impactor stages.

Table 3.8 also includes total mass and mass of each trace element collected using impactors and Method 5 trains both at the inlet and outlet of the baghouse. Although the particulate mass collection efficiency determined by the two sampling methods is almost the same, differences exist in particulate matter mass concentrations determined by two sampling methods. For example, inlet particulate mass concentrations are 1.09 g/m^3 with Method 5 and 1.57 g/m^3 with impactors. Similarly, Al concentration is 120,188 mg/m^3 with Method 5 and 86,000 mg/m^3 with impactors.

For an ideal comparison of concentrations (particulate mass as well as trace element) measured by the impactor and EPA Method 5, the following parameters must be considered:

sampling time and duration,

variations in particulate matter emission rates during sampling, and

stratification of particulate matter in the duct.

Method 5 samplings are performed by traversing the entire duct (16 points) during the 24-h sampling period. However, the low-pressure impactor samples are collected at a single point and sampling times are much shorter to avoid overloading of the stages. Therefore, particulate matter stratification in the duct may also account for the mass as well as the trace element concentration differences determined by the two sampling methods.

6.3. Trace Element Balances

The EPA Method 5 data are generally used to perform a trace element balance around the control device to provide a gross check on sampling and analytical accuracies. For calculation purposes, it is assumed that the trace element mass entering the inlet of the control device either leaves through the outlet or the ash hopper. A 0% imbalance indicates closure around the control device; that is, mass input exactly matches mass output. The elemental hopper ash rate is the elemental concentration of hopper ash multiplied by the hopper ash flow rate which is obtained by subtracting total inlet mass from total outlet mass. Therefore, the imbalance calculation may be expressed as

$$\text{Elemental imbalance} = \frac{\begin{array}{c}(\text{Concentration entering hopper}) - \\ (\text{Concentration leaving hopper and flue gas})\end{array}}{(\text{Concentration entering})} \qquad (7)$$

Table 3.11 shows typical percent imbalances for trace elements obtained at a baghouse. A 0% imbalance indicates perfect agreement between the measured input and output masses of a particular trace element, or perfect mass balance closure. Positive or negative imbalance values generally indicate the presence of error in trace element concentrations or total mass flow rates. Trace element imbalances can also occur because of a "hidden stream," for example, an element escaping as vapor from the outlet or an inefficient sample collection method. Imbalances are more likely caused by the stratification of hopper ash, and thus the samples collected and analyzed are not representative of the test period.

Table 3.11. Trace Metal Mass Balances around the Baghouse[a]

Element (ug/m^3)	Inlet	Outlet	Hopper Ash	Percent Imbalance
Al	120,188	55	104,599	+13
Sb	2.26	0.011	1.1	+51
As	14.02	0.009	8.0	+43
Ba	2,570	1.0	2,345	+ 9
Br	5.35	9.30	10.0	-260
Ca	70,914	249	50,535	+28
Cs	1.44	0.0007	0.97	+33
Cr	63	0.19	34	+46
Co	17	0.008	12.1	+29
Eu	0.78	0.003	1.03	-32
Ga	29	0.06	15.0	+48
Hf	6.7	0.003	9.2	-37
I	2.1	0.05	2.68	-30
Fe	20,516	13.4	30,666	-49
La	49	0.02	43	+12
Mg	146,063	9.4	18,501	+87
Mn	551	0.32	351	+36
Hg	0.0007	0.00023	<0.04	+67
Ni	53.9	0.26	28	+48
K	4,971	2.25	2,707	+45
Rb	37	1.63	23	+33
Sm	6.1	0.003	5.79	+ 5
Sc	6.7	0.004	7.0	- 5
Se	15.5	1.021	7.0	+48
Ag	2.21	0.027	<0.2	+99
Na	1,808	13	995	+46
Sr	615	0.33	705	-15
Ta	1.48	0.0008	1.0	+32
Th	12.1	0.004	12.1	- 0.8
Ti	4,893	2.08	3,543	+28
U	5.7	0.004	3.68	+35
V	94	0.05	51	+46
Zn	1,321	1.1	114	+91
Zr	218	0.33	229	- 5

[a]EPA Method 5 samples were analyzed for calculations that includes volatile trace element concentrations. Trace elements showing concentrations less than the detection limit (see Table 3.8) are not included. Neutron activation analysis method was used. (Ensor, et al, 1981).

The mass balance results (Table 3.11) indicate that less then 15% of the mass inputs from coal of Al, Ba, Cd, La, Sm, Sc, Th, and Zr leave as stack emmissions in the baghouse outlet. Thus, the bulk of these elements apparently is retained in the power plants. On the other hand, elements such as Sb, As, Br, Cr, Ni, Hg, Ga, Mg, K, Se, Na, Ag, V, and Zn show large portions unaccounted for and may be present as vapors. Such a state for some of these elements is confirmed (see Table 3.2). Imbalances for Ga, Fe, Mg, Pt, K, and Na are probably accounted for by analytical inaccuracies or sampling artifacts.

6.4. Impactor Data Interpretation

6.4.1. NAA Data

At the NAA laboratory, the total number of sample spots corresponding to deposits below the impactor jets on each sample aliquot are manually counted before irradiation. Upon analysis, trace element data are presented as total micrograms or nanograms as determined by using NAA. Then the total mass of each trace element on an impactor stage is calculated by multiplying the total mass as determined by NAA by the ratio of the total number of sample spots (number of impactor jet holes) on an impactor stage to the number on the aliquot. The calculated total trace element mass is divided by the sampling volume to obtain micrograms per cubic meter.

Trace element impactor data are reduced with an acceptable method for a mass data computer program (Markowski et al., 1980; McCain et al., 1980). The impactor stages are characterized by D_{50}s (the particle diameter with 50% collection efficiency). The cumulative trace element mass concentration is plotted against D_{50} and then fitted with a series of overlapping parabolas. The slope of the curves at the selected sizes is used to compute a differential particle size distribution curve dM/dD. The penetration for discrete particle diameters is obtained by dividing the outlet dM/dDs by the inlet dM/dDs. The advantage of this procedure is that differently designed impactors can be used at the inlet and outlet.

6.4.2. Size Distribution Data

For a typical inlet, size distribution data of a few trace elements are shown in Figure 3.10. Consideration of data in Figure 3.10 indicates that most of the trace elements included show a bimodal size distribution. The large and small particle modes have a mean diameter of approximately 4 and 0.08 μm, respectively. The size distribution curves above 10 μm are not included because of losses of particles to the walls of the impactor. In general, the size distribution curves for individual elements show a flatter profile (see Fig. 3.10) indicating a larger proportion of their mass in submicrometer particle ranges, probably because of their volatility during coal combustion.

The trace element size-dependent penetration through a particulate control device is the ratio of the outlet to inlet concentration as represented by differential size distribution curves. A summary of size-dependent elemental penetration through the baghouse is given in Figure 3.11 (Ensor et al., 1981). Most elemental penetration is less than 5% (i.e., greater than 95% efficiency for all particle sizes) and is quantitatively similar to the mass penetration. An obvious exception is Se which shows a penetration ten times greater than that

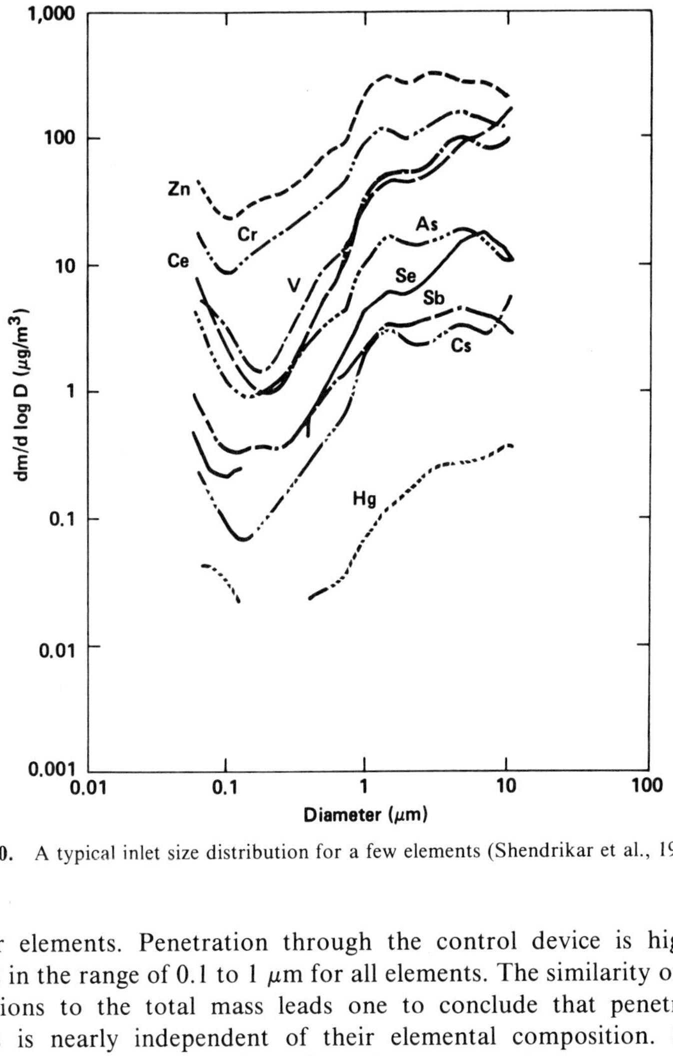

Figure 3.10. A typical inlet size distribution for a few elements (Shendrikar et al., 1983).

of other elements. Penetration through the control device is highest for particles in the range of 0.1 to 1 μm for all elements. The similarity of element penetrations to the total mass leads one to conclude that penetration of particles is nearly independent of their elemental composition. Selenium behavior may be due to condensation of vaporous selenium compounds to form particles within the control device due to a drop in flue-gas temperature. This is a reasonable conclusion because the collection efficiency of other elements of the same physical diameter is very high, and measurements taken with an electrical mobility analyzer at the same time indicates modification of the submicrometer size distribution to a greater extent than expected by only filtration processes (Ensor et al., 1981).

6.4.3. Enrichment Analysis of Particle Size Data

It is now generally recognized that elements volatilized during coal combustion eventually become enriched in the smaller particles. Such enrichment depends on many factors such as the properties of the coal, combustion

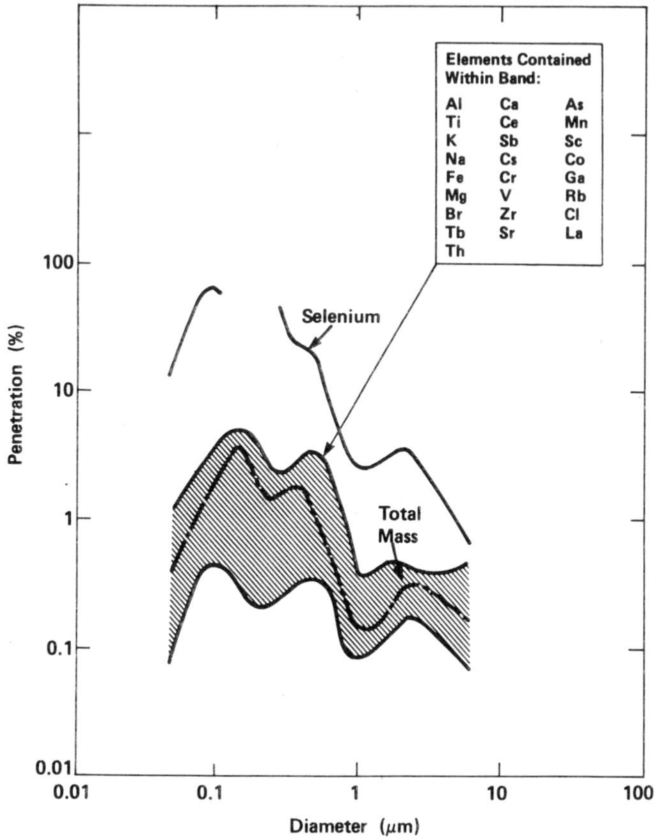

Figure 3.11. Summary of trace element penetrations through the baghouse (Shendrikar et al., 1983).

conditions, excess oxygen, residence times, and particulate control. However, within certain limitations, it is possible to make generalizations regarding enrichment profiles.

In the simplest case, elements may be divided into two groups on the basis of their concentration dependence upon particle size: (1) those that show *no* enrichments in the submicrometer particle range, and (2) those that *are* enriched. A recent study (Smith, 1980) has shown that the elements Mn, Br, V, Se, Zn, Pb, Hg, and S, among others, are usually volatilized during coal combustion to a significant extent. While elements like Si, Al, Fe, Ca, Sr, La, Sm, Eu, Tb, Dy, Yb, Sc, Zr, Ta, Na, Th, As, and In are either not volatilized or show only a minor trend, depending on combustion conditions or may be related to geochemistry of the mineral matter in the coal.

Two methods of enrichment computation are commonly used to gain additional insight into mechanisms of aerosol formation or modification. These are single ratio computations for relative enrichment and double ratios for absolute enrichment. The ratios indicate enrichment if they are greater than 1 and depletion if less than 1 when compared to stable matrix elements in the process stream. (This criterion applies only to the double ratio enrichment calculations.)

6.4.3.1. Single Ratio Enrichment. The single ratio enrichment is computed by dividing the concentration of the element of interest by the concentration of a reference element such as Fe, Sc, Al, Ti, and Eu in the same sample. The reference element ideally should meet the following criteria:

1. It follows total mass in its size-dependent behavior.
2. It is stable at the temperatures encountered in a boiler.
3. The analytical method that is used is sensitive, so precise and accurate data are available.

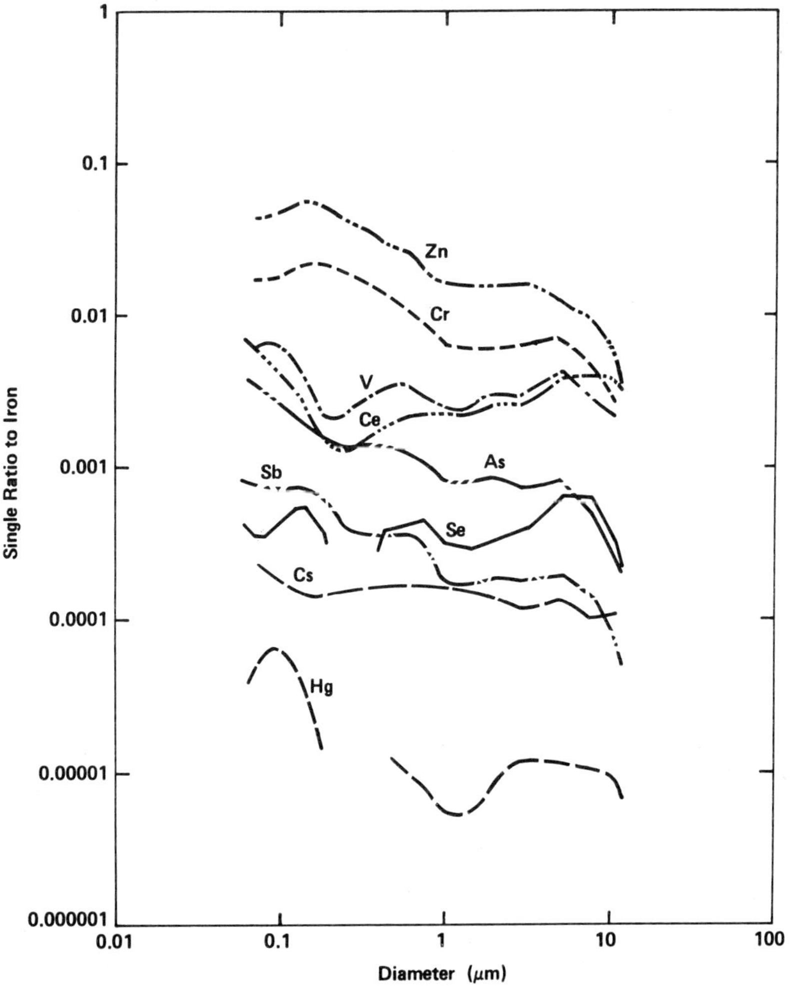

Figure 3.12. Single ratio enrichment (Ensor et al., 1981).

The single ratio analysis is useful in determining relative enrichment of the elements of interest for a particle size range. A phenomenon such as heterogeneous condensation may be verified by submicrometer single ratio enrichment resulting from the large surface-to-mass ratio in the small particle size range. Figure 3.12 shows the single ratio to iron for a few inlet elemental size distributions at a baghouse. The trace elements such as Zn, Cr, V, As, Se, Sb, Hg, and Cl show enrichment relative to iron, probably because of volatization–condensation mechanisms. Arsenic, Cu, Cr, Se, V, Sb, Cs, Ni, Zn, and Hg show an increasing single ratio with a decrease in size. This indicates that these elements are more concentrated in the smaller size range because the smaller particles have a larger surface-to-volume ratio than larger particles and therefore offer relatively more surface area for condensation of material that is volatilized during combustion.

The following conclusions could be made from the single ratio enrichment data presented in Figure 3.12:

1. A flatter curve indicates no change in enrichment as a function of particle size.
2. An elemental curve with a positive slope indicates that the element is more enriched in the larger particle size range with respect to the reference element.
3. A negative elemental curve slope indicates enrichment in the smaller particle size range with respect to the reference element than in the larger particle size range.

6.4.3.2. Double Ratio Enrichment. Double ratio enrichment was first reported by Gordon and Zoller (1973) to describe trace element concentration patterns in atmospheric aerosols. They chose to use aluminum for normalization which was also used later by Kaakinen (1974) for emission sources. The double ratio enrichment is defined as the ratio of an outlet stream's elemental mass to its reference element mass, divided by the ratio of total elemental mass in all outlet streams to the total reference element mass in all outlet streams.

Mathematically, double ratio enrichment may be expressed by

$$R_{ij} = \frac{M_{ij}/A_j}{\sum_{j=1}^{n} M_{ij} \Big/ \sum_{j=1}^{n} A_j} \tag{8}$$

where M_{ij} is the mass flow rate of element i leaving outlet streams j ($j = 1, 2, 3, \ldots n$ outlet streams), and A_j is the mass flow rate of the reference element in the jth outlet stream. The summation term in the denominator represents the total mass of element i and of the reference element leaving the various outlet streams. This technique of using all outlet streams is considered much less sensitive to process variability than using the inlet coal as a reference.

The double ratio enrichment calculation can provide enrichment ratios independent of specific coal composition and material matrices. However, it may be necessary to use the selected reference elemental concentration in the coal to calculate the flow rate of the outlet bottom ash stream. Here again, the reference element should ideally meet all the criteria mentioned for the single ratio enrichment calculation.

The elemental particle-size-dependent enrichment through the baghouse is computed with the double ratio method using Eq. (8) normalized to the reference element, iron. The results obtained from impactor data (University of Washington impactors Mark III–IV at the outlet and Mark 10 at the inlet) are presented in Figure 3.13. A solid line across the 1 on the *Y* axis is drawn to indicate clearly size-dependent enrichments, depletion with reference to

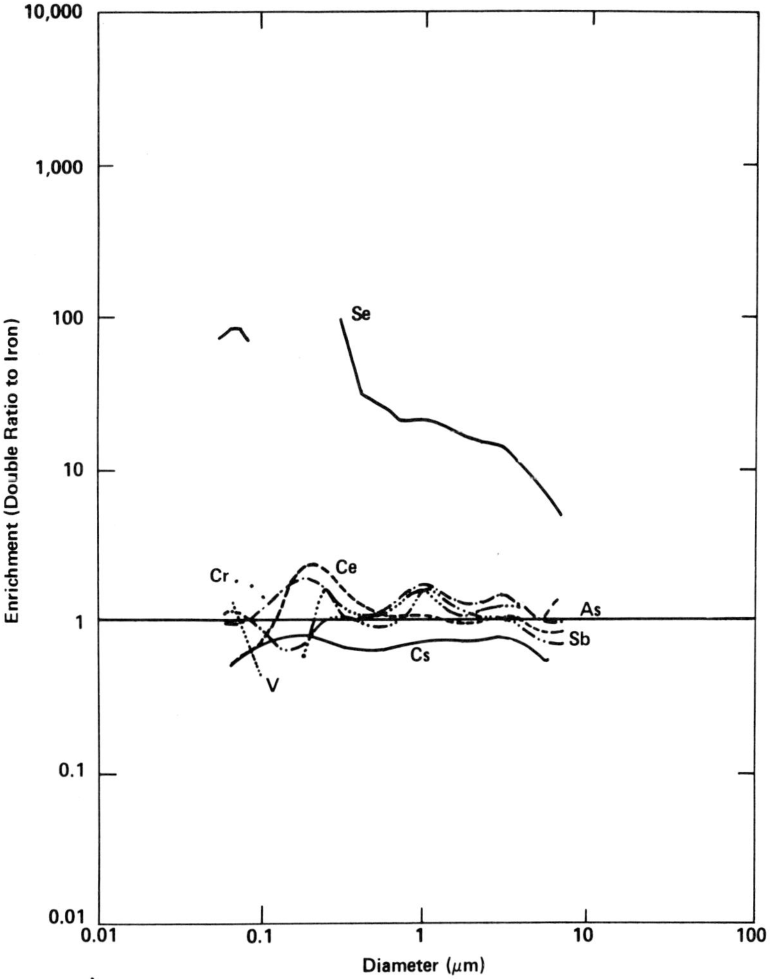

Figure 3.13. Double ratio enrichment (Ensor et al., 1981).

iron, or both. The double ratio enrichment curves for many elements are fairly flat and equal to about 1, indicating that they penetrate the baghouse in about the same way as the reference element.

7. CONCLUSIONS AND RECOMMENDATIONS FOR FUTURE RESEARCH

The emission of particulate matter and gaseous pollutants due to coal combustion has been an active concern for some time. In fact, nearly all advances in the field of control technology have centered around removal or control of particulate matter and obvious gaseous pollutants such as SO_x, NO_x, and POM. Our interest in emissions of trace materials from coal-fired power plants is relatively new, only about 10–15 yr. Data acquired to date indicate that fossil-fuel combustion of this type is a significant anthropogenic source of trace elements, many of which are potentially harmful to humans, animals, and to our biological system.

Control technology should be concerned with chemical behavior in addition to physical and engineering principles. There are still several areas to the problem of trace element emissions where research efforts are needed, and some of these are pointed out below:

1. Detailed and quantitative investigations of chemical, physical, and morphological inclusions of trace elements in coals of various types and sources. Data acquired to date on this subject matter still lacks the desired degree of quantitativeness for valid extrapolations. Such types of research efforts may help to predict the behavior of some of the trace elements of concern during coal combustion or assist in developing an overall control strategy.
2. The sample collection from any of the process streams within a facility is important. It is only through representative sample collection and accurate physical and chemical analysis that valid effects of trace element emissions can be evaluated. This area requires reseach efforts including (1) evaluations of sample collection efficiency of the sampling devices and neutrality of sample collection substrates such as filters, probe liners, and chemical solutions in impingers; (2) investigation of physical and chemical changes of trace element samples due to sampling device material (e.g., nozzle, probe, cyclone, filters); (3) improvement of present methods or devising new methods for sampling efficiently emitted volatile trace elements such as Hg, Se, As, and Sb; and (4) thorough evaluation of the integrity of size-dependent trace element samples collected from the stack gases. The problems and effects of overloading of impactors' stages, bounce-off phenomenon, wall losses, and accuracy of size segregation, particularly in the lower impactor stages, still need to be throroughly examined. Similarly, stability of sample-collection substrates requires investigation as does the compositional homogeneity of the sample on an entire impactor stage.

3. Sample storage effects for solid, liquid, and gaseous samples that are collected should be investigated. Maintenance of sample integrity should be considered essential for any kind of extrapolations. At present, no systematic study exists in the literature that shows solid sample stability or instability during storage periods. However, in related matters, some investigators have observed changes in fly ash samples due to storage. A detailed study of the stability of trace elements in liquid samples is also needed. Investigations of methods to stabilize solid, liquid, and gaseous samples for trace elements are urgently needed.

4. The role of analytical chemistry in the analysis of samples for trace elements also is an important one. In fact, sampling and analysis effects are interrelated; hence, these two must be considered in terms of mutual compatibility because sample analysis data are only as good as the sample on which the analyses are performed.

Trace element analysis methods as applied to samples from fossil-fuel combustion are still in a developmental stage. The methods discussed earlier in this chapter contribute significantly to understanding the behavior of trace elements through the coal combustion processes; however, increased efforts are still needed to validate the obtained data in terms of precision and accuracy. These two analytical parameters are sample-matrix-dependent and hence must be established using actual field samples in addition to NBS-SRM or EPA standard samples, or standard samples from the U.S. Geological Survey. Importantly, a thorough understanding of the limitations and assets of analytical methods used is essential for obtaining data that can be extrapolated with utmost confidence.

In spite of rapid developments in the field of analytical chemistry in the past few decades, we still do not have a method capable of analyzing samples of all kinds and that always produces precise and accurate data for all the trace elements of interest. Such a fact must be understood while developing an analytical protocol for the program. Research efforts are urgently needed to develop analytical methods capable of analyzing samples of fly ash *in situ* in the flue-gas environment. Such methods may contribute significantly to understanding trace element behavior in different process streams and also show some insights into the nature and mechanisms of fly ash particle progressive enrichments with respect to volatile elements such as As, Se, Hg, Cd, Cr, and Zn downstream from the boiler.

The carcinogenic and toxic potentials of trace elements are known to be dependent upon the chemical nature. For example, elemental mercury is toxic, whereas mercurous chloride is not considered to be toxic. Freshly prepared nickel oxide is potent, whereas aged nickel oxide is not. Iron in the form of dextran is toxic, but its elemental form is of nutritional value. A few of the particle surface characterization methods discussed earlier in this chapter are providing some interesting information on the fly ash surface composition and chemical nature of trace elements present. But these do not show the desired degree of quantitativeness. Efforts are needed to establish parameters such as specificity, selectivity, sensitivity, accuracy, and precision of these methods.

REFERENCES

Adams, F., Van Espen, P., and Maenhout, W. (1983). "Aerosol composition of Chacaltaya, Bolivia, as determined by size-fractionated sampling." *Atmospher. Environ.* **17,** 1521–1536.

Amdur, M., and Corn, M. (1963). "The irritant potency of zinc ammonium sulfate of different particle sizes." *Am. Ind. Hyg. Assoc. J.,* 24, 326.

Andren, A.W., and Klein, D.H. (1975). "Selenium in coal-fired power plant emissions." *Environ. Sci. Technol.* **9,** 856.

Babcock and Wilcox Corp. (1978) *Steam—Its generation and use.* 39th Ed. Babcock and Wilcox, New York.

Barnes, R.M., Ed. (1978) *Applied inductively coupled plasmas emission spectroscopy.* Franklin Institute, Philadelphia, Pa.

Barnes, R.M., Ed. (1979). *Applied emission spectroscopy.* Academic Press, New York, 147 pp.

Benes, P., and Rajam, I. (1969). "Radiochemical study of adsorption of trace elements IV: adsorption and deadsorption of bi-valent mercury on polyethylene." *Coll. Czech. Chem.* **341,** 1375.

Biermann, A.H., and Ondov, J.M. (1980). "Application of surface-deposition models to size-fractioned coal fly ash." *Atmospher. Environ.* **14,** 289–295.

Billings, C.E., and Matson, W.R. (1972). "Mercury emissions from coal combustion." *Science* **176,** 1232.

Billings, C.E., and Wilder, J. (1971). *Handbook of fabric filter technology.* NTIS PB 200-648 (July). U.S. Department of Commerce, Springfield, Va.

Bolton, N.E., Van Hook, R.I., Fulkerson, W., Emory, J.R., Lyon, W.S., Andren, A.W., and Carter, J.A. (1973). *Trace element measurements at the coal-fired allen steam plant.* ORNL-NSF-EP-43. Oak Ridge National Laboratory, Oak Ridge, Tenn.

Cahill, T.A. (1975). "Ion-Excited X-Ray Analysis of Environmental Samples." In J. F. Ziegler, Ed., *New Uses of Ion Accelerators.* Plenum Press, New York, pp. 1–71.

Cato, C.A., and Venezia, R.A. (1976). "Trace element and organic emission from industrical boilers." Presented at the 69th APCA annual meeting, Portland, Ore.

Cleland, J.G., and Kingsbury, G.L. (1977). *Multimedia Environmental Goals for Environmental Assessment.* EPA-600/7-77-136a and EPA-600/7-77-1366 (November). U.S. Environmental Protection Agency, Washington, D.C.

Cohen, J.J., and Montan, D.M. (1967). "Theoretical considerations, design and evaluation of a cascade impactor." *Am. Ind. Hyg. Assoc. J.* **28,** 95–104.

Conley, M.K., Sotera, J.J., and Kahn, H.L. (1981) "Reduction of interferences in furnace atomic absorption spectroscopy. Instrumentation Laboratory Report No. 149. Waltham, MA.

Cowherd, C. (1980) "The technical basis for a size-specific particulate standard." *J. Air Pollut. Control Assoc.* **30**(9), 971–982.

Damle, A.S., Ensor, D.S., and Ranade, M.B. (1982). "Coal combustion aerosol formation mechanisms—a review." *Aerosol Sci. Technol.* **1,** 119–133.

Diehl, R.C., Hattman, E.A., Schultz, H., and Haren, R.T. (1972). *Fate of mercury in the combustion of coal.* Bureau of Mines Technical Progress Report 54. Washington, D.C.

Duke, K.M., Davis, N.E., and Dennis, A.J. (1977). *IERL-RTP procedures manual: level 1 environmental assessment and biological tests for pilot studies.* EPA-600/7-77-043. U.S. Environmental Protection Agency, Washington, D.C.

DuPont, *Kapton polyimide film, type H—summary of properties,* Bulletin H-1D. E.I. du Pont de Nemours & Co., Inc., Film Department, Wilmington, Del.

Dzubay, T.G., Hines, L.E., and Stevens, R.K. (1976). "Particle bounce errors in cascade impactors. *Atmospher. Environ.* **10,** 229–234.

Ensor, D.S., and Hooper, R.G. (1976) "Cascade impactor measurement." In D. Van Osdell, Ed., *Proceedings of seminar on in-stack particle sizing for particulate control devices evaluation.* EPA-600/2-77-60. U.S. Environmental Protection Agency, Washington, D.C., p. 314.

Ensor, D.S., Cahill, T., and Sparks, L.E. (1975). "Elemental analysis of fly ash from combustion of a low sulfur coal." Proceedings of the 68th APCA Meeting, Boston, Mass., June 15, Paper 75-33-17.

Ensor, D.S., Cowen, S., Hooper, R., and Markowski, G.R. (1979a). *Evaluation of the George Neal no. 3 electrostatic precipitator.* EPRI-780-1. Electric Power Research Institute.

Ensor, D.S., Hooper, R.G., Markowski, G.R., and Carr, R.C. (1979b). "Evaluation of performance and particle size dependent efficiency of baghouses." *Proceedings: Advances in Particle Sampling and Measurement.* EPA-600/7-79-065. Asheville, N.C., pp. 314–336.

Ensor, D.S., Cowen, S.J., Shendrikar, A.D., Markowski, G.R., and Woffenden, G.J. (1981). *Kramer station fabric filter evaluation.* EPRI-CS-1669. Project 1130-1. Electric Power Research Institute, Palo Alto, Calif.

EPA (1977). *Control strategy preparation manual for particulate matter.* EPA-450/2-77-023. U.S. Environmental Protection Agency, Research Triangle Park, N.C.

EPA (1979). *Methods for chemical analysis of water and wastes.* EPA-600/4-79-020, March. U.S. Environmental Protection Agency, Research Triangle Park, N.C.

Federal Register (1971). "Standards of performance for new stationary sources." *Fed. Reg.* **36,** (247), 24876–24895.

Felix, L.G., Clinard, G.I., Lacey, G.E., and McCain, J.D. (1977). *Inertial cascade impactor substrate media for flue gas sampling.* EPA-600/7-77-060. U.S. Environmental Protection Agency, Washington, D.C.

Filby, R.H., Shah, K.R., Hunt, M.L., Khalil, S.R., and Saulter, C.A. (1978). *Solvent-refined coal process—trace elements.* Contract Ex-76-C-01-496. U.S. Department of Energy, Washington, D.C.

Finkelman, R.B. (1970). "Determination of trace element sites in the Waynesburg by SEM analysis of accessory minerals." *Scanning Electr. Microsc.* **1,** 60.

Fisher, G.L., Chang, D.P.Y., and Brummer, M. (1976). "Fly ash collected from the electrostatic precipitators—microcrystalline structure and the mystery of the spheres." *Science* **192,** 553.

Flagan, R.C., and Friedlander, S.K. (1978). "Particle formation in pulverized coal combustion—a review." In D.T. Shaw, Ed., *Recent developments in aerosol science.* Wiley, New York, Ch. 2.

Flagan, R.C., and Taylor, D.D. (1980). "Laboratory studies of submicron particles from coal combustion." *Proceedings of the 18th Symposium on Combustion,* Combustion Institute, Pittsburgh, Pa., pp. 1227–1237.

Flagen, R.C., and Friedlander, S.K. (1976) "Particle formation in pulverized coal combustion." Presented at the Symposium on Aerosol Science and Technology at the 82nd National Meeting of the American Institute of Chemical Engineers, Atlantic City, N.J. (Aug. 29 to Sept. 1).

Flegal, C.A., Starkovich, J.A., Maddalone, R.F., Kroft, M.L., Zee, C.A., and Lin, C. (1975). *Procedures for Process Measurements of Trace Inorganic Material.* TRW Systems, Inc., EPA Contract 68-02-1393 (July).

Francis, W. (1961). *Coal, its formation and composition.* Edward Arnold, Ltd., London.

Gluskoter, H.J., Ruch, R.R., Niller, W.H., Cahill, R.W. Dreher, G.B., and Kuhn, J.K. (1977).

Trace elements in coal: occurrence and distribution. EPA Report 600/7-77-064 (June). U.S. Environmental Protection Agency, Washington, D.C.

Gluskoter, H.J., Shimp, N.F., and Ruch, R.R. (1981). "Coal analysis, trace elements and mineral matter." In M.A. Elliott, Ed., *Chemistry of coal utilization.* Wiley, New York, Ch. 7.

Goldberg, A.J. (1972). *A survey of emissions and control for hazardous and other pollutants.* U.S. Environmental Protection Agency, Air Pollution Technology Branch, Research Triangle Park, N.C.

Gordon, G.E., and Zoller, W.H. (1973). "Normalization and interpretation of atmospheric trace element concentration patterns." Presented at 1st Annual NSF Trace Contaminant Conference, Oak Ridge, Tenn. (Aug. 8–10).

Gould, R.F. (1986). *Adsorption from aqueous solutions,* American Chemical Society Publication, Washington, D.C.

Hamersma, J.W., Reynolds, S.L., and Maddalone, R.F. (1976). *IERL-RTP procedures manual: level 1 environmental assessment.* EPA-600/2-76-100a. U.S. Environmental Protection Agency, Washington, D.C.

Harris, D.B. (1977). *Procedures for cascade impactor calibration and operation in process streams.* EPA-600/2-77-004. U.S. Environmental Protection Agency, Washington, D.C.

Headlee, A.J.W., and Hunter, P.B. (1955). "The inorganic elements in the coals." In *Characteristics of mineable West Virginia coals.* West Virginia Geological Survey, Vol. XIIIA. Morgantown, W.Va.

Henry, W.M., and Knapp, K.T. (1980). "Compound forms of fossil fuel fly ash emissions." *Environ. Sci. Technol.* **14,** 450–456.

Hering, S.V., Flagan, R.C., and Friedlander, S.K. (1978). "Design and evaluation of new low-pressure impactor. 1." *Environ. Sci. Technol.* **12,** 667–673.

Hering, S.V., Friedlander, S.K., Collins, J.J., and Richards, L.W., (1979). "Design and evaluation of a new low-pressure impactor. 2." *Environ. Sci. Technol.* **13,** 184–188.

Huelett, L.D., and Weinberger, A.J. (1980). "Some etching studies of the microstructures and composition of large aluminosilicate particles in fly ash from coal-burning power plants." *Environ. Sci. Technol.* **14,** 965–970.

Kaakinen, J.W. (1974). "Trace element in a pulverized coal-fired power plant." Ph.D. Disseration, University of Colorado, Boulder, Colo.

Kaakinen, J.W., Jorden, R.M., Lawasani, M.H., and West, R.F. (1975). "Trace-element behavior in a coal-fired power plant." *Environ. Sci. Technol.* **9,** 862.

Klein, D.H., Andren, A.W., Carter, J.A., Emery, J.F., Feldman, C., Fulkerson, W., Lyon, W.S., Ogle, J.C., Talme, Y., Van Hook, R.I., and Bolton, N. (1975). "Pathways of thirty-seven trace elements through coal-fired power plant." *Environ. Sci. Technol.* **9,** 973.

Knapp, K.T. (1982). "Elemental composition of sized profiles emitted from stationary sources." In *Proceedings of the International Symposium on Recent Advances in Particulate Science and Technology,* Madras, India (Dec. 8–10).

Linton, R.W., Loh, A., Natusch, D.F.S., Evans, Jr., C.A., and Williams, P. (1976). "Surface predominance of trace elements in airborne particles." *Science* **191,** 852–854.

Lundgren, D.A., and Balfour, W.D. (1980). *Use of limitations of in-stack impactors.* EPA-600/2-80-048. U.S. Environmental Protection Agency, Washington, D.C.

Maddalone, R.F., and Quinlivan, S.C. (1976). *Technical manual for inorganic sampling and analysis.* Document No. 29416-6038-RU-00. TRW Systems, Inc. Redondo Beach, Calif.

Maddalone, R.F., Jackson, B., and Yu, C. (1981). *Scrubber-generated particulate matter—literature survey.* EPRI-CS-1739, Project 982-11. Electric Power Research Institute. Palo Alto, Calif.

Magee, E.M. (1976). *Evaluation of pollution control in fossil fuel conversion processes.* Report PB 255-842. U.S. Environmental Protection Agency, Research Triangle Park, N.C.

Malte, P.C. (1977). "Inorganic pollutants from pulverized coal combustion." Presented at the 1977 fall meeting western states section, Combustion Institute, Stanford,, Calif. (Oct.).

Mann, R.E., Magee, R.A., Celling, R.V., Fusch, M.R., and Mesici, F.G. (1978). *Trace element in fly ash.* EPA Contract No. 08-01-8788. Research Triangle Park, N.C.

Markowski, G.R., Ensor, D.S., Drehsen, M.E., and Shendrikar, A.D. (1980). *Fine particle sampling, analysis and data reduction procedures, and manual for low pressure impactor and electrical aerosol analyzer.* EPRI Project 1410-3. Electric Power Research Institute. Palo Alto, Calif.

Marple, V.A., and Liu, B.Y.H. (1974). "Charactereistics of laminar jet impactors." *Environ. Sci. Technol.* **8,** 648.

Matsunoga, K., Konishi, S., and Nishimura, M. (1979). "Possible errors caused prior to measurement of mercury in natural water with special reference to sea water." *Environ. Sci. Technol.* **13,** 63–65.

McCain, J.D., Clinard, B., Felix, L., and Johnson, J. (1979). "A data reduction system for cascade impactors." *Proceedings, advances in particle sampling and measurement.* EPA-600/7-79-065. Asheville, N.C., May 1978.

McDonald, J.R. (1978). *Mathematical model of electrostatic precipitation* (Revision 1). Volumes I and II. EPA-600/7/77-111a and b, June. U.S. Environmental Protection Agency, Washington, D.C.

McElroy, M.W., Carr, R.C., Ensor, D.S., and Markowski, G.R. (1982). "Size distribution of fine particles from coal combustion." *Science* **215,** 13–19.

McQuaker, N.R., and Sandberg, D.K. (1982). "The determination of mercury source emissions in the presence of high levels of SO_2." *J. Air Pollut. Control Assoc.* **32,** (6), 634–636.

Mezey, E.J., et al. (1976). *Fuel contaminants, Vol. 1 of Chemistry.* EPA-600/2-76-177a (Battelle Columbus Laboratories). Columbus, Ohio.

Mims, C.A., Neville, M, Quamm, R.J., and Sarofim, A.F. (1979). "Laboratory studies of trace elements transformation during coal combustion." Paper 78 presented at the 87th AICHE meeting, August. Boston.

Moffitt, Jr., A.E., and Kupel, R.E. (1970). "A rapid method employing impregnated charcoal and atomic absorption spectrometry for the determination for the mercury in atmospheric, biological, and aquatic samples." *Atomic Absorp. Newslett.* **9,** 113.

Moore, C.B. (1981). "Geological and inorganic minerals." *Anal. Chem.* **53,** 39R.

MRI (1974). *Inertial cascade impactor model 1502 instructional manual.* IM-174. Meteorology Research Inc., Altadena, Calif.

Murthy, J. (1975). "Determination of mercury in coals by peroxide digestion and cold vapor atomic absorption spectrometry." *Atomic Absorp. Newslett.* 14, 151.

Natusch, D.F.S., and Wallace, J.R. (1976). "Determination of airborne particle size distributions: calculation of cross-sensitivity and discreteness effects in cascade impaction." *Atmospher. Environ.* **10,** 315–324.

Natusch, D.F.S., Wallace, J.R., and Evans, Jr., G.A. (1974). "Toxic trace elements—preferential concentration in respirable particles." *Science* **193,** 202.

Neville, M., Quann, R.J., Haynes, B.S., and Sarofim, A.F. (1980). "Vaporization and condensation of mineral matter during pulverized coal combustion." Presented at 18th Symposium on Cumbustion, Waterloo, Canada, August.

Ode, W.H. (1963). "Coal analysis and mineral matter." In H.H. Lowoy, Ed., *Chemistry of coal utilization,* Supplementary Volume. John Wiley & Sons, Inc., New York.

O'Gorman, J.V., and Walker, Jr., P.L. (1972). *Mineral matter and trace elements in U.S. coals.* Report No. 61, Interim Report No. 2. Office of Coal Research, U.S. Department of the Interior, Research and Development, Washington, D.C.

Ondov, J.M., Zoller, W.H., Olmez, I., Aras, N.K., Gorden, G.E., Rancitelli, L.A., Abel, K.H.,

Filby, R.H., and Shah, K.R. (1974). *Four-laboratory comparative instrumental nuclear analysis of NBS coal and fly ash standard reference material.* Special Publication 422. National Bureau of Standards. Washington, D.C.

Ondov, J.M., Zoller, W.H., Olmez, I., Aras, N.K., Gordon, G.E., Rancitell, L.A., Abel, K.H., Filby, R.H., Shah, K.R., and Ragaini, R.C. (1975). "Elemental concentrations in the National Bureau of Standards environmental coal fly ash Standard Reference Materials." *Anal. Chem.* **47,** 1102.

Ondov, J.M., Ragaini, R.C., and Biermann, A.H. (1978). "Elemental particle-size emissions from coal-fired power plants: use of an inertial cascade impactor." *Atmospher. Environ.* **12,** 1175–1185.

Ondov, J.M., Ragaini, R.C., and Biermann, A.H. (1979). "Elemental emissions from a coal-fired power plant. Comparison of a venturi wet scrubber system with a cold-side electrostatic precipitator." *Environ. Sci. Technol.* **13,** 598–607.

Peele R. (1966) *Mining engineer's handbook.* Vol. 2, 3rd Ed. John Wiley & Sons, Inc., New York.

Perry J.H., Chilton, C.H., et al. (1969). *Chemical engineer's handbook.* 4th Ed. McGraw-Hill Book Company, New York.

Pilat, M.J., Ensor, D.S., and Bosch, J.C. (1970). "Source test cascade impactor." *Atmospher. Environ.* **4,** 671–679.

Pilat, M.J., Raemhild, G.A., Powell, E.B., Fiorette, G.M., and Meyer, D.F. (1978). *Development of cascade impactor system for sampling 0.02 to 20-micron diameter particles.* FP-844, Vol. 1. University of Washington, Seattle, Wash.

Ramsden, A.R. (1969). "A microscopic investigation into the formation of fly ash during the combustion of pulverized bituminous coal." *Fuel (London)* **48,** 121–237.

Robertson, O.E. (1968). "Adsorption of trace elements in sea-water on various container surfaces." *Anal. Chem. Acta* **42,** 533.

Robinson, R.L. (1978). "Elemental analysis, plasmas revive emission spectroscopy." *Science* **198,** 369.

Ruch, R.R., Gluskoter, H.J., and Shimp, N.F. (1963). *Occurrences and distributions of potentially volatile trace elements in coal.* An Interim Report, Illinois State Geological Survey, April. Champaign, Ill.

Sarofim, A.F., Howard, J.B., and Padia, A.S. (1977). "The physical transformation of the mineral matter in pulverized coal under simulated coal combustion conditions." *Combust. Sci. Technol.* **16,** 187–204.

Shendrikar, A.D. (1982). "Critical review—sample storage of emitted particulate matter from power plants." Report submitted to the Electric Power Research Institute as part of a contract managed by Gordon A. Enk Associates, Inc., Markelyhouse, Medusa, N.Y.

Shendrikar, A.D., and West, P.W. (1974). "A study of adsorption characteristics of traces of Chromium-III and VI on selected surfaces." *Anal. Chem. Acta* **72,** 91.

Shendrikar, A.D., and West, P.W. (1975). "The rate of loss of selenium from aqueous solutions stored in various containers." *Anal. Chem. Acta* **74,** 189.

Shendrikar, A.D., Dharmarajan, V., Merrick, H.W., and West, P.W. (1976). "Adsorption losses of barium, beryllium, cadmium, managanese, lead, and zinc from dilute solutions." *Anal. Chem. Acta* **84,** 409.

Shendrikar, A.D., Damle, A.D., Gutknecht, W.F., and Briden, F.E. (1983). "Mercury atmosphere generation and media collection efficiency evaluations for the SASS impinger system." Paper presented at the APCA Specialty Conference, Chicago, Ill., March.

Shendrikar, A.D., and Ensor, D.S. (1981). "The role of analytical chemistry in the measurements of combustion aerosols." Paper presented at the AIChE meeting, Detroit, August 17–18.

Shendrikar, A.D., Filby, R., Markowski, G.R., and Ensor, D.S. (1984). "Trace element loss onto polyethylene container walls from impinger solutions from flue gas sampling." *J. Air Pollut. Control. Assoc.* **34,** 233–236.

Shendrikar, A.D., Ensor, D.S., Cowen, S.J., Woffindend, G.J., and McElroy, M.W. (1983). "Size-dependent penetration of trace elements through a utility baghouse." *Atmos. Environ.* **17,** 1411–1421.

Smith, R.D. (1980). "The trace element chemistry of coal during combustion and the emissions from coal-fired power plants." *Prog. Energy Combust. Sci.* **6,** 53–114.

Smith, R.D., and Street, G.B. (1978). "Mass spectrometric study of the volatile phase species of brominated polymeric sulfur nitride and tetrasulfur tetranitride." *Inorganic Chem.* **17,** 938.

Smith, R.D., Cambell, J.A., and Nelson, K.K. (1979). "Characterization and formation of submicron particles in coal-fired plants." *Atmospher. Environ.* **13,** 607–617.

Smith, W.B., and Wilson, R.R. (1978). *Development and laboratory evaluation of a five-stage cyclone system.* EPA-600/7-78-008. U.S. Environmental Protection Agency, Research Triangle Park, N.C.

Starik, E. (1959). *Principles of radiochemistry.* U.S. Atomic Energy Commission. Washington, D.C.

Swain, D.J. (1976). *Trace elements in coal—recent contributions to geochemistry and analytical chemistry.* A.E. Tugarinao, Ed. John Wiley & Sons, Inc., New York.

Taggart (1945). *Handbook of mineral dressing.* John Wiley & Sons, Inc., New York.

Task Group on Lung Dynamics (1966), *Health Phys.* **12,** 1973.

Tzavaras, J., and Shendrikar, A.D. (1984). "An ICP analytical scheme to screen environmental samples for metals." *American Lab.,* July, pp. 59–65.

Van Espen, P., Adam, F., and Maerhaut, W. (1981). "Analysis of size-fractionated air particulate matter by energy-dispersive X-ray fluorescence spectrometry." *Bull. Soc. Chem. Belge* **90,** 305–315.

Walt, J.D. (1968). *The occurrence, origin, identity, distribution, and estimation of the mineral species in British coals.* British Coal Utilization Research Assoc., Leatherhead, Surry, England.

Watson, J.G. (1981). *The state-of-the-art of receptor models relating ambient suspended particle matter to sources.* EPA-600/2-81-039, March. U.S. Environmental Protection Agency, Research Triangle Park, N.C., 88 pp.

Watters, R.L., and Norris, J.A. (1978). "Applied inductively coupled plasma emission spectroscopy." Conference Proceedings. Franklin Institute Press, Philadelphia, Pa., pp. 65–81.

Wolfe, G., and Flocchini, R. (1977). "Collection surfaces of cascade impactors." In T. Dzubay, Ed., *X Ray fluorescence analysis of environmental samples.* Ann Arbor Science, Ann Arbor, Mich.

Wroblewski, S., Spittler, T.M., and Harrison, P.R. (1974). "Mercury concentration in the atmosphere in Chicago. A new ultrasensitive method employing amalgamation." *J. Air Pollut. Control Assoc.* **24,** 778–781.

Yung, S.S., Calvert, S., and Barbarika, H. (1977). *Venturi scrubber performance model.* EPA-600/2-77-172 (NTIS PB 271-515), August. U.S. Environmental Protection Agency, Research Triangle Park, N.C.

4

SMELTING OPERATIONS AND TRACE METALS IN AIR AND PRECIPITATION IN THE SUDBURY BASIN

Walter H. Chan
Maris A. Lusis

Air Resources Branch
Ontario Ministry of the Environment
Toronto, Ontario

1. INTRODUCTION

This chapter will consider the atmospheric fate of the trace metal emissions from two large copper and nickel smelters—INCO Limited and Falconbridge Limited—at Sudbury, Ontario. The work to be described was carried out, largely during the period mid-1978 to mid-1980, as part of the Ontario Ministry of the Environment's Sudbury Environmental Study—a multidisciplinary study into the atmospheric, aquatic, and terrestrial impacts on the Sudbury Basin of the smelting operations—an overview of which is given elsewhere (Ontario Ministry of the Environment, 1982). The atmospheric program included detailed measurements of the nature and quantity of the emissions, and their subsequent dispersion, chemical transformation, and wet and dry deposition (see Chan (1982) and references cited therein), and yielded information on air and precipitation concentrations of trace metals in the Sudbury Basin, the impact of the local smelter sources on air and precipitation levels of trace metals, as well as the portion of the emissions removed within the study area by wet and dry deposition processes.

It is noted that many trace metals are emitted from the Sudbury smelters in varying amounts. This study focused on eight trace metals, namely Fe, Cu, Ni, Pb, Zn, Cr, Cd, and Al, because of their emitted quantities and toxicity. Although the study was designed to examine the contributions from both the INCO and the Falconbridge sources, because of the scarcity of proper sampling sites around the Falconbridge source and the lower emission rates of the Falconbridge source compared to INCO, more emphasis is put on the results of the INCO smelter. The discussion is applicable to emission conditions during the study period; however, it is expected that the general conclusions can be extrapolated to other similar sources and emission configurations.

2. EMISSIONS: SOURCES AND CHARACTERISTICS

Figure 4.1 shows the study area, including the major sources—the two smelters (INCO Limited at Copper Cliff, Falconbridge Limited in the town of Falconbridge), and a nickel refinery–iron ore recovery plant complex operated by INCO (at Copper Cliff). Table 4.1 gives an overview of the sulfur dioxide, sulfuric acid, total particulate, and major metal emissions, based on the available measurements during the period 1973–1981 (Ozvacic, 1982). This

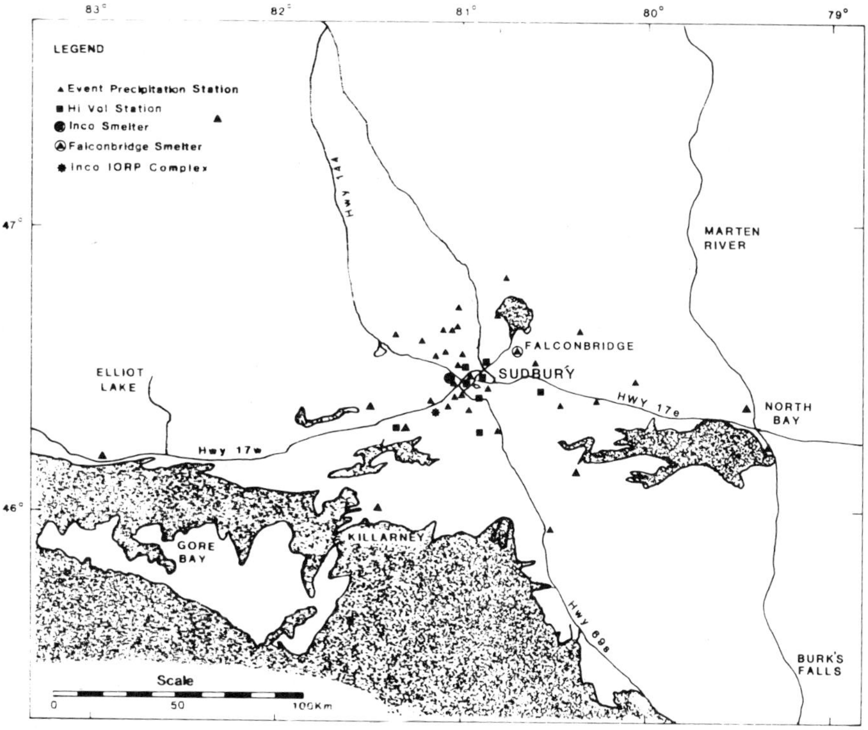

Figure 4.1. Locations of monitoring sites around Sudbury.

table shows that there has been a considerable year-to-year variability in the emission rates, and therefore, when considering the atmospheric concentration and deposition rate results, one should keep in mind that they apply primarily to the emissions during the study period.

The most significant source in the Sudbury area is the 381-m stack at the INCO smelter, which has also been the object of the most extensive testing activities. This stack emits particulates and gases from pyrometallurgical smelting processes at the Copper Cliff smelter, where nickel and copper of appropriate purity are made, in the form of nickel Bessemer matte and blister copper, from suitably prepared ore concentrates. About 80% of the total Sudbury sulfur emissions, as well as a large portion of the particulates, are attributable to the 381-m stack. Table 4.2 gives detailed information on yearly metal emissions from this source during the 1973–1980 period, estimated from stack testing results (Ozvacic, 1982). Iron, Cu, Ni, Pb, and As were the metals emitted in the largest amounts. Note that the relatively low values in 1978 and 1979 were due to a scheduled shutdown and a strike during this period. INCO was shut down for maintenance during July 17–August 27, 1978, and because of a strike during September 16, 1978–June 7, 1979.

Table 4.1. Yearly Emissions (in t) of Major Pollutants in the Sudbury Basin for the Period 1973–1981[a]

Source	Variation	Sulfur Dioxide[b]	Sulfuric Acid	Total Particulate	Iron	Copper	Nickel	Lead	Arsenic
INCO 381-m stack[c]	Maximum	1,185,449	14,541	14,494	1,454	350	342	298	201
	Average	885,667	7,270	11,417	990	245	228	184	114
	Minimum	383,000	3,241	6,491	201	70	53	88	70
INCO 194-m stack	Maximum	104,390							
	Average	54,568	1,664	2,380	643	171	226	6	4
	Minimum	18,980							
INCO Smelter	Maximum	16,000		755	90	312	40	0.8	0.1
(low level)	Average	12,000	88	586	70	242	31	0.6	0.1
	Minimum	6,000		283	34	117	15	0.3	0.1
Falconbridge	Maximum	274,000							
93-m stack	Average	173,000	438	865	98	11	9.6	13.4	6.4
	Minimum	88,000							
INCO[d]									
Two 45-m stacks	Average			(4,073)	(2,354)				
Total	Average	1,125,235	9,460	15,248	1,801	669	500	204	125

[a] Basis: 365 days × 24 h/day production.

[b] Except for INCO Smelter (low level), all sulfur dioxide emissions were obtained by mass balance calculations. Annual low level ammonia emissions from INCO's iron ore recovery plant in 1977 was 4,219 t.

[c] Oxides of nitrogen emissions expressed as NO_2 = 3,281 t; hydrogen chloride emissions = 530 t. Emissions of particulates, iron, copper, and nickel from the 381-m stack are yearly values calculated from correlations with sulfur dioxide emissions. Emissions of sulfuric acid, lead, and arsenic are average measured values.

[d] Emissions ceased in April, 1980. Emission rates are not included in average.

Table 4.2. **Average Measured Emissions of Metals and Other Particulate Pollutants from the INCO 381-m Stack (in kg/h)[a]**

Year	Fe	Cu	Ni	Pb	Al	As	Zn	Cr	Se	Cd	Bi	Be	Mn
1973	228	65	48	21	18	11	6	6	6	4	3	2	1
1974	258	55	55	29	6	10	7	15	6	5	3		2
1975	64	27	15	23	4	23	11	0.5	6	0.8	2		0.2
1976	74	25	22	12	3	15	7	0.2		1	1		0.1
1977	142	41	33	18	9	8	11	2		1	5		0.4
1978	80	20	20	10			1						
1979	31	13	12										
1980	147	47	44	34	19	10	6	2	3	2			0.3
Average	128	37	31	21	10	13	7	4	5	2	3	2	0.7

Year	Sn	Sb	Sr	Ag	V	Li	Si	Mg	B	Ca	Te	Co	Mb	Ti	K	Na
1973	0.8	0.6	0.4	0.3	0.2	0.1	109	15	10	9	0.3					
1974							19	5		5		2.0	4			
1975							14	1		3		0.3				
1976			0.1	0.1	0.1	0.6						0.1	0.1	1	4	20
1977					1.0		32	6		7		1.0			3	2
1978																
1979												0.3				
1980								4				2.0				
Average	0.8	0.6	0.3	0.2	0.4	0.3	43	6	10	6	0.3	0.9	2	1	3	11

[a] Emission rates of mercury were measured in 1980 at 7 g/h.

For the interpretation of atmospheric deposition results as well as health assessment studies, it is necessary to have information on the particle size distribution for the trace metals, in addition to their emission rates. Particle size distribution measurements have been made both within the 381-m stack (Ozvacic and McDonald, 1982) and in the plume near the smelter using Andersen impactors mounted on a helicopter (Chan et al., 1983a). The in-plume studies are considered to be of greater relevance here (since particle size distributions are modified when the hot gases are mixed with ambient air on exiting from the stack), and results of mass median diameter determinations from a number of helicopter flights are shown in Table 4.3 (Chan et al., 1983a). Note that metals such as Fe, Cu, Ni, Al, Mg, and Mn are found predominantly in the coarse category (with mass median diameters greater than 2.5 μm), while Pb, Zn, Se, and As occur in fine particulates (with mass median diameters less than 2.5 μm). Results for Cd and Cr seem to fluctuate from run to run.

The 194-m stack at the INCO Iron Ore Recovery Plant is the second largest source of industrial emissions in the Sudbury area and receives emissions from a complex consisting of an iron ore recovery plant, a nickel refinery, and three adjacent sulfuric acid plants. This stack emits comparable amounts of Fe, Cu, and Ni to the 381-m stack at the smelter, but considerably smaller amounts of other trace metals. In-plume particle size distribution measurements have also been made at the 194-m stack, and the results have been found to depend quite strongly on the operating conditions of the nickel refinery portion of the complex (Chan et al., 1982a). Generally, mass median diameters of Fe, Cu,

Table 4.3. Mass Median Diameter (μm) of Particulate in the INCO 381-m Stack Plume by Andersen Sampling

Trace Metal	1979 8 Sept.	1980 31 Jan.	1980 4 Feb.	1980 5 Feb.[a]	1980 7 Feb.
Fe	5.2	>9	>9	>9	>9
Cu	>9	>9	>9	>9	2.8
Ni	>9	>9	>9	>9	>9
Pb	1.0	1.05	0.92	1.7	1.15
Zn	1.1	1.1	0.96	6.6	2.1
Al[b]	3.2	2.1	8.2	>9	5.6
Cr		3.8	1.0		
Cd	3.1	1.5	1.9	1.1	3.4
Mn	4.2				
Se				1.25	1.4
As		0.98	0.92	0.96	1.0

[a] Emissions on this date were atypical. Particulate fallout near the stack was observed.

[b] Probably contamination due to Al sampler.

Ni, and Al were found to be lower in this plume than those from the 381-m stack, whereas those of Pb, Zn, and As were found to be somewhat larger, but most of the above metals could be classified in the coarse category.

The two 45-m stacks noted in Table 4.1 were operational during most of the present study period (until April 1980) and received emissions from a pelletizing plant associated with the iron ore recovery process (currently shut down). The INCO low-level emissions are another source of trace metal emissions of potential importance because they escape various process fume collection steps in the smelter and are emitted relatively close to the ground on building roofs, windows, and ventilators. Owing to the scarcity of measurements, there is a great deal of uncertainty about the values shown in Table 4.1 for this source (Ozvacic, 1982).

The Falconbridge 93-m stack receives off-gases from the Falconbridge smelter operations, which include smelting and converting, as well as gases from an adjacent sulfuric acid plant. Compared to the INCO smelter, it is a relatively small source of trace metal emissions, although the low emission height, combined with local topographical and meteorological features, leads to frequent fumigations (and hence dry deposition) fairly close to the smelter. Particle size distribution measurements (Chan et al., 1983a) have been carried out at this source also; the in-plume measurement results are shown in Table 4.4. In general, the trace metals emitted by the Falconbridge 93-m stack seem to be either in smaller, or comparable, particle size ranges to those from the INCO 381-m and 194-m stacks.

It may be noted that during the airborne plume particle sampling experiments, a large body of data was also collected on ratios of trace metals and SO_2 concentrations in the emissions. These results (shown later in Table 4.14) allowed trace metal emission rates to be estimated from the known SO_2 rates during study periods when no metal emission measurements were

Table 4.4. Mass Median Diameter (μm) by Andersen Sampling in the Falconbridge Smelter Plume, 1979

Trace Metal	1 Mar.	2 Mar.	3 Mar.	1 Sept.	2 Sept.
Fe	>9	>9	8.65	7.1	7.0
Cu	>9	7.35	8.20	5.9	4.4
Ni	>9	>9	>9	9	6.6
Pb	0.92	1.01	0.87	0.95	
Zn	<4.7	1.90	1.00	0.84	
Al[a]		5.40		7.2	2.9
Mn				5.5	
Cd	0.7	1.09	0.93	5.3	2.3
As	0.89	0.67	0.69		

[a] Probably contamination due to Al sampler.

available. Sulfur dioxide emission rates are continuously monitored at some of the sources and can be estimated from a mass balance at others.)

3. AIR MEASUREMENTS AND DRY DEPOSITION RATES

3.1. Air Concentrations and Dry Deposition of Trace Metals

To determine trace metal dry deposition rates in the present study, it was decided to make air concentration measurements, from which the dry deposition could be inferred using deposition velocities and the equation

$$\text{flux} = (\text{deposition velocity}) \times (\text{air concentration})$$

3.1.1. Experimental

Air concentrations were measured using high-volume samplers at eight locations out to about 50 km from the INCO smelter (Fig. 4.1), mainly to the south and southeast of the smelter, such that upwind and downwind (background and smelter-impacted) data could be obtained on many occasions. Twenty-four hour sampling was done (from 0000 to 2400 EST), using Whatman 41 cellulose filters, which are a good collection medium due to their low trace metal blank values. However, this sampling technique suffers from the disadvantage that there is a possibility of copper contamination of the sample from the exhaust of the HiVol sampler motor (King and Toma, 1975). Evidence of copper contamination was in fact found in the samples, although the other trace metals of interest seemed to be unaffected. Thus, the values reported here for copper should be regarded as qualitative only.

Samples were analyzed for trace metals by flame and flameless atomic absorption spectroscopy. In order to cut down the analysis load (since three samples were taken per week at each location), filters were analyzed only on days when simultaneous background and smelter-impacted samples could be identified. This was done by close inspection of local meteorological data (including continuous wind measurements at two meteorological towers (ground level and 114 m), and pilot balloon data and visual plume observations whenever available), as well as continuous SO_2 data from nearby monitoring stations operated by the Ministry of Environment's Northeastern Region. Samples thus identified as being potentially influenced by smelter emission, as well as those suitable for use as background samples, were then submitted for trace metal analysis.

The HiVol air sampling program was operated from July 1978 to May 1980. For a complete discussion of the network, see Chan et al. (1982b).

To estimate values of deposition velocities, information is needed on particle size distributions of the trace metals. The in-plume sampling work at a number of the major emission sources has already been alluded to in Section 2, and the information shown in Tables 4.3 and 4.4 was used for calculation of

Table 4.5. Summary of Mass Median Diameters (μm) of Particulates in Background Air in the Sudbury Area[a]

Trace Metal	August 1978		April 1979		August 1979		March 1980	
	MMD	Av	MMD	Av	MMD	Av	MMD	Av
Fe	8.0, 5.5, 9.0, 6.0	7.1	7.2, 6.0, 7.4, 5.5	6.5	9.0, 4.0, 7.0, 2.7	5.7	9.0, 2.6, 5.0, 2.1	4.7
Cu	3.8, 2.8, 7.4, 2.3	4.1	3.0, 2.6, 5.4, 2.0	3.3	8.0, 4.0, 2.7, 1.6	4.1	1.15, 2.0, 2.2, 1.5	1.7
Ni	>9.0, —, >9.0, >9.0	>9	7.0, —, >9.0, 1.7	—	8.2, —, 6.2, —	7.2	6.4, —, 8.0, 2.4	5.6
Pb	1.1, —, *3.3*, 0.7	0.9	0.7, 1.1, 1.5, 1.4	1.2	0.56, 0.68, 0.84, 0.8	0.7	0.56, 0.44, 0.42, 0.55	0.5
Zn	0.4, 0.64, *7.2*, —	0.52	<0.4, 1.1, 1.7, 0.4	0.8	<0.4, 0.74, 0.4, 0.74	0.5	1.1, 0.92, <0.4, 0.7	0.7
Mn	—, 3.9, —, —	3.9	4.7, 3.7, *0.68*, —	4.2	2.3, 2.3, *0.36*, 1.55	2.05	2.6, 1.6, —, —	2.1
Al	>9.0, 5.3, >9.0, 8.0	—	6.0, 5.8, 5.5, 4.0	5.3	7.4, 4.2, 5.3, *<0.4*	5.63	6.6, 3.5, 2.5, 1.1	3.4
Cd	3.0, —, 1.6, >9.0	—	1.9, 1.15, 1.1, 1.05	1.3	0.64, 1.5, 0.84, 0.76	0.9	0.60 *2.5*, 0.66, 0.90	1.2
Cr	—	—	2.8	2.8	—	—	—	—
As	—, 0.55, —, —	0.55	1.0, 1.15, 0.85, 0.9	1.0	—, —, —, —	—	—, —, —, —	—
Se	—, 0.68, —, —	0.7	1.0, 1.35, 0.9, —	1.1	—, —, —, —	—	—, —, —, —	—

[a] Italicized values are suspicious and are not included in calculations.

deposition velocities at samplers being impacted by the INCO or Falconbridge sources. In addition, measurements were made (using similar particle sizing techniques) near ground level at four locations in the Sudbury Basin (Ash Street, Burwash, Lake Panache, and Happy Valley) and at various times of the year. These represent data clear of any smelter plume for most of the sampling period and were used to determine deposition velocities at samplers considered to be measuring background air. See Table 4.5 for a summary of the data, and Chan et al. (1982b) for a discussion of the experimental methods used.

3.1.2. *Deposition Velocity Calculations*

Although the approach taken here to determine the dry deposition rate is currently considered to be most promising for routine monitoring (Hicks et al., 1980), information on deposition velocities is scarce, especially for fine particulate matter, and consequently there is a large amount of uncertainty in our trace metal dry deposition rate estimates. Several approaches to calculating the deposition velocity were explored, but the one finally adopted made use of the deposition velocity–particle size relationship published by McMahon and Denison (1979), which is based on various field and laboratory measurements (see Fig. 2 of their paper). For a particular trace metal, the deposition velocity corresponding to each size range of particles trapped on a given stage of the impactor was multiplied by the corresponding fraction of

Table 4.6. Summary of Deposition Velocities V_d (cm/s)

Trace Metal	Background[a]	Smelter Source[b]
Fe	1.05	1.78
Cu	0.72	1.32
Ni	1.38	1.78
Pb	0.23	0.34
Zn	0.25	0.38
Mn	0.85	
Al	0.87	1.23
Cd	0.25	0.49
Cr	0.95	0.82
As	0.14	(0.22)
Se	0.25	0.19

[a] Inferred from particle sizing data obtained at ground level in August 1978, April and August 1979 and March 1980.

[b] Inferred from particle sizing data obtained in the INCO and Falconbridge plumes in September 1979 and March 1980.

the total mass (for that metal) in that particular size range (Chan et al., 1984). In other words, the mass mean deposition velocity was determined for each trace metal and particle sizing experiment. Representative results from a number of experiments are shown in Table 4.6. These were used in subsequent calculations. Note that trace metals found in coarse particles, such as iron, have relatively large deposition velocities, while submicron particles have values severalfold smaller.

Results for plume and background deposition velocities are somewhat different because of the difference in particle size distributions. The values seem to be in the general range recently reported in the literature (McMahon and Dennison, 1979; Sehmel, 1980; Hicks, 1983).

3.2. Results

3.2.1. Air Concentration of Trace Metals in the Sudbury Basin

A statistical summary of the two years' data from all HiVol stations in the network is given in Table 4.7. The data may be taken as representative of average long-term trace metal concentrations in the Sudbury Basin. Note that the data for copper contain a contribution from the sampler exhaust and thus are not representative of the ambient air. There is a great deal of variability in the data, as can be seen from the concentration range for each parameter in Table 4.7. This is expected in the vicinity of large point sources, where occasional large excursions of concentration occur due to plume fumigations. A limited number of samples were also analyzed for As, Se, and Mn, which have arithmetic mean concentrations ($\mu g/m^3$) of 0.003, 0.003, and 0.008, respectively. The values for the first two species are probably underestimates because they may be associated with volatile compounds.

When the air data are inspected for individual stations at different distances from the smelters, there is some indication of a decrease in concentration with distance from the source (even in the long-term data), which becomes quite obvious when attention is focused only on those samples which are identified as being potentially impacted by the smelter. For example, Table 4.8

Table 4.7. Statistical Summary of Sudbury Environmental Study HiVol Results ($\mu g/m^3$)

Statistic	Fe	Ni	Cu	Pb	Zn	Cd	Al
Sample size	1245	1235	1246	1246	1244	1236	854
Maximum	8.675	0.732	2.306	0.563	0.387	0.032	2.222
Minimum	0.010	0.001	0.003	0.002	0.005	0.000	0.045
Arithmetic mean	0.525	0.021	0.426	0.066	0.037	0.002	0.189
Arithmetic standard deviation	0.765	0.056	0.370	0.069	0.039	0.003	0.255
Geometric mean	0.286	0.005	0.286	0.037	0.024	0.001	0.101
Geometric standard deviation	3.130	4.860	2.675	3.615	2.593	4.187	3.258

Table 4.8. Background and INCO Plume Air Concentrations ($\mu g/m^3$)

	BACKGROUND		INCO		
Trace Metal	Mean, *A*	Std. Dev.	Mean, *B*	Std. Dev.	Ratio, *B/A*
Fe	0.239	0.262	0.698	0.686	2.93
Ni	0.002	0.003	0.052	0.051	25.41
Cu	0.314	0.212	0.533	0.297	1.70
Pb	0.034	0.033	0.078	0.063	2.32
Zn	0.024	0.023	0.042	0.036	1.75
Cd	0.001	0.002	0.002	0.002	1.95
Al	0.128	0.145	0.182	0.175	1.42

compares background and plume-sector concentrations taken around the INCO smelter. (For the data shown in Table 4.8, there was usually one or two upwind, or background, measurements, and several measurements downwind at various distances for the smelter on each sampling day used to arrive at the average.) The stratification process made use of local wind and continuous SO_2 monitor information, and the final results were substantiated by the observed elevated concentrations of smelter-originated species of those samples that were under the plume shadow (Chan et al., 1982b). The first column of this table lists the parameters of interest, while the second and fourth columns show the average background and plume-sector concentrations. The third and fifth columns give the standard deviations of the corresponding concentrations, while the sixth column shows the ratio of plume-sector and background values (always greater than unity). If a paired-*t* test is used to determine whether or not there is a significant difference between the two sets of concentrations, it is found that for all parameters the result is positive at the 99% confidence level. Results for the Falconbridge smelter were similar. Again, it should be noted that the values for copper are probably artificially elevated due to sampler self-contamination. The fact that the in-plume copper concentration is significantly different from that of the large background (due to HiVol exhaust) suggests that the smelters' contribution of copper to these samples is large.

The dependence of trace metal concentrations on distance downwind of the smelter is best illustrated by computing the additional plume-sector concentration (by subtracting background from plume-sector concentrations) at various distances from the source. Tables 4.9 and 4.10 show the results of such a calculation for the INCO and Falconbridge smelters respectively, for a number of trace metals, together with the standard deviation and (in parentheses) number of values used for each metal and downwind distance. In most cases, there is quite a pronounced decrease with distance from the source

Table 4.9. Additional Concentration ($\mu g/m^3$) Dependence on Distance from the INCO Source

Trace Metal	4 km (representing 0–4 km)	10 km	21 km	31 km	39 km
Fe	1.225 ± 1.716 (69)	0.447 ± 0.389 (64)	0.352 ± 0.405 (41)	0.057 ± 0.186 (54)	0.268 ± 0.332 (16)
Ni	0.124 ± 0.146 (65)	0.045 ± 0.050 (63)	0.023 ± 0.026 (41)	0.017 ± 0.030 (51)	0.010 ± 0.017 (15)
Pb	0.101 ± 0.069 (69)	0.028 ± 0.037 (64)	0.019 ± 0.032 (41)	0.008 ± 0.025 (51)	0.005 ± 0.019 (14)
Zn	0.042 ± 0.050 (68)	0.013 ± 0.020 (64)	0.012 ± 0.021 (41)	0.004 ± 0.021 (54)	0.017 ± 0.019 (16)
Cd	0.003 ± 0.004 (68)	0.0007 ± 0.0016 (64)	0.0005 ± 0.0009 (41)	0.0001 ± 0.0009 (52)	0.0005 ± 0.0006 (16)
Al	0.119 ± 0.216 (68)	0.067 ± 0.148 (63)	0.084 ± 0.139 (40)	−0.011 ± 0.098[a] (54)	0.096 ± 0.162 (14)

[a] The negative value is an artifact of the present calculation method.

Table 4.10. Additional Concentration ($\mu g/m^3$) Dependence on Distance from the Falconbridge Source

Trace Metal	1 km (representing 0–1 km)	15 km	21 km	37 km
Fe	0.522 ± 0.673 (56)	0.310 ± 0.291 (50)	0.133 ± 0.330 (104)	0.045 ± 0.153 (30)
Ni	0.036 ± 0.042 (58)	0.009 ± 0.012 (47)	0.004 ± 0.009 (106)	0.003 ± 0.013 (30)
Pb	0.017 ± 0.041 (57)	0.066 ± 0.077 (50)	0.019 ± 0.046 (104)	0.015 ± 0.045 (30)
Zn	0.010 ± 0.038 (59)	0.014 ± 0.024 (50)	0.006 ± 0.023 (106)	0.003 ± 0.019 (28)
Cd	0.003 ± 0.004 (57)	0.0008 ± 0.0023 (50)	0.0005 ± 0.0015 (105)	0.0002 ± 0.0017 (29)
Al	0.070 ± 0.094 (38)	0.042 ± 0.104 (27)	0.034 ± 0.114 (63)	0.026 ± 0.058 (17)

for this subset of samples. By fitting the average additional concentration to an empirical expression of the form

$$\text{conc} = A + B/R$$

where conc = additional concentration ($\mu g/m^3$), A, B = constants, and R = distance from the source (km),

a reasonable fit to the data format parameters was obtained of the additional concentration dependence on the inverse distance from the sources. The results are summarized in Table 4.11. Except in the case of Falconbridge for Zn and Pb, all parameter concentrations follow a reasonably well-defined inverse distance dependence. The poor fit in the Zn and Pb cases may be due to some anomalous outliers which are obtained very close to the source.

Table 4.11. Dependence of Additional Air Concentration on Distance from Source Using Grouped Data Points, conc ($\mu g/m^3$) = $A + B/R$ (km)

	INCO			Falconbridge		
Trace Metal	Correlation Coefficient	*A*	*B*	Correlation Coefficient	*A*	*B*
Fe	.97	0.0464	4.6460	.87	0.1425	0.3849
Ni	1.00	−0.0016	0.4982	.99	0.0035	0.0329
Pb	.99	−0.0060	0.4180	−.30	0.0337	−0.0159
Zn	.91	0.0045	0.1418	.31	0.0074	0.0032
Cd	.97	−0.0001	0.0104	.99	0.0004	0.0026
Al	.55	0.0444	0.2935	.95	0.0322	0.0382

3.2.2. *Dry Deposition Rates of Smelter Emissions in the Sudbury Basin*

Dry deposition rates of smelter constituents were estimated by combining additional plume sector concentration data (Table 4.11) with the deposition velocities in Table 4.6. For comparison purposes, background dry deposition rates in the Basin, due to sources other than the smelters (long range transport, windblown dust, traffic, etc.), can be estimated in a similar way, but using background deposition velocity and air concentration values. The formula used for calculating the dry deposition rate of the smelter emissions was:

$$F = \int_{R=0}^{40\ \text{km}} C_{AA}(R) V_d \, dA_{\text{plume}}(R)$$

where $F =$ total dry deposition per unit time out to R km from the smelter,

$C_{AA}(R) =$ average additional plume concentration (a function of distance; see Table 4.11),

$V_d =$ deposition velocity (Table 4.6), and

$A_{\text{plume}}(R) =$ downwind area within the plume sector (in this case, the average sector angle was about 60°).

The calculations were made out to 40 km from each of the smelters, the approximate distance covered by the HiVol network. Tables 4.12 and 4.13 show the results for both the INCO and Falconbridge smelters for a number of metals. The first column shows the metals concerned, while the second and third columns show the number of sampling days and actual samples on which the calculations are based. The fourth column shows the estimated daily average metal emission rates during the sampling days. This was done by prorating the available emission measurements according to the SO_2 emission rate based on thc average observed metal-to-SO_2 ratios (see Table 4.14 and the discussion in Sec. 2).

Column 5 of Tables 4.12 and 4.13 shows the dry deposition rate of smelter emissions. However, since only samples from plume impingement days were submitted for analysis, the results in Column 5 are overestimates for the overall daily average deposition over the year, since on many days of the year there was no dry deposition (when the plumes were aloft). They therefore were prorated by multiplying by the portion of total sampling days on which impingement occurred (0.69 and 0.63 for INCO and Falconbridge in our set of data), and the results are shown in Column 6.

Column 7 shows that the percentage of emissions deposited from the INCO smelter is small, generally less than 15%. Values for Falconbridge are considerably greater, nearly total deposition within the Sudbury Basin being implied for the smelter-related iron, and 30 and 60% for Ni and Al, respectively. This is presumably due to the lower average emission height for

Table 4.12. Dry Deposition of INCO Emissions[a]

Parameters	NN[b]	N[c]	Daily Emissions (kg)	Dry Deposition (kg)	Prorated Dry Deposition[d] (kg)	% Deposited	Background Deposition (kg)	Total[e] (%)
Fe	97	274	1,626	361.34	247.88	15.2	1,090.12	19
Ni	91	265	798	30.21	20.72	2.6	11.68	64
Pb	97	269	480	3.74	2.57	0.53	33.68	7.1
Zn	96	273	126	3.23	2.22	1.76	26.04	7.9
Cd	95	271	37	0.02	0.01	0.03	1.07	0.9
Al	95	269	406	52.89	36.28	8.9	484.05	7.0

[a] Calculations were made out to 40 km from the smelter.

[b] NN = number of days included in the calculations.

[c] N = number of samples.

[d] Prorated by the fraction of sampling days with impingement, factor = 0.69 (see text).

[e] (INCO deposition)/(INCO deposition + background deposition) × 100 = % total.

Table 4.13. Dry Deposition of Falconbridge Emissions[a]

Parameter	NN[b]	N[c]	Daily Emissions (kg)	Dry Deposition (kg)	Prorated Dry Deposition[d] (kg)	% Deposited	Background Deposition (kg)	Total[e] (%)
Fe	107	269	100	199.36	124.80	125	934.76	12
Ni	107	270	14	6.47	4.05	29	12.00	25
Pb	108	270	37	7.82	4.9	13	41.07	11
Zn	108	272	5.7	1.9	1.19	21	33.33	3.5
Cd	108	270	12	0.02	0.01	0.09	1.07	0.9
Al	66	174	32	29.52	18.48	58	385.02	4.6

[a] Calculations were made out to 40 km from the smelter.

[b] NN = number of days included in calculations.

[c] N = number of samples.

[d] Prorated by the fraction of sampling days with impingement, factor = 0.63 (see text).

[e] (Falconbridge deposition)/(Falconbridge deposition + background deposition) × 100 = % total.

Table 4.14. Summary of Particulate-to-SO_2 (M/SO_2) Ratios[a]

Parameter (M)	INCO 381-m Stack	N	INCO IORP Stack[b]	Falconbridge Stack	N
Fe	$(3.55 \pm 2.41) \times 10^{-4}$	23	4.14×10^{-3}	$(3.44 \pm 2.67) \times 10^{-4}$	23
Cu	$(1.81 \pm 1.64) \times 10^{-4}$	30	9.85×10^{-4}	$(7.86 \pm 6.39) \times 10^{-5}$	28
Ni	$(1.04 \pm .66) \times 10^{-4}$	28	2.74×10^{-3}	$(4.82 \pm 4.82) \times 10^{-5}$	28
Pb	$(2.06 \pm .63) \times 10^{-4}$	22	2.04×10^{-4}	$(1.26 \pm .55) \times 10^{-4}$	28
Zn	$(4.48 \pm 4.40) \times 10^{-5}$	26	1.47×10^{-4}[c]	$(1.96 \pm .42) \times 10^{-5}$	2
Al	$(1.74 \pm 1.26) \times 10^{-4}$	5	[d]	$(1.10 \pm .09) \times 10^{-4}$	3
Cr	$(4.6 \pm 4.5) \times 10^{-5}$	6	1.00×10^{-4}[c]	3.59×10^{-6}	1
Cd	$(1.63 \pm 1.87) \times 10^{-5}$	17	1.30×10^{-5}[c]	$(4.08 \pm 5.06) \times 10^{-5}$	19

[a] Low level emission ratios assumed to be the same as that of the smelter stack.

[b] Single measurement.

[c] Based on incomplete cycle data, that is, individual or combination of background flue, blow, charge, and reduction phases.

[d] Assumed to be the same as that of the INCO 381-m stack.

the Falconbridge source as compared to those at INCO. Also, it may be noted that metals in the coarse particle size range, such as Fe, Ni, and Al, are dry-deposited more efficiently than fine particulates because of their higher deposition velocities. Note that no results are shown for copper due to the sampling problems alluded to earlier, but on the basis of emission rate and deposition velocity data, similar values to those for nickel might be expected. The much higher percentage deposition of Pb and Zn in the case of Falconbridge may also be related to the poorer fit of the concentration dependence on distance from the source (Table 4.11).

Column 8 shows the background deposition out to 40 km from each smelter (values are slightly different for INCO and Falconbridge because the geographical areas concerned do not entirely overlap, and the sampling days used in the calculations are also somewhat different). By comparing Columns 6 and 8, the percentage local source contribution to the total trace metal dry deposition in the Sudbury Basin can be calculated (Col. 9). Note that the contribution of the smelters is relatively small (10% or less) for most of the metals examined, with the exception of nickel and iron, which are estimated to receive on average 64 and 25% (for Ni) and 19 and 12% (for Fe) of the total dry deposition from the INCO and Falconbridge smelters, respectively. Copper is also expected to have a major smelter contribution (similar to Ni), on the basis of arguments presented earlier. It is interesting to note that the uncertainty in the smelter dry deposition impact (Col. 9) is smaller than that in the individual deposition rates (Cols. 6 and 8), because the smelter impact is obtained from a *ratio* of deposition rates and is largely determined by (measured) concentrations rather than (estimated) deposition velocities (which are similar for the plume and background and hence cancel out in the calculation).

4. PRECIPITATION MEASUREMENTS AND WET DEPOSITION RATES

4.1. Experimental

Precipitation samples were collected daily (at 0800 EST) on every precipitating day from mid-1978 to mid-1980. Typically, the network consisted of 25 stations in the summer and 15 in the winter. These stations were distributed primarily within a radius of 50 km from Sudbury (see Fig. 4.1).

The precipitation collector used in this network consisted of a large diameter polyethylene bucket with a polyethylene bag insert. Two models of the sampler were used. The earlier (before May 1979) consisted of a black high-density polyethylene bucket, 42 cm o.d. by 47 cm high, and a standard, food-grade, linear polyethylene bag insert. The later version consisted of a 44.5 cm o.d. 56.6 cm high, green polyethylene bucket with a custom-made polyethylene bag insert. This bag had an extra seal running from the midpoint of one side diagonally towards the bottom of the other side. At the end of the seal was an opening which allowed precipitation to funnel into the compartment below the seal. This design was implemented to reduce the effect of evaporation of the collected sample. It was used in the summer and autumn periods only, whereas the standard bag without the seal was used in the winter and spring. The effective collection area of both the 1978 and 1979 samplers was 1410 cm^2.

Extensive laboratory testing was carried out on the bags prior to their use. It was confirmed that no detectable adsorption or desorption of ions or metals in the precipitation sample took place on the bag walls over a 24-h period (Chan et al., 1983).

In cases when precipitation was collected, the bag was removed, a bottom corner was cut, and the sample was transferred to storage bottles. One of these was spiked with 0.5 mL of 5% HNO_3 for trace metal analysis by atomic absorption spectroscopy (AA).

In 1980, the method of analysis for all trace metals except Fe and Al changed from AA to inductively coupled plasma spectroscopy (ICP). However, those samples with metal concentrations near the detection limits of ICP were reanalyzed by flameless AA.

4.2. Results

4.2.1. Precipitation Concentration of Trace Metals in Sudbury Basin

Table 4.15 summarizes metal concentration of precipitation samples collected in the vicinity of Sudbury from mid-1978 to mid-1980. As indicated by the magnitude of the standard deviation, the variability of the observed concentration is extremely high. For instance, the highest Fe concentration was close to 4 mg/L, and the lowest was below detection limit (in milligrams per liter,

Table 4.15. Results of Event Precipitation Samples Collected around Sudbury, conc (mg/L)

Statistic	Fe	Cu	Ni	Pb	Zn	Al	Cr	Cd
Sample size	2046	2046	2046	2046	2042	2012	2044	2042
Maximum	3.940	2.470	0.932	0.500	1.670	2.310	0.0810	0.0300
Minimum	0.0005	0.0005	0.0005	0.0005	0.0005	0.0005	0.0003	0.0001
Arithmetic mean	0.103	0.035	0.016	0.019	0.021	0.054	0.0007	0.0007
Arithmetic standard deviation	0.220	0.132	0.059	0.027	0.065	0.108	0.0028	0.0019
Geometric mean	0.043	0.005	0.002	0.012	0.010	0.021	0.0004	0.0003
Geometric standard deviation	3.636	5.912	5.613	2.465	2.971	4.563	2.201	3.731
Volume-weighted mean	0.075	0.020	0.010	0.014	0.015	0.045	0.0006	0.0005

detection limits were as follows: Fe, 0.001; Cu, 0.001; Ni, 0.001; Pb, 0.001; Zn, 0.001; Al, 0.001; Cr, 0.0005; and Cd, 0.0001). Values corresponding to one half of the detection limits were used in the calculations when the reported concentrations were at these limits. Most of the reported minimum values in Table 4.15 are in this category. Aside from the variations due to source emissions, the observations may also reflect the concentration dependence on precipitation volume (the concentration tends to be higher for the lower volume events). The last entries are the volume-weighted mean concentrations over the period of the study.

In order to gain some insight into the origin of the pollutants observed in the precipitation samples, correlations between the various precipitation concentration pairs were calculated using the raw data from which the statistical summary in Table 4.15 was obtained. Only correlation coefficients that are significant at the 99% confidence level are retained in Table 4.16. The number of data pairs used is indicated in parentheses for each metal pair. As can be seen, Fe, Cu, Ni, Pb, and Cd, which are characteristic of smelter operation, are highly correlated with each other. Other species, which are emitted in lower quantities, do not show such a distinct pattern.

Tables 4.17 and 4.18 summarize concentration correlations of the chemical constituents measured in the INCO 381-m stack and the Falconbridge stack plumes with instruments mounted on a helicopter (these have been referred to earlier). Again, only statistically significant correlations at the 99% confidence level are given in these tables. No Al and Cr (Zn also, in the case of

Table 4.16. Correlation[a] of Precipitation Constituent Concentrations in Sudbury

	Fe	Cu	Ni	Pb	Zn	Al	Cr	Cd
Fe	1.000 (2046)							
Cu	.654 (2046)	1.000 (2046)						
Ni	.776 (2046)	.924 (2046)	1.000 (2046)					
Pb	.558 (2046)	.709 (2046)	.685 (2046)	1.000 (2046)				
Zn					1.000 (2042)			
Al	.570 (2012)					1.000 (2012)		
Cr							1.000 (2044)	
Cd	.525 (2042)	.683 (2042)	.656 (2042)	.655 (2042)				1.000 (2042)

[a] Significant at the 99% confidence level.

Table 4.17. INCO 381-m Stack Plume Constituent Correlation[a,b]

Fe	1.00 (20)							
Cu	.846 (20)	1.00 (21)						
Ni	.994 (20)	.847 (21)	1.00 (21)					
Pb	.590 (20)		.606 (21)	1.00 (21)				
Zn				.718 (21)	1.00 (21)			
Cd	.765 (20)	.673 (20)	.753 (20)	.826 (20)		1.00 (20)		
As	.596 (20)	.613 (21)	.633 (21)	.559 (21)	.564 (21)	.731 (20)	1.00 (21)	
Se	.832 (20)	.811 (21)	.824 (21)			.769 (20)	.703 (21)	1.00 (21)
	Fe	Cu	Ni	Pb	Zn	Cd	As	Se

[a] September 1979 and February 1980 studies.
[b] Significant at the 99% confidence level.

Table 4.18. Falconbridge 93-m Stack Plume Constituent Correlation[a,b]

Fe	1.00 (23)						
Cu	.858 (23)	1.00 (28)					
Ni	.806 (23)	.928 (28)	1.00 (28)				
Pb	.832 (23)	.870 (28)	.739 (28)	1.00 (28)			
Cd					1.00 (19)		
As	.852 (22)	.924 (24)	.834 (24)	.946 (25)		1.00 (24)	
Se	.632 (16)	.703 (18)		.933 (18)		.858 (16)	1.00 (18)
	Fe	Cu	Ni	Pb	Cd	As	Se

[a] March 1979 and September 1979 studies.
[b] Significant at the 99% confidence level.

Falconbridge) data are shown because of the unquantifiable amounts on the filters. In both smelter cases, As and Se are included, and they are highly correlated with other smelter-related species. It is of interest to note that Fe, Cu, Ni, Pb, and Cd show high cross-correlations with each other in the case of the INCO plume. However, in the Falconbridge plume case, high correlations of Cd with Fe, Cu, Ni, and Pb do not exist. A comparison of Table 4.16 with Tables 4.17 and 4.18 suggests that the precipitation concentrations of trace metals in the Sudbury area include a contribution of the smelter sources, especially that which is characteristic of INCO.

4.2.2. *Wet Deposition Rates of Smelter Emissions in the Sudbury Basin*

Using a method similar to that described for the HiVol sample analysis, precipitation samples were stratified to distinguish between background and plume-affected samples. Examples of the resulting data are given in Table 4.19 for INCO.

In Table 4.19, the first column lists the parameters of interest. The second and fourth columns correspond to the mean concentrations of the background and INCO plume-affected samples. The scatter of the data is shown as the standard deviations in Columns three and five. The last column gives the ratios of the average plume sample to background concentrations. Plume-affected precipitation concentrations are manyfold higher than the background values and the student-t test results indicate that these differences are statistically significant at the 99% confidence level. Similar results were also found in the case of the Falconbridge smelter.

Using the same treatment applied to the HiVol data, after subtracting the background contributions from the observed sample concentrations downwind of the smelters, the dependence of the additional precipitation concentration of smelter constituents as a function of distance from the source can be established. Tables 4.20 and 4.21 give the results of the INCO and Falcon-

Table 4.19. Background and INCO Plume-Affected Precipitation Concentration (mg/L) in Sudbury Area

Trace Metal	Background		INCO		Ratio,
	Mean, A	Std. Dev.	Mean, B	Std. Dev.	B/A
Fe	0.05187	0.04904	0.18996	0.20789	3.66
Cu	0.00324	0.00362	0.11106	0.19318	34.23
Ni	0.00164	0.00191	0.05015	0.08299	30.64
Pb	0.01154	0.00957	0.02783	0.02820	2.41
Zn	0.01018	0.00788	0.01962	0.01886	1.93
Al	0.04394	0.05431	0.07209	0.11099	1.64
Cr	0.00038	0.00022	0.00075	0.00106	1.97
Cd	0.00026	0.00035	0.00119	0.00129	4.55

Table 4.20. Additional Precipitation Concentration (mg/L) Dependence on Distance from the INCO Source

Trace Metal	5 km (representing 3–6 km)	10 km (9–11 km)	15 km (14–16 km)	20 km (17–24 km)	30 km (27–32 km)	40 km (35–50 km)
Fe	.1953 ± .2755 (166)	.1059 ± .2280 (68)	.1528 ± .2079 (28)	.1111 ± .2364 (73)	.0224 ± .0571 (54)	.0086 ± .0165 (5)
Cu	.1552 ± .2758 (168)	.0574 ± .0916 (68)	.0735 ± .0869 (28)	.0698 ± .1259 (73)	.0171 ± .0331 (54)	.0051 ± .0038 (5)
Ni	.0691 ± .1059 (168)	.0333 ± .0714 (68)	.0313 ± .0314 (28)	.0331 ± .0703 (73)	.0072 ± .0144 (54)	.0010 ± .0010 (5)
Pb	.0206 ± .0297 (168)	.0102 ± .0176 (68)	.0157 ± .0159 (28)	.0157 ± .0232 (73)	.0056 ± .0137 (53)	.0069 ± .0073 (4)
Zn	.0120 ± .0223 (167)	.0075 ± .0150 (68)	.0070 ± .0136 (28)	.0092 ± .0309 (73)	.0112 ± .0299 (52)	.0034 ± .0095 (5)
Al	.0294 ± .0764 (167)	.0224 ± .0731 (68)	.0585 ± .1917 (28)	.0280 ± .1019 (72)	.0009 ± .0407 (53)	.0065 ± .0174 (5)
Cr	.0006 ± .0021 (164)	.0003 ± .0009 (66)	.0001 ± .0004 (27)	.0001 ± .0004 (71)	.0004 ± .0016 (49)	.0001 ± .0001 (5)
Cd	.0012 ± .0021 (167)	.0005 ± .0012 (68)	.0007 ± .0010 (28)	.0010 ± .0020 (72)	.0005 ± .0009 (52)	.0001 ± .0001 (5)

Table 4.21. Additional Precipitation Concentration (mg/L) Dependence on Distance from the Falconbridge Source

Trace Metal	8 km (representing 6–11 km)	17 km (17–19 km)	23 km (21–24 km)	32 km (28–33 km)	40 km (38–41 km)	50 km
Fe	.0885 ± .1863	.0372 ± .1016	.0357 ± .1726	−.0025 ± .0362[a]	−.0021 ± .1104[a]	.0271 ± .0400
	(46)	(35)	(73)	(20)	(18)	(6)
Cu	.0228 ± .0394	.0069 ± .0144	.0128 ± .0274	.0071 ± .0216	.0011 ± .0025	.0037 ± .0036
	(47)	(36)	(73)	(19)	(19)	(6)
Ni	.0089 ± .0140	.0028 ± .0057	.0040 ± .0093	.0048 ± .0141	−0.0000 ± .0013[a]	−.0001 ± .0003[a]
	(47)	(35)	(72)	(19)	(19)	(5)
Pb	.0097 ± .0158	.0109 ± .0150	.0045 ± .0107	.0036 ± .0045	.0069 ± .0098	.0058 ± .0121
	(47)	(34)	(74)	(20)	(19)	(6)
Zn	.0168 ± .0576	.0079 ± .0169	.0050 ± .0172	−.0005 ± .0067[a]	.0024 ± .0119	.0242 ± .0305
	(46)	(35)	(72)	(20)	(19)	(6)
Al	.0732 ± .1757	.0226 ± .0524	.0141 ± .0435	−.0135 ± .0425[a]	.0149 ± .0588	.0054 ± .0129
	(47)	(35)	(72)	(17)	(19)	(5)
Cr	.0004 ± .0008	.0002 ± .0005	.0002 ± .0008	.0000 ± .0003	.0001 ± .0003	.0001 ± .0003
	(45)	(35)	(73)	(20)	(19)	(6)
Cd	.0003 ± .0006	.0002 ± .0003	.0002 ± .0005	.0001 ± .0004	.0001 ± .0002	.0005 ± .0005
	(46)	(35)	(72)	(20)	(19)	(6)

[a] The negative value is an artifact of the present calculation method.

bridge sources using grouped data points corresponding to samples collected over certain distance intervals. For each parameter–distance combination, an entry is given of the average additional plume concentration and one standard deviation with the number of data points indicated in the brackets. The farthest distance interval has the least data points. In these tables, a considerable amount of scatter is noted, as reflected by the large standard deviation at each distance interval. By using the same empirical expression for the HiVol data, a reasonable fit to the data was obtained of the dependence of the additional plume concentration on the inverse distance from the sources. Table 4.22 gives the correlation coefficients together with the regression coefficients. Aluminum in INCO's case and Zn and Cd in Falconbridge's case did not yield good fits. This point should be borne in mind when it comes to the interpretation of the deposition results. This is a direct result of the bias introduced at the farthest distance interval.

Wet deposition due to both the background and the local sources was calculated using the following expression:

$$\mathrm{DEP}_{\mathrm{av}} = D_{\mathrm{av}} \int_{R=0}^{40\ \mathrm{km}} (C_{\mathrm{add}}(R)\,dA_{\mathrm{plume}}(R));$$

where

$\mathrm{DEP}_{\mathrm{av}}$ = average areal deposition through wet scavenging,
D_{av} = average precipitation depth (mm),
$C_{\mathrm{add}}(R)$ = average additional concentration due to interaction of plume and precipitation (in mg/L unit),
$A_{\mathrm{plume}}(R)$ = area encompassed by the plume sector, and
R = distance from the source (km).

Table 4.22. Dependence of Additional Precipitation Concentration on Distance from Source Using Grouped Data Points, conc (mg/L) = $A + B/R$ (km)

	INCO			Falconbridge		
Trace Metal	Correlation Coefficient	A	B	Correlation Coefficient	A	B
Fe	.81	0.02738	0.90914	.90	−0.00862	0.77551
Cu	.92	0.00313	0.75641	.91	−0.00010	0.18121
Ni	.93	0.00170	0.34688	.86	−0.00038	0.07421
Pb	.75	0.00713	0.06716	.63	0.00455	0.04672
Zn	.54	0.00631	0.02629	.26[a]	0.00618	0.06159
Al	.31[a]	0.01649	0.09842	.92	−0.01538	0.68845
Cr	.80	0.00006	0.00247	.91	0.00001	0.00327
Cd	.71	0.00034	0.00444	−.01[a]	0.00023	−0.00003

[a] Poor fit due to bias in results obtained at the farthest distance interval.

Table 4.23. Wet Deposition of INCO Emissions, R = 40 km

Parameter	Emissions (kg)	Wet Deposition (kg)	% Scavenged	Background Deposition (kg)	% Total Deposition[a]
Fe	714	484	67.8	1,938	20.0
Cu	260	272	104.6	121	69.2
Ni	351	126	35.9	61	67.4
Pb	211	70	33.2	431	14.0
Zn	55	51	92.7	380	11.8
Al	178	142.2	79.9	1,642	8.0
Cr	52	1.2	2.3	14.2	7.8
Cd	16.4	3.7	22.6	9.7	27.6

[a](INCO deposition)/(INCO deposition + background deposition) × 100 = % total deposition.

An average precipitation depth of 7.4 and 8.9 mm and plume-sector width of 64° and 68° were used for INCO and Falconbridge, respectively, in the calculations. These parameters were determined from the observations at each source. Because of the large uncertainties in the Falconbridge data due to network design, only results of the INCO plume are given in Table 4.23. In Table 4.23, parameters of interest are listed under the first column. The second column gives the corresponding emissions during an average precipitating day over the precipitation period. Emission rates of trace metals were estimated during the study period from the SO_2 emission rate and the data in Table 4.14. Wet deposition of these parameters over the same interval is given in Column 3. The fraction of the emission scavenged by precipitation was calculated by dividing the deposition by the emissions and is summarized in Column 4. Background deposition was determined by using the mean background precipitation concentration (Table 4.19) and the average precipitation depth (7.4 mm) for the study period for the area defined by a radius of 40 km. The last column is a summary of the INCO contribution to the total deposition in the area within a 40-km radius from the source.

From Table 4.23, it is seen that except for Cr, the scavenging of trace metals is quite rapid and approaches total washout for Cu and Zn. Similar results have been reported by other workers (Wiebe and Whelpdale, 1974; Muller and Kramer, 1974) and confirmed in a more recent intensive study of scavenging from the INCO plume (Lusis et al., 1983). It can be seen that the deposition due to the INCO source corresponds to 8 to 28% of the total with the exception of Cu and Ni which contribute as much as 70%. The latter results are not surprising considering the fact that below-plume concentrations of these species are very high with respect to background values (see Table 4.19).

It is worth pointing out that, farther downwind from the source than the 40-km radius of the calculation, the plume-sector concentration will be lower, and the relative contribution to the total deposition due to the INCO source will be smaller than those given in Table 4.23.

5. TOTAL METAL DEPOSITION IN THE SUDBURY BASIN

Data from Sections 3 and 4 can be combined to yield an overall picture of the relative importance of dry and wet deposition from INCO. Also shown is the ratio of the total deposition due to INCO and that due to background sources. The same exercise cannot be repeated for Falconbridge as the precipitation results are not adequate for this purpose because of the relatively poor network coverage. Two assumptions were made: (1) precipitation occurs on every third day, and (2) dry deposition of Cu is similar to that of Ni. These results are summarized in Table 4.24.

In Table 4.24, parameters of interest are listed in Column 1. Column 2 gives the relative importance of wet to dry deposition of the INCO emissions,

Table 4.24. A Comparison of Wet and Dry Deposition and Relative Contribution Due to INCO And Background[a]

Parameter	$\frac{(\text{Wet})_{\text{INCO}}}{(\text{Dry})_{\text{INCO}}}$	$\frac{(\text{Wet and Dry})_{\text{Smelter}}}{(\text{Wet and Dry})_{\text{Total}}}$
Fe	0.45	0.23
Cu	3.0[b]	0.7[b]
Ni	1.39	0.69
Zn	5.26	0.12
Pb	6.24	0.13
Al	0.90	0.09
Cd	6.17	0.23

[a] Assuming precipitation occurs, on average, every third day. Results are average values for areas within 40 km of the source.

[b] Assuming similar dry deposition to that for Ni.

and Column 3 gives the relative contribution of INCO to the total deposition in the Sudbury Basin. It is noted that for trace metals in large particles (Fe, Ni, Al), wet and dry deposition is similar and particles in the sub-micron size range (Zn, Pb, Cd) are primarily deposited by precipitation. As far as total deposition is concerned, the major smelter impact is on Cu and Ni. Of the other substances examined, INCO contributes less than 25%.

6. CONCLUSIONS

A daily precipitation network and a daily HiVol sampling network were operated in the Sudbury area from mid-1978 to mid-1980 by the Ontario Ministry of the Environment to quantify the relative contribution of local smelter emissions from the smelter sources to precipitation quality, air quality, and wet and dry deposition in the Sudbury Basin. Because of network design, the data analysis yielded useful information primarily for the INCO source. The following conclusions are reached.

1. By meteorological stratification of the data, it was possible to separate the smelter contribution from the general background. The downwind precipitation and air concentrations of smelter-related constituents were significantly higher than those upwind, by as much as an order of magnitude for Ni and Cu.

2. At a distance of 40 km from the source, except for Cr, for the parameters examined, 23 to 100% of the INCO emissions were scavenged during precipitation events. It was estimated that for metals in the fine particle size range (Pb, Zn, Cd), less then 2% of the emissions were dry-deposited.

3. In general, the wet deposition within 40 km of Sudbury due to the INCO source corresponded to less than 20% of the total wet deposition for most of the substances examined, with the notable exception of Cu and Ni, which contributed as much as 70%. The smelter contribution to total dry deposition within 40 km was greatest for nickel, making up 64% of the total. The contribution of the other metals examined was generally less than 20%.

4. For trace metals in large particles (Fe, Ni, Al), wet and dry deposition was similar. Particles in the submicron size range (Zn, Pb, Cd) were primarily deposited by precipitation. As far as total (wet plus dry) deposition is concerned, the major smelter impact was on Cu and Ni. For most of the other substances examined, INCO contributed about 20% or less.

ACKNOWLEDGMENTS

The authors would like to thank the technical and scientific staff of the Sudbury Environmental Study for their assistance in data collection and analysis. Competent typing by the Word Processing Unit is gratefully acknowledged.

REFERENCES

Chan, W.H. (1982). *Sudbury Environmental Study—Atmospheric Research Program: A Synopsis.* Air Resources Branch Report #ARB-27-82-ARSP. Ontario Ministry of the Environment, Toronto, Ontario.

Chan, W.H., Vet, R.J., Lusis, M.A., and Skelton, G.B. (1982a). *Size distribution and emission rate measurements of particulates in INCO 381 m chimney and iron ore recovery plant stack plumes, 1979–80.* Air Resources Branch Report #ARB-TDA-62-80. Ontario Ministry of the Environment, Toronto, Ontario.

Chan, W.H., Tang, A.J.S., Lusis, M.A., Vet, R.J., and Ro, C.U. (1982b). *An analysis of the impact of smelter emissions on atmospheric dry deposition in the Sudbury area: Sudbury Environmental Study Airborne Particulate Matter Network results.* Air Resources Branch Report #ARB-012-81-ARSP. Ontario Ministry of the Environment, Toronto, Ontario.

Chan, W.H., Vet, R.J., Lusis, M.A., and Skelton, G.B. (1983a). "Airborne particulate size distribution measurements in nickel smelter plumes." *Atmos. Environ.* **17,** 1173–1181.

Chan, W.H., Tomassini, F., and Loescher, B. (1983b). "An evaluation of sorption properties of precipitation constituents on polyethylene surfaces." *Atmos. Environ.* **17,** 1779–1785.

Chan, W.H., Vet, R.J., Ro, C.U., Tang, A.J.S., and Lusis, M.A. (1984). "Impact of INCO smelter emissions on wet and dry deposition in the Sudbury area." *Atmos. Environ.,* 1001–1008.

Hicks, B.B. (1983). "Measurement techniques: Dry deposition." In *Proceedings of symposium on monitoring and assessment of airborne pollutants with special emphasis on long-range transport and deposition of acidic materials and workshop on air and precipitation monitoring networks, 30 August–3 September, 1982.* NRCC No. 20642. Ottawa, Can.

Hicks, B.B., Wesley, M.L., and Durham, J.L. (1980). *Critique of methods to measure dry deposition. Workshop summary.* Environmental Sciences Research Report. Research Triangle Park, N.C.

King, R.B., and Toma, J. (1975). *Copper emissions from a high-volume air sampler.* NASA Technical Memorandum TX-71693.

Lusis, M.A., Chan, W.H., Tang, A.J.S., and Johnson, N.D. (1983). "Scavenging rates of sulfur and trace metals from a smelter plume." In *Proceedings of the 4th international conference on precipitation scavenging, dry deposition and resuspension, Santa Monica, Nov. 29–Dec. 3, 1982,* H.R. Pruppacher, R. G. Semonin, and W.G.N. Slinn, Eds. Elsevier, New York.

McMahon, D.J., and Dennison, P.J. (1979). "Empirical atmospheric deposition parameters–A survey." *Atmos. Environ.* **13,** 571–585.

Muller, E.F., and Kramer, J.R. (1977), "Precipitation scavenging in central and northern Ontario." In *Precipitation scavenging,* R.E. Semonin and R.W. Beadle, Eds. Technical Information Centre, ERDA, pp. 590–601. Springfield, Va.

Ontario Ministry of the Environment (1982). *Sudbury Environmental Study, 1973–1980 synopsis.* Toronto, Ontario.

Ozvacic, V. (1982). *Emissions of sulphur oxides, particulates and trace elements in Sudbury Basin.* Air Resources Branch Report #ARB-ETRD-09-82. Ontario Ministry of the Environment, Toronto, Ontario.

Ozvacic, V. and McDonald, J. (1982). Determination of the extent of particulate stratification at the INCO 381 m chimney at Copper Cliff, Ontario. Air Resources Branch Report #ARB-TDA-58-80. Ontario Ministry of the Environment, Toronto, Ontario.

Sehmel, G.A. (1980). "Particle and gas dry deposition: A review." *Atmos. Environ.* **14,** 983–1011.

Wiebe, H.A., and Whelpdale, D.M. (1977). "Precipitation scavenging from a tall stack plume." In *Precipitation scavenging,* R.G. Semonin and R.W. Beadle, Eds. Technical Information Centre, ERDA, pp. 118–126. Springfield, Va.

5

EMISSIONS AND AIR QUALITY RELATIONSHIPS FOR ATMOSPHERIC TRACE METALS

Glen R. Cass

Environmental Engineering Science Department
and Environmental Quality Laboratory
California Institute of Technology
Pasadena, California

Gregory J. McRae

Department of Chemical Engineering
Carnegie-Mellon University
Pittsburgh, Pennsylvania

1. INTRODUCTION

Atmospheric particulate matter samples taken in urban and rural locations can be analyzed routinely for more than 40 trace elements. With the increasing use of automated X-ray fluorescence and neutron activation analyses (Dzubay, 1977; Cooper, 1973), the cost of trace metals determination in airborne particulate samples has been greatly reduced. As a result, large volumes of data are being acquired that contain considerable chemical resolution, including concentration data on toxic trace elements like lead, arsenic, cadmium, and nickel.

While air quality data that define toxic metals pollutant loading in the atmosphere are becoming available, the emission sources responsible for the release of these materials often are not obvious. Trace metals emissions arise from more than 60 distinctly different source types in a large urban area. There are, for example, autos burning leaded fuel, electric-arc steel furnaces, Kraft recovery boilers, and secondary lead smelters. The U.S. Environmental Protection Agency (1977) emission factor manual (AP-42) presents a fairly comprehensive list of the source types that might be present, and there are many of them. This proliferation of sources and the fact that trace metals often are only a minor fraction of the mass emissions from each source obscure the relative importance of the contributors to atmospheric metals levels.

The purpose of this chapter is to illustrate methods available for quantifying the sources that contribute trace metals emissions to the atmosphere. Receptor-oriented air quality models will be described that can identify the contribution of individual emission source types to the aerosol mass loading observed at community air monitoring sites. It will be shown how trace metals emission inventories can be used to improve receptor model reliability. A comprehensive emission inventory procedure will be developed for each of the individual trace metals released to the atmosphere. The consistency of these inventories will be tested based on comparison of the relative abundance of trace metals estimated from emissions data versus that measured in the atmosphere. Examples will be given based on data available for Los Angeles, California and for Houston, Texas.

2. SOURCE APPORTIONMENT BY CHEMICAL MASS BALANCE METHODS

Trace metals emissions undergo atmospheric dilution and transport, resulting in the pollutant concentrations observed at community air monitoring sites. Knowledge of the relative importance of each of many emission source types to air quality at receptor air monitoring stations is a prerequisite to the design of efficient pollutant abatement strategies. Mathematical air quality models can be constructed based on observed meteorological data that will track the contributions from many emission sources as they undergo atmospheric chemical reaction and pollutant transport. These fluid mechanical dispersion models are elegant, but the data requirements in practical applications can become so large that engineers are deterred from using them to solve complex multiple source urban air quality problems (see Cass and McRae, 1981b).

In the case of particulate emissions, an alternative to the use of fluid mechanical transport models is available. It is possible to attack the source contribution identification problem in reverse order, proceeding from measured air quality at a receptor site backward to the responsible emission sources. The unique metals content of the emissions from each source type is viewed as a tracer for the presence of material from that source in an ambient aerosol sample (Hidy and Friedlander, 1970; Miller et al., 1972; Friedlander, 1973; Kowalczyk et al., 1977). A chemical mass balance then is constructed on the concentration c_i of each element $i = 1,2, \ldots ,n$ in the ambient sample based on a linear combination of the chemical element profiles of the major emission sources present:

$$c_i = \sum_j f_{ij} a_{ij} s_j \tag{1}$$

where s_j is the mass concentration of material from sources $j = 1,2, \ldots ,m$ observed at the receptor site, a_{ij} is the fraction of chemical species i in the particulate emissions from source j, and f_{ij} is the coefficient of fractionation, representing any modification to the source emission profiles (the a_{ij}'s) due to gravitational settling or other atmospheric processes that occur between the source and the receptor point. Fractionation often is neglected.

Equation (1) defines a series of expressions, each of which constitutes a mass balance on a single chemical element in an aerosol sample. Given the chemical composition of the ambient sample and the source emission profiles, Eq. (1) may be solved for the source contributions s_j. In cases where the number of chemical elements exceeds the number of source types, one linear combination that fits the ambient sample can be found by least squares regression:

$$\mathbf{s} = [A^{\mathrm{T}} W A]^{-1} A^{\mathrm{T}} W \mathbf{c} \tag{2}$$

where $\mathbf{s}$ is a vector of estimated source contributions to the ambient sample, $\mathbf{c}$ is a vector of the concentrations of species $i = 1,2, \ldots ,n$ measured at the monitoring site, A is the matrix $f_{ij}\, a_{ij}$ appearing in Eq. (1), and W is a diagonal matrix of weighting factors. The weighting factors are selected to reflect the accuracy with which the concentration of a particular chemical species is measured. A common choice for these weighting factors is $1/\sigma_i^2$, where σ_i is the standard deviation of a single determination of the concentration of species i in an ambient sample (Kowalczyk et al., 1977).

3. EMISSION INVENTORY ASSISTED RECEPTOR MODELS

In the early stages of receptor model development, investigators had chemically resolved source signatures available for only 5 to 10 source types and did not face two important problems: (1) a number of source profiles exceeding the number of elements measured, or (2) *extreme* multicollinearity, with many industrial sources indistinguishable from one another on the basis of chemical composition. As the number of source profiles increases, these two problems can become quite serious.

If a chemical element balance application is to be useful under such circumstances, some method must be used to reduce the number of source signatures applied to the mass balance equations to well below the number of actual types of sources in an air basin. There are perhaps four ways to accomplish this reduction: intuition, "interactive" chemical mass balance approaches, factor analysis, and emission inventory assisted chemical mass balance approaches. The intuitive approach generally results in *a priori* selection of only those profiles that belong to sources that are obviously important, such as the automobile. In the interactive chemical mass balance method (Heisler, 1982), one utilizes all available source profiles in the initial mass balance calculations, discarding sources with predicted negative source contributions one at a time beginning with those having the largest relative uncertainty in the value of s_j. Then, once all predicted source contributions are positive, the analyst continues to remove sources one at a time, again beginning with those having the largest relative uncertainty in the value of s_j, until the source combination yielding the best fit to the ambient data is found. Factor analysis often is used to estimate the minimum number of important source types present by synthesizing source profiles that in combination will reproduce the chemical composition of a series of samples (Henry, 1977; Hopke et al., 1980; Alpert and Hopke, 1980; Hopke, 1981). Each of the above methods has serious drawbacks: Intuitive selection of the important source profiles is arbitrary, the interactive approach depends on the judgment and skill of the analyst when deciding which sources to delete, and factor analysis can result in a collection of synthetic source profiles that look nothing like the chemical composition of the effluent from any real emission sources.

The fourth alternative for reducing the dimension of the chemical mass balance problem is to use an emission inventory assisted approach (Cass and McRae, 1981a; 1983). Emission inventory data are used to systematically assess the ability of a limited number of source types to complete a mass balance on selected chemical species measured in ambient samples. Extraneous sources are eliminated from consideration in two ways. First, sources shown to constitute a negligible fraction of the mass emissions of all easily traced elements are eliminated from the source matrix. If the air sampling site of interest is not in the immediate vicinity of such a source, then the source will not be detected, even if present. Secondly, with the assistance of mass emission inventory data, source profiles for a variety of activities that are linked together physically, so that their emissions occur in known proportions with the same spatial distribution, can be combined into a single emissions-weighted average profile. For example, all highway traffic emissions from catalyst-equipped autos and light trucks, leaded gasoline fueled autos and trucks, diesel vehicles, tire dust, and brake dust can be combined into a single profile that often will be easily identified by its lead content. Likewise, the many dissimilar sources at a steel mill, petroleum refinery, or entire industrial complex might be combined into a single profile for the facility as a whole. In some cases, an equivalent to these composite profiles could be obtained experimentally by atmospheric sampling, such as in a highway tunnel. But the composite based on an emission-weighted average of several single source profiles has the advantage that the mass associated with the composite, once identified at a receptor site, can be separated easily into the impact of each single source within the group. This is important for control strategy evaluation purposes if each source in the composite group is to be controlled by different techniques and to a different degree. The emission inventory assisted approach can be used to resolve receptor modeling problems that resist solution by other methods because it brings additional data to bear on the problem, not just a different mathematical data reduction procedure.

Obviously, for an emissions inventory assisted chemical element balance analysis to work, the emissions inventory and ambient monitoring data must be accurate. Further, there is no guarantee that any particular air basin will necessarily have a population of sources that can be reduced to a manageable number by this method. Therefore, the first step in this approach is to run consistency checks on the available data and to determine if the problem can be formulated so that a mass balance can be achieved on key chemical elements without the source matrix becoming degenerate. In some cities this series of checks will succeed, whereas in others it will fail. One procedure for making this determination is as follows.

4. TRACE METALS EMISSIONS

A comprehensive inventory of trace metals emissions is constructed for each mobile, stationary, and fugitive source type in the air basin. An energy balance is constructed on the region's fuel supply. Fuel combustion data next are combined with information on the level of industrial process and fugitive source activity, and an inventory of total suspended particulate emissions to the atmosphere is computed by standard methods (Cass et al., 1982; U.S. Environmental Protection Agency, 1972; Taback et al., 1979). This inventory should be based, if possible, on source tests performed within the air basin of interest. If fuel switching is practiced on a seasonal basis, the inventory should be computed from fuel use during the time span of the ambient sampling program.

Next, the particulate inventory is subdivided into fine and coarse particle fractions using data on the size distributions of the particulate matter emitted from the sources of interest. Then the size-resolved mass emissions inventory is converted into a separate inventory for each chemical compound or element of interest by multiplying the mass emissions rate from each source type by its fractional chemical composition within the particle size range of interest. An example of the size distribution of the chemical composition of the particulate emissions from a petroleum refinery fluid catalytic cracking unit is given in Table 5.1. Similar data on the size distribution of the chemical composition of the emissions from more than 60 types of important particulate pollutant sources have been accumulated and catalogued in several useful reference works (Watson, 1979; Taback et al., 1979; Cass and McRae, 1981a).

An element-by-element emission inventory for fine particulate matter in the Los Angeles area can be used to illustrate the power of the size-resolved source fingerprint method for trace metals emissions estimation. The example chosen is based on Cass and McRae (1981a, 1983) and involves 70 separate pollutant source types and 39 source chemical composition profiles.

Table 5.1 Source Profile, Petroleum–FCC Units/CO Boiler[a]

Size Range (Micron)	0–99	0–1	1–3	3–10	10–99	0–10
Weight Percent (%)	100.00	51.00	6.00	4.00	39.00	61.00
Species						
Arsenic	0.05[b]	0.05	0.05	0.05	0.05	0.05
Calcium	0.55[c]	0.55	0.55	0.55	0.55	0.55
Cesium	1.00	2.00	1.00	1.00	1.00	1.836
Gallium	0.05	0.05	0.05	0.05	0.00	0.05
Iron	0.55	1.00	1.00	0.500	0.55	0.967
Lanthanum	0.55	0.55	0.55	5.550	0.55	0.878
Lead	0.05	0.05	0.05	0.05	0.05	0.05
Molybdenum	0.05	0.05	0.05	0.05	0.05	0.05
Neodymium	0.55	0.55	0.55	0.55	0.55	0.55
Nickel	0.05	0.05	0.05	0.05	0.05	0.05
Praeseodymium	0.55	0.55	0.55	0.55	0.55	0.55
Silicon	10.00	0.05	20.00	20.00	20.00	3.320
Strontium	0.05	0.05	0.05	0.05	0.05	0.05
Titanium	0.55	0.55	0.55	0.55	0.55	0.55
Sulfates	30.00	50.00	6.00	7.00	7.00	42.852
Total Carbon	2.00	4.00	0.00	0.00	0.00	3.344
(Volatile Carbon)	1.500	3.00	0.00	0.00	0.00	2.508
Other	53.400	39.900	68.950	63.450	68.450	44.302
Totals	100.00	100.00	100.00	100.00	100.00	100.00

[a] Reprinted with permission from Taback et al. (1979).

[b] Elements detected at concentrations below 0.1% but above 0.0% are shown as 0.05%.

[c] Elements detected at concentrations between 1.0% and 0.1% are shown as 0.55%.

The geographic area studied is the South Coast Air Basin that surrounds Los Angeles, as shown in Figure 5.1. An energy budget within that region during the year 1976 first was constructed by the methods of Cass et al. (1978; 1982). This fuel-burning survey was combined with industrial process and fugitive source emission data provided by Taback et al. (1979) and by Grisinger et al. (1981). An inventory of total particulate emissions then was constructed as shown in Table 5.2. Information on the size distribution and chemical composition of emissions from 39 different source types was used to divide the total particulate burden into inventories for each of 37 different chemical species within the fine particle size range (particle diameter, $d_p \leq 10\ \mu m$). A complete description of this emission inventory procedure is given by Cass and McRae (1981a).

The Los Angeles fine particle trace element inventory is summarized in Table 5.3 and for some of the key elements studied in Figure 5.2 and Table 5.4. The crustal elements Si, Al, Fe, and Ca are among the largest contributors to particulate air quality. They are emitted to the Los Angeles atmosphere principally in the form of fugitive dust. Primary aerosol carbon and primary sulfates also account for a significant fraction of the particulate emissions from combustion sources. Lead is the most abundant of the toxic trace metals, emitted principally from combustion of leaded gasoline. Nickel is emitted at the rate of about 0.5 t/day, largely from combustion of residual fuel oil. Other toxic trace elements like As and Cd are emitted at the rate of a few tens of kilograms daily. Fifty-eight percent of the mass of the Los Angeles fine particulate emissions can be accounted for by the elements listed in Table 5.3. Most of the remaining material consists of oxygen and hydrogen, with the metals often present as oxides.

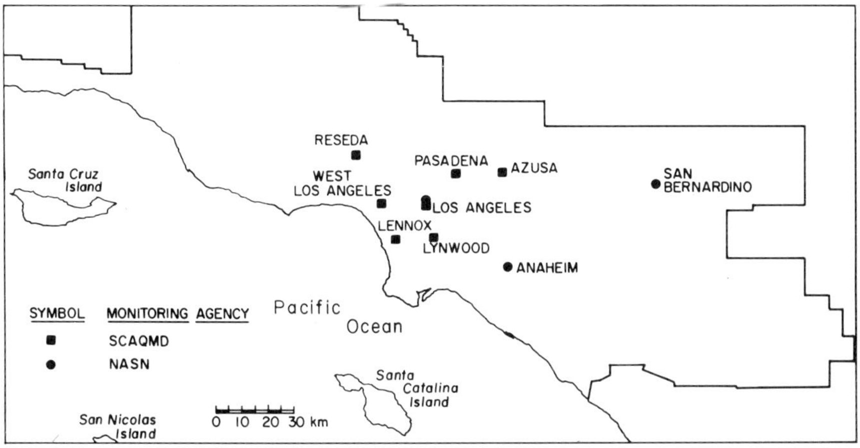

Figure 5.1. Location of monitoring sites within the South Coast Air Basin.

Table 5.2. Fine Particle Emission Inventory[a], Los Angeles Area, 1976 .

	Profile	Estimated Fuel Usage (10^9 BTU/day)	Emission Factor	Emissions (kg/day)	Mass Fraction <10 μm (%)	Fine Particle Emissions (kg/day)
Stationary Sources						
Fuel combustion						
Electric utilities						
Natural gas	27	227.45	1.081 kg/10^9 BTU	246.	95.0	233.6
Residual oil (0.50% S)	64	993.42	21.619 kg/10^9 BTU	21477.	97.0	20832.4
Landfill and digester gas	27	0.82	1.081 kg/10^9 BTU	1.	95.0	0.8
Refinery fuel						
Natural gas	27	93.03	9.080 kg/10^9 BTU	845.	95.0	802.5
Refinery gas	27	395.95	9.080 kg/10^9 BTU	3595.	95.0	3415.5
Residual oil	1	32.97	21.619 kg/10^9 BTU	713.	87.0	620.1
Nonrefinery industrial fuel						
Natural gas	27	421.64	7.567 kg/10^9 BTU	3191.	95.0	3031.0
LPG	27	2.74	7.567 kg/10^9 BTU	21.	95.0	19.7
Residual oil	1	53.42	21.619 kg/10^9 BTU	1155.	87.0	1004.8
Distillate oil	2	42.74	23.520 kg/10^9 BTU	1005.	98.0	985.1
Digester gas (IC engines)	50	6.30	20.430 kg/10^9 BTU	129.	99.0	127.4
Coke oven gas	27	37.53	7.567 kg/10^9 BTU	284.	95.0	269.8
Residential commercial						
Natural gas	51	1181.92	8.071 kg/10^9 BTU	9539.	95.0	9062.3
LPG	51	18.08	8.071 kg/10^9 BTU	146.	100.0	145.9
Residual oil	1	22.19	21.619 kg/10^9 BTU	480.	100.0	479.7
Distillate oil	2	22.19	21.619 kg/10^9 BTU	480.	98.0	470.1
Subtotal						41500.8

Mobile Sources						
Highway vehicles						
Catalyst autos and light trucks	53	368.78	2.137 kg/10^9 BTU	788.	100.0	788.1
Noncatalyst autos and light trucks	54	1255.16	37.814 kg/10^9 BTU	47463.	60.0	28477.6
Medium and heavy gasoline vehicles	54	228.17	43.596 kg/10^9 BTU	9947.	60.0	5968.4
Diesel vehicles	52	125.52	64.200 kg/10^9 BTU	8058.	96.0	7736.0
Civil aviation						
Jet aircraft	55	44.56	U.S. EPA (1977)	733.	100.0	733.0
Aviation gasoline	54	1.29	9.08 g/LTO cycle	28.	60.0	16.8
Commercial shipping						
Residual oil-fired ships boilers	1	29.41	85.386 kg/10^9 BTU	2511.	87.0	2184.7
Diesel ships	52	17.43	49.102 kg/10^9 BTU	856.	96.0	821.6
Railroad						
Diesel oil	52	19.94	81.837 kg/10^9 BTU	1632.	96.0	1566.6
Military						
Gasoline	54	6.03	43.596 kg/10^9 BTU	263.	60.0	157.7
Diesel oil	52	17.81	78.564 kg/10^9 BTU	1399.	96.0	1343.3
Jet fuel	55	16.71	U.S. EPA (1977)	659.	100.0	659.0
Residual oil (bunker fuel)	1	0.27	83.386 kg/10^9 BTU	23.	87.0	19.6
Miscellaneous						
Off-highway vehicles	52	39.73	78.564 kg/10^9 BTU	3121.	96.0	2996.5
Subtotal						53468.9

[a] Reprinted with permission from Cass and McRae (1983). Copyright 1983 American Chemical Society.

Table 5.2. *(continued)*

	Profile	Emissions (kg/day)	Mass Fraction <10 μm (%)	Fine Particle Emissions (kg/day)
Stationary Sources				
Industrial Process Point Sources				
Petroleum industry				
Refining	28	1943.	61.0	1185.2
Paving and roofing materials	19	527.	98.0	516.5
Other (calcining—mineral)	28	454.	61.0	276.9
Organic solvent use				
Surface coating	22	772.	96.0	741.1
Printing	56	9.	99.0	8.9
Storage loss	56	9.	99.0	8.9
Other	56	12.	99.0	11.9
Chemical plants	24	1952.	96.0	1873.9
Metallurgical industry				
Metals—general	65	1716.	100.0	1716.0
Primary metals	65	4041.	100.0	4041.0
Secondary metals				
Nonferrous metals	69	953.	95.0	905.3
Other	70	799.	95.0	759.0
Metal fabrication				
Nonferrous metals	69	1153.	95.0	1095.3
Other	67	409.	100.0	409.0
Mineral industry				
Glass furnaces	17	708.	98.0	693.8
Rock, stone, other	33	10433.	10.0	1043.3
Waste burning at point sources	5	209.	100.0	209.0

Wood and paper burning	5	263.	100.0	263.0
Food and agriculture				
Food and kindred	56	6637.	100.0	6637.0
Grain mill and bakery	29	944.	29.0	273.8
Vegetable oil	56	236.	100.0	236.0
Other	29	427.	29.0	123.8
Miscellaneous industrial				
Iron and steel foundry	70	272.	100.0	272.0
Nonferrous metals	69	182.	95.0	172.9
Other	67	663.	100.0	663.0
Unspecified	67	3514.	100.0	3514.0
Subtotal				27650.7
Fugitive Sources				
Road and building construction	57			176625.0
Agricultural tilling	57			23633.0
Refuse disposal sites	57			746.0
Livestock feedlots	59			3234.0
Unpaved road travel	57			57216.0
Paved road travel	58			165400.0
Forest fires (seasonal)	60			5790.0
Structural fires	35			398.4
Fireplaces	63			1244.0
Cigarettes	39			1990.0
Agricultural burning	61			1244.0
Tire attrition	38			2409.2
Brake lining attrition	41			7712.0
Sea salt	62			49753.0
Subtotal				497394.6

Table 5.3. Summary of Particulate Emissions Ranked by Emission Rate,[a,b] Los Angeles Area, 1976

Species	Symbol	Fine Particulate Emissions (kg/day)	Ratio of Element Emissions to Lead
Silicon	Si	89129.1	10.163
Total carbon	TC	89069.8	10.156
Aluminum	Al	35497.0	4.048
Chlorine	Cl	32469.4	3.702
Sulfates	SO_4^2	24711.1	2.818
Sodium	Na	23630.4	2.695
Iron	Fe	21149.5	2.412
Calcium	Ca	10490.0	1.196
Lead	Pb	8769.9	1.0
Magnesium	Mg	8519.2	0.971
Potassium	K	6791.2	0.774
Bromine	Br	3067.1	0.350
Titanium	Ti	2205.4	0.251
Nitrates	NO_3	2197.0	0.251
Zinc	Zn	693.3	0.079
Manganese	Mn	594.2	0.068
Nickel	Ni	572.7	0.065
Chromium	Cr	295.5	0.034
Copper	Cu	248.2	0.028
Barium	Ba	193.0	0.022
Vanadium	V	160.2	0.018
Selenium	Se	148.5	0.017
Cobalt	Co	118.6	0.013
Cesium	Cs	102.3	0.012
Arsenic	As	34.9	0.004
Cadmium	Cd	24.9	0.003
Molybdenum	Mo	20.7	0.002
Lanthanum	La	16.2	0.001
Strontium	Sr	12.7	0.001
Neodymium	Nd	10.1	0.001
Praeseodymium	Pr	10.1	0.001
Zirconium	Zr	9.1	0.001
Rubidium	Rb	7.9	0.001
Tin	Sn	6.9	0.0008
Silver	Ag	3.6	0.0004
Antimony	Sb	3.4	0.0004
Bismuth	Bi	0.1	0.0000
	Elements Listed	360983.	(58%)
	Other Emissions	259032.	(42%)
	Total Emissions	620015.	(100%)

[a] Reprinted with permission from Cass and McRae (1983). Copyright 1983 American Chemical Society.

[b] Aerodynamic diameter $d_p < 10\ \mu m$.

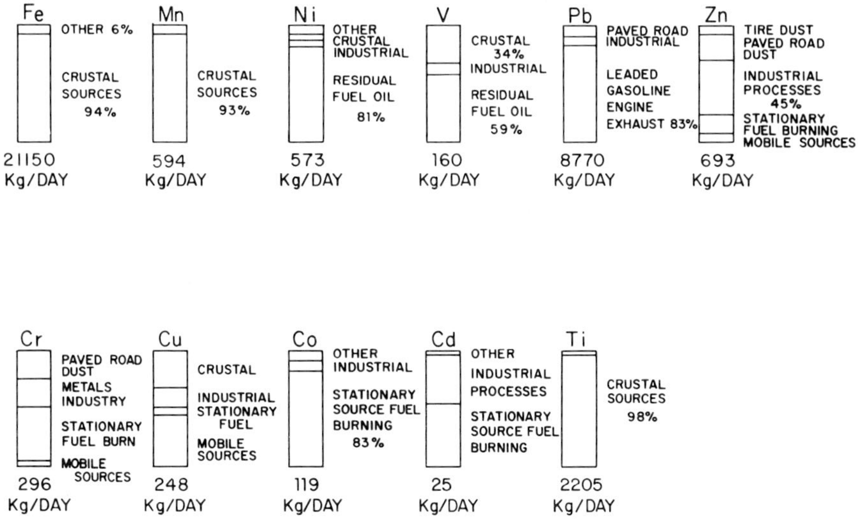

Figure 5.2. Summary of fine particle trace metals emissions in the Los Angeles area, 1976. Reprinted with permission from Cass and McRae (1983). Copyright 1983 American Chemical Society.

5. CONFIRMATION TESTS

Next, a consistency check is performed on the emission inventory and ambient monitoring data. The total mass emission rates for fine particle Fe, Si, Al, and so forth, are rank ordered. The relative abundance of each element in the fine particle emissions inventory is compared to the relative abundance of elements in long-term average ambient monitoring data at stations within the urban plume of the city. Results of such a comparison for elements traditionally monitored in the Los Angeles atmosphere are shown in Figure 5.3. With the exception of Cu and Co, the agreement between the relative abundance of elements in both emissions and air quality data is quite good. The mismatch for copper in that case is known to be due to contamination of the samples by copper worn from the high-volume sampler motors (Radke et al., 1977), whereas the mismatch for cobalt is probably due to slight cobalt contamination of the source signature for natural gas combustion taken at a source that periodically burns fuel oil.

Consistency checks also can be performed on the ambient data separately to assure that the ambient data are of high quality. The ambient data can be reviewed to see if lead and bromine concentrations are highly correlated (both usually arise from the same source, leaded auto exhaust). Crustal elements can be checked for high cross-correlation.

Table 5.4. Emissions of Selected Trace Metals,[a] Los Angeles Area, 1976

	Profile	Emissions (kg/day)	Elements							
			Al	Cd	Cr	Fe	Pb	Ni	V	Zn
Stationary Sources										
Fuel combustion										
Electric utilities										
Natural gas	27	233.58	0.0	0.1	1.3	1.3	0.1	1.3	0.0	1.3
Residual oil (0.50% S)	64	20832.44	0.0	4.6	27.1	235.4	13.3	441.6	70.8	20.8
Landfill and digester gas	27	0.84	0.0	0.0	0.0	0.0	0.0	0.0	0.0	0.0
Refinery fuel										
Natural gas	27	802.48	0.0	0.4	4.4	4.4	0.4	4.4	0.0	4.4
Refinery gas	27	3415.46	0.0	1.7	18.8	18.8	1.7	18.8	0.0	18.8
Residual oil	1	620.12	0.0	0.0	3.4	19.6	0.0	3.4	3.4	0.0
Nonrefinery industrial fuel										
Natural gas	27	3031.02	0.0	1.5	16.7	16.7	1.5	16.7	0.0	16.7
LPG	27	19.70	0.0	0.0	0.1	0.1	0.0	0.1	0.0	0.1
Residual oil	1	1004.75	0.0	0.0	5.5	31.8	0.0	5.5	5.5	0.0
Distillate oil	2	985.14	0.0	0.5	5.3	0.0	5.4	0.5	0.0	5.4
Digester gas (IC engines)	50	127.42	0.0	0.0	0.0	0.0	0.0	0.1	0.0	0.0
Coke oven gas	27	269.79	0.0	0.1	1.5	1.5	0.1	1.5	0.0	1.5
Residential/commercial										
Natural gas	51	9062.31	0.0	4.5	49.8	49.8	4.5	0.0	0.0	49.8
LPG	51	145.92	0.0	0.1	0.8	0.8	0.1	0.0	0.0	0.8
Residual oil	1	479.73	0.0	0.0	2.6	15.2	0.0	2.6	2.6	0.0
Distillate oil	2	470.13	0.0	0.2	2.5	0.0	2.6	0.2	0.0	2.6
Subtotals		41500.8	0.0	13.8	139.9	395.5	29.8	496.8	82.4	122.2

Mobile Sources										
Highway vehicles										
Catalyst autos and light trucks	53	788.08	0.9	0.0	0.0	0.9	0.0	0.1	0.0	0.6
Noncatalyst autos and light trucks	54	28477.57	12.2	0.0	0.0	71.2	6008.8	0.0	0.0	6.0
Medium and heavy gasoline vehicles	54	5968.38	2.6	0.0	0.0	14.9	1259.3	0.0	0.0	1.3
Diesel vehicles	52	7736.05	26.3	0.0	0.0	102.1	7.3	0.0	0.8	17.8
Civil aviation										
Jet aircraft	55	733.00	0.0	0.0	0.0	0.0	0.0	0.0	0.0	0.0
Aviation gasoline	54	16.80	0.0	0.0	0.0	0.0	3.5	0.0	0.0	0.0
Commercial shipping										
Residual oil-fired ships boilers	1	2184.75	0.0	0.0	12.0	69.2	0.0	12.0	12.0	0.0
Diesel ships	52	821.61	2.8	0.0	0.0	10.8	0.8	0.0	0.1	1.9
Railroad										
Diesel oil	52	1566.56	5.3	0.0	0.0	20.7	1.5	0.0	0.2	3.6
Military										
Gasoline	54	157.73	0.1	0.0	0.0	0.4	33.3	0.0	0.0	0.0
Diesel oil	52	1343.26	4.6	0.0	0.0	17.7	1.3	0.0	0.1	3.1
Jet fuel	55	659.00	0.0	0.0	0.0	0.0	0.0	0.0	0.0	0.0
Residual oil (bunker fuel)	1	19.59	0.0	0.0	0.1	0.6	0.0	0.1	0.1	0.0
Miscellaneous										
Off-highway vehicles	52	2996.49	10.2	0.0	0.0	39.6	2.8	0.0	0.3	6.9
Subtotals		53468.9	65.0	0.0	12.1	348.2	7318.7	12.2	13.6	41.2

[a] Aerodynamic diameter $d_p < 10$ μm.

Table 5.4. *(Continued)*

	Profile	Emissions (kg/day)	Elements							
			Al	Cd	Cr	Fe	Pb	Ni	V	Zn
Stationary Sources										
Industrial process point sources										
Petroleum industry										
Refining	28	1185.23	0.0	0.0	0.0	11.5	0.6	0.6	0.0	0.0
Paving and roofing materials	19	516.46	0.0	0.3	0.0	10.3	0.3	2.8	0.0	2.8
Other (calcining—mineral)	28	276.94	0.0	0.0	0.0	2.7	0.1	0.1	0.0	0.0
Organic solvent use										
Surface coating	22	741.12	0.0	0.0	0.0	4.1	0.0	0.0	0.0	0.0
Printing	56	8.91	0.0	0.0	0.0	0.0	0.0	0.0	0.0	0.0
Storage loss	56	8.91	0.0	0.0	0.0	0.0	0.0	0.0	0.0	0.0
Other	56	11.88	0.0	0.0	0.0	0.0	0.0	0.0	0.0	0.0
Chemical plants	24	1873.92	0.0	0.9	0.0	1.1	0.0	0.9	0.0	0.9
Metallurgical industry										
Metals—general	65	1716.00	0.0	2.1	20.0	100.7	38.9	2.9	2.9	1.8
Primary metals	65	4041.00	0.0	4.8	47.0	237.1	91.6	6.9	6.9	4.2
Secondary metals										
Nonferrous metals	69	905.35	0.0	0.0	0.5	34.8	113.2	5.0	0.0	95.1
Other	70	759.05	7.3	0.7	0.2	51.2	129.0	0.2	0.0	49.3
Metal fabrication										
Nonferrous metals	69	1095.35	0.0	0.0	0.5	42.1	136.9	6.0	0.0	115.0
Other	67	409.00	0.0	0.1	0.4	3.2	0.8	0.3	0.1	0.2
Mineral industry										
Glass furnaces	17	693.84	0.0	0.0	3.8	3.8	3.8	0.3	0.0	0.0
Rock, stone, other	33	1043.30	0.0	0.0	0.0	5.7	0.0	0.5	0.0	0.0
Waste burning at point sources	5	209.00	0.0	0.1	0.1	2.4	0.1	1.1	0.0	0.9

Wood and paper burning	5	263.00	0.0	0.1	0.1	3.0	0.1	1.4	0.0	1.1
Food and agriculture										
Food and kindred	56	6637.00	0.0	0.0	0.0	0.0	0.0	0.0	0.0	0.0
Grain mill and bakery	29	273.76	0.0	0.0	0.0	0.1	0.0	0.0	0.0	0.0
Vegetable oil	56	236.00	0.0	0.0	0.0	0.0	0.0	0.0	0.0	0.0
Other	29	123.83	0.0	0.0	0.0	0.1	0.0	0.0	0.0	0.0
Miscellaneous industrial										
Iron and steel foundry	70	272.00	2.6	0.3	0.1	18.4	46.2	0.1	0.0	17.7
Nonferrous metals	69	172.90	0.0	0.0	0.1	6.6	21.6	1.0	0.0	18.2
Other	67	663.00	0.0	0.1	0.7	5.2	1.3	0.5	0.1	0.3
Unspecified	67	3514.00	0.0	0.6	3.6	27.7	7.0	2.6	0.5	1.8
Subtotals		27650.7	9.9	10.1	77.1	571.8	591.6	33.4	10.5	309.3
Fugitive Sources										
Road and building construction	57	176625.00	14483.3	0.0	0.0	5652.0	35.3	7.1	10.6	8.8
Agricultural tilling	57	23633.00	1937.9	0.0	0.0	756.3	4.7	0.9	1.4	1.2
Refuse disposal sites	57	746.00	61.2	0.0	0.0	23.9	0.1	0.0	0.0	0.0
Livestock feedlots	59	3234.00	265.2	0.0	0.0	103.5	0.6	0.1	0.2	0.2
Unpaved road travel	57	57216.00	4691.7	0.0	0.0	1830.9	11.4	2.3	3.4	2.9
Paved road travel	58	165400.00	13893.6	0.0	66.2	11445.7	777.4	19.8	38.0	181.9
Forest fires (seasonal)	60	5790.00	83.4	0.0	0.0	11.0	0.0	0.0	0.0	0.0
Structural fires	35	398.40	0.0	0.0	0.0	10.2	0.0	0.0	0.0	0.0
Fireplaces	63	1244.00	0.3	0.0	0.0	0.0	0.1	0.0	0.0	1.5
Cigarettes	39	1990.00	0.0	1.0	0.0	0.0	0.0	0.0	0.0	0.0
Agricultural burning	61	1244.00	5.6	0.0	0.1	0.7	0.0	0.0	0.0	0.0
Tire attrition	38	2409.20	0.0	0.0	0.0	0.0	0.0	0.0	0.0	24.1
Brake lining attrition	41	7712.00	0.0	0.0	0.0	0.0	0.0	0.0	0.0	0.0
Sea salt	62	49753.00	0.0	0.0	0.0	0.0	0.0	0.0	0.0	0.0
Subtotals		497394.6	35422.1	1.0	66.3	19834.1	829.8	30.3	53.7	220.6

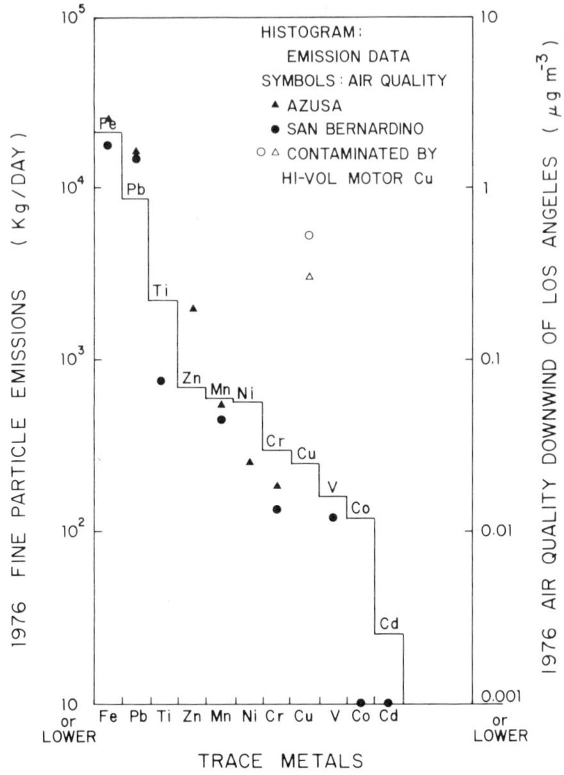

Figure 5.3. Comparison of emissions and air quality data in the Los Angeles area. Reprinted with permission from Cass and McRae (1983). Copyright 1983 American Chemical Society.

6. SOURCE PROFILE SELECTION FOR RECEPTOR MODEL CALCULATIONS

The mass emission inventory for each chemical element or compound is examined next, as shown in Figure 5.2. Elements are sought that serve as nearly unique tracers for one or a very few sources. These are the elements for which a nearly complete mass balance can be constructed using only a few source profiles. In the Los Angeles example shown in Figure 5.2, soil-like crustal materials dominate Fe and Mn emissions, Pb is a nearly unique tracer for the automobile, and fuel oil dominates Ni emissions. In contrast, metals like Zn, Cd, and Cr arise from small contributions from a large number of industrial and fugitive sources (see Table 5.4) that are grouped together in Figure 5.2. Zinc alone is emitted from more than 40 different source types, and it is clear that a mass balance equation cannot be written successfully for Zn in Los Angeles unless an exceptionally large number of source profiles are used in the data reduction process. Since the number of elements used in the solution of Eq. (1) and (2) must equal or exceed the number of source profiles present, the inclusion of an accurate mass balance equation for an element like Zn would substantially increase the dimension of the source assignment problem.

On the basis of the emission inventory findings summarized in Figure 5.2, a near mass balance can be achieved on Fe, Mn, Ni, V, and Pb in Los Angeles that can probably identify the relative importance of crustal material (soil or road dust), fuel oil fly ash, and leaded highway vehicle exhaust. The emission inventory further can be used to help quantify additional source contributions by linking together sources that have the same spatial and temporal distribution of emissions as one of the readily traced sources given above. Leaded automobile exhaust can be tracked by our receptor model. Those emissions are accompanied by other roadway emissions including diesel engine exhaust, exhaust from automobiles burning unleaded fuel, tire wear, and brake-lining dust. A composite source profile for all highway emissions is constructed in Table 5.5 based on an emissions inventory weighted average of each of these source types. If that composite profile is used in the chemical mass balance equations in place of a leaded auto exhaust profile, then five source types can be tracked simultaneously rather than just leaded auto exhaust alone.

Table 5.5. Source Profiles for Fine Aerosol Emissions from Highway Vehicles[a]

	Gasoline Autos and Trucks (Leaded Fuel)	Automobile (Unleaded)	Diesel Engine	Tire Tread	Brake Lining	Highway Composite
1976 Mass Emissions ($d_p < 10$ μm) (tons/day)	34.75 (60.7%)	0.79 (1.39%)	7.74 (13.6%)	6.04 (10.6%)	7.71 (13.6%)	56.71
Chemical Composition (%)						
Aluminum	0.043	0.12	0.34			0.074
Bromine	8.2		0.031			4.98
Calcium		0.17	0.84		5.5	0.86
Chlorine	5.4		1.69			3.51
Copper	0.004	0.024	0.73			0.10
Iron	0.25	0.11	1.32			0.333
Lead	21.1		0.095			12.8
Magnesium					8.25	1.12
Manganese		0.015	0.027			0.0039
Nickel		0.015				0.0002
Potassium		0.044				0.001
Silicon	0.075	0.51	0.17		15.4	2.17
Sodium			0.37			0.05
Vanadium			0.01			0.00136
Zinc	0.021	0.08	0.23	1.0		0.151
Sulfates	0.213	50.0	4.2			1.40
Nitrates			0.72			0.1
Carbon	54.5	39.0	70.0	87.0	28.3	56.2

[a] Reprinted with permission from Cass and McRae (1983). Copyright 1983 American Chemical Society.

7. RECEPTOR MODEL RESULTS

The emission inventory assisted chemical mass balance method has been applied to Los Angeles (Cass and McRae, 1981a, 1983). The source profiles for fuel oil fly ash, highway aerosol, crustal emissions, plus secondary sulfates and nitrates given in Table 5.6 were fit to the chemical elements for which a near mass balance could be assured: Pb, Fe, Mn, Ni, SO_4^{2-}, and NO_3^- at South Coast Air Quality Management District (SCAQMD) monitoring sites and Pb, Fe, Mn, V, SO_4^{2-} and NO_3^- at National Air Surveillance Network (NASN) sites. As seen in Figure 5.4, the material balance clearly accounts for about 80% of the total suspended particulate matter (TSP) concentrations at most high-volume sampler sites in that city, even if only a few chemical elements are available for use in the model.

Table 5.6. The Chemical Composition of Selected Sources of Atmospheric Fine Particulate Matter[a]

	Percentages (%)					
Chemical Component[b]	Soil Dust[c]	Road Dust[d]	Fuel Oil Fly Ash[e]	Highway Aerosol[f]	Ammonium Sulfate	Ammonium Nitrate
SO_4^{2-}		0.62	31.9	1.40	72.7	
NO_3^-		0.42	4.50	0.1		77.5
Fe	3.2	6.92	1.13	0.333		
Mn	0.11	0.137	0.052	0.0039		
Ni	0.004	0.012	2.12	0.0002		
Pb	0.02	0.47	0.064	12.8		
Cr		0.04	0.128			
Cu	0.008	0.032	0.042	0.10		
Zn	0.005[g]	0.11	0.101	0.151		
V	0.006	0.023	0.339	0.0014		
Cd			0.022			
Co	0.002		0.047			
Sn						
Ti	0.4	0.67	0.008			

[a] Reprinted with permission from Cass and McRae (1983). Copyright 1983 American Chemical Society.

[b] Chemical components listed are those under study in the SCAQMD and NASN ambient air quality data base. For complete source composition profiles, see Appendix B of Cass and McRae (1981).

[c] Resuspended soil dust samples taken in the Los Angeles area (Friedlander, 1973).

[d] Road dust fine particle samples taken in Portland, Ore. (Watson, 1979)

[e] Direct average of elemental percentages obtained by Taback et al. (1979) in South Coast Air Basin utility boiler tests 11, 12, 13, 21, 22, 24, 32, and 33 as reported in "Total" columns of Tables 4-26 to 4-33.

[f] Composite of gasoline and diesel powered highway vehicle exhaust, plus tire and brake dust (see Table 5.5).

[g] Given as $<0.01\%$.

8. IDENTIFICATION OF APPLICATIONS WHERE CHEMICAL MASS BALANCE MODELS MAY NOT PERFORM WELL

By contrast, the emission inventory assisted method applied to a city such as Houston, Texas, shows that not all urban areas can be modeled as readily as Los Angeles. The 1980 particulate emission inventory provided for the Houston area by the U.S. Environmental Protection Agency's National Emission Data System and by Environmental Research and Technology (Heisler et al., 1981) was associated with local soil dust composition data (Stevens, 1982) and with source profiles assembled by Taback et al. (1979), Watson (1979), and Cass and McRae (1981a). An element-by-element fine particle emission inventory was created by the methods described previously. In Figure 5.5, the available emission inventory data for Houston are compared to the average of ambient fine particle concentration data obtained by Dzubay et al. (1982) during September 1980 at a monitoring site on the University of Houston campus. It is seen that emissions and air quality data do not match as closely as in Los Angeles but that the general trend from abundant to rare elements is evident. The Houston emission inventory may be incomplete or inaccurate. Atmospheric Ca, Si, and Fe concentrations do not match the total inventory in relative abundance, nor do they closely match the composition of local soil dust. Manganese is present at a level consistent with stated road and soil dust emissions, but low compared to the total emission inventory. Aluminum concentrations by X-ray fluorescence are close to their abundance in both soil and the total inventory. Bromide and Cl are depleted relative to the inventory, as expected if they are being displaced from the aerosol by reaction with acidic sulfates or nitric acid. Airborne sodium exceeds the abundance in the inventory, perhaps because sea salt is excluded as a fugitive source in the local emission inventory. Phosphorus and Ba are both elevated in the atmosphere and rare in source emissions. They probably could be used to identify their source(s) if locally obtained source profiles were available for all sources. A phosphate fertilizer plant is present in Houston, but no corresponding source signature is available.

The information given in Figure 5.6 shows why receptor modeling problems in Houston will be tough to solve given available data. In contrast to Los Angeles, very few elements act as nearly unique tracers for one or two sources. Lead sources other than leaded auto exhaust are evident. Bromine would be a good tracer for leaded gasoline use, but it is not conserved in the aerosol phase. Aluminum would be the best soil dust tracer, but crustal emissions still account for only 71% of the local Al emissions. There are an enormous number of industrial sources of the other conventionally crustal elements, such as Si, Ca, Fe, and Mn. Most of the ambient vanadium data are below the detection limit, and the vanadium content of local soil dust is not known.

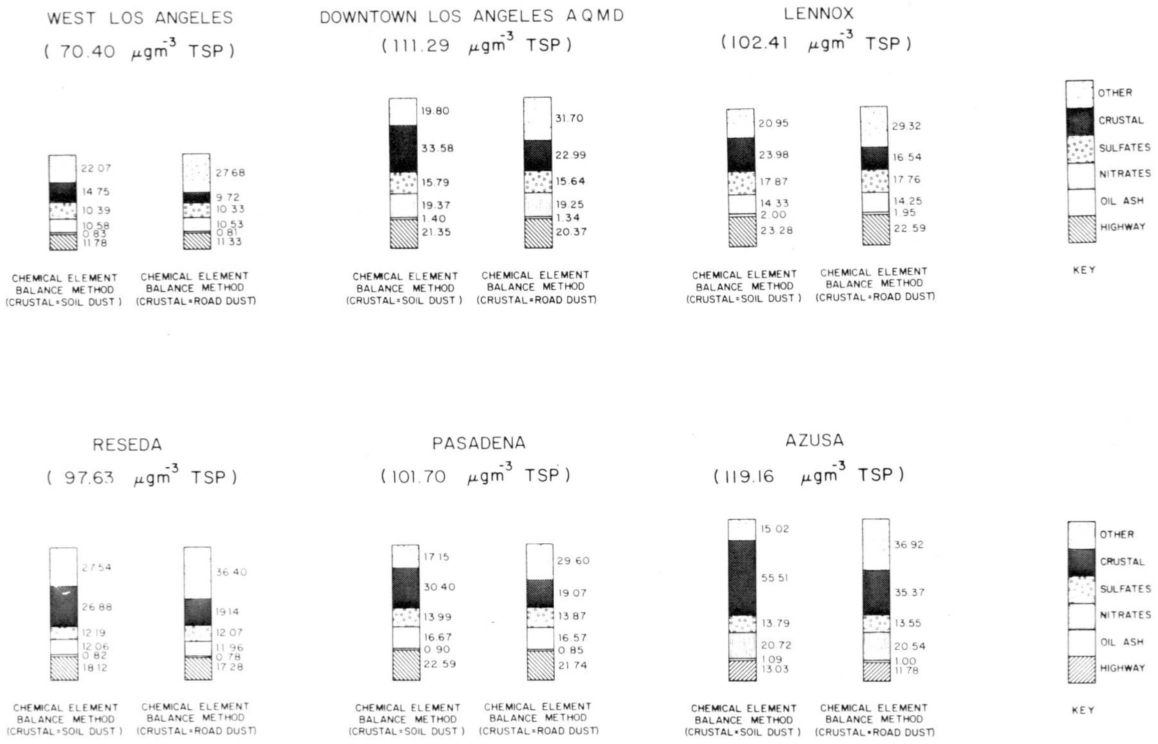
WEST LOS ANGELES
(70.40 μgm-3 TSP)
22.07
14.75
10.39
10.58
0.83
11.78
27.68
9.72
10.33
10.53
0.81
11.33
CHEMICAL ELEMENT BALANCE METHOD (CRUSTAL=SOIL DUST)
CHEMICAL ELEMENT BALANCE METHOD (CRUSTAL=ROAD DUST)
DOWNTOWN LOS ANGELES AQMD
(111.29 μgm-3 TSP)
19.80
33.58
15.79
19.37
1.40
21.35
31.70
22.99
15.64
19.25
1.34
20.37
CHEMICAL ELEMENT BALANCE METHOD (CRUSTAL=SOIL DUST)
CHEMICAL ELEMENT BALANCE METHOD (CRUSTAL=ROAD DUST)
LENNOX
(102.41 μgm-3 TSP)
20.95
23.98
17.87
14.33
2.00
23.28
29.32
16.54
17.76
14.25
1.95
22.59
CHEMICAL ELEMENT BALANCE METHOD (CRUSTAL=SOIL DUST)
CHEMICAL ELEMENT BALANCE METHOD (CRUSTAL=ROAD DUST)
OTHER
CRUSTAL
SULFATES
NITRATES
OIL ASH
HIGHWAY
KEY
RESEDA
(97.63 μgm-3 TSP)
27.54
26.88
12.19
12.06
0.82
18.12
36.40
19.14
12.07
11.96
0.78
17.28
CHEMICAL ELEMENT BALANCE METHOD (CRUSTAL=SOIL DUST)
CHEMICAL ELEMENT BALANCE METHOD (CRUSTAL=ROAD DUST)
PASADENA
(101.70 μgm-3 TSP)
17.15
30.40
13.99
16.67
0.90
22.59
29.60
19.07
13.87
16.57
0.85
21.74
CHEMICAL ELEMENT BALANCE METHOD (CRUSTAL=SOIL DUST)
CHEMICAL ELEMENT BALANCE METHOD (CRUSTAL=ROAD DUST)
AZUSA
(119.16 μgm-3 TSP)
15.02
55.51
13.79
20.72
1.09
13.03
36.92
35.37
13.55
20.54
1.00
11.78
CHEMICAL ELEMENT BALANCE METHOD (CRUSTAL=SOIL DUST)
CHEMICAL ELEMENT BALANCE METHOD (CRUSTAL=ROAD DUST)
OTHER
CRUSTAL
SULFATES
NITRATES
OIL ASH
HIGHWAY
KEY

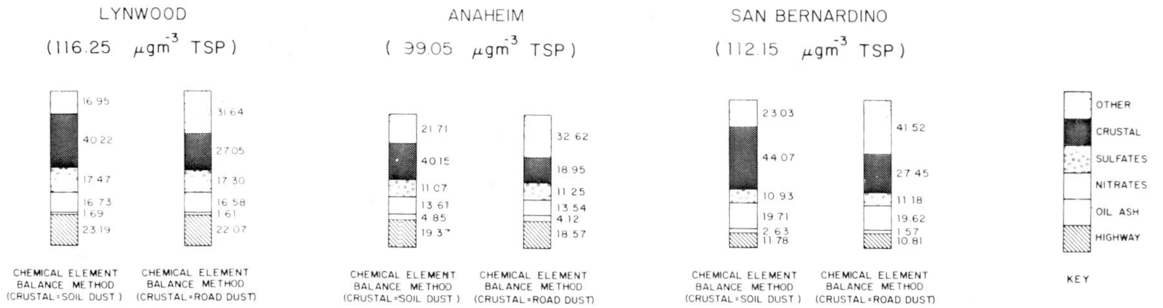

Figure 5.4. Chemical mass balance model results in the Los Angeles area, 1976. Reprinted with permission from Cass and McRae (1983). Copyright 1983 American Chemical Society.

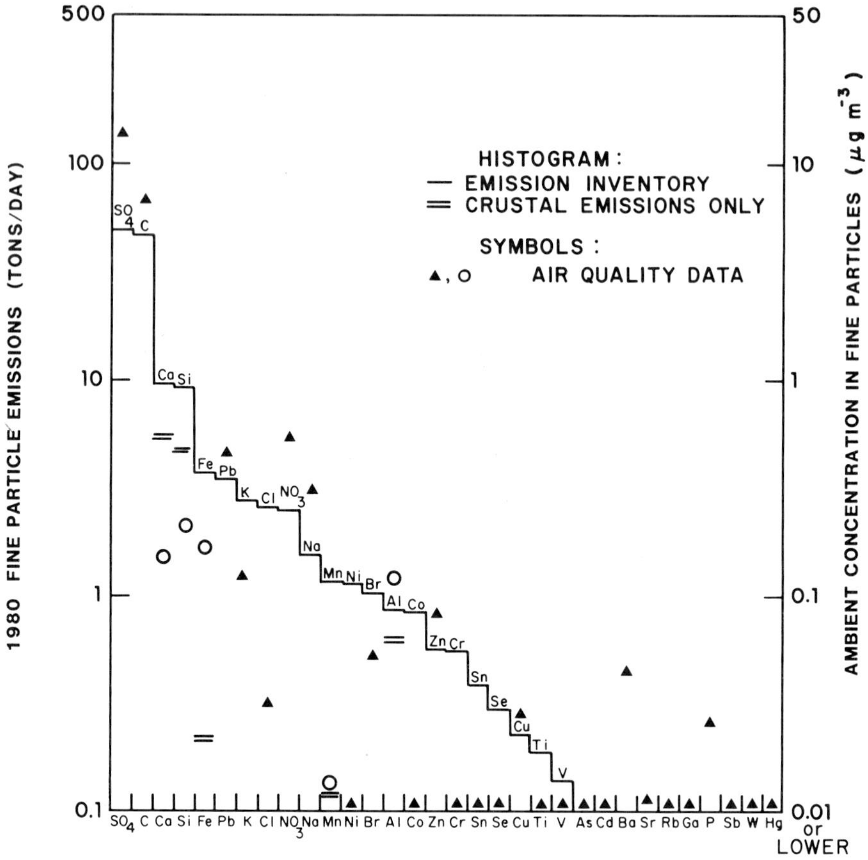

Figure 5.5. Comparison of emissions and air quality data in the Houston area.

If one wishes to proceed with a chemical mass balance analysis of the Houston data set, the next step is to assemble composite emission profiles for the steel mill, pulp mill, petroleum refineries, and so on, until enough sources are included to nearly complete a mass balance on the abundant easily detected elements. When this is done it is found that the ferroalloy furnaces, steel mill, pulp mill, and fluid catalytic cracking unit signatures are in many cases nearly linear combinations of one another. This renders the chemical mass balance problem nearly degenerate. The emissions inventory assisted chemical mass balance method thus indicates that the present data base for Houston is not a highly promising candidate for mass balance based receptor models. This situation might be improved by a better emission inventory and locally determined source profiles.

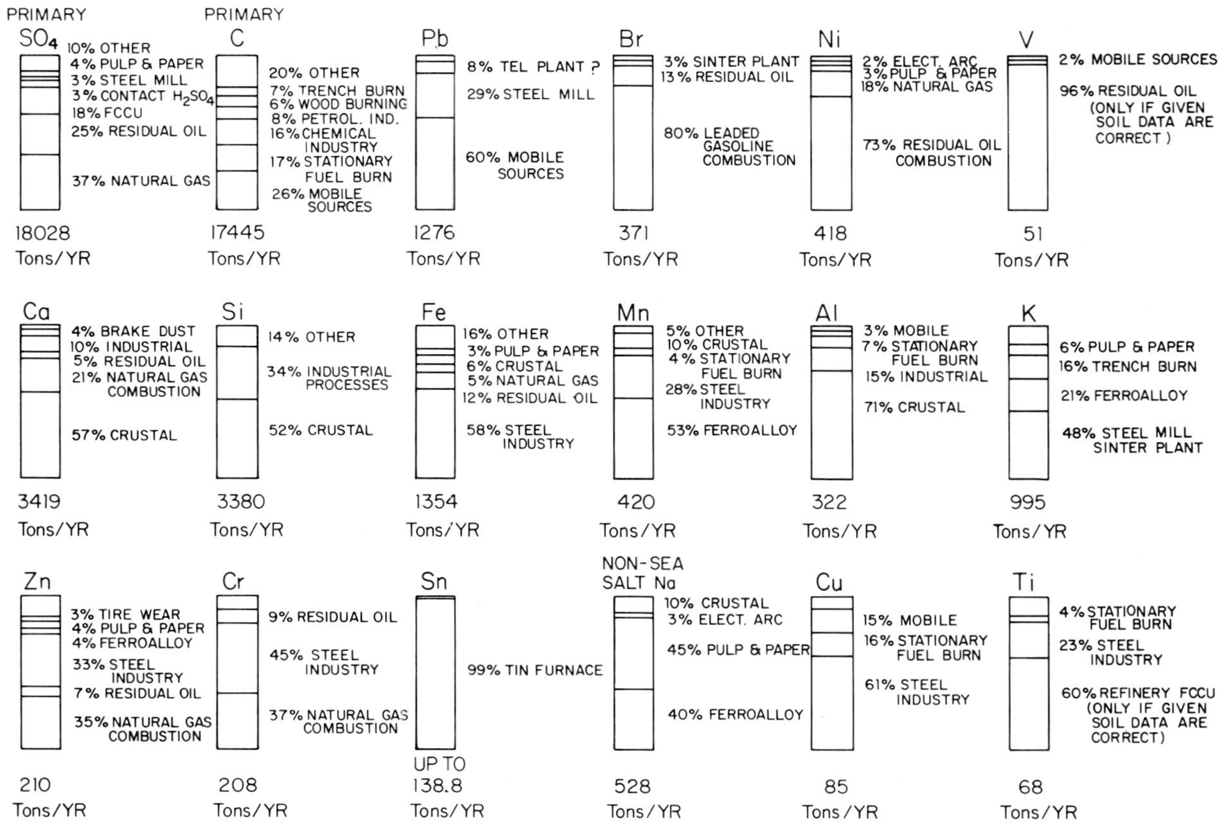

Figure 5.6. Summary of fine particle trace metals emissions in the Houston area, 1980.

REFERENCES

Alpert, D.J., and Hopke, P.K. (1980). "A quantitative determination of the sources in the Boston urban aerosol." *Atmos. Environ.* **14,** 1137–1146.

Cass, G.R., Boone, P.M., and Macias, E.S. (1982) "Emissions and air quality relationships for atmospheric carbon particles in Los Angeles." In *Particulate carbon: Atmospheric life cycle,* G.T. Wolff and R.L. Klimisch, Eds. Plenum, New York.

Cass, G.R., McMurry, P.S., and Houseworth, J.E. (1978). *Methods for sulfate air quality management with applications to Los Angeles.* Environmental Quality Laboratory Report 16. California Institute of Technology, Pasadena, Calif.

Cass, G.R. and McRae, G.J. (1981a). *Source-receptor reconciliation of South Coast Air Basin particulate air quality data.* Final Report to the California Air Resources Board, Sacramento, Calif., under Agreement A9-014-031, NTIS PB-82-250093.

Cass, G.R., and McRae, G.J. (1981b). "Minimizing the cost of air pollution control." *Environ. Sci. Technol.* **15,** 748–757.

Cass, G.R., and McRae, G.J. (1983). "Source-receptor reconciliation of routine air monitoring data for trace metals: An emission inventory assisted approach." *Environ. Sci. Technol.* **17,** 129–139.

Cooper, J.A. (1973). *Urban aerosol trace element ranges and typical values.* Tech. Publications BNWL-SA-4690. Battelle Pacific Northwest Laboratories, Richland, Wash.

Dzubay, T.G., Stevens, R.K., and Lewis, C.W. (1982). "Visibility and aerosol composition in Houston, Texas." *Environ. Sci. Technol.* **16,** 514–525.

Dzubay, T.G. (1977) *X-Ray fluorescence analysis of environmental samples.* Ann Arbor Science Publishers, Ann Arbor, Mich.

Friedlander, S.K. (1973). "Chemical element balances and identification of air pollution sources." *Environ. Sci. Technol.* **7,** 235–240.

Grisinger, J.G., Ed. (1981). *Draft 1979 emission inventory for South Coast Air Basin.* 1982 AQMP Revision Working Paper No. 1. South Coast Air Quality Management District, El Monte, Calif.

Heisler, S.L. (1982). Personal communication. Environmental Research and Technology, Inc., Concord, Mass.

Heisler, S.L., Watson, J.G., Shah, J.J., Chow, J.C., Collins, J.C., and Whitney, J. (1981). *Recommendations for the design of aerosol characterization studies in Houston and El Paso, Texas.* Environmental Research and Technology, Concord, Mass.

Henry, R.C. (1977). "A factor model of urban aerosol pollution." Ph.D. Thesis. Oregon Graduate Center, Beaverton, Ore.

Hidy, G.M., and Friedlander, S.K. (1970). "The nature of the Los Angeles aerosol." In *Proceedings Second International Clean Air Congress,* H.M. Eglund and W.T. Berry, Eds. Academic Press, New York.

Hopke, P.K. (1981). "The application of factor analysis to urban aerosol source resolution." In *Atmospheric aerosol: Source/air quality relationships,* E.S. Macias and P.K. Hopke, Eds. American Chemical Society, Washington, D.C., pp. 21–49.

Hopke, P.K., Lamb, R.E., and Natusch, D.F.S. (1980). "Multi-elemental characterization of urban roadway dust." *Environ. Sci. Technol.* **14,** 164–172.

Kowalczyk, G.S., Choquette, C.E., and Gordon, G.E. (1977). "Chemical element balances and identification of air pollution sources in Washington, D.C." *Atmos. Environ.* **12,** 1143–1153.

Miller, M.S., Friedlander, S.K., and Hidy, G.M. (1972). "A chemical element balance for the Pasadena aerosol." *J. Colloid Interface Sci.* **39,** 165–176.

Radke, N.M., Cherniack, I., Witz, S., and MacPhee, R.D. (1977). *The effect of type of air sampler on composition of collected particulates: A comparison of brushless and brush-type Hi-Vol air samplers.* Technical Services Division Report. South Coast Air Quality Management District, El-Monte, Calif.

Stevens, R.K. (1982). Personal communication of data. U.S. Environmental Protection Agency, Research Triangle Park, N.C., letters forwarding Houston data to participants in the USEPA Quail Roost II Workshop.

Taback, H.J., Brienza, A.R., Macko, J., and Brunetz, N. (1979). *Fine particle emissions from stationary and miscellaneous sources in the South Coast Air Basin.* KVB Inc., Tustin, California, Report Number KVB 5806-783 (Profiles are contained in Appendix A).

U.S. Environmental Protection Agency (1977). *Compilation of air pollutant emission factors,* 3rd Ed. Document AP-42. U.S. Enviromental Protection Agency, Research Triangle Park, N.C.

U.S. Environmental Protection Agency (1972). "*Guide for compiling a comprehensive emission inventory.*" Report APTD-1135. U.S. Environmental Protection Agency, Research Triangle Park, N.C.

Watson, J.G. Jr. (1979). "Chemical element balance receptor model methodology for assessing the sources of fine and total suspended particulate matter in Portland, Oregon." Ph.D. Thesis. Oregon Graduate Center, Beaverton, Ore.

6

QUANTITATIVE SOURCE ATTRIBUTION OF METALS IN THE AIR USING RECEPTOR MODELS

Philip K. Hopke

Institute for Environmental Studies
Department of Civil Engineering
and
Nuclear Engineering Program
University of Illinois
Urbana, Illinois

1. INTRODUCTION

The advent of a national ambient air quality standard for total suspended particles (TSP) over a decade ago created the need to identify particle sources so that effective control strategies could be designed and implemented. These initial efforts at identification of particle sources focused on dispersion models of point sources and, in most cases, resulted in substantial reductions in TSP levels. However, as the increment of additional control needed to reach standard levels became smaller, the model uncertainties lead to difficulties in identifying the actual sources of the continuing problems. In addition, fugitive and other nonducted emissions are generally not treated or are poorly handled in these models. Thus, additional methods were required to identify and quantitatively apportion particle mass to sources.

These requirements have arisen at a time when new analytical methods have been developed that permit multielemental analysis of large numbers of airborne particle samples. Thus, large data bases on the composition of airborne particles are available for use in these receptor models, that use the measured properties of collected ambient samples to infer the sources of those particles. Although much of the thrust of these model developments has been aimed at identification of sources of particle mass, as part of the analysis these developments do elucidate the origins of the various measured elemental species observed in the samples. It is then possible to use receptor models to identify sources of metals and other toxic species and to quantitatively apportion the observed airborne concentrations among the various source types. This chapter will outline several of the applicable models and provide examples of their use in apportioning materials in a number of airsheds. A more complete description of receptor models is provided by Hopke (1985).

2. PRINCIPLE OF MASS CONSERVATION

All of the currently used receptor models are based on the concept of the conservation of mass and a mass balance analysis. For example, we can assume that the total lead concentration (in nanograms per cubic meter) measured at a site can be considered a linear sum of contributions from independent source types such as motor vehicles, incinerators, and smelters.

$$\mathrm{Pb}_T = \mathrm{Pb}_{\mathrm{auto}} + \mathrm{Pb}_{\mathrm{incin}} + \mathrm{Pb}_{\mathrm{smelt}} + \ldots \qquad (1)$$

However, a motor vehicle burning leaded gasoline emits particles containing materials other than lead. Therefore, the atmospheric concentration of lead from automobiles in ng/m^3, $\mathrm{Pb}_{\mathrm{auto}}$, can be considered to be the product of two cofactors: (1) the gravimetric concentration (ng/mg) of lead in automotive particulate emissons, $a_{\mathrm{Pb,auto}}$, and (2) the mass concentration (mg/m^3) of automotive particles in the atmosphere, f_{auto}.

$$\mathrm{Pb}_{\mathrm{auto}} = a_{\mathrm{Pb,auto}} f_{\mathrm{auto}} \tag{2}$$

The normal approach to obtaining a data set for receptor modeling is to determine a large number of elements in a number of samples. The mass balance equation can thus be extended to account for all m elements in the n samples as contributions from p independent sources

$$x_{ij} = \sum_{k=1}^{p} a_{\mathrm{ik}} f_{\mathrm{kj}}, \qquad i = 1,m, \quad j = 1,n \tag{3}$$

where x_{ij} is the ith elemental concentration measured in the jth sample, a_{ik} is the gravimetric concentration of the ith element in material from the kth source, and f_{kj} is the airborne mass concentration of particles from the kth source contributing to the jth sample. There are several different approaches to receptor model analysis that have been successfully applied including chemical mass balance (CMB), multivariate receptor models including target transformation factor analysis (TTFA), and computer-controlled scanning electron microscopy combined with a mass balance analysis. The basis for each of these methods will be presented in subsequent sections of this chapter along with examples of their application to the identification of sources of metals in the atmosphere.

3. CHEMICAL MASS BALANCE

The chemical mass balance (CMB), sometimes called the chemical element balance, solves Eq. (3) directly for each sample by assuming that the number of sources and their compositions at the receptor site are known. This approach was first independently suggested by Winchester and Nifong (1971) and by Miller et al. (1972). The ambient aerosol composition of a sample is then used in a multiple linear regression against source compositions to derive the mass contribution of each source to that particular sample. Miller et al. modified Eq. (3) explicitly to include changes in composition of the source material while in transit to the receptor

$$x_{ij} = \sum \alpha_{ik} \, a'_{ik} f_{kj} \tag{4}$$

where α_{ik} is the coefficient of fractionation so that if a'_{ik} were the composition of the particles as emitted by the source, $a_{ik} = \alpha_{ik} \, a'_{ik}$ is the composition of the particles at the receptor site. In practice, it is very difficult to determine the α_{ij} values and they are assumed to be unity ($a_{ik} = a'_{ik}$).

There were several early applications of this approach to urban aerosol mass apportionment including Pasadena, Calif. (Friedlander, 1973), Heidelberg, FRG (Bogen, 1973), Ghent, Belgium (Heindryckx and Dams, 1974), and Chicago, Ill. (Gatz, 1975). In all of these analyses, the quality of available source compositions severely limited the precision to which the ambient compositions could be reproduced.

Several major research efforts have subsequently resulted in substantially better source data. Studies at the University of Maryland have provided data on coal-fired power plants (Gladney, 1973; Gladney et al., 1976), oil-fired power plants (Mroz, 1976), incinerators (Greenberg, 1976; Greenberg et al., 1978; Greenberg et al., 1978), motor vehicle traffic (Ondov, 1974; Ondov, et al., 1982), soil (Thomae, 1977), and assorted industrial sources (Small, 1979). These studies led to much improved resolution of the particle sources in Washington, D.C. (Kowalczyk, et al., 1978; Kowalczyk, et al., 1982). In the first report of these studies, Kowalczyk, et al. (1978) used a weighted least-squares regression analysis to fit six sources with eight elements for ten ambient samples. In these analyses, the ambient elemental concentrations are weighted by the inverse of the square of the analytical uncertainty in that measurement. Subsequently, Kowalczyk et al. (1982) examined 130 samples using seven sources with 28 elements included in the fit. The average results for this extended data set are presented in Table 6.1 taken from Kowalczyk, et al. (1982). Although 28 elements were used in the fitting process, the fit did not change appreciably with varying numbers of elements included, with the exception of some of the key tracer elements such as Na, Pb, and V. The values of the larger to smaller ratios of the observed and predicted values are an indication of the amount of fluctuation in the fit. This elemental balance sheet allows the identification of the major sources of metals in the air. For example, vanadium and nickel primarily arise from oil-fired power plant emissions while zinc has a major incinerator source as well as a motor vehicles source. The reverse is true for lead with motor vehicles as the primary source and refuse incineration as a lesser but important source. In this way both sources of particulate mass and specific elements can be identified.

In 1979, Watson (1979) and Dunker (1979) independently suggested a mathematical formalism called effective variance weighting that includes the uncertainty in the measurement of the source composition profiles as well as the uncertainties in the ambient concentrations. As part of this analysis, a method has also been developed to permit the calculation of the uncertainties in the mass contributions.

The most extensive use of effective variance fitting has been made by Watson in his work on data from Portland, Ore. (Cooper et al., 1979; Watson, 1979; Watson et al., 1984). Since that study, a number of other applications of this approach have been made including Medford, Ore. (Cooper, 1979; DeCesar and Cooper, 1981), Philadelphia, Pa. (Chow et al., 1981), and at a number of locations in the U.S. Environmental Protection Agency's Inhalable Particulate Network (Chow et al., 1981).

A major study that combined ambient and source samples, conducted in Portland, Ore., was the Portland Aerosol Characterization Study (PACS) (Watson, 1979; Cooper et al., 1979; Core et al., 1981). In this study several important improvements in ambient and source sampling as well as the CMB modeling were developed as a result of the PACS program. The PACS program is a particularly good example of the use of the mass balance approach since they utilized 37 sources for which direct measurements were made during the time the ambient samples were being taken. Therefore, the composition and variation in source compositions are quite applicable to the aerosol particle mass apportionment problem being considered.

Another important innovation that was introduced into the PACS study was the attempt to obtain a representative yearly average without having to sample every day of the year. In their planning, they stratified the year into eight defined meteorological regimes. When conditions and time of year were appropriate, samples were taken. However, although many samples were obtained, only enough of the samples were analyzed to provide a meaningful average value for that meteorological regime. Ninety-four days per year were sampled with a 32-day subset selected for intensive chemical analysis. In this way, annual average values that are fairly representative can be obtained with a quite reasonable effort for both sampling and analysis.

Samples were taken of both total suspended particles and particles with aerodynamic diameter less than 2.5 μm. The average chemical composition of the downtown Portland aerosol is given in Figure 6.1. The aerosol mass source apportionment is given in Figure 6.2. Although there are a large number of sources identified in the resolution, there is a sizeable unidentified amount as well as several nonspecific sources including volatilizable carbon and nonvolatilizable carbon. Much of the unidentified portion is likely to be NH_4^+ and H_2O associated with the identified sulfate and nitrate. The sulfate in fine mode airborne particle samples may initially be acidic, but by the time gravimetric and elemental analyses are performed, it will generally be ammonium sulfate. Similarly the nitrate will be ammonium nitrate by the time the filter mass is determined. In a latter part of this section, there will be a closer examination of the ability of the mass balance approach to obtain such detailed resolutions.

One of the important uses of the mass balance results that were obtained in the PACS program was to calibrate the dispersion models being used for making air quality management decisions (Core et al., 1981, 1982). The Portland Airshed was modeled using GRID (Fabrick and Sklarew, 1975), a conservation of mass, advection–diffusion code that has been designed to model the complex terrain of the Portland area. The area is divided into 2-by-2-km grid cells. For each of the 5000 cells, it is necessary to know the topography, the wind field flows for each of the eight meteorological regimes, and the spatially and seasonally resolved point and area source emissions as well as the stack parameters for each point source.

The results of the dispersion model can be separated into the same source groups as the mass balance results so that direct comparisons can be made. In Figures 6.3, 6.4, and 6.5 the results of the GRID and mass balance models are compared for road dust, auto exhaust, and residual oil combustion. Road dust and residual oil impacts are in substantial disagreement with dust

Table 6.1. Average Results of Chemical Element Balance of 130 Samples from the Washington, D.C. Area for Summer 1976

Element	Predicted Contributions[a], ng/m^3										
	Soil	Limestone	Coal	Oil	Refuse	Motor Vehicle	Marine	Total Predicted	Observed[d]	Larger/Smaller[b]	Missing Values[c]
Na[e]	43	0.83	8.3	12	35	—	201	300	300 ± 20	1.00[f]	0
Mg[e]	74	101	27	3.8	5.4	32	26	270	440 ± 30	1.63[f]	3
Al[e]	812	9	517	0.4	6.1	—	<0.01	1340	1350 ± 110	1.16[f]	0
K[e]	154	6	67	0.4	47	13	7	295	400 ± 20	1.38[f]	0
Ca[e]	66	635	47	8.2	7.1	47	7.6	820	860 ± 40	1.09	0
Sc[e]	0.15	0.002	0.18	0.0002	0.0006	—	<0.0001	0.33	0.33 ± 0.03	1.19[f]	0
Ti[e]	52	0.83	31	0.03	1.0	—	<0.0001	85	110 ± 10	1.35[f]	0
V[e]	1.06	0.042	1.6	23	0.013	—	<0.0001	26	25 ± 2	1.07	0
Cr	0.81	0.023	0.84	0.062	0.21	—	<0.0001	2.0	14 ± 2	7.07[f]	5
Mn[e,g]	13	2.3	1.6	0.10	0.31	1.3	<0.0001	18	17 ± 2[g]	1.26[f]	0
Fe[e]	511	8.3	362	2.8	2.8	34	0.0002	920	1000 ± 60	1.19[f]	0
Co[e]	0.22	0.0002	0.25	0.16	0.003	0.05	<0.0001	0.68	0.83 ± 0.08	1.19[f]	0
Ni	0.46	0.042	1.04	4.0	0.07	0.34	<0.0001	6.0	17 ± 2	4.35[f]	1
Cu	0.23	0.009	2.2	0.89	0.71	3.0	0.0001	7.1	17 ± 2	2.68[f]	23
Zn[e]	1.14	0.04	2.6	1.6	51	7.3	<0.0001	64	85 ± 6	1.42[f]	0
Ga[e]	0.38	0.008	0.46	0.0001	—	—	<0.0001	0.85	1.29 ± 0.17	1.89[f]	42
As[e]	0.061	0.002	3.1	0.028	0.10	—	0.0001	3.32	3.25 ± 0.2	1.58[f]	0
Se	0.0009	0.0002	0.78	0.035	0.016	0.035	0.0001	0.87	2.5 ± 0.2	4.10	0
Br	0.097	0.013	2.1	0.054	0.66	167	1.25	171	139 ± 9	1.84	0
Rb[e]	1.19	0.0064	0.60	0.0001	—	—	0.0023	1.8	2.1 ± 0.2	1.54[f]	17
Sr[e]	3.65	1.29	3.5	0.09	0.027	—	<0.0001	8.6	10 ± 1	1.40[f]	32

Ag[e]	0.0007	0.0001	—	0.006	0.23	—	<0.0001	0.24	0.20 ± 0.01	1.44[f]	26
Cd[e]	0.0011	0.0001	0.13	0.0028	0.64	1.03	<0.0001	1.80	2.4 ± 0.02	1.89[f]	0
In	0.0007	0.0001	0.0023	<0.0001	0.0024	—	0.0004	0.0059	0.020 ± 0.001	5.12[f]	3
Sb[e]	0.0081	0.0004	0.13	0.007	0.89	0.60	<0.0001	1.6	2.1 ± 0.2	1.39[f]	0
I	0.058	0.0026	2.1	—	—	—	1.14	3.3	2.0 ± 0.1	2.41	12
Cs[e]	0.028	0.0011	0.039	0.0005	0.0025	—	<0.0001	0.07	0.17 ± 0.05	3.36[f]	2
Ba	7.1	0.02	4.7	2.0	0.30	6.4	0.0006	21	19 ± 2	1.42[f]	1
La[e]	0.75	0.014	0.31	0.016	0.0016	—	<0.0001	1.1	1.5 ± 0.1	1.42[f]	0
Ce[e]	0.98	0.024	0.62	0.014	0.007	—	<0.0001	1.6	2.0 ± 0.2	1.30[f]	1
Sm[e]	0.068	0.002	0.057	0.0012	0.0003	—	<0.0001	0.13	0.20 ± 0.02	1.59[f]	0
Eu[e,h]	0.015	0.0004	0.014	0.0005	0.0002	—	<0.0001	0.030	0.030 ± 0.003	1.32[f]	5
Yb	0.037	0.0011	0.030	0.0004	0.0009	—	<0.0001	0.070	0.034 ± 0.003	2.60[f]	19
Lu	0.0067	0.0004	0.0080	0.0001	0.0004	—	<0.0001	0.016	0.0056 ± 0.0006	3.34[f]	1
Hf[e]	0.031	0.0006	0.022	0.0008	0.0004	—	<0.0001	0.055	0.10 ± 0.01	1.80[f]	1
Ta[e,h]	0.041	0.0008	0.0077	0.0011	0.0018	—	<0.0001	0.052	0.036 ± 0.004	1.84[f]	39
W	0.014	0.0013	0.038	0.0004	0.0072	—	<0.0001	0.061	0.24 ± 0.02	4.76	23
Pb[e]	0.15	0.019	2.1	0.39	34	42.8	<0.0001	465	440 ± 20	1.28[f]	0
Th[e]	0.11	0.0036	0.10	0.0016	0.0008	—	<0.0001	0.22	0.25 ± 0.02	1.19[f]	1

[a] Contributions designated by "—" indicate that concentration of the element in particles from the source is not known.

[b] Larger/smaller is the average of the ratio of predicted/observed or vice versa, whichever is larger over all samples.

[c] Number of samples for which no value is available because peak is not strong enough in spectra for determination or concentration is not above filter blank.

[d] Uncertainty is standard deviation of mean value.

[e] Element fitted by least-squares procedure.

[f] Larger/smaller value included in average given below.

[g] For reasons described in text, all observed values reduced by factor of 0.69 prior to fitting. Actual average observed values was 26 + 3 ng/m.

[h] As noted in the text, three observed values were eliminated from fits.

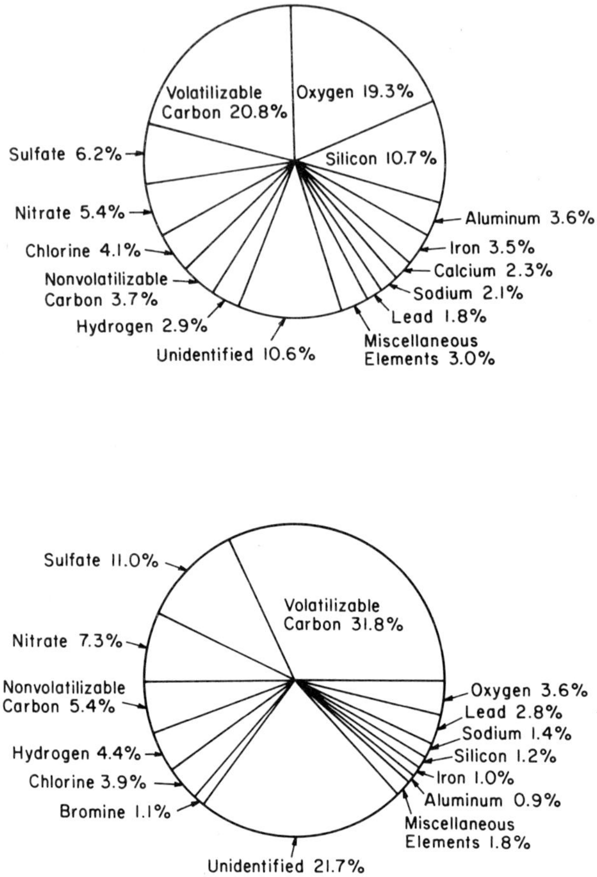

Figure 6.1. Average chemical composition of the total suspended particulate matter (top) and the fine (<2.5 μm) particulate matter (bottom) of the downtown Portland, Ore., aerosol. Reprinted with permission from Core et al. (1981). Copyright 1981 American Chemical Society.

substantially underpredicted by the dispersion model. For the auto exhaust, three sites agree within the uncertainty in the mass balance results. However, for one location (site 5) there is an underprediction by the GRID model. Core et al. (1982) indicated that these discrepancies can be readily understood. For the auto exhaust, a heavily travelled road was incorrectly assigned to an adjacent grid. When properly placed, the dispersion model value increases by 10%.

Several problems were found in modeling residual oil combustion emissions. First, the emissions had been assumed to be constant over the entire year. The dispersion model was thus modified to account for the actual monthly operating schedules for each meteorological regime. Second, the topographical assignments for each cell were examined. Each grid cell is given one of five altitude variations within a cell. Small changes in plume height

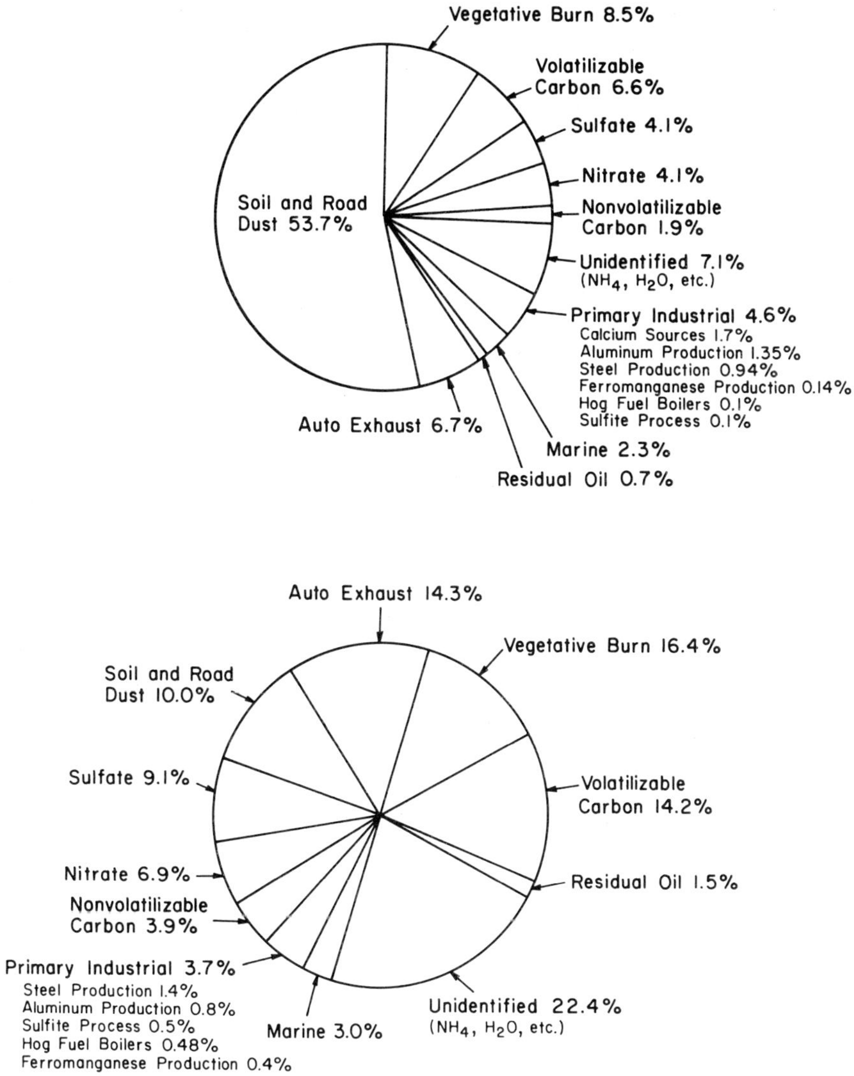

Figure 6.2. Source apportionment of the total suspended particulate matter (top) and fine (<2.5 μm) particulate matter (bottom) of the downtown Portland, Ore., aerosol as given in Core et al. (1982) and used with permission.

assignment can then substantially affect the predictions. Adjustments were made to the point source stack heights to better reflect the difference between the true stack and receptor heights. Finally, a single major source was modeled with an unrealistic operating schedule. After these corrections were made, quite good agreement was obtained.

The road dust was underpredicted based on the EPA generalized paved road emission factor. There are few unpaved roads, so it was felt that the

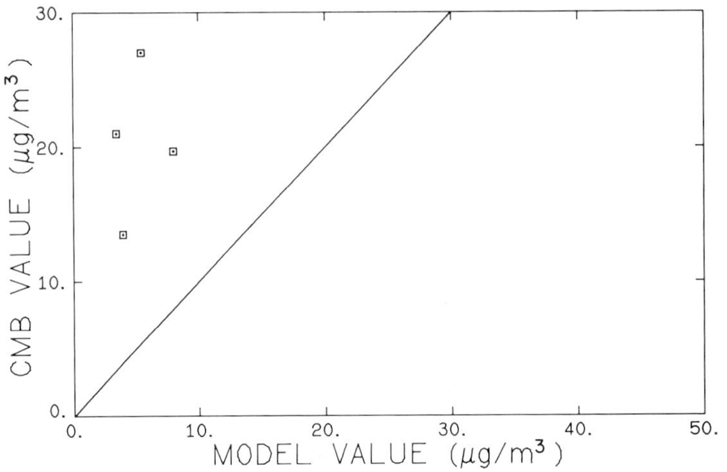

Figure 6.3. Comparison of initial results for road dust impacts in Portland, Ore., as derived by the CMB and dispersion models (Core et al., 1982) and used with permission.

paved road emission factor should be increased. Since the automotive emission factor appeared to be correct because of the generally good agreement seen in Figure 6.4, the road dust emission factor was scaled by the ratio of the mass balance result for road dust to that for auto exhaust. The road dust emission was also suggested to be larger in heavy industrial areas of Seattle, Wash. (Roberts et al., 1979) compared to commercial land use areas. This adjustment substantially increased the road dust contribution to the

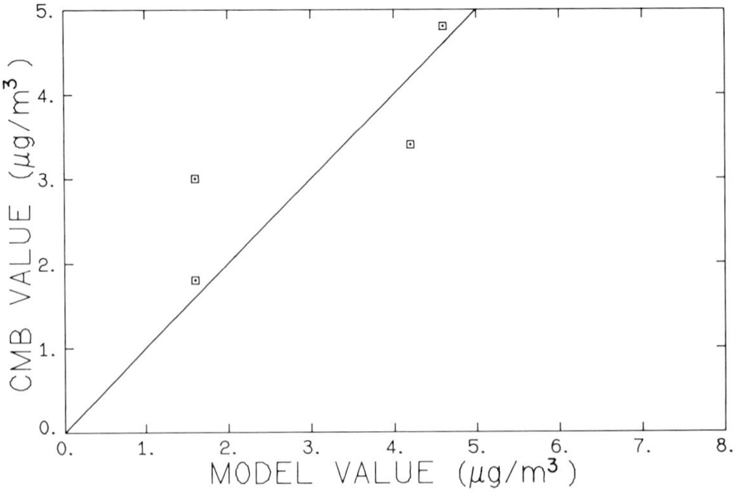

Figure 6.4. Comparison of initial results for motor vehicle tailpipe emissions in Portland, Ore., derived by the CMB and dispersion models (Core et al., 1982) and used with permission.

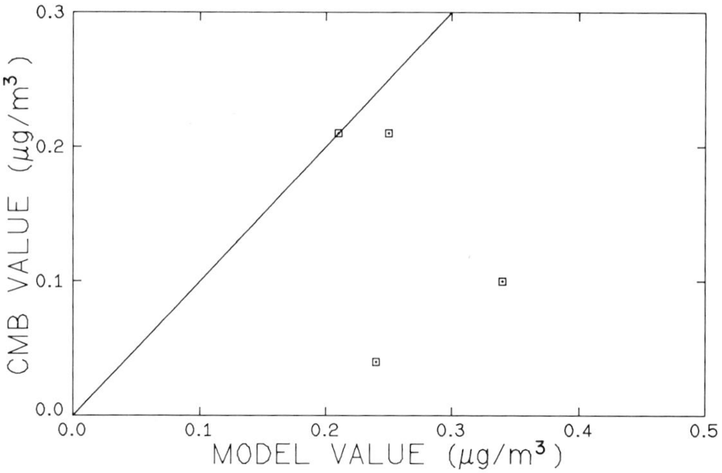

Figure 6.5. Comparison of initial results for residual oil impacts in Portland, Ore., as derived by the CMB and dispersion models (Core et al., 1982) and used with permission.

airborne particle loading. This adjustment brings the receptor and dispersional model results into agreement for road dust.

The test of these adjustments is then to compare dispersion model prediction of total suspended particulate matter (TSP) with the measured values. Figure 6.6 shows these comparisons before and after making the adjustments described above. Clearly, the use of the mass balance results to tune the dispersion model has made a noticeable improvement in the quality of the dispersion model predictions and hence its value as an air quality management tool.

The CMB method has also been directly applied to problems of identifying

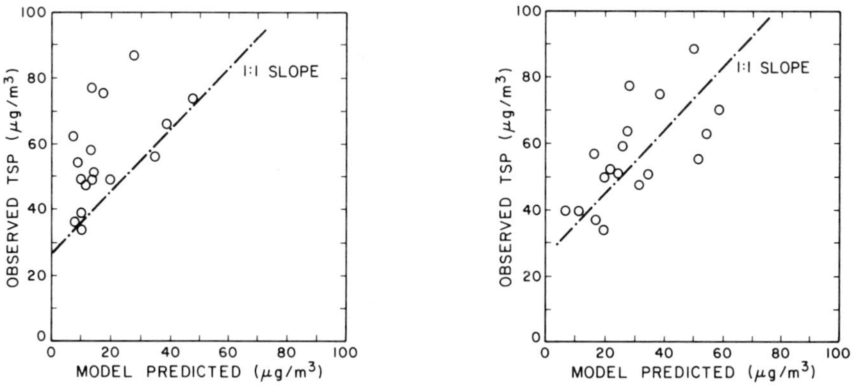

Figure 6.6. Comparison of the initial predictions of the Portland AQMA model for annual average TSP before (left) and after (right) CMB-suggested adjustments (Core et al., 1982) and used with permission.

the sources of particular metals. For example, Houck and Cooper (1983) have examined the source of lead in three airsheds with lead smelters. Particularly in East Helena, Mont., where they were able to obtain particle compositions for a number of specific sources with the plant, they were able to identify specific points where particle emission control would significantly alter the observed lead values. As seen in Figure 6.7, both the TSP and the lead can be apportioned to sources. While the major source of the TSP is road and soil dust (37 of the 79 $\mu g/m^3$ average TSP), it contributed less than 20% of the airborne lead The total lead value of 5.6 $\mu g/m^3$ was apportioned between eight sources, the largest of which were fugitive ore and residuals handling (35%) and blast furnace upsets (26%). The ability to identify areas from which lead emission is occurring offers a better opportunity to develop effective and economically efficient control strategies.

It must be made clear, however, that the CMB analysis works well in these examples because the programs were designed to measure both source and ambient samples during the same period. A much less detailed resolution of lead sources was all that was possible in Kellogg, Idaho (Houck and Cooper, 1983) when on-site samples could not be obtained. In a recent intercomparison study organized by the U.S. Environmental Protection Agency (Stevens and Pace, 1983) to examine receptor models, a set of ambient particulate elemental compositional data sets were analyzed by a number of investigators using similar CMB methods. The compositions of particles from sources in Houston were not available and were not measured during this program so that source composition profiles had to be obtained from literature sources. The lack of source data immediately raised problems in the use of the mass balance methods and comparison of results from different investigators. It is not always certain exactly which sources should be included in the analysis. Although emission inventories may be available for the region, it may be that the measured source composition for a coal-fired power plant in Maryland burning eastern bituminous coal is not a particularly good representation for a lignite-burning plant in Texas.

In addition, the motor vehicle profile may be undergoing changes in Pb and Br concentrations with time as the mix of new, catalyst-equipped and diesel cars and older leaded-fuel burning vehicles change. Even more of a problem are local sources of soil and road materials that may be quite different from locations where source measurements are available. If more than a qualitative or semiqualitative apportionment is desired, it is clearly essential to know the area that is to be modeled well. Specific features such as nearby point sources or other unusual activities must be known in order to assure that a reasonable resolution can be made. In spite of these problems, fairly reasonable agreement was obtained between groups for the general source categories that could be resolved (Dzubay and Stevens, 1983).

Another more fundamental problem that ultimately limits the resolving power of the CMB analysis has also been recently identified (Henry, 1983). It has been known for some time that problems arise in a multiple regression

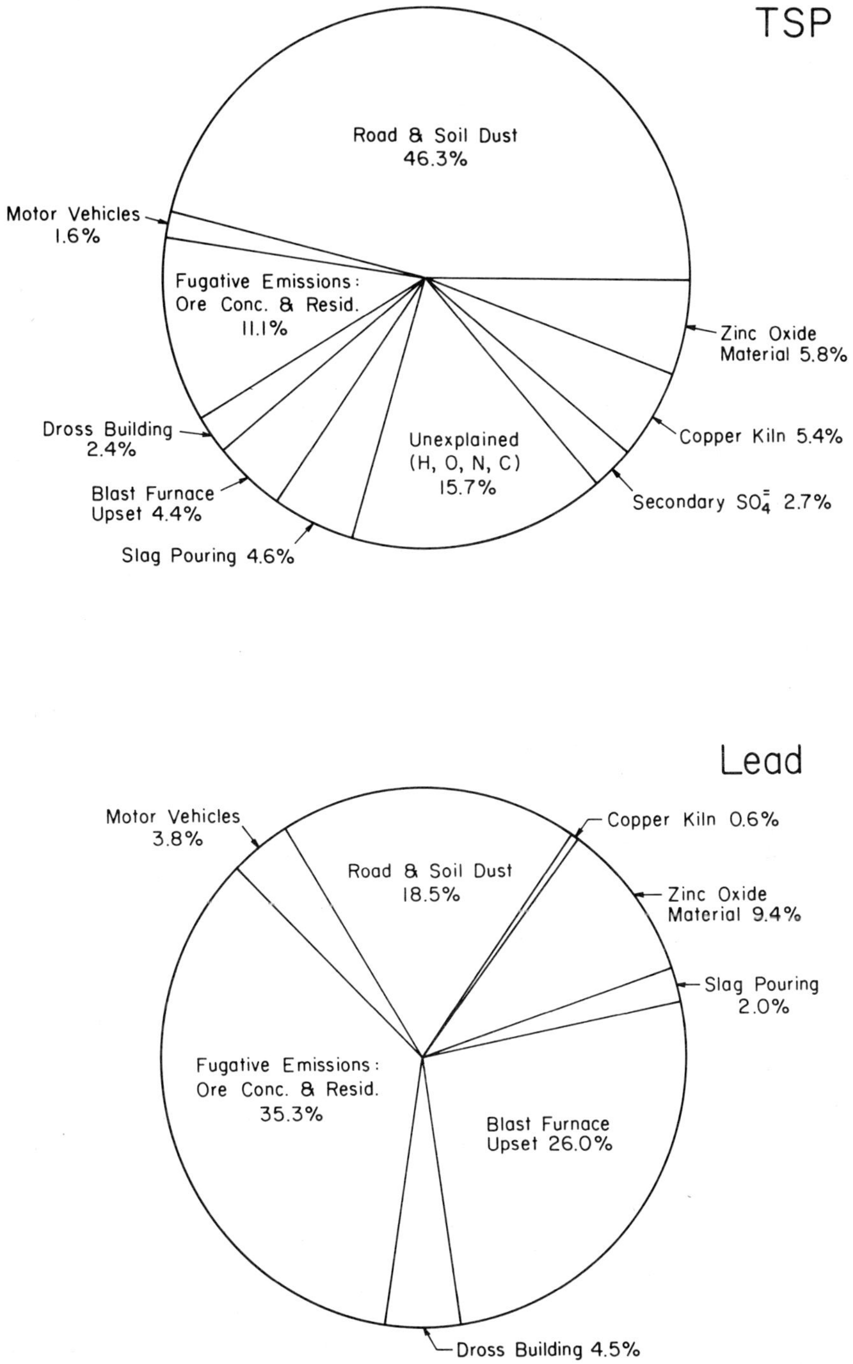

Figure 6.7. Apportionment of TSP and airborne particulate lead to sources in East Helena, Mont., as derived by CMB analysis (Houck and Cooper, 1983).

analysis when two or more of the sources being fit have similar elemental profiles. Problems of separating coal fly ash from soil have been frequently noted (Gatz, 1975; Kowalczyk et al., 1978, 1982). Henry (1983) has shown quite clearly that there are limits to how many sources can actually be resolved given a particular set of source profiles, and he has provided a procedure to identify estimatable from nonestimatable sources. He has also given an approach to identify estimatable combinations of sources having minimum variance. This important development provides a systematic approach to identifying what can and cannot be resolved in any particular CMB application. It should be reviewed by anyone who is planning to perform CMB analysis.

4. MULTIVARIATE RECEPTOR MODELS

Alternative approaches have been developed for identifying and quantitatively apportioning sources of airborne particles using multivariate statistical analysis. Factor analysis has been the principal method that has been applied to airborne particle composition data. Several different forms of factor analysis can be used including principal components analysis, classical factor analysis, and target transformation factor analysis.

Principal component and classical factor analyses are methods that try to simplify the description of a system by determining the minimum number of new variables necessary to reproduce the measured attributes of the system. The mathematical basis of these methods has been described by Hopke (1981, 1983, 1985).

The first receptor modeling applications of classical factor analysis were by Prinz and Stratmann (1968) and Blifford and Meeker (1967). Prinz and Stratmann examined both the aromatic hydrocarbon content of the air in 12 West German cities and data from Colucci and Begeman (1965) on the air quality of Detroit. In both cases they found three factor solutions and used a varimax rotation to give more readily interpretable results. Blifford and Meeker used a principal component analysis with both varimax and a nonorthogonal rotation to examine particle composition data collected by the National Air Sampling Network (NASN) during 1957–1961 in 30 U.S. cities. They were generally not able to extract much interpretable information from their data. Since there are a very wide variety of particle sources among these 30 cities and only 13 elements measured, it is not surprising that they were not able to provide much specificity to their factors. One interesting factor that they did identify was a unique copper factor for which they could not provide a convincing interpretation. It is likely that this factor represents the copper contamination from the brushes of the motors of the high-volume air samplers. This problem was identified to be a ubiquitous one by Hoffman and Duce (1971).

The factor analysis approach was then reintroduced by Hopke et al. (1976)

and Gaarenstroom et al. (1977) for their analysis of particle composition data from Boston, Mass., and Tucson, Ariz., respectively. In the Boston data for 90 samples at a variety of sites, six common factors were identified that were interpreted as soil, sea salt, oil-fired power plants, motor vehicles, refuse incineration, and an unknown manganese–selenium source. The six factors accounted for about 78% of the system variance. There was also a high unique factor for bromine that was interpreted to be fresh automobile exhaust. Large unique factors for antimony and selenium were found. These factors may possibly represent emission of volatile species whose concentrations do not covary with other elements emitted by the same source. Subsequent studies by Thurston and Spengler (1982) where other elements including sulfur were measured showed a similar result. They found that the selenium was strongly correlated with sulfur for the warm season (May 6 to Nov. 5). This result is in agreement with the Whiteface Mountain results (Parekh and Husain, 1981) and suggests that selenium is an indicator of long-range transport of coal-fired power plant effluents to the northeastern United States.

In the study of Tucson (Gaarenstroom et al., 1977), at each site whole filter data were analyzed separately. They found factors that are identified as soil, automotive, several secondary aerosols such as $(NH_4)_2SO_4$, and several unknown factors. They also discovered a factor that represented the variation of elemental composition in their aliquots of their neutron activation standard containing Na, Ca, K, Fe, Zn, and Mg. This finding illustrates one of the important uses of factor analysis; screening the data for noisy variables or analytical artifacts.

One of the valuable uses of this type of analysis is in screening large data sets to identify errors (Roscoe et al., 1981). With the use of atomic and nuclear methods to analyze environmental samples for a multitude of elements, very large data sets have been generated. Because of the ease in obtaining these results with computerized systems, the elemental data acquired are not always as thoroughly checked as they should be, leading to some, if not many, bad data points. It is advantageous to have an efficient and effective method to identify problems with a data set before it is used for further studies. Principal component factor analysis can provide useful insight into several possible problems that may exist in a data set including incorrect single values and some types of systematic errors. These uses are described by Dattner and Jenks (1981), Roscoe et al. (1982), and Hopke (1983).

Gatz (1978) applied a principal components analysis to aerosol composition data for St. Louis, Mo., taken as part of project METROMEX (Changnon et al., 1977; Ackerman et al., 1978). Nearly 400 filters collected at 12 sites were analyzed for up to 20 elements by ion-exchange X-ray fluorescence. Gatz used additional parameters in his analysis including day of the week, mean wind speed, percent of time with the wind from NE, SE, SW, or NW quadrants or variable, ventilation rate, and rain amount and duration. At several sites the inclusion of wind data permitted the extraction of additional factors that allowed identification of specific point sources. An important advantage of this form of factor analysis is the ability to include parameters such as wind speed and direction or particle size in the analysis.

Another approach to including the meteorological data that are available has been developed by Thurston and Spengler (1981). They separately analyze the elemental composition and the meteorological data. They then examine the correlations between the elemental factors and the meteorological factors. This approach assisted in the interpretation of their data from six different cities in the United States.

There have been a number of applications of principal components or factor analysis to particle source identification. Dattner (1978) has examined the sources of particles in a number of locations in Texas using X-ray fluorescence analysis on the Texas Air Control Board routine high-volume TSP samples. Sievering et al. (1980) have made extensive use of factor analysis in their interpretation of midlake aerosol composition and deposition data for Lake Michigan. Lewis and Macias (1980) have analyzed particle composition data from Charleston, W.Va. Their analysis was repeated by Hopke (1981) after separating their data set into one for fine particles and another for coarse particles. In this way a more detailed source resolution was obtained. Other studies have included examining the origins of particles in upstate New York (Parekh and Husain, 1981), the Arctic (Heidam, 1981), Vienna, Austria (Malissa et al., 1981), and Paris, France (Dutot et al., 1983).

A problem that exists with these forms of factor analysis is that they do not permit quantitative source apportionment of particle mass or of specific elemental concentrations. In an effort to find an alternative method that would provide information on source contributions when only the ambient particulate analytical results are available, Hopke and coworkers (Hopke, 1981; Hopke et al., 1980; Alpert and Hopke, 1980 and 1981; Chang et al., 1982; Liu et al., 1982; Severin et al., 1983) have developed target transformation factor analysis (TTFA). In this analysis, resolution similar to that obtained from a CMB analysis can be obtained. However, the CMB analysis can be made for a single sample if the source data is known while TTFA requires a series of samples with varying impacts by the same sources.

To illustrate these results and to demonstrate an alternative approach to obtaining source composition data, the results of Chang et al. (1982) on data from St. Louis, Mo., will be presented. As has been discussed above, one of the key problems in the application of receptor models to the source apportionment of aerosol mass is the determination of the composition of particles from a single source at the receptor site. There can be changes in the composition of particles from the point where they are emitted to the point where they are collected. New methods for determining these point source profiles are needed.

Rheingrover and Gordon have developed a method to identify samples strongly affected by single-point sources (1985). This approach, called wind trajectory analysis, can select samples from a large data base such as the one obtained in the Regional Air Pollution Study (RAPS) of St. Louis, Mo.

Samples that were heavily influenced by major sources of each element are identified first according to the following criteria:

1. Concentration of the element in question $X > X + Z_{cr}\,\sigma_X$, where X is the average concentration of that particular element for each station and size fraction (coarse or fine particle size fraction), Z_{cr} is typically set at about three for most elements, and σ_x is the standard deviation of the concentration of that element.

2. The standard deviation of the 6 or 12 hourly average wind directions for most samples, or minute averages for 2-h samples, taken during intensive periods is less than 20°.

Samples that are strongly affected by emissions from a source were identified through observations of clustering of mean wind directions for the sampling periods selected with angles pointing toward the source. Target transformation factor analysis (TTFA) for aerosol source apportionment offers a method of analysis for these data sets.

From May 1975 through April 1977, a total of about 35,000 individual ambient aerosol samples were collected at 10 selected sampling sites in the vicinity of St. Louis, Mo., as part of the Regional Air Pollution Study. The samples consisted of aerosol particles in coarse (2.4 to 20 μm) and fine (<2.4 μm) fractions deposited on membrane filters using automatic dichotomous air samplers (Nelson, 1979). The samples were analyzed for total mass by beta-gauge measurements and the concentrations of up to 27 elements were determined using energy dispersive X-ray fluorescence analysis (Goulding et. al., 1981). The normal sampling schedule consisted of 12-h sampling periods at eight of the stations and 6-h sampling periods at two of the stations. During an intensive study period in the summer of 1975, the schedule was modified to accommodate 6-h samples at most stations, with 2-h samples at three of the stations.

The RAPS data have been screened according to the criteria stated above. With wind trajectory analysis, specific emission sources could be identified even in cases where the sources were located very close together (Rheingrover and Gordon, 1985). A compilation was made of the selected impacted samples so that target transformation factor analysis can be employed to obtain elemental profiles of these specific sources at the receptor sites.

In matrix notation Eq. (3) can be rewritten as

$$X = AF$$

where X is the matrix of ambient aerosol compositions, A is the matrix of source composition profiles, and F is the matrix of mass contributions of the sources to the samples. The objectives of TTFA are (1) to determine p, the number of independent sources that contribute to the system, (2) to identify the components of matrix A, the elemental source profiles, and (3) to calculate F, the contribution of each source to each sample.

The number of sources is determined by performing an eigenvalue analysis on the matrix of correlations between the samples. A target transformation determines the degree of overlap between an input source profile and one of the calculated factor axes. The input source profiles, called test vectors, are developed from existing knowledge of the emission profiles of various sources or by an iterative technique from simple test vectors (Roscoe and Hopke, 1981a). The identified source profiles are then used in a simple weighted least-squares determination of the mass contributions of the sources (Severin et al., 1983).

To illustrate this method, the data set for fine particle zinc samples whose wind trajectories point toward a copper products plant in the East St. Louis, Ill., area will be analyzed. Some of the trajectories are shown in Figure 6.8.

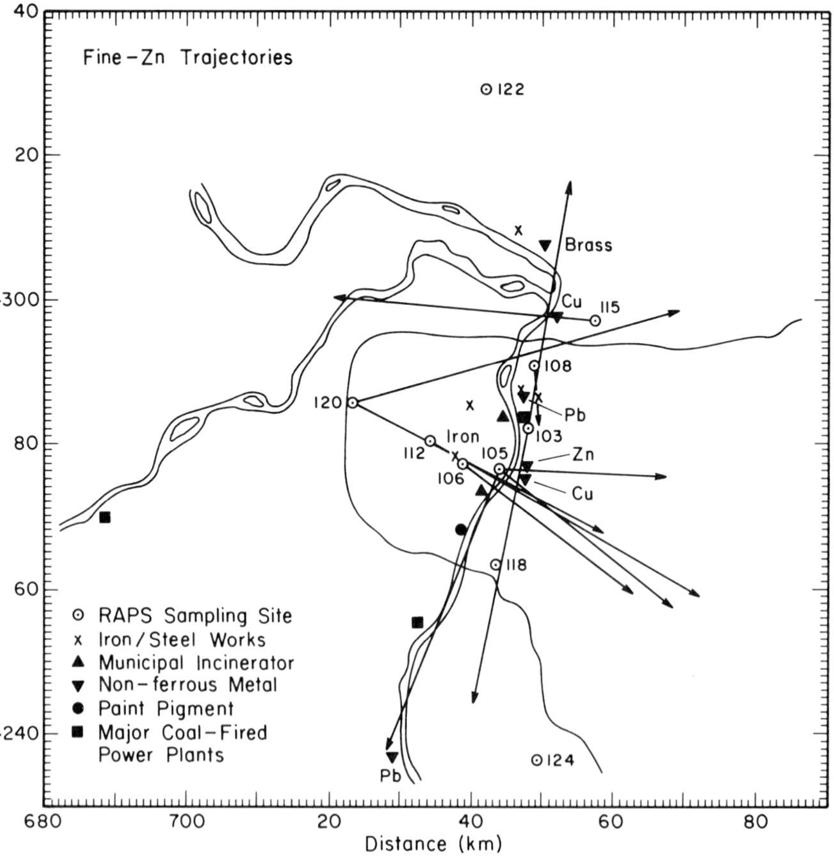

Figure 6.8. Major trajectories observed for fine Zn sources in metropolitan St. Louis, Mo. (Rheingrover and Gordon, 1985).

Table 6.2. Refined Source Concentration Profiles (mg/g) for Fine Particle Fraction Zn Trajectory

	Sulfate	Cu Plant	Motor Vehicle	Fly Ash/Soil
Al	0.0 ± 1.3	0.0 ± 5.0	0.0 ± 9.0	61.0 ± 15.0
Si	0.0 ± 1.2	0.0 ± 5.0	0.0 ± 8.0	195.0 ± 13.0
S	191.0 ± 1.0	47.5 ± 1.1	0.0 ± 1.9	17.0 ± 4.0
K	2.1 ± 1.5	8.5 ± 5.1	3.6 ± 9.0	26.0 ± 16.0
Ca	0.0 ± 0.8	6.5 ± 2.6	0.0 ± 5.0	37.0 ± 8.0
Ti	0.56 ± 0.25	0.0 ± 0.9	0.0 ± 1.6	4.3 ± 2.7
Mn	0.0 ± 0.6	4.0 ± 2.2	0.0 ± 4.0	9.0 ± 6.5
Fe	0.87 ± 1.17	2.5 ± 4.2	0.0 ± 8.0	68.0 ± 13.0
Ni	0.19 ± 0.08	2.35 ± 0.29	0.0 ± 0.5	0.7 ± 0.86
Cu	0.0 ± 0.7	108.0 ± 3.0	0.0 ± 5.0	0.0 ± 8.0
Zn	0.0 ± 4.0	175.0 ± 14.0	0.0 ± 30.0	0.0 ± 50.0
Se	0.08 ± 0.063	0.82 ± 0.23	0.45 ± 0.40	0.0 ± 0.7
Br	0.24 ± 0.34	1.2 ± 1.3	28.0 ± 2.2	4.3 ± 3.8
Cd	0.0 ± 0.18	5.03 ± 0.64	0.0 ± 1.2	0.0 ± 2.0
Sn	0.0 ± 0.7	21.2 ± 2.5	0.0 ± 5.0	0.0 ± 8.0
Pb	5.2 ± 1.3	101.0 ± 5.0	115.0 ± 9.0	8.5 ± 14.2

The details of the analysis have been given by Chang et al. (1982). The results for this analysis indicated motor vehicles, secondary sulfate, soil or fly ash, and the copper products plant as the major mass contributors. The calculated source profiles and the associated uncertainties that best reproduced this data set are listed in Table 6.2. The motor vehicle factor is the one with a strong dependence on lead and bromine. The bromine to lead ratio of 0.24 is typical of values reported in the literature (Dzubay et al., 1979; Chu and Macias, 1981). The concentration of lead in motor vehicle exhaust will vary from city to city and is dependent on the ratio of leaded gasoline to unleaded and diesel-powered vehicles. The values of 11.5% Pb is quite similar to that obtained in other studies of the RAPS data at other sites (Alpert and Hopke, 1981; Liu et al., 1982; Severin et al., 1983). The concentration of sulfur was not related to any other elements and presumably represents a secondary sulfate aerosol resulting from primary emissions of sulfur dioxide. The sulfur concentration of 19.1% is a bit lower than what would be expected when the material consists of only ammonium sulfate. The soil and fly ash factor could potentially represent both soil particles as well as materials originating from the combustion of coal. Because of the similarity in their elemental profiles, differentiating soil and coal fly ash is a problem often encountered in aerosol source resolution work. Reliable data for other elements, such as arsenic, might permit the resolution of the soil and coal fly ash contributions.

The copper products plant factor is associated with elements Cu, Zn, Pb, and S. Gatz (1978) noted the presence of a Zn/Pb factor and attributed it to smelters located primarily in the east St. Louis and Granite City areas. Rheingrover and Gordon (1985) indicate that the zinc smelter contributes primarily to the coarse samples while the copper products plant emits fine particles that are actually higher in zinc than copper. Therefore, this factor is attributed to a copper products plant. In this data set, secondary sulfate aerosols account for 68.3% of the mass of fine Zn trajectory samples. The copper products plant contributes 14.0%. Motor vehicle emissions account for 10.5% and the fly ash/soil represents the remaining 7.2%. The TTFA method thus represents a useful approach when source information for the area is lacking or suspect and if there is uncertainty as to the identification of all of the sources contributing to the measured concentrations at the receptor site.

5. MICROSCOPIC METHODS

Optical microscopy has been an important tool for particle identification for a long time. A variety of particles have unique morphological, crystallographic, or other features that make it possible to identify their source with a high degree of specificity (Crutcher, 1983). With specificity there is a problem with the number of individual particles that can be scanned and interpreted by a skilled observer. Thus, there are limits to the precision with which analyses can be made. There is also the problem in optical spectroscopy of the limits of visual resolution that require that detailed studies be made only on particles larger than 1 μm. Although it may be possible with polarized light to identify specific minerals, optical microscopy cannot directly provide elemental compositions of the particles. It is extremely useful for identifying sources of total particulate matter but may be less useful for sources of airborne metals, particularly if they are on fine particles.

An alternative approach that has recently been actively developed is the use of scanning electron microscopy. Recent advances in computers allowed the development of computer control of the scanning in conjunction with real-time image analysis to identify and categorize a large number of particles over a relatively short time.

In these methods a scanning electron microscope is used to determine particle size parameters as well as elemental composition information using the emitted X rays excited by the electron bombardment (Casuccio et al., 1982; Johnson et al., 1981). Figure 6.9 demonstrates the several kinds of information that can be obtained for a single particle. The secondary electron image of the particle (top) shows the particle morphology and surface texture. This qualitative information can often be helpful in identifying a particle's origin. The image and diameter analysis of those particles is also shown in Figure 6.9 (middle). The image analysis program defines the average size of the particle, typically converting the shape to an elipsoid of revolution of defined semimajor and semiminor axes. Finally, at the bottom of Figure 6.9 is the X-ray spectrum for this particle taken when the electron beam is at the

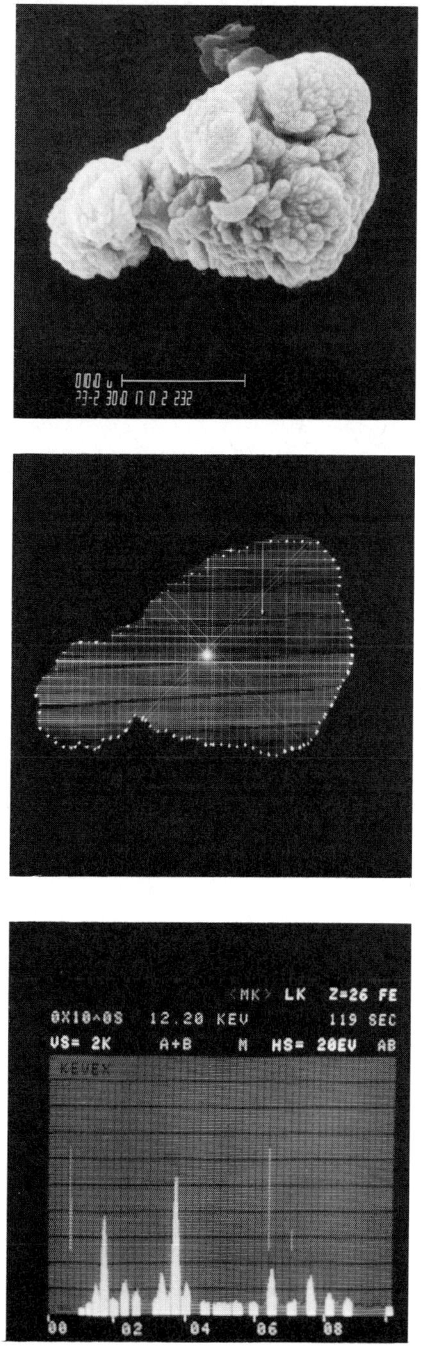

Figure 6.9. Example of a scanning electron microscopic examination of a particle. The top photograph shows the normal secondary electron image of a particle at a magnification of 700×. The image produced from the backscattered electron signal and the LeMont image analysis of this signal is shown in the middle view. The fluoresced X-ray spectrum produced by the electron bombardment at the image analyzed particle center is shown in the bottom photograph.

bright spot shown in the image analysis. Thus, information on the size, shape, and composition are directly available from the microscopic examination. These properties are then used to mathematically group the particles into collections attributable to a specific source and to estimate the mass contributions of that source to that sample. Johnson et al. (1983) use multivariate methods to make this apportionment while a CMB-like method is employed by Casuccio and coworkers (1982, 1983). Johnson and McIntyre (1983) have successfully applied their method to particle source apportionment in Syracuse, N.Y.

As an example, the use of their techniques to examine sources of airborne particles and lead in El Paso, Texas (Dattner et al., 1983) will be discussed. In

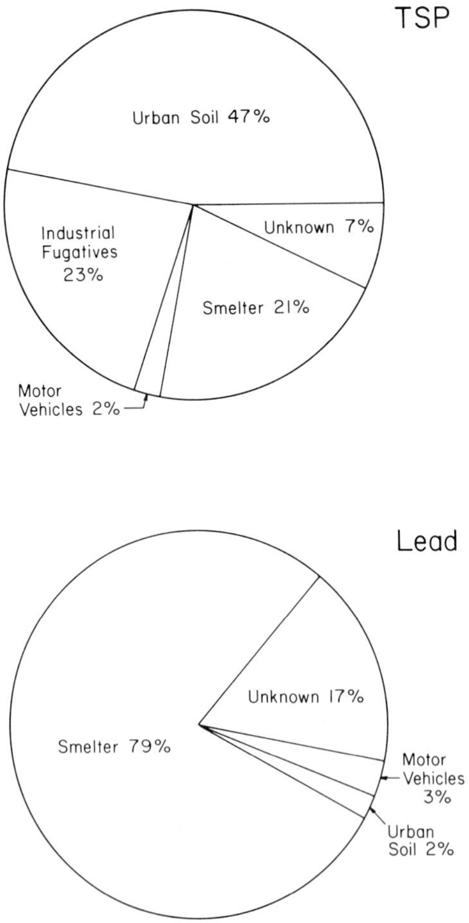

Figure 6.10. Apportionment of TSP and airborne particulate lead to sources in El Paso, Texas, as derived by computer-controlled scanning electron microscopy (Dattner et al., 1983).

El Paso, an air quality problem exists as shown by annual geometric mean value of total suspended particle (TSP) concentration and quarterly Pb concentrations significantly above the corresponding ambient air quality standards. In an effort to identify the sources of TSP and Pb, the Texas Air Control Board initiated a study whose primary approach utilized computer-controlled scanning electron microscopy. Both ambient filter samples and a large number of suspected source samples were taken in the El Paso area, particularly in the vicinity of a large smelter operation that was thought to be a major contributor to the observed TSP and lead levels. From the microscopic analysis and subsequent mass balance calculation, TSP and Pb were apportioned as shown in Figure 6.10. These results agree quite reasonably with two other apportionment studies (Radian, 1983; Trijonis, 1982) done for the same period primarily by examining the differences in lead levels with a period in 1980 when the smelter was closed because of a strike. They also agreed reasonably well with the tracer apportionment model developed and used by the Texas Air Control Board (Dattner et al., 1983). Thus, the microscopic methods have compared well with some of the receptor models based on the concentrations measured in the collected particles. It is likely that microscopic studies will increase in their importance and be more common as a method to examine the impacts of particle sources particularly containing metallic components.

6. SUMMARY

In this chapter several of the active areas of receptor modeling have been introduced. Their use as indicators of the sources of metals in the air can be very useful in developing air quality management strategies and can potentially become enforcement tools as well. Since receptor models must of necessity be retrospective in nature, another important use can be in the calibration and testing of the prognostic dispersion models so that prediction of changes in air quality can serve as a more reliable basis for management decisions. The field of receptor modeling has grown and developed rapidly during the last several years and can be expected to continue to do so as methods are improved and new applications discovered.

ACKNOWLEDGMENTS

I would like to thank Dr. David Johnson of SUNY—Syracuse for preparing the microscopy pictures (Fig. 6.9) especially for this chapter. My appreciation is also extended to John Core for his assistance in sorting out the PACS results and the preparation of the appropriate figure and to Mrs. Victoria Corkery for her help in preparing this manuscript.

REFERENCES

Ackerman, B., Changnon, S.A., Dzurisin, G., Gatz, D.L., Grosh, R.C., Hilberg, S.D., Huff, F.A., Mansell, J.W., Ochs, H.T., Peden, M.E., Schickedanz, P.T., Semonin, R.G., and Vogel, J.L. (1978). *Summary of METROMEX, Volume 2: Causes of precipitation anomalies.* Illinois State Water Survey Bulletin 63. Urbana, Ill., 395 pp.

Alpert, D.J., and Hopke, P.K. (1980). "A quantitative determination of sources in the Boston urban aerosol." *Atmos. Environ.,* **14,** 1137–1146.

Alpert, D.J., and Hopke, P.K. (1981). "A determination of the sources of airborne particles collected during the regional air pollution study." *Atmos. Environ.* **15,** 675–687.

Blifford, I.H., and Meeker, G.O. (1967). "A factor analysis model of large scale pollution." *Atmos. Environ.* **1,** 147–157.

Bogen, J. (1973). "Trace elements in atmospheric aerosol in the Heidelberg area, measured by instrumental neutron activation analysis." *Atmos. Environ.* **7,** 1117–1125.

Casuccio, G.S., and Janocko, P.B. (1983). *Identification of the sources of total suspended particulates and particulate lead in the El Paso area by quantitative microscopic analysis.* Final Report to the Texas Air Control Board, Austin, Texas.

Casuccio, G.S., Janocko, P.B., Lee, R.J., and Kelly, J.F. (1982). *The role of computer controlled scanning electron microscopy in receptor modeling.* Paper No. 82-21.4. Air Pollution Control Association, Pittsburgh, Pa., 19 pp.

Chang, S.N., Hopke, P.K., Rheingrover, S.W., and Gordon, G.E. (1982). *Target transformation factor analysis of wind-trajectory selected samples.* Paper No. 81-21.1. Air Pollution Control Association, Pittsburgh, Pa., 14 pp.

Changnon, S.A., Huff, R.A., Schickendenz, P.T., and Vogel, J.L. (1977). *Summary of METROMEX, Volume 1: Weather anomalies and impacts.* Illinois State Water Survey Bulletin 62. Urbana, Ill., 260 pp.

Chow, J.C., Shortell, V., Collins, J., Watson, J.G., Pace, T.G., and Burton, R. (1981). *A neighborhood scale study of inhalable and fine suspended particulate matter source contributions to an industrial area in Philadelphia.* Paper No. 81-14.1. Air Pollution Control Association, Pittsburgh, Pa., 16 pp.

Chow, J.C., Watson, J.G., and Shah, J.J. (1982). *Source contributions to inhalable particulate matter in major U.S. cities.* Paper No. 82-21.3. Air Pollution Control Association, Pittsburgh, Pa., 16 pp.

Chu, L.C., and Macias, E.S. (1981). "Carbonaceous urban aerosol—primary or secondary?" In *Atmospheric aerosol: Source/air quality relationships,* E.S. Macias and P.K. Hopke, Eds. (Symposium Series No. 167). American Chemical Society, Washington, D.C., pp. 251–268.

Colucci, J.M., and Begeman, C.R. (1965). "The automotive contribution to air-borne polynuclear aromatic hydrocarbons in Detroit. *J. Air Pollut. Control Assoc.* **15,** 113–122.

Cooper, J.A. (1979). *Medford aerosol characterization study: Application of chemical mass balance to identification of major aerosol sources in the Medford airshed.* Interim Report to the Oregon Department of Environmental Quality. Portland, Ore.

Cooper, J.A., Watson, J.G., and Huntzicker, J.J. (1979). *Summary of the Portland aerosol characterization study (PACS).* Paper No. 79-29.4. Air Pollution Control Association, Pittsburgh, Pa., 15 pp.

Core, J.E., Cooper, J.A., Hanrahan, P.L., and Cox, W.M. (1982). "Particulate dispersion model evaluation: A new approach using receptor models." *J. Air Pollut. Control Assoc.* **32,** 1142–1147.

Core, J.E., Hanrahan, P.L., and Cooper, J.A. (1981). "Air pollution control strategy development: A new approach using chemical mass balance methods." In *Atmospheric aerosol:*

Source/air quality relationships, E.S. Macias and P.K. Hopke, Eds. (Symposium Series No. 167). American Chemical Society, Washington, D.C., pp. 107–123.

Crutcher, E.R. (1983). "Light microscopy as an analytical approach to receptor modeling." In *Receptor models applied to contemporary pollution problems,* S.L. Dattner and P.K. Hopke, Eds. Air Pollution Control Association, Pittsburgh, Pa., pp. 266–284.

Dattner, S.L. (1978). *Preliminary analysis of the use of factor analysis of x-ray fluorescence data to determine the sources of total suspended particulate.* Draft Report, Texas Air Control Board, Austin, Texas.

Dattner, S.L., and Jenks, M.. (1981). *Identification of non-emission source related factors in sets of ambient particulate data.* Paper No. 81-64.1. Air Pollution Control Association, Pittsburgh, Pa., 13 pp.

Dattner, S.L., Mgebroff, S., Casuccio, G., and Janocko, P. (1983). *Identifying the sources of TSP and lead in El Paso using microscopy and receptor models.* Paper No. 83-49.3. Air Pollution Control Association, Pittsburgh, Pa., 14 pp.

DeCesar, R.T., and Cooper, J.A. (1981). *Medford aerosol characterization study.* Final Report to the Oregon Department of Environmental Quality. Portland, OR.

Dunker, A.M. (1979). *A method for analyzing data on the elemental composition of aerosols.* General Motors Research Laboratories Report GMR-3074 ENV-67. Warren, Mich.

Dutot, A.L., Elichegaray, C., and Vie le Sage, R. (1983). "Application de l'analyse des correspondances a l'etude de la composition physico-chemique de l'aerosol urbain." *Atmos. Environ.* **17,** 7–78.

Dzubay, T.G., and Stevens, R.K. (1983). "Intercomparison of results of several receptor models for apportioning houston aerosol." In *Receptor models applied to contemporary pollution problems,* S.L. Dattner and P.K. Hopke, Eds. Air Pollution Control Association, Pittsburgh, Pa., pp. 60–71.

Dzubay, T.G., Stevens, R.K., and Richards, L.W. (1979). "Composition of aerosols over Los Angeles freeways." *Atmos. Environ.* **13,** 653–659.

Fabrick, A.J., and Sklarew, R.C. (1975). Oregon/Washington Diffusion Modeling Study, Xonics, Inc.

Friedlander, S.K. (1973). "Chemical element balances and identification of air pollution sources." *Environ. Sci. Technol.* **7,** 235–240.

Gaarenstrom, P.D., Perone, S.P., and Moyers, J.P. (1977). "Application of pattern recognition and factor analysis for characterization of atmospheric particulate composition in southwest desert atmosphere." *Environ. Sci. Technol.* **11,** 795–800.

Gatz, D.F. (1975). "Relative contributions of different sources of urban aerosols: Application of a new estimation method to multiple sites in Chicago." *Atmos. Environ.* **9,** 1–18.

Gatz, D.F. (1978). "Identification of aerosol sources in the St. Louis area using factor analysis." *J. Appl. Met.* **17,** 600–608.

Gatz, D.F. (1983). *Source apportionment of rain water impurities in central Illinois.* Paper No. 83-28.3. Air Pollution Control Association, Pittsburgh, Pa., 16 pp.

Gladney, E.S. (1974). "Trace elemental emissions from coal-fired power plants: A study of the Chalk Point electric generating station." Ph.D. Thesis. University of Maryland, College Park, Md.

Gladney, E.S., Small, J.A. Gordon, G.E., and Zoller, W.H. (1976). "Composition and size distribution of in-stack particulate material at a coal-fired power plant." *Atmos. Environ.* **10,** 1071–1077.

Goulding, F.S., Jaklevic, J.M., and Loo, B.W. (1981). *Aerosol analysis for the regional air pollution study—Final Report.* EPA-600/S4-8L-006. U.S. Environmental Protection Agency, Research Triangle Park, N.C.

Greenberg, R.R. (1976). "A study of trace elements emitted on particles from municipal incinerators." Ph.D. Thesis. University of Maryland, College Park, Md.

Greenberg, R.R., Gordon, G.E., Zoller, W.H., Jacko, R.B. Neuendorf, D.W., and Yost, K.J. (1978). "Composition of particles emitted from the Nicosia municipal incinerator." *Environ. Sci. Technol.* **12,** 1329–1332.

Greenberg, R.R., Zoller, W.H., and Gordon, G.E. (1978). "Composition and size distribution of particles released in refuse incineration." *Environ. Sci. Technol.* **12,** 566–573.

Heidam, N.Z. (1981). "On the origin of the Arctic aerosol: A statistical approach." *Atmos. Environ.* **15,** 1421–1427.

Heindryckx, R., and Dams, R. (1974). "Continental, marine, and anthropogenic contributions to the inorganic composition of the aerosol of an industrial zone." *J. Radioanal. Chem.* **19,** 339–349.

Henry, R.C. (1983). "Stability analysis of receptor models that use least-squares fitting." In *Receptor models applied to contemporary pollution problems,* S.L. Dattner and P.K. Hopke, Eds. Air Pollution Control Association, Pittsburgh, Pa., pp. 141–157.

Hoffman, G.L., and Duce, R.A. (1971). "Copper contamination of atmospheric particulate samples collected with Gelman hurricane samplers." *Environ. Sci. Tech.* **5,** 1134–1136.

Hopke, P.K. (1981). "The application of factor analysis to urban aerosol source resolution." In *Atmospheric aerosol: Source/air quality relationships,* E.S. Macias and P.K. Hopke, Eds. (Symposium Series No. 167). American Chemical Society, Washington, D.C., pp. 21–49.

Hopke, P.K. (1983). "An introduction to multivariate analysis of environmental data." In *Analytical aspects of environmental chemistry,* D.F.S. Natusch and P.K. Hopke, Eds. John Wiley & Sons, Inc., New York, pp. 219–261.

Hopke, P.K. (1985) *Receptor modeling in environmental chemistry*, John Wiley & Sons, Inc., New York, 319 pp.

Hopke, P.K., Gladney, E.S., Gordon, G.E., Zoller, W.H., and Jones, A.G. (1976). "The use of multivariate analysis to identify sources of selected elements in the Boston urban aerosol. *Atmos. Environ.* **10,** 1015–1025.

Hopke, P.K., Lamb, R.E., and Natusch, D.F.S. (1980). "Multielemental characterization of urban roadway dust." *Environ. Sci. Technol.* **14,** 164–172.

Houck, J.E., and Cooper, J.A. (1983). "Receptor model source apportionment of lead in three airsheds with lead smelters." In *Receptor models applied to contemporary pollution problems,* S.L. Dattner and P.K. Hopke, Eds. Air Pollution Control Association, Pittsburgh, Pa., pp. 34–45.

Johnson, D.L., and McIntyre, B.L. (1983). "A particle class balance receptor model for aerosol apportionment in Syracuse, NY." In *Receptor models applied to contemporary pollution problems,* S.L. Dattner and P.K. Hopke, Eds. Air Pollution Control Association, Pittsburgh, Pa., pp. 238–248.

Johnson, D.L.. McIntyre, B., Fortmann, R., Stevens, R.K., and Hanna, R.B. (1981). "A chemical elemental comparison of individual particle analysis and bulk analysis methods." *Scanning Electr. Microsc.* **1,** 469–476.

Johnson, D.L., and Twist, J.P. (1983). "Statistical considerations in the employment of Sax results for receptor models." In *Receptor models applied to contemporary pollution problems,* S.L. Dattner and P.K. Hopke, Eds. Air Pollution Control Association, Pittsburgh, Pa., pp. 224–237.

Kowalczyk, G.S., Choquette, C.E., and Gordon, G.E. (1978). "Chemical element balances and identification of air pollution sources in Washington, D.C." *Atmos. Environ.* **12,** 1143–1153.

Kowalczyk, G.S., Gordon, G.E., and Rheingrover, S.W. (1982). "Identification of atmospheric particulate sources in Washington, D.C., using chemical element balances." *Environ. Sci. Technol.* **16,** 79–90.

Lewis, C.W., and Macias, E.S. (1980). "Composition of size-fractionated aerosol in Charleston, West Virginia." *Atmos. Environ.* **14,** 185–194.

Liu, B.Y.H., and Pui, D.Y.H. (1981). "Aerosol sampling inlets and inhalable particles." *Atmos. Environ.* **15,** 589–600.

Malissa, H., Puxbaum, H., and Wopenka, B. (1981). "Herkunftsanalyse des atmospharischen aerosol in wein." *Proceedings of the Second European Symposium on Physico-Chemical Behavior of Atmospheric Pollutants.* Varese, Italy.

Miller, M.S., Friedlander, S.K., and Hidy, G.M. (1972). "A chemical element balance for the Pasadena aerosol." *J. Colloid Interface Sci.* **39,** 165–176.

Mroz, E.J. (1976). "The study of the elemental composition of particulate emissions from an oil-fired power plant." Ph.D. Thesis. University of Maryland, College Park, Md.

Nelson, E. (1979). *Regional air pollution study: Dichotomous aerosol sampling system.* EPA-297310. U.S. Environmental Protection Agency, Research Triangle Park, N.C.

Ondov, J.M. (1974). "A study of trace element on particulates from motor vehicles." Ph.D. Thesis. University of Maryland, College Park, Md.

Ondov, J.M., Zoller, W.H., and Gordon, G.E. (1982). "Trace element emissions on aerosols from motor vehicles." *Envirion. Sci. Technol.* **16,** 318–328.

Parekh, P.P., and Husain, L. (1981). "Trace element concentrations in summer aerosols at rural sites in New York State and their possible sources." *Atmos. Environ.* **14,** 1717–1725.

Prinz, B., and Stratmann, H. (1968). "The possible use of factor analysis in investigating air quality." *Staub—Reinhalt luft 28,* 33—39.

Radian Corporation (1983). "A study to characterize ambient air quality and assess emission source contributions to ambient air pollution concentrations for El Paso County," Austin, Texas. Draft Final Report (DNC 82-144-771-04), March, 1983.

Rheingrover, S.G., and Gordon, G.E. (1985). "Wind-trajectory methods for determining compositions of particles from major air pollution sources." Submitted to *Environ. Sci. Technol.,* Nov. 1983.

Roberts, J., Watters, H., Austin, F., and Crooks, M. (1979). *Particulate emissions for paved roads in Seattle and Tacoma non-attainment areas.* Puget Sound Air Pollution Control Agency, Seattle, Wash.

Roscoe, B.A., and Hopke, P.K. (1981). "Comparison of weighted and unweighted target transformation rotations in factor analysis." *Computer Chem.* **5,** 1–7.

Roscoe, B.A., Hopke, P.K., Dattner, S.L., and Jenks, J.M. (1982). "The use of principal components factor analysis to interpret particulate compositional data sets." *J. Air Pollut. Control Assoc.* **32,** 637–642.

Severin, K.G., Roscoe, B.A., and Hopke, P.K. (1983). "The use of factor analysis in source determination of particulate emissions." *Particulate Sci. Technol.,* **1,** 183–192.

Sievering, H., Dave, M., Dolske, D., and McCoy, P. (1980). "Trace element concentrations over Midlake Michigan as a function of meteorology and source region." *Atoms. Environ.* **14,** 39–53.

Small, M. (1979). "Composition of particulate trace elements in plumes from industrial sources." Ph.D. Thesis. University of Maryland, College Park, Md.

Stevens, R.K., and Pace, T.G. (1983). "Status of source apportionment methods: Quail roost II." In *Receptor models applied to contemporary pollution problems,* S.L. Dattner and P.K. Hopke, Eds. Air Pollution Control Association, Pittsburgh, Pa., pp. 46–59.

Thomae, S.C. (1977). "Size and composition of atmospheric particles in rural areas near Washington, D.C." Ph.D. Thesis. University of Maryland, College Park, Md.

Thurston, G.D., and Spengler, J.D. (1981). "An assessment of fine particulate sources and their interaction with meteorological influences via factor analysis." Paper No. 81-64.4. Air Pollution Control Association, Pittsburgh, Pa., 16 pp.

Thurston, G.D., and Spengler, J.D. (1982). "Source contributions to inhalable particulate matter in metropolitan Boston, MA." Paper No. 82-21.5. Air Pollution Control Association, Pittsburgh, Pa., 16 pp.

Trijonis, J. (1982). *Analysis of ambient lead data near the ASARCO smelter in El Paso.* Sante Fe Research Corporation, Sante Fe, N.M.

Watson, J.G. (1979). "Chemical element balance receptor model methodology for assessing the source of fine and total suspended particulate matter in Portland, Oregon." Ph.D. Thesis. Oregon Graduate Center, Beaverton, Ore.

Watson, J.G., Cooper, J.A., and Huntzicker, J.J. (1984). "The effective variance weighting for least squares calculations applied to the mass balance receptor model," *Atmos. Environ.* **18,** 1347–1355.

Winchester, J.W., and Nifong, G.D. (1971). "Water pollution in Lake Michigan by trace elements from pollution aerosol fallout." *Water Air Soil Pollut.* **1,** 50–64.

7

TRACE METALS IN THE ATMOSPHERE OF REMOTE AREAS

G. Bruce Wiersma

Earth and Life Sciences
Idaho National Engineering Laboratory
Idaho Falls, Idaho

Cliff I. Davidson

Departments of Civil Engineering
and Engineering & Public Policy,
Carnegie-Mellon University, Pittsburgh, Pennsylvania

1. INTRODUCTION

This chapter discusses data on airborne trace element concentrations obtained in areas removed from the immediate influence of anthropogenic activities. The objective is to provide a summary of data, with limited interpretation, for general use by other investigators. Because many of the airborne trace element studies conducted before the mid-1970s have been summarized in a previous report (Rahn, 1976), this chapter focuses on data obtained or published during 1976–1983.

There are several reasons why remote area trace element data may be of interest. Of primary importance is the need for baseline values to which urban and industrial air pollutant concentrations can be compared. The influence of anthropogenic activities on airborne trace element levels can thus be studied. Second, trace elements may be used as indicators of long-range transport of other pollutant species. Examples of these other species include atmospheric acids such as aerosol sulfate (Rahn and Lowenthal, 1984) and gases that may impact global climate, such as CO_2 and CH_4 (Khalil and Rasmussen, 1983). Finally, there is concern over the impact of certain trace elements on biological systems in remote areas. For example, Patterson (1980) has shown that lead emitted from anthropogenic activities has increased lead levels in fauna and flora in remote ecosystems, resulting in possible changes in natural biochemical processes. Assessing such changes is assisted by acquisition of airborne concentration data.

Use of remote area trace element concentrations for these and other purposes requires high quality data. Unfortunately, such concentrations may be extremely small, making it difficult to obtain reliable information. The problem is often compounded by harsh sampling conditions; much of the data summarized here were obtained in the Arctic or Antarctic, in mountainous areas, or on board ships or aircraft. Limited availability of electric power may pose an additional constraint. The difficulties associated with remote area environmental sampling have been discussed by several investigators (e.g., Hoffman et al., 1976; Patterson and Settle, 1976; Boutron, 1979), all of whom identify sample contamination as the most important problem. Because techniques for trace element sampling, sample handling, and analysis have been improving with time, the data presented here, while highly imperfect in many instances, are generally more reliable than the results of earlier studies.

2. PRESENTATION OF AIRBORNE CONCENTRATIONS AND ENRICHMENT FACTORS IN REMOTE AREAS

The airborne concentrations of interest are presented in Table 7.1. Although not an exhaustive compilation, the listing includes a wide variety of data reported by many research groups investigating trace elements in remote areas. The elements are arranged in alphabetical order. For each element, data for the United States are first presented (alphabetically by state), followed by data for the rest of the world (alphabetically by country or geographical area). The approximate dates of sampling, if indicated in the original reference, are also listed. All concentrations are given in nanograms per cubic meter to two significant figures, unless the original study included data to only one significant figure.

Many of the studies, especially those conducted at high elevations, have reported concentrations adjusted to standard temperature and pressure conditions (STP). However, some of the investigations merely report concentrations applicable to field conditions during their sampling experiments. The data listed in Table 7.1 represent concentrations as reported in the original references. The problem in comparing values obtained at different temperatures and pressures is likely to be minor compared with overall inaccuracies in the data, caused by sampling and analytical difficulties as discussed later in this chapter.

For most of the concentration values, the crustal enrichment factor has been computed following Zoller et al. (1974):

$$EF_{crust} = \frac{X_{air} / Al_{air}}{X_{crust} / Al_{crust}} \qquad (1)$$

where X_{air} and Al_{air} are the airborne concentrations of any element X and of aluminum, respectively, and X_{crust} and Al_{crust} represent the corresponding concentrations in the earth's crust. The crustal composition data have been taken from Taylor (1964). Values of EF_{crust} near unity suggest that crustal erosion is the primary source of element X; values much greater than unity imply the importance of other sources, such as anthropogenic activities, volcanism, biological methylation, direct sublimation from the earth's crust, and aerosol production at the air–sea interface (Duce et al., 1975).

Values of EF_{crust} have been computed for all sets of data in Table 7.1 which include a value for the airborne concentration of aluminum. The calculation has not been performed, however, for concentrations that are upper or lower bounds. In many cases, values of EF_{crust} have been reported in the original publications using reference elements other than aluminum or using crustal composition data other than those of Taylor (1964). For consistency, these enrichment factors have been recalculated as described above. In a few instances where airborne aluminum data are lacking, values of EF_{crust} based on iron or scandium have been reported in the original studies; these values are reproduced in Table 7.1 with indication of the reference element used. Wherever possible, individual airborne concentration values for each sampling period have been used to calculate individual enrichment factors, and these enrichments have then been averaged to obtain the overall EF_{crust} value shown in the table. The enrichments have been computed from long term average airborne concentrations only for those data sets where individual concentration values have not been reported in the original reference. All enrichment factors are shown to two significant figures, except where only one significant figure is warranted.

Table 7.1. Airborne Concentrations and Crustal Enrichment Factors of Trace Elements in Remote Areas

Airborne Concentration (ng/m^3)	Crustal Enrichment Factor	Sampling Location	Reference
Aluminum (Al)			
27.	1.0	Alaska, Barrow, Dec. 1976–Apr. 1977	Rahn (1977)
1200.	1.0	Arizona, rural area Jan.–Dec. 1974	Moyers et al. (1977)
7.	1.0	Hawaii, Coast	Gordon et al. (1978)
49.	1.0	Hawaii, Mauna Loa, upslope wind, Jan.–Dec. 1980	Gordon et al. (1978)
9.	1.0	Hawaii, Mauna Loa, downslope wind no dust, Jan.–Dec. 1980	Gordon et al. (1978)
76.	1.0	Hawaii, Mauna Loa, downslope wind, Asian dust, Jan.–Dec. 180	Gordon et al. (1978)
250.	1.0	Michigan, Seney Nat. Wildlife Refuge, Jun. 1979	Alkezweeny et al. (1982)
930.	1.0	Montana, Colstrip, May–Sep. 1975	Crecelius et al. (1980)
160.	1.0	Montana, Glacier Nat. Park, Aug. 1981	Davidson et al. (1985)
71.	1.0	Tennessee, Smoky Mts. Nat. Park, Oct. 1979	Davidson et al. (1985)
20. (<2.5 μm)	1.0	Tennessee, Smoky Mts. Nat. Park, Sep. 1978	Stevens et al. (1980)
190. (>2.5 μm)	1.0	Tennessee, Smoky Mts. Nat. Park, Sep. 1978	Stevens et al. (1980)
44. (<2.5 μm)	1.0	Virginia, Shenandoah Valley, Jul.–Aug. 1980	Stevens et al. (1984)

310. (>2.5 μm)	1.0	Virginia, Shenandoah Valley, Jul.–Aug. 1980	Stevens et al. (1984)
110.	1.0	Washington, Olympic Nat. Park, Jul.–Aug. 1980	Davidson et al. (1985)
<0.30		Antarctica, Austral Winters 1975 & 1976	Cunningham & Zoller (1981)
0.83	1.0	Antarctica, Austral Summers 1971, 1975, 1976, & 1978	Cunningham & Zoller (1981)
70.	1.0	Atlantic Ocean, East	Duce et al. (1976)
59.	1.0	Bear Island, Nov. 1977–Mar. 1978	Rahn (1981)
500.	1.0	Bermuda, Summer 1975	Duce et al. (1976)
260.	1.0	Bolivia, Chacaltaya Mtn., Feb.–May 1976	Adams et al. (1977)
79.	1.0	Enewetak, Apr.–May 1979	Duce et al. (1983)
6.1	1.0	Enewetak, Jun.–Aug. 1979	Duce et al. (1983)
14.	1.0	Greenland, Kap Harald Moltke, Jun.–Aug. 1974	Flyger & Heidam (1978)
32.	1.0	Greenland, Thule, Feb.–Apr. 1978	Heidam (1981)
13.	1.0	Greenland, Prins Christianssund, Jan.–Feb. 1978	Heidam (1981)
94.	1.0	Nepal, Himalayas, Dec. 1979	Davidson et al. (1981a)
110.	1.0	Norway, Birkenes, Spring 1973	Semb (1978)
85.	1.0	Norway, Birkenes, Autumn 1973	Semb (1978)
68.	1.0	Norway, Skoganvarre, Spring 1973	Semb (1978)
43.	1.0	Spitsbergen, Dec. 1973–Mar. 1974	Larssen (1977)
50. (<2.5 μm)	1.0	USSR, Abastumani Mts., Jul. 1979	Stevens et al. (1984)
360. (>2.5 μm)	1.0	USSR, Abastumani Mts., Jul. 1979	Stevens et al. (1984)

Table 7.1. *(Continued)*

Airborne Concentration (ng/m^3)	Crustal Enrichment Factor	Sampling Location	Reference
Antimony (Sb)			
0.009	80.	Hawaii, Mauna Loa, upslope wind, Jan.–Dec. 1980	Gordon et al. (1978)
0.0021	96.	Hawaii, Mauna Loa, downslope wind, no dust, Jan.–Dec. 1980	Gordon et al. (1978)
0.0089	48.	Hawaii, Mauna Loa, downslope wind, Asian dust, Jan.–Dec. 1980	Gordon et al. (1978)
0.14	62.	Montana, Colstrip, May–Sep. 1975	Crecelius et al. (1980)
0.80	1400. (Sc)	New York, Whiteface Mtn., unpolluted air, Summer 1975	Parekh & Husain (1981)
0.0021		Antarctica, Austral Winters 1975 & 1976	Cunningham & Zoller (1981)
0.00045	220.	Antarctica, Austral Summers 1971, 1975, 1976, & 1978	Cunningham & Zoller (1981)
0.16	940.	Atlantic Ocean, East	Duce et al. (1976)
0.03	30.	Bermuda, Summer 1975	Duce et al. (1976)
0.51	800.	Bolivia, Chacaltaya Mtn., Feb.–May 1976	Adams et al. (1977)
0.0052	29.	Enewetak, Apr.–May 1979	Duce et al. (1983)
0.0024	180.	Enewetak, Jun.–Aug. 1979	Duce et al. (1983)
0.93	3400.	Norway, Birkenes, Spring 1973	Semb (1978)
0.35	2100.	Norway, Skoganvarre, Spring 1973	Semb (1978)

Arsenic (As)			
0.072	470.	Hawaii, Coast	Gordon et al. (1978)
0.15	140.	Hawaii, Mauna Loa, upslope wind, Jan.–Dec. 1980	Gordon et al. (1978)
0.020	100.	Hawaii, Mauna Loa, downslope wind, no dust, Jan.–Dec. 1980	Gordon et al. (1978)
0.49	290.	Hawaii, Mauna Loa, downslope wind, Asian dust, Jan.–Dec. 1980	Gordon et al. (1978)
<0.1		Michigan, Seney Nat. Wildlife Refuge, Jun. 1979	Alkezweeny et al. (1982)
1.8	88.	Montana, Colstrip, May–Sep. 1975	Crecelius et al. (1980)
1.9	640.	Montana, Glacier Nat. Park, Aug. 1981	Davidson et al. (1985)
2.0	360. (Sc)	New York, Whiteface Mtn., unpolluted air, Summer 1975	Parekh & Husain (1981)
1.0	520. (Sc)	New York, Whiteface Mtn., unpolluted air, Summer 1977	Parekh & Husain (1981)
<1.6		Tennessee, Smoky Mts. Nat. Park, Oct. 1979	Davidson et al. (1985)
2.2 (<2.5 μm)	5000.	Tennessee, Smoky Mts. Nat. Park, Sept. 1978	Stevens et al. (1980)
<1. (>2.5 μm)		Tennessee, Smoky Mts. Nat. Park, Sep. 1978	Stevens et al. (1980)
1. (<2.5 μm)	1000.	Virginia, Shenandoah Valley, Jul.–Aug. 1980	Stevens et al. (1984)
<2. (>2.5 μm)		Virginia, Shenandoah Valley, Jul.–Aug. 1980	Stevens et al. (1984)
2.3	960.	Washington, Olympic Nat. Park, Jul.–Aug. 1980	Davidson et al. (1985)
0.59 (<1 μm)	2000. (Fe)	China, Great Wall, Apr. 1980	Winchester et al. (1981)

Table 7.1. *(Continued)*

Airborne Concentration (ng/m^3)	Crustal Enrichment Factor	Sampling Location	Reference
<0.017		Antarctica, Austral Winters 1975 & 1976	Cunningham & Zoller (1981)
0.0084	460.	Antarctica, Austral Summers 1971, 1975, 1976, & 1978	Cunningham & Zoller (1981)
0.12	11.	Bermuda, Summer 1975	Duce et al. (1976)
1.6	280.	Bolivia, Chacaltaya Mtn., Feb.–May 1976	Adams et al. (1977)
2.		Germany, Mt. Kleiner Feldberg	Grosch et al. (1978)
1.7		Switzerland, Corviglia (Alps)	Grosch et al. (1978)
1.0		Switzerland, St. Moritz	Grosch et al. (1978)
<1.0 (<2.5 μm)		USSR, Abastumani Mts., Jul. 1979	Stevens et al. (1984)
0.3 (>2.5 μm)	40.	USSR, Abastumani Mts., Jul. 1979	Stevens et al. (1984)
Barium (Ba)			
4.0		California, High Sierra, Jun. & Aug. 1976, Aug. 1977	Elias & Davidson (1980)
2.3	1.8	Michigan, Seney Nat. Wildlife Refuge, Jun. 1979	Alkezweeny et al. (1982)
8.0	1.7	Montana, Colstrip, May–Sep. 1975	Crecelius et al. (1980)
10.	15.	Montana, Glacier Nat. Park, Aug. 1981	Davidson et al. (1985)

0.30	0.81	Tennessee, Smoky Mts. Nat. Park, Oct. 1979	Davidson et al. (1985)
1.1	1.9	Washington, Olympic Nat. Park, Jul.–Aug. 1980	Davidson et al. (1985)
0.020		Antarctica, Austral Winters 1975 & 1976	Cunningham & Zoller (1981)
0.019	4.6	Antarctica, Austral Summers 1971, 1975, 1976 & 1978	Cunningham & Zoller (1981)
0.53	1.3	Enewetak, Apr.–May 1979	Duce et al. (1983)
0.11	3.4	Enewetak, Jun.–Aug 1979	Duce et al. (1983)
1.0		Enewetak, Apr.–May 1979	Settle & Patterson (1982)
0.018		Enewetak, Jul.–Aug. 1979	Settle & Patterson (1982)
Bromine (Br)			
1.3	870.	Hawaii, Mauna Loa, upslope wind, Jan.–Dec. 1980	Gordon et al. (1978)
1.2	4400.	Hawaii, Mauna Loa, downslope wind, no dust, Jan.–Dec. 1980	Gordon et al. (1978)
1.1	480.	Hawaii, Mauna Loa, downslope wind, Asian dust, Jan.–Dec. 1980	Gordon et al. (1978)
2.1	280.	Michigan, Seney Nat. Wildlife Refuge, Jun. 1979	Alkezweeny et al. (1982)
2.8	99.	Montana, Colstrip, May–Sep. 1975	Crecelius et al. (1980)
12.	1700. (Sc)	New York, Whiteface Mtn., unpolluted air, Summer 1975	Parekh & Husain (1981)
13.	3600. (Sc)	New York, Whiteface Mtn., unpolluted air, Summer 1977	Parekh & Husain (1981)

Table 7.1. *(Continued)*

Airborne Concentration (ng/m^3)	Crustal Enrichment Factor	Sampling Location	Reference
18. (<2.5 μm)	30000.	Tennessee, Smoky Mts. Nat. Park, Sep. 1978	Stevens et al. (1980)
5. (>2.5 μm)	800.	Tennessee, Smoky Mts. Nat. Park, Sep. 1978	Stevens et al. (1980)
8. (<2.5 μm)	6000.	Virginia, Shenandoah Valley, Jul.–Aug. 1980	Stevens et al. (1984)
3. (>2.5 μm)	300.	Virginia, Shenandoah Valley, Jul.–Aug. 1980	Stevens et al. (1984)
6.3		Washington, Quillayute, onshore wind, Apr. 1975–Jun. 1975	Ludwick et al. (1977)
0.32		Antarctica, Austral Winters 1975 & 1976	Cunningham & Zoller (1981)
0.80	33000.	Antarctica, Austral Summers 1971, 1975, 1976, & 1978	Cunningham & Zoller (1981)
1.0	130.	Bolivia, Chacaltaya Mtn., Feb.–May 1976	Adams et al. (1977)
0.55 (<1 μm)	1400. (Fe)	China, Great Wall, Apr. 1980	Winchester et al. (1981)
20.	9000.	Enewetak, Apr.–May 1979	Duce et al. (1983)
20.	1.6×10^5	Enewetak, Jun.–Aug. 1979	Duce et al. (1983)
0.29	680.	Greenland, Kap Harald Moltke, Jun.–Aug. 1974	Flyger & Heidam (1978)
9.4	9700.	Greenland, Thule, Feb.–Apr. 1978	Heidam (1981)
3.9	9900.	Greenland, Prins Christianssund, Jan.–Feb. 1978	Heidam (1981)
4.7	1400.	Norway, Birkenes, Spring 1973	Semb (1978)

5.2	2000.	Norway, Birkenes, Autumn 1973	Semb (1978)
2.2	1100.	Norway, Skoganvarre, Spring 1973	Semb (1978)
2.8		Sweden, Sjoangen, annual ave.	Lannefors et al. (1983)
33.		Germany, Mt. Kleiner Feldberg	Grosch et al. (1978)
13.		Switzerland, Corviglia (Alps)	Grosch et al. (1978)
26.		Switzerland, St. Moritz	Grosch et al. (1978)
3. (<2.5 μm)	2000.	USSR, Abastumani Mts., Jul. 1979	Stevens et al. (1984)
1. (>2.5 μm)	90.	USSR, Abastumani Mts., Jul. 1979	Stevens et al. (1984)
Cadmium (Cd)			
0.37	5600.	Alaska, Barrow, Dec. 1976–Apr. 1977	Rahn (1977)
2.0	690.	Arizona, rural area, Jan.–Dec. 1974	Moyers et al. (1977)
0.022	1300.	Hawaii, Coast	Gordon et al. (1978)
2.2	3700.	Michigan, Seney Nat. Wildlife Refuge, Jun. 1979	Alkezweeny et al. (1982)
0.72	3100.	Montana, Glacier Nat. Park, Aug. 1981	Davidson et al. (1985)
0.15	870.	Tennessee, Smoky Mts. Nat. Park, Oct. 1979	Davidson et al. (1985)
0.54	2000.	Washington, Olympic Nat. Park, Jul.–Aug. 1980	Davidson et al. (1985)
<0.2		Antarctica, Austral Winters 1975 & 1976	Cunningham & Zoller (1981)
~0.049	~24000.	Antarctica, Austral Summers 1971, 1975, 1976, & 1978	Cunningham & Zoller (1981)

Table 7.1. *(Continued)*

Airborne Concentration (ng/m^3)	Crustal Enrichment Factor	Sampling Location	Reference
0.4	300.	Bermuda, Summer 1975	Duce et al. (1976)
0.064	190.	Bolivia, Chacaltaya Mtn., Feb.–May 1976	Adams et al. (1977)
0.0046	75.	Enewetak, Apr.–May 1979	Duce et al. (1983)
0.0025	180.	Enewetak, Jun.–Aug. 1979	Duce et al. (1983)
0.21		Norway, Birkenes, Sep.–Dec. 1974, Apr.–May 1975	Thrane (1978)
0.33	1600.	Norway, Birkenes, Autumn 1973	Semb (1978)
0.11	1100.	Spitsbergen, Dec. 1973–Mar. 1974	Larssen (1977)
Calcium (Ca)			
5.0	3.7	Alaska, Barrow, Dec. 1976–Apr. 1977	Rahn (1977)
790.	1.3	Arizona, rural area, Jan.–Dec. 1974	Moyers et al. (1977)
250.		California, High Sierra, Jun. & Aug. 1976, Aug. 1977	Elias & Davidson (1980)
260.		Colorado, Squaw Mtn., Apr. 1976	Winchester et al. (1979)
89.		Florida, Apalachicola Nat. Forest, May–Jul. 1973	Johansson et al. (1976)
1600.	450.	Hawaii, Coast	Gordon et al. (1978)
41.	1.7	Hawaii, Mauna Loa, upslope wind, Jan.–Dec. 1980	Gordon et al. (1978)

7.8	1.7	Hawaii, Mauna Loa, downslope wind, no dust, Jan.–Dec. 1980	Gordon et al. (1978)
190.	5.0	Hawaii, Mauna Loa, downslope wind, Asian dust, Jan.–Dec. 1980	Gordon et al. (1978)
190.	1.5	Michigan, Seney Nat. Wildlife Refuge, Jun. 1979	Alkezweeny et al. (1982)
390.	0.83	Montana, Colstrip, May–Sep. 1975	Crecelius et al. (1980)
270.	4.3	Montana, Glacier Nat. Park, Aug. 1981	Davidson et al. (1985)
150.		New Hampshire, Hubbard Brook Forest, Apr. 1976	Winchester et al. (1979)
280.		New Mexico, Jemez Mts., Apr. 1976	Winchester et al. (1979)
30.	0.84	Tennessee, Smoky Mts. Nat. Park, Oct. 1979	Davidson et al. (1985)
16. (<2.5 μm)	1.6	Tennessee, Smoky Mts. Nat. Park, Sep. 1978	Stevens et al. (1980)
320. (>2.5 μm)	3.3	Tennessee, Smoky Mts. Nat. Park, Sep. 1978	Stevens et al. (1980)
320.		Utah, Canyonlands, Jul. 1977–Dec. 1978	Flocchini et al. (1981)
520.		Utah, Henrieville, Jul. 1977–Sep. 1978	Flocchini et al. (1981)
290.		Utah, Zion Nat. Park, Oct. 1977–Dec. 1978	Flocchini et al. (1981)
35. (<2.5 μm)	1.6	Virginia, Shenandoah Valley, Jul.–Aug. 1980	Stevens et al. (1984)
300. (>2.5 μm)	1.9	Virginia, Shenandoah Valley, Jul.–Aug 1980	Stevens et al. (1984)
29.	0.53	Washington, Olympic Nat. Park, Jul.–Aug. 1980	Davidson et al. (1985)
63.		Washington, Quillayute, onshore wind, Apr. 1974–Jun. 1975	Ludwick et al. (1977)

Table 7.1. *(Continued)*

Airborne Concentration (ng/m^3)	Crustal Enrichment Factor	Sampling Location	Reference
1.9		Antarctica, Austral Winters 1975 & 1976	Cunningham & Zoller (1981)
0.55	1.3	Antarctica, Austral Summers 1971, 1975, 1976, & 1978	Cunningham & Zoller (1981)
150.	4.2	Atlantic Ocean, East	Duce et al. (1976)
200.	0.79	Bermuda, Summer 1975	Duce et al. (1976)
82.	0.60	Bolivia, Chacaltaya Mtn., Feb.–May 1976	Adams et al. (1977)
300.	8.1	Enewetak, Apr.–May 1979	Duce et al. (1983)
170.	84.	Enewetak, Jun.–Aug. 1979	Duce et al. (1983)
290.		Enewetak, Apr.–May 1979	Settle & Patterson (1982)
130.		Enewetak, Jul.–Aug. 1979	Settle & Patterson (1982)
9.1	1.3	Greenland, Kap Harald Moltke, Jun.–Aug. 1974	Flyger & Heidam (1978)
20.	1.2	Greenland, Thule, Feb.–Apr. 1978	Heidam (1981)
21.	3.2	Greenland, Prins Christianssund, Jan.–Feb. 1978	Heidam (1981)
80.	1.4	Norway, Birkenes, Spring 1973	Semb (1978)
74.	1.7	Norway, Birkenes, Autumn 1973	Semb (1978)
43.	1.2	Norway, Skoganvarre, Spring 1973	Semb (1978)
73.		Spitsbergen, Apr.–May 1979	Heintzenberg et al. (1981)

49.	2.2	Spitsbergen, Dec. 1973–Mar. 1974	Larssen (1977)
74.		Sweden, Sjoangen, annual ave.	Lannefors et al. (1983)
80. (<2.5 μm)	3.2	USSR, Abastumani Mts., Jul. 1979	Stevens et al. (1984)
330. (>2.5 μm)	1.8	USSR, Abastumani Mts., Jul. 1979	Stevens et al. (1984)
Cerium (Ce)			
0.075	2.1	Hawaii, Mauna Loa, upslope wind, Jan.–Dec. 1980	Gordon et al. (1978)
0.018	2.7	Hawaii, Mauna Loa, downslope wind, no dust, Jan.–Dec. 1980	Gordon et al. (1978)
0.11	2.0	Hawaii, Mauna Loa, downslope wind, Asian dust, Jan.–Dec. 1980	Gordon et al. (1978)
0.0042		Antarctica, Austral Winters 1975 & 1976	Cunningham & Zoller (1981)
0.0020	3.3	Antarctica, Austral Summers 1971, 1975, 1976, & 1978	Cunningham & Zoller (1981)
0.17	3.3	Atlantic Ocean, East	Duce et al. (1976)
0.6	1.6	Bermuda, Summer 1975	Duce et al. (1976)
0.089	1.6	Enewetak, Apr.–May 1979	Duce et al. (1983)
0.0048	1.0	Enewetak, Jun.–Aug. 1979	Duce et al. (1983)
Cesium (Cs)			
0.6	10.	Arizona, rural area, Jan.–Dec. 1974	Moyers et al. (1977)
0.052		California, High Sierra, Jun. & Aug. 1976, Aug. 1977	Elias & Davidson (1980)

Table 7.1. *(Continued)*

Airborne Concentration (ng/m^3)	Crustal Enrichment Factor	Sampling Location	Reference
0.096	2.8	Montana, Colstrip, May–Sep. 1975	Crecelius et al. (1980)
0.06	7. (Sc)	New York, Whiteface Mtn., unpolluted air, Summer 1975	Parekh & Husain (1981)
8.6×10^{-5}		Antarctica, Austral Winters 1975 & 1976	Cunningham & Zoller (1981)
0.00015	5.0	Antarctica, Austral Summers 1971, 1975, 1976, & 1978	Cunningham & Zoller (1981)
0.045	7.9	Bolivia, Chacaltaya Mtn., Feb.–May 1976	Adams et al. (1977)
0.013	4.7	Enewetak, Apr.–May 1979	Duce et al. (1983)
0.0012	5.4	Enewetak, Jun.–Aug. 1979	Duce et al. (1983)
Chlorine (Cl)			
34.		Florida, Apalachicola Nat. Forest, May–Jul. 1973	Johansson et al. (1976)
29.	370.	Hawaii, Mauna Loa, upslope wind, Jan.–Dec. 1980	Gordon et al. (1978)
68.	46.	Montana, Colstrip, May–Sep. 1975	Crecelius et al. (1980)
<10. (<2.5 μm)		Tennessee, Smoky Mts. Nat. Park, Sep. 1978	Stevens et al. (1980)
7. (>2.5 μm)	20.	Tennessee, Smoky Mts. Nat. Park, Sep. 1978	Stevens et al. (1980)
200.		Utah, Canyonlands, Jul. 1977–Dec. 1978	Flocchini et al. (1981)

160.		Utah, Henrieville, Jul. 1977–Sep. 1978	Flocchini et al. (1981)
63.		Utah, Zion Nat. Park, Oct. 1977–Dec. 1978	Flocchini et al. (1981)
10. (<2.5 μm)	140.	Virginia, Shenandoah Valley, Jul.–Aug. 1980	Stevens et al. (1984)
180. (>2.5 μm)	360.	Virginia, Shenandoah Valley, Jul.–Aug. 1980	Stevens et al. (1984)
2200.		Washington, Quillayute, onshore wind, Apr. 1974–Jun. 1975	Ludwick et al. (1977)
68.		Antarctica, Austral Winters 1975 & 1976	Cunningham & Zoller (1981)
6.6	5000.	Antarctica, Austral Summers 1971, 1975, 1976, & 1978	Cunningham & Zoller (1981)
39.	170.	Bolivia, Chacaltaya Mtn., Feb.–May 1976	Adams et al. (1977)
6.5 (<1 μm)	310. (Fe)	China, Great Wall, Apr. 1980	Winchester et al. (1981)
9500.	85000.	Enewetak, Apr.–May 1979	Duce et al. (1983)
7200.	1.1×10^6	Enewetak, Jun.–Aug. 1979	Duce et al. (1983)
5.3		England, S.W. Coast, onshore wind, Winter 1971–1972	Barnes & Eggleton (1977)
10.	450.	Greenland, Kap Harald Moltke, Jun.–Aug. 1974	Flyger & Heidam (1978)
154.	3000.	Greenland, Thule, Feb.–Apr. 1978	Heidam (1981)
1260.	61000.	Greenland, Prins Christianssund, Jan.–Feb. 1978	Heidam (1981)
60.	330.	Norway, Birkenes, Spring 1973	Semb (1978)
260.	1900.	Norway, Birkenes, Autumn 1973	Semb (1978)
10.	93.	Norway, Skoganvarre, Spring 1973	Semb (1978)
190.		Sweden, Sjoangen, annual ave.	Lannefors et al. (1983)

Table 7.1. *(Continued)*

Airborne Concentration (ng/m^3)	Crustal Enrichment Factor	Sampling Location	Reference
5. (<2.5 μm)	60.	USSR, Abastumani Mts., Jul. 1979	Stevens et al. (1984)
30. (>2.5 μm)	53.	USSR, Abastumani Mts., Jul. 1979	Stevens et al. (1984)
Chromium (Cr)			
3.1	2.1	Arizona, rural area, Jan.–Dec. 1974	Moyers et al. (1977)
0.2	20.	Hawaii, Coast	Gordon et al. (1978)
0.5	2.	Michigan, Seney Nat. Wildlife Refuge, Jun. 1979	Alkezweeny et al. (1982)
1.4	1.2	Montana, Colstrip May–Sep. 1975	Crecelius et al. (1980)
4.0	15. (Sc)	New York, Whiteface Mtn., unpolluted air, Summer 1975	Parekh & Husain (1981)
7.0	53. (Sc)	New York, Whiteface Mtn., unpolluted air, Summer 1977	Parekh & Husain (1981)
0.011		Antarctica, Austral Winters 1975 & 1976	Cunningham & Zoller (1981)
0.013	13.	Antarctica, Austral Summers 1971, 1975, 1976, & 1978	Cunningham & Zoller (1981)
0.2	2.	Atlantic Ocean, East	Duce et al. (1976)
0.5	0.8	Bermuda, Summer 1975	Duce et al. (1976)
0.57	1.8	Bolivia, Chacaltaya Mtn., Feb.–May 1976	Adams et al. (1977)

0.43 (<1 μm)	25. (Fe)	China, Great Wall, Apr. 1980	Winchester et al. (1981)
0.14	1.6	Enewetak, Apr.–May 1979	Duce et al. (1983)
0.037	5.0	Enewetak, Jun.–Aug. 1979	Duce et al. (1983)
0.09	5.0	Greenland, Kap Harald Moltke, Jun.–Aug. 1974	Flyger & Heidam (1978)
0.21	5.4	Greenland, Thule, Feb.1–Apr. 1978	Heidam (1981)
0.90	6.5	Norway, Birkenes, Spring 1973	Semb (1978)
1.2	12.	Norway, Birkenes, Autumn 1973	Semb (1978)
0.35	4.2	Norway, Skoganvarre, Spring 1973	Semb (1978)
2.6		Spitsbergen, Apr.–May 1979	Heintzenberg et al. (1981)
2.8		Sweden, Sjoangen, annual ave.	Lannefors et al. (1983)
Cobalt (Co)			
0.7	1.9	Arizona, rural area, Jan.–Dec. 1974	Moyers et al. (1977)
0.021	1.4	Hawaii, Mauna Loa, upslope wind, Jan.–Dec. 1980	Gordon et al. (1978)
0.0053	1.9	Hawaii, Mauna Loa, downslope wind, no dust, Jan.–Dec. 1980	Gordon et al. (1978)
0.029	1.3	Hawaii, Mauna Loa, downslope wind, Asian dust, Jan.–Dec. 1980	Gordon et al. (1978)
0.17	0.60	Montana, Colstrip, May–Sep. 1975	Crecelius et al. (1980)
0.00040		Antarctica, Austral Winters 1975 & 1976	Cunningham & Zoller (1981)
0.00060	2.4	Antarctica, Austral Summers 1971, 1975, 1976, & 1978	Cunningham & Zoller (1981)
0.04	2.	Atlantic Ocean, East	Duce et al (1976)

Table 7.1. *(Continued)*

Airborne Concentration (ng/m^3)	Crustal Enrichment Factor	Sampling Location	Reference
0.08	0.5	Bermuda, Summer 1975	Duce et al. (1976)
1.0	13.	Bolivia, Chacaltaya Mtn., Feb.–May 1976	Adams et al. (1977)
0.021	0.89	Enewetak, Apr.–May 1979	Duce et al. (1983)
0.0017	0.99	Enewetak, Jun.–Aug. 1979	Duce et al. (1983)
0.17	4.9	Norway, Birkenes, Spring 1973	Semb (1978)
0.16	6.1	Norway, Birkenes, Autumn 1973	Semb (1978)
0.049	2.4	Norway, Skoganvarre, Spring 1973	Semb (1978)
Copper (Cu)			
110.	140.	Arizona, rural area, Jan.–Dec. 1974	Moyers et al. (1977)
0.3	60.	Hawaii, Coast	Gordon et al. (1978)
3.8	23.	Michigan, Seney Nat. Wildlife Refuge, Jun. 1979	Alkezweeny et al. (1982)
2.0	3.2	Montana, Colstrip, May–Sep. 1975	Crecelius et al. (1980)
<6.9		Montana, Glacier Nat. Park, Aug. 1981	Davidson et al. (1985)
0.52	11.	Tennessee, Smoky Mts. Nat. Park, Oct. 1979	Davidson et al. (1985)
3. (<2.5 μm)	200.	Tennessee, Smoky Mts. Nat. Park, Sep. 1978	Stevens et al. (1980)

$<5.$ ($>2.5\ \mu m$)		Tennessee, Smoky Mts. Nat. Park, Sep. 1978	Stevens et al. (1980)
69.		Utah, Canyonlands, Jul. 1977–Dec. 1978	Flocchini et al. (1981)
13.		Utah, Henrieville, Jul. 1977–Sep. 1978	Flocchini et al. (1981)
15.		Utah, Zion Nat. Park, Oct. 1977–Dec. 1978	Flocchini et al. (1981)
5. ($<2.5\ \mu m$)	200.	Virginia, Shenandoah Valley, Jul.–Aug. 1980	Stevens et al. (1984)
6. ($>2.5\ \mu m$)	30.	Virginia, Shenandoah Valley, Jul.–Aug. 1980	Stevens et al. (1984)
5.6	76.	Washington, Olympic Nat. Park, Jul.–Aug. 1980	Davidson et al. (1985)
0.79		Washington, Quillayute, onshore wind, Apr. 1974–Jun. 1975	Ludwick et al. (1977)
0.079		Antarctica, Austral Winters 1975 & 1976	Cunningham & Zoller (1981)
0.059	110.	Antarctica, Austral Summers 1971, 1975, 1976, & 1978	Cunningham & Zoller (1981)
0.3		Arctic (Canadian), Late Winter 1980	Barrie et al. (1981)
0.9	20.	Atlantic Ocean, East	Duce et al. (1976)
2.	6.	Bermuda, Summer 1975	Duce et al. (1976)
1.3	7.4	Bolivia, Chacaltaya Mtn., Feb.–May 1976	Adams et al. (1977)
0.44 ($<1\ \mu m$)	55. (Fe)	China, Great Wall, Apr. 1980	Winchester et al. (1981)
0.072	1.9	Enewetak, Apr.–May 1979	Duce et al. (1983)
0.014	6.7	Enewetak, Jun.–Aug. 1979	Duce et al. (1983)

Table 7.1. *(Continued)*

Airborne Concentration (ng/m^3)	Crustal Enrichment Factor	Sampling Location	Reference
0.31	14.	Greenland, Thule, Feb.–Apr. 1978	Heidam (1981)
2.7	43.	Nepal, Himalayas, Dec. 1979	Davidson et al. (1981a)
5.0	66.	Norway, Birkenes, Spring 1973	Semb (1978)
6.9	120.	Norway, Birkenes, Autumn 1973	Semb (1978)
1.6	35.	Norway, Skoganvarre, Spring 1973	Semb (1978)
<1.9		Spitsbergen, Apr.–May 1979	Heintzenberg et al. (1981)
3.3	110.	Spitsbergen, Dec. 1973–Mar. 1974	Larssen (1977)
2.4		Sweden, Sjoangen, annual ave.	Lannefors et al. (1983)
7. (<2.5 μm)	200.	USSR, Abastumani Mts., Jul. 1979	Stevens et al. (1984)
6. (>2.5 μm)	30.	USSR, Abastumani Mts., Jul. 1979	Stevens et al. (1984)
Europium (Eu)			
0.016	1.2	Montana, Colstrip, May–Sep. 1975	Crecelius et al. (1980)
$9.\times10^{-6}$		Antarctica, Austral Winters 1975 & 1976	Cunningham & Zoller (1981)
2.0×10^{-5}	1.7	Antarctica, Austral Summers 1971, 1975 1976 & 1978	Cunningham & Zoller (1981)
0.0010	0.95	Enewetak, Apr.–May 1979	Duce et al. (1983)
9.0×10^{-5}	0.54	Enewetak, Jun.–Aug. 1979	Duce et al. (1983)

Gold (Au)			
9.2×10^{-5}		Antarctica, Austral Winters 1975 & 1976	Cunningham & Zoller (1981)
8.0×10^{-5}	2000.	Antarctica, Summer Austral Summers 1971, 1975, 1976, & 1978	Cunningham & Zoller (1981)
0.0015	280.	Bolivia, Chacaltaya Mtn., Feb.–May 1976	Adams et al. (1977)
Hafnium (Hf)			
0.044	1.3	Montana, Colstrip, May–Sep. 1975	Crecelius et al. (1980)
4.2×10^{-5}		Antarctica, Austral Winters 1975 & 1976	Cunningham & Zoller (1981)
9.0×10^{-5}	3.0	Antarctica, Austral Summers 1971, 1975, 1976, & 1978	Cunningham & Zoller (1981)
0.0039	1.4	Enewetak, Apr.–May 1979	Duce et al. (1983)
0.00045	2.2	Enewetak, Jun.–Aug. 1979	Duce et al. (1983)
Indium (In)			
0.00019		Antarctica, Austral Winters 1975 & 1976	Cunningham & Zoller (1981)
5.4×10^{-5}	54.	Antarctica, Austral Summers 1971, 1975, 1976, & 1978	Cunningham & Zoller (1981)
0.0078	44.	Bolivia, Chacaltaya Mtn., Feb.–May 1976	Adams et al. (1977)
Iodine (I)			
1.5	5000.	Hawaii, Mauna Loa, upslope wind, Jan.–Dec. 1980	Gordon et al. (1978)
2.0	37000.	Hawaii, Mauna Loa, downslope wind, no dust, Jan.–Dec. 1980	Gordon et al. (1978)

Table 7.1. *(Continued)*

Airborne Concentration (ng/m^3)	Crustal Enrichment Factor	Sampling Location	Reference
2.7	5800.	Hawaii, Mauna Loa, downslope wind, Asian dust, Jan.–Dec. 1980	Gordon et al. (1978)
0.18		Antarctica, Austral Winters 1975 & 1976	Cunningham & Zoller (1981)
0.080	16000.	Antarctica, Austral Summers 1971, 1975 1976, & 1978	Cunningham & Zoller (1981)
0.15	160.	Bolivia, Chacaltaya Mtn., Feb.–May 1976	Adams et al. (1977)
4.4	10000.	Enewetak, Apr.–May 1979	Duce et al. (1983)
1.7	73000.	Enewetak, Jun.–Aug. 1979	Duce et al. (1983)
0.58	3000.	Greenland, Thule, Feb.–Apr. 1978	Heidam (1981)
1.1	1500.	Norway, Birkenes, Spring 1973	Semb (1978)
0.48	1200.	Norway, Skoganvarre, Spring 1973	Semb (1978)
Iron (Fe)			
660.	0.80	Arizona, rural area, Jan.–Dec. 1974	Moyers et al. (1977)
310.		Colorado, Squaw Mtn., Apr. 1976	Winchester et al. (1979)
160.		Florida, Apalachicola Nat. Forest, May–Jul. 1973	Johansson et al. (1976)
8.	1.7	Hawaii, Coast	Gordon et al. (1978)
45.	1.3	Hawaii, Mauna Loa, upslope wind, Jan.–Dec. 1980	Gordon et al. (1978)

9.1	1.5	Hawaii, Mauna Loa, downslope wind, no dust, Jan.–Dec. 1980	Gordon et al. (1978)
60.	1.2	Hawaii, Mauna Loa, downslope wind, Asian dust, Jan.–Dec. 1980	Gordon et al. (1978)
120.	0.73	Michigan, Seney Nat. Wildlife Refuge, Jun. 1979	Alkezweeny et al. (1982)
410.	0.64	Montana, Colstrip, May–Sep. 1975	Crecelius et al. (1980)
240.	3.9	Montana, Glacier Nat. Park, Aug. 1981	Davidson et al. (1985)
100.		New Hampshire, Hubbard Brook Forest, Apr. 1976	Winchester et al. (1979)
175.		New Mexico, Jemez Mts., Apr. 1976	Winchester et al. (1979)
240.	1.7 (Sc)	New York, Whiteface Mtn., unpolluted air, Summer 1975	Parekh & Husain (1981)
260.	3.7 (Sc)	New York, Whiteface Mtn., unpolluted air, Summer 1977	Parekh & Husain (1981)
240.	4.9	Tennessee, Smoky Mts. Nat. Park, Oct. 1979	Davidson et al. (1985)
28. (<2.5 μm)	2.0	Tennessee, Smoky Mts. Nat. Park, Sep. 1978	Stevens et al. (1980)
120. (>2.5 μm)	0.88	Tennessee, Smoky Mts. Nat. Park, Sep. 1978	Stevens et al. (1980)
170.		Utah, Canyonlands, Jul. 1977–Dec. 1978	Flocchini et al. (1981)
180.		Utah, Henrieville, Jul. 1977–Sep. 1978	Flocchini et al. (1981)
150.		Utah, Zion Nat. Park, Oct. 1977–Dec. 1978	Flocchini et al. (1981)
54. (<2.5 μm)	1.8	Virginia, Shenandoah Valley, Jul.–Aug. 1980	Stevens et al. (1984)
160. (>2.5 μm)	0.74	Virginia, Shenandoah Valley, Jul.–Aug. 1980	Stevens et al. (1984)

Table 7.1. *(Continued)*

Airborne Concentration (ng/m^3)	Crustal Enrichment Factor	Sampling Location	Reference
310.	4.1	Washington, Olympic Nat. Park, Jul.–Aug. 1980	Davidson et al. (1985)
19.		Washington, Quillayute, onshore wind, Apr. 1974–Jun. 1975	Ludwick et al. (1977)
0.25		Antarctica, Austral Winters, 1975 & 1976	Cunningham & Zoller (1981)
0.68	1.2	Antarctica, Austral Summers, 1971, 1975, 1976, & 1978	Cunningham & Zoller (1981)
18.		Arctic, (European) Apr.–May 1979 & Mar. 1981	Heintzenberg (1982)
1.1		Arctic (European) Jul.–Sep. 1980	Heintzenberg (1982)
50.	1.0	Atlantic Ocean, East	Duce et al. (1976)
300.	0.9	Bermuda, Summer 1975	Duce et al. (1976)
180.	1.0	Bolivia, Chacaltaya Mtn., Feb.–May 1976	Adams et al. (1977)
7.9 ($<1\ \mu m$)	1.0 (Fe)	China, Great Wall, Apr. 1980	Winchester et al. (1981)
50.	0.89	Enewetak, Apr.–May 1979	Duce et al. (1983)
3.3	0.80	Enewetak, Jun.–Aug. 1979	Duce et al. (1983)
17.		Greenland, Dye 3, Jun.–Aug. 1979	Davidson et al. (1981b)
6.4	0.67	Greenland, Kap Harald Moltke, Jun.–Aug. 1974	Flyger & Heidam (1978)

32.	1.5	Greenland, Thule, Feb.–Apr. 1978	Heidam (1981)
11.	1.2	Greenland, Prins Christianssund, Jan.–Feb. 1978	Heidam (1981)
96.	1.2	Norway, Birkenes, Spring 1973	Semb (1978)
95.	1.6	Norway, Birkenes, Autumn 1973	Semb (1978)
70.	1.5	Norway, Skoganvarre, Spring 1973	Semb (1978)
64.		Spitsbergen, Apr.–May 1979	Heintzenberg et al. (1981)
42.	1.4	Spitsbergen, Dec. 1973–Mar. 1974	Larssen (1977)
88.		Sweden, Sjoangen, annual ave.	Lannefors et al. (1983)
70. (<2.5 μm)	2.0	USSR, Abastumani Mts., Jul. 1979	Stevens et al. (1984)
200. (>2.5 μm)	0.81	USSR, Abastumani Mts., Jul. 1979	Stevens et al. (1984)
Lanthanam (La)			
0.056	3.1	Hawaii, Mauna Loa, upslope wind Jan.–Dec. 1980	Gordon et al. (1978)
0.0065	2.0	Hawaii, Mauna Loa, downslope wind, no dust, Jan.–Dec. 1980	Gordon et al. (1978)
0.32	12.	Hawaii, Mauna Loa, downslope wind, Asian dust, Jan.–Dec. 1980	Gordon et al. (1978)
0.50	1.5	Montana, Colstrip, May–Sep. 1975	Crecelius et al. (1980)
<0.002		Antarctica, Austral Winters 1975 & 1976	Cunningham & Zoller (1981)
0.00078	2.6	Antarctica, Austral Summers 1971, 1975, 1976, & 1978	Cunningham & Zoller (1981)
0.092	1.6	Bolivia, Chacaltaya Mtn., Feb.–May 1976	Adams et al. (1977)
Lead (Pb)			
67.	370.	Arizona, rural area, Jan.–Dec. 1974	Moyers et al. (1977)

Table 7.1. *(Continued)*

Airborne Concentration (ng/m^3)	Crustal Enrichment Factor	Sampling Location	Reference
21.		California, High Sierra, Jun. & Aug. 1976, Aug. 1977	Elias & Davidson (1980)
26.		Florida, Apalachicola Nat. Forest, May–Jul. 1973	Johansson et al. (1976)
1.2	1100.	Hawaii, Coast	Gordon et al. (1978)
10.	270.	Michigan, Seney Nat. Wildlife Refuge, June 1979	Alkezweeny et al. (1982)
14.	99.	Montana, Colstrip, May–Sep. 1975	Crecelius et al. (1980)
4.6	260.	Montana, Glacier Nat. Park, Aug. 1981	Davidson et al. (1985)
19.	1700.	Tennessee, Smoky Mts. Nat. Park, Oct. 1979	Davidson et al. (1985)
97. (<2.5 μm)	32000.	Tennessee, Smoky Mts. Nat. Park, Sep. 1978	Stevens et al. (1980)
14. (>2.5 μm)	470.	Tennessee, Smoky Mts. Nat. Park, Sep. 1978	Stevens et al. (1980)
30.		Utah, Canyonlands, Jul. 1977–Dec. 1978	Flocchini et al. (1981)
72.		Utah, Henrieville, Jul. 1977–Sep. 1978	Flocchini et al. (1981)
48.		Utah, Zion Nat. Park, Oct. 1977–Dec. 1978	Flocchini et al. (1981)
52. (<2.5 μm)	7800.	Virginia, Shenandoah Valley, Jul.–Aug. 1980	Stevens et al. (1984)
9. (>2.5 μm)	200.	Virginia, Shenandoah Valley, Jul.–Aug. 1980	Stevens et al. (1984)

2.2	130.	Washington, Olympic Nat. Park, Jul.–Aug. 1980	Davidson et al. (1985)
1.8		Washington, Quillayute, onshore wind Apr. 1974–Jun. 1975	Ludwick et al. (1977)
0.027	220.	Antarctica, Dec. 1974–Feb. 1975	Maenhaut et al. (1979)
1.0		Arctic (Canadian), Late Winter 1980	Barrie et al. (1981)
7.	700.	Atlantic Ocean, East	Duce et al. (1976)
7.	100.	Bermuda, Summer 1975	Duce et al. (1976)
3.7	93.	Bolivia, Chacaltaya Mtn., Feb.–May 1976	Adams et al. (1977)
5.6 (<1 μm)	2700. (Fe)	China, Great Wall, Apr. 1980	Winchester et al. (1981)
0.13	11.	Enewetak, Apr.–May 1979	Duce et al. (1983)
0.096	110.	Enewetak, Jun.–Aug. 1979	Duce et al. (1983)
0.23		Enewetak, Apr.–May 1979	Settle & Patterson (1982)
0.12		Enewetak, Jul.–Aug. 1979	Settle & Patterson (1982)
0.15		Greenland, Dye 3, Jun.–Aug. 1979	Davidson et al. (1981b)
0.20	94.	Greenland, Kap Harald Moltke, Jun.–Aug. 1974	Flyger & Heidam (1978)
8.0	1600.	Greenland, Thule, Feb.–Apr. 1978	Heidam (1981)
1.8	880.	Greenland, Prins Christianssund, Jan.–Feb. 1978	Heidam (1981)
0.86	60.	Nepal, Himalayas, Dec. 1979	Davidson et al. (1981a)
17.		Norway, Birkenes, Sep.–Dec. 1974, Apr.–May 1975	Thrane (1978)

Table 7.1. *(Continued)*

Airborne Concentration (ng/m^3)	Crustal Enrichment Factor	Sampling Location	Reference
24.	1800.	Norway, Birkenes, Autumn 1973	Semb (1978)
5.8	890.	Spitsbergen, Dec. 1973–Mar. 1974	Larssen (1977)
<4.9		Spitsbergen, Apr.–May 1979	Heintzenberg et al. (1981)
21.		Sweden, Sjoangen, annual ave.	Lannefors et al. (1983)
18. (<2.5 μm)	2400.	USSR, Abastumani Mts., Jul. 1979	Stevens et al. (1984)
1. (>2.5 μm)	20.	USSR, Abastumani Mts., Jul. 1979	Stevens et al. (1984)
Lithium (Li)			
8.9	30.	Arizona, rural area, Jan.–Dec. 1974	Moyers et al. (1977)
Lutetium (Lu)			
$<1.\times10^{-3}$		Antarctica, Austral Winters 1975 & 1976	Cunningham & Zoller (1981)
$7.\times10^{-6}$	1.4	Antarctica, Austral Summers 1971, 1975, 1976, & 1978	Cunningham & Zoller (1981)
Magnesium (Mg)			
162.	21.	Alaska, Barrow, Dec. 1976–Apr. 1977	Rahn (1977)
180.	0.53	Arizona, rural area, Jan.–Dec. 1974	Moyers et al. (1977)
410.	210.	Hawaii, Coast	Gordon et al. (1978)

28.	2.0	Hawaii, Mauna Loa, upslope wind, Jan.–Dec. 1980	Gordon et al. (1978)
5.1	2.0	Hawaii, Mauna Loa, downslope wind, no dust, Jan.–Dec. 1980	Gordon et al. (1978)
28.	1.3	Hawaii, Mauna Loa, downslope wind, Asian dust, Jan.–Dec. 1980	Gordon et al. (1978)
140.	2.8	Montana, Glacier Nat. Park, Aug. 1981	Davidson et al. (1985)
65.	3.2	Tennessee, Smoky Mts. Nat. Park, Oct. 1979	Davidson et al. (1985)
340.	11.	Washington, Olympic Nat. Park, Jul.–Aug. 1980	Davidson et al. (1985)
5.7		Antarctica, Austral Winters 1975 & 1976	Cunningham & Zoller (1981)
0.93	4.0	Antarctica, Austral Summers 1971, 1975, 1976, & 1978	Cunningham & Zoller (1981)
300.	15.	Atlantic Ocean, East	Duce et al. (1976)
300.	2.1	Bermuda, Summer 1975	Duce et al. (1976)
~18.	~0.40	Bolivia, Chacaltaya Mtn., Feb.–May 1976	Adams et al. (1977)
800.	38.	Enewetak, Apr.–May 1979	Duce et al. (1983)
560.	470.	Enewetak, Jun.–Aug. 1979	Duce et al. (1983)
<0.77		Greenland, Dye 3, Jun.–Aug. 1979	Davidson et al. (1981b)
23.	2.5	Greenland, Thule, Feb.–Apr. 1978	Heidam (1981)
110.	29.	Greenland, Prins Christianssund, Jan.–Feb. 1978	Heidam (1981)
15.	0.57	Nepal, Himalayas, Dec. 1979	Davidson et al. (1981a)
51.		Spitsbergen, Apr.–May 1979	Heintzenberg et al. (1981)

Table 7.1. *(Continued)*

Airborne Concentration (ng/m^3)	Crustal Enrichment Factor	Sampling Location	Reference
Manganese (Mn)			
1.1	3.5	Alaska, Barrow, Dec. 1976–Apr. 1977	Rahn (1977)
12.	0.87	Arizona, rural area, Jan.–Dec. 1974	Moyers et al. (1977)
0.16	2.0	Hawaii, Coast	Gordon et al. (1978)
0.65	1.1	Hawaii, Mauna Loa, upslope wind, Jan.–Dec. 1980	Gordon et al. (1978)
0.12	1.2	Hawaii, Mauna Loa, downslope wind, no dust, Jan.–Dec. 1980	Gordon et al. (1978)
0.90	1.0	Hawaii, Mauna Loa, downslope wind, Asian dust, Jan.–Dec. 1980	Gordon et al. (1978)
3.8	1.3	Michigan, Seney Nat. Wildlife Refuge, Jun. 1979	Alkezweeny et al. (1982)
9.3	0.87	Montana, Colstrip, May–Sep. 1975	Crecelius et al. (1980)
8.0	3.0 (Sc)	New York, Whiteface Mtn., unpolluted air, Summer 1975	Parekh & Husain (1981)
20.	15. (Sc)	New York, Whiteface Mtn., unpolluted air, Summer 1977	Parekh & Husain (1981)
3. (<2.5 μm)	6.	Virginia, Shenandoah Valley, Jul.–Aug. 1980	Stevens et al. (1984)
3. (>2.5 μm)	0.8	Virginia, Shenandoah Valley, Jul.–Aug. 1980	Stevens et al. (1984)
0.0067		Antarctica, Austral Winters 1975 & 1976	Cunningham & Zoller (1981)
0.014	1.5	Antarctica, Austral Summers 1971, 1975, 1976, & 1978	Cunningham & Zoller (1981)

0.5		Arctic (Canadian) Late Winter 1980	Barrie et al. (1981)
0.3	0.4	Atlantic Ocean, East	Duce et al. (1976)
1.1	1.7	Bear Island, Nov. 1977–Mar. 1978	Rahn (1981)
3.	0.5	Bermuda, Summer 1975	Duce et al. (1976)
2.6	0.86	Bolivia, Chacaltaya Mtn., Feb.–May 1976	Adams et al. (1977)
0.58 (<1 μm)	3.7 (Fe)	China, Great Wall, Apr. 1980	Winchester et al. (1981)
0.79	0.86	Enewetak, Apr.–May 1979	Duce et al. (1983)
0.063	1.1	Enewetak, Jun.–Aug. 1979	Duce et al. (1983)
0.21	1.3	Greenland, Kap Harald Moltke, Jun.–Aug. 1974	Flyger & Heidam (1976)
0.69	1.9	Greenland, Thule, Feb.–Apr. 1978	Heidam (1981)
0.32	2.1	Greenland, Prins Christianssund, Jan.–Feb. 1978	Heidam (1981)
8.6	6.5	Norway, Birkenes, Spring 1973	Semb (1978)
9.9	10.	Norway, Birkenes, Autumn 1973	Semb (1978)
2.1	2.7	Norway, Skoganvarre, Spring 1973	Semb (1978)
1.5		Spitsbergen, Apr.–May 1979	Heintzenberg et al. (1981)
1.1	2.2	Spitsbergen, Dec. 1973–Mar. 1974	Larssen (1977)
6.7		Sweden, Sjoangen, annual ave.	Lannefors et al. (1983)
190. (<2.5 μm)	320.	USSR, Abastumani Mts., Jul. 1979	Stevens et al. (1984)
38. (>2.5 μm)	9.1	USSR, Abastumani Mts., Jul. 1979	Stevens et al. (1984)

Table 7.1. *(Continued)*

Airborne Concentration (ng/m^3)	Crustal Enrichment Factor	Sampling Location	Reference
Mercury (Hg)			
0.60	3300. (Sc)	New York, Whiteface Mtn., unpolluted air, Summer 1975	Parekh & Husain (1981)
1.3		Switzerland, Corviglia (Alps)	Grosch et al. (1978)
2.7		Switzerland, St. Moritz	Grosch et al. (1978)
3.4		Norway, Birkenes, Sep.–Dec. 1974, Apr.–May 1975	Thrane (1978)
Nickel (Ni)			
3.2	2.9	Arizona, rural area, Jan.–Dec. 1974	Moyers et al. (1977)
0.85	3.8	Michigan, Seney Nat. Wildlife Refuge, Jun. 1979	Alkezweeny et al. (1982)
0.57	0.67	Montana, Colstrip, May–Sep. 1975	Crecelius et al. (1980)
1. (<2.5 μm)	50.	Tennessee, Smoky Mts. Nat. Park, Sep. 1978	Stevens et al. (1980)
1. (>2.5 μm)	6.	Tennessee, Smoky Mts. Nat. Park, Sep. 1978	Stevens et al. (1980)
1. (<2.5 μm)	20.	Virginia, Shenandoah Valley, Jul.–Aug. 1980	Stevens et al. (1984)
1. (>2.5 μm)	4.	Virginia, Shenandoah Valley, Jul.–Aug. 1980	Stevens et al. (1984)
0.13		Washington, Quillayute, onshore wind, Apr. 1974–Jun. 1975	Ludwick et al. (1977)

0.56	2.3	Bolivia, Chacaltaya Mtn., Feb.–May 1976	Adams et al. (1977)
0.22 (<1 μm)	20. (Fe)	China, Great Wall, Apr. 1980	Winchester et al. (1981)
0.08	6.	Greenland, Kap Harald Moltke, Jun.–Aug. 1974	Flyger & Heidam (1978)
0.13	4.4	Greenland, Thule Feb.–Apr. 1978	Heidam (1981)
0.70		Spitsbergen, Apr.–May 1979	Heintzenberg et al. (1981)
1.5		Sweden, Sjoangen, annual ave.	Lannefors et al. (1983)
<1.0 (<2.5 μm)		USSR, Abastumani Mts., Jul. 1979	Stevens et al. (1984)
<1.0 (>2.5 μm)		USSR, Abastumani Mts., Jul. 1979	Stevens et al. (1984)
Potassium (K)			
530.	1.7	Arizona, rural area, Jan.–Dec. 1974	Moyers et al. (1977)
180.		California, High Sierra, Jun. & Aug. 1976, Aug. 1977	Elias & Davidson (1980)
130.		Colorado, Squaw Mtn., Apr. 1976	Winchester et al. (1979)
51.		Florida, Apalachicola Nat. Forest, May–Jul. 1973	Johansson et al. (1976)
120.	68.	Hawaii, Coast	Gordon et al. (1978)
24.	1.9	Hawaii, Mauna Loa, upslope wind, Jan.–Dec. 1980	Gordon et al. (1978)
4.5	2.0	Hawaii, Mauna Loa, downslope wind, no dust, Jan.–Dec. 1980	Gordon et al. (1978)
34.	1.8	Hawaii, Mauna Loa, downslope wind, Asian dust, Jan.–Dec. 1980	Gordon et al. (1978)

Table 7.1. *(Continued)*

Airborne Concentration (ng/m^3)	Crustal Enrichment Factor	Sampling Location	Reference
83.	1.3	Michigan, Seney Nat. Wildlife Refuge, Jun. 1979	Alkezweeny et al. (1982)
280.	1.2	Montana, Colstrip, May–Sep. 1975	Crecelius et al. (1980)
51.		New Hampshire, Hubbard Brook Forest, Apr. 1976	Winchester et al. (1979)
85.		New Mexico, Jemez Mts., Apr. 1976	Winchester et al. (1979)
120.	1.6 (Sc)	New York, Whiteface Mtn., unpolluted air, Summer 1975	Parekh & Husain (1981)
150.	4.2 (Sc)	New York, Whiteface Mtn., unpolluted air, Summer 1977	Parekh & Husain (1981)
40. (<2.5 μm)	7.9	Tennessee, Smoky Mts. Nat. Park, Sep. 1978	Stevens et al. (1980)
110. (>2.5 μm)	2.2	Tennessee, Smoky Mts. Nat. Park, Sep. 1978	Stevens et al. (1980)
260.		Utah, Canyonlands, Jul. 1977–Dec. 1978	Flocchini et al. (1981)
160.		Utah, Henrieville, Jul. 1977–Sep. 1978	Flocchini et al. (1981)
120.		Utah, Zion Nat. Park, Oct. 1977–Dec. 1978	Flocchini et al. (1981)
61. (<2.5 μm)	5.5	Virginia, Shenandoah Valley, Jul.–Aug. 1980	Stevens et al. (1984)
130. (>2.5 μm)	1.6	Virginia, Shenandoah Valley, Jul.–Aug. 1980	Stevens et al. (1984)
55.		Washington, Quillayute, onshore wind, Apr. 1974–Jun. 1975	Ludwick et al. (1977)

1.3		Antarctica, Austral Winters 1975 & 1976	Cunningham & Zoller (1981)
0.61	2.9	Antarctica, Austral Summers 1971, 1975, 1976, & 1978	Cunningham & Zoller (1981)
90.	5.1	Atlantic Ocean, East	Duce et al. (1976)
200.	1.6	Bermuda, Summer 1975	Duce et al. (1976)
47.	1.2	Bolivia, Chacaltaya Mtn., Feb.–May 1976	Adams et al. (1977)
23. (<1 μm)	5.6 (Fe)	China, Great Wall, Apr. 1980	Winchester et al. (1981)
270.	15.	Enewetak, Apr.–May 1979	Duce et al. (1983)
160.	170.	Enewetak, Jun.–Aug. 179	Duce et al. (1983)
300.		Enewetak, Apr.–May 1979	Settle & Patterson (1982)
120.		Enewetak, Jul.–Aug. 1979	Settle & Patterson (1982)
3.8	1.1	Greenland, Kap Harald Moltke, Jun.–Aug. 1974	Flyger & Heidam (1978)
15.	1.8	Greenland, Thule, Feb.–Apr. 1978	Heidam (1981)
20.	5.8	Greenland, Prins Christianssund, Jan.–Feb. 1978	Heidam (1981)
140.	4.8	Norway, Birkenes, Spring 1973	Semb (1978)
35.	1.6	Norway, Birkenes, Autumn 1973	Semb (1978)
32.		Spitsbergen, Apr.–May 1979	Heintzenberg et al. (1981)
76.		Sweden, Sjoangen, annual ave.	Lannefors et al. (1983)
130. (<2.5 μm)	10.	USSR, Abastumani Mts., Jul. 1979	Stevens et al. (1984)
100. (>2.5 μm)	1.1	USSR, Abastumani Mts., Jul. 1979	Stevens et al. (1984)

Table 7.1. *(Continued)*

Airborne Concentration (ng/m^3)	Crustal Enrichment Factor	Sampling Location	Reference
Rubidium (Rb)			
2.0	1.5	Arizona, rural area, Jan.–Dec. 1974	Moyers et al. (1977)
0.54		California, High Sierra, Jun. & Aug. 1976, Aug. 1977	Elias & Davidson (1980)
1.2	1.2	Montana, Colstrip, May–Sep. 1975	Crecelius et al. (1980)
0.0030		Antarctica, Austral Winters 1975 & 1976	Cunningham & Zoller (1981)
<0.004		Antarctica, Austral Summers 1971, 1975, 1976, & 1978	Cunningham & Zoller (1981)
0.67	2.3	Bolivia, Chacaltaya Mtn., Feb.–May 1976	Adams et al. (1977)
0.22	2.7	Enewetak, Apr.–May 1979	Duce et al. (1983)
0.060	12.	Enewetak Jun.–Aug. 1979	Duce et al. (1983)
5.5		Enewetak, Apr.–May 1979	Settle & Patterson (1982)
0.06		Enewetak, Jul.–Aug. 1979	Settle & Patterson (1982)
Samarium (Sm)			
0.073	1.1	Montana, Colstrip, May–Sep. 1975	Crecelius et al. (1980)
<0.0004		Antarctica, Austral Winters 1975 & 1976	Cunningham & Zoller (1981)

0.00010	1.7	Antarctica, Austral Summers 1971, 1975, 1976, & 1978	Cunningham & Zoller (1981)
0.014	1.2	Bolivia, Chacaltaya Mtn., Feb.–May 1976	Adams et al. (1977)
Scandium (Sc)			
0.011	0.84	Hawaii, Mauna Loa, upslope wind, Jan.–Dec. 1980	Gordon et al. (1978)
0.0027	1.1	Hawaii, Mauna Loa, downslope wind, no dust, Jan.–Dec. 1980	Gordon et al. (1978)
0.020	0.98	Hawaii, Mauna Loa, downslope wind, Asian dust, Jan.–Dec. 1980	Gordon et al. (1978)
0.13	0.52	Montana, Colstrip, May–Sep. 1975	Crecelius et al. (1980)
0.05	1.0 (Sc)	New York, Whiteface Mtn., unpolluted air, Summer 1975	Parekh & Husain (1981)
0.03	1.0 (Sc)	New York, Whiteface Mtn., unpolluted air, Summer 1977	Parekh & Husain (1981)
3.7×10^{-5}		Antarctica, Austral Winters 1975 & 1976	Cunningham & Zoller (1981)
0.00018	0.84	Antarctica, Austral Summers 1971, 1975 1976, & 1978	Cunningham & Zoller (1981)
0.019	1.0	Atlantic Ocean, East	Duce et al. (1976)
0.06	0.45	Bermuda, Summer 1975	Duce et al. (1976)
0.030	0.63	Bolivia, Chacaltaya Mtn., Feb.–May 1976	Adams et al. (1977)
0.016	0.78	Enewetak, Apr.–May 1979	Duce et al. (1983)
0.00094	0.63	Enewetak, Jun.–Aug. 1979	Duce et al. (1983)
0.019	0.62	Norway, Birkenes, Spring 1973	Semb (1978)
0.009	0.49	Norway, Skoganvarre, Spring 1973	Semb (1978)

Table 7.1. *(Continued)*

Airborne Concentration (ng/m^3)	Crustal Enrichment Factor	Sampling Location	Reference
Selenium (Se)			
0.027	910.	Hawaii, Mauna Loa, upslope wind, Jan.–Dec. 1980	Gordon et al. (1978)
0.013	2400.	Hawaii, Mauna Loa, downslope wind, no dust, Jan.–Dec. 1980	Gordon et al. (1978)
0.028	610.	Hawaii, Mauna Loa, downslope wind, Asian dust, Jan.–Dec. 1980	Gordon et al. (1978)
<1.0		Michigan, Seney Nat. Wildlife Refuge, Jun. 1979	Alkezweeny et al. (1982)
0.27	480.	Montana, Colstrip, May–Sep. 1975	Crecelius et al. (1980)
0.70	4700. (Sc)	New York, Whiteface Mtn., unpolluted air, Summer 1975	Parekh & Husain (1981)
0.84	11000 (Sc)	New York, Whiteface Mtn., unpolluted air, Summer 1977	Parekh & Husain (1981)
1.4 (<2.5 μm)	1.2×10^5	Tennessee, Smoky Mts. Nat. Park, Sep. 1978	Stevens et al. (1980)
0.2 (>2.5 μm)	2000.	Tennessee, Smoky Mts. Nat. Park, Sep. 1978	Stevens et al. (1980)
1.2 (<2.5 μm)	45000.	Virginia, Shenandoah Valley, Jul.–Aug. 1980	Stevens et al. (1984)
<1. (>2.5 μm)		Virginia, Shenandoah Valley, Jul.–Aug. 1980	Stevens et al. (1984)
0.0069		Antarctica, Austral Winters 1975 & 1976	Cunningham & Zoller (1981)
0.0063	12000.	Antarctica, Austral Summers 1971, 1975, 1976, & 1978	Cunningham & Zoller (1981)

0.3	7000.	Atlantic Ocean, East	Duce et al. (1976)
0.19	630.	Bermuda, Summer 1975	Duce et al. (1976)
0.10	1200.	Bolivia, Chacaltaya Mtn., Feb.–May 1976	Adams et al. (1977)
<0.10 (<1 μm)		China, Great Wall, Apr. 1980	Winchester et al. (1981)
0.15	3700.	Enewetak, Apr.–May 1979	Duce et al. (1983)
0.11	48000.	Enewetak, Jun.–Aug. 1979	Duce et al. (1983)
0.026	3100.	Greenland, Kap Harald Moltke, Jun.–Aug. 1974	Flyger & Heidam (1978)
0.48	6900.	Norway, Birkenes, Spring 1973	Semb (1978)
0.66	13000.	Norway, Birkenes, Autumn 1973	Semb (1978)
0.17	4100.	Norway, Skoganvarre, Spring 1973	Semb (1978)
<1. (<2.5 μm)		USSR, Abastumani Mts., Jul. 1979	Stevens et al. (1984)
<1. (>2.5 μm)		USSR, Abastumani Mts., Jul. 1979	Stevens et al. (1984)
Silicon (Si)			
3900.	0.95	Arizona, rural area, Jan.–Dec. 1974	Moyers et al. (1977)
190.		Colorado, Squaw Mtn., Apr. 1976	Winchester et al. (1979)
450.	0.53	Michigan, Seney Nat. Wildlife Refuge, Jun. 1979	Alkezweeny et al. (1982)
2700.	0.85	Montana, Colstrip, May–Sep. 1975	Crecelius et al. (1980)
59.		New Hampshire, Hubbard Brook Forest, Apr. 1976	Winchester et al. (1979)

Table 7.1. *(Continued)*

Airborne Concentration (ng/m^3)	Crustal Enrichment Factor	Sampling Location	Reference
350.		New Mexico, Jemez Mts., Apr. 1976	Winchester et al. (1979)
38. (<2.5 μm)	0.55	Tennessee, Smoky Mts. Nat. Park, Sep. 1978	Stevens et al. (1980)
580. (>2.5 μm)	0.87	Tennessee, Smoky Mts. Nat. Park, Sep. 1978	Stevens et al. (1980)
1200.		Utah, Canyonlands, Jul. 1977–Dec. 1978	Flocchini et al. (1981)
1400.		Utah, Henrieville, Jul. 1977–Sep. 1978	Flocchini et al. (1981)
940.		Utah, Zion Nat. Park, Oct. 1977–Dec. 1978	Flocchini et al. (1981)
120. (<2.5 μm)	0.77	Virginia, Shenandoah Valley, Jul.–Aug. 1980	Stevens et al. (1984)
810. (>2.5 μm)	0.76	Virginia, Shenandoah Valley, Jul.–Aug. 1980	Stevens et al. (1984)
21. (<1 μm)	0.5 (Fe)	China, Great Wall, Apr. 1980	Winchester et al. (1981)
43.	0.90	Greenland, Kap Harald Moltke, Jun.–Aug. 1974	Flyger & Heidam (1978)
140.	1.3	Greenland, Thule, Feb.–Apr. 1978	Heidam (1981)
38.	0.82	Greenland, Prins Christianssund, Jan.–Feb. 1978	Heidam (1981)
200. (<2.5 μm)	1.2	USSR, Abastumani Mts., Jul. 1979	Stevens et al. (1983)
940. (>2.5 μm)	0.76	USSR, Abastumani Mts., Jul. 1979	Stevens et al. (1983)

Silver (Ag)			
0.14	1700.	Montana, Glacier Nat. Park, Aug. 1981	Davidson et al. (1985)
0.022	370.	Tennessee, Smoky Mts. Nat. Park, Oct. 1979	Davidson et al. (1985)
0.12	1300.	Washington, Olympic Nat. Park, Jul.–Aug. 1980	Davidson et al. (1985)
~0.001		Antarctica, Austral Winters 1975 & 1976	Cunningham & Zoller (1981)
<0.003		Antarctica, Austral Summers 1971, 1975, 1976, & 1978	Cunningham & Zoller (1981)
0.0040	83.	Enewetak, Apr.–May 1979	Duce et al. (1983)
0.0048	1200.	Enewetak, Jun.–Aug. 1979	Duce et al. (1983)
0.002		Greenland, Dye 3, Jun.–Aug. 1979	Davidson et al. (1981b)
Sodium (Na)			
910.	120.	Alaska, Barrow, Dec. 1976–Apr. 1977	Rahn (1977)
280.	0.81	Arizona, rural area, Jan.–Dec. 1974	Moyers et al. (1977)
3000.	1500.	Hawaii, Coast	Gordon et al. (1978)
110.	7.8	Hawaii, Mauna Loa, upslope wind, Jan.–Dec. 1980	Gordon et al. (1978)
10.	3.9	Hawaii, Mauna Loa, downslope wind, no dust, Jan.–Dec. 1980	Gordon et al. (1978)
24.	1.1	Hawaii, Mauna Loa, downslope wind, Asian dust, Jan.–Dec. 1980	Gordon et al. (1978)
90.	0.34	Montana, Colstrip, May–Sep. 1975	Crecelius et al. (1980)
360.	14.	Montana, Glacier Nat. Park, Aug. 1981	Davidson et al. (1985)

Table 7.1. *(Continued)*

Airborne Concentration (ng/m^3)	Crustal Enrichment Factor	Sampling Location	Reference
80.	1.0 (Sc)	New York, Whiteface Mtn., unpolluted air, Summer 1975	Parekh & Husain (1981)
96.	2.4 (Sc)	New York, Whiteface Mtn., unpolluted air, Summer 1977	Parekh & Husain (1981)
30.	1.5	Tennessee, Smoky Mts. Nat. Park, Oct. 1979	Davidson et al. (1985)
260.	8.2	Washington, Olympic Nat. Park, Jul.–Aug. 1980	Davidson et al. (1985)
40.		Antarctica, Austral Winters 1975 & 1976	Cunningham & Zoller (1981)
5.1	22.	Antarctica, Austral Summers 1971, 1975, 1976, & 1978	Cunningham & Zoller (1981)
2000.	100.	Atlantic Ocean, East	Duce et al. (1976)
2000.	14.	Bermuda, Summer 1975	Duce et al. (1976)
36.	0.85	Bolivia, Chacaltaya Mtn., Feb.–May 1976	Adams et al. (1977)
5600.	270.	Enewetak, Apr.–May 1979	Duce et al. (1983)
4000.	3600.	Enewetak, Jun.–Aug. 1979	Duce et al. (1983)
130.	14.	Greenland, Thule, Feb.–Apr. 1978	Heidam (1981)
790.	200.	Greenland, Prins Christianssund, Jan.–Feb. 1978	Heidam (1981)
400.	12.	Norway, Birkenes, Spring 1973	Semb (1978)
480.	20.	Norway, Birkenes, Autumn 1973	Semb (1978)
310.	16.	Norway, Skoganvarre, Spring 1973	Semb (1978)
820	66.	Spitsbergen, Dec. 1973–May 1974	Larssen (1977)

Strontium (Sr)			
3.0	0.56	Arizona, rural area, Jan.–Dec. 1974	Moyers et al. (1977)
1.8		California, High Sierra, Jun. & Aug. 1976, Aug. 1977	Elias & Davidson (1980)
<0.15		Antarctica, Austral Winters 1975 & 1976	Cunningham & Zoller (1981)
0.031	8.5	Antarctica, Austral Winters 1971, 1975, 1976, & 1978	Cunningham & Zoller (1981)
1.1	0.92	Bolivia, Chacaltaya Mtn., Feb.–May 1976	Adams et al. (1977)
4.0		Enewetak, Apr.–May 1979	Settle & Patterson (1982)
2.2		Enewetak, Jul.–Aug. 1979	Settle & Patterson (1982)
0.08	1.3	Greenland, Kap Harald Moltke, Jun.–Aug. 1974	Flyger & Heidam (1978)
0.31	2.1	Greenland, Thule, Feb.–Apr. 1978	Heidam (1981)
0.52	8.5	Greenland, Prins Christianssund, Jan.–Feb. 1978	Heidam (1981)
Sulfur (S)			
1200.	320.	Arizona, rural area Jan.–Dec. 1974	Moyers et al. (1977)
230.		Colorado, Squaw Mtn., Apr. 1976	Winchester et al. (1979)
270.		Florida, Apalachicola Nat. Forest, May–Jul. 1973	Johansson et al. (1976)
560.	710.	Michigan, Seney Nat. Wildlife Refuge, Jun., 1979	Alkezweeny et al. (1982)
550.	190.	Montana, Colstrip, May–Sep. 1975	Crecelius et al. (1980)
310.		New Hampshire, Hubbard Brook Forest, Apr. 1976	Winchester et al. (1979)

Table 7.1. *(Continued)*

Airborne Concentration (ng/m^3)	Crustal Enrichment Factor	Sampling Location	Reference
220.		New Mexico, Jemez Mts., Apr. 1976	Winchester et al. (1979)
3700. (<2.5 μm)	59000.	Tennessee, Smoky Mts. Nat. Park, Sep. 1978	Stevens et al. (1980)
200. (>2.5 μm)	330.	Tennessee, Smoky Mts. Nat. Park, Sep. 1978	Stevens et al. (1980)
440.		Utah, Canyonlands, Jul. 1977–Dec. 1978	Flocchini et al. (1981)
510.		Utah, Henrieville, Jul. 1977–Sep. 1978	Flocchini et al. (1981)
380.		Utah, Zion Nat. Park, Oct. 1977–Dec. 1978	Flocchini et al. (1981)
4500. (<2.5 μm)	32000.	Virginia, Shenandoah Valley, Jul.–Aug. 1980	Stevens et al. (1984)
260. (>2.5 μm)	270.	Virginia, Shenandoah Valley, Jul.–Aug. 1980	Stevens et al. (1984)
29.		Antarctica, Austral Winters 1975 & 1976	Cunningham & Zoller (1981)
76.	30000.	Antarctica, Austral Summers 1971, 1975, 1976, 1978	Cunningham & Zoller (1981)
900.		Arctic (European), Apr.–May 1979 & Mar. 1981	Heintzenberg (1982)
90.		Arctic (European), Jul.–Sep. 1980	Heintzenberg (1982)
95.	210.	Bolivia, Chacaltaya Mtn., Feb.–May 1976	Adams et al. (1977)
310.	7700. (Fe)	China, Great Wall, Apr. 1980	Winchester et al. (1981)

490.		England, S.W. Coast, onshore wind, Winter 1971–1972	Barnes & Eggleton (1977)
31.		Greenland, Dye 3, Jun.–Aug. 1979	Davidson et al. (1981b)
27.	610.	Greenland, Kap Harald Moltke, Jun.–Aug. 1974	Flyger & Heidam (1978)
250.	2500.	Greenland, Thule, Feb.–Apr. 1978	Heidam (1981)
79.	1900.	Greenland, Prins Christianssund, Jan.–Feb. 1978	Heidam (1981)
50. (<1 μm)		South America ave. of 8 remote sites, Jul.–Aug. 1976 & Jan.–Mar. 1977	Lawson and Winchester (1979)
690.		Spitsbergen, Apr.–May 1979	Heitzenberg et al. (1981)
800.		Sweden, Sjoangen, annual ave.	Lannefors et al. (1983)
1600. (<2.5 μm)	10000.	USSR, Abastumani Mts., Jul. 1979	Stevens et al. (1984)
80. (>2.5 μm)	70.	USSR, Abastumani Mts., Jul. 1979	Stevens et al. (1984)
Tantalum (Ta)			
0		Hawaii, Mauna Loa, downslope wind, no dust, Jan.–Dec. 1980	Gordon et al. (1978)
0.0016	0.87	Hawaii, Mauna Loa, downslope wind, Asian dust, Jan.–Dec. 1980	Gordon et al. (1978)
0.015	0.66	Montana, Colstrip, May–Sep. 1975	Crecelius et al. (1980)
6.2×10^{-5}		Antarctica, Austral Winters 1975 & 1976	Cunningham & Zoller (1981)
7.2×10^{-5}	3.7	Antarctica, Austral Summers 1971, 1975, 1976, & 1978	Cunningham & Zoller (1981)
0.0012	0.65	Enewetak, Apr.–May 1979	Duce et al. (1983)

Table 7.1. *(Continued)*

Airborne Concentration (ng/m^3)	Crustal Enrichment Factor	Sampling Location	Reference
0.00012	0.86	Enewetak, Jun.–Aug. 1979	Duce et al. (1983)
Terbium (Tb)			
0.02	2.	Montana, Colstrip, May–Sep. 1975	Crecelius et al. (1980)
Thorium (Th)			
0.010	1.7	Hawaii, Mauna Loa, upslope wind, Jan.–Dec. 1980	Gordon et al. (1978)
0.0020	1.9	Hawaii, Mauna Loa, downslope wind, no dust, Jan.–Dec. 1980	Gordon et al. (1978)
0.017	1.9	Hawaii, Mauna Loa, downslope wind, Asian dust, Jan.–Dec. 1980	Gordon et al. (1978)
0.15	1.4	Montana, Colstrip, May–Sep. 1975	Crecelius et al. (1980)
5.0×10^{-5}		Antarctica, Austral Winters 1975 & 1976	Cunningham & Zoller (1981)
0.00016	1.7	Antarctica, Austral Summers 1971, 1975, 1976, & 1978	Cunningham & Zoller (1981)
0.016	2.0	Atlantic Ocean, East	Duce et al. (1976)
0.05	0.9	Bermuda, Summer 1975	Duce et al. (1976)
0.030	1.6	Bolivia, Chacaltaya Mtn., Feb.–May 1976	Adams et al. (1977)
0.017	1.9	Enewetak, Apr.–May 1979	Duce et al. (1983)

0.0011	1.6	Enewetak, Jun.–Aug. 1979	Duce et al. (1983)
Titanium (Ti)			
100.	1.2	Arizona, rural area, Jan.–Dec. 1974	Moyers et al. (1977)
20.		Florida, Apalachicola Nat. Forest, May–Jul. 1973	Johansson et al. (1976)
5.8	1.7	Hawaii, Mauna Loa, upslope wind, Jan.–Dec. 1980	Gordon et al. (1978)
1.0	1.6	Hawaii, Mauna Loa, downslope wind, no dust, Jan.–Dec. 1980	Gordon et al. (1978)
7.3	1.4	Hawaii, Mauna Loa, downslope wind, Asian dust, Jan.–Dec. 1980	Gordon et al. (1978)
8.7	0.51	Michigan, Seney Nat. Wildlife Refuge, Jun. 1979	Alkezweeny et al. (1982)
34.	0.53	Montana, Colstrip, May–Sep. 1975	Crecelius et al. (1980)
3.1	0.64	Tennessee, Smoky Mts. Nat. Park, Oct. 1979	Davidson et al. (1985)
<6.0 (<2.5 μm)		Tennessee, Smoky Mts. Nat. Park, Sep. 1978	Stevens et al. (1980)
18. (>2.5 μm)	1.3	Tennessee, Smoky Mts. Nat. Park, Sep. 1978	Stevens et al. (1980)
<10. (<2.5 μm)		Virginia, Shenandoah Valley, Jul.–Aug. 1980	Stevens et al. (1984)
17. (>2.5 μm)	0.79	Virginia, Shenandoah Valley, Jul.–Aug. 1980	Stevens et al. (1984)
4.7	0.61	Washington, Olympic Nat. Park, Jul.–Aug. 1980	Davidson et al. (1985)
0.18		Antarctica, Austral Winters 1975 & 1976	Cunningham & Zoller (1981)
0.11	2.0	Antarctica, Austral Summers 1971, 1975, 1976, & 1978	Cunningham & (1981)

Table 7.1. *(Continued)*

Airborne Concentration (ng/m^3)	Crustal Enrichment Factor	Sampling Location	Reference
16.	0.88	Bolivia, Chacaltaya Mtn., Feb.–May 1976	Adams et al. (1977)
<1.0		Greenland, Dye 3, Jun.–Aug. 1979	Davidson et al. (1981b)
0.85	0.88	Greenland, Kap Harald Moltke, Jun.–Aug. 1974	Flyger & Heidam (1978)
2.5	1.1	Greenland, Thule, Feb.–Apr. 1978	Heidam (1981)
1.0	1.1	Greenland, Prins Christianssund, Jan.–Feb. 1978	Heidam (1981)
8.6	1.1	Norway, Birkenes, Spring 1973	Semb (1978)
6.8	1.1	Norway, Birkenes, Autumn 1973	Semb (1978)
<5.2		Spitsbergen, Apr.–May 1979	Heintzenberg et al. (1981)
5.0		Sweden, Sjoangen, annual ave.	Lannefors et al. (1983)
<10. (<2.5 μm)		USSR, Abastumani Mts., Jul. 1979	Stevens et al. (1984)
9. (>2.5 μm)	0.4	USSR, Abastumani Mts., Jul. 1979	Stevens et al. (1984)
Tungsten (W)			
<0.005		Antarctica, Austral Winters 1975 & 1976	Cunningham & Zoller (1981)
0.0019	130.	Antarctica, Austral Summers 1971, 1975, 1976, & 1978	Cunningham & Zoller (1981)
0.43	170.	Bolivia, Chacaltaya Mtn., Feb.–May 1976	Adams et al. (1977)

Vanadium (V)			
0.62	14.	Alaska, Barrow, Dec. 1976–Apr. 1977	Rahn (1977)
0.14	12.	Hawaii, Coast	Gordon et al. (1978)
0.075	0.93	Hawaii, Mauna Loa, upslope wind, Jan.–Dec. 1980	Gordon et al. (1978)
0.014	0.95	Hawaii, Mauna Loa, downslope wind, no dust, Jan.–Dec. 1980	Gordon et al. (1978)
0.12	1.0	Hawaii, Mauna Loa, downslope wind, Asian dust, Jan.–Dec. 1980	Gordon et al. (1978)
0.75	1.8	Michigan, Seney Nat. Wildlife Refuge, Jun. 1979	Alkezweeny et al. (1982)
1.4	0.92	Montana, Colstrip, May–Sep. 1975	Crecelius et al. (1980)
<4. (<2.5 μm)		Tennessee, Smoky Mts. Nat. Park, Sep. 1978	Stevens et al. (1980)
<5. (>2.5 μm)		Tennessee, Smoky Mts. Nat. Park, Sep. 1978	Stevens et al. (1980)
<10. (<2.5 μm)		Virginia, Shenandoah Valley, Jul.–Aug. 1980	Stevens et al. (1984)
<20. (>2.5 μm)		Virginia, Shenandoah Valley, Jul.–Aug. 1980	Stevens et al. (1984)
~0.0009		Antarctica, Austral Winters 1975 & 1976	Cunningham & Zoller (1981)
0.0016	1.2	Antarctica, Austral Summers 1971, 1975, 1976, & 1978	Cunningham & Zoller (1981)
0.1		Arctic (Canadian), Late Winter 1980	Barrie et al. (1981)
0.15	1.3	Atlantic Ocean, East, Summer 1975	Duce et al. (1976)
1.0	11.	Bear Island, Nov. 1977–Mar. 1978	Rahn (1981)
1.5	1.8	Bermuda, Summer 1975	Duce et al. (1976)

Table 7.1. *(Continued)*

Airborne Concentration (ng/m^3)	Crustal Enrichment Factor	Sampling Location	Reference
0.23	0.92	Bolivia, Chacaltaya Mtn., Feb.–May 1976	Adams et al. (1977)
0.53 (<1 μm)	25. (Fe)	China, Great Wall, Apr. 1980	Winchester et al. (1981)
0.17	1.3	Enewetak, Apr.–May 1979	Duce et al. (1983)
0.042	5.5	Enewetak, Jun.–Aug. 1979	Duce et al. (1983)
0.12	5.2	Greenland, Kap Harald Moltke, Jun.–Aug. 1974	Flyger & Heidam (1978)
0.28	5.3	Greenland, Thule, Feb.–Apr. 1978	Heidam (1981)
2.2	12.	Norway, Birkenes, Spring 1973	Semb (1978)
2.6	19.	Norway, Birkenes, Autumn 1973	Semb (1978)
1.1	10.	Norway, Skoganvarre, Spring 1973	Semb (1978)
0.29		Spitsbergen, Apr.–May 1979	Heintzenberg et al. (1981)
0.70	9.9	Spitsbergen, Dec. 1973–Mar. 1974	Larssen (1977)
4.5		Sweden, Sjoangen, annual ave.	Lannefors et al. (1983)
Zinc (Zn)			
15.	650.	Alaska, Barrow, Dec. 1976–Apr. 1977	Rahn (1977)
110.	110.	Arizona, rural area, Jan.–Dec. 1974	Moyers et al. (1977)
7.2		Colorado, Squaw Mtn., Apr. 1976	Winchester et al. (1979)
16.		Florida, Apalahicola Nat. Forest, May–Jul. 1973	Johansson et al. (1976)

0.8	130.	Hawaii, Coast	Gordon et al. (1978)
0.4	10.	Hawaii, Mauna Loa, upslope wind, Jan.–Dec. 1980	Gordon et al. (1978)
0.12	16.	Hawaii, Mauna Loa, downslope wind, no dust, Jan.–Dec. 1980	Gordon et al. (1978)
0.44	6.8	Hawaii, Mauna Loa, downslope wind, Asian dust, Jan.–Dec. 1980	Gordon et al. (1978)
6.2	30.	Michigan, Seney Nat. Wildlife Refuge, Jun. 1979	Alkezweeny et al. (1982)
Jan.–Dec. 1979	(1982)		
6.5	8.2	Montana, Colstrip, May–Sep. 1975	Crecelius et al. (1980)
9.0	45.	Montana, Glacier Nat. Park, Aug. 1981	Davidson et al. (1985)
11.		New Hampshire, Hubbard Brook Forest, Apr. 1976	Winchester et al. (1979)
3.2		New Mexico, Jemez Mts., Apr. 1976	Winchester et al. (1979)
20.	100. (Sc)	New York, Whiteface Mtn., unpolluted air, Summer 1975	Parekh & Husain (1981)
39.	390. (Sc)	New York, Whiteface Mtn., unpolluted air, Summer 1977	Parekh & Husain (1981)
4.3	71.	Tennessee, Smoky Mts. Nat. Park, Oct. 1979	Davidson et al. (1985)
9. ($<2.5\ \mu m$)	500.	Tennessee, Smoky Mts. Nat. Park, Sep. 1978	Stevens et al. (1980)
$<4.$ ($>2.5\ \mu m$)		Tennessee, Smoky Mts. Nat. Park, Sep. 1978	Stevens et al. (1980)
13.		Utah, Canyonlands, Jul. 1977–Dec. 1978	Flocchini et al. (1981)
17.		Utah, Henrieville, Jul. 1977–Sep. 1978	Flocchini et al. (1981)
13.		Utah, Zion Nat. Park, Oct. 1977–Dec. 1978	Flocchini et al. (1981)

Table 7.1. ***(Continued)***

Airborne Concentration (ng/m^3)	Crustal Enrichment Factor	Sampling Location	Reference
11. (<2.5 μm)	290.	Virginia, Shenandoah Valley, Jul.–Aug. 1980	Stevens et al. (1984)
6. (>2.5 μm)	20.	Virginia, Shenandoah Valley, Jul.–Aug. 1980	Stevens et al. (1984)
8.9	94.	Washington, Olympic Nat. Park, Jul.–Aug. 1980	Davidson et al. (1985)
7.8		Washington, Quillayute, onshore wind, Apr. 1974–Jun. 1975	Ludwick et al. (1977)
0.077		Antarctica, Austral Winters 1975 & 1976	Cunningham & Zoller (1981)
0.035	50.	Antarctica, Austral Summers 1971, 1975, 1976, & 1978	Cunningham & Zoller (1981)
1.0		Arctic (Canadian), Late Winter 1980	Barrie et al. (1981)
5.	80.	Atlantic Ocean, East	Duce et al. (1976)
6.	10.	Bermuda, Summer 1975	Duce et al. (1976)
2.4	8.1	Bolivia, Chacaltaya Mtn., Feb.–May 1976	Adams et al. (1977)

6.3 (<1 μm)	570. (Fe)	China, Great Wall, Apr. 1980	Winchester et al. (1981)
0.21	3.6	Enewetak, Apr.–May 1979	Duce et al. (1983)
0.12	33.	Enewetak, Jun.–Aug. 1979	Duce et al. (1983)
<1.3		Greenland, Dye 3, Jun.–Aug. 1979	Davidson et al. (1981b)
0.18	15.	Greenland, Kap Harald Moltke, Jun.–Aug. 1974	Flyger & Heidam (1978)
1.7	62.	Greenland, Thule, Feb.–Apr. 1978	Heidam (1981)
2.8	240.	Greenland, Prins Christianssund, Jan.–Feb. 1978	Heidam (1981)
29.	300.	Norway, Birkenes, Spring 1973	Semb (1978)
36.	500.	Norway, Birkenes, Autumn 1973	Semb (1978)
4.7	81.	Norway, Skoganvarre, Spring 1973	Semb (1978)
3.2		Spitsbergen, Apr.–May 1979	Heintzenberg et al. (1981)
12.	330.	Spitsbergen, Dec. 1973–Mar. 1974	Larssen (1977)
24.		Sweden, Sjoangen, annual ave.	Lannefors et al. (1983)
15. (<2.5 μm)	350.	USSR, Abastumani Mts., Jul. 1979	Stevens et al. (1984)
6. (>2.5 μm)	20.	USSR, Abastumani Mts., Jul. 1979	Stevens et al. (1984)

Many of the studies referenced in Table 7.1 include data sets for polluted as well as relatively unpolluted air; only data for the latter have been reported here. A few of the original references list separate data for Whatman cellulose filters as well as for one or more types of membrane filters. Only the cellulose filter data have been reported for such studies, which generally represent a larger compilation than data for the membrane filters.

3. DISCUSSION

The airborne concentrations in Table 7.1 show wide ranges of values. This reflects the variety of remote sites represented, which are influenced to varying degrees by anthropogenic activities, crustal erosion, sea spray, and other processes. The smallest concentrations for each element are generally those reported by Cunningham and Zoller (1981) for the South Pole (in the case of Pb, reported by Maenhaut et al., 1979), consistent with the great distance from this site to sources of aerosol. Many of the elements have, in general, smaller concentrations in the Southern Hemisphere than in the Northern Hemisphere, due to smaller anthropogenic emission rates as well as less exposed land area.

Of particular interest are the airborne concentrations of lead, since the presence of this element in many remote areas has been shown to be influenced by anthropogenic activities. Patterson (1980) estimates that over 99% of global lead emissions, totaling some 400,000 t annually, is derived from anthropogenic sources. Nriagu (1978) estimates that over 96% is anthropogenic; in the mid-1970s, approximately 60% of this lead was emitted from combustion of leaded gasoline, with the remainder from a variety of stationary sources. Recent decreases in the use of leaded gasoline may have altered these values.

The airborne lead data listed in Table 7.1 are consistent with the findings of these authors. Remote areas that are most likely to be influenced by automotive and industrial emissions, such as those only a moderate distance from populated areas in industrialized countries, show the greatest airborne lead concentrations. Most of the remote sites in the United States and Europe are in this category. Concentrations are typically in the range 10–100 ng/m^3. Remote continental locations in South America and Asia, and sampling sites over the Atlantic Ocean, show concentrations of $\approx$ 1–20 ng/m^3. Finally, lead levels in the Arctic and Antarctic, and over a remote area of the Pacific Ocean (Enewetak Atoll), are mostly below 1 ng/m^3 and can reach <0.1 ng/m^3.

Also of considerable interest are the data for sulfur. The values in Table 7.1 refer to particulate sulfur collected by filters and impactors, although the influence of SO_2 as an artifact may also have been significant in some of the studies. The data generally reflect total particulate sulfur concentration uncorrected for the influence of sea spray or soil dust. Airborne concentrations of SO_4^{2-} given in many of the original references have been converted to the equivalent concentrations of sulfur for the listings of Table 7.1.

The concentrations of sulfur range from 20 to 30 ng/m^3 during seasonal minima in the polar regions, to >1000 ng/m^3 in remote continental regions of the United States. Cunningham and Zoller (1981) have suggested that sulfate at the South Pole, which is the dominant aerosol species at this location, results mainly from nucleation of sulfur-containing gases emitted from a wide variety of natural and anthropogenic sources in more temperate latitudes. Virtually all of the nucleation and transport occurs in the troposphere; there is little evidence of stratospheric injection. The influence of sulfate derived from seaspray is also believed to be minor at the South Pole. In contrast, sulfate in remote areas of eastern United States results mainly from nucleation of sulfur-containing gases emitted from anthropogenic activities. For example, Stevens et al. (1980) have reported that fine particle aerosol in Smoky Mountains National Park during September 1978 was dominated by acid sulfates. High airborne concentrations of particulate sulfur in the northeastern United States have been attributed to nucleation of SO_2 emitted from the combustion of fossil fuels (Leaderer et al., 1979).

Besides lead and sulfur, variations in concentration of several orders of magnitude are seen for many of the other elements. For some elements such as Cd, Cu, Ni, and Zn, global anthropogenic emissions exceed emissions from natural sources (Nriagu, 1979; Nriagu, 1980; Nriagu and Davidson, 1980; Schmidt and Andren, 1980), and hence the variation may in part reflect varying anthropogenic influence. For most of the elements, however, the wide concentration ranges are probably indicative of different concentrations of soil dust, sea spray, and other natural aerosol.

Table 7.2 shows the geometric mean values and standard deviations of the crustal enrichments for each element as listed in Table 7.1. The sets of enrichments for chlorine and manganese each include a maximum value approximately an order of magnitude greater than the second largest enrichment in each set; the sets of enrichments for Ca, Mg, K, and Na each include two values that are substantially greater than the remaining enrichments. These excessively large values have not been included in calculating the geometric means. For the dichotomous sampler data in Table 7.1, new enrichment factors have been computed on the basis of total airborne concentrations in both size ranges before performing these calculations. Only aluminum-based enrichments in Table 7.1 have been used.

Table 7.2 also shows the geometric mean enrichments for remote crustal and marine areas as reported by Rahn (1976). No attempt has been made to divide the sites listed in Table 7.1 into these two categories, since there may not be clear distinctions for many of the sites characterized by both continental and marine influence. Overall, there is general agreement between the enrichments summarized in this chapter and those reported by Rahn (1976).

Figure 7.1 illustrates the geometric mean enrichment factors computed in the current study, as shown in the third column of Table 7.2. Values one standard deviation above and below the geometric mean are also indicated. The elements are arranged in order of increasing geometric mean value.

Table 7.2. Geometric Mean and Standard Deviation of Crustal Enrichment Factors, Based on the Reference Element Aluminum

	Based on Table 7.1			Taken from Rahn (1976)[a]	
Element	No. of Values	Geometric Mean	Standard Deviation	Geometric Mean (Remote Continental)	Geometric Mean (Remote Marine)
Aluminum	29	1.0	—	1.0	1.0
Antimony	12	211.	5.5	500.	2000.
Arsenic	13	190.	3.6	200.	1000.
Barium	8	2.5	2.5	~2.	
Bromine	18	1900.	7.0	300.	10^4
Cadmium	14	1100.	4.7	2000.	5000.
Calcium	25	1.8	2.0	1.5	8.
Cerium	8	2.0	1.5	2.	2.
Cesium	6	5.5	1.6	2.	20.
Chlorine	14	650.	13.	40.	10^4–10^5
Chromium	15	3.6	2.6	6. (N) 1. (S)	20. (N) 1. (S)
Cobalt	14	1.9	2.5	1.5 (N) 0.9 (S)	4. (N) 0.9 (S)
Copper	20	25.	3.6	20.	150.
Europium	4	1.0	1.6	1.5	1.5
Gold	2	750.	4.0		
Hafnium	4	1.9	1.5	~2.	~2.
Indium	2	49.	1.2	30.	150.
Iodine	10	5100.	5.9	300.	2000.
Iron	26	1.3	1.7	1.5 (N) 1. (S)	2.5 (N) 1. (S)
Lanthanam	6	2.8	2.2	2.	2.
Lead	22	320.	4.1	2000. (N) 80. (S)	2000. (N) 150. (S)
Lithium	1	30.			
Lutetium	1	1.4			
Magnesium	16	3.5	4.1	0.7	10.–100.
Manganese	23	1.5	2.1	2. (N) 1. (S)	3. (N) 1. (S)
Nickel	8	3.6	2.3	50.	100.
Potassium	19	2.3	1.9	1.5	6.
Rubidium	5	2.7	2.5	~1.5	~10.–100.
Samarium	3	1.3	1.3	1.5	1.5
Scandium	12	0.71	1.3	0.8	0.8
Selenium	16	3500.	3.7	1000.	6000.
Silicon	9	0.84	1.3	0.7	0.7
Silver	5	610.	3.5		
Sodium	20	11.	6.8	0.4	100.–1000.
Strontium	6	2.2	3.2	1.	5.–20.

Table 7.2. *(Continued)*

Element	Based on Table 7.1 No. of Values	Geometric Mean	Standard Deviation	Taken from Rahn (1976)[a] Geometric Mean (Remote Continental)	Geometric Mean (Remote Marine)
Sulfur	11	1300.	4.7		
Tantalum	5	1.0	2.1		
Terbium	1	2.0			
Thorium	10	1.6	1.3	1.5	1.5
Titanium	18	0.93	1.6	1.2	1.2
Tungsten	2	150.	1.2		
Vanadium	20	3.3	3.1	1.5	15. (N) 1.5 (S)
Zinc	27	50.	4.0	80.	400.

[a](N) and (S) refer to data collected in the Northern and Southern Hemispheres, respectively.

The figure summarizes characteristics of remote-area enrichment factors which have been discussed in many of the original references of Table 7.1. The elements Cu, Li, In, Zn, W, As, Sb, Pb, Ag, Cl, Au, Cd, S, Br, Se, and I have large geometric mean enrichments ranging from 25 to >5000; these are often termed the anomolously enriched elements (e.g., Duce et al., 1976). The elements Cs and Na have moderate values of 5.5 and 11, respectively. All of the other elements have enrichments near unity, and may be classified as predominantly soil-derived.

In general, the anomolously and moderately enriched elements have much greater standard deviations (mean value 4.6, excluding elements with only 1 or 2 data points) than the crustal elements (mean value 2.0, with similar exclusion). This most likely signifies the multiplicity of sources influencing elements in the former category. Table 7.1 shows that the enriched elements have values of EF_{crust} which vary over one to three orders of magnitude, while values for the crustal elements range over factors of 10 or less in many instances. These results would not be substantially different if the enrichment factors in Table 7.1 were based on bulk seawater rather than crustal composition, since many elements have similar marine and crustal abundances.

The difficulty in attempting to define characteristics of global background aerosol is apparent from the wide variations in concentration and enrichment from site to site. Besides spatial inhomogeneities, large temporal variations in the background aerosol have been illustrated by trace element analysis of filter samples collected sequentially (i.e. Barrie et al., 1981; Heidam, 1983) and by real-time measurements of condensation nuclei which show rapid fluctuations in concentration (Flyger and Heidam, 1978; Murphy and Bodhaine, 1980).

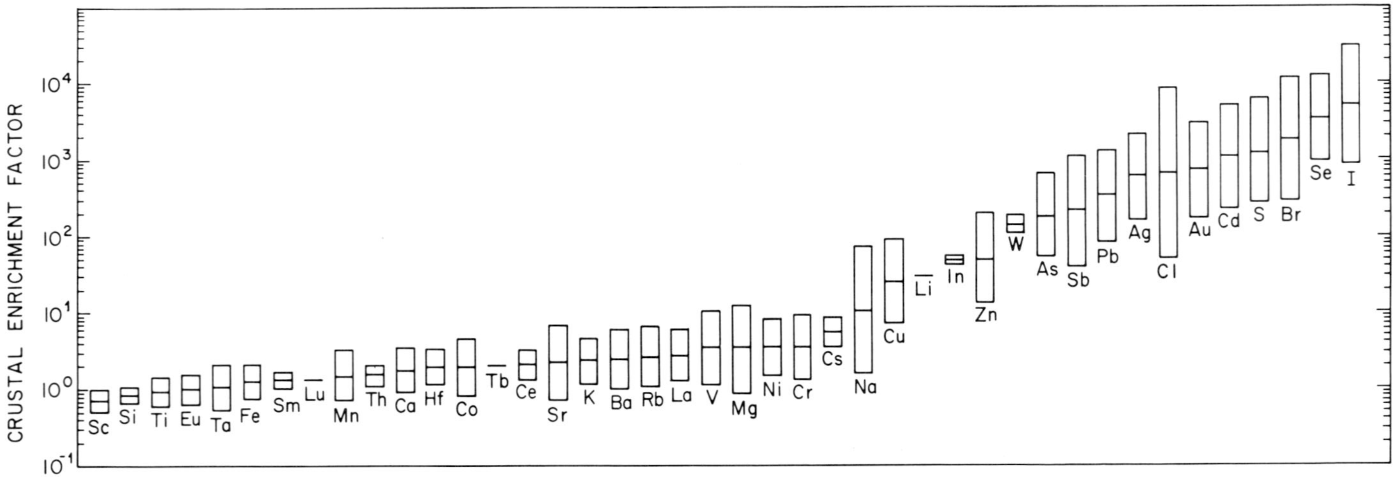

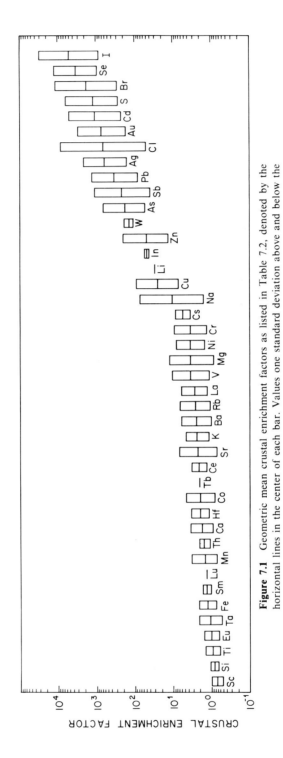

Figure 7.1 Geometric mean crustal enrichment factors as listed in Table 7.2, denoted by the horizontal lines in the center of each bar. Values one standard deviation above and below the geometric mean are shown by the vertical extent of the bar.

The complexity of the background aerosol thus makes estimation of global minimum concentrations of particular species difficult.

Several potential problems with the data presented in this chapter must be acknowledged. These problems include difficulties with the sampling and analytical work leading to the airborne concentrations in Table 7.1, as well as weaknesses in the use of enrichment factors.

Most trace element sampling has been conducted with filters composed of cellulose, glass fibers, Teflon, polycarbonate, or other material. In most cases, the efficiency of particle capture and retention is satisfactory (Spurny et al., 1969; Stafford and Ettinger, 1972), although elements with an appreciable vapor-phase component or predominantly associated with submicron particles may have low collection efficiencies for some filter types at commonly encountered air flow rates. Artifact formation of sulfate may occur on certain types of filter media (Pierson et al., 1980; Stevens et al., 1980), while artifact loss of Cl and I may pose additional problems (Klockow et al., 1979). For those analytical techniques requiring dissolution of the particles in liquid media, complete extraction of the particles from the filter may be difficult to achieve. Impurities in the filter media, or in the reagents used to extract the particles, may be significant relative to the small amounts of material sampled from the air. Contamination-controlled laboratory facilities are essential to minimize contamination during sample preparation (Patterson and Settle, 1976).

Numerous problems are inherent in trace element analytical techniques. Most of the data in Table 7.1 have been obtained using energy dispersive X-ray fluorescence spectrometry, atomic absorption spectrophotometry, proton-induced X-ray emission, or instrumental neutron activation analysis. These methods may suffer, to varying degrees, from lack of sensitivity needed for remote area samples, and from potential interferences from species other than those under investigation. The analytical results are also critically dependent upon the accuracy of the calibration standards. Contamination during analysis may occur if samples are improperly handled. These problems are discussed in many of the references listed in Table 7.1.

The use of enrichment factors to identify crustally derived elements has several potential problems, as summarized by Rahn (1976). Of primary concern is the variation in crustal composition from one site to another. Use of average composition values, such as those of Taylor (1964), may provide misleading results if the crustal composition near the aerosol sampling site is substantially different from the average composition data used to calculate enrichment.

A second related problem concerns the use of rock as opposed to soil for the reference material. Although most crustal aerosol is derived from erosion of soil, more complete data sets are available for the elemental composition of rock, and hence the latter reference material is generally used. Rahn (1976) reports that the chemical composition of crustal aerosol resembles a combination of characteristics of both rock and soil, implying that soil composition data might not be the obvious choice even if more complete soil data were available.

A third problem involves physical and chemical fractionation which may occur during formation of soil-derived aerosols. Physical fractionation refers to differences in the size distributions of these aerosols compared with the sizes of particles composing the parent crustal material. Several investigators have demonstrated that crustal aerosols found several meters above the ground are derived preferentially from the smallest particles present in the soil (Schutz and Jaenicke, 1974; Gillette, 1976). Because the chemical composition of most soils varies with particle size, such physical fractionation is likely to result in chemical fractionation. Thus the elemental composition of the aerosol may be different from that of the bulk soil. Miller et al. (1972) have confirmed evidence of chemical fractionation for aerosols derived from a variety of soils.

Finally, it must be recognized that the use of enrichment factors provides little information on sources of the anomolously enriched aerosols, which are likely to be those of greatest interest. Large enrichment factors may result from a variety of natural as well as anthropogenic processes, as discussed earlier. For identifying the types and locations of aerosol sources, more sophisticated techniques such as element ratios are needed (e.g., Rahn and Lowenthal, 1984).

4. SUMMARY

Airborne concentration data for 43 trace elements in remote areas throughout the world have been summarized. The data have been used to calculate crustal enrichment factors, based on aluminum, to assist with interpretation.

The elements Cu, Li, In, Zn, W, As, Sb, Pb, Ag, Cl, Au, Cd, S, Br, Se, and I have large geometric mean crustal enrichment factors. These elements are listed in order of increasing enrichments, which range from 25 for Cu to >5000 for I. The elements Cs and Na have moderate enrichments of 5.5 and 11, respectively. The remaining 24 elements have enrichments near unity, implying the earth's crust as a major source.

For all of the elements, the airborne concentrations vary greatly among the remote sites for which data are available. These variations reflect differences in characteristics of the sites: The influences of crustal erosion, sea spray, volcanism, biogenic emissions, and anthropogenic emissions vary considerably from site to site. Values of the crustal enrichment factor also show wide variations for the highly enriched elements. However, elements with geometric mean crustal enrichments near unity show little variation in the enrichment factor in spite of wide variations in airborne concentration; this suggests that the relative importance of crustal erosion, compared with noncrustal sources, varies much less than the absolute mass contribution from crustal erosion among the remote sites considered.

The data presented in this chapter suggest that the elemental composition of global background aerosol is complex, and may vary greatly over small distance and time scales. It may thus be difficult to arrive at meaningful estimates of global minimum airborne concentrations of particular chemical species.

ACKNOWLEDGMENTS

The authors gratefully acknowledge the U.S. National Man and Biosphere Program for supporting part of the preparation of this chapter. The manuscript was prepared by Gloria Blake.

REFERENCES

Adams, F., Dams, R., Guzman, L., and Winchester, J.W. (1977). "Background aerosol composition on Chacaltaya Mountain, Bolivia." *Atmos. Environ.* **11,** 629–634.

Alkezweeny, A.J., Laulainen, N.S., and Thorp, J.M. (1982). "Physical, chemical and optical characteristics of a clean air mass over Northern Michigan." *Atmos. Environ.* **16,** 2421–2430.

Barnes, R.A., and Eggleton, A.E.J. (1977). "The transport of atmospheric pollutants across the North Sea and English Channel." *Atmos. Environ.* **11,** 879–892.

Barrie, L.A., Hoff, R.M., and Daggupaty, S.M. (1981). "The influence of mid-latitudinal pollution sources on haze in the Canadian Arctic." *Atmos. Environ.* **15,** 1407–1419; data summarized by Heidam (1984).

Boutron, C. (1979). "Reduction of contamination problems in sampling of Antarctic snows for trace element analysis." *Geochim. Cosmochim. Acta* **106,** 127–130.

Crecelius, E.A., Lepel, E.A., Laul, J.C., Rancitelli, L.A., and McKeever, R.L. (1980). "Background air particulate chemistry near Colstrip, Montana." *Environ. Sci. Tech.* **14,** 422–428.

Cunningham, W.C., and Zoller, W.H. (1981). "The chemical composition of remote area aerosols." *J. Aerosol Sci.* **12,** 367–384.

Davidson, C.I., Grimm, T.A., and Nasta, M.A. (1981a). "Airborne lead and other elements derived from local fires in the Himalayas." *Science* **214,** 1344–1346.

Davidson, C.I., Chu, L., Grimm, T.C., Nasta, M.A., and Qamoos, M.P. (1981b). "Wet and dry deposition of trace elements onto the Greenland ice sheet." *Atmos. Environ.* **15,** 1429–1437.

Davidson, C.I., Wiersma, G.B., Goold, W.D., Mathison, T.P., Brown, K.W., and Reilly, M.T. (1985). "Airborne trace elements in Great Smoky Mountains, Olympic, and Glacier National Parks." *Environ. Sci. Technol.* **19,** 27–35.

Duce, R.A., Hoffman, G.L., and Zoller, W.H. (1975). "Atmospheric trace metals at remote Northern and Southern Hemisphere sites: Pollution or natural?" *Science* **198,** 59–61.

Duce, R.A., Ray, B.J., Hoffman, G.L., and Walsh, P.R. (1976). "Trace metal concentration as a function of particle size in marine aerosols from Bermuda." *Geophys. Res. Lett.* **3,** 339–342; some of the data also taken from Gordon et al. (1978), which includes a more complete listing of the Duce et al. (1976) data.

Duce, R.A., Arimoto, R., Ray, B.J., Unni, C.K., and Harder, P.J. (1983). "Atmospheric trace elements at Enewetak Atoll: 1, Concentrations, sources, and temporal variability." *J. Geophys. Res.* **88,** 5321–5342.

Elias, R.W., and Davidson, C.I. (1980). "Mechanisms of trace element deposition from the free atmosphere to surfaces in a remote High Sierra canyon." *Atmos. Environ.* **14,** 1427–1432.

Flocchini, R.G., Cahill, T.A., Ashbaugh, L.L., and Eldred, R.A. (1980). "Seasonal behavior of particulate matter at three rural Utah sites." *Atmos. Environ.* **15,** 315–320.

Flyger, H., and Heidam, N.Z. (1978). "Ground level measurements of the summer tropospheric aerosol in Northern Greenland." *J. Aerosol Sci.* **9,** 157–168.

Gillette, D.A. (1976) "Production of fine dust by wind erosion of soil: Effect of wind and soil texture." In *Atmosphere-surface exchange of particulate and gaseous pollutants (1974), proceedings of the symposium, Sep. 4–6, 1974, Richland, Wash.,* CONF-740921. National Technical Information Service, Springfield, Va., pp. 591–609.

Gordon, G.E., Moyers, J.L., Rahn, K.A., Gatz, D.F., Dzubay, T.G., Zoller, W.H., and Corrin, M.H. (1978). "Atmospheric trace elements: Cycles and measurements." Report on the National Science Foundation Atmospheric Chemistry Workshop. Panel on Trace Elements, National Center for Atmospheric Research, Boulder, Colo.

Grosch, M., Grosch, W., Wolf, G., and Kreyling, H. (1977). "Quantitative analysis of mercury, arsenic, and bromine in atmospheric aerosols." *Atmos. Environ.* **12,** 1235–1237.

Heidam, N.Z. (1981). "On the origin of the Arctic aerosol: A statistical approach." *Atmos. Environ.* **15,** 1421–1427.

Heidam, N.Z. (1983). *"Studies of the aerosol in the Greenland atmosphere."* Report MST LUFT-A73. Riso National Laboratory, Copenhagen, Denmark, 169 pp.

Heidam, N.Z. (1985) "Trace metals in the Arctic aerosol." In J.O. Nriagu and C.I. Davidson, Eds., *Toxic metals in the atmosphere,* Wiley, New York pp. 267–293.

Heintzenberg, J. (1982). "Size-segregated measurements of particulate elemental carbon and aerosol light absorption at remote Arctic locations." *Atmos. Environ.* **16,** 2461–2469.

Heintzenberg, J., Hansson, H.-C., and Lannefors, H. (1981). "The chemical composition of Arctic haze at Ny Alesund, Spitsbergen." *Tellus* **33,** 162–171.

Hoffman, E.J., Hoffman, G.L., and Duce, R.A. (1976). "Contamination of atmospheric particulate matter collected at remote shipboard and island locations." In *Accuracy in trace analysis: Sampling, sample handling, and analysis. Proceedings of the 7th IMR Symposium, Oct. 7–11, 1974, Gaithersburg, MD.* (*Nat. Bur. Stand. Spec. Publ.* **422,** 377–388).

Johansson, T.B., Van Grieken, R.E., and Winchester, J.W. (1976). "Elemental abundance variation with particle size in north Florida aerosols." *J. Geophys. Res.* **81,** 1039–1046.

Khalil, M.A.K., and Rasmussen, R.A. (1983). "Sources, sinks, and seasonal cycles of atmospheric methane." *J. Geophys. Res.* **88,** 5131–5144.

Klockow, D., Jablonski, B., and Niessner, R. (1979). "Possible artifacts in filter sampling of atmospheric sulfuric acid and acidic sulphates." *Atmos. Environ.* **13,** 1665–1676.

Lannefors, H., Hansson, H.-C., and Granat, L. (1983). "Background aerosol composition in southern Sweden—Fourteen micro and macro constituents measured in seven particle size intervals at one site during one year." *Atmos. Environ.* **17,** 87–101.

Larssen, S. (1977). Data collected in Spitsbergen, published by Gordon et al. (1978).

Lawson, D.R., and Winchester, J.W. (1979). "Atmospheric sulfur aerosol concentrations and characteristics from the South American continent." *Science* **205,** 1267–1269.

Leaderer, B.P., Holford, T.R., and Stolwijk, J.A.J. (1979). "Relationship between sulfate aerosol and visibility." *J. Air Poll. Control Assoc.* **29,** 154–157.

Ludwick, J.D., Fox, T.D., and Garcia, S.R. (1977). "Elemental concentrations of Northern Hemispheric air at Quillayute, Washington." *Atmos. Environ.* **11,** 1083–1087.

Maenhaut, W., Zoller, W.H., Duce, R.A., and Hoffman, G.L. (1979). "Concentration and size distribution of particulate trace elements in the South Polar atmosphere." *J. Geophys. Res.* **84,** 2421–2431.

Miller, M.S., Friedlander, S.K., and Hidy, G.M. (1972). "A chemical element balance for the Pasadena aerosol." *J. Coll. Interface Sci.* **39,** 165–176.

Moyers, J.L., Ranweiler, L.E., Hopf, S.B., and Korte, N.E. (1977). "Evaluation of particulate trace species in southwest desert atmosphere." *Environ. Sci. Tech.* **11,** 789–795.

Murphy, M.E., and Bodhaine, B.A. (1980). "The Barrow, Alaska automatic condensation nuclei counter and four wavelength nephelometer: Instrument details and four years of observations." National Oceanic and Atmospheric Administration, Boulder, Colo. Technical Memorandum ERL ARL-90. 101 pp.

Nriagu, J.O. (1978). "Lead in the atmosphere." In J.O. Nriagu, Ed., *The biogeochemistry of lead in the environment,* Vol. 1. Elsevier, Amsterdam, pp. 137–183.

Nriagu, J.O. (1979). "Copper in the atmosphere and precipitation." In J.O. Nriagu, Ed., *Copper in the environment,* Vol. 1. Wiley, New York, pp. 43–75.

Nriagu, J.O. (1980). "Cadmium in the atmosphere and precipitation." In J.O. Nriagu, Ed., *Cadmium in the environment,* Vol. 1. Wiley, New York pp. 71–114.

Nriagu, J.O. and Davidson, C.I. (1980). "Zinc in the atmosphere." In J.O. Nriagu, Ed., *Zinc in the environment,* Vol. 1. Wiley, New York, pp. 113–159.

Parekh, P.P., and Husain, L. (1981). "Trace element concentrations in summer aerosols at rural sites in New York State and their possible sources." *Atmos. Environ.* **15,** 1717–1725.

Patterson, C.C. (1980). "An alternative perspective—lead pollution in the human environment: Origin, extent, and significance." In National Academy of Sciences, *Lead in the human environment,* National Academy Press, Washington, D.C., pp. 265–349.

Patterson, C.C., and Settle, D. (1976). "The reduction of orders of magnitude errors in lead analyses of biological materials and natural waters by evaluating and controlling the extent and sources of industrial lead contamination introduced during sample collecting, handling, and analysis." In *Accuracy in trace analysis: sampling, sample handling and analysis. Proceedings of the 7th IMR Symposium, Oct. 7–11, 1974, Gaithersburg, MD. (Nat. Bur. Stand. Spec. Publ.* **422,** 321–351).

Pierson, W.R., Brachaczek, W.W., Korniski, T.J., Truex, T.J., and Butler, J.W. (1980). "Artifact formation of sulfate, nitrate, and hydrogen ion on backup filter: Allegheny Mountain experiment." *J. Air Poll. Control Assoc.* **30,** 30–34.

Rahn, K.A. (1976). *The chemical composition of the atmospheric aerosol.* Technical Report. Graduate School of Oceanography, University of Rhode Island, Kingston, R.I., 265 pp.

Rahn, K.A. (1977). Data collected in Barrow, Alaska, published by Gordon et al. (1978).

Rahn, K.A. (1981). "The Mn/V ratio as a tracer of large-scale sources of pollution aerosol for the Arctic." *Atmos. Environ.* **15,** 1457–1464.

Rahn, K.A., and Lowenthal, D.H. (1984). "Elemental tracers of distant regional pollution aerosols." *Science,* **223,** 132–139.

Schmidt, J.A., and Andren, A.W. (1980). "The atmospheric chemistry of nickel." In J.O. Nriagu, Ed., *Nickel in the environment.* Wiley, New York, pp. 93–135.

Schutz, L., and Jaenicke, R. (1974). "Particle number and mass distributions above 10^{-4} cm radius in sand and aerosol of the Sahara Desert." *J. Appl. Meteor.* **13,** 863–870.

Semb, A. (1978). "Deposition of trace elements from the atmosphere in Norway." SNSF project, Norway, FR 13/78, 28 pp. Norwegian Institute for Air Research, Lillestrøm, Norway.

Settle, D.M., and Patterson, C.C. (1982). "Magnitudes and sources of precipitation and dry deposition fluxes of industrial and natural leads to the North Pacific at Enewetak." *J. Geophys. Res.* **87,** 8857–8869.

Spurny, K.R., Lodge, J.P., Jr., Frank, E.R., and Sheesley, D.C. (1969). "Aerosol filtration by means of nuclepore filters." *Environ. Sci. Tech.* **3,** 453–464.

Stafford, R.G., and Ettinger, H.J. (1972). "Filter efficiency as a function of particle size and velocity." *Atmos. Environ.* **6,** 353–362.

Stevens, R.K., Dzubay, T.G., Shaw, R.W., Jr., McClenny, W.A., Lewis, C.W., and Wilson, W.E. (1980). "Characterization of the aerosol in the Great Smoky Mountains." *Environ. Sci. Tech.* **14,** 1491–1498.

Stevens, R.K., Dzubay, T.G., Lewis, C.W., and Shaw, R.W., Jr. (1984). "Source apportionment methods applied to the determination of the origin of ambient aerosols that affect visibility in forested areas." *Atmos. Environ.* **18,** 261–272.

Taylor, S.R. (1964). "Abundance of chemical elements in the continental crust: A new table." *Geochim. Cosmochim. Acta* **28,** 1273–1285.

Thrane, K.E. (1978). "Background levels in air of lead, cadmium, mercury and some chlorinated hydrocarbons measured in South Norway." *Atmos. Environ.* **12,** 1155–1161.

Winchester, J.W.,, Ferek, R.J., Lawson, D.R., Pilotte, J.O., and Thiemens, M.H. (1979). "Comparison of aerosol sulfur and crustal element concentrations in particle size fractions from continental U.S. locations." *Water, Air, Soil Pollut.* **12,** 431–440.

Winchester, J.W., Wei-xiu, L., Luxin, R., Ming-Xing, W., and Maenhaut, W. (1981). "Fine and coarse aerosol composition from a rural area in North China." *Atmos. Environ.* **15,** 933–937.

Zoller, W.H., Gladney, E.S., and Duce, R.A. (1974). "Atmospheric concentrations and sources of trace metals at the South Pole." *Science* **183,** 198–200.

8

TRACE METALS IN THE ARCTIC AEROSOL

Niels Z. Heidam

National Agency of Environmental Protection
Air Pollution Laboratory
Riso^e National Laboratory
Roskilde, Denmark

1. INTRODUCTION

The Arctic atmosphere has in recent years been found to contain a polluted aerosol, particularly during the winter. This pollution consists mainly of soot, sulfur, and a large number of trace metals. The Arctic winter aerosol may be responsible for the formation of Arctic haze, which degrades the usually very great visibility, and it may significantly affect both the atmospheric radiation budget and the acidity of precipitation.

Problems relating to the possible environmental effects, such as atmospheric fluxes, wet and dry deposition, and radiative effects of airborne pollutants in the Arctic have been discussed by several workers (Flyger et al., 1980; Shaw and Stamnes, 1980; Davidson et al., 1981; Rahn, 1982; Patterson et al., 1982).

The polluted aerosol is thought to originate from very long-range atmospheric transport from midlatitude source regions. During the winter the aerosol accumulates because of the sparsity of precipitation in the Arctic,

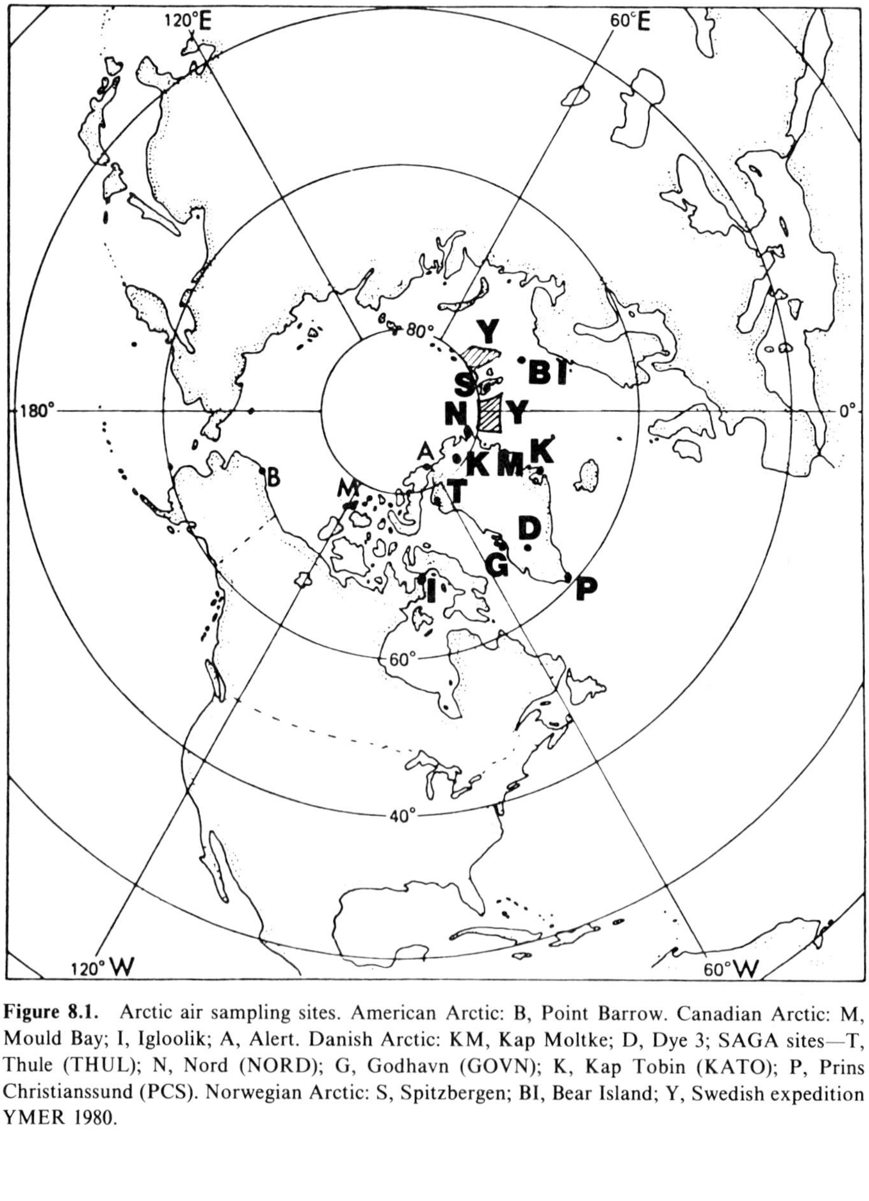

Figure 8.1. Arctic air sampling sites. American Arctic: B, Point Barrow. Canadian Arctic: M, Mould Bay; I, Igloolik; A, Alert. Danish Arctic: KM, Kap Moltke; D, Dye 3; SAGA sites—T, Thule (THUL); N, Nord (NORD); G, Godhavn (GOVN); K, Kap Tobin (KATO); P, Prins Christianssund (PCS). Norwegian Arctic: S, Spitzbergen; BI, Bear Island; Y, Swedish expedition YMER 1980.

which increases normal atmospheric residence time by almost an order of magnitude (Rahn and McCaffrey, 1980; Rahn, 1981a; Carlson, 1981).

This deplorable picture does, however, need some correction. First, the annual observed pollution levels are very low, often one or two orders of magnitude less than levels in rural areas in industrialized countries. Second, a certain, but perhaps diminishing, fraction of the aerosol has natural origins, such as eroded windblown dust or sea spray. To some extent the acidity of precipitation also has natural causes (Winkler, 1983).

A study of the trace metal contents in the Arctic aerosol may help to make these distinctions and clarify the picture.

The enhanced levels in the winter aerosol have been explained by Rahn et al. (1977) as an effect of the position of the polar front. The polar front is a zone of intensive mixing between temperate and polar air masses. The mixing processes in the Iceland and Atlantic Lows developing at the polar front are accompanied by strong precipitation and lead to intense scavenging. The polar front acts, therefore, as an effective barrier to meridional mixing and aerosol transport (Winkler, 1983). In winter the polar front lies south of populated and industrialized regions, which therefore become part of a northern air mass. In summer, on the other hand, the Arctic atmosphere is isolated effectively from lower latitudes by the polar front in its northern position, and the atmospheric circulation is in essence confined to the Arctic itself. In addition, precipitation scavenging is more pronounced in summer than in winter, so that aerosol residence times in the Arctic atmosphere are at a maximum during winter.

The overall effect is that the Arctic atmosphere behaves as a reservoir that becomes progressively dirtier, albeit at very low levels, throughout the winter and that it is cleaned during summer, when air pollution reaches a minimum.

Measurements at Point Barrow, Alaska (Rahn et al., 1977, 1980), in the Canadian Arctic (Barric et al., 1981), at Spitzbergen (Heintzenberg et al., 1981) and the surrounding ocean (Lannefors et al., 1983b)), and in Greenland (Heidam, 1981, 1984) have been shown to be in agreement with this reservoir model. The locations of these and other sampling sites are shown on the map in Figure 8.1.

2. TRACE METAL CONCENTRATIONS

The seasonal levels of trace metal concentrations in the Arctic aerosol are shown in Table 8.1, which contains results from a number of studies in various parts of the Arctic (see Fig. 8.1). As can be seen, there is in general very good agreement between results from different arctic locations, and the large differences between summer and winter levels are evident.

A comparison with measurements at a rural station in Sweden shows that in winter the arctic concentrations are typically a factor of 2 to 4 lower than in rural Sweden and that this difference increases considerably in summer (Lannefors et al., 1983a).

Table 8.1. Average Trace Metal Concentrations (in ng/m^3) in the Arctic Aerosol[a]

Atomic No./ Element	KM N. Greenland, Summer 1974[b]	D Greenland, Icecap, Summer 1979[c]	Y Arctic Ocean, Summer 1980[d]	S Spitzbergen, Late Winter 1979[e]	M,I Canadian Arctic, Late Winter 1980[f]	B Alaska, 1976–1977[g]
12/Mg		<0.35–1.7		51		
13/Al	14					30
19/K	3.8		2.6	32		
20/Ca	9.1		2.7	73		
22/Ti	0.85	<0.6–1.5	0.16	<5.2		
23/V	0.12			0.29	0.1	0.2–0.6
24/Cr	0.09			2.6		
25/Mn	0.21			1.5	0.5	1.1
26/Fe	6.4	10–30	1.1	64		
28/Ni	0.08		<0.06	0.7		
29/Cu			0.041	<1.9	0.3	
30/Zn	0.18	<0.85–2.1	0.14	3.2	1.0	
47/Ag		<0.001–0.004				
82/Pb	0.20	0.1–0.2	0.14	<4.9	1.0	

[a]Site designations refer to the map in Figure 8.1.
[b]From Flyger and Heidam (1978).
[c]From Davidson et al. (1981).
[d]From Lannefors et al. (1983b), fine fraction.
[e]From Heintzenberg et al. (1981).
[f]From Barrie et al. (1981).
[g]From Rahn (1981b).

Annual geometric means and standard deviations for a number of elements are listed in Table 8.2. These results were obtained over a year (1979–1980) at five stations in Greenland in a Danish Study of Aerosols in the Greenland Atmosphere (SAGA) (Heidam, 1983, 1984). The SAGA stations THUL, NORD, GOVN, KATO, and PCS are located as shown on the map in Figure 8.1. THUL should not be confused with the Thule air base at Dundas, about 100 km to the south. At these stations samples were taken continuously at a rate of two per week. Elemental contents were determined by PIXE.

In general, the levels on the western coast of Greenland fall off towards the north, GOVN displaying the highest and THUL the lowest annual levels. On the eastern coast the levels at NORD and KATO are rather similar, lying between THUL and GOVN. Notable exceptions are the metals Cr, Cu, Zn, and Pb, which are most abundant at NORD. It is noteworthy that the southern station PCS exhibits some of the lowest levels for these metals.

The polar pack ice is permanent at NORD and prevails over most of the year at THUL and KATO, so these stations are considered to be true arctic stations. The high levels of Cl at GOVN and PCS are, however, evidence of a strong marine influence at these sites.

The geometric standard deviations, which range from 2.0 to 3.5 in Table 8.2, indicate that the annual mean concentrations are not very representative of the levels as they occur throughout the year. To illustrate the seasonal variation, Table 8.3 shows the monthly geometric means at station KATO. For most of the elements maximum values occur in winter or late spring when they may be 10 or 20 times larger than the minimum summer values. Similar annual cycles are also found at other stations as illustrated by Figures 8.2 and 8.3, which show monthly geometric means for a selection of metals.

Historically, the observations at Point Barrow of the very large seasonal variation of V, a tracer for heavy fuel oil combustion, were the first indication that air pollution might also be a problem in the remote Arctic (Rahn et al., 1977). It is evident from Figure 8.3 that the problem concerns the whole Arctic and involves several other tracers, too.

3. CRUSTAL ENRICHMENTS

Enrichment factors are measures of the excess concentrations of constituents relative to a specific source. The definition of enrichment factors is based on the assumption that constituents from the source are found in the sample in the same proportion as they occur in the source. If the source is the earth's crust, then crustal enrichments are around 1, whereas components that derive from other sources will have high crustal enrichments. Low crustal enrichments (less than 1) signify depletion with respect to the crustal source.

Table 8.2. SAGA 1979–1980: Geometric Means (in ng/m^3) and Standard Deviations[a]

Atomic No./ Element	THUL[b]		NORD		GOVN		KATO		PCS[c]	
	Conc	Dev.	Conc	Dev.	Conc	Dev.	Conc	Dev.	Conc	Dev.
13/Al	1.91	2.42	8.39	3.18	18.2	2.97			6.39	2.42
14/Si	19.0	2.64	161	2.80	260	2.61	69.1	3.04		
16/S	93.7	3.00	178	3.27	205	1.76	111	2.41	154	1.78
17/Cl	21.5	3.22			504	3.80	91.6	5.94	1173	2.12
19/K	4.15	2.31	15.9	2.83	33.5	2.14	11.0	2.79	31.3	1.84
20/Ca	2.71	2.54	19.2	3.10	50.9	2.36	11.3	2.81	35.2	1.89
22/Ti	0.24	2.46	1.29	3.51	3.56	3.46	0.78	3.07	0.57	3.15
24/Cr	0.03	1.95	0.16	2.52	0.14	2.30			0.06	1.81
25/Mn	0.13	2.33	0.50	2.99	0.77	2.92	0.21	2.83	0.19	2.86
26/Fe	4.16	2.74	24.2	3.36	53.0	3.18	12.5	3.09	14.0	2.68
28/Ni					0.17	2.12				
29/Cu	0.07	2.63	0.30	3.91	0.21	2.27	0.12	2.48	0.09	2.58
30/Zn	0.68	3.08	1.97	3.69	1.52	2.81	0.59	3.10	1.41	2.17
35/Br	2.02	3.20	3.73	3.23	4.05	2.26	1.02	3.05	4.95	1.98
38/Sr	0.06	2.66	0.25	2.95	1.00	2.17	0.20	3.04	0.85	1.68
82/Pb	1.82	2.57	4.98	3.50	1.28	2.48	0.80	2.44	0.92	3.87
Total suspended particles	1177	2.18							4841	1.68

[a] Total number of samples: THUL, 51; NORD, 99; GOVN, 98; KATO, 76; PCS, 19 (Heidam, 1984).

[b] THUL: February–August 1980 only.

[c] PCS: September 1979–January 1980 only.

Table 8.3. Concentrations of Aerosol Constitutents at 901 KATO, 1979–1980, Monthly Geometric Means (in ng/m^3)

Atomic No./Element	Sept.	Oct.	Nov.	Dec.	Jan.	Feb.	Mar.	Apr.	May	June	July
Av no. of results above detection limits	7.92	7.06	8.58	4.43	7.82	7.72	2.00	1.00	8.52	8.50	6.35
14 Si	215	58.9	58.5	153	106	102	236	33.9	72.0	26.3	33.3
16/S	145	72.1	134	105	117	205	611	106	103	63.3	82.4
17/Cl	1389	124	114	107	192	262		76.1	34.0	37.8	80.9
19/K	46.9	9.85	12.4	13.6	11.9	18.9	25.7	16.2	8.71	4.39	6.81
20/Ca	41.4	10.3	13.7	12.0	12.6	18.8	26.5	9.59	8.91	5.21	7.35
22/Ti	3.95	1.07	0.99	0.66	0.65	1.13	1.12	0.72	0.82	0.34	0.44
25 Mn	0.66	0.19	0.20	0.19	0.24	0.46	0.59	0.20	0.22	0.08	0.13
26/Fe	61.6	10.4	13.7	11.0	12.3	21.5	21.9	13.3	12.4	4.91	4.32
29/Cu	0.17	0.15	0.08	0.09	0.16	0.25	0.50	0.13	0.09	0.07	0.07
30/Zn	0.73	0.88	0.56	1.26	0.56	2.09	3.79	0.64	0.34	0.17	0.35
35/Br	3.18	0.71	0.99	0.97	0.69	2.91	5.67	3.58	1.41	0.42	0.69
38/Sr	0.82	0.20	0.25	0.30	0.22	0.37	0.32	0.18	0.14	0.10	0.21
82/Pb	0.51	0.60	0.76	0.84	1.21	1.62	3.21	1.27	0.95	0.78	0.37

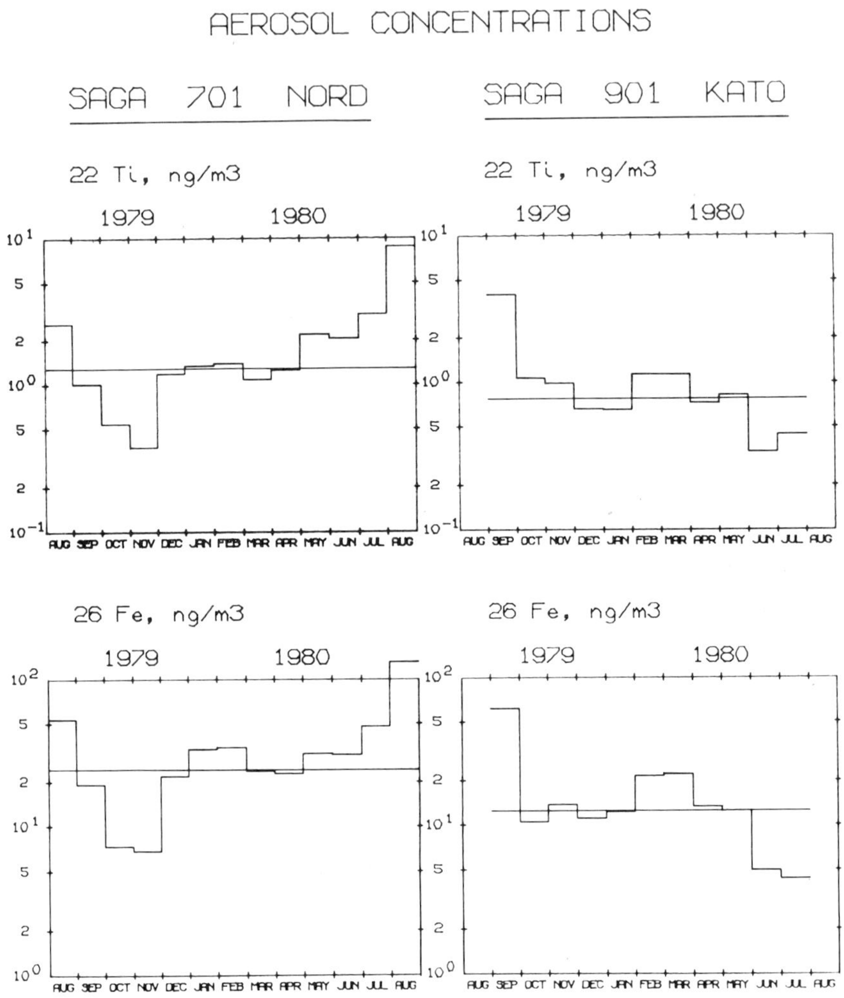

Figure 8.2. Monthly geometric mean concentrations of metals in the aerosol in Greenland.

For elemental aerosol concentrations c_k the crustal enrichment factor of element k is defined as

$$E_k = \frac{c_k / c_0}{e_k / e_0}$$

where subscript zero designates a reference element specific to the earth's crust and e is the abundance in the globally averaged crust (Mason, 1966).

As demonstrated by Schütz and Rahn (1982) the basic assumption about enrichment factors (i.e., that the ratios of crustal abundancies are preserved from source to sampling site) is well founded for particles less than 10 μm in

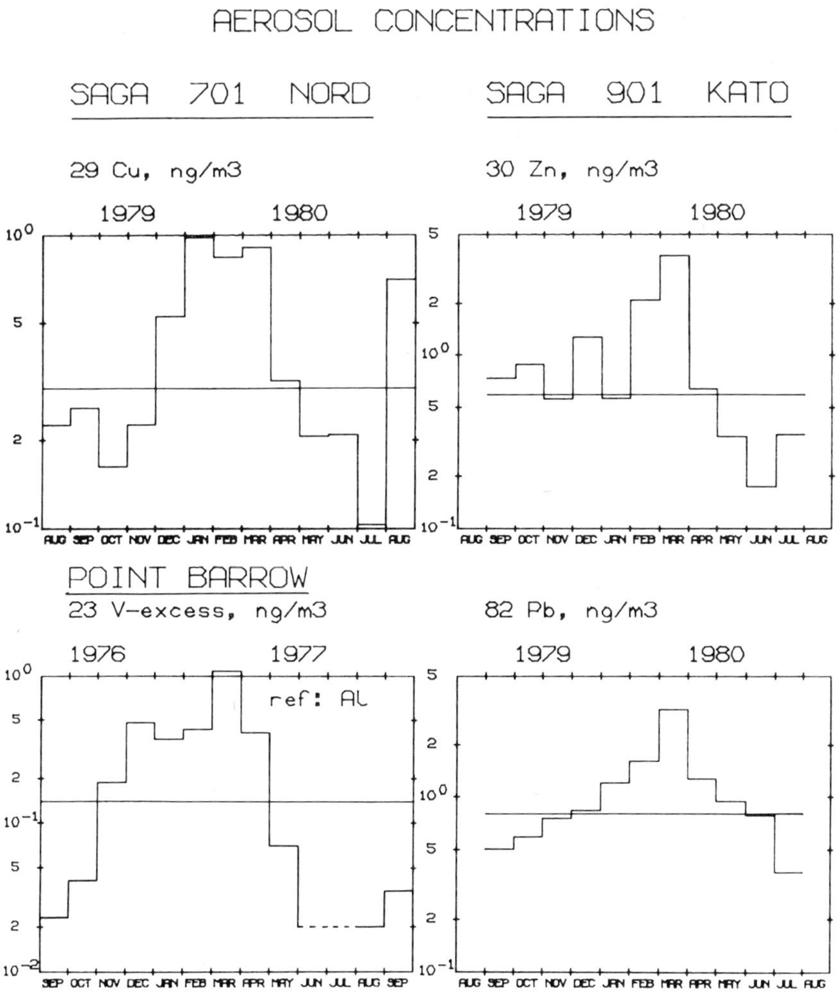

Figure 8.3. Monthly geometric mean concentrations of metals in the aerosols in Greenland and Alaska (the latter after Rahn and McCaffrey, 1980).

diameter. At remote locations long-range transported particles will most likely be in this size range.

Results on crustal enrichments in the arctic aerosol have been published by Flyger and Heidam (1978) for a summer period at the KHM site, by Rahn et al. (1977, 1980) for the seasonal variation of excess V at the Barrow site, and by Heidam (1983) for a range of elements measured over a year at the SAGA sites in Greenland. The crustal reference element used in these three studies were Si, Al, and Ti, respectively. They are all excellent reference elements (Rahn, 1976).

For the extensive results from the SAGA study, which will be used to illustrate the main points, Ti was preferred over the more universally used Al for reasons of detectability by PIXE analysis. The total geometric means and standard deviations for Ti enrichments relative to the average global crust are shown in Table 8.4.

In general the enrichments at NORD and GOVN are smaller than at THUL and KATO. The significance of this geographically diagonal grouping of Greenland stations, if any, is not understood. The variations among stations tend to be larger for the highly enriched elements. Apart from that the most characteristic feature is the large differences in enrichments among groups of aerosol constituents.

The crustal elements Al, Si, K, Ca, Mn, and Fe, which are all metals or semimetals, are characterized by low enrichment factors from 1.0 to 3.0 at all stations, except that Al appears to be somewhat depleted. That may, however, have been an effect of fairly high detection limits. The metals Ni and Sr also fall in this group of nonenriched elements.

In the group of metals consisting of Cr, Cu, and Zn there are considerable enrichments, 5–180, so they do not appear to be crustally derived.

Finally, there is a group of highly enriched elements S, Cl, Br, and Pb. The enrichments range from 120 to 15,000, and the geographic spread is considerable in this group.

Similar groupings were found by Flyger and Heidam (1978), and they have also been observed in rural Sweden by Lannefors et al. (1983a), who, incidentally, have also used Ti as a crustal reference.

A roughly similar grouping of elements is obtained from Table 8.4 if they are arranged according to the geometric standard deviations which measure the scale of relative variations over the year.

The nonenriched metals have low standard deviations from 1.25 to 2.5, whereas the enriched metals and the highly enriched elements vary with factors from 2.5 to 6 with a tendency for larger variations for the larger enrichment factors.

To illustrate the annual variations the monthly geometric mean enrichment factors at the KATO site are shown in Table 8.5. As for the concentrations in Table 8.3, maximum values tend to occur in winter or late spring, but as expected it is only for the enriched elements that variations remain large.

The annual variations are also shown in Figures 8.4 and 8.5 for a selection of metals at various Arctic sites.

The enrichments of crustal metals in Figure 8.4 show that to a large extent they are just random residuals as is to be expected from excess concentrations. In some cases, notably Mn, there is however still a systematic behavior left. That indicates that the description by means of enrichment factors is not exhaustive, either because there are additional noncrustal sources or because the reference element Ti is not a complete descriptor of the crustal sources for the Arctic aerosol.

The annual courses for enriched metals are shown in Figure 8.5 and are, as expected, quite similar to the concentrations (see Figure 8.3). For both Cu

Table 8.4. Annual Geometric Means and Standard Deviations for Ti-Enrichments, SAGA, 1979–1980

	THUL		NORD		GOVN		KATO	
Atomic No./Element	Mean	Dev.	Mean	Dev.	Mean	Dev.	Mean	Dev.
13/Al	0.43	1.57	0.35	1.34	0.28	1.78		
14/Si	1.27	1.66	1.99	1.73	1.16	2.40	1.41	2.24
16/S	6649	2.66	2349	5.96	979	4.30	2418	2.80
17/Cl	3052	4.12			4799	4.33	3985	3.92
19/K	2.96	1.86	2.10	2.01	1.60	2.37	2.40	1.87
20/Ca	1.38	2.26	1.81	1.88	1.74	1.79	1.76	1.81
24/Cr	5.91	2.08	5.58	2.62	1.77	1.92		
25/Mn	2.59	1.73	1.80	2.07	1.00	1.43	1.22	1.60
26/Fe	1.53	1.41	1.66	1.28	1.31	1.25	1.41	1.25
28/Ni					2.74	2.31		
29/Cu	22.6	2.47	18.9	4.96	4.68	2.50	11.9	2.91
30/Zn	180	3.09	96.3	5.33	26.9	4.31	47.5	3.79
35/Br	14905	2.93	5100	5.58	2005	4.21	2318	2.64
38/Sr	3.01	2.20	2.27	3.12	3.30	2.17	3.07	2.06
82/Pb	2586	2.45	1310	5.70	121	3.77	347	3.60

Table 8.5. Ti-Enrichment Factors at 901 KATO, 1979–1980, Monthly Geometric Means

Atomic No./ Element	1979				1980						
	Sept.	Oct.	Nov.	Dec.	Jan.	Feb.	Mar.	Apr.	May	June	July
No. of samples	8.00	8.00	9.00	5.00	9.00	8.00	2.00	1.00	9.00	9.00	8.00
14/Si	0.86	0.96	1.11	3.69	2.59	1.43	3.34	0.75	1.40	1.23	1.24
16/S	621	1578	2716	2693	3049	3066	9197	2484	2145	3185	4311
17/Cl	11893	5070	3461	4471	3203	4615	3475	3567	1687	2872	5040
19/K	2.02	1.82	2.52	3.50	2.50	2.84	3.88	3.81	1.81	2.22	2.91
20/Ca	1.27	1.35	1.99	2.21	1.88	2.02	2.86	1.61	1.32	1.88	2.23
25/Mn	0.78	0.90	1.11	1.35	1.68	1.87	2.41	1.25	1.27	1.10	1.16
26/Fe	1.37	1.31	1.44	1.47	1.66	1.67	1.72	1.62	1.34	1.28	1.18
29/Cu	3.39	14.3	7.75	18.2	16.3	17.5	35.4	14.7	8.77	16.3	15.9
30/Zn	11.6	79.0	41.7	120	54.1	116	211	55.3	26.0	29.5	61.3
35/Br	1414	1470	2075	2180	1862	4516	8874	8726	3039	2223	2172
38/Sr	2.69	2.58	3.54	4.34	3.04	3.78	3.38	2.91	1.99	2.99	3.87
82/Pb	43.4	267	307	431	550	482	965	595	392	789	391

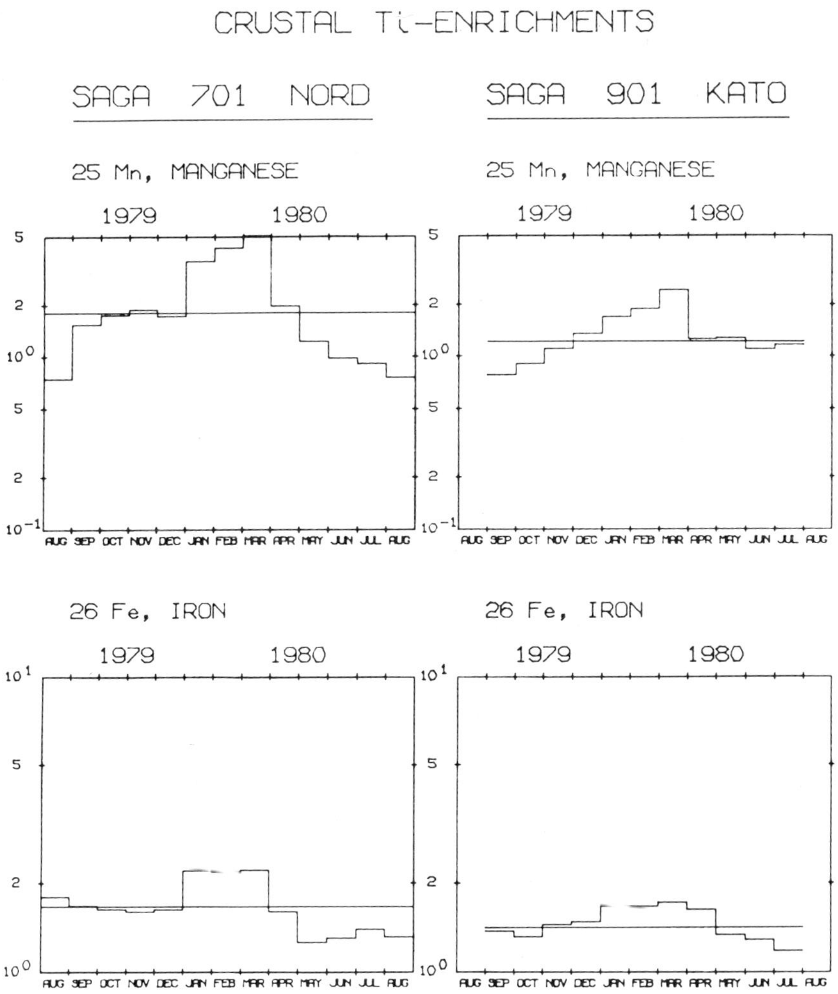

Figure 8.4. Monthly geometric mean enrichments of metals in the aerosol in Greenland.

and Zn enrichments build up in northeastern Greenland during the winter to peak at highly enriched levels in March, whereupon they decrease rapidly in April and May. The low summer levels are, however, still considerably enriched.

The enrichment of Pb, the tracer for motor vehicle exhausts, exhibits a similar, but even more dramatic, behavior in North Greenland. In the autumn the enrichment increases rapidly by a factor close to 50 and remains at highly enriched levels until spring, when it decreases to low, but still enriched, levels. That behavior is similar to the observations for V by Rahn et al. (1977, 1980) at Barrow in 1976–1977, also shown in Figure 8.5.

For these enriched metals the annual courses of the enrichment factors are

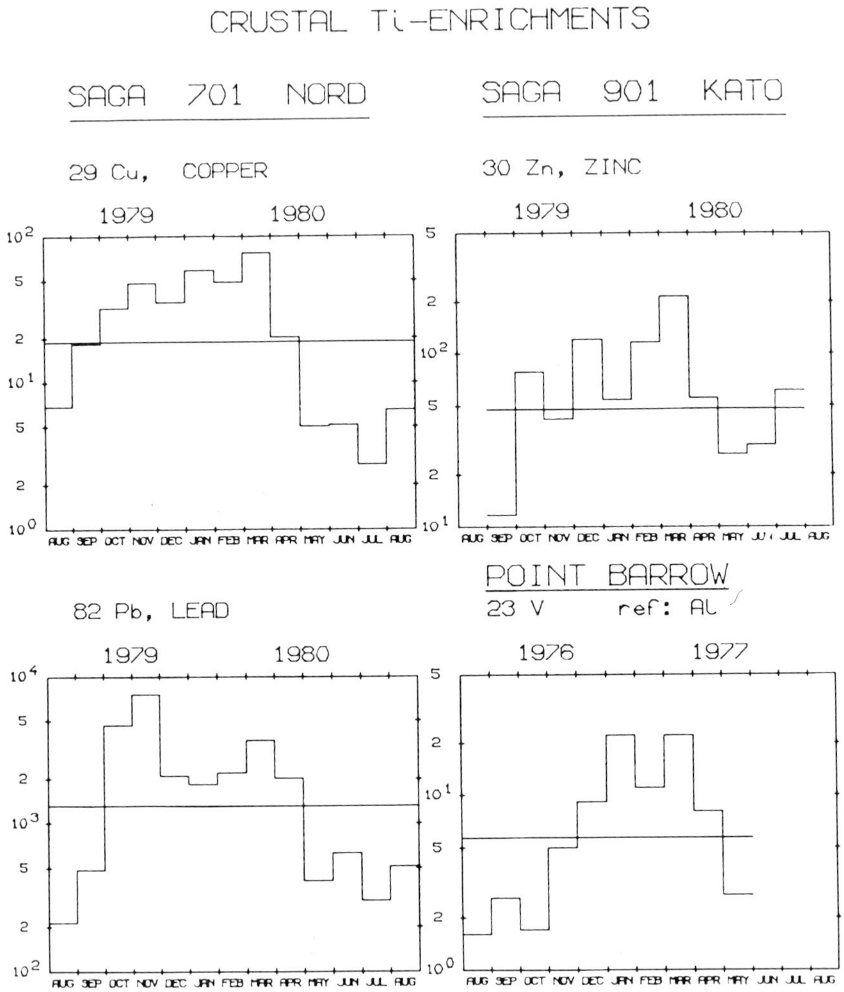

Figure 8.5. Monthly geometric mean enrichments of metals in the aerosols in Greenland and Alaska (the latter after Rahn and McCaffrey, 1980).

in accordance with the reservoir model of the Arctic atmosphere. Thus the excess metal contents, tracers of far larger aerosol masses, appear to be of very distant origin.

4. SOURCE-RELATED AEROSOL COMPONENTS

The grouping of trace metals in the Arctic aerosol according to the magnitude of their enrichments indicates that the aerosol may be composed of a few components that can be traced to various source types. In order to investigate this question it is necessary to study the correlations among the trace elements. That has been done for the Greenland data by factor analysis (Harman, 1976; Heidam, 1982, 1983, 1984).

Factor analysis is a statistical technique in which the correlation matrix is diagonalized and the ensuing eigenvector solution is used to build a linear model of the variations of the variables. By this method the large set of intercorrelated variables are replaced by a much smaller set of uncorrelated stochastic variables—the factors. The factors are in principle unobservable with zero mean and unit variance. The couplings between factors and variables are equal to the correlation coefficients, and their varying magnitudes often permit a physically meaningful interpretation of the factors. The factor model also allows the unobservable factors to be estimated by, for example, least squares criteria.

As an example of the starting point for such an analysis, the correlation matrix for 13 trace elements detected regularly at KATO is shown in Table 8.6. The aerosol constituents are to a good approximation log–normally distributed, and therefore, the logarithms of the concentrations are used as the basic variables. This matrix shows a widespread significant coupling between the elements, only Cl and Pb seem somewhat decoupled. That can be seen from the last row of percentages of significant nonzero correlation coefficients.

The factor models for the composition of the aerosol at the SAGA sites are summarized in Table 8.7. For each station and each factor the more important elements loaded on that factor are listed. The elements have been divided into three categories, namely, prominent, weak, and minor members, according to the strength of coupling to the factors. Minor members are not listed, but in cases where they collectively are of importance, the factor has been labeled as a minor member collector (MMC). As the appearance of minor members in a factor can be ascribed partly to the stochastic nature of the data, an MMC factor is not only of a physical, but also to some degree of a purely statistical nature. The total variance of the standardized variables is equal to the number of variables (n), and the sum of the variances contributed by the factors is equal to the total variance explained by the model. The percentage of total variance explained is also listed in Table 8.7.

Table 8.7 shows that the atmospheric aerosols in Greenland can be described by the same four or five different and statistically independent components. In view of the separation of the sampling locations by distances of up to 1600 km, that is a remarkable result. It means, at least on an annual basis, that the tropospheric aerosol in Greenland has a uniform composition except for minor local variations. In view of the general similarity of results of aerosol composition measurements in the Arctic, this concept of one aerosol may possibly be extended to the whole Arctic region.

By combining the knowledge of the origins of elements (Rahn, 1976) with the coupling of elements to the factors, it is deduced that the Greenland aerosol is composed of at most two naturally occurring components and of at most three anthropogenic components.

Table 8.6 Correlation coefficients of the Log Concentrations of 13 Constituents of the Aerosol at KATO, 1979–1980

	14/Si	16/S	17/Cl	19/K	20/Ca	22/Ti	25/Mn	26/Fe	29/Cu	30/Zn	35/Br	38/Sr	82/Pb
14/Si	1.00	0.40	0.28	0.60	0.59	0.74	0.77	0.76	0.55	0.37	0.44	0.50	0.32
16/S	0.40	1.00	0.42	0.67	0.68	0.49	0.63	0.53	0.45	0.55	0.63	0.63	0.48
17/Cl	0.28	0.42	1.00	0.83	0.83	0.64	0.54	0.62	0.21	0.28	0.67	0.90	−0.02
19/K	0.60	0.67	0.83	1.00	0.97	0.83	0.83	0.85	0.48	0.47	0.80	0.96	0.26
20/Ca	0.59	0.68	0.83	0.97	1.00	0.85	0.84	0.86	0.44	0.41	0.78	0.97	0.26
22/Ti	0.74	0.49	0.64	0.83	0.85	1.00	0.91	0.98	0.46	0.30	0.62	0.79	0.21
25/Mn	0.77	0.63	0.54	0.83	0.84	0.91	1.00	0.94	0.59	0.46	0.69	0.77	0.37
26/Fe	0.76	0.53	0.62	0.85	0.86	0.98	0.94	1.00	0.52	0.35	0.67	0.80	0.27
29/Cu	0.55	0.45	0.21	0.48	0.44	0.46	0.59	0.52	1.00	0.76	0.46	0.41	0.55
30/Zn	0.37	0.55	0.28	0.47	0.41	0.30	0.46	0.35	0.76	1.00	0.43	0.43	0.41
35/Br	0.44	0.63	0.67	0.80	0.78	0.62	0.69	0.67	0.46	0.43	1.00	0.78	0.39
38/Sr	0.50	0.63	0.90	0.96	0.97	0.79	0.77	0.80	0.41	0.43	0.78	1.00	0.18
82/Pb	0.32	0.48	−0.02	0.26	0.26	0.21	0.37	0.27	0.55	0.41	0.39	0.18	1.00
Percent[a] ABS(R) $>R_0(R_0)$	100	100	85	100	100	92	100	100	92	100	100	92	77

[a] For hypothesis $R = 0$, $R_0(R_0) = 0.248$ is significant at 5%.

Table 8.7. The Components of the Arctic Aerosol[a]

Station	Crust	Marine	Combustion	Metal	Engine Exhausts	Sums
601 THUL[b]						
Minor member collector[c]		MMC	MMC			
Prominent members	Al,Si,Ti, Fe	Cl,TSP	S,Br	K,Ca,Cu, Zn,Sr		
Weak members	S,K,Ca, Cr,Mn,Sr,		Al,Cr, Fe,Pb,TSP	Ti,Cr,Mn, Fe,Pb		
Variance ($n=16$)	4.5	1.8	2.8	4.6		13.7
Percentage	28.2	11.1	17.5	28.8		85.6
701 NORD						
Minor member collector[c]				MMC		
Prominent members	Al,Si,K, Ca,Ti,Mn, Fe		S,Br	Cu,Zn	Pb	
Weak members	Cr,Sr		Cr,Sr, Zn,Pb	S,Cr,Mn, Sr		
Variance ($n=14$)	6.8		2.1	2.5	1.0	12.3
Percentage	48.6		15.0	17.5	6.9	88.0
801 GOVN						
Minor member collector[c]				MMC	MMC	
Prominent members	Al,Si,K, Ca,Ti,Cr, Mn,Fe,Ni, Sr	Cl	S,Br	Zn	Pb	
Weak members	Cu	K,Ca,Br, Sr	Zn,Pb	Cu,Ni	−Si,Cu	
Variance ($n=16$)	8.1	1.8	2.0	1.4	1.1	14.5
Percentage	50.7	11.1	12.6	9.0	7.0	90.4
901 KATO						
Minor member collector[c]				MMC	MMC	
Prominent members	Si,Ti,Mn, Fe	Cl,K,Ca, Br,Sr	Pb	Cu,Zn		
Weak members	K,Ca,Cu, Sr	S,Ti,Mn, Fe	S,Cu,Br	S		
Variance ($n=13$)	3.4	4.7	1.4	1.8		11.4
Percentage	26.2	36.4	10.8	14.0		87.4

[a] Factor model survey, SAGA, 1979–1980 (Heidam, 1984).

[b] Feb.–Aug. 1980

[c] If minor members are collectively important, factor is labeled MMC.

The naturally occurring crustal component consists of the metals Al, Si, K, Ca, Ti, Mn, Fe, and Sr. It has been observed at all stations and accounts for 25 to 50% of the total variance of the logarithmized concentrations.

The second natural component is marine with Cl as the most prominent member, but couplings to S, Br, and the metals K, Ca, Ti, Mn, Fe, and Sr occur with various strengths and in various combinations. At THUL total suspended particulate TSP is a prominent member. The marine component accounts for 10 to 35% of the total variance. It has not been observed at NORD, which is permanently locked in the polar pack ice.

The anthropogenic combustion component comprises S and Br and in various combinations also the metals Cr, Cu, Zn, and Pb. That composition points to an origin in various combustion processes in both power plants, incinerators, and motor vehicles in distant source areas. It accounts for 10 to 15% of the total variance and has been observed at all stations. A marine contribution of S and Br is possible but difficult to reconcile with the metal content. That interpretation is discarded finally for reasons given later. The presence in the Arctic aerosol of combustion products in the form of graphitic soot of distant origin has previously been reported at Barrow, Alaska, and at Spitzbergen (Rosen et al., 1981; Heintzenberg, 1982).

The second anthropogenic component, observed at all stations, has been labeled metallic as it comprises (in various combinations) Cr, Ni, Cu, and Zn, that is, the enriched metals, and occasionally S, K, Ca, Ti, Mn, Fe, Sr, and Pb. The component accounts for 10 to 30% of the variance. The appearance of metals at remote locations are sometimes ascribed to volcanism (Duce et al., 1975). However, the worldwide volcanic emissions of Cu and Zn are estimated to constitute but a tiny fraction of emissions from all natural and anthropogenic sources (Nriagu, 1979), and the systematic seasonal variations which are evident in Figure 8.3 do not support the hypothesis of volcanic origin.

The last aerosol component identified is related to engine exhausts. It consists almost solely of the highly enriched Pb with Br as a minor member and accounts for 5 to 10% of the variance. This component was not found at KATO and THUL.

The factor models in Table 8.7 are quite good; they reproduce all correlations correctly, and they explain 85 to 90% of both the total variances and the individual element variances.

On an annual basis the mass concentration contributions can be found in Table 8.1. At NORD the crustal elements contribute about 55% of the mass detected by PIXE, followed by a contribution of about 43% from the combustion elements, mainly S. The elements allocated primarily to the metallic factor contribute a mere 0.5%, and Pb accounts for the remaining 1%. Usually, however, the mass concentration detected by PIXE amounts to only 10 to 20% of the total aerosol mass concentration, see the THUL column in Table 8.1.

When considering this general composition of the aerosol in Greenland it should be noted that the various components, despite their common descriptive names, may vary considerably with respect to their composition of prominent and weak members. In particular, the classification of the combustion component at KATO remains tentative at this stage. Despite these differences in composition, the similarities in aerosol composition do suggest that the atmospheric aerosol components observed in various parts of Greenland originate from common or similar sources and are subject to common or similar large-scale meteorological transport mechanisms.

5. SEASONAL VARIATIONS

It is evident that a study of the seasonal variations of the chemical composition of the Arctic aerosol is greatly simplified by the factor models. The problem is reduced to a study of a few statistically independent components and their variations, which are considered to be the main causes for the variations of the individual trace constituents.

The large differences in aerosol levels between summer and winter are, as described earlier, seen as an effect of the position of the polar front. The average summer and winter positions differ by the exclusion and inclusion, respectively, of major land areas in the northern air mass. The potential source areas are the northeastern United States of America, Europe, and the Soviet Union.

The factors are, as mentioned, stochastic variables with zero mean and unit variance, and they can be estimated by a least squares criterion. The factor variations for the Greenland aerosols have been described by Heidam (1984). In most cases, however, the factors are so well correlated with some of their prominent member elements that the variations of the log concentrations of these elements are a very good substitute for the factors, albeit on another scale.

The seasonal variation of the crustal factor is very well represented by the variations of the log concentrations of Ti and Fe in Figure 8.2. This is especially true at NORD where the correlations between these two metals and the factor were 0.98.

The annual cycle of the crustal factor at NORD is characterized by an autumn minimum, a stable winter level (i.e., with small scatter), and a maximum in summer. The autumn minimum in October and November probably reflects both the absence of remote crustal influences because the atmospheric circulation is still dominantly local and a quenching of local sources due to increasing snowcover. According to Rahn and McCaffrey (1980) the polar front moves south rather abruptly in late autumn, and the increase in December, associated with a large scatter towards the mean value in winter, indicates that the blocking of injections of midlatitude air into the Arctic disappears. Throughout the winter these injections sustain the levels in the atmospheric reservoir of the Arctic. The dust concentration increases in

May concurrent with the spring in the midlatitude source areas. During the summer the crustal component increases somewhat, and this also applies to the scatter. This behavior is consistent with a shift from a steady remote influence to indigenous sources that make their presence felt occasionally.

The behavior at KATO is somewhat different, thus the autumn minimum occurs later, but the interpretation of a long-range transported crustal component in winter and a component of indigenous origin in summer is still valid. At THUL and GOVN the crustal components behave similarly, but at GOVN the pattern is delayed about a month.

The winter levels are relatively stable at all stations until July, when an increase to more unstable summer levels occurs, probably signaling a shift to local influence.

The marine component exhibits, as expected, winter minima and summer maxima. At KATO, the only site where this component contains trace metals, there is no pronounced cycle.

The seasonal variation of the combustion factor is well represented at KATO by the log concentration of Pb in Figure 8.3; the correlation with the factor was 0.87. The results for V at Barrow also shows the typical behavior of the combustion component. It has a very pronounced and well-defined annual cycle. It rises steadily from low levels in summer to high levels in winter and reaches a maximum in March. The annual course of this component finally rules out the possibility mentioned in the preceding section that it is a marine component; the behavior of a marine component is in fact the inverse with a winter minimum.

The well-defined annual cycle of the combustion component points to a remote origin, reflecting large-scale circulation of air masses. Local activity and its accompanying local emission are very small in winter, except for heating of the living quarters. It is concluded that local influence, if any, can manifest itself only in the summer values, but they are so low that they might as well represent general background values (Flyger and Heidam, 1978).

The increase in combustion levels is effective about a month earlier than for the crustal factors. Evidently the meteorological processes responsible for transport to the Arctic differ for these two aerosol components, indicating that the source regions are also different.

The combustion component exhibits quite similar behavior at the other Greenland sites. This latter fact is the main reason for classifying the KATO factor as a combustion component, despite a somewhat diverging composition (see Table 8.7). It is remarkable that at THUL and NORD the monthly means are in exact phase and that the decisive shift from low summer to high winter values occurs a month later at GOVN and KATO. The winter maxima appear at all stations in March and April.

For the early winter increase in both the crustal and combustion components of the aerosol there is thus a time lag of about a month at the more southerly stations compared to the northern stations. The aerosol winter thus behaves qualitatively as the climatic winter, progressing from north to south. Evidently the observed time lag reflects the southward progress of the influence of the polluted reservoir rather than direct transport from midlatitudes, which would be expected to manifest itself first at the more southerly sites.

The seasonal variations of the metallic factors are quite well represented by the log concentrations of Cu and Zn in Figure 8.3; the correlations with the factors were 0.86 and 0.90. The metallic factors scatter around the annual means in the summer and autumn, reach well-defined maxima from January to March, and fall rapidly in April towards quite well-defined minimum values in the spring and early summer, May to July. This behavior does not reveal any manifest source areas for the metallic component, but the well-defined extrema in the first half-year indicate that it is controlled by some large-scale atmospheric circulation, presumably of remote origin.

For the second half of the year the possibility that it stems from local refuse burning cannot be ruled out. Aerosols from refuse incinerators have been found to contain considerable amounts of Zn, Cu, and Pb (Greenberg et al., 1978a, b). Refuse burning takes place occasionally on a small scale, and the expected behavior would be an erratic meandering around zero as is in fact observed in the second half-year.

It is, again, remarkable that the monthly winter means of the metallic component at THUL are in phase with the component at NORD. That is decisive indication of a large-scale phenomenon. At KATO a quite similar behavior is observed, but with a time lag of a month.

The engine exhaust components at NORD and GOVN appear to be of local origin (Heidam, 1984), but they contribute very little to the total variance (see Table 8.7).

A closer study of the concentration of Pb at NORD, or its enrichment in Figure 8.5, reveals, however, a pronounced systematic behavior, which is not seen in the factor values. It is reminiscent of both the combustion and metallic factors. A similar result was found at THUL (scc Table 8.7), and it therefore appears that Pb at NORD has several sources, both distant and local. It is noteworthy that the enrichment reaches a maximum in autumn in the period where the crustal component passes through a minimum (compare Figs. 8.2 and 8.5).

It can therefore be concluded that the seasonal variations of the chemical composition of the Arctic aerosol to a large extent can be explained in terms of the variations in the basic source-related components of the aerosol and that the variations of these components are in good general agreement with the reservoir model of the Arctic troposphere. This general agreement does not, however, lead to a distinction between the potential source regions. A closer scrutiny of the results is needed if this distinction is to be obtained.

6. TRANSPORT AND SOURCES

The relative importance of sources in northeastern America versus Eurasian sources has been discussed by Rahn (1981a, b) who considered both meteorological and compositional evidence. Among the latter he considered in particular the Mn_x/V_x ratio, where x means excess or noncrustal concentrations. This ratio appears to be a little higher in North American than in Eurasian aerosols. Rahn concludes that although Eurasia appears to be a major source compared to North America, the question of dominant source areas has not been finally settled.

For the Greenland results another approach has been used (Heidam, 1984). Meteorological evidence was combined with compositional evidence in that the individual factor values at the NORD site were studied in relation to the synoptic meteorological situations existing at the time of the individual samples. The analyses were carried out in episodic cases when the concentrations of elements belonging to one of the components are enhanced, corresponding to factor values larger than perhaps 1, or when the factor values change abruptly between small or even negative values and large positive ones.

The average synoptic situation in the Arctic in January is shown in Figure 8.6 (Liljequist, 1970). The circulation is dominated by the North Atlantic cyclone and by the continental anticyclones over central Asia and northwestern Canada. They are connected through a high-pressure ridge over the Polar Sea with a bulge towards Greenland.

The Iceland Low, which dominates the Atlantic sector, is actually a huge trough extending to the west into northeastern Canada and to the east into the Barents Sea. The eastern leg roughly marks the most frequent northern route taken by the Atlantic Lows before they die out in the Kara Sea east of Novaya Zemlya. There is also a more southerly route terminating in the same area.

These moving cyclones drive a great sweeping circulation around the Barents and Kara Seas that carries with it air masses that have previously passed over the European continent. As can be seen from the map there is vigorous circulation within the Arctic. Winds from northeastern directions are dominant in northeast Greenland. In cases when the Polar ridge retreats from Greenland these winds may effectively penetrate the region.

A study of the individual sample values of the factors revealed that the combustion factor was nonepisodic and rarely exhibited sudden shifts to values larger than 1; in general, it changed smoothly from sample to sample. From these observations it is concluded that in winter the Arctic troposphere constitutes a well-mixed aerosol reservoir of combustion products, where concentrations are largely independent of local wind directions. As the combustion factor displayed the same smooth course at all the SAGA stations it can also be concluded that the routes of injections to the reservoir only rarely pass Greenland.

In the few episodes observed in the winter 1979–1980, it was found from the synoptic weather maps that both North American and European sources may have contributed to the combustion component of the North Greenland aerosol. In general, however, it does not seem possible to single out any

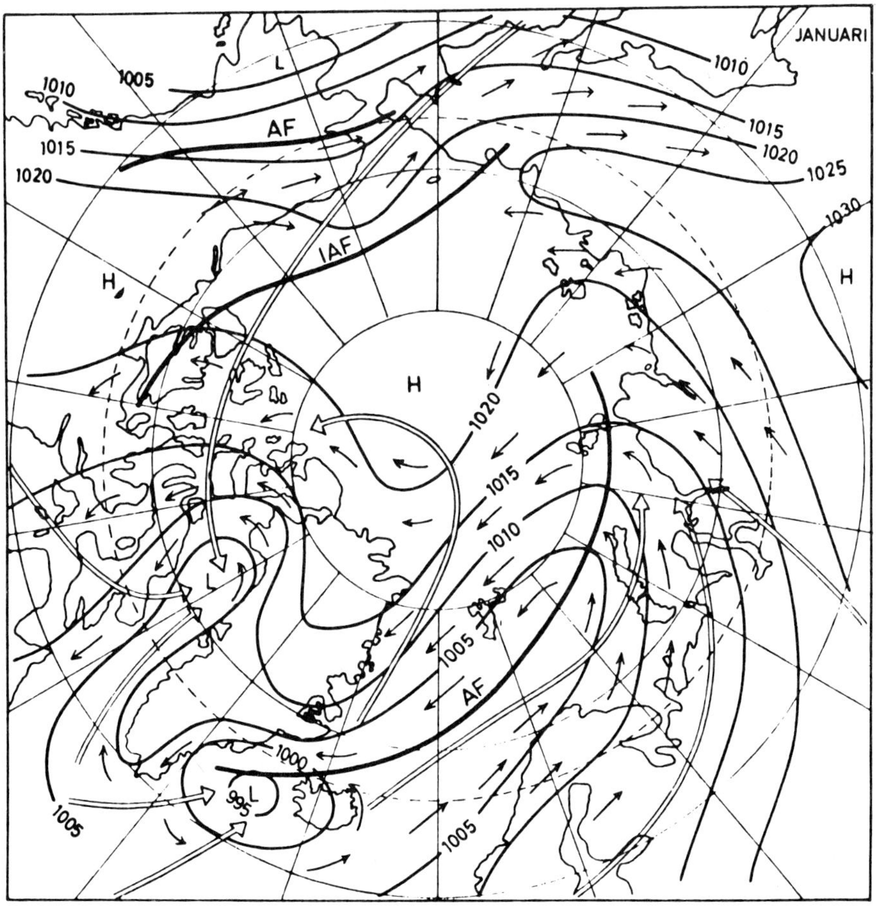

Figure 8.6. Average distributions of surface pressures, winds, fronts, and cyclone trajectories in January. Winds are marked by short arrows, cyclone trajectories by long double arrows, and fronts by heavy lines. AF, arctic front; IAF, inner arctic front. Data from 1931–1960 or 1954–1960 (Liljequist, 1970).

particular or dominant source area. This is in accord with the reservoir model in which the origins of the injections are masked by mixing in the high Arctic troposphere.

The metallic factor is statistically independent of the combustion factor on an annual basis. The two factors do, however, also differ in their detailed behavior. Thus the metallic component is highly episodic in contrast to the combustion factor.

In the early winter the metallic episodes are sharp but short, 3 to 7 days corresponding to only one or two samples. Some of these episodes were characterized by the combined effect of an Atlantic Low and a High centered over Scandinavia and northwestern USSR. That resulted in a northward flow from Europe, in particular Great Britain. In other cases a strong cyclonic flow from North America was moving at 500 millibar towards North Greenland along the east coast.

In the winter period from January to March the metallic episodes are less clearly distinguished, but generally they are of a much longer duration, from weeks up to a month, and the major transport must occur here. Thus, during an episode in the latter half of January 1980 and in a month-long episode from the end of February and into March 1980, the incoming Lows from the Atlantic moved north to the Barents Sea region, exactly as in Figure 8.6. At times this pattern could be seen simultaneously both at the surface and at 500 mb. The Barents Sea Lows, reinforced at times by polar Highs, a Siberian High, or both, drove a strong flow from eastern Russia and western Siberia out over the Kara Sea and the Arctic Ocean directly towards northeastern Greenland. Occasionally there were also indications of an altitude flow from North America.

The occurrence of enhanced levels of metals, notably Cu and Zn, in these situations strongly indicates that the Ural region is the primary source area. In the southern Urals some of the richest mineral deposits in the world are found, and the region constitutes a vast complex of heavy industries ranging from mines over smelters to all kinds of processing plants (Oxford Atlas, 1967). Thus the episodicity of the metallic factor seems to derive mainly from the flows directed over the source areas in the Urals into North Greenland.

The difference in character between the combustion and metallic factors in northern Greenland thus appears to be caused by different meteorological pathways.

The metallic component disappeared almost completely in May (see Fig. 8.3). From 15 May to 18 June there was a prolonged antiepisode of this component with factor estimates from -1.3 to -2.0. This period was characterized by a trough in the 500-mb surface, which resided west of Greenland, and at times also a ridge southeast of Greenland. The result was an altitude current from northern Canada across the Greenland ice cap.

For the engine exhaust component at NORD single large factor values coincided on a few occasions with one of the above mentioned episodes. In these instances this otherwise local component may also have had a distant origin.

The crustal factor, which contains the largest number of trace metals, rarely attained large positive values in winter, as is also evident from Figure 8.2. The small scatter in this period indicates, as for the other components, that the presence of the component is largely a consequence of its accumulated level in the reservoir.

The only winter episode observed was, however, strong and clear with factor values 1.8–2.0 during a week late in December 1979. This period followed immediately after a metal episode with possible origin in North America and was itself followed by the first metal episode of the Ural type.

The altitude maps show that a cyclonic flow from America was directed into the Barents Sea where it was picked up by a polar trough that disappeared with this episode. At the surface a pronounced Low (965 mb) stretching from the Pole to Spitzbergen directed a strong flow of polar air to North Greenland during these days. The explanation for this crustal episode could therefore be that the polar front by late December had moved so far south that the flow from America now could pick up soil dust from the still dry areas in southern United States and carry it into the Arctic reservoir, where there just happened to be a strong circulation directed straight towards North Greenland. On the other hand, there is also the possibility that the surface flow picked up dust from arid Peary Land northwest of NORD. That the subsequent Ural episode carried no erosion component is perhaps explainable as a result of wet or snow-covered soil in this and adjacent areas.

7. CONCLUSIONS

The Arctic aerosol, its seasonal variations, and its chemical composition have been studied by several groups in various parts of the Arctic for both short and extended periods of time. The results show that the aerosol all over the Arctic has a very pronounced annual cycle and that it is geographically quite homogeneous with respect to the chemical composition as expressed by the contents of trace elements and trace metals.

The elemental composition of the Greenland aerosol has been studied by means of factor analysis of the logarithmic concentrations. The analyses show that the atmospheric aerosol can be described by 4–5 different source-related and statistically independent components.

Two of these components have a natural origin as they consist of a crustal component originating from erosion processes and a marine component produced by sea spray. They account for 40 to 60% of the total logarithmic variance. In winter the crustal component, which comprises a large number of nonenriched trace metals, appears to have a very distant origin, but the transport routes rarely pass directly over Greenland. In summer it probably has indigenous sources. In autumn neither of these sources are of importance and the crustal component passes through a minimum. The marine component is at maximum in summer, and its annual course can be attributed to mesoscale transport mechanisms inasmuch as it seems controlled by the icecover in the Greenland waters.

The remaining two or three aerosol components are of anthropogenic nature, and they account for 25 to 45% of the total variance. One of these has been termed the combustion component because it comprises some highly enriched trace elements that are related to a variety of combustion processes. It displays a very pronounced cycle with a winter maximum and a summer minimum equivalent to an almost total absence. For these reasons it is believed to derive exclusively from very distant industrial source areas, but transport is only rarely directed straight towards Greenland. The second anthropogenic component has a metallic composition and exhibits a maximum in midwinter and a late winter minimum, which also indicates that some large-scale atmospheric transport is involved. It contains a number of enriched trace elements, notably Cu and Zn. For this component the transport routes do appear to be directed towards northern Greenland from primary source areas in the Urals. The last component identified is related to engine exhaust. At some, but not all, sites it seems to be largely of local origin but it may also contain traffic aerosols of distant origin.

The annual variations of concentration levels and aerosol composition have been explained in terms of a reservoir model. According to this model the high Arctic troposphere in winter constitutes a mixed aerosol reservoir which is subject to a series of injections of pollutants from lower latitudes. Because of the sparse precipitation these pollutants accumulate in the reservoir. In spring, precipitation scavenging sets in and the polar front migrates north from latitudes south of the major source areas and isolates with increasing efficiency the Arctic from midlatitude influences. In autumn, the polar front moves south again, allowing an increasing number of injections from populated and industrialized regions to penetrate into the Arctic.

In accordance with the model, the enhanced winter aerosol levels appear at the northern sites about a month earlier than at the more southerly stations. The model can also explain why the crustal component appears later in winter than the anthropogenic components. As the polar front moves south in the autumn it incorporates industrialized regions with wet soils, and it is not until later that the more southerly and dry regions are also incorporated in the northern air mass.

On the other hand, the identification of source areas is difficult because the origin of individual injections is masked by the mixing in the reservoir. Consequently, aerosol component episodes with enhanced concentration levels are rare; the levels just increase smoothly during winter. In some cases it has, nevertheless, been possible to link episodes with source areas by a study of concurrent synoptic weather maps. The sample of episodes available showed that combustion products may arrive in Greenland from both North American and European sources, and that the major source for the metallic component probably is the Ural region with its vast complex of mines, smelters, and other metal processing plants.

ACKNOWLEDGMENTS

The SAGA project has in various stages been partially funded by the Nordic Ministerial Council through its Expert Group on Air Pollution.

REFERENCES

Barrie, L.A., Hoff, R., and Daggupaty, S.M. (1981). "The influence of mid-latitudinal pollution sources on haze in the Canadian Arctic." *Atmos. Environ.* **15**, 1407–1419.

Carlson, T.N. (1981). "Speculations on the movement of polluted air to the Arctic." *Atmos. Environ.* **15**, 1473–1477.

Davidson, C.I., Chu, L., Grimm, T.C., Nasta, M.A., and Qamoos, M.P. (1981). "Wet and dry deposition of trace elements onto the Greenland ice sheet." *Atmos. Environ.* **15,** 1429–1437.

Duce, R.A., Hoffman, G.L., and Zoller, W.H. (1975). "Atmospheric trace metals at remote northern and southern hemisphere sites: Pollution or natural?" *Science* **187,** 59–61.

Flyger, H., and Heidam, N.Z. (1978). "Ground level measurements of the summer tropospheric aerosol in northern Greenland." *J. Aerosol Sci.* **9,** 157–168.

Flyger, H., Heidam, N.Z., Hansen, K., Rasmussen, L., and Megaw, W.J. (1980). "The background levels of the summer tropospheric aerosol and trace gases in Greenland." *J. Aerosol Sci.* **11,** 95–110.

Greenberg, R.R., Zoller, W.H., and Gordon, G.E. (1978a). "Composition and size distributions of particles released in refuse incineration." *Environ. Sci. Technol.* **12,** 566–573.

Greenberg, R.R., Gordon, G.E., Zoller, W.H., Jacks, R.B., Neuendorf, D.W., and Yost, K.J. (1978b). "Composition of particles emitted from the Nicosia municipal incinerator." *Environ. Sci. Technol.* **12,** 1329–1332.

Harman, H.H. (1976). *Modern Factor Analysis.* 3rd rev. ed. The University of Chicago Press, Chicago, Ill.

Heidam, N.Z. (1983). *Studies of the Aerosol in the Greenland Atmosphere. SAGA I: Observations 1979–1980, a Base Report.* Air Pollution Laboratory Report MST LUFT A-73.

Heidam, N.Z. (1984). "The components of the Arctic aerosol." *Atmos. Environ.* **18,** 329–343.

Heintzenberg, J. (1982). "Size segregated measurements of particular elemental carbon and aerosol light absorption at remote Arctic locations." *Atmos. Environ.* **16,** 2461–2469.

Heintzenberg, J., Hansson, H.C., and Lannefors, H. (1981). "The chemical composition of arctic haze at Ny-Ålesund, Spitzbergen." *Tellus* **33,** 162–171.

Lannefors, H., Hansson, H.-C., and Granat, L. (1983a). "Background aerosol composition in southern Sweden." *Atmos. Environ.* **17,** 87–101.

Lannefors, H., Heintzenberg, J., and Hansson, H.-C. (1983b). "A comprehensive study of physical and chemical parameters of the Arctic summer aerosol." *Tellus* **35B,** 40–54.

Liljequist, G.H. (1970). *Klimatologi* (in Swedish). Generalstabens Litografiska Anstalt, Stockholm, p. 324ff.

Mason, B. (1966). *Principles of Geochemistry.* 3rd. ed. John Wiley & Sons, Inc., New York.

Nriagu, J.O. (1979). "Global inventory of natural and anthropogenic emissions of metals to the atmosphere." *Nature* **279,** 409–411.

Oxford Economic Atlas of the World (1967). 4th ed. Oxford University Press. Oxford, U.K.

Patterson, E.M., Marshall, B.T., and Rahn, K.A. (1982). "Radiative properties of the Arctic aerosol." *Atmos. Environ.* **16,** 2967–2977.

Rahn, K.A. (1976). "The chemical composition of the atmospheric aerosol." Technical Report. Graduate School of Oceanography, University of Rhode Island, Kingston, R.I.

Rahn, K.A. (1981a). "Relative importances of North America and Eurasia as sources of Arctic aerosol." *Atmos. Environ.* **15,** 1447–1455.

Rahn, K.A. (1981b). "The Mn/V ratio as a tracer of large-scale sources of pollution aerosol for the Arctic." *Atmos. Environ.* **15,** 1457–1464.

Rosen, H., Novakov, T., and Bodhaine, B.A. (1981). "Soot in the Arctic." *Atmos. Environ.* **15,** 1371–1374.

Shaw, G.E., and Stamnes, K. (1980). "Arctic haze: perturbation of the radiation budget." *Ann. N.Y. Acad. Sci.* **338,** 533–539.

Schutz, L., and Rahn, K.A. (1982). "Trace element concentrations in erodible soils." *Atmos. Environ.* **16,** 171–176.

Winkler, P. (1983). "Acidity of aerosol particles and of precipitation in the North Polar region and over the Atlantic." *Tellus* **35B,** 25–30.

9

CHEMICAL ELEMENTS AS TRACERS OF POLLUTANT TRANSPORT TO A RURAL AREA

Liaquat Husain

Wadsworth Center for Laboratories and Research
New York State Department of Health
Albany, New York

1. INTRODUCTION

Combustion of fossil fuel releases large amounts of sulfur oxides and nitrogen oxides into the atmosphere. Air mass movement can transport these oxides hundreds or even thousands of kilometers. In the atmosphere, these oxides may be removed by physical processes such as precipitation or interaction with the earth's surface. Chemical reactions may also convert these oxides into sulfuric or nitric acid or their salts, and these may subsequently be deposited either by precipitation or dry deposition. These acids, of course, lower the pH of rain and snow. The pH of precipitation in the Adirondacks ranges from 3.7 to 4.8 compared to pure rainwater pH, which is 5.6 to 5.8. Two thirds of the enhanced acidity of precipitation in the northeastern United States comes from SO_4^{2-}, with nitrates accounting for the remainder.

Acid precipitation is held responsible for many ponds and lakes which now have pH ~5 and even lower. Fish populations in hundreds of Adirondack lakes have been severely depleted or completely eradicated during the recent decades of increased acidity. Acid rain injures forests by dissolving certain essential metals from sensitive tree roots in unbuffered soils. It also directly affects public health as the inhalation of the acidic particles aggravates existing respiratory ailments or initiates new ones. Sulfate ion deposited in unbuffered environments can enhance the leaching of toxic metals in ground water.

An understanding of the causes and effects of acid rain is essential for developing an effective strategy for its control. Emissions of SO_2 from individual states in the United States have been calculated (Benkovitz, 1982). Approximately two thirds of the emissions in the 31 states east of the Mississippi River originate in the Midwest. Precipitation acidity is, however, greater in the Northeast, and due to lower buffering capacity of the soil there, effects are more severe.

Sulfur dioxide emission sources are distributed nonuniformily over vast areas. The existing meteorologic conditions at the emission points determine whether SO_2 is removed rapidly or transported long distances. The chemical fate of SO_2 also depends on many variables such as humidity, solar radiation, temperature, concentrations of oxidants, and other chemically reactive species. Because of such a large number of variables, many of which are not accurately measured or even known, needed accuracy is not possible in mathematically relating emissions with impacts at distant receptors. Under these circumstances the most viable approach appears to be a receptor apportionment method in which extensive data obtained at deposition sites are used to identify contributions from various sources.

At a rural site particulate matter originates primarily from three sources: re-entrainment from soils, local fossil-fuel combustion in automobile and space heating, and, under appropriate conditions, transport from distant urban/industrial centers. At rural Whiteface Mountain, New York, SO_4^{2-} and trace metal concentrations can increase tenfold during pollution episodes lasting from 1 to 3 days (Husain and Samson, 1979; Dutkiewicz et al., 1983; Husain et al, 1984). This indicates that local sources are generally small and that transport from distant sources can have a major influence at this site. To properly understand pollutant transport, local contributions should be accurately characterized and the effects of varied meteorological conditions studied in detail. Extensive daily measurements of SO_4^{2-} and trace metals have been made at the summit of Whiteface Mountain spanning several years and all seasons. In this chapter the available SO_4^{2-} and trace element data are used to characterize seasonal variation, background levels, and the relationship between wind direction and observed concentrations at this site. Further, the usefulness of these trace elements as "fingerprints" of SO_4^{2-} source regions is discussed, including a detailed test case for the promising Mn/V ratio technique.

2. SAMPLING AND ANALYSIS

Although samples were collected at several sites, the primary site was at the summit of Whiteface Mountain (44°23′N, 73°51′W, 1.5 km above sea level), which is isolated from significant urban and industrial areas. The nearest cities are (distances in km): Montreal, 110; Albany, 175; Syracuse, 225; Boston, 300; New York, 370; and Buffalo and Toronto, 400. Particulate air samples were collected daily from July 1978 to December 1979. In addition, samples were collected continuously every 6 h during July and August 1981 at Mayville (MAY), Brewerton (BRW), and Whiteface Mountain (WFC), and every 24 h at Oneonta (ONT) and West Harverstraw (WHV) (see Fig. 9.9). Meteorologic data (Parekh and Husain, 1982) have shown that summer pollution episodes at WFC are associated with air masses from the Midwest which frequently pass through MAY and BRW. The 6-h sample collection was designed to intercept air masses at these locations as they traveled eastward. The ONT and WHV sites were chosen to identify air masses from metropolitan New York, northern New Jersey, and central Pennsylvania. Airborne particles were collected on 25 × 20 cm Whatman 41 filter papers using high volume samplers equipped with Sierra mass flow controllers which maintained flow rate at 75 m^3/h (±3%) (Dutkiewicz et al., 1983). A portion of filter paper was ashed in a low temperature asher, and the residue was treated with HF, evaporated to dryness, and taken up in 2% HNO_3. Concentrations of Na, Mg, Al, K, Ca, Fe, Zn, and Pb were determined by using both flame and flameless atomic absorption spectrophotometry (Husain et al., 1980, 1984). Soluble sulfates were extracted from another portion of the filter and [SO_4^{2-}] (where [X] means concentration of X) determined using either methyl thymol blue method with an automated Technicon Autoanalyzer (Lazrus et al., 1966) or pyrolysis–microcoulometry (Canelli and Husain, 1982) and ion chromatography (Husain et al., 1984).

3. CHARACTERIZATION OF SO_4^{2-} AND ELEMENTAL CONCENTRATIONS

3.1. Seasonal Variations

Concentrations of total suspended particulates (TSP), Na, Mg, Al, K, Ca, Fe, Zn, Pb, and SO_4^{2-} ions were determined in daily samples collected at WFC for all but 13 days in 1979. This is quite satisfactory in view of the severe weather conditions frequenting this location. For brevity the daily data (Husain et al., 1980) will not be presented here. In general, above average concentrations were observed during March through October and below average during remaining months. During summer, [SO_4^{2-}] as high as 45 $\mu g/m^3$ were recorded during episodes that lasted 1 to 3 days. Trace metals concentrations were also elevated during such episodes although these elevations were not necessarily proportional to SO_4^{2-} increases.

Monthly average concentrations are shown in Figure 9.1. The seasonal changes are striking: A distinct summer peak is observed in each case. Seasonal [SO_4^{2-}] were calculated from the monthly data ($\mu g/m^3$): summer (June–August), 7.3 ± 3.4; autumn (September–November), 3.0 ± 1.1; winter (December–February), 1.3 ± 0.3; and spring (March–May) 3.6 ± 0.7. Thus, summer [SO_4^{2-}] were 5.6 times higher than winter and approximately twice that in spring and in autumn. Similar summer increases in [SO_4^{2-}] were observed at Narragansett, RI (Rahn et al., 1982). Seasonal patterns in other parts of the United States are not as well defined. The reasons for the observed changes are not fully understood. Increased photochemical oxidation of SO_2 to SO_4^{2-} in summer, and to a lesser degree in autumn and spring, can only account for part of the increase. The predominant wind directions at WFC are from north to northwest in winter and southwest during summer. Since major SO_2 emission sources are located southwest of the site, increased long-range transport must also significantly contribute to the observed seasonal changes. Thus, a combination of the increased photochemical reactivity and changes in regions through which the air passes before reaching this site probably accounts for much of the seasonal change in [SO_4^{2-}].

Concentrations of metals were lowest in winter (1/3 to 1/2 of the annual average), intermediate in spring and autumn, and highest in summer (1.3 to 1.9 times the annual average). The changes in photochemical reactivity should not play any significant role in the concentrations of metallic ions. Wind directions and meteorologic conditions (stagnation of air) that result in buildup of total suspended particles (TSP) (Husain and Samson, 1979) probably cause a major portion of the observed seasonal changes. Winter snowcover must also significantly reduce the availability of soil particles for re-entrainment. Subtle differences in the behavior of lithophilic elements Mg, Al, K, Ca, and Fe and anthropogenic elements Zn and Pb were also observed (Fig. 9.1). Whereas the lithophilic elements and TSP exhibited a strong

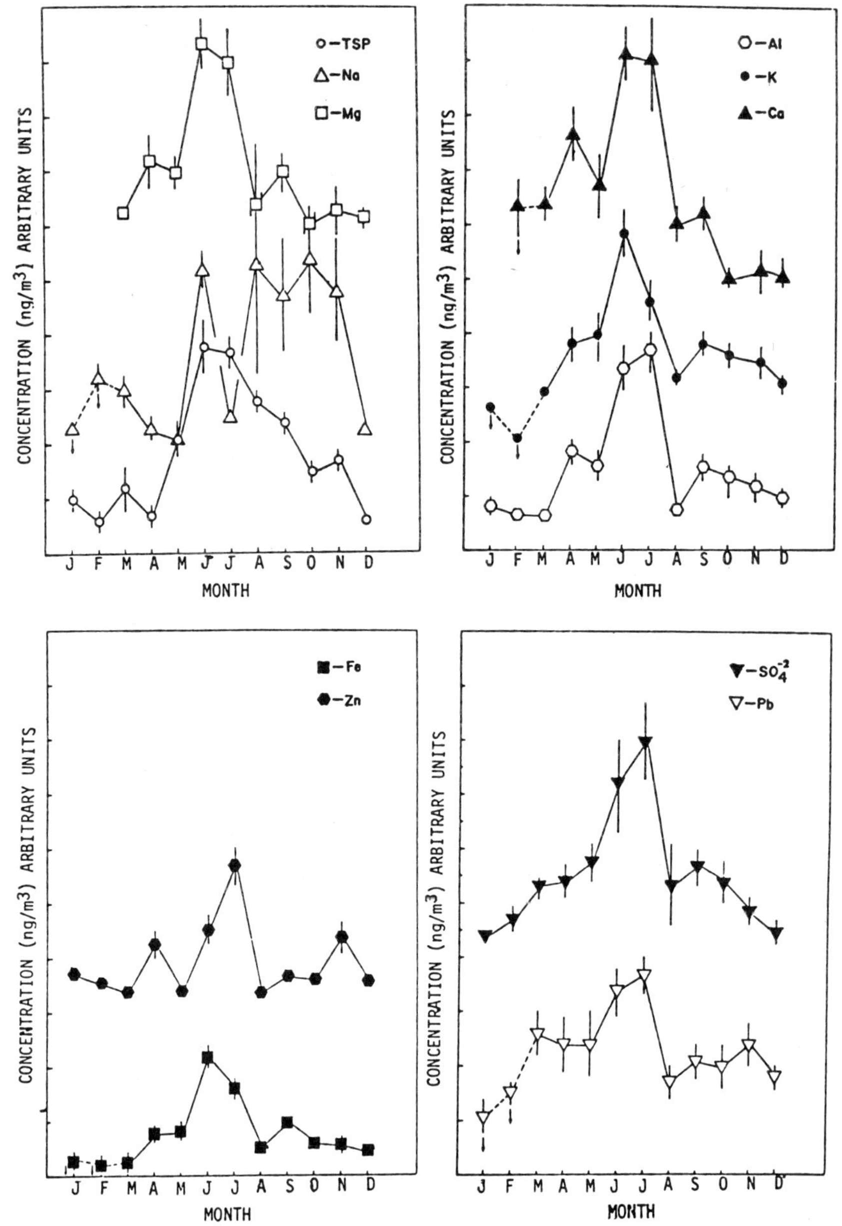

Figure 9.1. Distribution patterns of TSP, Na, Mg, Al, K, Ca, Fe, Zn, Pb, and SO_4^{2-} in air particulates of Whiteface Mountain, N.Y. Whatman 41 filters contained high and variable Na blanks. This results in relatively high uncertainties in [Na]. Although Na data are included in this figure, they are excluded from discussion in the text.

seasonal pattern similar to the SO_4^{2-} pattern, Zn and Pb showed smaller variations. The principal source of Pb in ambient air is automotive exhaust, and thus long-range transport probably plays a smaller role in observed [Pb].

3.2. Composition of Background Aerosols

Airborne particles at a rural site originate primarily from the three above-mentioned sources. While particles from each source are ubiquitous, their relative proportions vary. Ample evidence shows that SO_4^{2-} and trace elements are highly concentrated when stagnant air masses from the Midwest move into New York State (Samson, 1978; Husain and Samson, 1979; Dutkiewicz et al., 1983; Husain et al., 1983, 1984) (Sec. 4). On the other hand, northerly windflows yield very low concentrations, suggesting small contributions not only from local sources but also from areas north of WFC. Because wind flows from the north infrequently, a background concentration derived from a small number of days may not be representative. Another approach to calculate background concentrations includes only data when concentrations were unusually low. Figure 9.2 shows the distribution of the daily [SO_4^{2-}] in 1979. In approximately 50% of the samples, [SO_4^{2-}] was <2 $\mu g/m^3$. Of 166 days when [SO_4^{2-}] was <2 $\mu g/m^3$, [SO_4^{2-}] exceeded 1 $\mu g/m^3$ on only 55 days, and only on 10 days did it approach 2 $\mu g/m^3$ (Fig. 9.2 inset). A weighted mean of 0.8 $\mu g/m^3$ is calculated from these data. This value perhaps represents the most accurate and unbiased estimate of background [SO_4^{2-}] in the Adirondacks because it is based on 166 daily measurements from all four seasons in one year. This value is slightly lower than that deduced by Altschuller (1976) and Chung (1978), who deduced a background value based

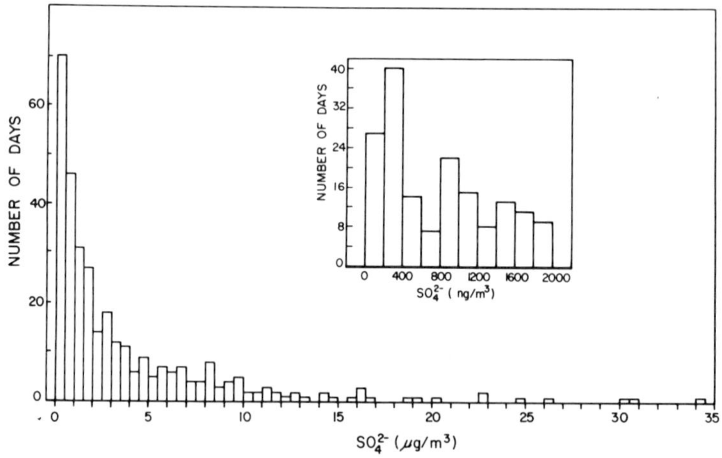

Figure 9.2. Frequency distribution of daily [SO_4^{2-}] at Whiteface Mountain, N.Y.

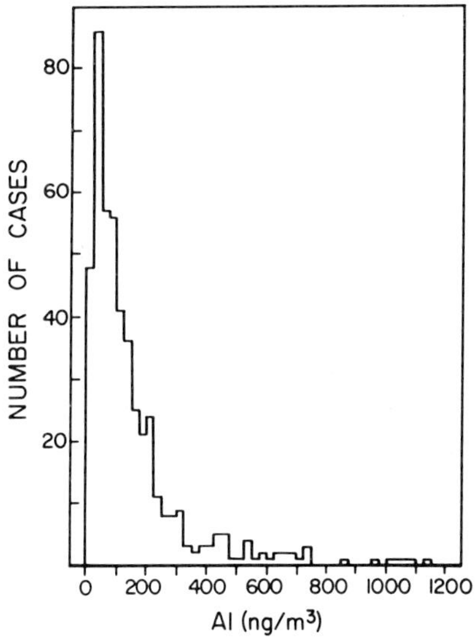

Figure 9.3. Frequency distribution of the daily Al concentrations.

on their lowest observed values, ~1 $\mu g/m^3$, which also was the detection limit of their discontinuous measurements.

A similar approach can determine background concentrations of trace metals which exhibited large (up to 100-fold) variations. Daily [Al] were between 50 and 100 ng/m^3 for approximately 25% of the time and less than 50 ng/m^3 for about 10% of the time (Fig. 9.3). In only 15% of the observations did [Al] exceed 250 ng/m^3. The low concentrations as represented by the peak at 50 to 100 ng/m^3 and the very high concentration (>300 ng/m^3) which appear at the tail end of the distribution probably represent differences in principal sources of atmospheric particles. The highest [Al] were associated with the highest [SO_4^{2-}], which were measured primarily during summer (Fig. 9.3), and the high [Al] were also most likely transported from distant sources. On the other hand, low [Al] probably originated from local sources. Particulate elemental concentrations associated with [SO_4^{2-}] ≤ 2 $\mu g/m^3$ may also be taken as the background elemental concentrations. Accordingly the geometric mean of [Al] on days when [SO_4^{2-}] were < 2 $\mu g/m^3$ is 77 ng/m^3.

Frequency distributions for Fe, Zn, and Pb were similar to that for Al, except that high concentrations were even less frequent for Zn and Pb. For [Mg], [K], and [Ca] the frequency distributions were less sharply peaked. The background concentrations were calculated in a manner similar to that for [Al].

Background concentrations at WFC are compared in Table 9.1 with those reported for five other remote locations of the world. Concentrations of the lithophilic elements Mg, Al, K, and Ca were similar to those measured at Jungfraujoch, Switzerland; Twin Gorges, Canada; and North Cape, Norway,

Table 9.1 Background Elemental Concentrations (ng/m^3) in Aerosols at Rural and Remote Sites

Species	White face Mt.[a]	Jungfraujoch[b]	Twin Gorges[c]	North Cape[c]	Chacaltaya[d]	Amundsen–Scott Station[e]
Mg	18	10	16	70	29	1.0
Al	77	51	66	43	253	1.1
K	49	20	54	48	80	0.92
Ca	71	50	40	46	83	0.67
Fe	30	36	71	51	181	0.84
Zn	7	9.9	3.8	8.9	4.3	0.045
Pb	9	4.4		5.6	5.3	0.63
SO_4^{2-}	<800				516	202

[a]This work.

[b]Dams and DeLonge (1976).

[c]Rahn as quoted in Dams and DeLonge (1976).

[d]Adams et al. (1977).

[e]Zoller et al. (1974); Maenhaut and Zoller (1976).

but were lower than those measured at Chalcaltaya, Bolivia by factors of 2 to 6. Background Fe unexpectedly were lower at WFC than at any station except Amundsen-Scott, at the South Pole. The background [Zn] and [Pb] at WFC were within a factor of 2 of those observed in the Swiss, Canadian, Norwegian, and Bolivian background aerosols. Background concentrations of trace elements are ~20- to 100-fold lower at the South Pole. This is not so surprising in view of the snowcover there. However, background [SO_4^{2-}] may show less variation worldwide.

The differences in the background aerosols could be very site-specific. Crecelius et al. (1980), for instance, observed that several volatile elements (e.g., Zn, Pb, As, and Se) in Colstrip, Montana, were highly enriched relative to the average crustal abundances. However, enrichments were drastically reduced when normalized to local soil composition. Another example is the high [W] of 0.45 ng/m^3 in Bolivian aerosols, more than an order of magnitude greater than in the Swiss and Canadian aerosols. The rich wolframite–asiterite mineralization in the Silurian sediments of Chacaltaya presumably influences the [W] in the Bolivian aerosols (Schneider and Lehmann, 1977). Such background differences must be considered when using elements as tracers of atmospheric pollutions.

4. RELATIONSHIP BETWEEN SULFATE AND ELEMENTAL CONCENTRATIONS WITH WIND DIRECTIONS

The large number (~500) of daily measurements in 1978–1979 provides an excellent opportunity to study changes in concentrations of trace elements and SO_4^{2-} with wind-direction changes. To deduce windflow, air trajectories are often calculated using Heffter's atmospheric transport and dispersion model (Heffter, 1980). For each day, the model calculates the trajectories of air masses reaching the sampling location at 0200, 0800, 1400, and 2000 hours. Meteorologic data for North America averaged between 400 and 3000 m were used as input to the model to calculate trajectories for WFC. Examples of air trajectories will be discussed in Section 5.1. Trajectories were calculated for the entire July 1978 to December 1979 period.

The trajectories reaching WFC were divided into 12 sectors of 30° each (with 0° being north of the sampling site). Each trajectory reaching WFC was assigned a sector. If a trajectory crossed several sectors, the contribution from each sector was estimated. It was assumed that each of the four 6-h trajectories contributed equally to the observed concentrations. Parekh and Husain (1982) have discussed the procedure in detail.

The mean concentration $\overline{C_j}$ of M from sector j is defined as:

$$\overline{C_j} = \sum_{i=1}^{N} C_i f_{ij} \Bigg/ \sum_{i=1}^{N} f_{ij} \quad \text{or} \quad \overline{C_j} = \sum_{i=1}^{N} C_i f_{ij} \Bigg/ N_j \qquad (1)$$

where C_i is the concentration of the species on the ith day and f_{ij} is the fraction of the ith day that the trajectory passed through sector j. The denominator N_j is the total number of days that trajectories passed through sector j, and N is the total number of days data were collected.

Mean concentrations corresponding to the 12 sectors were calculated for SO_4^{2-} and trace metals using Eq. (1). Figure 9.4 shows the frequency of wind directions at WFC as well as mean [SO_4^{2-}] corresponding to the windflow in each 30° sector. Similar data for Mg, K, and Ca are shown in Figure 9.5, and for Al, Fe, Zn, and Pb in Figure 9.6. Data in Figure 9.4*b* clearly shows that the windflow is primarily from the west, northwest, and southwest and that windflows from easterly directions are negligible.

For all seven metals and sulfate, $\overline{C_j}$ peaked in the southwest quadrant and had a secondary smaller peak in the northeast quadrant. In general, the mean concentration of metals as a function of sector had a broader peak than the analogous spatial distribution of sulfate. For many of the metals this may be caused by contributions from crustal components, either long-range-transported, local, or both. For sulfate the maximum concentrations arrived from sector 8 and the minimum from sector 3. The ratio of maximum to minimum [SO_4^{2-}] was 5.9. For Fe, K, Al, Mg, and Ca the mean minimum concentrations were also observed from sector 3, but the maximum was from sector 9. The ratios of maximum to minimum for these five metals were 4.0, 2.8, 2.3, 2.0, and 1.8, respectively. While none of these metals had a spatial distribution as sharply peaked as that of SO_4^{2-}, Ca was particularly flat. This may be due to a local source of Ca, for example, from salting the roads during winter. For Zn,

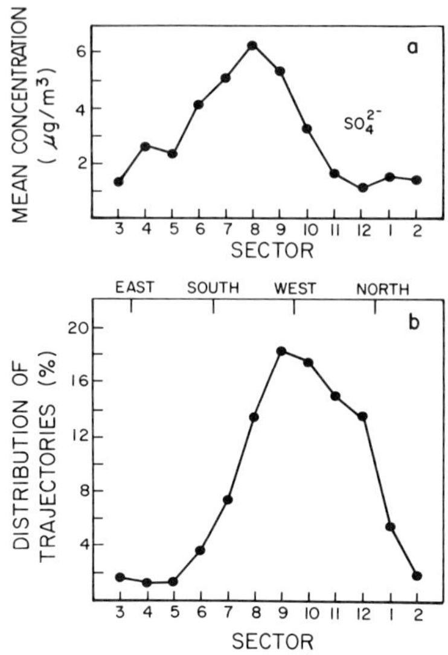

Figure 9.4.(*a*) Mean concentration of sulfate as a function of wind direction (sector). (*b*) Distribution of air trajectories as a function of sector.

the maximum concentration was from sector 8 and the minimum from sector 4. The ratio of maximum to minimum was 2.8. Lead was unique in that the maximum concentration was associated with sector 7, although the concentrations from sectors 8 and 9 were also relatively high.

The main feature of the spatial distribution in each case is the peak in the southwest quadrant. Emissions from the industrial and metropolitan areas lying to the southwest of the Adirondacks are most likely responsible for this peak. The secondary peak in the Northeast quadrant may arise from emissions in metropolitan Montreal region and industrial areas along the St. Lawrence River.

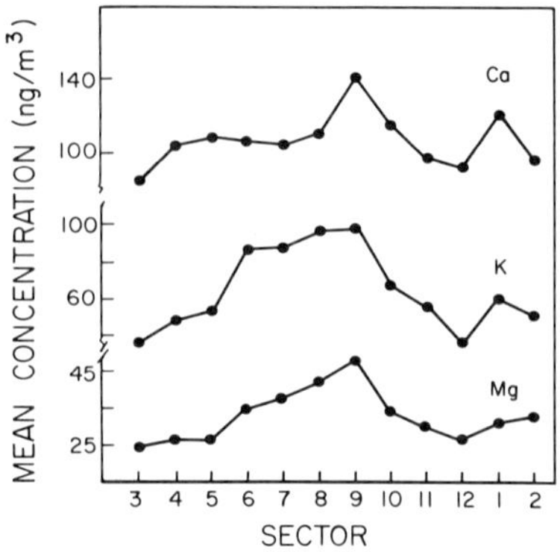

Figure 9.5. Mean concentrations of Ca, K, and Mg as a function of sector.

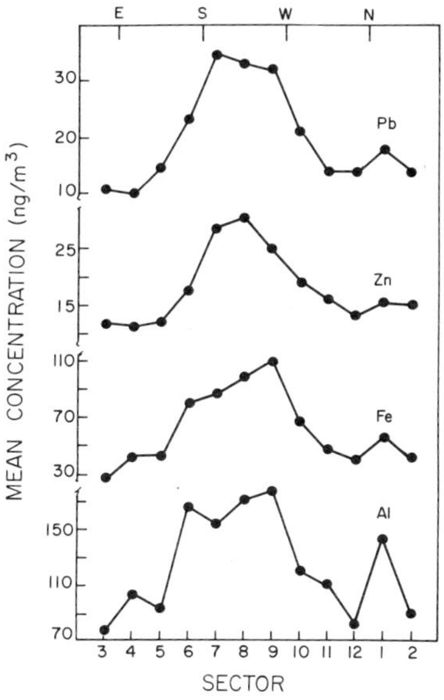

Figure 9.6. Mean concentration of Pb, Zn, Fe, and Al as a function of sector.

To compare contributions from various directions, the frequency of windflow from each direction must also be taken into account. For example, mean [SO_4^{2-}] from sectors 2, 3, and 4 was ~ 2μg/m^3 or approximately one third that of sector 8 (Fig. 9.4*a*). However, the contributions of easterly windflows from sectors 2, 3, and 4 would be far smaller because easterlies rarely influence WFC (Fig. 9.4*b*). The relative contributions (%M) for SO_4^{2-} and trace metals from individual sectors may be calculated as:

$$\%M = \frac{\overline{C}_j N_j}{\sum_{J=1}^{12} (\overline{C}_j N_J)} 100 \qquad (2)$$

The results are summarized in Table 9.2. Contributions from sectors 8, 9, and 10 predominate, although sectors 7, 11, and 12 also contribute significantly.

A limitation of this analysis is whether the partitioning of trajectories is proper when they pass through several sectors. For example, the trajectories on August 2, 1981 discussed in Section 5.1 pass through the high SO_2 emission areas in the Ohio River Valley and also along Lake Erie before looping to WFC. The above analysis would assign a portion each to sectors 8 and 9. Summer trajectories frequently pass through sectors 8, 9, and 10. Therefore, it may be more meaningful to determine total contributions by combining windflows from these sectors. The data in Table 9.2 yields net contributions from sectors 8–10 of 56, 58, 58, 56, 62, 59, 61, and 67% for Mg,

Table 9.2. Relative Contributions (%) from Each Sector

Sector No.	Sector (deg)	Mg	Al	K	Ca	Fe	Zn	Pb	SO_4^{2-}
1	0–30°	5	5	5	6	4	4	4	3
2	30–60°	2	1	1	2	1	1	1	<1
3	60–90°	1	<1	<1	1	<1	<1	<1	<1
4	90–120°	1	1	1	1	<1	<1	<1	<1
5	120–150°	<1	<1	<1	1	<1	<1	<1	<1
6	150–180°	4	5	4	4	4	3	4	4
7	180–210°	8	9	9	7	9	10	11	10
8	210–240°	16	18	18	14	19	20	20	24
9	240–270°	24	24	24	23	27	22	25	27
10	270–300°	16	16	16	18	16	17	16	16
11	300–330°	13	13	11	13	10	12	9	7
12	330–360°	10	8	9	12	7	9	8	5

Al, Ca, Fe, Zn, Pb, and SO_4^{2-}, respectively. For SO_4^{2-} the northeast and Mid-Atlantic states (sectors 2–7) contribute 17%, and northern areas in Canada (sectors 11, 12, 1, and 2) contribute 16%. Trace metals closely follow SO_4^{2-}.

5. IDENTIFICATION OF POLLUTION SOURCES

5.1. Chemical Elements as Tracers of Regional Pollution Sources

The data presented in Section 4 indicate that approximately 67% of the [SO_4^{2-}] at WFC are attributable to the transport of air masses from southwest of the site. These air masses also pass through areas of lower SO_2 emissions within New York State. Because these areas are much closer to Whiteface Mountain, they can have a larger relative effect and their contribution must be considered. Neither present transport models nor the data discussed in Section 4 can adequately accomplish this. This difficulty has been a serious hurdle in answering the acid-rain source question. If SO_2 or its product, SO_4^{2-}, could be traced in the atmosphere, it would further considerably our efforts to determine the impact of regional "individual" SO_2 sources. Artificial tracers such as SF_6 or $^{13}C^2H_4$ are not suitable for studies of the transport of SO_2 or SO_4^{2-} because of large differences in chemical and physical removal processes. Radioactive tracers such as ^{35}S could be very useful in sulfate transport studies but the public perception of radioactivity hazards, real or imagined, would prohibit such investigations. These considerations limit us to the application of chemicals either naturally present in source regions or emitted in industrial operations.

Since principal SO_2 sources are combustion of coal and residual oil, trace elements emitted during combustion may fingerprint the aerosols if their composition is unique. This could be significant since the Midwest relies heavily on coal, and the Northeast on oil. The primary source of V in air is the combustion of residual oil used in oil-fired power plants. V emissions from oil-fired plants are tenfold higher compared to coal-fired plants. Figure 9.7 shows that the [V] (Homolya, 1983) in northeastern states are much higher than in midwestern states. Emissions from Wisconsin, Michigan, Illinois, Indiana, Kentucky, Ohio, and West Virginia total 3×10^6 kg/yr compared to 12×10^6 kg/yr for the smaller area of the northeastern states. Therefore, V seems well-suited to distinguish midwestern aerosols from those produced in the Northeast.

The concentration of an element in air is greatly affected by atmospheric processes, such as atmospheric mixing and precipitation scavenging, and hence

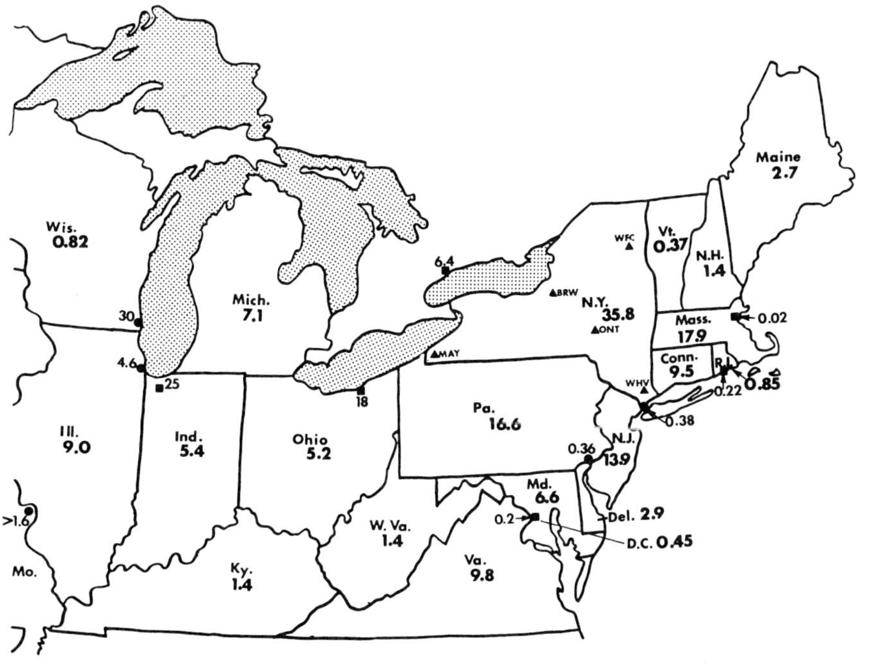

Figure 9.7. Vanadium emissions (10^5kg/yr) and Mn/V for states from Midwest to Northeast. Boldface numerals, calculated V emissions for 1978 (Homolya, 1983). ●, uncorrected Mn/V. ■, Mn/V with crustal corrections where [Al] were known. ▲, sampling sites in New York State: BRW, Brewerton; MAY, Mayville; ONT, Oneonta; WFC, Whiteface Mountain; WHV, West Haverstraw. Data were obtained from the literature as follows: Chicago, Ill., Milwaukee, Wisc., and northwestern Indiana (Winchester and Nifong, 1971), Cleveland, Ohio (King et al., 1976), Philadelphia, Pa., and Washington, D.C. (Lee et al., 1972), Boston, Mass. (Hopke et al., 1976), St. Louis, Mo. (Dzubay and Stevens, 1975), New York, N.Y. (Kleinman et al., 1974), Toronto, Ont., Can. (Paciga and Jervis, 1976), and Narragansett, R.I. (Rahn et al., 1982). From Husain et al. (1984). Reprinted with permission from Pergamon Press, Ltd.

cannot be used singly to distinguish regional sources of air masses. Ratios of two elements which are present at a known ratio in a source region and are affected equally by atmospheric processes provide a much better basis for such comparisons. Rahn et al. (1982) made the empirical observation that Mn/V ratios in the Northeast (<0.2) are much smaller compared to the Midwest (>1). We have calculated Mn/V ratios (corrected for crustal contributions and written simply as Mn/V from here on) for these regions (Fig. 9.7) from the data available in literature and they are, in fact, sharply different. Differences principally in [V] apparently yield the distinctive Mn/V thus enhancing their use as tracers. Mn/V in the Midwest vary from ~2 to 30, tenfold or larger than on the East Coast, ~0.2 (Fig. 9.7). These values are qualitatively different. Extensive data have been obtained to investigate the usefulness of Mn/V as a tracer (Rahn et al., 1982; Husain et al., 1983, 1984). Other possible tracers, especially Se and As, hold promise. The V/Se ratios in the crust, coal-fired, and oil-fired fly ash are ~1,000, ~20, and 1,500–42,000, respectively—values that are sufficiently source-specific to merit investigation. Similarly, V/As ratios from these sources are also vastly different: crust, 55; coal-fired fly ash, ~10; and oil-fired fly ash, ~500–4,000.

Carefully designed experiments must be implemented to determine the usefulness of these elements as tracers of atmospheric transport. Our investigations from 1978 to 1982 have shown that conventional daily aerosol sampling has limited usefulness in scrutinizing source-specific tracers. Weekly or semiweekly sampling is even less useful and may lead to erroneous conclusions. Significant meteorologic changes, that is, air trajectory shifts, often occur over a period of a few hours. Therefore, sampling durations must be minimized to maximize resolution of aerosols from a given region. Furthermore, to add confidence to source-specific sampling it is highly desirable to intercept (by sampling) an air mass at several widespread locations. Because of the multiplicity of sources, only experiments carefully designed in light of these considerations can lead to the development of dependable atmospheric tracers.

5.2. Mn/V Ratio as a Tracer

The usefulness of Mn/V ratio as a fingerprint of the midwestern and northeastern aerosols was tested by Husain et al. (1984) who conducted an experiment in New York State in 1981. As described in Section 2, samples were collected at five strategically located sites every 6 h so that a pollution episode could be sampled several times as the air mass moved across the state. A dozen or so 6-h samples, which were collected during each of three episodes when [SO_4^{2-}] exceeded 30 $\mu g/m^3$, allowed detailed examination. The results of this extensive study have been published elsewhere (Husain et al., 1984). With permission from Pergamon Press, discussion of one test case is reproduced below with minor modifications.

5.2.1. August 1–4, 1981 Episode

At MAY, [SO_4^{2-}] began increasing on July 31, reached 34 $\mu g/m^3$ by August 2, and remained high until the morning of August 5 (Fig. 9.8). High concentrations were also measured at the other four sites. Air masses were arriving from the north early on July 31, when [SO_4^{2-}] were low (Fig. 9.9). Later in the day, [SO_4^{2-}] began to increase at MAY, when the air masses began to cross into western Pennsylvania before looping back into MAY. Air masses reaching BRW and WFC were still from the north, and [SO_4^{2-}] at BRW were near background level. By August 1, New York State was receiving air masses from Pennsylvania, and by August 2, the air circulation pattern had shifted so that air masses were looping westward into Ohio before returning to the state. The highest [SO_4^{2-}] coincided with the air masses coming initially from Ohio (on Aug. 2–3) and then from Indiana and Michigan (Aug. 4). The average [SO_4^{2-}]

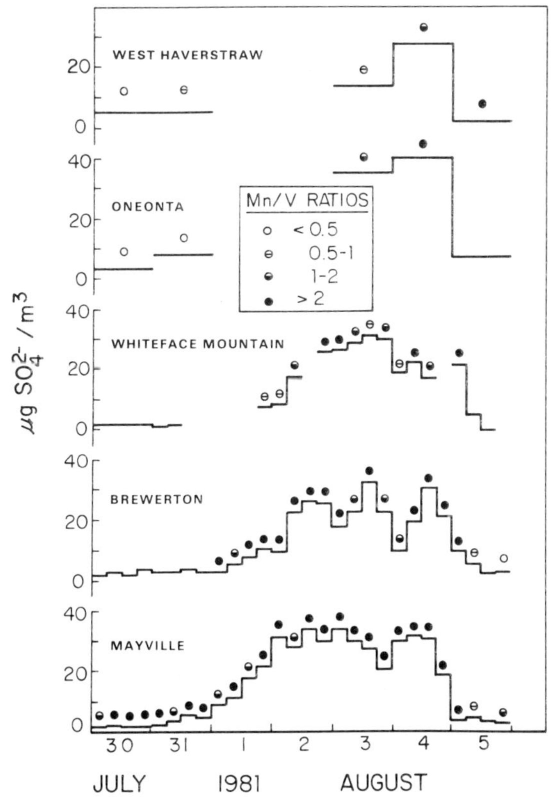

Figure 9.8. [SO_4^{2-}] in New York State, July 30–August 5, 1981, collected every 6 h at MAY, BRW, and WFC and every 24 h at ONT and WHV. Mn/V during the August 1–4 episode yielded a midwestern signature at all five sites. From Husain et al. (1984). Reprinted with permission from Pergamon Press, Ltd.

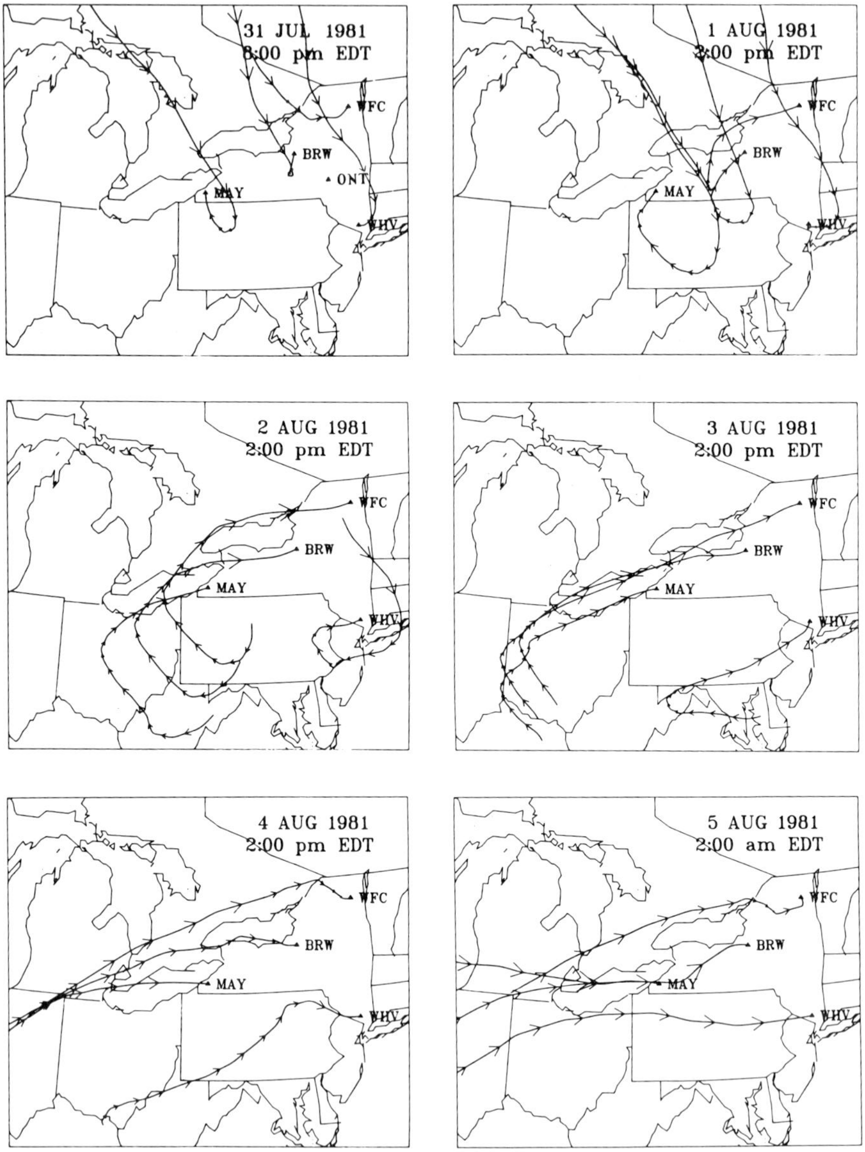

Figure 9.9. Selected air trajectories to sampling sites, July 31–August 5, 1981. From Husain et al. (1984). Reprinted with permission from Pergamon Press, Ltd.

on August 2 were (in $\mu g/m^3$): 30.8 at MAY, 20.7 at BRW, and 17.3 at WFC. The decreasing concentrations at BRW and WFC may be due partly to further mixing over the longer transport distance. They may also reflect regional differences in the emissions of SO_2 and SO_4^{2-}, since air masses reaching these sites on August 2 appeared to be from different areas in Ohio and Pennsylvania.

On August 3, the average [SO_4^{2-}] (in $\mu g/m^3$) were similar: 28.1 at MAY, 23.7 at BRW, and 29.1 at WFC. This similarity probably stemmed from the influence of the same air mass on all three sites.

On August 4, the average [SO_4^{2-}] (in $\mu g/m^3$) were 27.9 at MAY, 20.3 at BRW, and 19.5 at WFC. The higher [SO_4^{2-}] at MAY most likely reflects its nearness to the midwestern source. As the air masses travel eastward, passing through low SO_2 emission areas, mixing and removal processes appear to lower [SO_4^{2-}], as observed at BRW and WFC. On August 2–4, the average decreases in [SO_4^{2-}] at BRW compared to MAY were 32, 15, and 27%, respectively. Decreases at WFC compared to MAY were 44, 0, and 30%, respectively. An average decrease in [SO_4^{2-}] between MAY and WFC was ~25%. This is significantly smaller than a 60% decrease observed in monthly average concentrations.

During this episode air masses influencing MAY, BRW, WFC, and probably ONT were quite different from those influencing WHV. On August 3, [SO_4^{2-}] at WHV (13.5 $\mu g/m^3$) was less than half of the concentrations at the other sites. Air masses reaching WHV had traveled through Maryland, Virginia, West Virginia, southern Pennsylvania, and New Jersey. The lower [SO_4^{2-}] at WHV compared to rural sites in western, central, and northern New York provides striking testimony on the sources of SO_4^{2-}. If SO_2 emissions in New York State were important contributors, the highest [SO_4^{2-}] should have been observed at WHV (much closer to the New York City–New Jersey metropolitan area) and the lowest at MAY—exactly opposite of what was observed.

The change in [SO_4^{2-}] at WHV on August 4 leaves no doubt that transport from the Midwest was responsible for the high [SO_4^{2-}] at all sites. On August 4, WHV was influenced by the air masses which had passed through the Ohio River Valley, and [SO_4^{2-}] at this site had doubled to 27.9 $\mu g/m^3$, a concentration very similar to those at the other four sites when they were influenced by the midwestern air masses during this episode.

Examination of the 6-h data provides additional support for the midwestern origin of SO_4^{2-}. Only at MAY and BRW were the air masses on August 1 arriving from Pennsylvania (Fig. 9.9), a high-emission state. On the same day [SO_4^{2-}] at these sites began to increase. The increase occurred earlier at MAY than at BRW and was larger at MAY (15 vs. 6.5 $\mu g/m^3$, Fig. 9.8). The first three samples for August 1 could not be collected at WFC, but subsequent samples showed lower [SO_4^{2-}] than at MAY. On August 5 [SO_4^{2-}] suddenly decreased at all five sites following widespread precipitation.

At BRW, WFC, ONT, and WHV available data show marked increases in Mn/V during August 1–4. The increases in Mn/V were related to decreased [V], particularly at ONT and WHV. For example, during July 16–31, average [V] on nonprecipitation days at ONT and WHV were 27.3 and 29.5 ng/m^3, respectively. On August 3 and 4, a three- to ninefold decrease in [V] was measured at these sites (ONT, 4.4. and 3.2; WHV, 9.3 and 6.1 ng/m^3). Comparison of [SO_4^{2-}] and air trajectories above showed that high [SO_4^{2-}] at all sites were due to transport from the neighboring midwestern states. Mn/V

(Fig. 9.8), which were almost invariably larger than 1 at all sites during this episode, are consistent with this interpretation.

On August 1, Mn/V at MAY in the first three samples ranged from 1.1 to 2.0 with an average of 1.5. Air masses during this time were from western Pennsylvania. On August 2 and 3, Mn/V averaged 4.0 when air masses were principally from Ohio. On August 4, Mn/V increased markedly—to 11; the air masses were now mainly from Indiana and southern Michigan. At MAY, therefore, Mn/V appears to yield a regional signature: 1 from northwest Pennsylvania, 4 from Ohio, and ~11 from Indiana and southern Michigan. Mn/V appears to decrease with transport eastward as observed daily averages on August 2–4 were MAY, 3.4, 4.7, and 10.9; BRW, 3.8, 2.5, and 3.7; and WFC, 1.8, 1.5, and 1.8. The most direct comparison can be made from the August 3 data when all 6-h sites were influenced by the same air mass. A two- and threefold decrease in the ratio from MAY to BRW to WFC is seen in the observed values of 4.7, 2.5, and 1.5 at MAY, BRW, and WFC, respectively.

Taken together, air trajectories, Mn/V, and [SO_4^{2-}] show that during August 2–4, SO_4^{2-} aerosols were transported into New York State from the nearby midwestern states and Pennsylvania.

5.3. Reliability and Limitations of Mn/V as a Tracer

In air masses reaching New York State, Mn/V was usually effective in distinguishing aerosols produced in the Midwest from those produced in the Northeast. The Mn/V were invariably > 1 for samples associated with air masses from the Midwest. The usefulness of this technique was rather clearly demonstrated in the July 24–26 episode (Husain et al., 1984) when MAY was influenced by a midwestern air mass system and BRW and WFC by air masses passing through oil-consuming areas. The Mn/V at MAY bore the midwestern signature, whereas the BRW and WFC samples yielded values typically observed on the East Coast, ~0.2.

A limitation of this technique appears in its application in areas where large amounts of V are emitted. [V] and [SO_4^{2-}] for summer 1981 samples collected at MAY, ONT, and WHV are plotted in Figure 9.10. (WFC and BRW data were similar to those for MAY.) At MAY [V] were between 1 and 10 ng/m^3, with an average of ~ 3. The range was much greater (between 1 and 70 ng/m^3) at ONT and WHV. The highest [SO_4^{2-}] at these sites were distinctly associated with low [V], and low [SO_4^{2-}] were associated with high [V].

The low [V] at MAY reflects the low V emissions in the Midwest. Since [V] in urban areas in the East can be at least tenfold higher (Rahn et al., 1982; Hopke et al., 1976; Lioy et al., 1980), a small amount of mixing can drastically alter the Mn/V of an incoming midwestern air mass as it travels eastward.

Since [V] associated with the midwestern air masses are only a few ng/m^3, one questions the validity of applying the Mn/V technique in areas where local

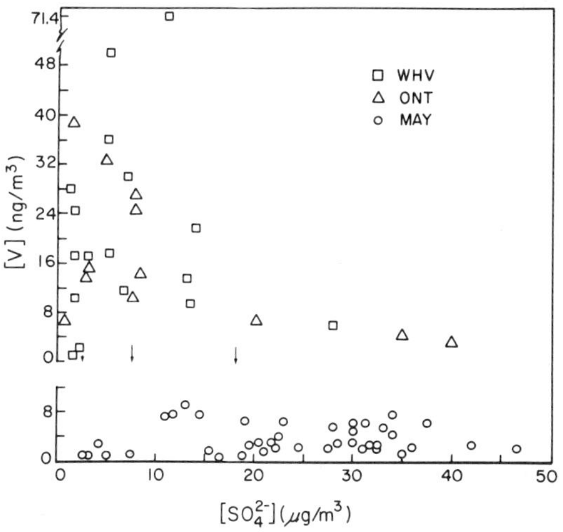

Figure 9.10. [V] vs. [SO_4^{2-}] at MAY, ONT, and WHV, July 16–August 5, 1981. From Husain et al. (1984). Reprinted with permission from Pergamon Press, Ltd.

emissions may result in [V] tenfold or higher. This possibility was investigated by Husain et al. (1983) using samples collected in July 1976 at four locations in New York and one in New Jersey. The sampling site at Albany was ~15 km from an oil-fired power plant. The [V] in Albany samples were much higher (average, ~30 ng/m^3) as expected. During an episode when all five sampling sites were influenced by midwestern air masses, Mn/V at all sites except Albany bore the midwestern signature. Samples from Albany, however, retained the East Coast signature. This observation is neither unexpected nor difficult to explain. Because the midwestern air masses contain only a few ng/m^3 of V, small contamination from local aerosols with [V] of up to 100 ng/m^3 would result in ratios symbolic of East Coast signature.

Whereas Mn and V are associated with primary particle emissions, [SO_4^{2-}] is produced from the oxidation of SO_2 in the atmosphere. Therefore, a quantitative relationship between Mn/V and SO_4^{2-} was neither expected nor was observed. At this time, the Mn/V technique should be taken only as a qualitative tracer. Another limitation was indicated by a distinct decrease in Mn/V accompanying air masses as they moved eastward from MAY to WFC. Caution is advised, therefore, in applying Mn/V as an aerosol tracer at much larger distances from the emission sources.

5.4. Identification of Pollution Source Types by Electron Microscopy

Coal-fired sources and oil-fired sources of sulfate-bearing particles can be distinguished by electron microscopical (EM) techniques. Scanning EM with energy-dispersive X-ray spectroscopy (EDXRS) has revealed that oil fly ash is morphologically and chemically distinct from coal fly ash. Supramicrometer oil fly ash is honeycombed and contains high concentrations of vanadium, whereas coal fly ash is typically smooth Si, Fe, or Al spheres. Electron and X-ray diffraction have shown that mullite, quartz, magnetite, hematite, and glass are major mineral components of coal fly ash.

Preliminary analyses of microparticles collected at WFC during the summer of 1983 have yielded the most direct evidence yet of a coal-burning source of high-SO_4^{2-} air masses reaching the Northeast (Webber et al., 1984). In a sample collected during a brief episode (24 μg SO_4^{2-} / m^3) on June 23, submicrometer coal fly ash (Fig. 9.11) was identified by high-voltage EM with EDXRS and electron diffraction. These magnetite and glass spheres were not

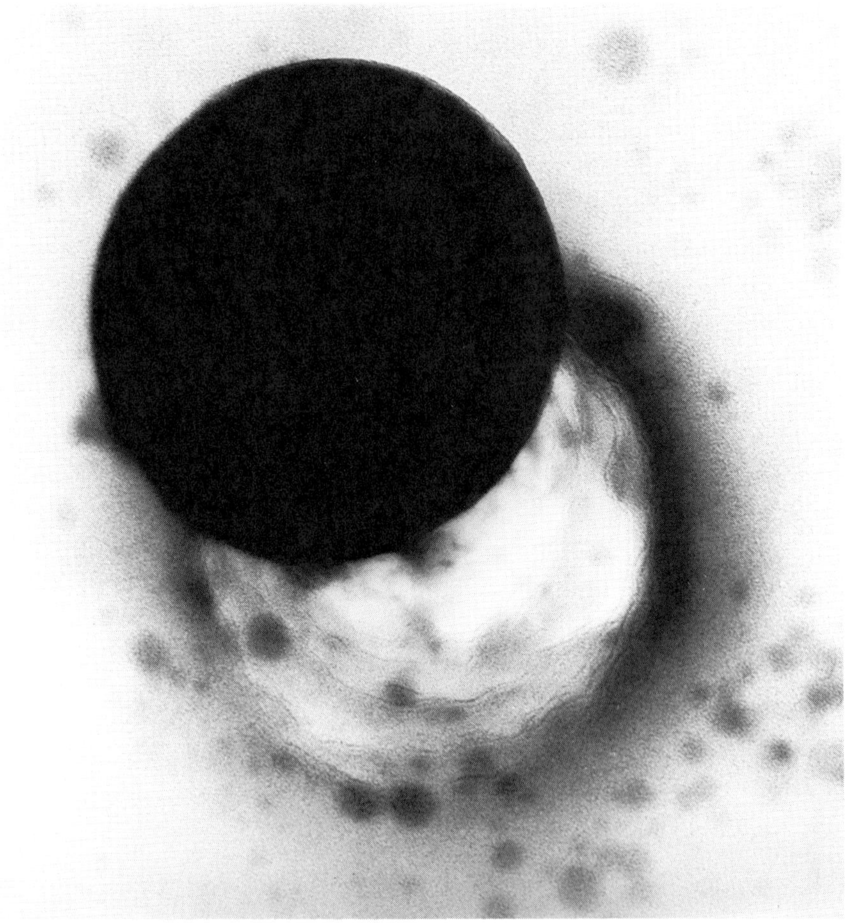

Figure 9.11. Transmission electron micrograph of a typical coal fly ash particle collected during a high-sulfate episode at Whiteface Mountain, N.Y. The 0.25-μm sphere is lodged on the replicated edge of an incomplete Nuclepore filter pore.

seen 12 h before or 12 h after the episode. The lack of supramicrometer coal fly ash during the episode pointed to a distant source, and a midwestern source of this coal fly ash was indicated by meteorologic conditions and a midwestern Mn/V ratio.

An earlier sample collected when [SO_4^{2-}] was only 2.7 $\mu g/m^3$ and [V] was high (12.8 ng/m^3) revealed no coal fly ash, but some amorphous submicrometer particles did yield weak V X-ray peaks. These were most likely oil fly ash particles from an oil-fired power plant ~200 km upwind. These V-bearing particles were not seen during the episode.

6. SUMMARY

Concentrations of trace metals and SO_4^{2-} have been determined in daily samples of suspended particulate matter collected from July 1978 to December 1979 at Whiteface Mountain, N.Y. Large variations (up to 100-fold) were observed. Concentrations were highest in summer, intermediate in autumn and spring, and lowest in winter. Comparison of daily concentrations with wind direction shows that high summer concentrations were due to the increased transport of air masses from southwest of the site—the industrial Midwest. Concentrations of SO_4^{2-} and trace elements deduced in background aerosols from daily measurements are only a fraction of those measured during pollution episodes.

In addition to the large emissions in the industrial Midwest, smaller operations in the Northeast also introduce SO_4^{2-} and trace elements into the atmosphere. In order to distinguish emissions from the two sources, chemical elements were considered as possible fingerprints of pollution source regions. Since V emission on the East Coast are distinctly higher than in the Midwest and crustal-corrected Mn/V ratios in the Midwest (≥ 1) are suggested to be at least tenfold greater than in the Northeast (≤ 0.2), air-transport experiments using five sampling sites and 6-h sampling were conducted in New York State. Results showed that in rural areas Mn/V is a reliable tracer of Midwestern and East Coast aerosols but also revealed limitations of the technique. Average [V] in the Midwestern air masses were only a few ng/m^3. In Northeastern areas where [V] are tenfold or higher, Mn/V cannot be generally used to fingerprint Midwestern aerosols. The second limitation was a Mn/V decrease with distance from the source. Consequently, the Mn/V technique cannot quantify the transport of SO_4^{2-} components, but it can qualitatively fingerprint source regions if applied judiciously.

Identification of pollution source types by electron microscopy appears promising. Analysis of microparticles collected during a SO_4^{2-} episode showed the presence of coal fly ash. Mn/V ratio yielded a midwestern signature and the meteorologic conditions also suggested air transport from the Midwest. A combination of trace elements and electron microscopic techniques may provide the most powerful tool to identify distant pollution sources.

ACKNOWLEDGMENT

The author wishes to express his appreciation to V. A. Dutkiewicz for several stimulating discussions and suggestions on organization of this chapter; J. S. Webber for commenting on the manuscript; and P. P. Parekh for assistance in trajectory calculations.

REFERENCES

Adams, F., Dams, R., Guzman, L., and Winchester, J.W. (1977). "Background aerosol composition on Chacaltaya Mountain, Bolivia." *Atmos. Environ.* **11,** 629.

Altschuller, A. P. (1976). "Regional transport and transformation of sulfur dioxide to sulfate in the U.S." *J. Air Pollut. Contr. Assoc.* **26,** 318.

Benkovitz, C. M. (1982). "Compilation of an inventory of anthropogenic emissions in the United States and Canada." *Atmos. Environ.* **16,** 1551.

Canelli, E., and Husain L. (1982). "Determination of total particulate sulfur in air samples at Whiteface Mountain, New York by pyrolysis microcoulometry." *Atmos. Environ.* **16,** 945.

Chang, Y.S. (1978). "Ground level ozone, sulfate and total suspended particulates in Canada." *Proceedings of the 1978 International Clean Air Conference* (E.T. White et al., eds.) Brisbane, Australia, May 15–19. *Ann Arbor Science, Michigan,* **91.**

Crecelius, E.A., Lepel, E.A., Laid, J.C., Rancitelli, L.A., and McKeever, R.L. (1980). "Background air particulate chemistry near Colstrip, Montana." *Environ. Sci. Technol.* **14,** 422.

Dams, R., and DeJonge, J. (1976). "Chemical composition of Swiss aerosols from the Jungfraujoch." *Atmos. Environ.* **10,** 1079.

Dutkiewicz, V.A., Halstead, J.A., Parekh, P.P., Khan, A., and Husain, L. (1983). "Anatomy of an episode of high sulfate concentration at Whiteface Mountain, New York." *Atmos. Environ.* **17,** 1475.

Dzubay, T.G., and Stevens, R.K. (1975). "Ambient air analysis with dichotomous samples and x-ray fluorescence spectrometer." *Environ. Sci Technol.* **9,** 663.

Heffter, J.L. (1980). "Air Resources Laboratory atmospheric transport and dispersion model (ARL-ATAD). NOAA Technical Memo ERL ARL-81. Air Resources Laboratories, Boulder, Colo.

Homolya, J. (1983). "Anthropogenic emissions." In *The Acid Deposition Phenomenon and its Effects,* Vol. 1 of *Atmospheric Sciences.* Public Review Draft EPA-600 :8-83-016A. Office of Research and Development, Washington, D.C.

Hopke, P.K., Galdney, E.S., Gordon, G.E., Zoller, W.H., and Jones, A.G. (1976). "The use of multivariate analysis to identify sources of selected elements in the Boston urban aerosol." *Atmos. Environ.* **10,** 1015.

Hopke, P.K., Lamb, R.E., and Natusch, D.F.S. (1980). "Multielemental characterization of urban roadway dust." *Environ. Sci. Technol.* **14,** 164.

Husain, L., and Samson, P.J. (1979). "Long range transport of trace elements." *J. Geophys. Res.* **84,** 1237.

Husain, L., Dutkiewicz, V.A., and Parekh, P.P. (1980). "Sources of ozone and sulfate in northeastern United States." Annual Progress Report (C00-4501-4) to the US Department of Energy Contract No. DE-AC02-77EV04 501.A000. Washington, D.C.

Husain, L., Webber, J.S., and Canelli, E., (1983) "Erasure of midwestern Mn/V signature in an area of high vanadium concentration." *J. Air Pollut. Contr. Assoc.* **33,** 1185.

Husain, L., Webber, J.S., Canelli, E., Dutkiewicz, V.A., and Halstead, J.A. (1984). "Mn/V ratio as a tracer of aerosol sulfate transport." *Atmos. Environ.* **18,** 1059.

King, R.R., Fordyce, J.S., Antoine, A.C., Liebecki, H.F., Neustadter, H.E., and Sidik, J.M. (1976). "Elemental composition of airborne particulates and source identification: An extensive one year survey." *J. Air Poll. Contr. Assoc.* **26,** 1073.

Kleinman, M.T., Kneip, T.J., and Eisenbud, M., (1974) "Meteorological influences on airborne trace metals and suspended particulates." In *Trace Substances in Environmental Health VII* (ed., D.D. Hemphill). University of Missouri, Columbia, Mo., p. 147.

Lazrus, A.L., Hill, K.C., and Lodge, J.P. (1966). "A new colorimetric micro determination of sulfate ion." Proceedings of Technicon Symposia *Automation in Analytical Chemistry.* Mediad Inc., New York.

Lee, R.E., Garanson, S.S., Environe, R.E., and Morgan, G.B. (1972). "National air surveillance cascade impactor network II size distribution measurements of trace metal components." *Environ. Sci. Technol.* **6,** 1025.

Lioy, P.J., Mallon, R.P., and Kneip, T.J. (1980). "Long term trends in total suspended particulates, vanadium, manganese and lead at near street level and elevated sites in New York, City." *J. Air Poll. Contr. Assoc.* **30,** 153.

Maenhaut, W., and Zoller, W.H., (1976) "Determination of the chemical composition of the South Pole aerosols by instrumental neutron activation analysis." Proceedings of the 1976 International Conference. *Modern Trends in Activation Analysis.* Munich, FRG.

Paciga, J.J., and Jervis, R.E. (1976). "Multielement size characterization of urban aerosols." *Environ. Sci. Technol.* **10,** 1124.

Parekh, P.P., and Husain, L. (1982). "Ambient sulfate concentrations and windflow patterns at Whiteface Mountain, New York." *Geophys. Res. Lett.* **9,** 79.

Rahn, K.A., Lowenthal, D.H., and Lewis, N.F. (1982). "Elemental tracers and source areas of pollution aerosol in Narragansett, R.I." Technical Report. University of Rhode Island, Narragansett, R.I.

Samson, P.J. (1978). "Ensemble trajectory analysis of summertime sulfate concentrations in New York State." *Atmos. Environ.* **12,** 1889.

Schneider, H.J., and Lehmann, B.L. (1977). "Contributions to a new genetic concept on the Bolivian tin province." In *Time and Strata Bound Ore Deposits* (Klemn, D.D., and Schneider, H.J., eds.). Springer-Verlag, Berlin.

Webber, J.S., Dutkiewicz, V.A., and Husain, L. (1985). "Evidence of submicrometer coal fly ash during an episode of high sulfate at Whiteface Mountain, New York" *Atmos. Environ.* **19,** 285.

Winchester, J.W., and Nifong, G.D. (1971). "Water pollution in Lake Michigan by trace elements from pollution aerosol fallout." *Water Air Soil Pollut.* **1,** 50–64.

Zoller, W.H., Gladney, E.S., and Duce, R.A. (1974). "Atmospheric concentrations and sources of trace metals at the South Pole." *Science* **183,** 198.

10

CHEMICAL SPECIATION AND REACTION PATHWAYS OF METALS IN THE ATMOSPHERE

Roy M. Harrison

Department of Chemistry
University of Essex
Colchester, England

1. INTRODUCTION

Until recently, very little has been known of the chemical speciation and reaction pathways of metals in air. The analytical problems are intense and even now no fully satisfactory speciation methodology is available. This has hampered not only the identification of discrete chemical species of metals, but also the elucidation of chemical reaction pathways, since this is dependent upon precise characterization of the starting materials and end products of such reaction processes.

An understanding of chemical speciation is important for two prime reasons. First, the toxicity, both by ingestion and inhalation, of a metal is dependent upon the chemical form. In an extreme example, the toxicology of inorganic mercury and alkylmercury are very different. In a less obvious case, the differing particle sizes and chemical species of lead in the air of a primary smelter can have important industrial hygiene implications (O'Neill et al., 1982). While the particle size influences the point of deposition in the respiratory system, the chemical form determines the rate and efficiency of absorption of an inhaled or ingested metal. In particular, gastrointestinal absorption, which may be the prime route of absorption of *airborne* metal in the case of particles of 2–10 μm (which deposit primarily in the tracheobronchial region and enter the stomach via the muco-ciliary escalator mechanism), is critically dependent upon the chemical form of the metal and the nutritional status of the exposed individual.

The second major significance of chemical speciation is that it strongly influences the environmental pathways and fate of a metal. Large, insoluble particles may travel only short distances and be relatively immobile after deposition, while smaller, water-soluble particles may travel over long distances and be of greater environmental mobility in the aquatic or terrestrial environment.

For these and other reasons the study of chemical speciation and environmental reaction pathways is of considerable importance. This chapter seeks to highlight areas of substantial knowledge and to indicate subjects where information is seriously lacking.

2. SPECIATION METHODS

2.1. Particulate Metal Compounds

There has been a considerable volume of published work (e.g., Hamilton and Adie, 1982; Butler et al., 1976) involving scanning electron microscopy or electron microprobe with X-ray microanalysis (SEM/XES) to identify individual metal-rich particles in air samples collected by filtration. This method, however, has some important limitations. With a scanning electron microscope the smallest individual particle that can be analyzed is $\sim$0.5 μm

(Butler et al., 1976), which is larger than many of the important metal-bearing particles (the mass median aerodynamic diameter for lead in air is often <0.3 μm). The elemental analyses generated by X-ray microanalysis are only semiquantitative, and only elements of atomic number ≥ 11 are detected; thus, for example, oxygen and nitrogen are not analyzed and it may not be possible to discriminate between an elemental metal and a metal oxide, or between a metal sulfide and sulfate. Ter Haar and Bayard (1971) gave a range of very specific structural assignments for lead compounds in urban air, which have been questioned by Heidel and Desborough (1975), who doubted the ability of the electron microprobe to give the precise elemental data necessary for such assignments. Probably the major limitation of single particle techniques such as the SEM/XES method is the near impossibility of studying a statistically significant number of particles. A monodisperse aerosol of particles of 1-μm diameter and 2-g/cm^3 density contains 10^7 particles/m^3 at an atmospheric loading of 100 $\mu g/m^3$, and a polydisperse aerosol of similar mass median diameter would contain many more. Detailed examination of only relatively small numbers of particles is possible, and hence it is unlikely that results will represent adequately the great diversity of particles in an ambient aerosol.

The use of a transmission electron microscope offers certain advantages. Smaller particles, with a lower limit of about 0.1 μm, may be examined, and, as well as X-ray analytical data, it is possible to generate electron diffraction patterns from which crystalline materials may be identified (Bloch et al., 1979, Yakowitz et al., 1972). While the structural assignments resulting from this method may be viewed with far greater confidence than in the SEM/XES technique, the impossibility of examining a very large number of particles is a considerable drawback.

Adams and co-workers (1981) have reported the application of laser microprobe mass spectrometry (LAMMA) to the analysis of atmospheric particles collected on a filter or impactor stage. A high intensity laser pulse of resolution about 1 μm is used to volatilize and ionize discrete particles, the ionic fragments being fractionated by a time-of-flight mass spectrometer, generating both positive and negative ion mass spectra. In some cases, the spectra from individual particles correspond to those of simple inorganic salts, but in many cases complex patterns arise from the analysis of multicomponent particles (Adams et al., 1981). Again, this is a single-particle method, with the associated drawbacks.

The most important macroscopic technique used in air particulate speciation studies is X-ray powder diffraction (XRD). Crystalline phases give characteristic diffraction patterns which allow their identification in quite complex mixtures. The technique may be applied to whole air samples collected by filtration or impaction and thus does not suffer from the problems of statistical representativeness which afflict single-particle methods. The main drawback is the fact that amorphous and poorly crystalline phases do not give clear diffraction patterns and are hence not identified even when present in some abundance in the sample.

XRD signals from metallic pollutants in airborne particle samples are frequently obscured by natural mineral materials such as α-quartz and calcite. This problem may be overcome by size fractionation during air sampling using a dichotomous sampler (O'Connor and Jaklevic, 1981) or a cascade impactor (Biggins and Harrison, 1979a,b). Alternatively, after air sampling, the particles may be fractionated by density separation (Fukasawa et al., 1980; Biggins and Harrison, 1979a), the metallic constituents of interest generally being in the most dense fraction. The latter method, however, involves the use of organic solvents, and some chemical changes may occur as a result of interactions with the solvent; for this reason size-fractionated sampling is generally preferable.

Davis (1981) and Davis and Cho (1977) have reported a sophisticated technique for quantification of XRD data. Other workers, however, have not quantified their analyses. The sensitivity of XRD is such that a crystalline phase must comprise at least 1–5% of the mixture before it can be readily identified. Thus, available data are generally qualitative, but refer only to fairly major components of the atmospheric aerosol.

2.2. Gaseous Metal Compounds

Most attention in this field has to date focused upon mercury and lead. A variety of methods have been used, and these inevitably have to be matched to the possible molecular and elemental forms present, rather than being generally applicable to a range of elements.

In the case of mercury, a number of chemical forms may be present in air in the vapor phase. These include volatile inorganic salts (e.g., $HgCl_2$), monoalkyl derivatives (e.g., CH_3HgCl), dialkyl derivatives (e.g., $(CH_3)_2Hg$), and elemental vapor (Hg^o). Selective collectors have been developed by Trujillo and Campbell (1975), Henriques et al. (1973), and Braman and Johnson (1974). These are summarized in Table 10.1. After thermal desorption, the mercury collected in the various traps is analyzed by cold vapor atomic absorption or, in the case of Braman and Johnson (1974), by a specially designed d.c. discharge causing emission at 257.3 nm.

Lead is known to exist in air as vapor-phase tetraalkyllead compounds (TAL) derived from gasoline additives, typically accounting for about 2–8% of total lead in an urban area where leaded gasoline is burned (Birch et al., 1980). There is circumstantial evidence of the existence of vapor-phase trialkyllead (e.g., $(CH_3)_3PbCl$) in the atmosphere, but no direct measurements have been reported.

Many early measurements purporting to determine TAL in air were in fact made by techniques subject to interference and hence liable to give unreliable data (Harrison and Perry, 1977). The commonly used iodine monochloride procedure is subject to interference from inorganic lead (Harrison and Perry, 1977), but it has been substantially amended by Hancock and Slater (1975) to

Table 10.1. Selective Collectors for Vapor Phase Mercury Species

Species	Collector	Reference
Inorganic vapor	HCl-treated Chromosorb W	Braman and Johnson (1974)
Monoalkylmercury	Carbosieve B	Trujillo and Campbell (1975)
	$KMnO_4$ scrubber	Henriques et al. (1973)
	NaOH-treated Chromosorb W	Braman and Johnson (1974)
Dialkylmercury	Carbosieve B	Trujillo and Campbell (1975)
	Au filter	Henriques et al. (1973)
	Au-coated glass beads	Braman and Johnson (1974)
Elemental mercury	Silvered Chromosorb P	Trujillo and Campbell (1975)
	Au coated with Au–Si alloy	Henriques et al. (1973)
	Silvered glass beads	Braman and Johnson (1974)

overcome this interference. The latter method responds to all vapor-phase alkyllead, thus including trialkyllead vapor if present in the air.

The most specific analytical procedure for TAL involves gas chromatographic separation coupled with a specific detector, such as flame atomic absorption (Ebdon et al., 1982), flameless atomic absorption (De Jonghe et al., 1980), microwave plasma (Reamer et al., 1978), or mass spectrometer (Nielsen et al., 1981). In the latter method, interferences may arise and it is essential to establish the normal isotope ratios for lead in lead alkyl fragments to confirm the absence of interfering substances. Any of these methods will give detection of specific tetraalkyllead compounds at picogram levels. Combination with an adsorption tube (Nielsen et al., 1981) or freeze-out technique (De Jonghe et al., 1981) of air sampling provides a sensitive, specific procedure for tetraalkyllead in ambient air.

3. RESULTS OF SPECIATION STUDIES

3.1. Particulate Metal Compounds

The greatest interest has focused upon the compounds of lead present in ambient air polluted by vehicle exhaust or by industrial emissions of the metal.

The most comprehensive data from XRD analysis upon automotive lead in air are due to Biggins and Harrison (1979a), a selection of whose data appears in Table 10.2. While PbBrCl, $PbBrCl \cdot 2NH_4Cl$, and α-$2PbBrCl \cdot NH_4Cl$ are known primary vehicle exhaust emissions, the other compounds are not, and their routes of formation are discussed in Section 4.1. The major component in most samples was found to be $PbSO_4 \cdot (NH_4)_2SO_4$, which has also been observed in air in the United States by O'Connor and Jaklevic (1981).

Table 10.2. Selected Results of Speciation of Automotive Lead in Air[a]

Site	Date	Compounds identified
Central Lancaster, A6 road	May 17–19, 1977	$PbSO_4 \cdot (NH_4)_2SO_4$; $PbSO_4$
	Jan. 30–Feb. 6, 1978	$PbSO_4 \cdot (NH_4)_2SO_4$; α-$2PbBrCl \cdot NH_4Cl$; $PbBrCl \cdot 2NH_4Cl$; NH_4Cl
	June 20–July 7, 1978	$PbSO_4 \cdot (NH_4)_2SO_4$; α-$2PbBrCl \cdot NH_4Cl$; PbBrCl
	Dec. 1–4, 1978	$PbSO_4 \cdot (NH_4)_2SO_4$; α-$2PbBrCl \cdot NH_4Cl$; PbBrCl; $(NH_4)_2SO_4$
Shap, M6 Motorway	Sept. 19–21, 1978	$PbSO_4 \cdot (NH_4)_2SO_4$
Red Scar, Preston, M6 Motorway	Oct. 26–31, 1977	$PbSO_4 \cdot (NH_4)_2SO_4$
	Nov. 14–21, 1977	$PbSO_4 \cdot (NH_4)_2SO_4$; $PbBrCl \cdot (NH_4)_2BrCl$
	Nov. 21–28, 1977	$PbSO_4 \cdot (NH_4)_2SO_4$; $PbBrCl \cdot (NH_4)_2BrCl$
Preston, A6 road	Aug. 1–8, 1977	$PbSO_4 \cdot (NH_4)_2SO_4$; PbBrCl
	Aug. 8–15, 1977	$PbSO_4 \cdot (NH_4)_2SO_4$
	Oct. 4–11, 1977	$PbSO_4 \cdot (NH_4)_2SO_4$; $PbSO_4$
London, A5 road	March 21–28, 1978	$PbSO_4 \cdot (NH_4)_2SO_4$
	Apr. 4–11, 1978	$PbSO_4 \cdot (NH_4)_2SO_4$; $(NH_4)_2SO_4$

[a] From Biggins and Harrison (1979a).

Lead compounds arising from smelting operations are tabulated in Table 10.3 which includes two phases, PbO and Pb° identified in smelter emissions, but not in ambient air outside the smelter. The smelter involved in this work produced Pb, Zn, and Cd, and in addition to the lead phases, ZnO, α-ZnS, and CdO were identified in the stack discharges (Harrison and Williams 1983). Table 10.4 shows the phases identified within the atmosphere of the

Table 10.3. Metal Phases in Stack Discharges from Smelting Operations

Compound	Stack/Ambient	Reference
PbS	Stack/Ambient	Harrison and Williams (1983); Foster and Lott (1980)
$PbSO_4$	Stack/Ambient	Harrison and Williams (1983); Foster and Lott (1980)
$PbO \cdot PbSO_4$	Stack/Ambient	Harrison and Williams (1983); Foster and Lott (1980)
β-ZnS	Stack/Ambient	Harrison and Williams (1983)
PbO (litharge)	Stack	Harrison and Williams (1983)
Pb°	Stack	Harrison and Williams (1983)
ZnO	Stack	Harrison and Williams (1983); Foster and Lott (1980)
CdO	Stack	Harrison and Williams (1983)

Table 10.4. Metal Phases in the Internal Atmosphere of a Primary Zinc–Lead Smelter[a]

β-ZnS	PbS	CdO
ZnO	PbO (litharge)	Cd°[b]
	PbO (massicot)	$Cd(OH)_2$[b]
	$PbSO_4$	
	Pb°	
	$PbO \cdot PbSO_4$	

[a] From Harrison et al. (1981a).
[b] Detected in floor dusts.

smelter itself in the work of Harrison, Williams, and O'Neill (1981a). Since, for reasons of industrial hygiene, workplace air is vented through stacks outside the smelter, these phases will also exist in outside ambient air, although obviously at much lesser concentrations.

Table 10.5 summarizes other metallic compounds observed in ambient air samples. These include natural mineral materials such as calcite and gypsum, as well as the products of atmospheric reactions of anthropogenic emissions, such as the frequently observed metal ammonium sulfates.

Table 10.5. Miscellaneous Metal Phases Identified in Ambient Air

Compound	Analytical Method	Reference
NaCl	XRD	Biggins and Harrison (1979b)
	STEM/electron diffraction	Bloch et al. (1979)
Na_2SO_4	XRD	Biggins and Harrison (1979b)
	STEM/electron diffraction	Bloch et al. (1979)
$CaCO_3$ (calcite)	XRD	Biggins and Harrison (1979b)
	STEM/electron diffraction	Bloch et al. (1979)
$CaSO_4 \cdot 2H_2O$ (gypsum)	XRD	Biggins and Harrison (1979b)
	STEM/electron diffraction	Bloch et al. (1979)
$MgCO_3 \cdot CaCO_3$ (dolomite)	XRD	Davis (1981)
Fe_2O_3 (hematite)	XRD	Davis (1981)
$FeO \cdot Fe_2O_3$ (magnetite)	XRD	Davis (1981)
$CaSO_4 \cdot \frac{1}{2}H_2O$ (hemihydrate gypsum)	XRD	Fukasawa et al. (1980)
$MgSO_4 \cdot 7H_2O$	XRD	Fukasawa et al. (1980)
$Fe_2(SO_4)_3 \cdot 3(NH_4)_2SO_4$	XRD	Biggins and Harrison (1979c)
$CaSO_4 \cdot (NH_4)_2SO_4 \cdot H_2O$	XRD	Harrison and Sturges (1983)
$ZnSO_4 \cdot (NH_4)_2SO_4 \cdot 6H_2O$	XRD	O'Connor and Jaklevic (1981)
K_2SO_4	STEM/electron diffraction	Bloch et al. (1979)
TiO_2	STEM/electron diffraction	Bloch et al. (1979)

3.2. Gaseous Metal Phases

Johnson and Braman (1974) have reported that mercury in air exists primarily (> 90%) in the vapor phase. Volatile mercury in air can arise from a range of sources: Elemental vapor is produced naturally by degassing of the earth's mantle, whereas organic forms are released from aquatic systems as a result of formation in bioalkylation processes.

In the Tampa, Fl. area, Johnson and Braman (1974) found mercury present in all of the forms analyzed (i.e., particulate; inorganic vapor; methylmercury; dimethylmercury; and elemental vapor). Total mercury levels ranged from 1.8 to 298 ng/m^3 with higher levels at nighttime than during the day. From 54 speciation measurements, the volatile mercury accounted for 96% of the total and was on average distributed as follows: inorganic vapor, 25%; methylmercury compounds, 21%; elemental vapor, 49%; dimethylmercury, 1%. The distribution between species was, however, highly variable from sample to sample.

The tetraalkyllead compounds that are observed in the atmosphere reflect their usage in gasoline and changes in relative abundance which occur during atmospheric transport. Table 10.6 shows measurements made by de Jonghe et al. (1981) in terms of relative abundance at different sites. The gasoline station samples show a composition which relates to that in the gasoline itself, but with enrichment of the more volatile methyl species relative to the ethyl and some influence of emissions from vehicles on nearby streets. Samples taken at greater distance from major sources (e.g., the residential and rural areas) show further depletion of ethyl derivatives relative to methyl, presumably as a result of the lower atmospheric stability of the former compounds (Harrison and Laxen, 1978).

Table 10.6. Speciation of the Atmospheric Tetraalkyllead Concentrations at Different Measuring Sites (Average Composition)[a]

	% of Total Tetraalkyllead Content				
Site	Me_4Pb	Me_3EtPb	Me_2Et_2Pb	$MeEt_3Pb$	Et_4Pb
Gasoline station A	18 ± 10	12 ± 9	5 ± 4	2 ± 2	66 ± 21
Gasoline station B	46 ± 11	31 ± 5	12 ± 5	7 ± 6	5 ± 3
Car-repair workshop	53 ± 5	27 ± 4	11 ± 2	1.8 ± 0.5	7 ± 4
Underground tunnel	69 ± 10	13 ± 4	5 ± 3	2 ± 2	12 ± 10
Highway crossing	35 ± 14	25 ± 3	13 ± 3	4 ± 3	24 ± 12
Central-city street	61 ± 7	13 ± 2	5 ± 1	1.2 ± 0.4	19 ± 6
Residential area[b]	76 ± 16	15 ± 10	6 ± 2	nd[c]	nd
Rural area	85 ± 12	20 ± 6	nd	nd	nd

[a] From De Jonghe et al. (1981).

[b] Two values for $MeEt_3Pb$ and Et_4Pb above the detection limit not included.

[c] nd = not detected.

Although alkyl derivatives of several other metals and metalloids are found in environmental waters and sediments (e.g., Sn, As), little, if any, attention has been given to their probable presence in the atmosphere.

4. CHEMICAL PATHWAYS IN THE ATMOSPHERE

While rather little is known of chemical speciation, it is extremely difficult to infer chemical changes within the atmosphere. This is an area requiring considerably increased research effort in terms both of environmental measurements and laboratory studies.

4.1. Inorganic Lead

The inorganic compounds of lead identified in a study of an area without major industrial sources of lead emissions are shown in Table 10.2. It is well known from studies of vehicle exhaust (Hirschler et al., 1957; Habibi, 1973) that the major compound emitted is PbBrCl, with lesser amounts of α-$2PbBrCl\cdot NH_4Cl$, β-$2PbBrCl\cdot NH_4Cl$, and $PbBrCl\cdot 2NH_4Cl$ emitted in urban driving modes. Indeed, three of these four compounds are found at the central Lancaster urban site (Table 10.2).

The other observed compounds of lead must arise from atmospheric chemical transformations of primary lead compounds, and Biggins and Harrison (1979a) used laboratory experiments to investigate possible formation mechanisms. It was found that direct photolysis of PbBrCl did not proceed with tropospheric solar UV light wavelengths, reaction with sulfur dioxide was negligible, and with ammonia in water-saturated air a very slow conversion to Pb(OH)Br occurred. The main reactions were with sulfates. The reaction that appears to be most important in the atmosphere is with ammonium sulphate:

$$2PbBrCl + 2(NH_4)_2SO_4 \longrightarrow PbSO_4\cdot(NH_4)_2SO_4 + PbBrCl\cdot(NH_4)_2BrCl$$

This leads to the formation of $PbSO_4\cdot(NH_4)_2SO_4$, the most commonly observed atmospheric compound of lead in the work of Biggins and Harrison (1979a). Both products of the above reaction appear in two samples from Red Scar, Preston, in Table 10.2. In the atmosphere, the reaction is believed to arise from coagulation of primary vehicle exhaust PbBrCl, typically of size ~ 0.015 μm, and thus subject to rapid coagulation processes (Chamberlain et al., 1979) with the ambient submicrometer aerosol, in which $(NH_4)_2SO_4$ is often the most abundant single component (Harrison and Pio, 1983a). Ammonium sulfate is hygroscopic and exists as solution droplets in air at relative humidities > 80% and often at lower humidities due to hysteresis effects (Charlson et al., 1978). Thus, liquid-phase reactions may readily proceed.

Loss of bromine to the gas phase upon aging of vehicle exhaust aerosol is a well-known phenomenon (Harrison and Sturges, 1983a). This may be explained by the presence of acid sulfates, H_2SO_4 or NH_4HSO_4, in air, with the consequent reactions leading to loss of bromine as HBr:

$$PbBrCl + H_2SO_4 \longrightarrow PbSO_4 + HCl + HBr$$

and

$$6PbBrCl + 2NH_4HSO_4 \longrightarrow 2PbSO_4 + 2PbBrCl \cdot NH_4Br + \alpha\text{-}2PbBrCl \cdot NH_4Cl + HCl + HBr$$

As well as explaining the observed bromine loss from aerosols, these reactions account for the formation of $PbSO_4$, observed in some ambient air samples (Table 10.2).

There are no comprehensive theories of the atmospheric chemistry of lead emitted from smelters. The PbS reported in Table 10.3 is from loss of primary ore minerals, generally by wind. $PbO \cdot PbSO_4$ and $PbSO_4$ arise from the sintering of lead sulfide ores (Harrison and Williams, 1983) and are also emitted from lead blast furnaces (Foster and Lott, 1980). Formation of PbO must arise from oxidation of lead metal and salts in the smelting processes and Pb^{o} formation is due to addition of sawdust to molten lead in the decopperizing operation (Harrison et al., 1981a).

The subsequent chemical changes affecting industrial lead emissions have not been the subject of investigation, although the observation of similar compounds in both stack gases and ambient air close to a primary smelter (Table 10.3) suggests that no rapid chemical changes occur in the air.

It has been reported that lead sulfate is the most frequently observed crystalline (i.e., identifiable by XRD) compound of lead in street dusts (Olsen and Skogerboe, 1975; Biggins and Harrison, 1980). The latter workers, however, found that $PbSO_4$ accounted for, at most, only a few percent of total lead and could be formed by water leaching of $PbSO_4 \cdot (NH_4)_2SO_4$ deposited from the atmosphere. Further interaction of $PbSO_4$ with the dust, however, causes conversion of lead to noncrystalline forms more typical of a soil or sediment (Harrison et al., 1981b).

4.2. Tetraalkyl Lead (TAL)

Both homogeneous and heterogeneous reaction pathways are possible for tetraalkyllead compounds. Edwards and co-workers (1974, 1975) postulated absorption of TAL on the surface of aerosol particles, with possible subsequent decomposition. From measurements of adsorbed TAL in air

(Harrison and Laxen, 1977) and from laboratory studies (Harrison and Laxen, 1978), it has been shown that this route is only of minor significance in the atmosphere.

The major reaction route for TAL in air is homogeneous and involves direct photolytic breakdown, or reaction with ozone, triplet atomic oxygen (O^3P), or hydroxyl radicals (Harrison and Laxen, 1978). The appropriate rate constants and typical estimated breakdown rates for summer and winter conditions for both tetramethyllead and tetraethyllead are shown in Table 10.7. Under most atmospheric conditions, it is the reaction with hydroxyl which is the most rapid process, as is the case for most hydrocarbons in air. Using more sophisticated kinetic techniques, Nielsen et al. (1982) have redetermined the TAL–OH rate constants, finding an almost identical value for the tetramethyllead reaction, but a smaller rate for tetraethyllead. They predict slower atmospheric decomposition rates for TAL than were calculated by Harrison and Laxen (1978), but the difference arises primarily from different assumed concentrations of OH radical in the two studies. Unfortunately, there are at present no fully satisfactory means of determining this variable, and hence entirely reliable predictions of breakdown rates are not possible.

The mechanism of breakdown of TAL is open to doubt. While Harrison and Laxen (1978) postulated hydrogen abstraction as the initial step, that is,

$$(CH_3)_3PbCH_3 + OH \longrightarrow (CH_3)_3PbCH_2 + H_2O$$

Nielsen et al. (1982) believe cleavage of the Pb–C bond to be more probable. More importantly, it is not yet known whether stable intermediate species exist in the breakdown of TAL to inorganic Pb^{2+}, or whether once attack on TAL occurs, the full cycle of decay rapidly ensues.

Table 10.7. Estimated Upper Limit Rates of Tetramethyllead (TML) and Tetraethyllead (TEL) Decay at Midday in a Moderately Polluted Irradiated Atmosphere[a]

			Decay Rate (% h^{-1})			
	Concentration of Reactive Species		TML		TEL	
Decay Path	Summer	Winter	Summer	Winter	Summer	Winter
OH attack	$(1\text{–}3)\times10^{-7}$ ppm	$(1\text{–}2)\times10^{-8}$ ppm	8–21	1–1.5	51–88	7–13
Photolysis	$z \sim 40°$	$z \sim 75°$	8	2	26	7
O_3 attack	100–200 ppb	40 ppb	1–2	0.5	9–17	4
$O(^3P)$ attack	10^{-8} ppm	10^{-9} ppm	<0.1	<<0.1	0.1	<0.1
Particulates	200 $\mu g/m^3$	200 $\mu g/m^3$			0.03	0.03
Total			16–29	3–4	67–93	17–23

[a] From Harrison and Laxen (1978).

4.3. Marine Aerosol

The major components of marine aerosol are NaCl and $MgCl_2$; there is a minor quantity of Na_2SO_4 and also many other elements at trace levels. The only observed atmospheric chemical processes affecting metals are those involving NaCl, which may react with strong acids. Thus, NaCl reacts with particulate H_2SO_4 or gaseous HNO_3 to form Na_2SO_4 and $NaNO_3$, respectively, with concomitant loss of HCl to the gas phase. These reactions may be inferred from size-discriminated ion balance studies (Harrison and Pio, 1983b).

The chemistry of trace metals in seawater is extremely complex, but is progressively becoming clearer. A rather similar pattern of chemical speciation of trace metals in marine aerosol to that in seawater may occur, but chemical fractionation processes, commonly involving trace metal enrichment upon formation of the aerosol, may be expected to cause some alteration in the pattern of speciation.

4.4. Other Metals

The rather frequent observation of metal ammonium sulfates in ambient air (Table 10.5) probably implies that the types of chemical reaction process undergone by airborne inorganic lead (Sec. 4.1) are important for other metals also.

Thus calcium ammonium sulfate may arise from the coagulation of particles containing ammonium sulfate and calcium sulfate, the latter possibly arising from atmospheric reaction of the rather abundant calcium carbonate with sulphuric acid:

$$CaSO_4 + (NH_4)_2SO_4 \longrightarrow CaSO_4 \cdot (NH_4)_2SO_4$$

In the case of ferric ammonium sulfate, the most probable starting material is ferric oxide, Fe_2O_3, which occurs commonly in the environment as an iron corrosion product:

$$Fe_2O_3 + 3H_2SO_4 \longrightarrow Fe_2(SO_4)_3 + 3H_2O$$
$$Fe_2(SO_4)_3 + 3(NH_4)_2SO_4 \longrightarrow Fe_2(SO_4)_3 \cdot 3(NH_4)_2SO_4$$

Alternatively a one-step process involving reaction with ammonium bisulfate is possible (Biggins and Harrison, 1979c):

$$Fe_2O_3 + 6NH_4HSO_4 \longrightarrow Fe_2(SO_4)_3 \cdot 3(NH_4)_2SO_4 + 3H_2O$$

Similar reactions may be postulated for many metal sulfates or oxides in the atmosphere. The metal ammonium sulfates appear to offer relatively stable crystalline products for these processes.

5. CONCLUSIONS

Although intensive research in certain areas has advanced knowledge of chemical speciation and reaction pathways, it is clear that across the field as a whole our knowledge is extremely sparse. For this reason, much research is still required, most urgently in the area of aerosol speciation techniques. Only with the help of better tools will a fuller knowledge of speciation become available, and this will surely lead to a clearer understanding of chemical pathways.

REFERENCES

Adams, F., Bloch, P., Natusch, D.F.S., and Surkyn, P. (1981). "Microscopical analysis for source identification in air chemistry and air pollution." *Proceedings of the International Conference on Environmental Pollution*. Thessaloniki, Greece, pp. 122–142.

Biggins, P.D.E., and Harrison, R.M. (1979a). "Atmospheric chemistry of automotive lead." *Environ. Sci. Technol.* **13,** 558–565.

Biggins, P.D.E., and Harrison, R.M. (1979b). The identification of specific chemical compounds in size-fractionated atmospheric particulates collected at roadside sites." *Atmos. Environ.* **13,** 1213–1216.

Biggins, P.D.E., and Harrison, R.M. (1979c). "Characterization and classification of atmospheric sulphates." *J. Air Pollut. Contr. Assoc.* **29,** 838–840.

Biggins, P.D.E., and Harrison, R.M. (1980). "Chemical speciation of lead compounds in street dust." *Environ. Sci. Technol.* **14,** 336–339.

Birch, J., Harrison, R.M., and Laxen, D.P.H. (1980). "A specific technique for 24–48 hour analysis of tetraalkyl lead compounds in air." *Sci. Total Environ.* **14,** 31–42.

Bloch, P., Adams, F., Van Landuyt, J., and Van Goethem, L. (1979). "Morphological and chemical characterization of individual aerosol particles in the atmosphere." Proceedings of the Symposium *Physico-Chemical Behaviour of Atmospheric Pollutants*. Ispra, Italy. E.E.C., pp. 307–321.

Braman, R.S., and Johnson, D.L. (1974). "Selective absorption tubes and emission technique for determination of ambient forms of mercury in air." *Environ. Sci. Technol.* **8,** 996–1003.

Butler, J.D., McMurdo, S.D., and Stewart, C.J. (1976). "Characterization of aerosol particulate by scanning electron microscope and X-ray energy fluorescence analysis." *Int. J. Environ. Stud.* **9,** 93–103.

Chamberlain, A.C., Heard, M.J., Little P., and Wiffen, R.D. (1979). "The dispersion of lead from motor exhausts." *Phil. Trans. Roy. Soc. London A.* **290,** 577–89.

Charlson, R.J., Covert, S.D., Larson, T.V., and Waggoner, A.P. (1978). "Chemical properties of tropospheric sulfur aerosols." *Atmos. Environ.* **12,** 39–54.

Davis, B.L., and Cho, N.K. (1977). "Theory and application of X-ray diffraction analysis to high-volume filter samples." *Atmos. Environ.* **11,** 73–85.

Davis, B.L. (1981). "Quantitative analysis of crystalline and amorphous airborne particulates in the Provo-Orem vicinity, Utah." *Atmos. Environ.* **15,** 613–618.

De Jonghe, W., Chakraborti, D., and Adams, F. (1980). "Graphite furnace atomic absorption spectrometry as a metal-specific detection system for tetraalkyllead compounds separated by gas–liquid chromatography." *Anal. Chim. Acta* **115,** 89–101.

De Jonghe, W.R.A., Chakraborti, D., and Adams F. (1981). "Identification and determination of individual tetralkyllead species in air." *Environ. Sci. Technol.* **15,** 1217–1222.

Ebdon, L., Ward, R.W., and Leathard, D.A. (1982). "Development and optimisation of atom cells for sensitive coupled gas chromatography–flame atomic absorption spectrometry." *Analyst* **107,** 129–143.

Edwards, H.W., and Rosenvold, R.J. (1974). "Uptake of tetraethyllead by atmospheric dust components." *Trace Contaminants in the Environment.* Lawrence Berkeley Lab. Publ. LBL-3217, pp. 59–63. Berkeley, Calif.

Edwards, H.W., Rosenvold, R.J., and Wheat, H.G. (1975). "Sorption of organic lead vapour on atmospheric dust particles." In *Trace Subst. Environmental Health.,* Vol. IX, Hemphill D.D., ed. University of Missouri Press, Columbia, Mo., pp. 197–205.

Foster, R.L., and Lott, P.F. (1980). "X-ray diffractometry examination of air filters for compounds emitted by lead smelting operations." *Environ. Sci. Technol.* **14,** 1240–1244.

Fukasawa, T., Iwatsuki, M., Kawakubo, S., and Miyazaki, K. (1980). "Heavy liquid separation and X-ray diffraction analysis of airborne particulates." *Anal. Chem.* **52,** 1784–1787.

Johnson, D.L., and Braman, R.S. (1974). "Distribution of atmospheric mercury species near ground." *Environ. Sci. Technol.* **12,** 1003–1009.

Habibi, K. (1973). "Characterization of particulate matter in vehicle exhaust." *Environ. Sci. Technol.* **7,** 223–234.

Hamilton, R., and Adie, G. (1982). "Size, shape and elemental associations in an urban aerosol." *Sci. Total Environ.* **23,** 393–402.

Hancock, S., and Slater, A. (1975). "A specific method for the determination of trace concentrations of tetramethyl- and tetraethyl-lead vapours in air." *Analyst* **100,** 422–429.

Harrison, R.M., and Perry, R. (1977). "The analysis of tetraalkyl lead compounds and their significance as urban air pollutants." *Atmos. Environ.* **11,** 847–852.

Harrison, R.M., and Laxen, D.P.H. (1977). "Organolead compounds adsorbed upon atmospheric particulates: a minor component of urban air." *Atmos. Environ.* **11,** 201–203.

Harrison, R.M., and Laxen, D.P.H. (1978). "Sink processes for tetraalkyllead compounds in the atmosphere." *Environ. Sci. Technol.* **12,** 1384–1392.

Harrison, R.M., Williams, C.R., and O'Neill, I.K. (1981a). "Characterization of airborne heavy metals within a primary zinc-lead smelting works." *Environ. Sci. Technol.* **15,** 1197–1204.

Harrison, R.M., Laxen, D.P.H., and Wilson, S.J. (1981b). "Chemical associations of lead, cadmium, copper and zinc in street dusts and roadside soils." *Environ. Sci. Technol.* **15,** 1378–1383.

Harrison, R.M., and Pio, C.A. (1983a). "Major ion composition and chemical associations of inorganic atmospheric aerosols." *Environ. Sci. Technol.* **17,** 169–174.

Harrison, R.M., and Pio, C.A. (1983b). "Size differentiated composition of inorganic atmospheric aerosols of marine and polluted continental origin." *Atmos. Environ.* **17,** 1733–1738.

Harrison, R.M., and Sturges, W.T. (1983a). "The measurement and interpretation of Br/Pb ratios in airborne particles." *Atmos. Environ.* **17,** 311–328.

Harrison, R.M., and Sturges, W.T. (1983b). Unpublished data.

Harrison, R.M., and Williams, C.R. (1983). "Physico-chemical characterization of atmospheric trace metal emissions from a primary zinc-lead smelter." *Sci. Total Environ.* **31,** 129–140.

Heidel, R.H., and Desborough, G.A. (1975). "Limitations on analysis of small particles with an electron probe: pollution studies." *Environ. Pollut.* **8,** 185–191.

Henriques, A., Isberg, J., and Kjellgren, D. (1973). "Collection and separation of metallic mercury and organo-mercury compounds in air." *Chemica Scripta* **4,** 139–142.

Hirschler, D.A., Gilbert, L.F., Lamb, F.W., and Niebylski, L.M. (1957). "Particulate lead compounds in vehicle exhaust gas." *Ind. Eng. Chem.* **49,** 1131–1142.

Nielsen, T., Egsgaard, H., Larsen, E., and Schroll, G. (1981). "Determination of tetramethyllead and tetraalkyllead in the atmosphere by a two-step enrichment method and gas-chromatographic–mass spectrometric isotope dilution analysis." *Anal. Chim. Acta* **124,** 1–13.

Nielsen, O.J., Nielsen, T., and Pagsberg, P. (1982). "Direct spectrokinetic investigation of the reactivity of OH with tetraalkyllead compounds in gas phase. Estimate of lifetimes of tetraalkyllead compounds in ambient air." Riso National Laboratory Report-R-463, 17 pp.

O'Connor, B.H., and Jaklevic, J.M. (1981). "Characterization of ambient aerosol particulate samples from the St. Louis area by X-ray powder diffractometry." *Atmos. Environ.* **15,** 1681–1690.

Olsen, K.W., and Skogerboe, R.K. (1975). "Identification of soil lead compounds from automotive sources." *Environ. Sci. Technol.* **9,** 227–230.

O'Neill, I.K., Harrison, R.M., and Williams, C.R. (1982). "Characterization of airborne heavy metal particulate in a zinc–lead smelter. Potential importance of gastro-intestinal absorption in occupational exposure." *Trans. Inst. Min. Metall.* **91,** C84–90.

Reamer, D.C., Zoller, W.H., and O'Haver, T.C. (1978). "Gas chromatograph–microwave plasma detector for the determination of tetraalkyllead species in the atmosphere." *Anal. Chem.* **50,** 1449–1453.

Ter Haar, G., and Bayard, M.A. (1971). "Composition of airborne lead particles." *Nature* **232,** 553–554.

Trujillo, P.E., and Campbell, E.E. (1975). "Development of a multistage air sampler for mercury." *Anal. Chem.* **47,** 1629–1634.

Yakowitz, H., Jacobs, M.H., and Hunneyball, P.D. (1972). "Analysis of urban particulates by means of combined electron microscopy and X-ray microanalysis." *Micron.* **3,** 498–505.

11

CHARACTERIZATION OF TRACE METAL COMPOUNDS IN THE ATMOSPHERE IN TERMS OF DENSITY

Akiyoshi Sugimae

Environmental Pollution Control Center, Osaka Prefecture
Nakamichi, Higashinari-ku
Osaka, Japan

1. INTRODUCTION

There has been increasing concern about the health effects, damage to vegetation and materials, and reduced visibility caused by atmospheric particulates generated by anthropogenic activities. Bertine and Goldberg

(1971) and Zoller et al. (1974) indicated that volatile elements might be selectively introduced into the atmosphere from a variety of sources such as industrial emissions, fuel combustions, incineration, and transportation.

Volatile elements that vaporize during high temperature processes and later condense either homogeneously or heterogeneously would be distributed over submicron-sized particles. In homogeneous nucleation, particles (or nuclei) are formed in the vapor phase by molecular clustering without the aid or intervention of foreign nuclei. The term *heterogeneous nucleation* refers to the situation where foreign particles or gas ions (clusters of molecules) are present in the vapor and act as nuclei for the subsequently growing particles. Both homogeneous self-nucleation and heterogeneous nucleation on foreign nuclei may occur. Whether homogeneous or heterogeneous nucleation occurs depends on the relative amount of surface area provided by foreign nuclei. When there is a sufficient amount of particle surface area necessary to accommodate the condensable vapors, the vapors deposit on the existing particles. The mechanism would give rise to surface predominance of the volatile elements, and this has been observed for coal fly ash and steel furnace dust (Campbell et al., 1978a; Keyser et al., 1978; Linton et al., 1976, 1977; Van Craen et al., 1982, 1983). If foreign nuclei and gaseous ions are present only in very low concentrations, particle formation proceeds by molecular clustering of condensable species (i.e., by homogeneous nucleation). The homogeneous nucleation may be of special importance in the early phases of particle formation and would yield a very large number of extremely small particles. Flagan and Friedlander (1978) explored the contribution of homogeneous nucleation to fine-particle formation in pulverized coal combustion and reported that a very large number of extremely small particles, probably much smaller than 0.01 μm in diameter, would result from homogeneous nucleation. These particles may coagulate with other particles, or they may diffuse to the surfaces of the much larger particles.

Fine particles, which exhibit violent Brownian motion in the atmosphere, have great opportunities to collide with other particles. Particle growth caused by coagulation during atmospheric transport provides the mechanism whereby the physicochemical properties could change. The coagulation rate constants estimated by Fuchs (1964), Husar et al. (1972), and Hirsch et al. (1981) indicated that urban particles coagulated on almost every collision. Chemical reactions and mixing with atmospheric background particles would also be expected to alter the properties.

The processes involved are extremely complex, and the exact chemical and physical mechanisms are not yet established. The physical processes such as nucleation, condensation, absorption, adsorption, and coagulation are primarily responsible for determining such physicochemical properties as number concentration, density distribution, size distribution, optical properties, and solubility of the formed particles.

The nature and behavior of atmospheric particulates depend on their chemical nature and also on their physical characteristics. In recent years, increasing emphasis has been placed on the physicochemical characterization of trace elements in the atmosphere. It is assumed that particles derived from a given source exhibit distinguishable size and density characteristics, though during atmospheric transport and exposure the physicochemical characteristics could change to some extent. In this sense, each particle still bears a lot of information about the process by which it was formed.

Whitby et al. (1972) found the particle size distributions in Pasadena to be bimodal with a saddle at 2–3 μm indicating two types of particles with different formation histories and little subsequent interconversion. There is now an overwhelming amount of evidence that two modes are usually observed in the size distributions of ambient aerosols (Durham et al., 1975; Haff and Jaenicke, 1980; Noonkester, 1981; Sheih et al., 1983; Sverdrup et al., 1975; Whitby et al., 1974, 1975; Whitby, 1977; Willeke and Whitby, 1975). The fine and coarse modes are usually chemically quite different. It is well established that mechanical processes such as abrasion and wind erosion release particles of predominantly coarse particles and that volatilization followed by condensation and coagulation yields particles of very small sizes. Size distribution curves of elements suspended in the atmosphere can be divided into the following broad classes: predominantly coarse-particle, preferentially fine-particle, and mixed fine- and coarse-particle distributions (Adams et al., 1983; Dzubay and Stevens, 1975; Flocchini et al., 1976; Fujimura et al., 1978; Gladney et al., 1974; Hardy et al., 1976; Kadowaki, 1979; Paciga and Jervis, 1976; Tanaka et al., 1980; Wangen, 1981). However, these size distributions are mainly based on sampling by cascade impactors. Particle size measurements in this manner cannot account for variation in particle density. The densities of atmospheric particulates are assumed to remain constant for all particles.

There is little known about the densities of atmospheric particulates. However, the particle density is also related to the mode of origin, composition, and interaction with the surrounding vapors and with each other. It is generally accepted that the density of an agglomerate which includes void spaces is less than that of individual component particles (Silverman et al., 1971). Particles originating from a particular combustion process are either cenospheres or plerospheres with low apparent densities (Fisher et al., 1976; Campbell et al., 1978b; Smith, 1980). On the other hand, the density of particles formed by comminution, attrition, or disintegration will resemble that of the parent material. Thus, atmospheric particulates are formed from materials of various densities.

Each particle thus has an individual history in the atmosphere with regard to the origin, growth, interaction, and decay. Urban and natural particulates are polydispersed with respect to particle density as well as particle size. Because of the agglomerated nature of atmospheric particulates and the attendant large variations of particle densities, it is necessary to define the quantity related to the aerodynamic behavior of the particle. Indeed, practically all models of respiratory tract deposition of inhalable particles have employed normalizing techniques based on the aerodynamic equivalent diameter, which is defined as the diameter of a unit density sphere that has the same terminal settling velocity as the given particle. In the field measurements on particle size distributions, results are usually interpreted in terms of the aerodynamic behavior of unit density spheres. However, particle motion in the atmosphere is characterized by the product of the density multiplied by the diameter squared. In any estimation of the particle motion in the atmosphere, it is essential to take into account some measure of particle density as well as particle size. It is anticipated that fine and coarse particles in the atmosphere are usually physicochemically quite different from each other. Some attention should be made where there are obvious differences in the density of particles. For example, Watson et al. (1983) assumed the densities of 1.7 and 2.6 g/cc for the fine particles and the coarse particles, respectively, when volume median diameters were converted to approximate aerodynamic diameters.

It is apparent from the above discussion that our knowledge of atmospheric particulates is fragmentary. In order to fingerprint the pollution sources and to determine the extent of the anthropogenic contribution, a fundamental study of the physicochemical properties of atmospheric particulates is required. Also, for the studies of source characteristics, atmospheric transport processes, and removal rates, an accurate knowledge of density distribution and size distribution is essential. Extensive data on elemental density distributions of street dusts and roadside dusts have only recently become available for use in elemental source tracing (Hopke et al., 1980; Linton et al., 1980). There are few published data on the elemental constituents as a function of density of atmospheric particulates (Sugimae, 1983a, 1983b, 1984).

This paper presents the results of density distributions of Si, Fe, Mn, Zn, Cu, and other elements in total suspended particulates, inhalable particulates, size-fractionated suspended particulates, suspended dusts (collected on the air-inlet filter of an air conditioning system), and dustfalls. To evaluate the contribution of windblown soil to local atmospheric particulates, elemental density distributions for soils taken near the sampling station were also examined.

2. EXPERIMENTAL

2.1. Sampling Procedures

Atmospheric particulates were sampled either by filtration from the suspension in the atmosphere or by collection of deposited particles as they fall out under gravitational influence, known as dustfall. Samples of suspended particulates were collected on the rooftop (about 10 m above street level) of the Environmental Pollution Control Center in a metropolitan area of Osaka using both a conventional high-volume sampler (hi-vol) and a low-volume sampler (low-vol). The hi-vol was used to collect samples of total suspended particulates (TSPs) on 8 × 10 in. filters. The sampler is fitted in an aluminum shelter which prevents direct settling of dusts onto the filter. Particles with a terminal settling velocity in air greater than approximately 30 cm/s are excluded from the shelter. The maximum theoretical particle size has been calculated to be 100 μm (particle density, 1.0 g/cc) (Jutze and Foster, 1967). However, gravitational and wind velocity effects can interfere with sampling of suspended particulates from ambient air. Collection efficiencies for large particles are quite sensitive to small changes in wind velocity, approaching angle, and flow rate (Wedding et al., 1977). The upper limit of particle size actually collected by the sampler is believed to be 15–20 μm aerodynamic diameter. The Japanese Environment Agency specifies the use of low-vol for the long-term monitoring of suspended particulate levels. Coarse particles are preliminarily separated by three cyclones installed in the sampler. The low-vol collects inhalable particulates (IPs) on 48-mm diameter filters at a flow rate of 20 L/min and has a D_{50} (aerodynamic diameter at an effectiveness of 50%) of 8 μm.

The polystyrene filter (Microsorban type 99/97, Delbag-Luftfilter GMBH, Berlin, Germany), as well as the glass-fiber filter (Toyo GB-100R), was used as the collection media in this sampling program. Samples routinely analyzed for trace elements in this laboratory have been obtained by drawing air through the glass-fiber filter for periods varying from a few hours to several days. The polystyrene filter has collection efficiencies, capacities, and loading characteristics similar to the more commonly used glass-fiber filter, yet it has much lower blank concentrations of trace elements than do glass-fiber filters (Dams et al., 1972). Suspended particulate samples were collected on the polystyrene filter for later determination of elemental density distributions.

Samples in which suspended particulates was fractionated according to size were also collected using a modified Andersen cascade impactor. The impactor was housed in a proper shelter allowing sufficient air access, yet protecting the filters from rain. The flow rate of 1 cfm through the impactor was maintained, and the sampling period was 10 days. The particle size fractions were collected in the ranges of equivalent aerodynamic diameter (>11, 3.3–11, 1.1–3.3, and 0.43–1.1 μm) by impaction on Delbag polystyrene filters about 50 cm^2 in area mounted on circular stainless steel disks. Particles of equivalent aerodynamic diameter <0.43 μm were collected by the fifth stage, that is, the polystyrene filter.

Batches of suspended dusts were also collected from the air-inlet filters of the air conditioning systems of a building and homogenized.

Particulates that fell as a result of their own settling velocity and that were carried down with precipitation were collected in deposit gauges. The gauge is an open-top polyethylene cylinder of 105-mm diameter and vertical sides of 215-mm height. The sample collection usually continues for a month. After collection, water-insoluble dustfall was filtered off on a polystyrene filter.

To evaluate the contribution of windblown soil to local atmospheric particulates, soil samples were taken near the sampling station.

2.2. Analytical Procedures and Instrumentation

The density fractions were determined by means of the heavy-liquid density gradient technique, which has been described elsewhere (Sugimae, 1983a, 1983b, 1984). The basis of the method is that the components of atmospheric particulates, differing in particle density, will be separated into distinct bands in a density gradient in a suitable heavy liquid (Harrison and Parker, 1977). Similar methods have been developed for use in the mineral processing industry and for characterization of soil clay. Their potential for environmental analysis has only recently been realized (Fukasawa et al., 1980; Pilkington and Warren, 1979). A number of investigators have applied the technique to the separation of Pb compounds in street dusts (Biggins and Harrison, 1978, 1979a, 1980; Linton et al., 1980; Olson and Skogerboe, 1975) and to the multielement characterization of roadway dusts (Hopke et al., 1980). However, the application of this technique to density fractionation of actual atmospheric particulates has been very limited. This is probably because of the difficulty in removing the suspended particulates from the filter loaded with sample. The use of polystyrene filter for the collection medium or the impaction surface was a good choice because the suspension of captured particulates for the subsequent density separation was prepared by dissolving the filter material in dichloromethane (density, 1.3 g/cc).

A sample disk of 48-mm diameter punched out from a hi-vol filter, half a low-vol filter or half an impaction surface of Andersen cascade impactor was placed in a 10-mL centrifuge tube, and 10 mL of dichloromethane was added to dissolve the filter material. In the case of powder samples such as soils, the powder is directly dispersed in the solvent. After centrifugation, the float was transferred and the residue was ultrasonically dispersed and centrifuged in a series of dichloromethane–diiodomethane mixtures such that the density of each successive mixture was incrementally raised by 0.2 g/cc. The final centrifugation was in diiodomethane (3.3 g/cc). Suspended particulates collected on the filter were thus successively classified into 12 density fractions from <1.3 to >3.3 g/cc.

A major problem of chemical analysis of the density fractions is that the amounts of suspended particulates derived are very small. There have been enormous improvements in the analytical techniques that can be used, in particular, the development of inductively coupled plasma–atomic emission spectrometric (ICP-AES) technique. The ICP-AES, because of high sensitivity, general lack of interferences, multielement capability, and simplicity of operation is an excellent technique for the simultaneous multielemental analysis of atmospheric particulates (Floyd et al., 1980; Long et al., 1979; McQuarker et al., 1979; Sugimae, 1979, 1981; Sugimae and Mizoguchi, 1982).

Table 11.1. Instrumentation and Working Conditions

Spectrometer	ARL Quantometric Analyser (1-m Paschen–Runge mounting, grating rules 1920 lines/mm, 0.52 nm/mm reciprocal linear dispersion, primary slit width 12 μm, secondary slit width 50 μm, HTV R-300W photomultiplier)
Readout	Digital readout of integrated signal
rf generator	Henry Radio 3000 PGC/27 (27.12 MHz, crystal controlled)
rf power	1.6 kW (forward), <10 W (reflected)
Plasma torch	Fused quartz torch
Nebulizer	Concentric glass nebulizer
Argon flow rate	Coolant gas 11.0 L/min, plasma gas 1.3 L/min, carrier gas 1.0 L/min; carrier gas is humidified with water vapor prior to its entry into the nebulizer.
Observation height in plasma	16 mm above load coil
Analytical lines	Cd(II) 226.50 nm, Cr(II) 283.56 nm, Cu(I) 324.75 nm, Fe(II) 259.94 nm, Mn(II) 257.61 nm, Ni(I) 231.60 nm, Pb(II) 220.35 nm, Si(I) 251.61 nm, V(II) 311.07 nm, Zn(II) 202.55 nm.

Iron, Mn, Zn, and Cu were successively determined in each density fraction, even in the case of suspended particulates fractionated into five particle size ranges by the modified Andersen cascade impactor, by ICP-AES following acid digestion in hot aqua regia. Silicon was determined by ICP-AES after alkaline fusion with sodium carbonate. The plasma operating parameters are shown in Table 11.1. The experimental methods have been described in detail elsewhere (Sugimae, 1979, 1981). Analyses were performed for many elements including Be, Cd, Cr, Cu, Fe, Mn, Ni, Pb, Si, V, and Zn. Not all the results from these elements were accepted as valid, however, and the elemental density distribution data presented in this paper are limited to those obtained for Si, Fe, Mn, Zn, and Cu.

3. RESULTS AND DISCUSSION

3.1. Elemental Density Distributions of Suspended Particulates

Concentration profiles as a function of particle density for Si, Fe, Mn, Zn, and Cu in TSPs collected on hi-vol filter and in IPs collected on low-vol filter are shown in Figure 11.1.

Possible chemical forms of these elements in the atmosphere include oxides, silicates, sulfates, nitrates, carbonates, and so forth. Soil-derived elements exist primarily as oxides (or silicates). Quartz is the major

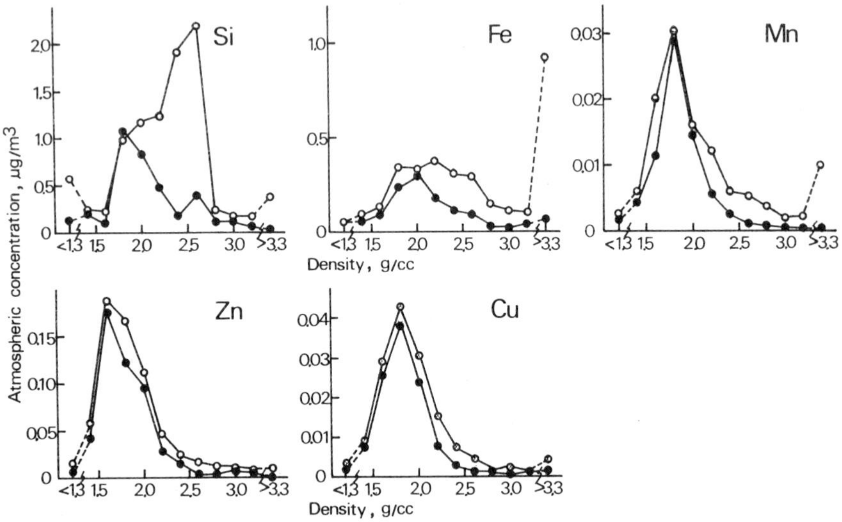

Figure 11.1. Density distributions of Si, Fe, Mn, Zn, and Cu in total suspended particulates and inhalable particulates. (1) Total suspended particulates. (2) Inhalable particulates.

atmospheric Si compounds. Iron oxide, principally as hematite (Fe_2O_3) or magnetite (Fe_3O_4), occurs in appreciable amounts in the atmosphere. Abrasional processes such as normal crustal weathering produce mostly coarse particles, while high temperature processes including industrial emission, incineration, fuel combustion, and automobile emission produce fine particles with diameters that are generally less than a micrometer. The submicron particles have significantly long atmospheric residence times. Additionally the increased area of exposed surface associated with fine particles leads to increased chemical reaction rates. Combustion-generated elements probably starting out as oxide fume or metal fume may be converted to other compounds during atmospheric transport and exposure (Biggins and Harrison, 1979b). $ZnSO_4$ and $ZnSO_4 \cdot (NH_4)_2SO_4 \cdot 6H_2O$, reported by Hemeon (1955), are examples of sulfates formed from atmospheric reactions of zinc oxide fume or zinc fume. The formation of sulfates in the atmosphere is attributed essentially to the existence of SO_2 discharged into the atmosphere from various sources. Jaklevic et al. (1980) reported that fine particles were predominantly composed of $ZnSO_4$ and $Zn(NH_4)_2(SO_4)_2$, whereas the coarse particles are apparently ZnO. The difference in the chemical composition indicates the independent behavior of the fine- and coarse-particle fractions. For approximately spherical particles, the ratio of surface area to volume is inversely proportional to the particle diameter. The fine particles could be converted to the sulfated form more readily due to their much higher surface to volume ratio.

The densities (g/cc) of the commonly recognized compounds which exist or are expected to exist in the atmosphere are as follows: SiO_2, 2.65; Fe_2O_3, 5.24; Fe_3O_4, 5.18; $Fe_2(SO_4)_3$, 3.094; MnO, 5.026; ZnO, 5.61; $ZnSO_4$, 3.54; CuO, 6.0; and $CuSO_4$, 3.61.

Common sense suggests that the compounds of heavy metals will be present primarily in association with particles of high density (>3.3 g/cc). Little is known about the elemental density distributions of suspended particulates. However, the bulk density can be calculated from the ratio of the particle mass to volume. Sverdrup et al. (1975) compared volume concentration distributions calculated from size distribution measurements by the Minnesota Aerosol Analyzing System (MAAS) to mass concentration distributions measured by the 8-stage Andersen cascade impactor. The results indicated that the average density of submicron particles in the atmosphere was 1.70 g/cc with uncertainty of ±0.17 g/cc. Durham et al. (1975) also found that the density ranged from 1.1 to 1.5 g/cc for submicron particles during an air pollution episode. The average density of submicron background particles was similarly inferred to be from 1.6 to 1.8 g/cc.

The calculation of density for larger particles could not be made because of incompatible collection efficiencies of the MAAS and the Andersen cascade impactor. However, it is noticeable that the average density for submicron particles during an air pollution episode is lower than those for background particles.

To make a detailed interpretation of data of the sort obtained, one would wish to know (1) density distributions of particles originating from various important types of sources and (2) the relative magnitude of sources of each element. At present, information on both points is far from complete. Regarding point (1), it is known that the density of a particle formed by attrition from a solid will be the same as that of the parent material and that volatilization followed by condensation and coagulation yields an agglomerate with density less than that of the individual component particles.

The density distribution profiles shown in Figure 11.1 suggest that there is a general tendency for the soil-derived elements in TSPs, typified by Si and Fe, to increase in concentration with relation to the density ranges of the parent materials (such as quartz, hematite, and magnetite).

The concentration of Si in TSPs is relatively high in the density range of 2.5–2.7 g/cc, corresponding to the density of SiO_2 (2.65 g/cc). Approximately 30% of Fe particles in TSPs collected by means of hi-vol is associated with particles of density higher than 3.3 g/cc, although it is not possible to identify a maximum for the higher density particles that 3.3 g/cc. However, the density distributions of Si and Fe in IPs collected on low-vol filter differ remarkably with those in TSPs. The density distribution pattern of Si is bimodal with one mode occurring in the density range of 1.7–1.9 g/cc and another in the density range of 2.5–2.7 g/cc. The density distribution of Fe particles also shows a definite peak in the density range of 1.9–2.1 g/cc. Iron particles with densities higher than 3.3 g/cc can account for only a few percent of total mass of Fe particles collected on low-vol filter. Comparison of the density distribution of Si and Fe between TSPs and IPs indicates that these elements are preferentially associated with lower density particles in IPs than in TSPs.

The differences between suspended particulate concentrations determined in parallel experiments by the hi-vol and the low-vol are quite large, amounting to 200–500%. The low-vol used in this study was limited to the collection of particles that have aerodynamic diameters no larger than 10 μm. Silicon and Fe are predominantly associated with coarse particles. The efficiency with which these coarse particles are drawn into the sampler is expected to be reduced as particle density is increased. It is noticeable that TSPs and IPs display recognizably different density distributions with respect to Si and Fe particles, which are predominantly of large size.

In the case of lithophilic elements, which occur in coarse particles and form stable oxides, the mechanisms responsible for the low density affiliation are not completely understood. However, the presence of Si and Fe particles with densities less than those of most likely parent materials is also remarkable. This is partially attributable to the presence of soil agglomerates with low apparent densities and partially to the presence of hollow spheres from a particular combustion process. It is well known that soil agglomerates of varying sizes are formed from soil particles. Clay particles remain largely agglomerated to other clay particles and to larger sand particles (Rahn, 1976). Indeed, Gillette et al. (1972) presented photographs of thin sections of soil creep samples collected during a sandstorm which showed a coating of submicron clay particles on the surface of the large sand particles.

Volatile (chalcophilic) elements such as Zn and Cu are concentrated in low density fractions which seem to consist of carbonaceous matter, judging from their black color and microscopic observation. The density distribution curves of Zn and Cu show only one peak in the range of 1.5–1.7 and 1.7–1.9 g/cc, respectively. It is notable that the density distributions of Zn and Cu do not differ appreciably between hi-vol and low-vol filter sample. Hi-vol and low-vol sampling gave approximately equivalent results for these elements. The densities of Zn and Cu compounds known to exist in the atmosphere are generally higher than 3.3 g/cc. Nevertheless, these high density particles of Zn and Cu can account for only a few percent of total mass of Zn and Cu. This is probably because these elements are capable of being volatilized during high temperature processes. And fine particles originating from nucleation of vapors grow by condensation and coagulation to form agglomerates with low densities.

In the case of Mn particles collected on hi-vol filter, the concentration goes through the maximum in the density range of 1.7–1.9 g/cc, and another small maximum at the density higher than 3.3 g/cc. Manganese is concentrated in the lower density particles in IPs. Gladney et al. (1974) reported that the size distribution for Mn showed both fine- and coarse-particle components, indicating that Mn particles originated from two or more processes. Manganese is much less volatile than the chalcophilic elements. However, low density components are probably introduced into the atmosphere after being volatilized. This process would account for the observed greater number of low density particles bearing Mn. Manganese has somewhat higher density particle concentrations in TSPs than in IPs. The gross atmospheric concentration obtained for Mn with hi-vol was somewhat higher than that obtained with low-vol. The presence of dense particles (>3.3 g/cc) in TSPs suggests that abrasional processes such as crustal weathering are also the source of atmospheric Mn particles.

The types of sampling device have marked effects on the elemental density distributions. In any case, the density distributions for the elements examined are generally lower than might be anticipated from *a priori* knowledge of the corresponding bulk density. The occurrence of low density particles is not surprising. Whytlaw-Gray and Patterson (1932) suggested that the density of agglomerate might be less than one-tenth the density of the parent material. The results support recent theories on the origins of suspended particulates based on the measurement of particle size distributions.

To evaluate the contribution of windblown soil to local suspended particulate, the density distributions of Si, Fe, Mn, Zn, and Cu in soils taken near the sampling station are shown in Figure 11.2. The maximum value for Si was found in the density range of 2.5–2.7 g/cc, which corresponds to the density of SiO_2. The concentrations of Si in the other density ranges were in decreasing order of particle size. However, all elements, with the exception of Si, showed two definite peaks in the density distributions in the density range of 2.5–2.7 g/cc and >3.3 g/cc. The presence of particles with density higher than 3.3 g/cc would be explained by the presence of these elements as oxides. The unusual feature of the low density of 2.5–2.7 g/cc is that it is so low—

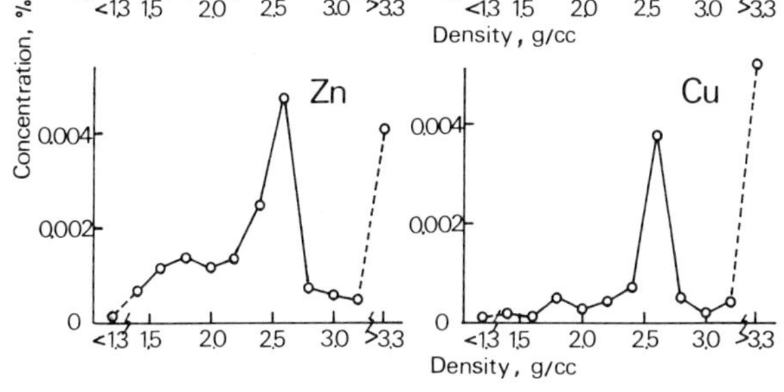

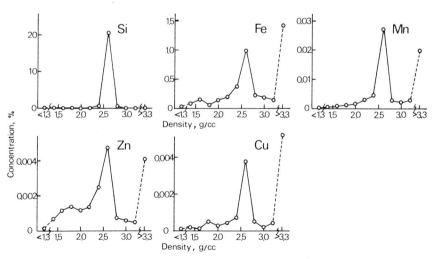

Figure 11.2. Density distributions of Si, Fe, Mn, Zn, and Cu in soil.

distinctly lower than the recognized chemical compounds in soil. We are unaware of any single crustal material, containing these elements with a density near 2.6 g/cc. Although the mechanism responsible for the low density is not yet understood, it is possible that agglomerated minerals play a major role. The presence of these low density particles may be explained by the presence of SiO_2 particles which are coated with submicron platelets containing these elements.

The soil particles introduced into the atmosphere as a result of wind turbulance are likely to give additional peaks in the density ranges of 2.5–2.7 and >3.3 g/cc. In urban areas, these dusts should be the source of coarse particles, most of which will rapidly fall out near their sources, but some of which should remain suspended. However, sampling effectiveness is dependent upon the size and density of the particles to be sampled. The density distributions shown in Figure 11.1 suggest that the low-vol as well as the hi-vol is not effective in capturing these large, high density soil particles from the atmosphere.

Figure 11.3 shows the typical density distributions of Si, Fe, Mn, Zn, and Cu in the dust collected from the air-inlet filter of an air conditioning system. These elements exhibit relatively high concentrations in particles with densities higher than 3.3 g/cc. Some elements are present at considerably higher concentrations in particles with densities of 2.5–2.7 g/cc. The Si, Fe, and Mn pattern in Figure 11.3 typifies this behavior, which may result from the predominance of soil-derived particles in the dust sample. This is probably because air velocity at the inlet of the air conditioning system is high enough to collect large, high density particles which will not be collected by means of hi-vol or low-vol.

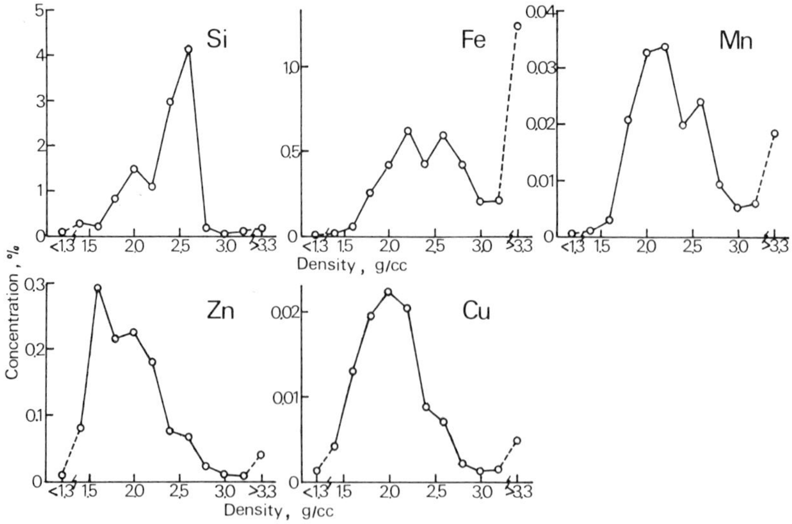

Figure 11.3. Density distributions of Si, Fe, Mn, Zn, and Cu in suspended dust deposited on the air-inlet filter of air conditioning system.

3.2. Elemental Density Distributions of Size-Fractionated Suspended Particulates

Chemical elements in suspended particulates are distributed in a way that reflects the sources of particulates and the atmospheric conversion processes. The composition of suspended particulates can vary with particle size as well as with particle density. In order to obtain information about these parameters, which determine the particle motion in the atmosphere, the size-fractionated particulates were subdivided according to the particle density and each size fraction was analyzed for Fe, Mn, Zn, and Cu.

A typical example of density distribution patterns for these elements in size-fractionated, suspended particulate samples, which were classified into five size classes from <0.43 to >11 μm by means of the modified Andersen cascade impactor, is shown in Figure 11.4. Characteristic differences can be seen among the density distributions for samples taken in the size ranges <0.43, 0.43–1.1, 1.1–3.3, 3.3–11, and >11 μm effective aerodynamic diameter at unit density. The maximum in elemental density distribution for the fractionated sample containing larger particulates is shifted toward higher density. The elemental density distribution appears to be somewhat broader with increasing particle size.

Density information can be combined with size separation as an effective means of source differentiation. As an example, consider Fe particles. The maximum occurs at the density of 1.6 g/cc for the suspended particulates with the size less than 0.45 μm, whereas the maximum occurs at the density of 2.4 g/cc for the particulates from the Andersen stage 1, representing particle sizes >11 μm. Iron, which is thought to originate primarily from mechanical processes such as windblown soil dust, is present in the coarse, high density

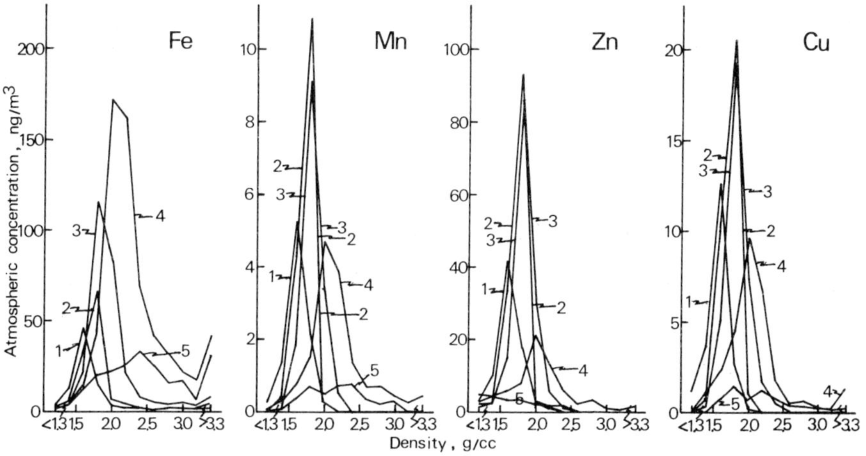

Figure 11.4. Density distributions of Fe, Mn, Zn, and Cu in size-fractionated suspended particulates. Particle size ranges: (1) <0.43 μm, (2) 0.43–1.1 μm, (3) 1.1–3.3 μm, (4) 3.3–11 μm, and (5) >11 μm.

particles. Manganese is associated with low density particles with intermediate particle size. Zinc and Cu, which originate as a result of combustion processes, are concentrated in the fine, low density particles.

It has been well established that the size distribution of suspended particulates often has two modes separated by a saddle at approximately 2 μm (Whitby et al., 1972). Separation at this particle diameter would need further research. The size range larger than 2 μm is called the mechanical particle range because particles in this range seem to result primarily from mechanical processes. The size range from about 0.1 to 2 μm is called the accumulation range because particulate material originating from coagulation of very small Aitken size particles or from the condensation of chemically formed vapors tend to accumulate in this range.

The elemental density distributions for the suspended particulates classified into two size fractions, one greater than 3.3 μm and another less than 3.3 μm, are shown in Figure 11.5, together with the density distributions for gross suspended particulates collected by means of the Andersen cascade impactor. The density distribution curves were constructed by summing up the values from each density range for a given size range. The point of 3.3 μm was set as close as possible to the minimum of the bimodal distribution which occurs at approximately 2 μm. Separation at particle diameter of about 3.5 μm is also needed for monitoring suspended particulates to protect human health, because a maximum deposition in the gas-exchange areas of the human respiratory system occurs with 3.5 μm particles (Miller et al., 1979). Indeed, the particle size of 3.5 μm is the cut-off point used in particulate sampling by the American Conference of Governmental Industrial Hygienists and the Department of Energy.

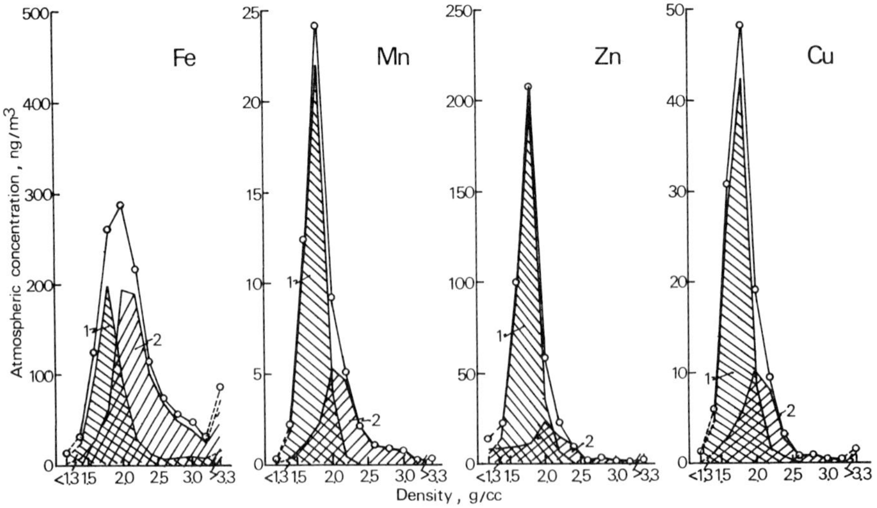

Figure 11.5. Density distributions of Fe, Mn, Zn, and Cu in suspended particulates with diameter smaller (1) and larger (2) than 3.3 μm.

Differences in the elemental density distributions for the coarse (>3.3 μm) and fine (<3.3 μm) particles in the atmosphere are evident. The fine particles mainly consisted of particles whose densities were lower than 2.1 g/cc, while the densities of coarse particles were in the broad range of <1.3 to >3.3 g/cc. Further inspection of the data in Figure 11.5 indicates that the elemental density distributions obtained for gross suspended particulates collected using the Andersen sampler are similar to those measured for a low-vol sample rather than those for a hi-vol sample. In addition, the sum of the concentrations of each element over all the size ranges provides an estimate of gross atmospheric concentration of an individual element collected by means of the Anderson cascade impactor. The elemental concentrations thus calculated yield values similar to the low-vol experiments. In a few cases, gross concentrations measured with Andersen cascade impactor were lower than the corresponding low-vol values. Although this could imply the presence of wall losses within the impactor (Cushing et al., 1979; Hu, 1971), the differences are not severe.

A difference in the elemental composition of each size range is also evident from Figure 11.5. Iron had 62% of its mass in particles greater than 3.3 μm, while 67%, 80%, and 72% of Mn, Zn, and Cu, respectively, occurred in particles less than 3.3 μm. Distributions of these elements can be classified as large (Fe), small (Zn) and intermediate-size (Mn and Cu) components. The observed size distributions are consistent with the known atmospheric injection mechanisms (Gladney et al., 1974). The results support previous studies which show extensive enrichment of certain trace elements in the smaller size fractions due to volatilization during combustion and subsequent condensation and coagulation.

3.3. Elemental Density Distributions of Dustfall

The observed particle density distributions of Si, Fe, Mn, Zn, and Cu in water-insoluble dustfall are shown in Figure 11.6. The elemental density distributions of dustfalls are completely different from those of suspended particulates. The observations presented above show that suspended particulates and dustfalls are of a quite different nature with respect to density. Silicon shows a definite peak in the density range of 2.5 to 2.7 g/cc, corresponding to the density of SiO_2. Iron, Mn, Cu, and Zn show definite peaks at the density ranges of 2.5–2.7 and >3.3 g/cc.

Gravitational fallout is an important removal mechanism for large particles. Dustfalls are probably made of soil materials to a large extent, and hence the elemental density distributions for dustfalls are expected to be similar to those for the surrounding soils. The elemental density distribution curves for soil, illustrated in Figure 11.2, indicate that the concentration of Si reaches a maximum at approximately 2.6 g/cc, whereas two peaks appear in the density distributions for Fe, Mn, Zn, and Cu at the density ranges of 2.5–2.7 g/cc and

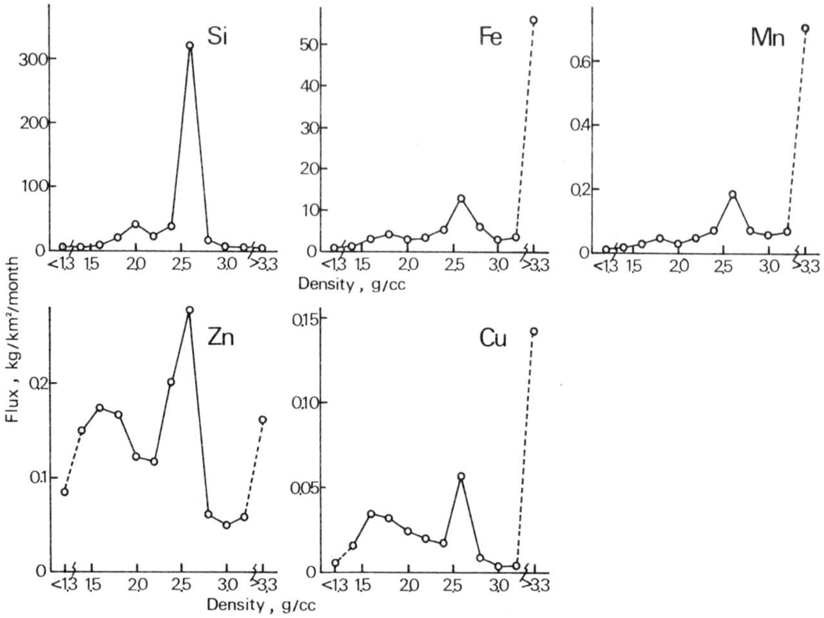

Figure 11.6. Density distributions of Si, Fe, Mn, Zn, and Cu in dustfall.

>3.3 g/cc. The particles associated with these density ranges of 2.5–2.7 g/cc (and >3.3 g/cc) can be attributed to the natural sources such as windblown soil dust.

The density distributions for Si, Fe, and Mn in the dustfall are essentially similar to those in the soil and reflect a strong contribution from windblown soil dust. However, further inspection of Figure 11.6 indicates that considerably fewer Fe and Mn particles with a density of 2.6 g/cc are in the dustfall than in the soil. It is to be noted that, for these elements, 70% or more of the total elemental mass are contained in the most dense fraction (>3.3 g/cc) of the dustfall. Some of these high density particles are probably due to rusting.

The concentrations of the chalcophilic elements Zn and Cu in the dustfall are also high in the density ranges of 2.5–2.7 and >3.3 g/cc. The association of these elements with the particles of 2.5–2.7 and >3.3 g/cc is expected of soil-derived particles. However, the results suggest that the elemental density distributions seem to depend on the contribution of particles produced by different mechanisms. The resolution into a two-component system is likely to be more appropriate in studying the evolution of the density distributions for Zn and Cu on the basis of the mechanisms of particle production. The density distributions of Zn and Cu indicate the presence of a low density component and suggest the contribution of high temperature, anthropogenic sources to the dustfall.

REFERENCES

Adams, F., and Van Espen, P. (1983). "Aerosol composition at Chacaltaya, Bolivia, as determined by size-fractionated sampling." *Atmos. Environ.* **17,** 1521–1536.

Bertine, K.K., and Goldberg, E.S. (1971). "Fossil fuel combustion and the major sedimentary cycle." *Science* **173,** 233–235.

Biggins, P.D.E., and Harrison, R.M. (1978). "Identification of lead compounds in urban air." *Nature* **272,** 531–532.

Biggins, P.D.E., and Harrison, R.M. (1979a). "Atmospheric chemistry of automotive lead." *Environ. Sci. Technol.* **13,** 558–565.

Biggins, P.D.E., and Harrison, R.M. (1979b). "Characterization and classification of atmospheric sulfates." *J. Air Pollut. Control Assoc.* **29,** 838–840.

Biggins, P.D.E., and Harrison, R.M. (1980). "Chemical speciation of lead compounds in street dusts." *Environ. Sci. Technol.* **14,** 336–339.

Campbell, J.A., Smith, R.D., and Davis, L.E. (1978a). "Application of X-ray photoelectron spectroscopy to the study of fly ash." *Appl. Spectrosc.* **32,** 316–319.

Campbell, J.A., Laul, J.C., Nielson, K.K., and Smith, R.D. (1978b). "Separation and chemical characterization of finely-sized fly-ash particles." *Anal. Chem.* **50,** 1032–1040.

Cushing, K.M., McCain, J.D., and Smith, W.B. (1979). "Experimental determination of sizing parameters and wall losses of five source-test cascade impactors." *Environ. Sci. Technol.* **13,** 726–731.

Dams, R., Rahn, R.A., and Winchester, J.W. (1972). "Evaluation of filter materials and impactor surfaces for nondestructive neutron activation analysis of aerosols." *Environ. Sci. Technol.* **5,** 441–448.

Durham, J.L., Wilson, W.E., Ellestad, T.G., Willeke, K., and Whitby, K.T. (1975). "Comparison of volume and mass distributions for Denver aerosols." *Atmos. Environ.* **9,** 717–722.

Dzubay, T.G., and Stevens, R.K. (1975). "Ambient air analysis with dichotomous sampler and X-ray fluorescence spectrometer." *Environ. Sci. Technol.* **9,** 663–668.

Fisher, G.L., Chang, D.P.Y., and Brummer, M. (1976). "Fly ash collected from electrostatic precipitators: microcrystalline structures and the mystery of the sphere." *Science* **192,** 553–555.

Flocchini, R.G., Cahill, T.A., Shadoan, D.J., Lange, S.J., Eldred, R.A., Feeney, P.J., Wolfe, G.W., Simmeroth, D.C., and Suder, J.K. (1976). "Monitoring California's aerosols by size and elemental composition." *Environ. Sci. Technol.* **10,** 76–82.

Floyd, M.A., Fassel, V.A., and D'Silva, A.P. (1980). "Computer-controlled scanning monochromator for the determination of 50 elements in geochemical and environmental samples by inductively coupled plasma-atomic emission spectrometry." *Anal. Chem.* **52,** 2168–2173.

Fragan, R.C., and Friedlander, S.K. (1978). "Particle formation in pulverized coal combustion—a review." In D.T. Shaw, Ed., *Recent Developments in Aerosol Science.* Wiley, New York, pp. 25–59.

Fuchs, N.A. (1964). *Mechanism of Aerosols.* Pergamon Press, New York.

Fujimura, M., Yano, N., and Hashimoto, Y. (1978). "Size distribution measurement and neutron activation analysis of sea salt particles in maritime atmosphere." *Nippon Kagaku Kaishi* **1978,** 456–461.

Fukasawa, T., Iwatsuki, M., Kawakubo, S., and Miyazaki, K. (1980). "Heavy-liquid separation and X-ray diffraction analysis of airborne particulates." *Anal. Chem.* **52,** 1784–1787.

Gillett, D.A., Blifford, I.H., and Fenster, C.R. (1972). "Measurements of aerosol size distributions and vertical fluxes of aerosols on land subjected to wind erosion." *J. Appl. Meteor.* **11,** 977–987.

Gladney, E.G., Zoller, W.H., Jones, A.G., and Gordon, G.E. (1974). "Composition and size distributions of atmospheric particulate matter in Boston area." *Environ. Sci. Technol.* **8,** 551–557.

Haff, W., and Jaenicke, R. (1980). "Results of improved size distribution measurements in the Aitken range of atmospheric aerosols." *J. Aerosol Sci.* **11,** 321–330.

Hardy, K.A., Akselsson, R., Nelson, J.W., and Winchester, J.W. (1976). "Elemental constituents of Miami aerosol as function of particle size." *Environ. Sci. Technol.* **10,** 176–182.

Harrison, R.M., and Parker, J. (1977). "Analysis of particulate pollutants." In R. Perry and R. Young, Eds., *Handbook of Air Pollution Analysis.* Chapman and Hall, London, pp. 84–156.

Hemeon, W.C.L. (1955). "The estimation of health hazards from air pollution." *A.M.A. Arch. Ind. Health* **11,** 397–402.

Hirsch, Y., Peleg, M., and Luria, M. (1981). "Characterization of suspended dust particles in the Jerusalem air." *Environ. Sci. Technol.* **15,** 1456–1460.

Hopke, P.K., Lamb, R.E., and Natusch, D.F.S. (1980). "Multielemental characterization of urban roadway dust." *Environ. Sci. Technol.* **14,** 164–172.

Hu, J.N. (1971). "An improved impactor for aerosol studies—modified Andersen sampler." *Environ. Sci. Technol.* **5,** 251–253.

Husar, R.B., Whitby, K.T., and Liu, B.Y.H. (1972). "Physical mechanisms governing the dynamics of Los Angeles smog aerosol." *J. Colloid Interface Sci.* **39,** 211–224.

Jaklevic, J.M., Kirby, J.A., Ramponi, A.J., and Thompson, A.C. (1980). "Chemical characterization of air particulate samples using X-ray absorption spectroscopy." *Environ. Sci. Technol.* **14,** 437–441.

Jutze, G.A., and Foster, K.I. (1967). "Recommended standard method for atmospheric sampling of fine particulate matter by filter media—high-volume sampler." *J. Air Pollu. Control Assoc.* **17,** 17–25.

Kadowaki, S. (1979). "Silicon and aluminum in urban aerosols for characterization of atmospheric soil particles in the Nagoya area." *Environ. Sci. Technol.* **13,** 1130–1133.

Keyser, T.R., Natusch, D.F.S., Evans, C.A., Jr., and Linton, R.W. (1978). "Characterizing the surfaces of environmental particles." *Environ. Sci. Technol.* **12,** 768–773.

Linton, R.W., Loh, A., Natusch, D.F.S., Evans, C.A., Jr., and Williams, P. (1976). "Surface predominance of trace elements in airborne particles." *Science* **191,** 852–854.

Linton, R.W., Williams, R., Evans, C.A., Jr., and Natusch, D.F.S. (1977). "Determination of the surface predominance of toxic elements in airborne particles by ion microprobe mass spectrometry and auger electron spectrometry." *Anal. Chem.* **49,** 1514–1521.

Linton, R.W., Natusch, D.F.S., Solomon, R.L., and Evans, C.A., Jr. (1980). "Physicochemical characterization of lead in urban dusts. A microanalytical approach to lead tracing." *Environ. Sci. Technol.* **14,** 159–164.

Long, S.J., Suggs, J.C., and Walling, J.F. (1979). "Lead analysis of ambient air particulates: interlaboratory evaluation of EPA lead reference method." *J. Air Pollu. Control Assoc.* **29,** 28–31.

McQuarker, N.R., Brown, D.F., and Kluckner, P.D. (1979). "Digestion of environmental materials for analysis by inductively coupled plasma–atomic emission spectrometry." *Anal. Chem.* **51,** 1082–1084.

Miller, F.J., Gardner, D.E., Graham, J.A., Lee, R.E. Jr., Wilson, W.E., and Backmann, J.D. (1979). "Size considerations for establishing a standard for inhalable particles." *J. Air Pollu. Control Assoc.* **29,** 610–615.

Noonkester, V. R. (1981). "Aerosol size spectra in a convective marine layer with stratus: results of airborne measurements near San Nicolas Island, California." *J. Appl. Met.* **20,** 1076–1080.

Olson, K.W., and Skogerboe, R.K. (1975) "Identification of soil lead compounds from automotive sources." *Environ. Sci. Technol.* **9,** 227–230.

Paciga, J.J., and Jervis, R.E. (1976). "Multielement size characterization of urban aerosols." *Environ. Sci. Technol.* **10,** 1124–1128.

Pilkington, E.S., and Warren, L.J. (1979). "Determination of heavy-metal distribution in marine sediments." *Environ. Sci. Technol.* **13,** 295–299.

Rahn, K.A. (1976). "Silicon and aluminum in atmospheric aerosols: crust–air fractionation?" *Atmos. Environ.* **10,** 597–601.

Sheih, C.M., Johnson, S.A., and DePaul, F.T. (1983). "Case studies of aerosol size distribution and chemistry during passages of a cold and a warm front." *Atmos. Environ.* **17,** 1299–1306.

Silverman, L., Billings, C.E., and First, M.W. (1971). *Particle Size Analysis in Industrial Hygiene.* Academic Press, New York, pp. 1–24.

Smith, R.D. (1980). "The trace element chemistry of coal during combustion and the emissions from coal-fired plants." *Prog. Energy Combust. Sci.* **6,** 53–119.

Sugimae, A. (1979). "Determination of trace elements in airborne particulate matter by inductively coupled plasma–optical emission spectrometry." *J. Japan Soc. Air Pollut.* **14,** 389–398.

Sugimae, A. (1981). "Simultaneous multielement analysis of airborne particulate matter by icp-oes." *ICP Inform. Newslett.* **6,** 619–644.

Sugimae, A., and Mizoguchi, T. (1982). "Atomic emission spectrometric analysis of airborne particulate matter by direct nebulization of suspensions into the inductively coupled plasma." *Anal. Chim. Acta* **144,** 205–212.

Sugimae, A. (1983a). "Physicochemical characterization of trace metal compounds in airborne particulate matter." *J. Japan Soc. Air Pollut.* **18,** 233–240.

Sugimae, A. (1983b). "Density distributions of trace metallic compounds in the atmosphere." *J. Japan Soc. Air Pollut.* **48,** 416–424.

Sugimae, A. (1984). "Elemental constituents of atmospheric particulates as function of particle density." *Nature* **307,** 145–147.

Sverdrup, G.M., Whitby, K.T., and Clark, W.E. (1975). "Characterization of California aerosols II. Aerosol size distribution measurements in the Mojave desert." *Atmos. Environ.* **9,** 483–494.

Tanaka, S., Darzi, M., and Winchester, J.W. (1980). "Short term effect of rainfall on elemental composition and size distribution of aerosol in north Florida." *Atmos. Environ.* **14,** 1421–1426.

Van Craen, M., Natusch, D.F.S., and Adams, F. (1982). "Quantitative surface analysis of steel furnace dust particles by secondary ion mass spectrometry." *Anal. Chem.* **54,** 1786–1792.

Van Craen, M.J., Denoyer, E.A., Natusch, D.F.S., and Adams, F. (1983). "Surface enrichment of trace elements in electric steel furnace dust." *Environ. Sci. Technol.* **17,** 435–439.

Wangen, L.E. (1981). "Elemental composition of size-fractionated aerosols associated with a coal-fired power plant plume and background." *Environ. Sci. Technol.* **15,** 1080–1088.

Watson, J.G., Chow, J.C., Shah, J.J., and Pace, T.G. (1983). "The effect of sampling inlets on the PM-10 and PM-15 to TSP concentration ratios." *J. Air Pollu. Control Assoc.* **33,** 114–119.

Wedding, J.B., McFarland, A.R., and Cermak, J.E. (1977). "Large particle collection characteristics of ambient aerosol samplers." *Environ. Sci Technol.* **11,** 387–390.

Whitby, K.T., Husar, R.B., and Liu, B.Y.H. (1972). "The aerosol size distribution of Los Angeles smog." *J. Colloid Interface Sci.* **39,** 177–204.

Whitby, K.T., Charlson, R.E., Wilson, W.E., and Stevens, R.K. (1974). "The size of suspended particle matter in air." *Science* **183,** 1098–1099.

Whitby, K.T., Clark, W.E., Marple, V.A., Sverdrup, G.M., Sem, G.J., Willeke, K., Liu, B.Y.H., and Pui, D.Y.H. (1975). "Characterization of California aerosols I. Size distributions of freeway aerosols." *Atmos. Environ.* **9,** 463–482.

Whitby, K.T. (1977). "The physical characteristics of sulfate aerosols." *Atmos. Environ.* **12,** 135–159.

Whytlaw-Gray, R., and Peterson, H.S. (1932). *Smoke.* Edward Arnold and Company, London.

Willeke, K., and Whitby, K.T. (1975). "Atmospheric aerosols: Size distribution interpretation." *J. Air Pollut. Control Assoc.* **25,** 529–534.

Zoller, W.H., Gladney, E.S., and Duce, R.A. (1974). "Atmospheric concentrations and sources of trace metals at the South Pole." *Science* **183,** 198–200.

12

THE SIZES OF AIRBORNE TRACE METAL CONTAINING PARTICLES

Cliff I. Davidson
James F. Osborn

Departments of Civil Engineering
and Engineering & Public Policy
Carnegie-Mellon University
Pittsburgh, Pennsylvania

1. INTRODUCTION

In a previous review paper, Lee and von Lehmden (1973) addressed trace metals in fossil fuels, industrial materials, and ambient air. A section of this article was concerned with airborne trace metal size distributions: The authors emphasized the role of particle diameter in influencing respiratory system deposition, atmospheric transport, and other processes. Critical gaps in our understanding of these processes were identified, including weaknesses in the fundamental models as well as lack of input data.

A wealth of new information has become available in the decade since this paper was published. Detailed models have been developed, allowing better prediction of particle behavior. Improved methods for determining the sizes of trace metal-containing aerosols have evolved. The current data base includes airborne size distributions measured near industrial sources, in ambient urban settings, and in remote regions.

A review of the literature has shown that detailed size data are available for several trace metals. Those metals having the greatest amount of information on size distributions include Pb, Cd, Cu, Fe, Mn, and Zn, each of which has been reported in at least 10 separate investigations. Other elements have generally received less attention; a list of references covering airborne concentrations and size distributions of a large number of metals is presented by Rahn (1976), while additional size distribution data for 38 elements have been summarized by Milford and Davidson (1985).

This chapter summarizes size distribution data from the literature for the above six metals. After the distributions are presented, several separate issues are discussed. First, the experimental techniques used to generate these distributions are reviewed to assess possible artifacts. Overall characteristics and consistencies in the spectra are then identified. Finally, as an example of the utility of the size data, the distributions are used with recently published information to estimate deposition in the human respiratory tract and dry deposition from the ambient atmosphere onto various surfaces.

Only distributions with size resolution sufficient to show detailed shapes of the spectra are discussed in this chapter. Data from dichotomous samplers, or from impactors and cyclones with less than four stages, have not been included.

2. PRESENTATION OF THE DATA

Figures 12.1–12.6 show distributions of the six metals of interest. The ordinate is the normalized mass distribution function $(\Delta C/C_T)/\Delta \log d_p$, where ΔC is the airborne mass concentration of the metal in a given size range (which extends from $d_{p\,min}$ to $d_{p\,max}$), C_T is the total concentration of the metal in all size ranges, and $\Delta \log d_p$ is the difference $\log d_{p\,max} - \log d_{p\,min}$. The aerodynamic diameter of a particle, d_p, represents the size of a unit density sphere whose aerodynamic transport characteristics are identical to those of the original particle. Note that the area under the curve between any two particle diameters is proportional to the fraction of airborne mass in that size interval. Data to construct these graphs have been obtained from differential or cumulative distribution plots, or from tables, in the original references.

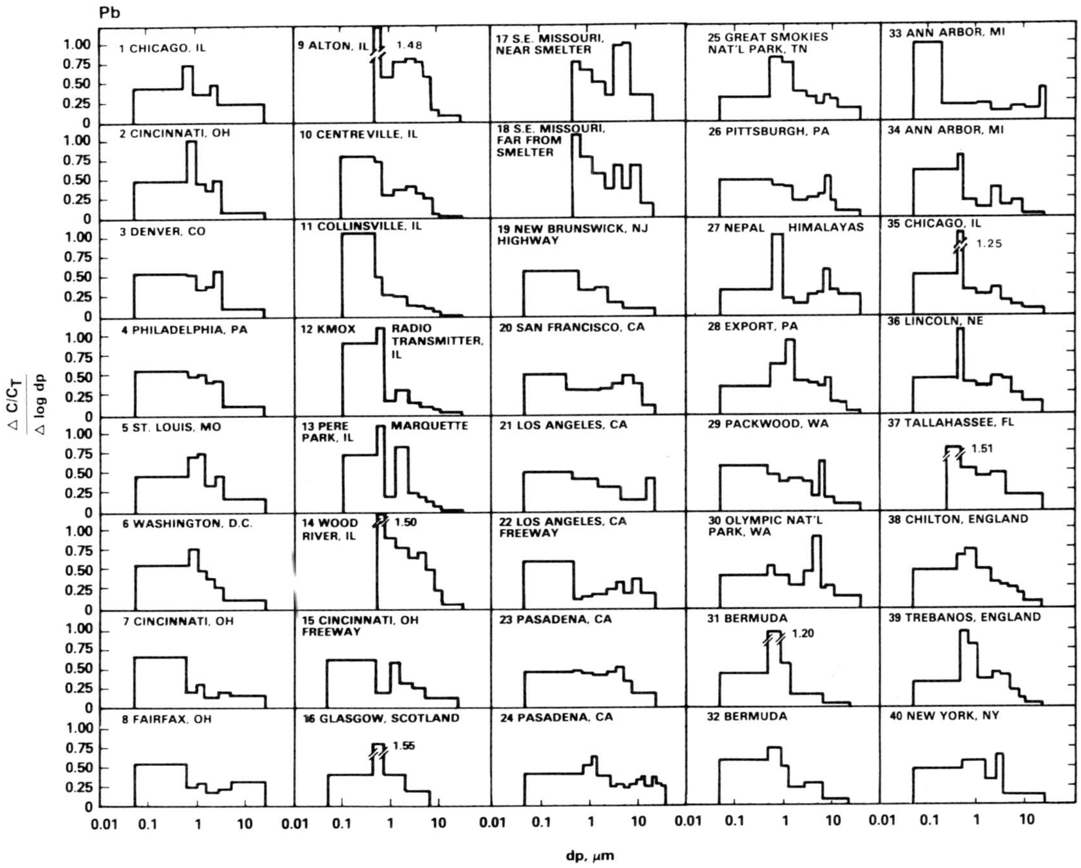

Figure 12.1. Distributions of airborne Pb mass with respect to particle size.

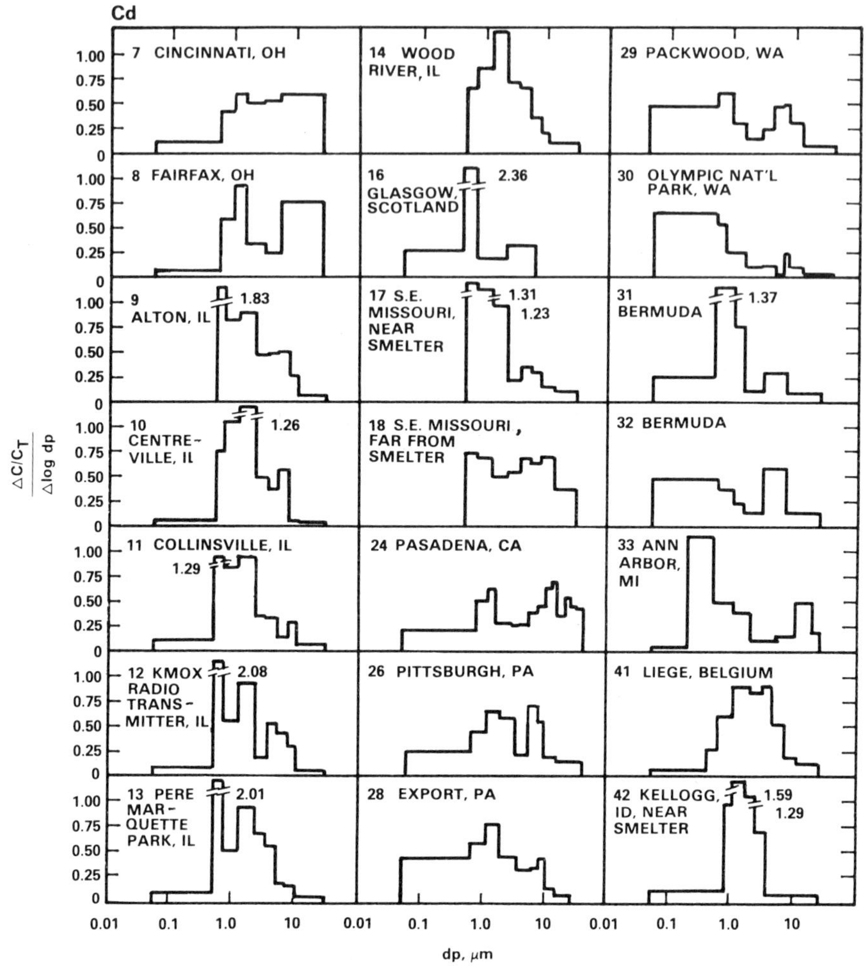

Figure 12.2. Distributions of airborne Cd mass with respect to particle size.

Personal communication with the authors was necessary in several instances to clarify or update the data.

All distributions have been plotted as though the particles are uniformly distributed in log d_p in each size range. The resulting histograms are thus only approximations to the true smooth-curve distributions which are unknown, although techniques for estimating the true size spectra have been reported in the literature (Esmen, 1977; Raabe, 1978). For consistency, overall minimum and maximum aerodynamic diameters of 0.05 and 25 μm, respectively, have been assumed after Hidy (1974) and Davidson (1977), except for a few cases where investigators have taken special measures to characterize the distribution endpoints. This style of graphing has been chosen to allow direct observation of those size ranges containing the bulk of the airborne mass, and

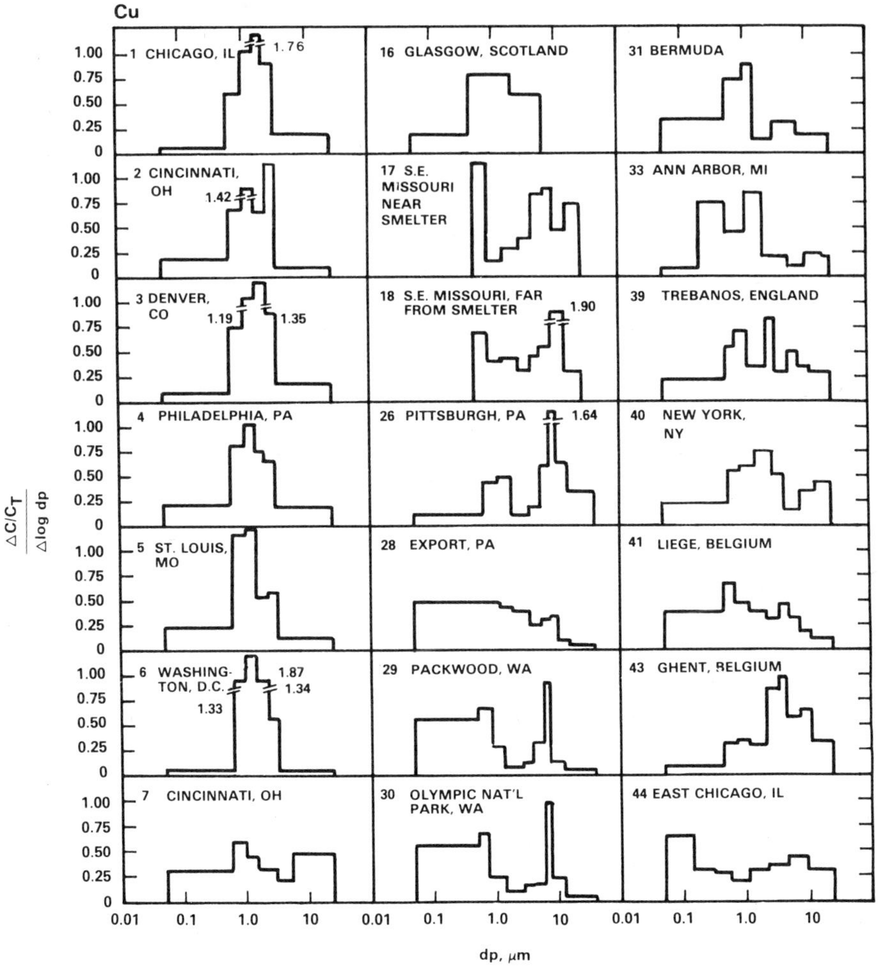

Figure 12.3. Distributions of airborne Cu mass with respect to particle size.

to emphasize differences in the shapes of the distributions. Such differences would be more difficult to identify in cumulative log probability plots.

Pertinent data associated with the graphs are given in Tables 12.1 and 12.2. Table 12.1 lists the authors, dates of sampling, location, and type of equipment used to collect the airborne particles. Table 12.2 gives the total airborne concentrations and mass median aerodynamic diameters (MMD) for each distribution. In some cases, the MMD values are slightly different than those listed in the original references, due to assumed minimum and maximum diameters in Figs. 12.1–12.6, which differ from those quoted in the references.

The figures and tables show wide variations in the shapes of the distributions as well as in values of MMD. The extent to which these

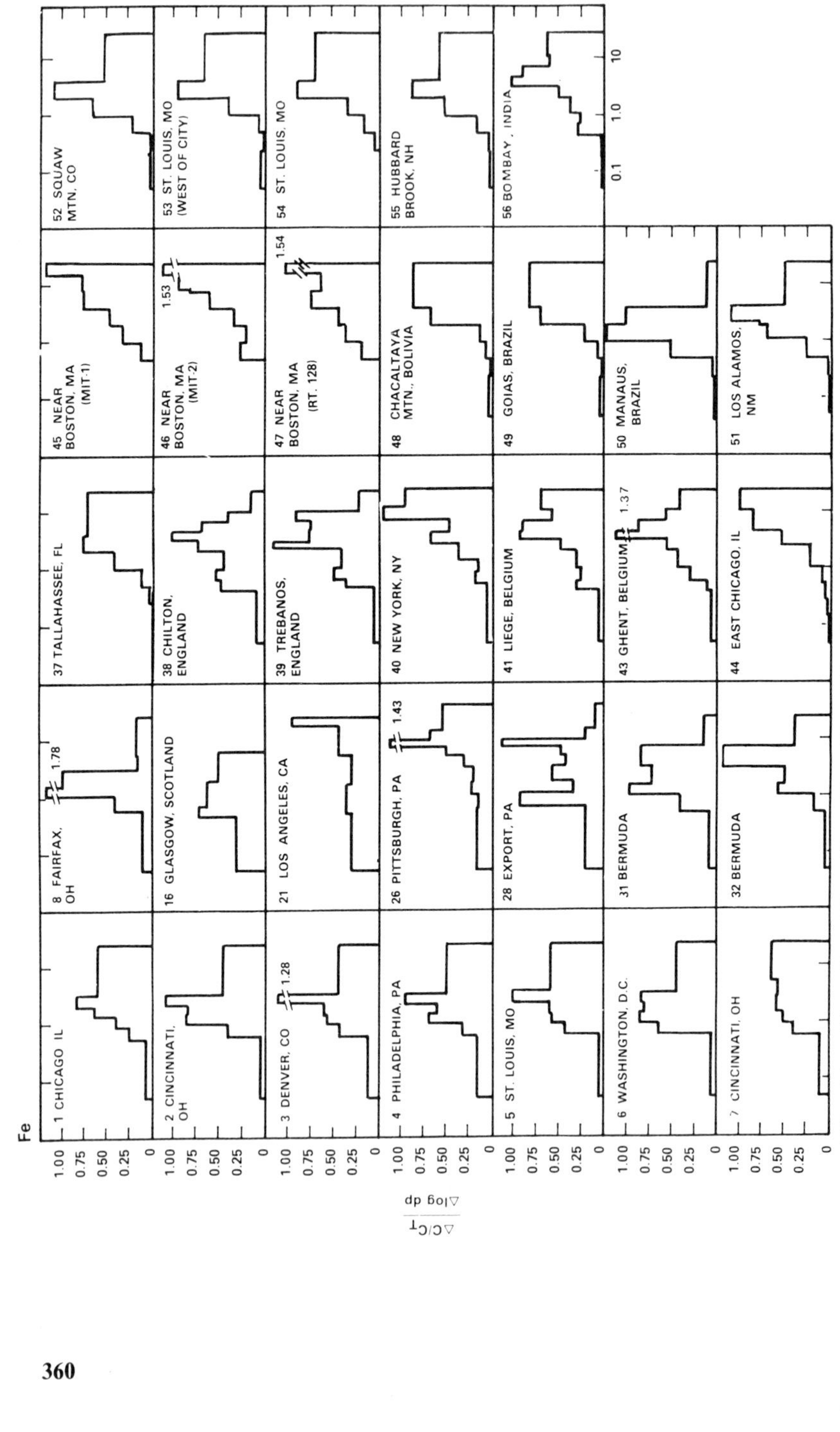

Figure 12.4. Distributions of airborne Fe mass with respect to particle size.

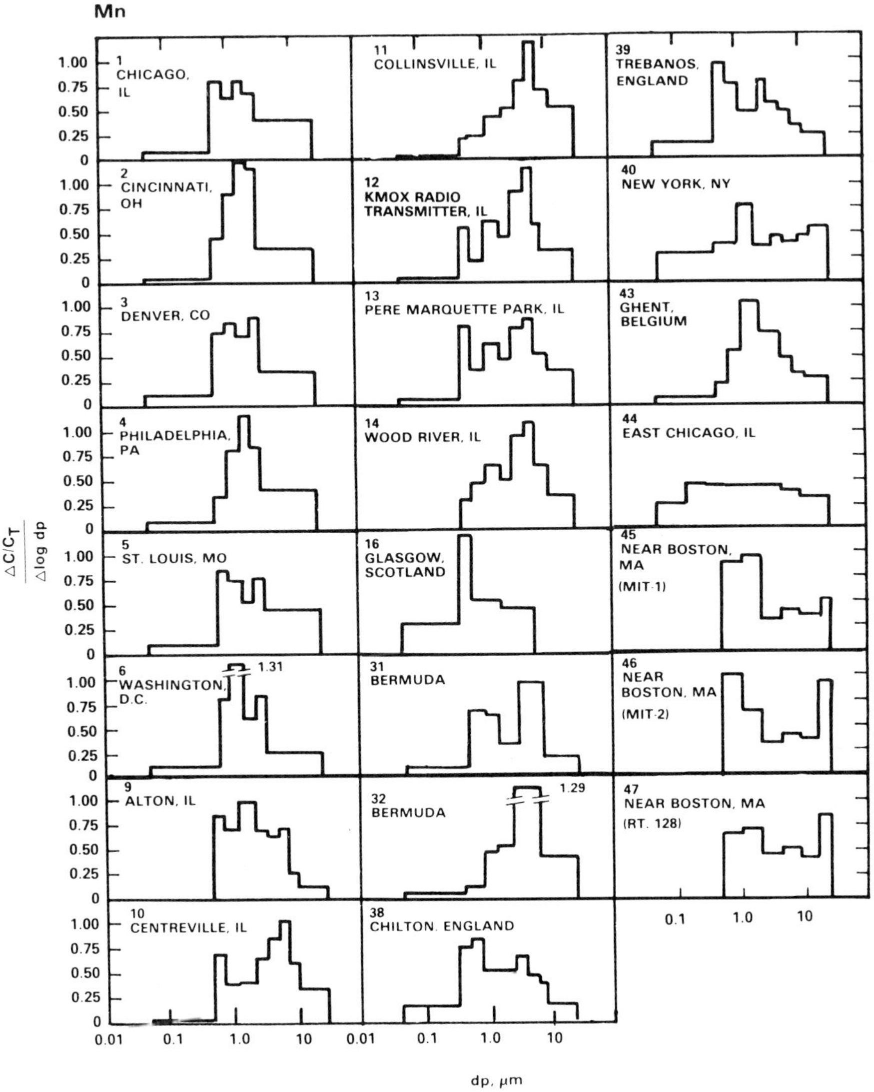

Figure 12.5. Distributions of airborne Mn mass with respect to particle size.

variations reflect true differences in ambient conditions as opposed to experimental artifacts is now discussed.

3. VARIATIONS IN THE DATA: REAL OR ARTIFACT?

Nearly all of the size distribution data in Figs. 12.1–12.6 were obtained with cascade impactors. These devices fractionate airborne particles according to

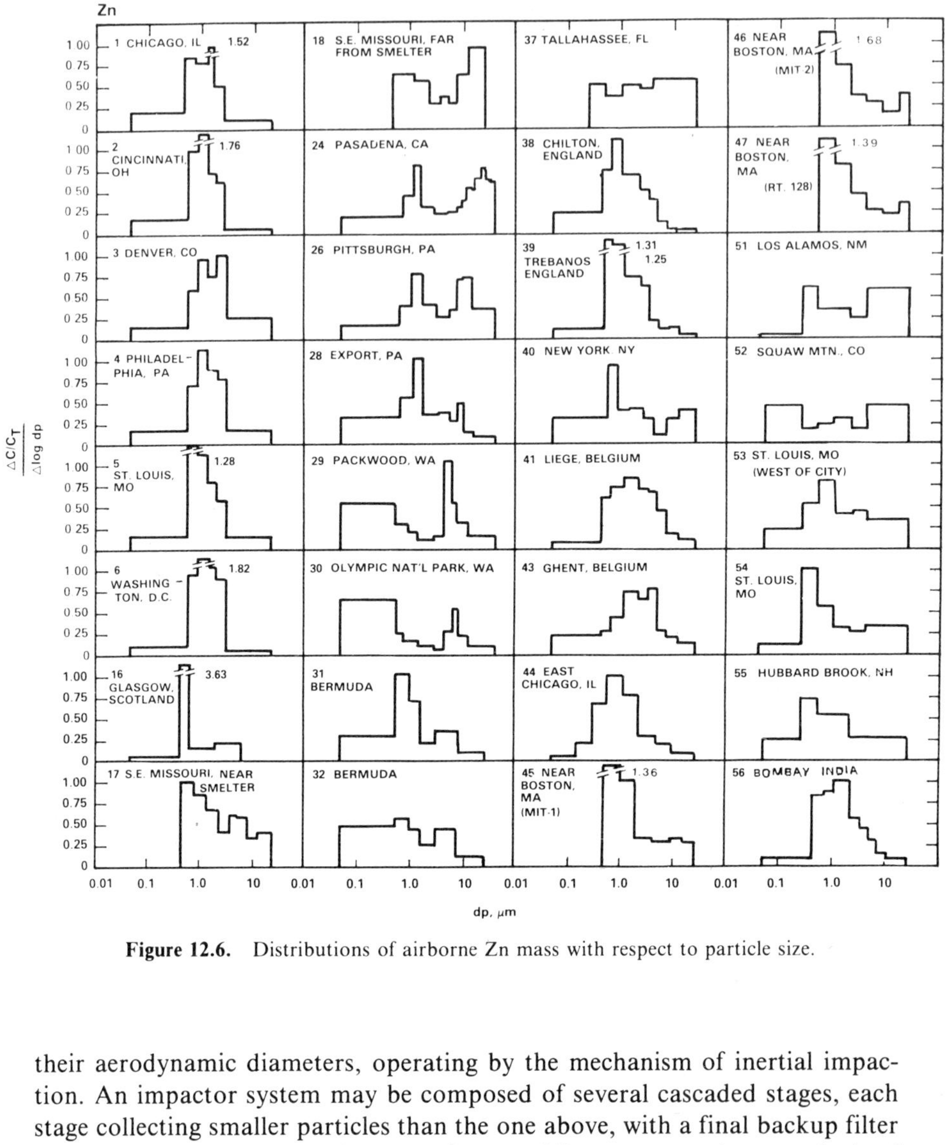

Figure 12.6. Distributions of airborne Zn mass with respect to particle size.

their aerodynamic diameters, operating by the mechanism of inertial impaction. An impactor system may be composed of several cascaded stages, each stage collecting smaller particles than the one above, with a final backup filter to collect very small particles. While providing a convenient method of fractionating aerosols for subsequent elemental analysis, impactors are highly imperfect.

One difficulty with impactor sampling concerns the upper end of the size spectrum. Obtaining a representative sample of large airborne particles requires careful measuring techniques: The air velocity at the impactor inlet must be approximately the same as the ambient wind speed (isokinetic sampling), and the inlet must be directionally aligned with the wind (Liu and Pui, 1981). Even with these precautions, particle loss within the inlets may

Table 12.1. Information Associated with the Airborne Mass Distributions of Figs. 12.1–12.6[a]

Graph No.	Reference	Dates of Sampling	Location of Sampling	Type of Sampler
1	Lee et al., 1972	Jan.–Dec. 1970 Average of 4 quarterly composited samples, representing a total of 21 sampling periods of 24 h each	Chicago, IL	Andersen impactor operated at 5–6 CFM
2	Lee et al., 1972	Mar.–Dec. 1970 Same averaging as Graph 1, total of 18 sampling periods	Cincinnati, OH	Same as Graph 1
3	Lee et al., 1972	Jan.–Dec. 1970 Same averaging as Graph 1, total of 21 sampling periods	Denver, CO	Same as Graph 1
4	Lee et al., 1972	Mar.–Dec. 1970 Same averaging as Graph 1, total of 20 sampling periods	Philadelphia, PA	Same as Graph 1
5	Lee et al., 1972	Jan.–Dec. 1970 Same averaging as Graph 1, total of 22 sampling periods	St. Louis, MO	Same as Graph 1
6	Lee et al., 1972	Jan.–Dec. 1970 Same averaging as Graph 1, total of 23 sampling periods	Washington, DC	Same as Graph 1
7	Lee et al., 1968	Sept. 1966 Average of 14 runs, 24 h each	Cincinnati, OH	Andersen impactor 1.2 m above the ground
8	Lee et al., 1968	Feb. 1967 Average of 3 runs, 4 days each	Fairfax, OH, suburb of Cincinnati	Same as Graph 7

Table 12.1. ***(Continued)***

Graph No.	Reference	Dates of Sampling	Location of Sampling	Type of Sampler
9	Peden, 1977	Summer 1975 Average of 4 runs, average 8 days each	Alton, IL, industrial area near St. Louis	Andersen impactor, no backup filter
10	Peden, 1977	Summer 1972 Average of 3 runs, average 10 days each	Centreville, IL, downwind of a zinc smelter	Andersen impactor with backup filter
11	Peden, 1977	Summer 1973 Average of 2 runs, average 5 days each	Collinsville, IL, industrial area near St. Louis	Same as Graph 10
12	Peden, 1977	Summer 1973 Average of 2 runs, average 6 days each	KMOX radio transmitter, IL, industrial area near St. Louis	Same as Graph 10
13	Peden, 1977	Summer 1972 Average of 9 runs, average 9 days each	Pere Marquette State Park, IL, upwind of St. Louis	Same as Graph 10
14	Peden, 1977	Summer 1975 Average of 4 runs, average 8 days each	Wood River, IL, industrial area near St. Louis	Same as Graph 9
15	Cholak et al., 1968	Apr. 1968 Average of several runs, 3 days each	3 sites: 10, 400, and 3300 m from Interstate 75, Cincinnati, OH	Andersen impactor
16	McDonald and Duncan, 1979	June 1975 One run of 15 days	Glasgow, Scotland	Casella impactor 30 m above the ground
17	Dorn et al., 1976	Winter, spring, summer 1972 Average of 3 runs, 27 days each	Southeast Missouri, 800 m from a lead smelter	Andersen impactor 1.7 m above the ground, no backup filter

18	Dorn et al., 1976	Winter, spring, summer 1972 Average of 3 runs, 14 days each	Southeast Missouri, 75 km from the lead smelter of Graph 17	Same as Graph 17
19	Daines et al., 1970	1968 Average of continuous 1-week runs over an 8-month period	3 sites: 9, 76, and 530 m from U.S. Route 1, New Brunswick, NJ	Cascade impactor 1.2 m above the ground
20	Martens et al., 1973	July 1971 One run of 4 days	9 sites throughout San Francisco, CA area	Andersen impactor
21	Lundgren, 1970	Nov. 1968 Average of 10 runs, 16 h each	Los Angeles, CA	Lundgren impactor
22	Huntzicker et al., 1975	May 1973 One run of 8 h	Shoulder of freeway near downtown Los Angeles, CA	Andersen impactor 2 m above the ground
23	Huntzicker et al., 1975	Feb. 1974 One run of 6 days	Pasadena, CA	Andersen impactor on roof of 4-story building
24	Davidson, 1977	May and July 1975 Average of 2 runs, 61 h each	Pasadena, CA	Modified Andersen impactor on roof of 4-story building
25	Davidson et al., 1980	Oct. 1979 One run of 120 h	Clingman's Dome, Great Smokies National Park, elev. 2024 m	2 Modified Andersen impactors 1.2 m above the ground
26	Davidson et al., 1981a	July–Sept. 1979 Average of 2 runs, 90 h each	Pittsburgh, PA	Modified Andersen impactor 4 m above the ground
27	Davidson et al., 1981b	Dec. 1979 One run of 52 h	Nepal Himalayas, elev. 3900 m	Modified Andersen impactor 1.2 m above the ground

Table 12.1. (*Continued*)

Graph No.	Reference	Dates of Sampling	Location of Sampling	Type of Sampler
28	Davidson et al., 1983	June 1980 One run of 72 h	Export, PA, 40 km east of downtown Pittsburgh	2 Modified Andersen impactors 1.2 m above the ground
29	Davidson et al., 1985	July 1980 One run of 34 h	Packwood, WA, rural site in Gifford Pinchot National Forest	Modified Andersen impactor 1.5 m above the ground
30	Davidson et al., 1985	July–Aug. 1980 One run of 92 h	Hurricane Ridge, Olympic National Park, elev. 1600 m	Modified Andersen impactor 1.5 m above the ground
31	Duce et al., 1976a	May–June 1975 One run of 112 h	Southeast coast of Bermuda	Sierra high volume impactor 20 m above the ground
32	Duce et al., 1976a	July 1975 One run of 79 h	Same as Graph 31	Same as Graph 31
33	Harrison et al., 1971	Apr. 1968 Average of 21 runs, 2 h each	Ann Arbor, MI	Modified Andersen impactor 20 m above the ground
34	Gillette and Winchester, 1972	Oct. 1968 Average of 15 runs, 24 h each	Ann Arbor, MI	Andersen impactor near ground level
35	Gillette and Winchester, 1972		Chicago, IL	Same as Graph 34
36	Gillette and Winchester, 1972		Lincoln, NE	Same as Graph 34
37	Johansson et al., 1976	June–July 1973 Average of 15 runs, average 50 h each	2 sites in Tallahassee, FL	Delron Battelle-type impactor on building roofs, no backup filter

38	Cawse et al., 1974	July–Dec. 1973	Chilton, England	Andersen impactor 1.5 m above the ground
39	Pattenden et al., 1974	May–Aug. 1973 Average of 4 runs, 1 month each	Trebanos, England	Same as Graph 38
40	Bernstein and Rahn, 1979	August 1976 Average of 4 runs, 1 week each	New York, NY	Andersen impactor on roof of 15-story building Pb distribution obtained with Cyclone sampling system
41	Rahn, 1976	Apr.–May 1972 One 3-week run	Liege, Belgium	Andersen impactor
42	U.S. Environmental Protection Agency, 1973, 1974	Apr. 1973–Jan. 1974 Average of 37 runs, 24 h each	Kellogg, ID, near a lead smelter	Modified Andersen impactor
43	Heindryckx, 1976	Sept. 1971–Nov. 1973 Average of 8 runs, 5–10 h each	Industrial area north of Ghent, Belgium	Andersen impactor
44	Nifong et al., 1972	July 1969–Feb. 1970 Average of 15 runs, 1–15 days each	3 sites in East Chicago, IL, industrial area	Andersen impactor
45	Gladney et al., 1974	Feb.–Mar. 1974 Average of 3 runs, 24 h each	Near Boston, MA, MIT campus	Scientific Advances impactor on roof of 2-story building, no backup filter
46	Gladney et al., 1974	Feb.–Mar. 1974 Average of 3 runs, 24 h each	Near Boston, MA, MIT campus	Scientific Advances impactor on roof of 19-story building, no backup filter

Table 12.1. (*Continued*)

Graph No.	Reference	Dates of Sampling	Location of Sampling	Type of Sampler
47	Gladney et al., 1974	Feb.–Mar. 1974 Average of 3 runs, 24 h each	Near Route 128 in a Boston, MA suburb	Scientific Advances impactor near ground level, no backup filter
48	Lawson and Winchester, 1979	Jan.–Mar. 1977 One run of 171 h	Chacaltaya Mtn., Brazil	Delron Battelle-type impactor
49	Lawson and Winchester, 1979	July 1976 Average of 3 runs, less than 72 h each	Goias, Brazil	Same as Graph 48
50	Lawson and Winchester, 1979	Jan.–Mar. 1977 Average of 3 runs, less than 72 h each	Manaus, Brazil	Same as Graph 48
51	Winchester et al., 1979	Apr. 1976 Average of 5 runs, 48 h each	Las Alamos, NM, elev. 2615 m	Same as Graph 48
52	Winchester et al., 1979	Apr. 1976 Average of 5 runs, 48 h each	Squaw Mountain, CO, elev. 3200 m	Same as Graph 48
53	Winchester et al., 1979	Apr. 1976 Average of 5 runs, 48 h each	St. Louis, MO, west of city	Same as Graph 48
54	Winchester et al., 1979	Apr. 1976 Average of 5 runs, 48 h each	St. Louis, MO	Same as Graph 48
55	Winchester et al., 1979	Apr. 1976 Average of 5 runs, 48 h each	Hubbard Brook, NH	Same as Graph 48
56	Sadasivan, 1981	Nov.–Dec. 1976 Average of 7 runs, 2–5 days each	Bombay, India	Andersen impactor 15 m above ground level

[a] All impactors and cyclones were operated with backup filters, unless otherwise indicated. The Andersen impactors refer to the standard 1 CFM samplers (28 L/min), while the Sierra high volume impactors are the slotted 40 CFM devices (1100 L/min). Other samplers listed in this table include the cyclone system of Bernstein and Rahn (15–210 L/min), the Lundgren impactor (82 L/min), the Castella impactor (17 L/min), the Scientific Advances impactor (12 L/min), and the Delron Battelle-type impactor (1 L/min). These airflows represent the manufacturers' rated values; actual flow rates were often different from these ratings.

Table 12.2. Total Airborne Concentration C_T (ng/m^3) and Mass Median Aerodynamic Diameter MMD (μm) for the Distributions of Fig. 12.1–12.6.

Graph No.	Reference and Location	Pb		Cd		Cu		Fe		Mn		Zn	
		C_T	MMD	C_T	MMD	C_T	MMD	C_T	MMD	C_T	MMD	C_T	MMD
1	Lee et al., 1972 Chicago, IL	3200.	0.67			100.	1.7	1100.	3.8	30.	2.0	450.	1.1
2	Lee et al., 1972 Cincinnati, OH	1800.	0.56			200.	1.3	1800.	2.6	170.	2.2	1700.	1.1
3	Lee et al., 1972 Denver, CO	1800.	0.42			400.	1.6	800.	2.7	20.	1.9	180.	1.7
4	Lee et al., 1972 Philadelphia, PA	1600.	0.42			100.	1.2	700.	2.4	50.	2.2	430.	1.3
5	Lee et al., 1972 St. Louis, MO	1800.	0.69			100.	1.0	1100.	3.4	30.	2.4	280.	1.2
6	Lee et al., 1972 Washington, DC	1300.	0.40			200.	1.3	600.	2.4	20.	1.4	350.	1.3
7	Lee et al., 1968 Cincinnati, OH	2800.	0.29	80.	3.2	190.	1.2	3100.	3.9				
8	Lee et al., 1968 Fairfax, OH	690.	0.42	20.	5.5			1200.	1.4				
9	Peden, 1977 Alton, IL	240.	1.9	1.2	1.3					62.	2.0		
10	Peden, 1977 Centreville, IL	620.	0.43	4.0	1.4					22.	4.1		
11	Peden, 1977 Collinsville, IL	670.	0.30	1.8	1.2					29.	5.5		
12	Peden, 1977 KMOX Radio, IL	600.	0.36	2.0	1.2					28.	4.2		
13	Peden 1977 Pere Marquette Park, IL	150.	0.51	1.8	1.3					16.	3.5		
14	Peden, 1977 Wood River, IL	270.	1.5	2.7	1.8					30.	4.3		

Table 12.2. *(Continued)*

Graph No.	Reference and Location	Pb		Cd		Cu		Fe		Mn		Zn	
		C_T	MMD	C_T	MMD	C_T	MMD	C_T	MMD	C_T	MMD	C_T	MMD
15	Cholak et al., 1968 Cincinnati, OH, near freeway	3500.	0.32										
16	McDonald and Duncan, 1979 Glasgow, Scotland	530.	0.51	2.2	0.54	20.	1.1	570.	0.88	12.	0.66	1300.	0.57
17	Dorn et al., 1976 S.E. Missouri, near smelter	1000.	3.8	25.	1.2	19.	4.9					170.	2.0
18	Dorn et al., 1976 S.E. Missouri, far from smelter	100.	2.0	3.7	3.2	9.5	6.1					71.	4.2
19	Daines et al., 1970 New Brunswick, NJ	3800.	0.34										
20	Martens et al., 1973 San Francisco, CA	890.	0.46										
21	Lundgren, 1970 Los Angeles, CA	590.	0.50					690.	2.0				
22	Huntzicker et al., 1975 Los Angeles, CA, near freeway	14000.	0.32										
23	Huntzicker et al., 1975 Pasadena, CA	3500.	0.72										
24	Davidson, 1977 Pasadena, CA	1200.	0.97	7.9	2.8							130.	2.7
25	Davidson et al., 1980 Great Smokies National Park, TN	14.	1.0										
26	Davidson et al., 1981a Pittsburgh, PA	590.	0.54	6.7	1.5	55.	5.8	4600.	8.1			2800.	2.6
27	Davidson et al., 1981b Nepal Himalayas	1.4	0.90										
28	Davidson et al., 1983 Export, PA	NA	1.1	NA	0.70	NA	0.59	NA	1.6			NA	1.1
29	Davidson et al., 1985 Packwood, WA	16.	0.40	1.1	0.55	10	0.42					13.	0.41

30	Davidson et al., 1985 Olympic National Park, WA	2.4	0.87	0.62	0.28	6.6	0.39					11.	0.28
31	Duce et al., 1976a Bermuda (May–June, 1975)	8.5	0.57	0.080	0.75	0.38	0.83	40.	2.1	0.44	2.1	3.8	0.82
32	Duce et al., 1976a Bermuda (July, 1975)	4.1	0.43	0.065	0.52			270.	4.1	2.3	4.3	1.8	0.57
33	Harrison et al., 1971 Ann Arbor, MI	2000.	0.16	61.	0.55	200.	0.96						
34	Gillette and Winchester, 1972 Ann Arbor, MI	NA	0.28										
35	Gillette and Winchester, 1972 Chicago, IL	NA	0.39										
36	Gillette and Winchester, 1972 Lincoln, NE	NA	0.52										
37	Johansson et al., 1976 Tallahassee, FL	240.	0.62					330.	5.2			15.	3.2
38	Cawse et al., 1974 Chilton, England	200.	0.57					370.	2.7	25.	1.4	160.	0.86
39	Pattenden et al., 1974 Trebanos, England	220.	0.74			38.	1.4	540.	3.3	18.	1.6	240.	0.95
40	Bernstein and Rahn, 1979 New York City, NY	1200.	0.57			67.	1.6	1400.	7.8	44.	1.5	230.	0.90
41	Rahn, 1976 Liege, Belgium			120.	2.0	210.	0.61	2900.	5.1			2800.	1.7
42	U.S. E.P.A., 1973, 1974 Kellogg, ID			1000.	1.5								
43	Heindryckx, 1976 Ghent, Belgium					NA	3.4	NA	4.1	NA	2.1	NA	1.5
44	Nifong et al., 1972 East Chicago, IL					57.	1.1	2000.	7.9	290.	1.2	690.	0.92
45	Gladney et al., 1974 Near Boston (MIT-1)							1500.	7.4	33.	1.7	260.	1.2
46	Gladney et al., 1974 Near Boston (MIT-2)							1100.	9.5	25.	2.0	430.	0.99
47	Gladney et al., 1974 Near Boston (Rt. 128)							760.	7.8	20.	3.1	100.	1.2

Table 12.2. *(Continued)*

Graph No.	Reference and Location	Pb		Cd		Cu		Fe		Mn		Zn	
		C_T	MMD	C_T	MMD	C_T	MMD	C_T	MMD	C_T	MMD	C_T	MMD
48	Lawson and Winchester, 1979 Chacaltaya Mtn., Bolivia							23.	6.8				
49	Lawson and Winchester, 1979 Goias, Brazil							1400.	6.3				
50	Lawson and Winchester, 1979 Manaus, Brazil							340.	1.8				
51	Winchester et al., 1979 Los Alamos, NM							180.	2.8			2.9	4.2
52	Winchester et al., 1979 Squaw Mountain, CO							310.	3.2			7.1	1.3
53	Winchester et al., 1979 St. Louis, MO (west of city)							430.	4.4			17.	0.88
54	Winchester et al., 1979 St. Louis, MO							1100.	4.8			77.	0.83
55	Winchester et al., 1979 Hubbard Brook, NH							100.	3.4			11.	0.76
56	Sadasivan, 1981 Bombay, India							2300.	4.7			340.	1.3

[a] Distributions for which values of C_T are not available are indicated by NA.

occur (Felix and McCain, 1981). Additional losses of large particles on the inner walls of impactors have been reported by McFarland et al. (1977), Cushing et al. (1979), and Chan and Lawson (1981).

Another problem with impactors is that particle bounceoff inside the units may occur. Large particles may make their way down to the lower impactor stages or onto the backup filter, biasing the distribution toward small particle sizes. Bounceoff may be especially severe in high volume impactors. The problem can be minimized by use of adhesive-coated collection substrates on each impactor stage (Wesolowski, 1973; Dzubay et al., 1976; Lawson, 1980). Because of the potential contamination and handling problems associated with adhesives, such precautions are not always implemented.

There is also evidence that glass fiber filters may decrease bounceoff relative to smoother surfaces such as cellulose (Sievering et al., 1978). However, Rao (1975) and Dzubay et al. (1976) have shown that glass fiber substrates may collect particles smaller than the rated cutoff diameters, due to filtration as the airstream penetrates the fibrous material. Additional studies have suggested that impactor sampling of marine aerosol with cellulose filter substrates may not be appreciably influenced by bounceoff; high humidities which moisten the cellulose filters, as well as the hygroscopic nature of marine aerosol, are believed responsible (Walsh et al., 1978).

Collection efficiency curves for typical impactors show that with proper operation, the devices generally fractionate aerosols into the approximate desired size ranges. But even with careful sampling and proper choice of substrate material, the fractionation is highly imperfect. Small particles may be collected to a certain extent on the upper impactor stages, while some of the large particles may not be impacted until reaching the lower stages. Such problems are the result of nonideal airflow through the impactor, particularly variations in the air velocity across each jet (Marple and Liu, 1974; Marple et al., 1974). Collection efficiency curves for typical impactors, generated with monodisperse particles in the laboratory, have been presented by May (1964), Flesch et al. (1967), Swartz et al. (1973), Knuth (1979), and in many of the studies referenced above.

Potential problems in conducting the analyses for trace metals must also be acknowledged. Of primary importance are difficulties with contamination control during sample collection, sample preparation for analysis, and operation of the analytical equipment. Although contamination is a potential problem for all trace metals, recent research by Patterson and co-workers (Patterson and Settle, 1976; Boutron and Patterson, 1983) have demonstrated that measurement of Pb concentrations in environmental samples is especially difficult. Contamination of air samples containing low levels of other trace metals has been discussed by Hoffman et al. (1976). These studies show that the small amounts of trace metals collected during impactor sampling require stringent field and laboratory procedures if reliable data are to be obtained. A second problem involves interferences from species other than those under investigation. For example, there is considerable literature on the difficulties of accurately measuring low concentrations of a metal in samples that contain high concentrations of other metals (Thompson et al., 1979; Waughman and Brett, 1980). Interferences caused by the medium containing the sample may also be significant (Wegscheider et al., 1977; Geladi and Adams, 1978). A third potential problem involves the generation of erroneous calibration data, such as through the use of inappropriate reference standards.

Most of the size distribution data in Figures 12.1–12.6 have been obtained using techniques that are far from ideal. Very few of the studies attempted to sample large particles accurately; most incorporated vertically oriented inlets. Some of the samplers used rainshields above the inlets, further reducing the collection efficiency of large particles. Impactor substrates included metal plates, Teflon film, polyethylene film, cellulose filters, and glass fiber filters, which in general were not coated with adhesive material. Several of the investigators made no attempt to calibrate their impactors, relying instead upon manufacturers' suggested cutoff diameters. Possible errors in air flow rates may have resulted in additional uncertainty in the cutoff diameters.

Of the 56 sets of data reported here, 8 did not incorporate backup filters. The importance of this problem can be readily demonstrated. The distributions of Figs. 12.1–12.6 have a d_p cutoff for the lowest impactor stage averaging 0.5 ± 0.15 μm. Table 12.3 lists the average mass fractions for $d_p <$ 0.5 μm for those distributions incorporating backup filters, providing an estimate of the amount of each trace metal missed by sampling without a backup filter. Note that Pb has the greatest fraction of mass below 0.5 μm, while Cd, Cu, and Zn have smaller but still significant fractions in this size range. Fe and Mn are generally associated with larger particles; sampling without backup filters is less important for these elements.

Table 12.3. Mass Fraction below 0.5 μm Aerodynamic Diameter and MMD for the Six Metals of Interest[a]

Metal	n	Mass Fraction <0.5 μm	MMD (μm)
Pb	35	0.51 ± 0.11	0.53 ± 0.23
Cd	17	0.25 ± 0.20	1.5 ± 1.3
Cu	19	0.28 ± 0.16	1.5 ± 1.2
Fe	29	0.087 ± 0.078	3.8 ± 2.0
Mn	18	0.12 ± 0.11	2.4 ± 1.3
Zn	26	0.27 ± 0.14	1.3 ± 0.82

[a] Only distributions in Figs. 12.1–12.6 which incorporated backup filters were used to calculate these values. The number of such distributions is denoted by n. Values shown are arithmetic averages and standard deviations.

Analytical difficulties also may have been important. Many of the studies did not employ adequate contamination control, either during field sampling or laboratory work. Possible artifacts generated by interfering species were not discussed in most of the references. Quality assurance procedures, such as interlaboratory exchanges of calibration standards or samples, were lacking in virtually all of the reported studies.

An additional problem concerns the small number of experiments conducted in many of the investigations. Although some of the distributions reflect the average of 20 or more separate sampling periods, Table 12.1 indicates that most of the spectra are based on very limited data. Many of the distributions reflect only a single experiment. The extent to which most of the distributions in Figs. 12.1–12.6 provide representative data must be questioned; Lodge (1976) has discussed in detail the issues of representativeness and accuracy in air pollution sampling.

It is clear that the distributions reported here must be considered, at best, merely rough estimates of the true size spectra. There are also obvious difficulties in comparing data collected with different sampling and analysis techniques, generated by independent investigators over a substantial period of time. In spite of these problems, the data show consistencies which suggest that one can infer general characteristics in the shapes of the distributions for each metal. The consistencies also suggest that the data can be used, with considerable caution, to improve our understanding of the inhalation risks and atmospheric deposition of trace metals. These topics are addressed next.

4. CHARACTERISTICS OF THE DISTRIBUTIONS

Several notable features may be observed in the distributions, despite variations from graph to graph within each figure. A most important characteristic, already demonstrated in Table 12.3, is the difference in fractions of small and large particles among the six elements. Lead is primarily submicron, found mostly in the smallest size range in each distribution. Significant amounts of submicron Cd, Cu, and Zn are also seen in many of the graphs. Mn is predominantly supermicron with appreciable material in the 1–10 μm range, while Fe is associated mostly with the largest size ranges. The mass median aerodynamic diameter (MMD) values listed in Table 12.3 follow a pattern opposite to that of the $d_p < 0.5$ μm mass fractions, increasing in the order Pb < Cd ~ Cu ~ Zn < Mn < Fe.

A second noteworthy characteristic is the bimodal shape of many of the distributions. One peak generally occurs below 1 μm, with the other peak above 5 μm. Similar bimodal spectra have also been reported for total mass size distributions of atmospheric particles, measured with optical techniques (National Academy of Sciences, 1979). However, Whitby et al. (1972) and Sievering et al. (1978) have cautioned that impactor data may not provide sufficient size resolution to identify the true detailed shapes of the distributions. Most of the graphs in Figs. 12.1–12.6 probably portray oversimplifications of more complex spectra. This may be particularly true for very small or very large particles, where size discrimination is lacking. It is also noteworthy that many of the distributions, including several of those that are bimodal, contain one predominant mode. For Pb, the predominant mode is almost always submicron; for Fe, the mode is supermicron. This is consistent with the suggestion of Natusch and Wallace (1974) that the size spectra of many trace elements can, for certain applications, be approximated as unimodal.

Variations in the nonideal characteristics among impactors are responsible for a third feature of Figs. 12.1–12.6: Differences among several distributions measured with a single sampling technique are not as significant as differences among distributions measured by different techniques. For example, Graphs 1 through 6 for Pb, Cd, Cu, Fe, Mn, and Zn represent data collected by the National Air Surveillance Network, using Andersen impactors operated at relatively high flow rates of 5–6 CFM (ft^3/min). Table 12.2 shows that many of the MMD values for these distributions are smaller than the Table 12.3 averages, especially for Fe and Mn; Dzubay et al. (1976) have commented that the high flow rates may have promoted particle bounceoff in these samplers, biasing the MMD values toward small particles. As another example, many of the distributions have been obtained with impactors which include one or more narrow size intervals. The narrow intervals are often characterized by large peaks in the mass distribution function. It is likely that these peaks may be somewhat exaggerated, since nonideal impaction characteristics have probably resulted in collection of particles outside the intervals.

Our knowledge of source emissions for each of the six metals can be used to understand consistencies in the shapes of the distributions. Pb is a volatile element, emitted primarily from the combustion of leaded gasoline. Lesser quantities are emitted from the production of nonferrous metals, coal combustion, waste incineration, and other industrial activities. These anthropogenic emissions contribute far more Pb to the atmosphere than natural sources (Nriagu, 1978; Patterson, 1980). Size distributions of Pb emitted from motor vehicles using leaded gasoline have been reported by several investigators (Hirschler et al., 1957; Hirschler and Gilbert, 1964; Mueller et al., 1964; Habibi, 1970, 1973; Ter Haar et al., 1972). Results show that steadily cruising vehicles emit primarily submicron Pb, although large amounts of supermicron Pb can be exhausted during accelerations. Based on these studies, one would expect most of the airborne Pb in the ambient atmosphere to be in the submicron range, consistent with the distributions of Figure 12.1.

Besides direct emissions from motor vehicles and stationary sources, the small amounts of supermicron material in Figure 12.1 may indicate Pb associated with soil dust, coal particles, and other coarse material. In some cases, this can reflect attachment of anthropogenically emitted submicron Pb to larger particles; such attachment may occur, for example, if Pb-containing particles deposit on soil that is subsequently resuspended. Successive deposition and resuspension processes may greatly alter the shape of the size distribution as material is transported through the environment. The supermicron material in Figure 12.1 may also indicate large fly ash particles that have a surface coating of condensed Pb (Linton et al., 1976; Van Craen et al., 1983). Other volatile species associated with large particles, including Cd and Zn, may have a similar surface predominance.

The most significant source category for Cd, Cu, and Zn is production of nonferrous metals. Lesser amounts are emitted from waste incineration, fossil-fuel combustion, and other industrial processes. Natural soil dust and volcanic emissions of these three metals are small relative to emissions from anthropogenic sources (Nriagu, 1979, 1980; Nriagu and Davidson, 1980), although there is evidence that sea spray may be an important contributor (Duce et al., 1976b; Cattrell and Scott, 1978). Size distributions of particles emitted from metal production, waste incineration, and fossil-fuel combustion are highly variable. Some processes may produce mostly submicron aerosol, while others produce large particles (Herring, 1971; Danielson, 1973; Davison et al., 1974; Natusch et al., 1974; Lee at al., 1975; Greenberg et al., 1978; Davidson et al., 1982a). Sea-spray aerosol is generally supermicron (Duce et al., 1976b; McDonald et al., 1982). Unlike Pb, which is predominantly submicron at most sites, the wide variety of distribution shapes in Fig 12.2, 12.3, and 12.6 probably reflects, in addition to sampling artifacts, the many different sources influencing airborne Cd, Cu, and Zn at each site.

To a significant extent, airborne Fe and Mn are associated with soil dust. This has been demonstrated by several investigators using the concept of enrichment factor EF:

$$\mathrm{EF} = \frac{C_{\mathrm{X,air}}/C_{\mathrm{Al,air}}}{C_{\mathrm{X,crust}}/C_{\mathrm{Al,crust}}} \tag{1}$$

where $C_{\mathrm{X,air}}$ and $C_{\mathrm{Al,air}}$ are the airborne concentrations of any element X and aluminum, respectively, and $C_{\mathrm{X,crust}}$ and $C_{\mathrm{Al,crust}}$ represent the corresponding concentrations in the earth's crust. Aluminum is generally used as a tracer for soil dust. Values of EF near unity imply the earth's crust as a source of that particular element, while large values suggest the importance of other sources such as volcanism, sea spray, biogenic processes, or anthropogenic activities (Zoller et al., 1974; Duce et al., 1975).

Average values of EF reported in the literature for Pb, Cd, Cu, Fe, Mn, and Zn for 29 urban areas are 3800, 940, 149, 2.2, 3.2, and 300, respectively (Rahn, 1976), showing the importance of soil dust for Fe and Mn. Note that human activities such as construction and farming may increase airborne soil dust concentrations without altering values of the enrichment factor. Thus the small average enrichments for Fe and Mn do not necessarily imply natural erosion.

Soil dust is characterized by a wide range of particle sizes, usually with a predominance of supermicron material (Patterson and Gillette, 1977). This is consistent with the graphs of Figs. 12.4 and 12.5, indicating most of the Fe and Mn in the uppermost size ranges. However, some of the distributions show significant amounts of submicron aerosol; this may reflect the influence of industrial sources. The National Academy of Sciences (1973, pp. 16–28) has reported substantial emissions of Fe and Mn from the production of ferrous metals. Enrichment factors greater than 10 for both Fe and Mn in industrial areas have been reported by Rahn (1976). Other natural sources such as volcanism, sea spray, and biogenic processes may also be significant.

5. PREDICTION OF PARTICLE DEPOSITION

As an example of the utility of particle size distributions, the data of Figs. 12.1–12.6 have been used to perform two types of calculations. First, the fractional deposition in each compartment of the human lung has been estimated, allowing assessment of the relative inhalation risk for each metal. Second, dry deposition velocities for transport from the atmosphere onto various surfaces have been calculated.

5.1. Deposition in the Lung

Trace metal deposition in the human respiratory tract has been studied previously by Natusch and Wallace (1974). These investigators have used the model of the International Committee on Radiological Protection (ICRP) with size distribution estimates for Pb, Fe, and other species to predict deposition in the nasopharyngeal, tracheobronchial, and pulmonary compartments of the lung. Their results suggest that Pb, predominantly submicron, is deposited to an appreciable extent in the pulmonary compartment. The ICRP model indicates a fractional mass deposition (mass deposited/mass inhaled) in the range 0.25–0.65 for submicron particles in this region, with much smaller fractions depositing in the other compartments (Task Group on Lung Dynamics, 1966). The deposition of Fe, on the other hand, is more pronounced in the nasopharyngeal region due to greater airborne mass in large particles.

Akselsson et al. (1976), Desaedeleer and Winchester (1975), and Desaedeleer et al. (1977) have compared the concentrations of several trace metals as a function of particle size in inhaled and exhaled air.. Their results suggest that deposition in the total respiratory tract is minimal for particles with $d_p \sim 0.5$ μm. The fractional mass deposition is less than 0.5 for this particle diameter. Slightly greater values are observed for smaller particles, while values as large as 0.96 are associated with particles of aerodynamic diameter greater than 4 μm.

New data for particle deposition in the lung have become available since these studies were published. Of particular importance is the work of

Lippmann (1977), Chan and Lippmann (1980), and Stahlhofen et al. (1980). These investigators have reported experimental data for monodisperse particle deposition in each compartment of the human lung, which can be used with published size distribution data to calculate respiratory tract deposition.

The results of these lung deposition studies, summarized by the U.S. Environmental Protection Agency (1982) have been used with the distributions of Figs. 12.1–12.6 to estimate the fractional mass deposition for the six metals of interest. In performing these calculations, the geometric mean particle diameter for each size interval has been used as the basis for estimating the mass deposition fraction f_i for that interval. Only lung deposition data for mouth breathing have been used to estimate the various f_i values. The final deposition fraction F for each compartment of the lung, for each distribution, has been computed from the relation:

$$F = \sum_{i=1}^{n} f_i \Delta C_i \Big/ \sum_{i=1}^{n} \Delta C_i \tag{2}$$

where ΔC_i is the airborne mass concentration of the metal in size range i. The values of F thus refer to the total mass of a metal deposited in a specific compartment of the lung divided by the total mass of the metal inhaled. The applicable respiration rates vary from 7.5 to 30 L/min, with tidal volumes generally in the range 1.0 to 1.5 L.

Results of the calculations are shown in Figure 12.7. The values of F for the individual distributions have been averaged for each metal and for each

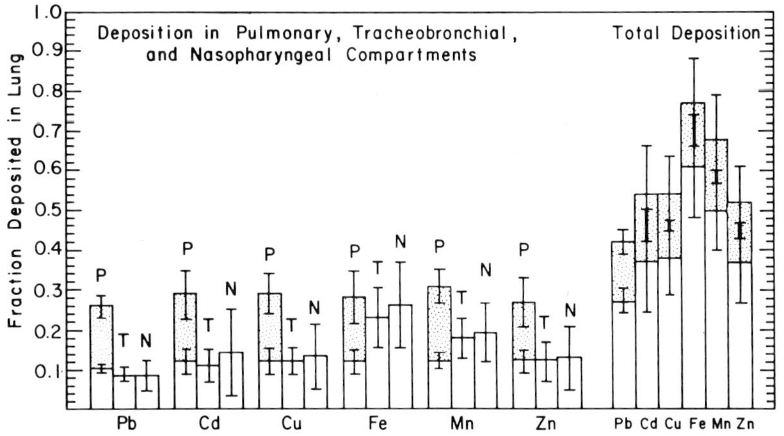

Figure 12.7. Mass deposition in the pulmonary (P), tracheobronchial (T), and nasopharyngeal (N) compartments of the human lung as fractions of the mass inhaled through mouth breathing. The total fractions for deposition in all three compartments are also shown. Values corresponding to the pulmonary compartment represent minimum and maximum deposition fractions, shown by clear and shaded areas, respectively (see text). The error bars indicate ± 1 standard deviation from the mean, based on deposition fractions calculated for each set of distributions in Figs. 12.1–12.6.

compartment of the lung to yield these results. Minimum and maximum pulmonary deposition fractions are based on the two pulmonary deposition curves which envelop the data, as reported by the U.S. Environmental Protection Agency (1982). Values for each of the three compartments have been summed to obtain the total deposition fractions.

Of interest is the surprisingly uniform average deposition fraction in the pulmonary compartment for all six metals. The minimum fraction is in the range 0.10–0.12, while the maximum is in the range 0.26–0.30. This uniformity is in contrast to the calculations of Natusch and Wallace (1974), who report a pulmonary deposition fraction of 0.32 for Pb but only 0.22 for Fe. The differences between the Figure 12.7 results and the earlier calculations reflect the relatively small deposition fractions for submicron particles reported in the recent lung deposition studies: the new pulmonary compartment data show deposition fractions below 0.2 for particles smaller than 0.5 μm, with a peak in the range 0.3–0.6 at approximately 3 μm. The ICRP model, in contrast, predicts a steadily increasing pulmonary deposition fraction as particle size decreases, reaching 0.4 at 0.1 μm. Chan and Lippmann (1980) state that the ICRP model overestimates pulmonary deposition by at least a factor of two for particles smaller than 0.5 μm. Such differences are likely to have a marked effect on the calculated pulmonary deposition of Pb, a predominantly submicron species.

The tracheobronchial deposition fractions in Figure 12.7 are in the range 0.081–0.12 for Pb, Cd, Cu, and Zn, with somewhat larger fractions for Mn and Fe. These values reflect the increasing tracheobronchial deposition with increasing particle size up to approximately 8 μm: the deposition fractions in Figure 12.7 correlate well with the values of MMD in Table 12.3 for the six metals. Similar trends are observed for deposition in the nasopharyngeal compartment, and for total deposition in all compartments of the respiratory tract. Note that deposition in the nasopharyngeal compartment is only slightly greater than that in the tracheobronchial compartment for each metal.

It is worthwhile to consider the limitations of the lung deposition calculations discussed above. Of primary importance are the inaccuracies associated with the size distribution data. For example, the lack of detailed information on the distribution of particles within each size interval has necessitated the use of the geometric mean particle diameter when estimating lung deposition. It is fortunate that the interval corresponding to the backup filter, which generally encompasses the widest size range, may not be as significant as the supermicron size ranges for the calculations of Figure 12.7; the lung deposition data for submicron particles show relatively small fractions depositing and show roughly constant deposition across a wide range of particle sizes. The error introduced by using aerodynamic diameters for these submicron particles is also likely to be small. However, the uncertainty caused by using the geometric mean particle diameter for the uppermost impactor stage may have been significant for the calculated tracheobronchial and nasopharyngeal deposition rates. Related to this problem is the poor sampling efficiency for the largest airborne particles in the distributions of Figs. 12.1–12.6, which may have resulted in calculated deposition fractions which are smaller than the true values.

Another problem concerns the enrichment of volatile species such as Pb, Cd, and Zn on the surfaces of particles, resulting from condensation of gases containing these species. Such surface coatings may be more readily available for absorption than material composing the bulk of each particle (Natusch and Wallace, 1974; Linton et al., 1976; Van Craen et al., 1983), complicating the interpretation of Figure 12.7.

A third problem concerns uncertainties in the lung deposition data. Especially important is the difficulty in generalizing the data, derived from a limited number of subjects, to a larger population. Because of these problems, the results of Figure 12.7 must be viewed cautiously.

5.2. Deposition from the Atmosphere onto Surfaces

The distributions of Figs. 12.1–12.6 have also been used with dry deposition models to estimate the deposition of each trace metal onto several types of surfaces. These calculations have been performed in a manner similar to the lung deposition estimates, using the geometric mean diameter for each size interval as the basis for estimating the dry deposition velocity v_{di} for that interval. The overall dry deposition velocity V_d for each distribution has been calculated from the relation

$$V_d = \sum_{i=1}^{n} v_{di}\Delta C_i \bigg/ \sum_{i=1}^{n} \Delta C_i \qquad (3)$$

Five different sets of v_{di} have been used for these calculations. The first two sets involve the model of Davidson et al. (1982b) for *Agrostis hyemalis* (hair grass) and *Andropogen virginicus* (broom sedge), species of wild grass. This model incorporates equations of particle transport by diffusion, interception, impaction, and sedimentation to predict dry deposition; data for the geometry of the vegetation and for windfield characteristics above and within the canopies have been applied. The third set of v_{di} values is derived from the model of Sehmel and Hodgson (1978). These investigators have developed empirical equations applicable to a wide variety of conditions, based on their wind tunnel data for monodisperse particle deposition onto several types of surfaces. The fourth set is based on the model of Slinn (1982) for a *Eucalyptus* forest. This investigator has used detailed equations of particle transport with measured windspeed profiles and assumed geometrical characteristics of the forest to estimate dry deposition. The final set of v_{di} values is based on sedimentation, providing an estimate of the minimum dry deposition expected.

Results of these calculations are shown in Table 12.4. The values of V_d for

Table 12.4. Dry Deposition Velocity (cm/s) for the Six Metals of Interest[a]

Metal	n	*Agrostis hyemalis*[b]	*Andropogen virginicus*[c]	Extrapolation from Wind Tunnel Data for Various Surfaces[d]	*Eucalyptus* Forest[e]	Sedimentation[f]
Pb	35	0.96 ± 0.40	0.20 ± 0.15	0.20 ± 0.092	0.64 ± 0.39	0.28 ± 0.20
Cd	17	1.5 ± 0.83	0.32 ± 0.30	0.30 ± 0.18	1.0 ± 0.85	0.45 ± 0.40
Cu	19	1.5 ± 0.65	0.29 ± 0.23	0.29 ± 0.14	1.0 ± 0.63	0.43 ± 0.33
Fe	29	2.7 ± 0.87	0.59 ± 0.30	0.54 ± 0.18	2.2 ± 0.89	0.85 ± 0.47
Mn	18	2.1 ± 0.63	0.41 ± 0.18	0.43 ± 0.14	1.5 ± 0.57	0.62 ± 0.30
Zn	26	1.5 ± 0.73	0.32 ± 0.31	0.29 ± 0.17	1.0 ± 0.79	0.44 ± 0.40

[a] Calculated using deposition models for several types of surfaces. These values were determined from the same distributions used in Table 12.3. Values shown are arithmetic averages and standard deviations.

[b] $h = 33$ cm, $d = 13$ cm, $z_o = 14$ cm, $u_* = 40$ cm/s (Davidson et al., 1982b).

[c] $h = 57$ cm, $d = 14$ cm, $z_o = 9$ cm, $u_* = 32$ cm/s (Davidson et al., 1982b).

[d] $z_o = 3$ cm, $u_* = 30$ cm/s (Sehmel and Hodgson, 1978).

[e] $h = 27.4$ m, $d = 21.6$ m, $z_o = 1.86$ m, $u_* = 75$ cm/s (Slinn, 1982).

[f] From Friedlander, 1977.

the individual distributions have been averaged for each metal and each surface type to yield these results. Values of the relevant parameters for each set of atmospheric and surface conditions are given in the table: h is the average height of the vegetation, d is the zero plane displacement of the canopy, z_o is the roughness height (i.e., momentum sink), and u_* is the friction velocity. All of the models are based on an adiabatic atmosphere.

For each of the five surfaces, the total deposition velocities increase in the order Pb < Cd ~ Cu ~ Zn < Mn < Fe, a pattern consistent with the data of Table 12.3. These calculations show that the largest particles in each distribution are responsible for most of the dry deposition mass.

For Pb, the large particles influencing deposition compose only a small fraction of the total airborne Pb mass. This suggests that the denominator of Eq. (3) should be reduced to include only the uppermost size ranges if a meaningful deposition velocity is to be calculated: the small V_d values for Pb in Table 12.4 do not reflect predominant deposition of small particles, but rather deposition of fewer large particles. This is because the predicted deposition velocities decrease rapidly with decreasing particle size below the maximum diameters assumed in Figures 12.1–12.6. A similar situation exists for many of the distributions of Cd, Cu, and Zn, where there is a relatively small fraction of large particles. Most of the airborne Fe and Mn is associated with large particles, however, and hence the dry deposition velocities of these metals in Table 12.4 may be more representative of the bulk of the airborne mass.

Table 12.4 shows much smaller deposition velocities for *Andropogen virginicus* than for *Agrostis hyemalis*. The difference is due to the greater leaf area index of *Agrostis*, as well as the presence of fine hairs which assist particle capture. The model of Sehmel and Hodgson (1978) provides results which are comparable to those of *Andropogen,* while sedimentation yields values approximately one third as large. It is interesting and somewhat surprising that deposition velocities for *Agrostis* are actually slightly greater than those of the *Eucalyptus* forest. Although this comparison is tenuous because of different assumptions in the two models, Slinn (1982) has calculated deposition velocities for grass and wheat as well as for *Eucalyptus,* and the results show values for the shorter canopies which are comparable to those of the forest. Relatively small windspeeds within the *Eucalyptus* canopy, as well as assumed characteristics of the vegetation geometry, are apparently responsible for keeping the deposition velocities small. In general, the results shown in Table 12.4 agree with the findings of similar calculations using subsets of the data of Figs. 12.1–12.6 with different deposition models (Davidson, 1980; Nriagu and Davidson, 1980).

Several difficulties with these dry deposition estimates, similar to problems encountered with the lung deposition calculations, must be considered. A major problem concerns inaccuracies in the measured size distribution data of Figs. 12.1–12.6. Difficulties in sampling the largest airborne particles, which greatly influence deposition, may have compromised the calculations. The deposition models must also be acknowledged as major sources of uncertainty. The models of Davidson et al. (1982b) and Slinn (1982) involve numerous assumptions and have never been tested experimentally. The model of Sehmel and Hodgson (1978) involves the tenuous extrapolation of experimental data for slightly rough artificial surfaces in a wind tunnel to much rougher natural vegetation exposed to the ambient atmosphere. As such, the results presented in Table 12.4 must be viewed with caution.

6. SUMMARY

This chapter has reviewed literature data on the sizes of airborne particles containing Pb, Cd, Cu, Fe, Mn, and Zn. Possible difficulties with the data have been discussed, and consistencies in the distributions have been identified. The data have then been used to perform two types of calculations. First, the deposition of each trace metal in the human respiratory tract has been estimated. Second, dry deposition of each metal from the ambient atmosphere onto various surfaces has been calculated.

Numerous problems with the size distribution data reported in this chapter must be acknowledged. Most of the measurements have been conducted with inertial impactors, which are subject to internal losses and particle bounceoff. Impactors operated without backup filters most likely have missed a significant fraction of the submicron aerosol. Nonisokinetic sampling and inlet losses have resulted in exclusion of large particles. Some of the distributions include size intervals subject to considerable uncertainty, because of errors in the air flow rates, use of inappropriate substrate material, and lack of calibration data. There also may be appreciable error associated with the analytical data, primarily because of difficulties in working with very small quantities of trace metals.

In spite of these problems, the data presented here show reasonable patterns. Lead is generally associated with submicron particles, while the distributions of Cd, Cu, and Zn contain submicron as well as supermicron aerosol. Fe and Mn are associated predominantly with particles larger than 1 μm aerodynamic diameter. These characteristics are consistent with our understanding of the sources of each metal in the atmosphere. Lead is derived primarily from the combustion of leaded gasoline, a process which produces mostly small particles. Cd, Cu, and Zn are emitted mainly from stationary sources, including nonferrous metal production, waste incineration, and fossil-fuel combustion. These processes produce a wide variety of particle sizes. Fe and Mn are normally associated with erosion of the earth's crust, a process which mainly produces large particles.

The distributions have been used with recent lung deposition data to estimate the mass fraction deposited in each compartment of the human respiratory tract as a function of mass inhaled. The average pulmonary compartment deposition for each of the six metals varies from 0.10 to 0.12, based on published estimates of the minimum expected deposition in this compartment using monodisperse particles. The corresponding average fraction based on maximum pulmonary deposition varies from 0.26 to 0.30. The small range in average pulmonary deposition among the six metals is in contrast to previously published estimates, which show much greater pulmonary deposition for submicron species (e.g., Pb) than for species associated with large particles (e.g., Fe). The relatively minor role of submicron particles in pulmonary deposition, as reflected in the recent inhalation studies, is primarily responsible; these recent studies show that particles in the range 1–5 μm, with a peak deposition at approximately 3 μm, have the greatest pulmonary deposition.

Average deposition fractions in the tracheobronchial compartment are similar to those in the nasopharyngeal region, ranging from 0.08 for Pb to 0.23 for Fe. Values for the six metals correlate well with the respective mass median aerodynamic diameters, since deposition in these compartments increases with increasing particle diameter. A similar trend is seen for deposition in the total respiratory tract. Based on maximum pulmonary values, the total respiratory tract deposition fraction varies from 0.42 for Pb to 0.77 for Fe.

Predicted dry deposition velocities for the six metals also correlate with mass median aerodynamic diameter. Values for deposition onto *Andropogen virginicus* (broom sedge) vary from 0.20 for Pb to 0.59 cm/s for Fe; values for *Agrostis hyemalis* (hair grass), a much denser vegetation, vary from 0.96 cm/s for Pb to 2.7 cm/s for Fe. Predicted deposition velocities for a *Eucalyptus* forest are slightly smaller than for *Agrostis.* For all surfaces examined, the deposition of each of the six trace metals is primarily influenced by the largest airborne particles present.

ACKNOWLEDGMENTS

This work was funded in part by National Institutes of Health Training Grant GMO-7477 and by Claude Worthington Benedum Foundation Grant 1-31335. The manuscript was prepared by Gloria Blake.

REFERENCES

Akselsson, K.R., Desaedeleer, G.G., Johansson, T.B., and Winchester, J.W. (1976). "Particle size distribution and human respiratory deposition of trace metals in indoor work environments." *Ann. Occup. Hyg.* **19,** 225–238.

Bernstein, D.M., and Rahn, K.A. (1979). "N.Y. summer aerosol study: Trace element concentration as a function of particle size." *Ann. N.Y. Acad. Sci.* **322,** 87–98.

Boutron, C.F., and Patterson, C.C. (1983). "The occurrence of lead in Antarctic recent snow, firn deposited over the last two centuries and prehistoric ice." *Geochim. Cosmochim. Acta* **47,** 1355–1368.

Cattrell, F.C.R., and Scott, W.D. (1978). "Copper in aerosol particles produced by the ocean." *Science* **202,** 429–430.

Cawse, P.A. (1974). *A Survey of Atmospheric Trace Elements in the U.K.* Report AERE-R7669. United Kingdom Atomic Energy Authority, Harwell, Oxfordshire, p. 81.

Chan, T.L., and Lawson, D.R. (1981). "Characteristics of cascade impactors in size determination of diesel particles." *Atmos. Environ.* **15,** 1237–1279.

Chan, T.L., and Lippmann, M. (1980). "Experimental measurements and empirical modelling of the regional deposition of inhaled particles in humans." *Am. Ind. Hyg. Assoc. J.* **41,** 399–409.

Cholak, J., Schafer L.J., and Yeager, D. (1968). "The air transport of lead compounds present in automobile exhaust gases." *Am. Ind. Hyg. Assoc. J.* **29,** 562–568.

Cushing, K.M., McCain, J.D., and Smith, W.B. (1979). "Experimental determination of sizing parameters and wall losses of five source-test cascade impactors." *Environ. Sci. Technol.* **13,** 726–731.

Daines, R.H., Motto, H., and Chilko, D.M. (1970). "Atmospheric lead: Its relationship to traffic volume and proximity to highways." *Environ. Sci. Technol.* **4,** 318–322.

Danielson, J.A. (1973). *Air Pollution Engineering Manual.* U.S. Environmental Protection Agency, Washington, D.C. Report 4-AP-40, Second edition.

Davidson, C.I. (1977). "The deposition of trace metal-containing particles in the Los Angeles area." *Powder Technol.* **18,** 117–126.

Davidson, C.I. (1980). "Dry deposition of cadmium from the atmosphere." In J.O. Nriagu, Ed., *Cadmium in the Environment,* Part I. Wiley, New York, pp. 115–139.

Davidson, C.I., Nasta, M.A., Reilly, M.T., and Suuberg, E.M. (1980). *Dry Deposition of Trace Elements in Great Smoky Mountains National Park.* Final Report to the U.S. Environmental Protection Agency, Contract No. LV-79-35. Carnegie-Mellon University, Pittsburgh, Pa, p. 24.

Davidson, C.I., Goold, W.D., Nasta, M.A., and Reilly, M.T. (1981a). "Airborne size distributions in an industrial section of Pittsburgh." 75th Annual Meeting, Air Pollution Control Association, Philadelphia, Pennsylvania, June 21–26, Paper 81-28.6.

Davidson, C.I., Grimm, T.C., and Nasta, M.A. (1981b). "Airborne lead and other elements derived from local fires in the Himalayas." *Science* **214,** 1344–1346.

Davidson, C.I., Santhanam, S., Stetter, J.R., Flotard, R.D., and Gebert, E. (1982a). "Characterization of airborne particles at a high-BTU coal gasification pilot plant." *Environ. Monit. Assess.* **1,** 313–335.

Davidson, C.I., Miller, J.M., and Pleskow, M.A. (1982b). "The influence of surface structure on predicted particle dry deposition to natural grass canopies." *Water, Air, Soil Pollut.* **18,** 25–43.

Davidson, C.I., Goold, W.D., and Wiersma, G.B. (1983). *Sources and Sinks of Airborne Trace Metals in the Olympic National Park Biosphere Reserve.* Final Report to the U.S. Environmental Protection Agency, Contracts No. V-1441-NAEX and V-4196-NAET. Carnegie-Mellon University, Pittsburgh, Pa, p. 44.

Davidson, C.I., Goold, W.D., Mathison, T.P., Wiersma, G.B., Brown, K.W., and Reilly, M.T. (1985). "Airborne trace elements in Great Smoky Mountains, Olympic, and Glacier National Parks." *Environ. Sci. Technol.* **19,** 27–35.

Davison, R.L., Natusch, D.F.S., Wallace, J.R., and Evans, C.A., Jr. (1974). "Trace elements in fly ash. Dependence of concentration on particle size." *Environ. Sci. Technol.* **8,** 1107–1113.

Desaedeleer, G.G., and Winchester, J.W. (1975). "Trace metal analysis of atmospheric aerosol particle size fractions in exhaled human breath." *Environ. Sci. Technol.* **9,** 971–972.

Desaedeleer, G.G., Winchester, J.W., and Akselsson, K.R. (1977). "Monitoring aerosol elemental composition in particle size fractions for predicting human respiratory uptake." *Nucl. Instr. Methods* **142,** 97–99.

Dorn, C.R., Pierce, J.O., Phillips, P.E., and Chase, G.R. (1976). "Airborne Pb, Cd, Zn and Cu concentration by particle size near a Pb smelter." *Atmos. Environ.* **10,** 443–446.

Duce, R.A., Hoffman, G.L., and Zoller, W.H. (1975). "Atmospheric trace metals at remote Northern and Southern Hemisphere sites: Pollution or natural?" *Science* **187,** 59–61.

Duce, R.A., Ray, B.J., Hoffman, G.L., and Walsh, P.R. (1976a). "Trace metal concentration as a function of particle size in marine aerosols from Bermuda." *Geophys. Res. Lett.* **3,** 339–342.

Duce, R.A., Hoffman, G.L., Ray, B.J., Fletcher, I.S., Wallace, G.T., Fasching, J.L., Piotrowicz, S.R., Walsh, P.R., Hoffman, E.J., Miller, J.M., and Heffter, J.L. (1976b). "Trace metals in the marine atmosphere: Sources and fluxes." In H.L. Windom and R.A. Duce, Eds., *Marine Pollutant Transfer.* D.C. Heath, Lexington, Mass., pp. 77–119.

Dzubay, T.G., Hines, L.E., and Stevens, R.K. (1976). "Particle bounce errors in cascade impactors." *Atmos. Environ.* **10,** 229–234.

Esmen, N.A. (1977). "An iterative impactor data analysis method." Paper presented at the 51st Colloid and Interface Science Symposium, June 19–22, Grand Island, N.Y.

Felix, L.G., and McCain, J.D. (1981). "Errors in recovered particle size distributions caused by impactor sampling with bent nozzles." 74th Annual Meeting, Air Pollution Control Association, Philadelphia, Pennsylvania, June 21–26, Paper 81-7.4.

Flesch, J.P., Norris, C.H., and Nugent, A.E., Jr. (1967). "Calibrating particulate air samplers with monodisperse aerosols: Application to the Andersen cascade impactor." *Am. Ind. Hyg. Assoc. J.* **28,** 507–516.

Geladi, P., and Adams, F. (1978). "The determination of cadmium, copper, iron, lead and zinc in aerosols by atomic-absorption spectrometry." *Anal. Chim. Acta* **96,** 229–241.

Gillette, D.A., and Winchester, J.W. (1972). "A study of the aging of lead aerosols—I. Observations." *Atmos. Environ.* **6,** 443–450.

Gladney, E.S., Zoller, W.H., Jones, A.G., and Gordon, G.E. (1974). "Composition and size distributions of atmospheric particulate matter in Boston area." *Environ. Sci. Technol.* **8,** 551–557.

Greenberg, R.R., Zoller, W.H., and Gordon, G.E. (1978). "The contribution of refuse incineration to urban aerosols." *Proceedings, Fourth Joint Conference on Sensing of Environmental Pollutants*, New Orleans, Louisiana, November 6–11, 1977. Paper #218.

Habibi, K. (1970). "Characterization of particulate lead in vehicle exhaust-experimental techniques." *Environ. Sci. Technol.* **4,** 239–248.

Habibi, K. (1973). "Characterization of particulate matter in vehicle exhaust." *Environ. Sci. Technol.* **7,** 223–234.

Harrison, P.R., Matson, W.R., and Winchester, J.W. (1971). "Time variations of lead, copper and cadmium concentrations in aerosols in Ann Arbor, Michigan." *Atmos. Environ.* **5,** 613–619.

Heindryckx, R. (1976). "Comparison of the mass-size function of the elements in the aerosols of the Gent industrial district with data from other areas: Some physico-chemical implications." *Atmos. Environ.* **10,** 65–71.

Herring, W.O. (1971). U.S. Environmental Protection Agency Report PB-201-739, APTD-0706.

Hidy, G.M. (1974). *Characterization of Aerosols in California.* Final Report, Air Resources Board Contract No. 358. Science Center, Rockwell International, Tulsa, Okla.

Hirschler, D.A., and Gilbert, L.F., (1964). "Nature of lead in automobile exhaust gas." *Arch. Environ. Health* **8,** 297–313.

Hirschler, D.A., Gilbert, L.F., Lamb, F.W., and Niebylski, L.M. (1957). "Particulate lead compounds in automobile exhaust gas." *Ind. Eng. Chem.* **49,** 1131–1142.

Hoffman, E.J., Hoffman, G.L., and Duce, R.A. (1976). "Contamination of atmospheric particulate matter collected at remote shipboard and island locations." In *Accuracy in Trace Analysis: Sampling, Sample Handling, and Analysis, Proceedings of the 7th IMR Symposium, Oct. 7–11, 1974, Gaithersburg, Md.* National Bureau of Standards Spec. Publ. 422, pp. 377–388.

Huntzicker, J.J., Friedlander, S.K., and Davidson, C.I. (1975). "Material balance for automobile-emitted lead in Los Angeles Basin." *Environ. Sci. Technol.* **9,** 448–457.

Johansson, T.B., Van Grieken, R.E., and Winchester, J.W. (1976). "Elemental abundance variation with particle size in North Florida aerosols." *J. Geophys. Res.* **81,** 1039–1046.

Knuth, R.H. (1979). *Calibration of a Modified Sierra Model 235 Slotted Cascade Impactor.* Report EML-360. Environmental Measurements Laboratory, U.S. Department of Energy, Washington, D.C.

Lawson, D.R. (1980). "Impaction surface coatings intercomparison and measurements with cascade impactors." *Atmos. Environ.* **14,** 195–199.

Lawson, D.R., and Winchester, J.W. (1979). "A standard crustal aerosol as a reference for elemental enrichment factors." *Atmos. Environ.* **13,** 925–930.

Lee, R.E., Jr., and von Lehmden, D.J. (1973). "Trace metal pollution in the environment." *J. Air Pollut. Control Assoc.* **23,** 853–857.

Lee, R.E., Jr., Patterson, R.K., and Wagman, J. (1968). "Particle-size distribution of metal components in urban air." *Environ. Sci. Technol.* **2,** 288–290.

Lee, R.E., Jr., Goranson, S.S., Enrione, R.E., and Morgan, G.B. (1972). "National air surveillance cascade impactor network. II: Size distribution measurements of trace metal components." *Environ. Sci. Technol.* **6,** 1025–1030.

Lee, R.E., Jr., Crist, H.L., Riley, A.E., and McLeod, K.E. (1975). "Concentration and size of trace metal emissions from a power plant, a steel plant, and a cotton gin." *Environ. Sci. Technol.* **9,** 643–647.

Linton, R.W., Loh, A., Natusch, D.F.S., Evans, C.A., Jr., and Williams, P. (1976) "Surface predominance of trace elements in airborne particles." *Science* **191,** 852–854.

Lippmann, M. (1977). "Regional deposition of particles in the human respiratory tract." In D.H.K. Lee, H.L. Falk, and S.D. Murphy, Eds., *Handbook of Physiology,* Section 9: *Reactions to Environmental Agents.* The American Physiological Society, Bethesda, Md., pp. 213–232.

Liu, B.Y.H., and Pui, D.Y.H. (1981). "Aerosol sampling inlets and inhalable particles." *Atmos. Environ.* **15,** 589–600.

Lodge, J.P., Jr. (1976). "Accuracy in air sampling." In *Accuracy in Trace Analysis: Sampling, Sample Handling, and Analysis, Proceedings of the 7th IMR Symposium, Oct. 7–11, 1974, Gaithersburg, Md.* National Bureau of Standards Spec. Publ. 422, pp. 311–320.

Lundgren, D.A. (1970). "Atmospheric aerosol composition and concentration as function of particle size and of time." *J. Air Pollut. Control Assoc.* **20,** 603–607.

Marple, V.A., and Liu, B.Y.H. (1974). "Characteristics of laminar jet impactors." *Environ. Sci. Technol.* **8,** 648–654.

Marple, V.A., Liu, B.Y.H., and Whitby, K.T. (1974). "Fluid mechanics of the laminar flow aerosol impactor." *Aerosol Sci.* **5,** 1–16.

Martens, C.S., Wesolowski, J.J., Kaifer, R., and John, W. (1973). "Lead and bromine particle size distributions in the San Francisco Bay area." *Atmos. Environ.* **7,** 905–914.

May, K.R. (1964). "Calibration of a modified Andersen bacterial aerosol sampler." *Appl. Microbiol.* **12,** 37–43.

McDonald, C., and Duncan, H.J. (1979). "Particle size distribution of metals in the atmosphere of Glasgow." *Atmos. Environ.* **13,** 977–980.

McDonald, R.L., Unni, C.K., and Duce, R.A. (1982). "Estimation of atmospheric sea salt dry deposition: Wind speed and particle size dependence." *J. Geophys. Res.* **87,** 1246–1250.

McFarland, A.R., Wedding, J.B., and Cermak, J.E. (1977). "Wind tunnel evaluation of a modified Andersen impactor and an all weather sampler inlet." *Atmos. Environ.* **11,** 535–539.

Milford, J.B., and Davidson, C.I. (1985). "The sizes of particulate trace elements in the atmosphere: A review." *J. Air Poll. Control Assoc.* (in press).

Mueller, P.K., Helwig, H.L., Alcocer, A.E., Gong, W.K., and Jones, E.E. (1964). "Concentration of fine particles and lead in car exhaust." *ASTM Special Technical Publication No. 352.* Philadelphia, Pa.

National Academy of Sciences (1973). *Manganese.* Publishing and Printing Office, National Academy of Sciences, Washington, D.C.

National Academy of Sciences (1979). *Airborne Particles.* University Park Press, Baltimore, Md., p. 41.

Natusch, D.F.S., and Wallace, J.R. (1974). "Urban aerosol toxicity: The influence of particle size." *Science* **186,** 695–699.

Natusch, D.F.S., Wallace, J.R., and Evans, Jr. C.A. (1974). "Toxic trace elements: Preferential concentration in respirable particles." *Science* **183,** 202–204.

Nifong, G.D., Boettner, E.A., and Winchester, J.W. (1972). "Particle size distributions of trace elements in pollution aerosols." *Am. Ind. Hyg. Assoc. J.* **33,** 569–575.

Nriagu, J.O. (1978). "Lead in the atmosphere." In J.O. Nriagu, Ed., *The Biogeochemistry of Lead in the Environment.* Vol. 1. Elsevier, Amsterdam, pp. 137–183.

Nriagu, J.O. (1979). "Copper in the atmosphere and precipitation." In J.O. Nriagu, Ed., *Copper in the Environment,* Vol. 1. Wiley, New York, pp. 43–75.

Nriagu, J.O. (1980). "Cadmium in the atmosphere and precipitation." In J.O. Nriagu, Ed., *Cadmium in the Environment,* Vol. 1. Wiley, New York, pp. 71–114.

Nriagu, J.O., and Davidson, C.I. (1980). "Zinc in the atmosphere." In J.O. Nriagu, Ed., *Zinc in the Environment,* Vol. 1. Wiley, New York, pp. 113–159.

Pattenden, N.J. (1974). *Atmospheric Concentrations and Deposition Rates of Some Trace Elements Measured in the Swanseal/Neath/Port Talbot Area.* Report AERE-R7729. United Kingdom Atomic Energy Authority, Harwell, Oxfordshire, p. 55.

Patterson, C.C. (1980). "An alternative perspective—lead pollution in the human environment: Origin, extent, and significance." In National Academy of Sciences, *Lead in the Human Environment.* National Academy Press, Washington, D.C., pp. 265–349.

Patterson, C.C., and Settle, D. (1976). "The reduction of orders of magnitude errors in lead analyses of biological materials and natural waters by evaluating and controlling the extent and sources of industrial lead contamination introduced during sample collecting, handling, and analysis." In *Accuracy in Trace Analysis: Sampling, Sample Handling, and Analysis, Proceedings of the 7th IMR Symposium, Oct. 7–11, 1974, Gaithersburg, Md.* National Bureau of Standards Spec. Publ. 422, pp. 321–351.

Patterson, E., and Gillette, D. (1977). "Commonalities in measured size distributions for aerosols having a soil-derived component." *J. Geophys. Res.* **82,** 2074–2082.

Peden, M.E. (1977). "Flameless atomic absorption determinations of cadmium, lead and manganese in particle size fractionated aerosols." In *Methods and Standards for Environmental Measurement, Proceedings of the 8th IMR Symposium, Sept. 20–24, 1976, Gaithersburg, Md.* National Bureau of Standards Spec. Publ. 464, pp. 367–377.

Raabe, O.G. (1978). "A general method for fitting size distributions to multi-component aerosol data using weighted least-squares." *Environ. Sci. Technol.* **12,** 1162–1167.

Rahn, K.A. (1976). *The Chemical Composition of the Atmospheric Aerosol.* Technical Report. Graduate School of Oceanography, University of Rhode Island, Kingston, R.I., 265 pp.

Rao, A.K. (1975). "An experimental study of inertial impactors." Ph.D. thesis. Mechanical Engineering Department, University of Minnesota, Minneapolis, Minn.

Sadasivan, S. (1981). "Trace elements in size separated atmospheric particulates at Trombay, Bombay, India." *Sci. Total Environ.* **20,** 109–115.

Sehmel, G.A., and Hodgson, W.H. (1978). *A Model for Predicting Dry Deposition of Particles and Gases to Environmental Surfaces.* Rep. PNL-SA-6721. Battelle Pacific Northwest Laboratories, Richland, Wash.

Sievering, H., Dave, M.J., McCoy, P.G., and Walther, K. (1978). "Cellulose filter high-volume cascade impactor aerosol collection efficiency: A technical note." *Environ. Sci. Technol.* **12,** 1435–1437.

Slinn, W.G.N. (1982). "Predictions for particle deposition to vegetative canopies." *Atmos. Environ.* **7,** 1785–1794.

Stahlhofen, W., Gebhart, J., and Heyder, J. (1980). "Experimental determination of the regional deposition of aerosol particles in the human respiratory tract." *Am. Ind. Hyg. Assoc. J.* **41,** 385–398.

Swartz, D.B., Denton, M.B., and Moyers, J.L. (1973). "On calibrating of cascade impactors." *Am. Ind. Hyg. Assoc. J.* **34,** 429–439.

Task Group on Lung Dynamics (1966). "Deposition and retention models for internal dosimetry of the human respiratory tract." *Health Phys.* **12,** 173–207.

Ter Haar, G.L., Lenane, D.L., Hu, J., and Brandt, M. (1972). "Composition, size, and control of automotive exhaust particulates." *Environ. Sci. Technol.* **22,** 39–46.

Thompson, M., Walton, S.J., and Wood, S.J. (1979). "Statistical appraisal of interference effects in the determination of trace elements by atomic-absorption spectrophotometry in applied geochemistry." *Analyst* **104,** 299–312.

U.S. Environmental Protection Agency (1973, 1974). Computer printouts of data supplied by G.G. Akland. U.S. Environmental Protection Agency, Environmental Monitoring and Support Laboratory, Research Triangle Park, N. C.

U.S. Environmental Protection Agency (1982). *Air Quality Criteria for Particulate Matter and Sulfur Oxides,* Volume Ill. Report EPA 600/8-82-029C. U.S. Environmental Protection Agency, Washington, D.C., pp. 11-1–11-74.

Van Craen, M.J., Denoyer, E.A., Natusch, D.F.S., and Adams, F. (1983). "Surface enrichment of trace elements in electric steel furnace dust." *Environ. Sci. Technol.* **17,** 435–439.

Walsh, P.R., Rahn, K.A., and Duce, R.A. (1978). "Erroneous elemental mass-size functions from a high-volume cascade impactor." *Atmos. Environ.* **12,** 1793–1795.

Waughman, G.J., and Brett, T. (1980). "Interference due to major elements during the estimation of trace heavy metals in natural materials by atomic absorption spectrophotometry." *Environ. Res.* **21,** 385–393.

Wedding, J.B., McFarland, A.R., and Cermak, J.E. (1977). "Large particle collection characteristics of ambient aerosol samplers." *Environ. Sci. Technol.* **11,** 387–390.

Wegscheider, W., Knapp, G., and Spitzy, H. (1977). "Statistical investigations of interferences in graphite furnace atomic absorption spectroscopy, Parts I, II, and III." *Z. Anal. Chem.* **283,** 9–14, 97–103, 183–190.

Wesolowski, J.J. (1973). "Ambient air aerosol sampling." In *Proceedings of the Second Joint Conference on Sensing of Environmental Pollutants, Dec. 10–12, 1973.* Washington, D.C., pp. 191–196.

Whitby, K.T., Husar, R.B., and Liu, B.Y.H. (1972). "The aerosol size distribution of Los Angeles smog." *J. Colloid Interface Sci.* **39,** 177–204.

Winchester, J.W., Ferek, R.J., Lawson, D.R., Pilotte, J.O., Thiemens, M.H., and Wangen, L.E. (1979). "Comparison of aerosol sulfur and crustal element concentrations in particle size fractions from continental U.S. locations." *Water, Air, Soil Pollut.* **12,** 431–440.

Zoller, W.H., Gladney, E.S., and Duce, R.A. (1974). "Atmospheric concentrations and sources of trace elements at the South Pole." *Science* **183,** 198–200.

13

METAL SOLUBILITY IN ATMOSPHERIC DEPOSITION

Donald F. Gatz
Lih-Ching Chu

Illinois Department of Energy and Natural Resources
State Water Survey Division
Champaign, Illinois

1. INTRODUCTION

Concern over potential toxicity of heavy metals has led to measurements of their input to terrestrial and aquatic ecosystems via atmospheric deposition (Andren and Lindberg, 1977; Eisenreich, 1980; Jeffries and Snyder, 1981; Lindberg et al., 1980; Schlesinger et al., 1974; Swanson and Johnson, 1980; Wiener, 1979). Effects of metals in ecosystems depend in part on their mobility and bioavailability (Hardy and Crecelius, 1981; Lindberg and Harriss, 1980). Mobility and bioavailability of metals have been characterized in terms of metal solubilities in water and various stronger extractants in samples of aerosols (Lindberg and Harrison, 1980), street dusts (Harrison et al., 1981), and roadside soils (Harrison et al., 1981). Since atmospheric deposition is a major mechanism by which airborne metals reach ecosystems, it is also important that we understand metal solubility in wet and dry deposition. Further, the mesoscale spatial distributions of soluble and insoluble forms of the same elements appear to be clues to the sources of the elements and the scavenging processes by which they are removed from the atmosphere by precipitation (Gatz, 1980).

Solubility of metals in many natural aqueous systems can be influenced by organic materials (Harrison et al., 1981; Slavek and Pickering, 1981; Wilber and Hunter, 1979), clay minerals (Slavek and Pickering, 1981; Wilber and Hunter, 1979), metal hydroxides (Harrison et al., 1981; Wilber and Hunter, 1979), pH (Cavallaro and McBride, 1978; Farrah and Pickering, 1977; Sadiq and Zaidi, 1981; Slavek and Pickering, 1981), and divalent ions (Garcia-Miragaya and Page, 1977; Griffin and Au, 1977; McBride, 1976). Of these, clay minerals and metal hydroxides would be part of the insoluble matter in precipitation samples. There have been pleas for separate analyses of soluble and insoluble impurities in precipitation (Lewis and Grant, 1978; Rattonetti, 1976), but the insoluble fraction is still rarely analyzed, and it appears from a recent review of literature on toxic substances in atmospheric precipitation (Galloway et al., 1980) that precipitation has not yet been systematically examined for the influence of insoluble materials on metal solubility.

Of 10 recent papers reporting heavy metal concentrations in precipitation (Andren and Lindberg, 1977; Betson, 1978; Dethier, 1979; Eisenreich, 1980; Hendry and Brezonik, 1980; Jeffries and Snyder, 1981; Navarre et al., 1980; Schlesinger et al., 1974; Swanson and Johnson, 1980; Wiener, 1979), none reported soluble and insoluble metals separately. None of these authors separated soluble from insoluble materials before analysis. In one case (Swanson and Johnson, 1980) separation by filtration was tested, but comparison of metal concentrations in 20 pairs of filtered and unfiltered subsamples showed no effect on measured concentrations, and filtration was abandoned. In other cases, separation of soluble and insoluble fractions was precluded by preacidification of the collection vessel for the purpose of avoiding wall losses from the precipitation samples. Depending on the specific analytical procedures used, these papers reported metal concentrations variously as soluble, total, or total acid leachable, or simply do not indicate what their measurements represent.

As pointed out by Rattonetti (1976), proper storage and treatment of natural water samples are necessary for accurate analyses, as well as to enable proper interpretations of the measurements. Rattonetti analyzed soluble and insoluble fractions of stream waters and rain and found that acidification of unfiltered samples artificially released metals from insoluble particles in the samples. Wall losses of dissolved metals were minimal in samples containing insoluble particles. Filtration was recommended to stabilize the distribution of metals between soluble and insoluble fractions. Acidification of the filtrate was recommended to avoid wall losses following filtration.

To help improve our knowledge in this area, we examined a set of atmospheric deposition samples for distributions of metal solubility in wet, dry, and bulk (wet + dry) precipitation, and for their relationship to concentrations of insoluble materials and sample pH.

2. EXPERIMENTAL METHODS

2.1. Sampling Methods

Both bulk and separate wet and dry samples were collected weekly (Tuesdays) except for a few 2- or 3-week periods. The wet and dry samples were collected in 29.4-cm diameter white high density polyethylene buckets using an Aerochem Metrics sampler. The bulk sample, open to the atmosphere continuously, was collected in the same type of bucket.

The collectors were located at a Commonwealth Edison Co. electric power substation in a residential area of Glen Ellyn, Ill., 42 km west of the Lake Michigan shoreline. The collector openings were positioned about 1.5 m above a grass surface, a few meters south of a seldom used crushed limestone parking lot. The site was bordered to the east and south by the asphalt parking lot of a three-story apartment building located 25 m to the east of the samplers. No obstacles were located higher than 45 degrees from the horizontal.

Rainfall was measured in a standard weighing bucket recording rain gauge with a 30.5-cm (12-in.) diameter top. In winter the rain-gauge collector bucket was charged with an antifreeze solution to prevent freezing of the accumulated sample. The antifreeze solution was covered with a thin layer of motor oil to prevent evaporation. An Alter-type windshield surrounded the rain gauge to lessen the effect of the wind on the catch. Openings and closings of the wet/dry collector were recorded on the rain-gauge chart with an event pen.

The preweighed sample containers were reweighed to the nearest gram to measure sample mass before shipment to our laboratory in Champaign, Ill., for analysis.

2.2. Laboratory Procedures

In the laboratory, sample processing was begun immediately upon arrival of the samples to minimize internal chemical changes. The first step was to weigh the containers as a check on possible leaks during shipment. Then, a 16-mL portion of the sample was removed and used for measuring pH and specific conductance. A volume of 250 mL of deionized water was added to the dry-side buckets and 50 mL to those wet and bulk collectors that were exposed during weeks when no precipitation fell. Samples to which water was added were allowed 24 h to equilibrate before mass, pH, and specific conductance were measured.

To separate the soluble (i.e., filterable) and insoluble (i.e., nonfilterable) materials, samples were filtered through preweighed 47-mm diameter, 0.4-μm pore diameter polycarbonate membranes. The portion of the filtrate designated for trace metals analysis was acidified to pH 2 with nitric acid to stabilize the metals in solution (U.S. EPA, 1979).

The filters containing the insoluble impurities were placed in a chamber maintained at a constant 47% relative humidity by a saturated lithium nitrate solution and were allowed to equilibrate for one week before being reweighed to determine insoluble mass. The filters were digested in small amounts of concentrated nitric acid and 30% stabilized hydrogen peroxide in Teflon*-lined stainless steel pressure vessels. Following digestion, the sample solutions were diluted to a 1.28*N* (8%) nitric acid concentration and stored in acid-washed linear polyethylene bottles. A reagent blank and filter blank were processed concurrently and analyzed with each batch of seven filters.

Flame atomic absorption spectrophotometry (AAS) was used to analyze the acidified portion of the filtrate for Zn and the insoluble portion of the sample for Al, Cu, Fe, and Zn. Flameless AAS was used for the analysis of soluble Cd, Cu, and Pb, and insoluble Cd and Pb. EPA Trace Metals Reference Samples of known concentrations were analyzed with the precipitation samples. Results showed accuracies mostly within 5% and precisions (two standard deviations) mostly less than 15% of the mean values, at concentrations of 10 times detection limits.

2.3. Laboratory Equipment

An Orion* model 811 pH meter with a Beckman† Futura microcombination electrode and a Metrohm‡ model 103 pH meter with microcombination electrode were used for pH measurements. Specific conductance was measured

*DuPont trademark
*Orion Research Inc., Cambridge, MA.
†Beckman Instruments, Inc., Irvine, CA.
‡Brinkman Instruments, Inc., Westbury, NY.

with a Yellow Springs Instrument Co.[§] manually balanced AC bridge with glass microelectrode. Instrumentation Laboratory[||] models 353 and 951 atomic absorption spectrophotometers were used for all flame atomic absorption measurements. Flameless AAS analyses were carried out on an Instrumentation Laboratory model 151 AAS fitted with Instrumentation Laboratory model 455 and Varian[*] model 63 carbon rod atomizers.

2.4. Reagents

All metal standard solutions were prepared from either a reagent grade soluble salt or metal dissolved in the minimum amount of acid and diluted to the desired concentration with polished deionized water. All other chemicals were reagent grade.

3. RESULTS

3.1. Data Summaries

Tables 13.1 and 13.2 give a partial summary of analytical results. Minimum and maximum concentrations, as well as median concentrations and standard errors are given for both soluble and insoluble fractions of all metals measured. Results for total insoluble mass (TIM) concentrations and for sample pH are given also. Examination of normal plots of log concentrations indicated that all concentrations were approximately log–normally distributed.

Partition coefficients, which describe the partitioning of metals between solid and liquid phase in atmospheric precipitation, were calculated using the definition of Eisenreich et al. (1981). Thus, the partition coefficient k_p is given by $k_p = (\mu g\ \text{metal/g solids})/(\mu g\ \text{metal/g water})$. The results are summarized in terms of observed median values in wet-only and bulk precipitation in Table 13.3. The table also gives observed values of k_p for the same four metals in lake water from Saginaw Bay, Lake Huron, for comparison. It appears that k_p is higher in Saginaw Bay, indicating a greater tendency for metals to adsorb onto solids in lake water than in precipitation. The available data do not permit a clear explanation of these differences, but the composition of suspended solids in lake water is no doubt different from those in precipitation, especially with regard to organic materials, and thus metal sorption characteristics can also be expected to differ.

Partition coefficients appear to be greater in wet-only than in bulk samples (Table 13.3), except for Pb. The reasons for these differences are not

§Yellow Springs Instrument Co., Yellow Springs, OH.
||Instrumentation Laboratory, Inc., Wilmington, MA.
*Varian Associates, Palo Alto, CA.

Table 13.1. Summary of Analytical Results for Wet Samples

Wet Sample	*N*	Concentrations (μg/L) Min	Median and Standard Error	Max
Soluble Zn	51	7.	36. ± 5.8	3,020
Insoluble Zn	51	0.02	1.1 ± 0.3	530
Soluble Cu	51	<0.4	3.09 ± 0.58	75.4
Insoluble Cu	51	0.059	0.77 ± 0.22	48.5
Soluble Cd	51	0.04	0.150 ± 0.026	6.79
Insoluble Cd	51	0.000	0.007 ± 0.002	6.04
Soluble Pb	51	0.9	9.54 ± 1.78	332.
Insoluble Pb	51	0.286	2.49 ± 0.57	377.
Insoluble Fe	51	3.05	88.3 ± 29.1	18,900
Insoluble Al	25	2.37	71.0 ± 25.2	2,616
Total insoluble mass	51	380	3,350 ± 290	267,000
pH[a]	49	3.61	4.43 ± 0.09	6.84

[a]pH units.

Table 13.2. Summary of Analytical Results for Bulk Samples

Bulk Sample	*N*	Concentrations (μg/L) Min	Median and Standard Error	Max
Soluble Zn	49	7.	43.9 ± 6.8	10,200
Insoluble Zn	48	0.25	12.6 ± 1.7	1,360
Soluble Cu	49	<0.4	3.43 ± 0.89	377
Insoluble Cu	47	0.00	4.69 ± 1.74	581
Soluble Cd	49	<0.02	0.242 ± 0.053	40.0
Insoluble Cd	48	0.001	0.044 ± 0.007	5.3
Soluble Pb	49	<0.7	5.0 ± 1.8	1,510
Insoluble Pb	48	0.747	32.5 ± 10.6	5,880
Insoluble Fe	49	37.0	606 ± 136	115,000
Insoluble Al	25	7.4	684 ± 243	23,700
Total insoluble mass	50	39.9	23,260 ± 3,690	2,167,000
pH[a]	42	3.73	5.95 ± 0.33	7.26

[a]pH units.

immediately apparent either, but again suggest a systematic difference in the composition of the solid materials between the two types of atmospheric deposition samples.

3.2. Distributions of Percent Soluble Metals

Frequency distributions of soluble fractions of Zn and Pb in wet, dry, and bulk samples from weekly sampling periods between December 1979 and

Table 13.3. Observed Partition Coefficients in Wet-Only and Bulk Precipitation Samples

	Median Partition Coefficients, k_p				Observed[a] range of k_p values in Saginaw Bay 1976–1978
Metal	N	Wet Samples	N	Bulk Samples	
Cu	43	39,800	54	24,400	100,000–156,000
Zn	43	9,500	55	6,300	277,000–412,000
Cd	49	14,300	54	4,600	224,000
Pb	48	60,600	49	104,000	418,000–438,000

[a]Annual values from Dolan and Bierman (1982).

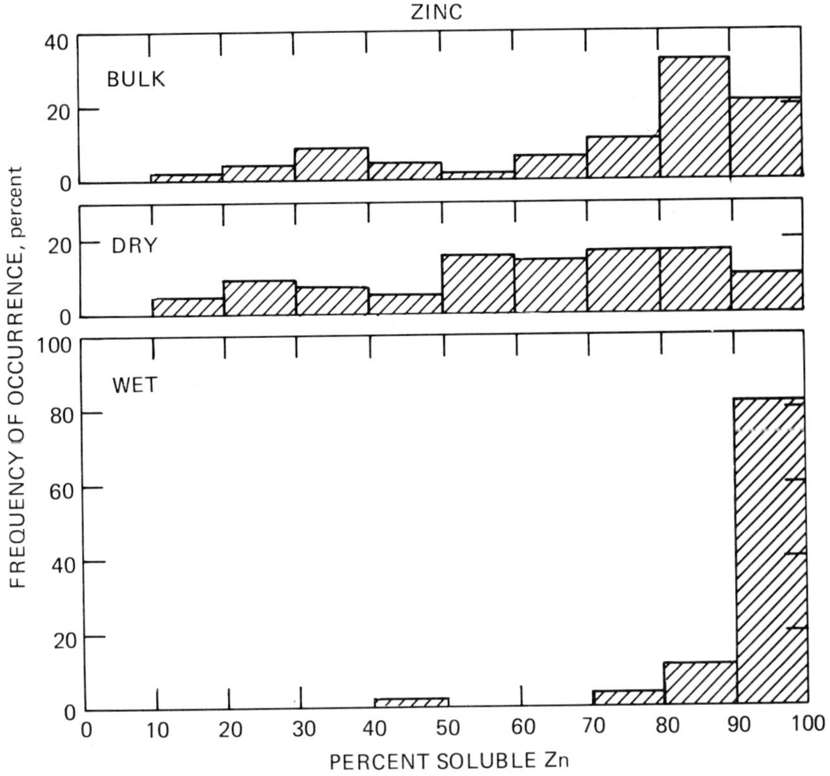

Figure 13.1. Frequency distributions of percent soluble Zn in wet, dry, and bulk samples.

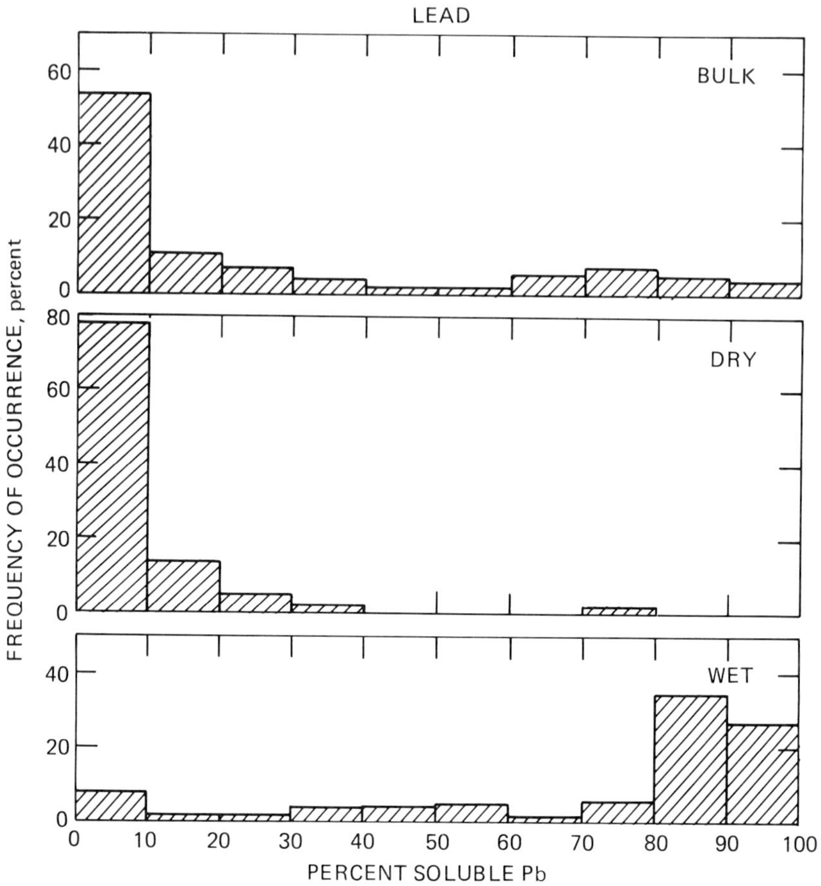

Figure 13.2. Frequency distributions of percent soluble Pb in wet, dry, and bulk samples.

March 1981 are shown in Figs. 13.1 and 13.2, respectively. Zinc was almost always very soluble in wet samples, but a wide range of solubilities occurred in the dry and bulk samples. Lead was also quite soluble in wet samples, but somewhat less than Zn. Most of the dry samples, however, had less than 10% soluble Pb.

Distributions of Cu and Cd solubilities (not shown) were intermediate between those of Zn and Pb. Table 13.4 shows that the median solubilities of Cd were very similar to those of Zn in wet, dry, and bulk samples, respectively. The corresponding median solubilities for Cu were intermediate between those of Zn and Pb.

Thus, all four metals examined were mostly soluble in wet samples, but exhibited a wide range of solubility in dry and bulk samples. The solubility distributions in the bulk samples were much more like those of the dry than the wet samples.

Table 13.4. Median Values for Total and Fe Insoluble Concentrations in Wet, Dry, and Bulk Samples, with Corresponding Median Metal Solubilities

	Percent Soluble				Total Insoluble Concentration[a] (mg/L)	Insoluble Fe Concentration (mg/L)
Sample Type	Zn	Cd	Cu	Pb		
Wet	96	95	90	83	3.3	0.086
Dry	67	66	30	5	(26.8)[b]	(0.68)[b]
Bulk	82	88	38	8	24.9	0.75

[a] Based on 49 samples.

[b] Concentration measured in diluted samples.

3.3. Metal Solubility as a Function of TIM Concentrations

To examine the possible influence of insoluble materials on metal solubility in atmospheric deposition, the soluble fractions of Zn and Pb were plotted against the concentration of total insoluble mass (TIM) (Fig. 13.3). For both metals, the wet and bulk samples show a consistent pattern with respect to TIM, with the wet samples on the whole having smaller TIM concentrations and greater solubilities. There were noticeable differences between Zn and Pb in their relationships to TIM concentrations. Both show a general tendency for solubility to decrease as TIM concentrations increase, but Zn was generally more soluble than Pb, at least at higher TIM concentrations. More specifically, for TIM concentrations below 50 mg/L, Zn was likely to be >80% soluble. For TIM concentrations >50 mg/L, Zn solubility was rather uniformly distributed between 20 and 90%. Lead, on the other hand, was likely to be >80% soluble only for TIM concentrations <3 mg/L. For TIM concentrations between 3 and 30 mg/L, Pb solubility was rather uniformly distributed between 0 and 90%. For TIM concentrations >30 mg/L, Pb was likely to be <20% soluble.

Results are not plotted for Cd and Cu, but examination of similar plots for these two metals showed that Cd behaved very similarly to Zn, and Cu similarly to Pb. Results for Cd and Cu are included in the summary of median metal solubilities and TIM concentrations given in Table 13.4. For each sample type, the median percent solubilities of Zn and Cd were very similar, and Cu was intermediate between those elements and Pb. Further, Table 13.4 shows that solubility varied inversely to TIM among the three types of samples. TIM was lowest in the wet samples, in which metal solubilities were greatest, and highest in dry samples, where solubilities were least.

3.4. Metal Solubility as a Function of Insoluble Fe Concentrations

One of the specific agents for removal of metal ions from solution, either by sorption at the surface or by inclusion in the metal hydroxide precipitate, is

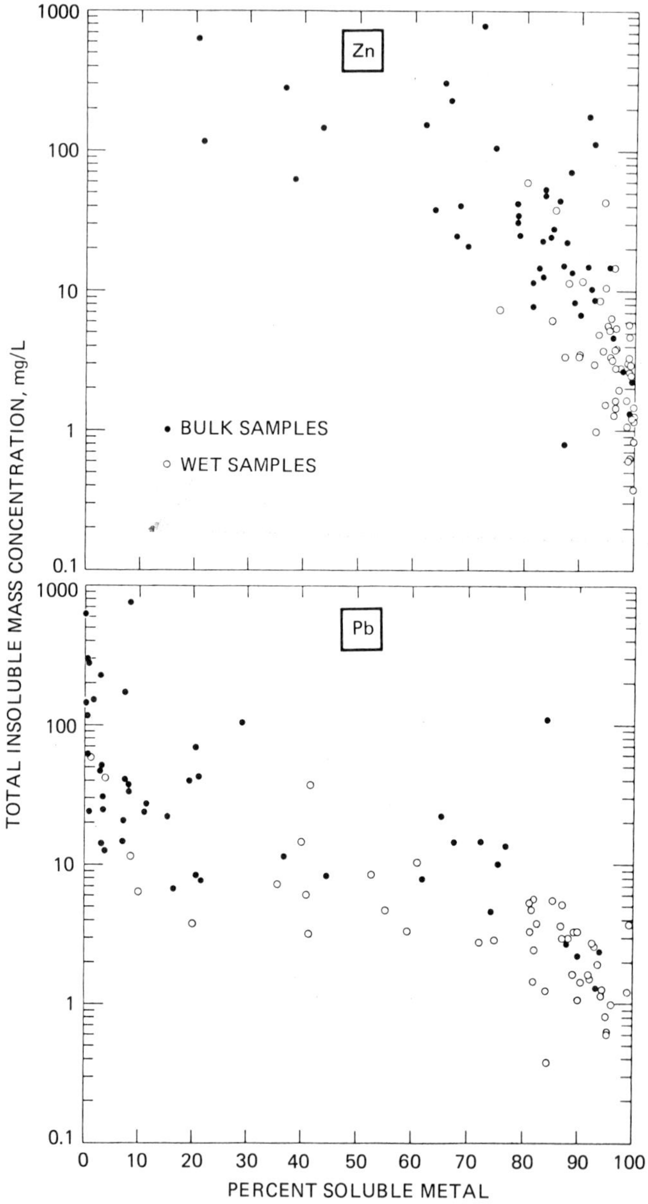

Figure 13.3. Metal solubility as a function of total insoluble mass concentration for wet and bulk samples.

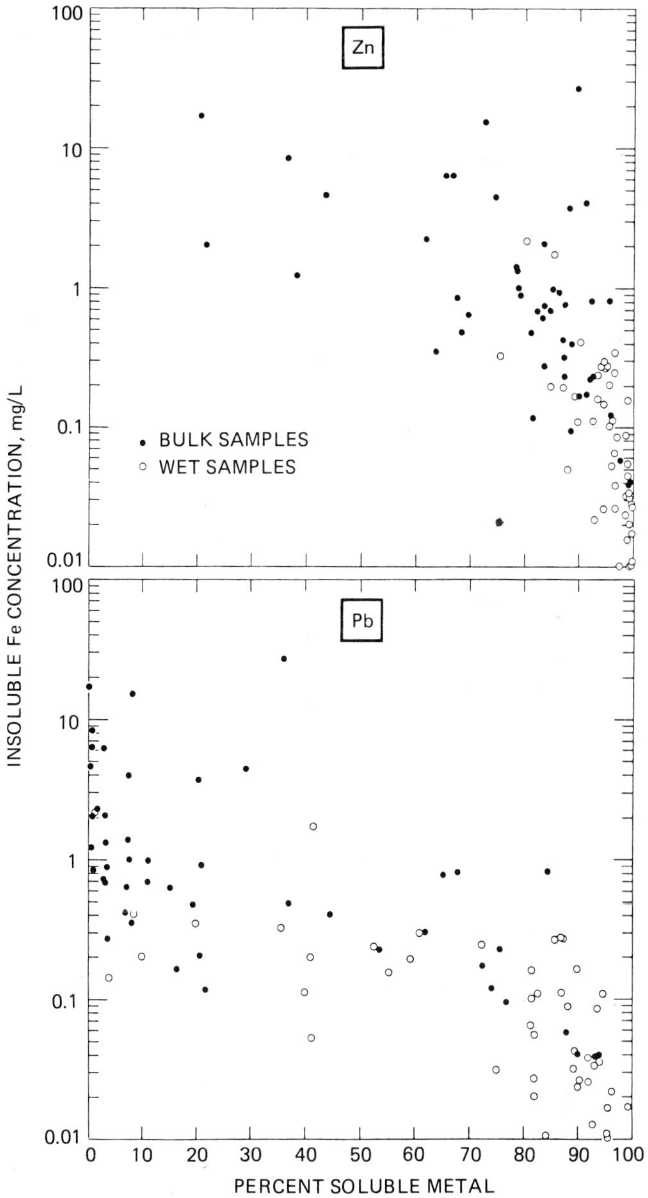

Figure 13.4. Metal solubility as a function of insoluble Fe concentration for wet and bulk samples.

metal hydroxides. As an indication of the possible effect of hydroxides on metal solubility, solubilities were plotted versus concentrations of insoluble Fe. Results for both Zn and Pb are shown in Figure 13.4. Both Zn and Pb behaved similarly with respect to insoluble Fe concentrations as they did with respect to TIM concentrations. That is, considering wet and bulk samples combined, there were only a few samples having Zn <50% soluble, while for Pb about half the samples had <50% of their Pb in soluble form. For both metals, the wet samples had generally higher solubilities than the bulk samples.

More specifically, for insoluble Fe concentrations <1 mg/L, Zn was likely to be >80% soluble, while above 1 mg/L Zn solubility was quite uniformly distributed between 20 and 90%.

Pb was likely to be >80% soluble only for insoluble Fe concentrations <0.1 mg/L, while the range of uniform solubility distribution was 0.1 to 1 mg/L. Above an insoluble Fe concentration of 1 mg/L, Pb was likely to be <10% soluble.

Again, the results for Cd and Cu are not plotted, but in this case also, the Cd results were very similar to those of Zn, and Cu results to those of Pb. As in the case of TIM, Table 13.4 shows an inverse relationship between median values of metal solubilities and insoluble Fe concentrations.

3.5. Metal Solubility as a Function of pH

Results for Zn and Pb are shown in Figure 13.5. At first glance, Figure 13.5 appears to show almost random variability of metal solubility with pH, but differences between sample types may be discerned with study of the data. Zinc in wet samples was almost all >80% soluble, regardless of pH. In the bulk samples, however, solubility was quite variable at pH >4.5, but Zn was mostly >70% soluble at pH <4.5.

Pb in wet samples showed quite a strong tendency to increase in solubility with decreasing pH. The same general pattern occurred in the bulk samples, but again pH 4.5 was the dividing line between regimes. At pH >4.5, Pb was mostly <20% soluble, but between pH 4.0 and 4.5 the distribution of solubility was again more or less uniform from very low to very high values.

3.6. The Sample Volume Effect

The effect of rainfall amount, or sample volume, on concentrations of almost any material in precipitation is well known. Concentrations tend to decrease, often exponentially (Hales and Dana, 1979), with rainfall amount. There is also an effect of sample volume on pH, as shown in Figure 13.6. Two separate regimes appear. For rainfall >1 cm, the pH values were predominantly between 4 and 5. On the other hand, for rainfall <1 cm, the pH was relatively

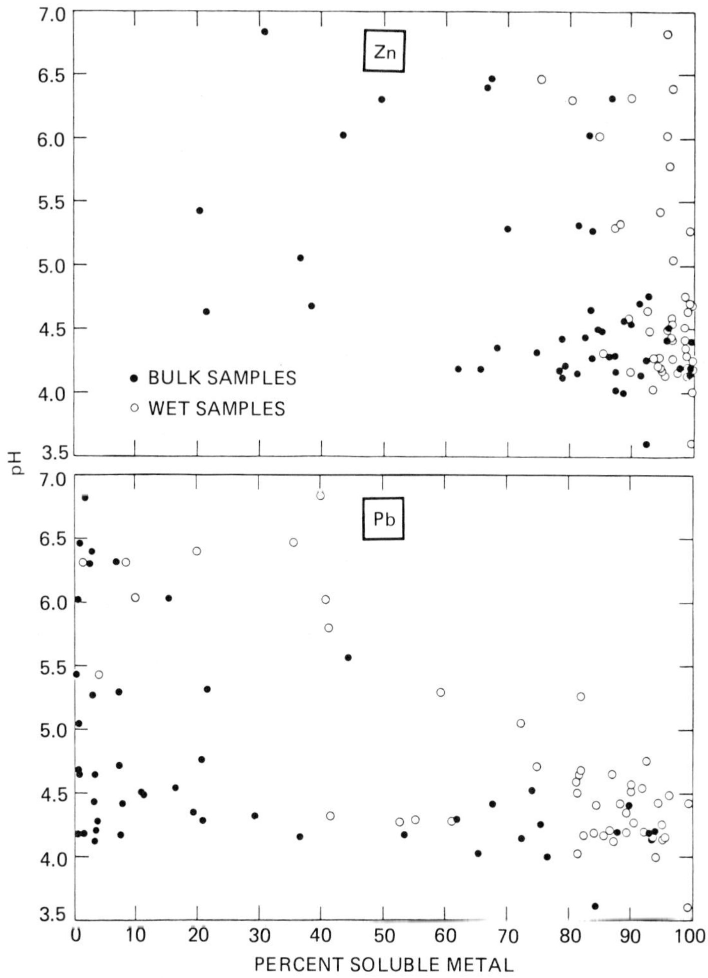

Figure 13.5. Metal solubility as a function of sample pH.

uniformly distributed between pH 4 and 7 for both wet and bulk samples. The reasons for this result are not known, but could be related to differing removal mechanisms for alkaline and acidic materials from the atmosphere.

4. DISCUSSION

The systematic differences in metal solubility between wet, dry, and bulk samples need an explanation. Figs. 13.1 and 13.2 and Table 13.4 demonstrate (1) that the metals were generally very soluble in wet samples, but that solubility was distributed more uniformly in the dry and bulk samples, and (2)

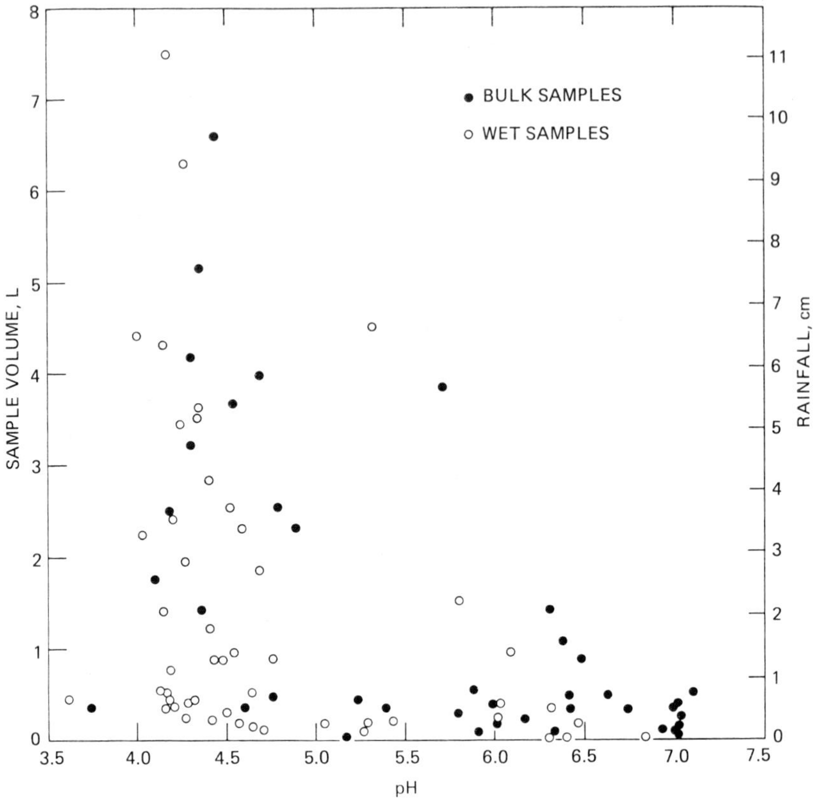

Figure 13.6. Variation of sample pH with rainfall.

that, in a given type of sample, Zn and Cd were the most soluble, followed in order by Cu and Pb.

The reason for the systematic differences in metal solubility between wet and dry or bulk samples appears to lie in the enhanced concentrations of insoluble materials in the dry and bulk samples and the fact that insoluble materials tend to remove ions from solution through some sort of sorption process. As was shown in Figure 13.3, metal solubility decreased with increasing concentrations of TIM.

One component of the insoluble material that may sorb metal ions on its surface or incorporate them by coprecipitation is the hydrous metal oxides, such as hydrous Fe oxide. The similarity of the metal solubility relationships to TIM and insoluble Fe concentrations in Figs. 13.3 and 13.4, respectively, indicates that sorption or coprecipitation involving hydrous metal oxides is very likely to be involved in determining metal solubility in atmospheric precipitation.

Another component of the insoluble material that may sorb metal ions is aluminosilicate clay minerals. Indeed, the observed order of solubility (Zn $\simeq$

Cd > Cu > Pb, Table 13.3) is opposite to observed metal ion affinities for montmorillonite, kaolinite, and illite, as reported by Farrah and Pickering (1977).

Local soils contain the clay minerals montmorillonite, kaolinite, and illite in mass fractions totaling about 30%. Thus, the clays are the dominant Al-bearing minerals, and the insoluble Al concentration in precipitation should be a good index of the clay mineral concentration. However, plots (not shown) of the insoluble Al concentration in 25 randomly chosen samples from the sample set versus soluble fractions of Zn and Pb showed only a weak tendency for lower solubilities with higher concentrations of insoluble Al.

The results also illustrate the key role of meteorological factors in precipitation chemistry. It is well known that concentrations of most materials in rain vary inversely with rainfall amount (Hales and Dana, 1979). The variation of pH with rainfall is not so well known, but Figure 13.6 shows a rather strong relationship for our site in which pH was rather uniformly distributed in rainfall <1 cm, but confined to a rather narrow range of pH (between 4 and 5) with heavy rains. Thus, heavy rains tend to have low concentrations of insoluble materials and low pH, both of which tend to enhance metal solubility. This suggests that two sites having equal total metal deposition but different distributions of rainfall amount could have different amounts of *soluble* metals deposited. The site with the greater frequency of heavy rains would tend to have a higher soluble metal deposition.

These results also have important implications regarding sample *collection* methods. If one wants to measure the soluble metal composition of the rain *as it hits the ground,* it is clear that only a wet-only sample will give the proper result, because the insoluble materials that accumulate in a bulk sample during dry periods will very likely remove some portion of the soluble metals from solution before they can be measured.

These results also have important implications of the methods used to prepare precipitation samples for metal analyses. Measurements of *total* metal content of precipitation are the most appropriate for comparison of metal concentrations in precipitation against those in air, or in mass budgets where comparison is made to emissions, since both airborne concentrations and emissions are generally given in terms of total (i.e., not just soluble) metals. To measure total metal concentrations in precipitation, the analyst must be certain that both soluble and insoluble fractions of the metal in question are being measured. Some methods, such as neutron activation analysis of liquid samples or of residue left after evaporation of a precipitation sample, can measure totals, but analytical methods that require solids to be dissolved before measurement, such as flame AAS, should be used on separate samples of the soluble and insoluble fractions of precipitation samples. The soluble metals may be determined directly in the sample filtrate, but the insoluble metals must be determined in an appropriately dissolved portion of the insoluble materials.

These results are also important in other areas of precipitation chemistry.

For example, to understand the chemistry of metals in precipitation, one clearly needs a complete picture of the distribution of the metal between soluble and insoluble phases, as well as measurements of other species that affect metal solubility in both soluble and insoluble fractions. A similarly complete picture is also needed for other studies involving metals in precipitation, such as scavenging of airborne metals by precipitation, the use of metals in precipitation as source tracers, and effects of increased precipitation acidity on metal mobility and bioavailability in ecosystems.

The results given here do not agree with those of Swanson and Johnson (1980), which found no differences in metal concentrations between pairs of filtered and unfiltered samples. No explanation of the earlier results is apparent, unless the samples were collected in a location where insoluble particles did not reach the sampler. The present results show that one cannot assume that there will be no insoluble matter in precipitation samples.

5. SUMMARY AND CONCLUSIONS

Both soluble and insoluble metals were measured in weekly wet, dry, and bulk precipitation samples collected in suburban Chicago during 59 sampling periods from December 1979 to March 1981.

Distributions of percent-soluble metals show differences in solubility between metals, Zn = Cd > Cu > Pb, and differences for all metals between sample types, wet > bulk = dry.

Metal solubility in precipitation is similar to that of other natural waters with respect to the effects of pH and insoluble materials. Solubility decreases as pH and the concentrations of total mass and insoluble Fe increase. Thus, metals are less soluble in dry and bulk samples than in wet-only samples because dry and bulk samples contain more insoluble matter. Only a weak relationship was found between concentrations of insoluble Al (as an index of clay mineral concentrations) and soluble fractions of Zn and Pb.

These results have important implications for precipitation sampling and sample preparation methodology, as well as for the understanding of metal chemistry in precipitation, airborne metals scavenging processes, metals as tracers for sources of acidity, and effects of acid precipitation on ecosystems.

ACKNOWLEDGMENTS

Collection and analysis of the samples used in this study was funded by the U.S. EPA Nationwide Urban Runoff Program, through a subcontract with the Northeastern Illinois Planning Commission (NIPC). Donald L. Hey was Principal Investigator and Gary C. Schaefer was Project Coordinator for NIPC. Samples were analyzed using equipment obtained under Contract DEAS0276EVA01199, with the U.S. Department of Energy. Data analysis and interpretation were funded in part by Grants ATM 77-24294 and ATM 80-14893 from the National Science Foundation, Atmospheric Chemistry Program. The Commonwealth Edison Co. and Glenbard West High School allowed us the use of their facilities for sampling and sample shipping, respectively. Samples were collected by Jerry Dudgeon of the DuPage County Regional Planning Commission, and by John Hudson and Laura Streitberger, students at Glenbard West High School. At the Water Survey, Michael Slater was responsible for sample preparation; the samples were analyzed by Barbara Keller under the general direction of Mark Peden; and Randall K. Stahlhut and Stacy Craft provided assistance with programming and data processing, and Jean Dennison typed the manuscript.

REFERENCES

Andren, A.W., and Lindberg, S.E. (1977). "Atmospheric input and origin of selected elements in Walker Branch Watershed, Oak Ridge, Tennessee." *Water, Air, Soil Pollut.* **8,** 199–215.

Betson, R.P. (1978). "Bulk precipitation and streamflow quality relationships in an urban area." *Water Resour. Res.* **14,** 1165–1169.

Cavallaro, N., and McBride, M.B. (1978). "Copper and cadmium adsorption characteristics of selected acid and calcareous soils." *Soil Sci. Soc. Am. J.* **42,** 550–556.

Dethier, D.P. (1979). "Atmospheric contributions to stream water chemistry in the North Cascade Range, Washington." *Water Resour. Res.* **15:** 787–794.

Dolan, D.M., and Bierman, V.J. (1982). "Mass balance modeling of heavy metals in Saginaw Bay, Lake Huron." *J. Great Lakes Res.* **8**(4), 676–694.

Eisenreich, S.J. (1980). "Atmospheric input of trace metals to Lake Michigan." *Water, Air, Soil Pollut.* **13,** 287–301.

Eisenreich, S.J., Looney, B.B., and Thornton, J.D. (1981). "Airborne organic contaminants in the Great Lakes ecosystem." *Environ. Sci. Technol.* **15,** 30–38.

Emmel, R.H., Sotera, J.J., and Stux, R.L. (1977). Standard Conditions for Flame Operation, Vol. 1 of Atomic Absorption Methods Manual. Instrumentation Laboratory, Lexington, Mass.

Farrah, H., and Pickering, W.F. (1977). "Influence of clay-solute interactions on aqueous heavy metal ion levels." *Water, Air, Soil Pollut.* **8,** 189–197.

Galloway, J.N., Eisenreich, S.J., and Scott, B.C., Eds., (1980). *Toxic Substances in Atmospheric Deposition, A Review and Assessment.* Workshop Report, Jekyll Island, Georgia, November 1979. Published by the National Atmospheric Deposition Program, Colorado State University, Fort Collins, Colo., 146 pp.

Garcia-Miragaya, J., and Page, A.L. (1977). "Influence of exchangeable cation on the sorption of trace amounts of Cd by montmorillonite." *Soil Sci. Am. J.* **41,** 718–721.

Gatz, D.F. (1980). "Associations and mesoscale spatial relationships among rainwater constituents." *J. Geophys. Res.* **85,** 5588–5598.

Griffin, R.A., and Au, A.K. (1977). "Lead adsorption by montmorillonite using a competitive Langmuir equation." *Soil Sci. Soc. Am. J.* **41,** 880–882.

Hales, J.M., and Dana, M.T. (1979). "Precipitation scavenging of urban pollutants by convective storm systems." *J. Appl. Meteorol.* **36,** 294–316.

Hardy, J.T., and Crecelius, E.A. (1981). "Is atmospheric particulate matter inhibiting marine primary productivity?" *Environ. Sci. Technol.* **15,** 1103–1105.

Harrison, R.M., Laxen, D.P.H., and Wilson, S.J. (1981). "Chemical associations of lead, cadmium, copper, and zinc in street dusts and roadside soils." *Environ. Sci. Technol.* **15,** 1378–1383.

Hendry, C.D., and Brezonik, P.L. (1980). "Chemistry of precipitation at Gainesville, Florida." *Environ. Sci. Technol.* **14,** 843–849.

Jeffries, D.S., and Synder, W.R. (1981). "Atmospheric deposition of heavy metals in central Ontario." *Water, Air, Soil Pollut.* **15,** 127–152.

Lewis, M.W., and Grant, M.C. (1978). "Sampling and interpretation of precipitation for mass balance studies." *J. Geophys. Res.* **14,** 1098–1104.

Lindberg, S.E., and Harriss, R.C. (1980). "Trace metal solubility in aerosols produced by coal combustion." In *Environmental and Climatic Impact of Coal Utilization,* J.J. Singh and A. Deepak, Eds. Academic Press, New York, pp. 589–608.

Lindberg, S.E., Harriss, R.C., and Turner, R.L. (1982). "Atmospheric deposition of metals to forest vegetation." *Science* **215,** 1609–1611.

McBride, M.B. (1976). "Exchange and hydration properties of Cu^{2+} on mixed-ion Na^{+}–Cu^{2+} smectites." *Soil Sci. Soc. Am. J.* **40,** 452–456.

Navarre, J.-L., Ronneau, C., and Priest, P. (1980). "Deposition of heavy elements on Belgian agricultural soils." *Water, Air, Soil Pollut.* **14,** 207–213.

Peden, M.E., Skowron, L.M., and McGurk, F.F. (1979). "Precipitation sample handling, analysis and storage procedures." Research Report 4 to U.S. Dept. of Energy, Contract No. EY-76-S-02-1199. Illinois State Water Survey, Champaign, Ill.

Rattonetti, A. (1976). "Stability of metal ions in aqueous environmental samples." In *Accuracy in Trace Analysis: Sampling, Sample Handling and Analysis,* Proceedings of the 7th IMR Symposium. National Bureau of Standards Special Publication 422, Gaithersburg, Md.

Sadiq, M., and Zaidi, T.H. (1981). "The adsorption characteristics of soils and removal of cadmium and nickel from wastewaters." *Water, Air, Soil Pollut.* **16,** 293–299.

Schlesinger, W.H., Reiners, W.A., and Knopman, D.S. (1974). "Heavy metal concentrations and deposition in bulk precipitation in montane ecosystems of New Hampshire, USA." *Environ. Pollut.* **6,** 39–47.

Slavek, J., and Pickering, W.F. (1981). "The Effect of pH on the retention of Cu, Pb, Cd, and Zn by clay-fulvic acid mixtures." *Water, Air, Soil Pollut.* **16,** 209–221.

Sotera, J.J., Bancroft, M.F., Smith, S.B., and Corum, T.L. (1981). *Flameless Operations,* Vol. 2 of Atomic Absorption Methods Manual. Instrumentation Laboratory, Lexington, Mass.

Swanson, K.A., and Johnson, A.H. (1980). "Trace metal budgets for a forested watershed in the New Jersey pine barrens." *Water Resour. Res.* **16,** 373–376.

U.S. EPA. (1979). "Methods for chemical analysis of water and wastes." U.S. EPA Report EPA-600/4-79-020. Environmental Monitoring and Support Laboratory, Cincinnati, Ohio.

Wiener, J.G. (1979). "Aerial inputs of cadmium, copper, lead, and manganese into a freshwater pond in the vicinity of a coal-fired power plant." *Water, Air, Soil Pollut.* **12,** 343–353.

Wilber, W.G., and Hunter, J.V. (1979). "Distribution of metals in street sweepings, stormwater solids, and urban aquatic sediments." *J. Water Pollut. Control Fed.* **51,** 2810–2822.

14

IMPACT OF ATMOSPHERIC INPUTS ON THE HYDROSPHERIC TRACE METAL CYCLE

Wim Salomons

Delft Hydraulics Laboratory
Haren Branch
c/o Institute for Soil Fertility
Haren
The Netherlands

1. INTRODUCTION

The influence of increased metal levels due to human activities is found in all parts of the hydrological cycle (e.g., soils, rivers, lakes, estuaries, and the ocean). In this respect the atmosphere plays a unique role. Metals in the atmosphere have a short residence time, estimated to vary between days and weeks (Hidy, 1973). However, within this short time span, they are able to travel large distances and influence all parts of the hydrological cycle (Fig. 14.1). In addition, transition metals act as catalysts in the atmosphere (e.g., the oxidation of sulfur dioxide) (Graedel and Weschler, 1981; Lindberg, 1981).

This review will focus on the influence the atmospheric trace metals have on the first inner ring of Figure 14.1 in particular. The complex interactions

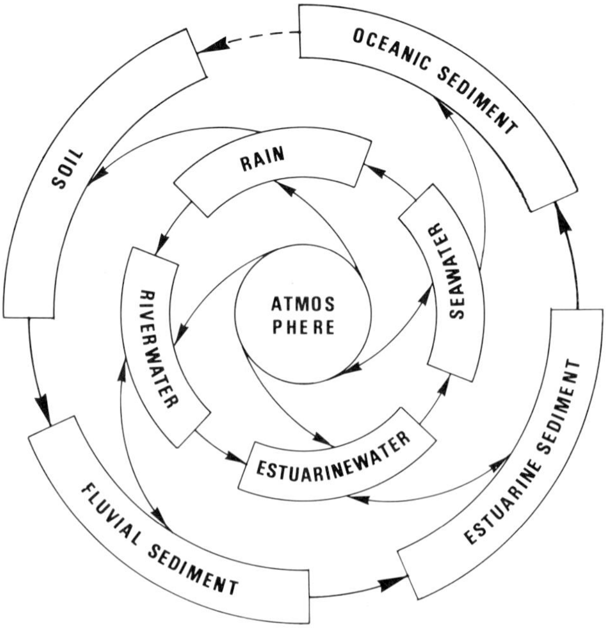

Figure 14.1. The hydrological cycle.

affecting trace metals during their transport in the hydrological cycle and their interactions with solid phases are treated elsewhere in detail (Salomons and Förstner, 1984). In fact, this chapter is partly based on this book. Furthermore, the arctic aerosol and its influence on the composition of arctic precipitation is omitted from this review. This aspect is treated in a Chapter 8 of this volume and elsewhere in more detail (Salomons and Forstner, 1984).

Several sources contribute to atmospheric trace metals. Most of these sources, like forest fires, windblown dust, vegetation, and seasalt sprays, influence the composition of the lower troposphere. The upper boundary of the troposphere, the tropopause, is found at 11–17 km above the earth's surface. Volcanic eruptions may inject particulate metals through the tropopause boundary to the stratosphere. Particles in the stratosphere will be subject to a global distribution, whereas the particles in the troposphere will be transported in the zonal circulation before returning to the earth's surface (Cawse, 1982). Compilations of the amounts of trace metals brought into the atmosphere have been made by Nriagu (1979), Lantzy and Mackenzie (1979), and Weisel (1981). Some estimates of anthropogenic and natural sources of trace metals emitted to the atmosphere are presented in Table 14.1. The data are order-of-magnitude estimates, since some are based on broad assumptions or isolated literature data which have been extrapolated to global emissions. Nevertheless, these data provide an estimate of the relative importance of the various sources.

To illustrate the impact of atmospheric trace metals on the hydrological cycle, a number of selected areas where the atmospheric input is dominant will be discussed. These include (1) terrestrial ecosystems, (2) ocean areas far away from the input of rivers and other point sources, (3) the polar regions, and (4) lakes.

The impact of the hydrological cycle on the composition of the atmospheric aerosol will be discussed in Section 5.2.

Table 14.1. Atmospheric Emissions of Trace Metals ($\times 10^9$ g/yr)[a]

	Natural Sources					
Metal	Windblown Dust	Forest Fires	Volcanogenic Particles	Vegetation	Seasalt Sprays	Industrial Sources
Cd	0.1	0.012	0.52	0.2	0.4	7.3
Cu	12	0.3	3.6	2.5	0.08	56
Ni	20	0.6	3.8	1.6	0.04	47
Pb	16	0.5	6.4	1.6	5	449
Zn	25	2.1	7.0	9.4	10	314

[a] Data are based on Nriagu (1979), except for the Zn, Pb, and Cd inputs from sea spray which are from Weisel (1981).

2. FLUXES AND TRENDS IN ATMOSPHERIC POLLUTION

2.1. Natural and Anthropogenic Emissions of Trace Metals

Three approaches for determining the natural and anthropogenic input of trace metals into the atmosphere have been reviewed by Galloway et al. (1980, 1982):

1. Compare the actual metal emission rates from natural and anthropogenic processes. This is the approach which has been used by Lantzy and Mackenzie (1979).
2. Compare the ratios of atmospheric concentrations to those in the natural sources contributing to it. This approach was used by Duce et al. (1975) in comparing the metal concentrations in the atmosphere with those in crustal material.
3. Determine the temporal trend in the composition of metals in atmospheric deposition. One of these techniques is the use of lake sediments as a historical record for trace metal emissions.

In Table 14.2, natural (soil dust, volcanic dust, and volcanic emanation) and anthropogenic emission rates are presented. Also given is the mobilization factor, which is defined as the ratio of the flux from anthropogenic sources to that from natural sources.

The second approach is the comparison of trace metals in the atmosphere with those in one of its natural sources: crustal material. In this comparison aluminum is used as a reference element. The enrichment factor EF (crust) is defined as:

$$\mathrm{EF(crust)} = (\mathrm{X/Al})\mathrm{air}/(\mathrm{X/Al})\mathrm{crust}$$

where (X/Al)air and (X/Al)crust refer, respectively, to the ratio of the concentrations of metal X to that of Al in the atmosphere and in average crustal material. Data for EF(crust) values for Bermuda and Eniwetok atoll are presented in Figure 14.2. Values of EF(crust) near 1.0 suggest that continental weathering is the likely source of the particles (Duce et al., 1975, Rahn, 1976). Values of EF (crust) larger than about 4.0 are called (anomalously) enriched elements and have some source other than crustal weathering, which may be anthropogenic.

However, several other sources apart from anthropogenic inputs and crustal material contribute to the atmospheric trace metals, for example, low temperature volatilization. Particles released by vegetation and coal and fly ash have relative concentrations of trace metals similar to average crustal material (Bertine and Goldberg, 1971). Enrichment factors must therefore be used cautiously in attempts to ascertain atmospheric trace metal sources (Duce et al., 1976).

Table 14.2. Natural and Anthropogenic Emissions of Trace Metals in the Atmosphere and the Mobilization factor[a]

	Emissions (10^8 g/yr)		
	Natural	Anthropogenic	Mobilization Factor
Ag	0.6	50	83
As	28(210)	780	3.3
Cd	2.9	55	19
Co	70	50	0.71
Cr	580	940	1.6
Cu	190	2600	13
Hg	0.4(250)	110	0.44
Mn	6100	3200	0.53
Mo	11	510	45
Ni	280	980	3.5
Pb	40	4000	100
Sb	9.8	380	39
Se	4.1(30)	140	4.7
Sn	52	430	8.3
V	650	2100	3.2
Zn	360	8400	23

[a] Based on Lantzy and Mackenzie (1979) and Galloway et al. (1980). The values in brackets refer to vapor emissions of the volatile species (As, Hg, and Se) from land and sea. The data on Pb are from Ng and Patterson (1981).

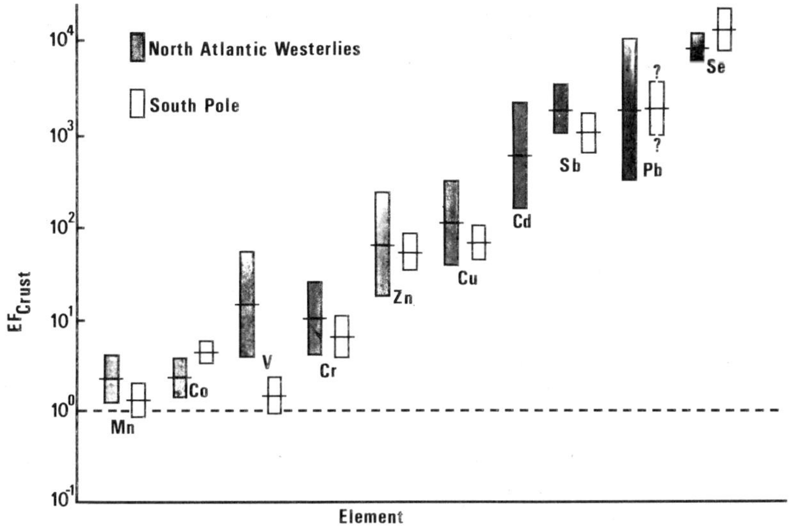

Figure 14.2. EF (crust) values for metals in the atmosphere at Bermuda and the South Pole (Duce et al., 1975).

Table 14.3. Expected Enrichment for Trace metals in the Atmosphere and in Atmospheric Deposition[a]

	Expected Enrichment in Atmospheric Deposition[b]			
Techniques	Low	Moderate	Large	No Data
Mobilization factor[c]	Co, Mn, Hg	As, Cr, Ni, Se V	Ag, Cd, Cu, Mo, Pb, Sb, Se, Sn, Zn	Be, Te Tl
Enrichment factor	Co, Mn Ni	Cr, V	Cd, Cu, Pb, Sb, Se, Zn	Ag, As, Be, Hg Mo, Sn Te, Tl
Historical factor	Co, Mn, Ni, Be	Cr, V, Cu, Ag, Zn, Se	As, Cd, Pb, Sb	Mo, Sn, Te, Tl, Hg

[a] From Galloway et al. (1982).

[b] Low = < 2 × enrichment; moderate = 2 to 4 ×; large = > 4 ×.

[c] The MF is based on the comparison of global emission rates. On a reduced scale, such as for the United States, the relative order will change; for example, Hg, As, and Se would be expected to be in categories of higher enrichment.

A comparison of the enrichment of atmospheric trace metals as derived from the mobilization factor, the crustal enrichment factor, and dated lake sediments has been made by Galloway et al. (1982). Their results are presented in Table 14.3. The results show that the elements Ag, Cd, Cu, Pb, Se, V, As, and Cr have highly elevated concentrations in the atmosphere. For the eastern United States the rates of atmospheric deposition of Ag, Cd, Cu, Pb, Sb, V, and Zn are strongly influenced, if not controlled, by anthropogenic processes (Galloway et al., 1980).

2.2. Metal Concentrations in Urban, Rural, and Remote Atmospheres

Trace metal concentrations at different localities may vary by orders of magnitude in the atmosphere. Low concentrations are found at remote sites and over the oceans, whereas the highest concentrations are observed at urban sites. Annual average concentrations at a rural site in the United Kingdom (Wraymires) are compared in Table 14.4 with levels recorded at background European (Jungfrau), remote South Pole, and city atmospheres (Cawse, 1981). The results show the strong maritime influence (high concentrations of Cl, Na, K, and Mg) for the nonurban site in the UK compared with the other nonurban sites. A comparison with city atmospheres shows that urbanization and industry have resulted in order-of-magnitude increases in air concentrations of Br, Co, Cr, Cs, Ni, Pb, Sb, and Zn (Cawse, 1981).

Table 14.4. Concentrations of Elements in Air at Nonurban and Urban Locations (ng/kg air)

	Non urban			Urban	
Element	Wraymires, Lake District, UK, 1980	Jungfrau Central Europe Inland Background	S. Pole Remote Site	20 Sites Average in UK	Central Swansea
Al	105	42	0.67	6.4	370
As	2.4	0.19	0.006	6.4	15
Br	17	1.1	1.1		320
Ca	310		0.40		
Cd	<1	0.4		2.8	
Ce	0.19				0.80
Cl	1450	5.9	2.1		4600
Co	0.11	0.037		1.4	4.5
Cr	1.2	0.29	<0.03	14	6.1
Cs	0.024	0.012	0.024		0.27
Cu	16	0.72	0.51	19	57
Fe	96	29		680	940
Hg	0.047	0.024			
I	1.6	0.22	0.068		
In	0.027	0.0008			<0.7
K	600	16	0.56		
Mg	160	8.2	0.59		
Mn	4.0	1.2		31	25
Mo	<0.4	0.24			
Na	710	18	2.7		1960
Ni	<3			13	66
Pb	38	3.6		340	500
Sb	0.79	0.16		7.3	4.0
Sc	0.031	0.0063			0.16
Se	0.81	0.034	0.005		2.7
Ti	<7	2.0	0.08	39	
V	3.7	0.24		17	21
Zn	19	8.1	0.027	260	310
Reference	Cawse, 1981	Dams and De Jonge 1976	Maenhaut and Zoller, 1977	McInnes 1979	Pattenden, 1974

Note: 1.0 m^3 air at 15°C, 760 mm Hg (Standard Cubic Metre) = 1.226 kg

Galloway et al. (1982) made an inventory of all available data on atmospheric metal concentrations in rural, remote, and urban areas. To show the relative contribution of anthropogenic sources on various elements, the ratios of urban/remote and rural/remote are presented in Figure 14.3. The ratio of urban/remote follows the order Zn>Pb>Cu>Mn>Co>Sb>Cr>As>Ag>Ni>V>Se>Hg>Cd. The ratios for rural/remote follow a similar order, but are about one order of magnitude lower. An exception is cobalt, which shows very little enrichment. It seems probable that this is due to the high MMD (mass median diameter) and consequently high washout ratio near its sources.

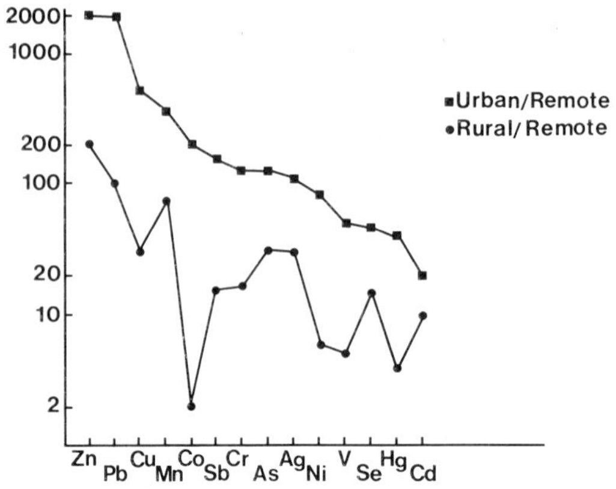

Figure 14.3. The ratio of metal concentrations in urban/remote and rural/remote atmospheres (Galloway et al., 1980, 1982).

2.3. Trends in Atmospheric Pollution

It is known that the atmosphere has been contaminated with lead since about 4500 B.P. (Fig. 14.4) when technology for smelting lead sulfide ores and cupellation of silver from lead were developed in Southwest Asia (Patterson 1971). Until 2600 B.P. the annual production of lead was about 200 t/yr, while it rose to 10,000 after the discovery of the use of silver coinage in the eastern region of the Mediterranean (Patterson 1972). During the period 2100 to 1800 B.P., the Romans mined and smelted lead at a rate of about 80,000 t/yr and during this period the concentrations in the atmosphere probably increased about fivefold above natural levels (Patterson, 1972). Around 1940 there was a sharp increase in atmospheric lead concentrations due to lead emissions from automobiles (Murozimi et al., 1969). The present input of lead into the atmosphere is estimated at 400,000 t/yr (Ng and Patterson, 1981).

Information on long-term trends and seasonal fluctuations in trace metal concentrations in the atmosphere are available for the United Kingdom. The seasonal fluctuation in absolute concentrations of elements in air is shown in Fig. 14.5. Bromine showed the greatest increase in winter, to almost three times its concentrations in summer. Increases for As, Cr, Pb, Sb, V, and Zn are 11- to 2-fold compared with the summer period. This variation is attributed to increased combustion of fossil fuels and the persistence of inversion layers in winter (Cawse, 1981). The long-term records at the various stations (Fig. 14.5) show that downward trends are evident for several elements over the last 8–10 yr. Downward trends are found for Al, Br, Co, Cr, Mn, Pb, Sc, V, and Zn for all sites. The downward trend is most marked for zinc. A decrease in industrial emissions owing to the economic recession is considered the most likely cause of lower trace element concentrations in air (Cawse, 1981). The deposition of trace metals from the atmosphere depends

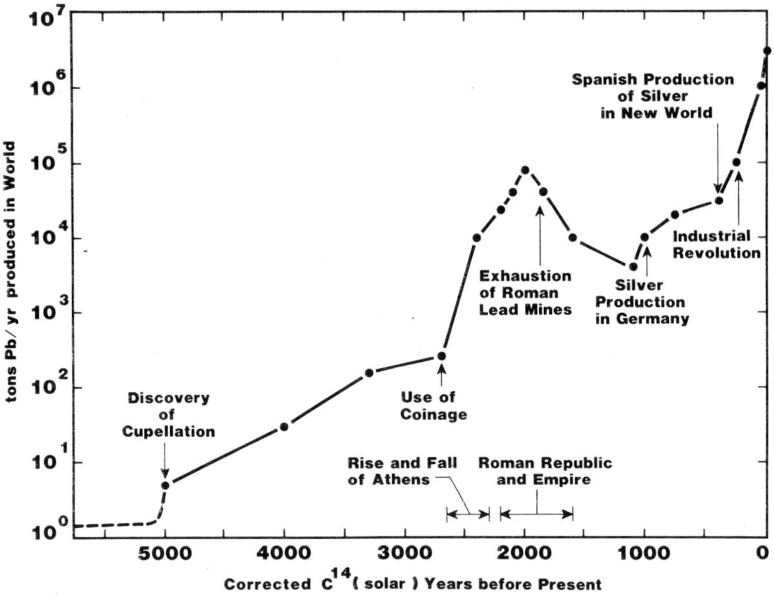

Figure 14.4. Historical development in the use of lead during the last 5500 yr (Settle and Patterson, 1980).

on dry versus wet deposition and on local atmospheric conditions. Trends for trace metals in urban air are presented in Figure 14.29.

3. COMPOSITION OF THE ATMOSPHERIC AEROSOL RELEVANT FOR PROCESSES IN THE HYDROLOGICAL CYCLE

Information on the speciation of trace metals in solids can be obtained with selective leaching techniques (Salomons and Förstner, 1984). These techniques have recently been used to determine the mode of occurrence of trace metals in urban particulates (Harrison et al., 1981). Standard Reference Material 1648, Urban Particulate Matter, was analyzed with a sequential extraction procedure by Lum et al. (1982) (Fig. 14.6). The results show high proportions of soluble metals, especially for Zn, Pb, and Cd. Cobalt and Cr are more tightly bound in the residual fraction. Lindberg and Harris (1983) studied the relationship between solubility and particle size in detail (Fig. 14.7).

All elements exhibit a trend of increasing mean solubility as particle size decreases (see also Fig. 14.26). This high solubility of trace metals in atmospheric particulates is a general phenomenon (Wallace et al., 1977; Crecelius, 1980; Gatz et al., 1983; Hodge et al., 1978; Ochs and Gatz, 1980) and shows that deposition on wet surfaces (e.g., leaves) and in surface waters will result in a release of trace metals to the ecosystem.

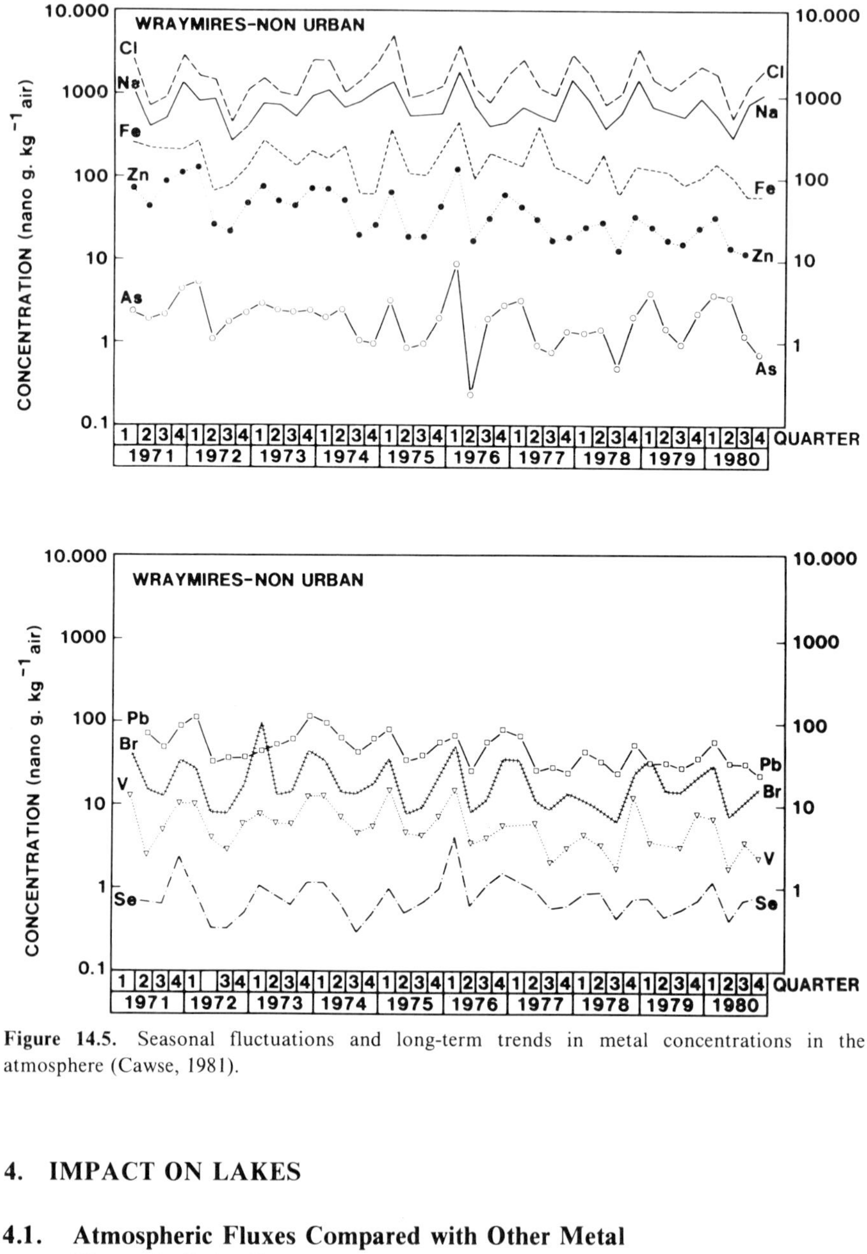

Figure 14.5. Seasonal fluctuations and long-term trends in metal concentrations in the atmosphere (Cawse, 1981).

4. IMPACT ON LAKES

4.1. Atmospheric Fluxes Compared with Other Metal Fluxes in Lake Systems

Sources for trace metals in lakes are the atmosphere, riverine inputs, and various waste discharges (Fig. 14.8). Metals are introduced in lakes both in solution and in particulate form. The allochtonous particles will partially settle to the bottom, whereas the dissolved trace metals are subject to removal processes by adsorption, uptake by biota, and incorporation in authigenous

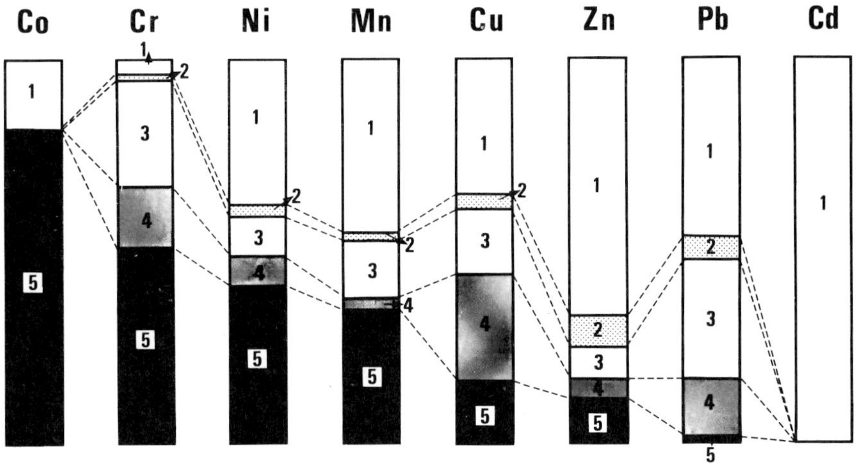

Figure 14.6. Mode of occurrence of trace metals in Standard Reference Material 1648 (Urban particulate matter). (Drawn after Table 2 in Lum et al., 1982). 1, Exchangeable fraction; 2, surface oxide and carbonate-bound fraction; 3, associated with Fe–Mn oxides; 4, organically bound fraction; 5, residual fraction.

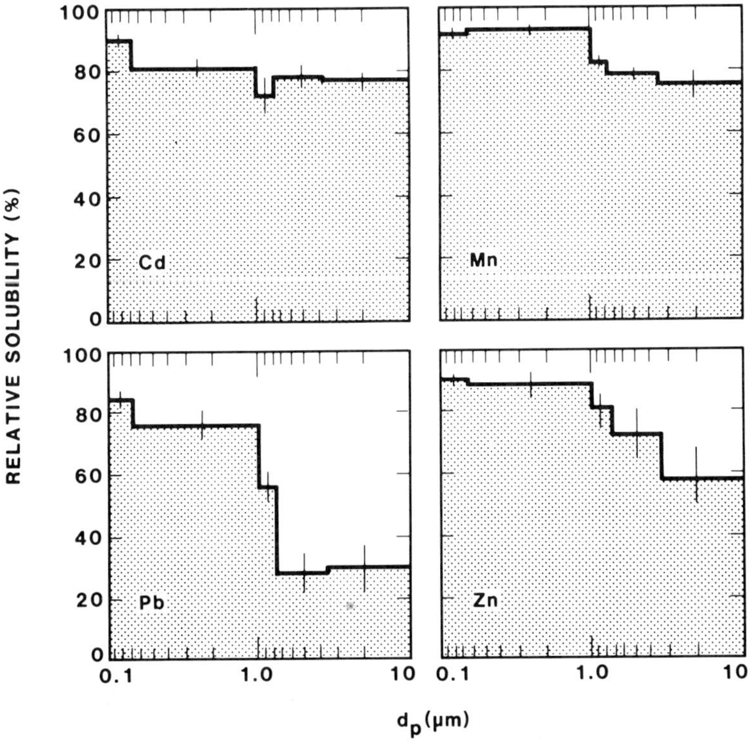

Figure 14.7. Relationship between relative solubility (water soluble concentrations X 100/water soluble plus dilute acid leachable concentrations) and particle diameters for cadmium, manganese, lead, and zinc (Lindberg and Harris, 1983).

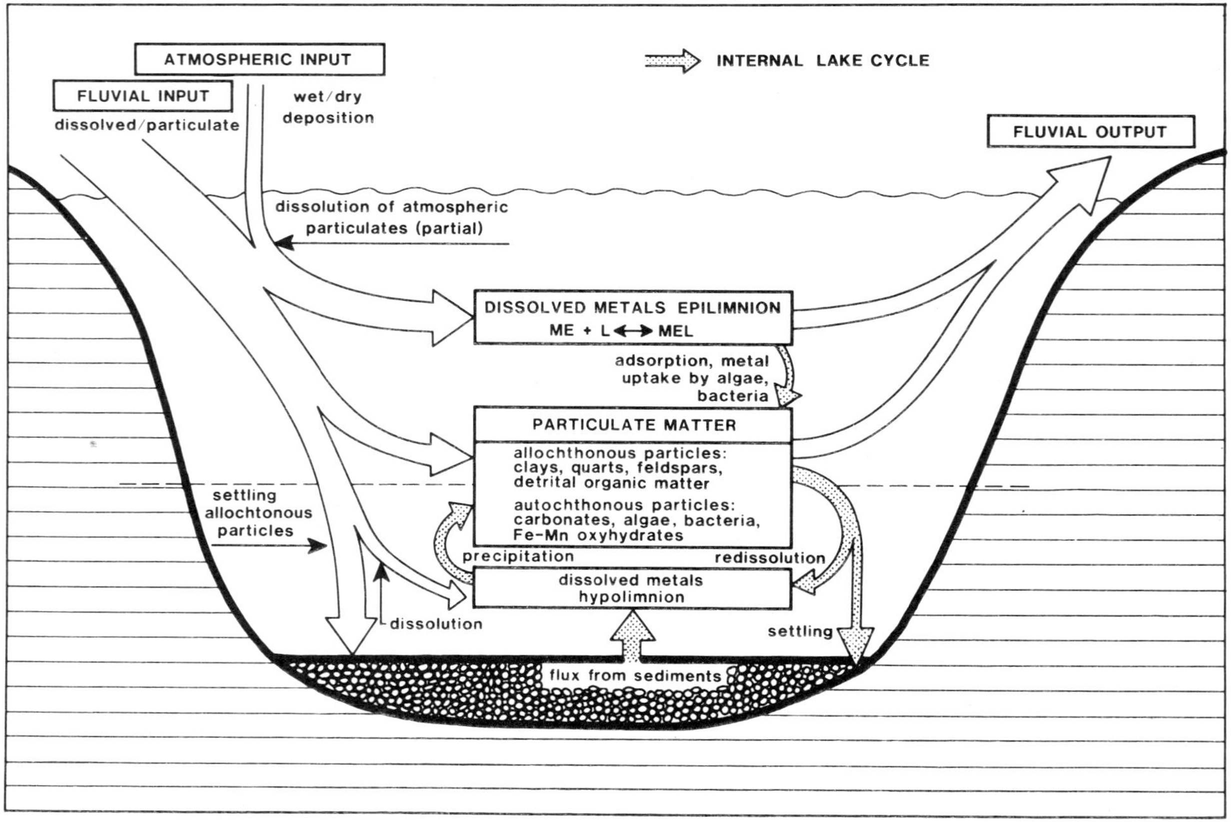

Figure 14.8. Processes affecting trace metals in lakes.

phases. In the oxic epilimnion, there is a strong interaction between the carbon cycles and the metal cycle. The production of algal material causes an increase in pH and reprecipitation of calcium carbonate, thus causing removal of trace metal by their incorporation in algal tissues and in new mineral phases and by providing active surfaces for adsorption. During their fall through the water column, a partial dissolution of the trace metals may take place in the same way as in the oceans.

In the lakes with an anoxic hypolimnion there is a strong interaction between the metal and the redox cycle. In anoxic waters, a strong redissolution of iron and manganese takes place, and the metals diffuse upward and precipitate at the oxic/anoxic interface in the lake. The resulting hydrous iron and manganese particles are able to remove dissolved metals. Detailed examples of the complicated processes affecting trace metals in lakes are published elsewhere (Salomons and Förstner, 1984). In this chapter we will discuss atmospheric inputs in relation to other trace metals inputs in lakes and the interaction between acid and metal loadings of lakes.

Natural and anthropogenic fluxes for Lake Erie have been compared by Nriagu et al. (1979). The present flux of metals to the sediments is calculated as the product of sedimentation rates and metal concentrations in the top 1.0-cm layer. The anthropogenic component was derived by subtracting the precolonial flux (below the ambrosia horizon, ~1850 A.D.) from the total elemental flux to the surficial sediments. Because of the higher overall sedimentation rates, the average anthropogenic flux of metals into the eastern Basin sediments is higher than those entering the other two lake basins. As shown in Table 14.5, the anthropogenic input of zinc, lead, and cadmium into the western Basins exceeds that for precolonial sediments fivefold, whereas in the central and eastern Basins the anthropogenic fluxes of these metals are approximately two to three times greater than the natural inputs. Metal pollution in Lake Erie originates from tributary inflow, coastal runoff, industrial and domestic effluent discharges, atmospheric fallout, shipping operations, and the recreational use of the lake. The inventory of the sources and sinks of metals in the lake agrees, give or take one order of magnitude, with the loading rates recorded in the sediments (Table 14.5; Nriagu et al., 1979). It is interesting to note that direct atmospheric inputs account for 8, 34, and 13% of the Cu, Pb, and Zn, respectively, delivered annually to the lake. The annual contributions of Cu, Pb, and Zn from sewage effluents are 18%, 15%, and 11%, respectively.

Anthropogenically derived changes in the sedimentary flux of metals has been discussed by Rippey et al. (1982) for dated cores from Lough Neagh, Northern Ireland, and from comparison with other lacustrine situations. The alteration of a lake's sedimentary regime, as a result of deforestation or other similar agricultural change in the catchment, has been recorded as increased sedimentary K, Mg, and Na concentrations, and by increased accumulation rates. The Mg profile of cores B41 from Lough Neagh (Fig. 14.9) indicates that a small but detectable increase in the erosion rate occurs at 52–54 cm

Table 14.5. Inventory of Sources and Sinks of Heavy Metals in Lake Erie and Their Residence Times[a]

	Flow Rate ($\times 10^{-3}$)(kg/yr)[b]			
Source	Cadmium	Copper	Lead	Zinc
Detroit River (import from upper Lakes)		1660	630	5220
Tributaries, USA		100	52	271
Tributaries, Ontario		31	19	140
Sewage discharges	5.5	448	283	759
Dredged spoils	4.2	42	56	175
Atmospheric inputs	39	206	645	903
Shoreline erosion	7.9	190	221	308
Total, all sources		2477	1906	7776
Export, Niagara river and Welland Canal		1320	660	4400
Retained in sediments		1157	1246	3776
Retained in % of input		50%	65%	35%
Residence time (days)		104	180	152

[a] From Nriagu et al. (1979).

[b] Unless otherwise noted.

(corresponding to 1670 A.D.) and stabilizes at 44–46 cm. Above 24–26 cm (~1880 A.D.) Mg concentrations decrease again, suggesting a change in the erosion regime. There is, however, no evidence for any major land-use changes in the last 100 yr. The increase in erosion rates above 52–54 cm not only produces increasing Mg concentrations but also relates to increasing Cu and Pb concentrations (Fig. 14.9). Above 24–26 cm (around 1880), Cu, Pb, Zn, and Hg concentrations increase toward the surface and suggest that contamination of the sedimentary material is occurring. Phosphorous, on the other hand, behaves differently to trace metals (Fig. 14.9) and appears to reflect the increasing P loadings to the lake over the last 100 yr or so as a result of sewered population increase and the introduction of phosphate-rich detergents. Calculations of recent net sedimentary and background fluxes show that background atmospheric flux could be a significant component for Cu, Zn, Hg, and Pb in this area (Table 14.6).

The importance of the atmospheric input of trace metals in lakes is clearly shown by the data in Table 14.6. A characteristic example depicts the atmospheric input of trace metals in lake Michigan compared with a number of other sources in Fig. 14.10 (Eisenreich, 1980).

The atmospheric loading is especially important for lead (60%) and is a significant source for zinc (33%); for copper it contributes 13% and for

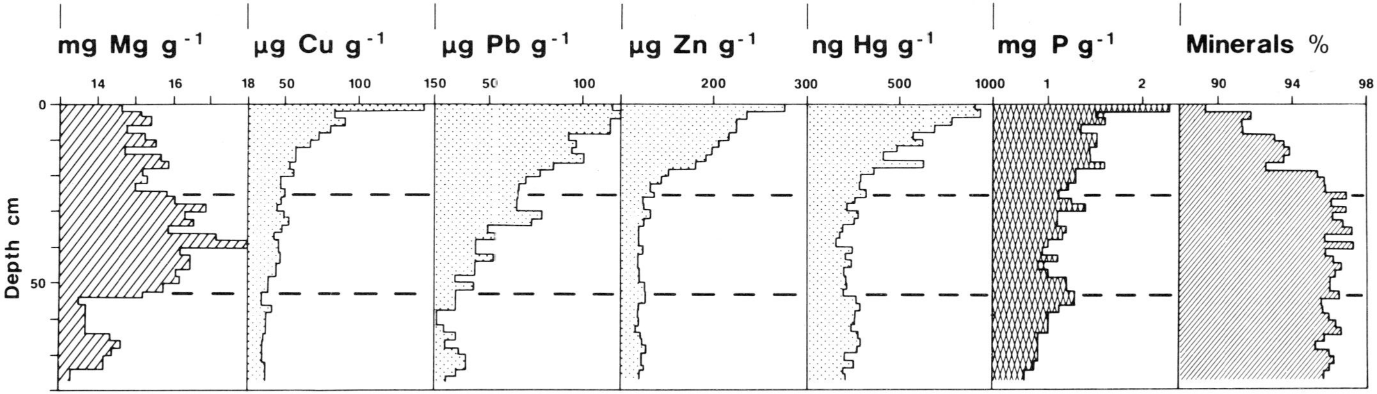

Figure 14.9. Metal concentrations along sediment cores in Lough Neagh, North Ireland (Rippey et al., 1982).

Table 14.6. Selection of Data on the Sedimentary Fluxes in Lakes (mg/m^2 yr)

Lake	Cr	Ni	Cu	Zn	Cd	Hg	Pb	V
Lake Huron[a]								
Anthropogenic			2–14	7–57		0–0.05	8–42	
Natural			4–8	12–17		0.01–0.04	5–8	
Lake Ontario[a]								
Anthropogenic			22–33	109–217	0.9–3	0.2–1.4	63–169	
Natural			13–17	29–48	0.3–0.4	0.01–0.05	8–12	
Lake Erie[b]								
Anthropogenic			1.5–120	10–362	0.1–5.4		2–118	
Natural			5.9–78	6.8–148	0.1–1.6		3.8–43	
Lake Neagh[c]								
Anthropogenic	84	58	61	164	1.8	0.5	72	
Natural	61	56	15	47	1.2	0.08	12	
Atmospheric	3	8	23	75		0.3	27	
Woodhull Lake[d]								
Anthropogenic	11		6.3	65	1.2		56	12.6
Natural	1.1		2.7	11	0.05		5.4	5.4
Atmospheric	3.6		13	50	1.2		25	16
Windermere[e]								
Anthropogenic			18	169		0.49	99	
Natural			12	75		0.05	26	
Atmospheric			32	120		<0.24	55	

[a] From Kemp and Thomas (1976).
[b] From Nriagu et al. (1979).
[c] From Rippey et al. (1981).
[d] From Galloway and Likens (1979).
[e] From Hamilton-Taylor (1979).

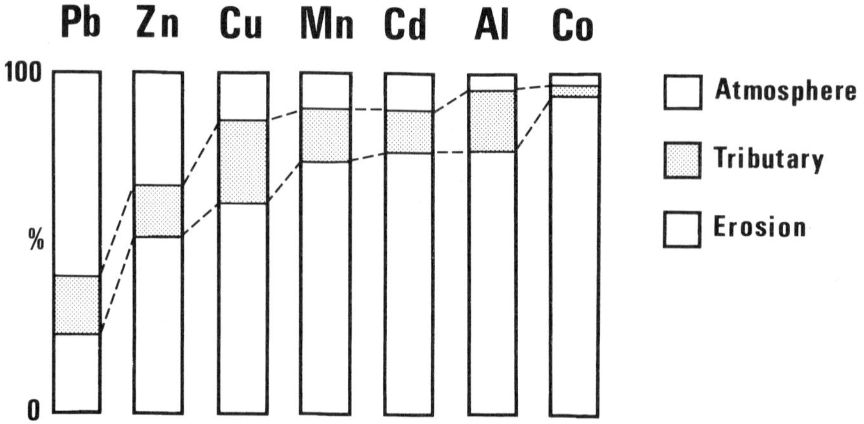

Figure 14.10. Relative importance of the input of trace metals in Lake Michigan by various sources. Based on Eisenreich (1980).

cadmium and manganese, 11%. The atmosphere is a minor source for aluminum, iron, and cobalt. This correlates with the data of Sievering et al. (1980, 1981) who have studied the input of atmospheric trace metals over Lake Michigan in detail (see also Winchester and Nifong, 1971; Gatz, 1975). More than 90% of calcium and magnesium are lake derived, and aluminum is almost entirely soil derived. It is estimated that over 95% of the lead, 75% of the zinc and manganese, and more than 50% of iron are of anthropogenic origin (Sievering et al., 1980).

Sievering et al. (1981) compared the dry deposition loading with wet precipitation and surface runoff (Table 14.7). They found that the atmospheric input by dry deposition contributes 75% of the total loading for Pb and 50% for the total loading of zinc to the lake ecosystem. The interaction between the dry deposition and the lake waters is highly complex due to processes in the surface microlayer (see Sec. 5.2.2). Moreover, part of the particles are highly soluble and contribute in this way to the dissolved trace metal loading of the lake.

The first profiles of dissolved and particulate trace metal concentrations in deep lakes have recently been obtained for the Bodensee (Germany) (Sigg et al., 1982). The results for cadmium, lead, and copper are presented in Fig. 14.11. The results are similar to profiles obtained for ocean systems: a removal of trace metals in the surface layers by biological activity and partial decomposition during settling of the authigenous particles to the bottom (compare with Figs. 14.21 and 14.23). However, for the Bodensee, the high atmospheric input results in high levels of copper and lead in surface layers. Although the input of trace metals in lakes are one or two orders of magnitude higher than that of the oceans (Table 14.8), the metal concentrations are of the same order of magnitude. Therefore apparently highly efficient removal processes are operating in lakes (Sigg et al., 1982).

Table 14.7. Annual Loading Estimate to Southern Lake Michigan by Dry Deposition Compared to Wet Deposition and Surface Runoff[a]

	Loading (t/yr)		
Element	Dry Deposition	Precipitation	Total Lake Runoff
Al	550		
Ca	2000		
Fe	1100	950	1450
Mg	700		
Mn	60	55	450
Pb	500	90	100
Zn	200	50	180

[a] From Sievering et al. (1981).

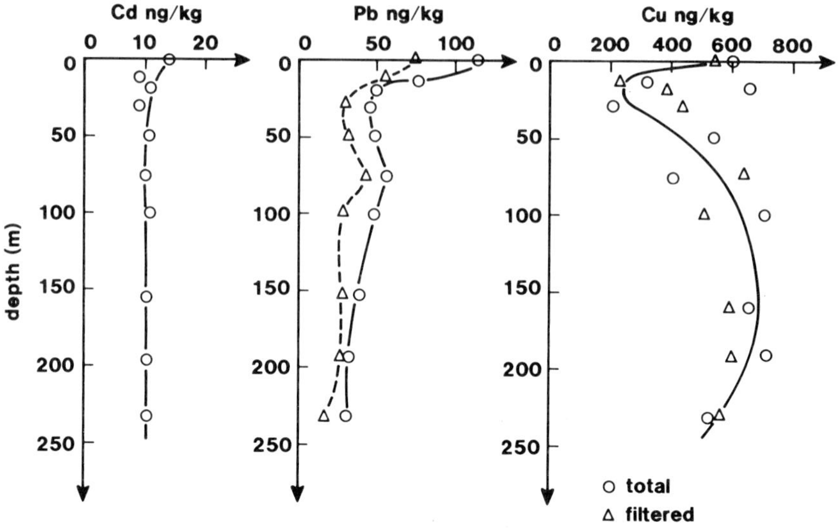

Figure 14.11. Profiles for trace metals in the Bodensee (Sigg et al., 1982).

Table 14.8. Inputs, Deposition Rates, and Concentrations of Heavy Metals in the Bodensee Compared with the North Atlantic Ocean[a]

	Cu	Pb	Cd	Zn
Bodensee				
Atmospheric input (ng/cm^2 yr)	714	11000	20	8400
Deposition (ng/cm^2 yr)	6500	9500	100	36000
Concentrations in the lake (ng/L)	300–800	50–100	6–20	1000–4000
North Atlantic				
Atmospheric input (ng/cm^2 yr)	25	310		130
Deposition (ng/cm^2 yr)	234	330		1040
Concentrations in the ocean (ng/L)	100	3	10	100
	(30–300)	(1–15)	(1–120)	(10–600)

[a]From Sigg et al. (1982).

4.2. Trace Metals and Acidic Rain

The acidic deposition from the atmosphere affects the pH of soft water lakes. Thousands of lakes and streams in southern Scandinavia and eastern North America and Canada are affected by it; fish populations are lost and other vital ecosystem components and processes are disturbed. With regard to heavy metal behavior, three processes can be distinguished:

the increased leaching of trace metals from soils in the drainage basin causing elevated metal concentrations in the receiving waters;

the changes in the distribution of trace metals over the particulate and dissolved phase in lakes (e.g., more trace metals will be leached from atmospheric particulates entering the lake); and

the active leaching of trace metals from sediments in the lake.

Changes in pH affect the distribution of trace metals over the dissolved and particulate phase; however, the actual changes in the lake depend to a large extent on its buffering capacity. Kramer (1976) used the carbonate alkalinity of lakes to access their susceptibility to acid input. In Fig. 14.12 this

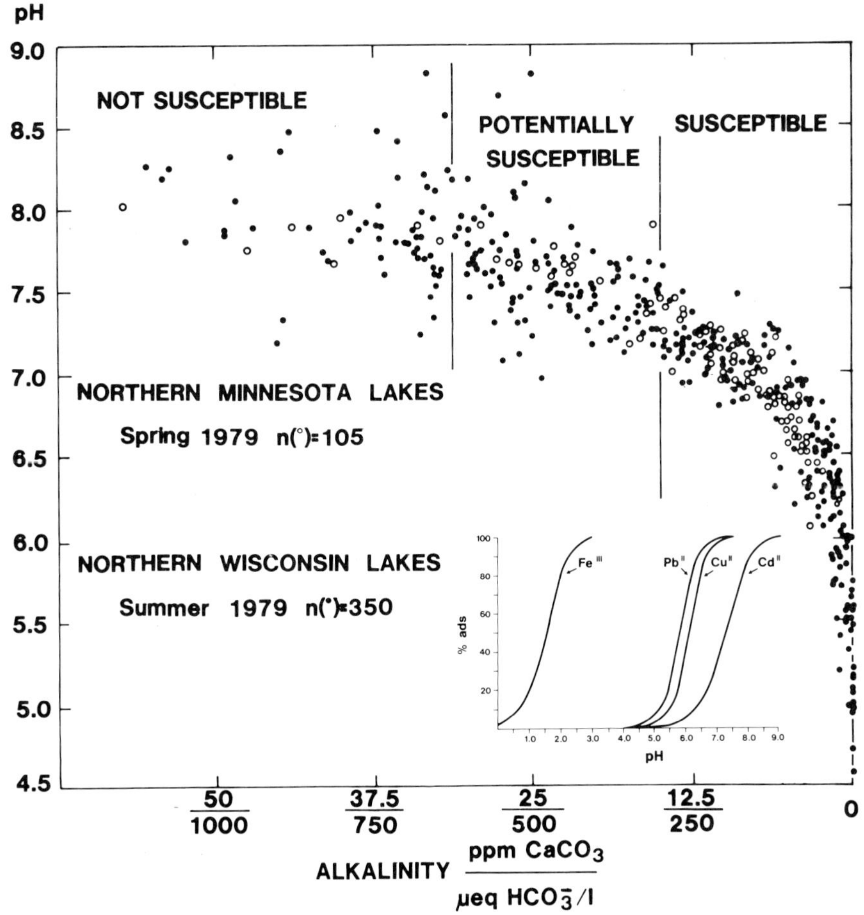

Figure 14.12. A, Relationship between pH and alkalinity in 105 northern Minnesota lakes and 350 northern Wisconsin lakes (Glass et al., 1980), and B, the adsorption of trace metals on silica as a function of the pH.

classification is superimposed on the relationship between pH and alkalinity for Minnesota and northern Wisconsin lakes (Glass et al., 1980).

Also shown in Fig. 14.12 is the relationship between adsorption of trace metals on silica and the pH. This shows that a small shift in pH can cause a large shift in adsorption equilibria. Research on acid mine drainage and trace metal behavior has shown that pH is important in predicting the geochemical behavior of trace metals.

Cameron and Ballantyne (1975) studied the fate of dissolved lead and zinc originating from oxidizing mineral deposits during their flow in a river and lake system in a permafrost region of northern Canada. With an increase in pH the dissolved metal concentrations decrease and those in the sediments increase. As is expected from Figure 14.13, the reduction for lead is stronger

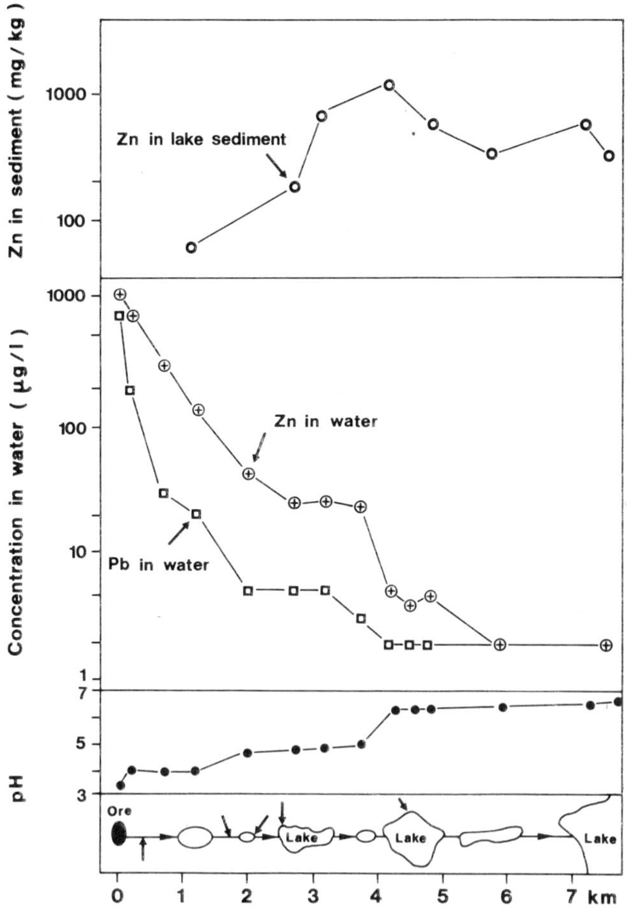

Figure 14.13. Decay of zinc and lead concentrations in lake and stream water, Northwest Territories, Canada, and corresponding increase of Pb and Zn in lake sediments (from Cameron and Ballantyne, 1975; after Rose et al., 1979).

Table 14.9. Metal Concentrations ($\mu g/L$) in Acidifying and Nonacidified Lakes in Scandinavia and in North America[a]

	Al	Cd	Mn	Zn
Lakes with pH 6.0–7.8	<50	<0.1	<100	<30
Lakes with pH 4.1–5.3	up to 600	0.6	400	120

[a] From Haines (1981).

than for zinc. Similar observations on the distribution of trace metals over the dissolved and particulate phases as a function of the pH have been observed for the IJsselmeer, Netherlands (Salomons and Mook, 1980). They found that an increase of 0.5 pH unit was sufficient to cause an almost complete removal of dissolved cadmium from river water entering the lake. In acidifying lakes the reverse may be expected. Haines (1981) made a survey of the literature on reported metal concentrations and the pH in lakes. Some of his results are presented in Table 14.9.

In acidified lakes, metal concentrations are significantly higher than in nonacidified lakes. Large changes are found for the dissolved Mn and Al concentrations in particular. In Swedish lakes a pronounced correlation is found between dissolved metals levels and pH (Dickson, 1980; Borg, 1983). Some of the observed correlations are presented in Figure 14.14.

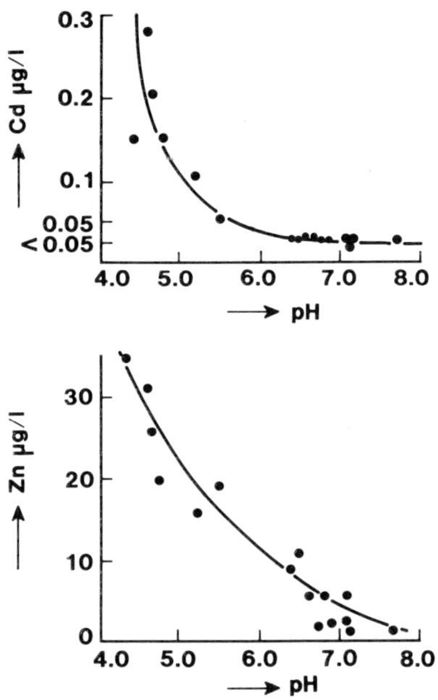

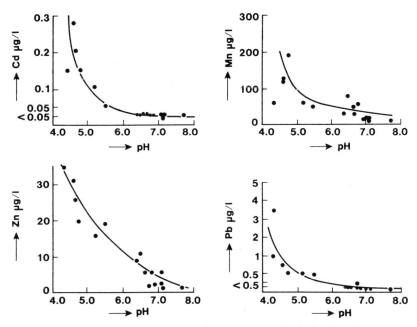

Figure 14.14. Metals in 16 lakes on the Swedish west coast with similar deposition but with a different pH (Dickson, 1980).

Davis et al. (1982) studied the influence of acidic precipitation on the release of lead from sediments. Compared with the results on pure mineral particles (Fig. 14.12), the adsorption edges were shifted to lower pH values. Significant desorption only occurred at pH values <3.0 in sediments from Woods Lake and <2.0 in sediment from Lake Sagamore.

The input of acidity in lakes is not a continuous process, especially in spring where the melting of snow results in large influxes. Johannessen and Henriksen (1978) showed, with laboratory and lysimeter experiments, that 50–80% of the pollutant load is released with the first 30% of the meltwater. The average concentrations of pollutants in this fraction is 2–2.5 times the concentration in the snowpack itself. This phenomenon has been attributed to ion separation during melting (Hultberg, 1977) or to a freeze-concentration process (Johannessen and Henriksen, 1978).

Norton et al. (1981), in a study of New England lakes, observed a decrease in zinc concentrations from near-recent to present sediment in lakes with pH values <5.5, which was explained by a leaching of zinc from the sediments by acidic deposition. Three processes were suggested for this release.

Ion exchange:

$$ZnX_2 + 2H^+ \rightarrow Zn^{2+} + 2HX \text{ (where X is a cation-exchange surface)}$$

Solution of a host (solid solution phase):

$$(Zn,Mn)O(OH) + H^+ \rightarrow Mn^{2+} \text{ (dominant)} + Zn^{2+} + e^- + H_2O$$

Congruent solution of a Zn-phase:

$$ZnS + 2H^+ = Zn^{2+} + H_2S$$

Figure 14.15 (from Reuther et al., 1981) presents data from two lakes in Southern Norway and shows the different behavior of zinc at different pH values. Lake Hovvatn had, before lime treatment in 1980–1981, a pH of 4.4 and had been barren of fish since the 1940s. Upon comparison with Lake Langtjern (pH 4.95), it is evident that zinc is remobilized from the upper sediment layers at that degree of acidity, mainly from the easily reducible fractions. A significant decrease of the reducible fractions is also observed for cadmium, cobalt, and nickel, whereas lead and copper are affected by these changes only to a small, insignificant degree. The fact that these effects are not discerned in Lake Langtjern indicated that an increase in pH of half a point (0.5) may be enough to reduce the mobility of these elements (see also insert in Fig. 14.12).

Increased inputs of trace metals in lakes from the atmosphere and their subsequent incorporation in bottom deposits will mainly affect benthic organisms. However, the combined input of metals and acids will affect both benthic and pelagic organisms because the acid will cause a lowering of the

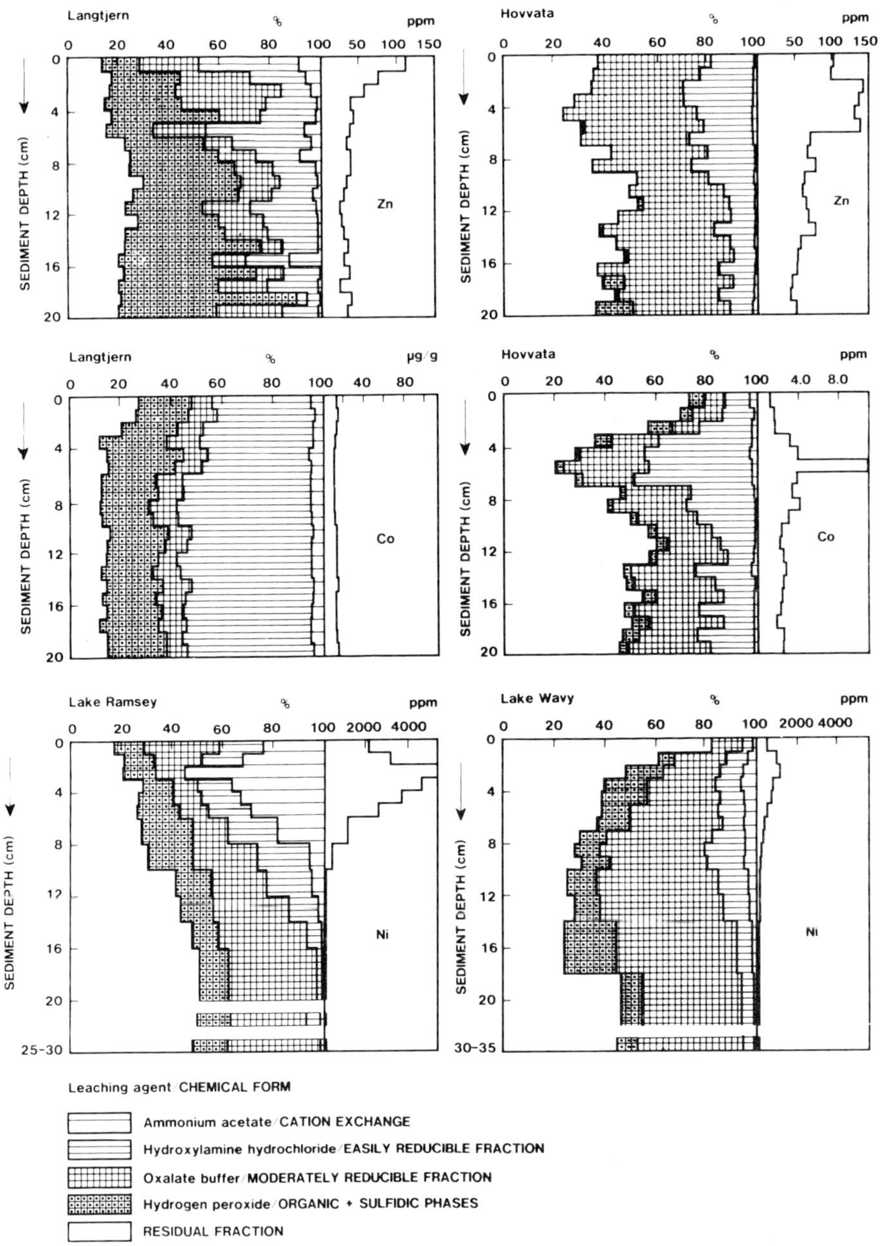

Figure 14.15. Historical evolution of chemical forms of trace metals in lake sediments of areas affected by acid precipitation. Above zinc and cobalt in sediments of Hovvatn and Langtjern, Norway. Below nickel in sediment cores from lake Ramsey and lake Wavy in the Sudbury mining and smelting area, Ontario (Reuther et al., 1981).

pH and metals will remain in solution and/or will be solubilized from the sediment.

5. IMPACT ON THE MARINE ENVIRONMENT

5.1. Introduction

The oceans are the last part of the hydrological cycle, and here trace metals are removed from the hydrological cycle and incorporated into the sediments. There they spend several hundred millions of years before taking part in the next hydrological cycle.

Roughly three sources contribute to the input of trace metals into the world oceans: (1) rivers, (2) the atmosphere, and (3) hydrothermal inputs from active ridges.

River inputs are retained to a large extent in the estuaries and in the coastal zone (Salomons and Förstner, 1984). The atmospheric inputs are less known than the fluvial inputs, especially with regard to the complex processes at the sea–air interface and the origin of the oceanic aerosol. Little information is available on the influence of active ridges on the supply of trace elements to the world oceans.

The processes affecting trace metals in the oceans can be schematized with box models (Broecker, 1974; Lerman, 1979). A four-box model which includes the sediments is shown in Figure 14.16.

The four boxes in the model include:

1. The Surface Mixed Layer. This box receives trace metals from a number of sources: the riverine input, the atmospheric input, and input from upwelling water.

The rate of upwelling is about 4 months/yr, which is 40 times the water discharge of the rivers (0.1 month/yr). For metals which have higher concentrations in the deep water compared with the surface water, upwelling is a major source. In the surface layer, biological processes take place which result in the formation of particulate matter. Trace metals are incorporated in the particulate matter or are adsorbed on it.

Metals are removed from the surface layer by the sinking particles and through downwelling of water.

2. The Deep Layer. This layer is subject to a continuous passing through of particulate matter, causing adsorption of some trace metals on it. On the other hand, part of the biogenic particulate matter decomposes releasing trace metals. Since the oceanic stirring time is about 1600 yr and the residence time of the large particulates only days to weeks and those of the small particles about 50–100 yr, the particles continuously provide the deep layer with trace metals whereas for some trace metals scavenging takes place.

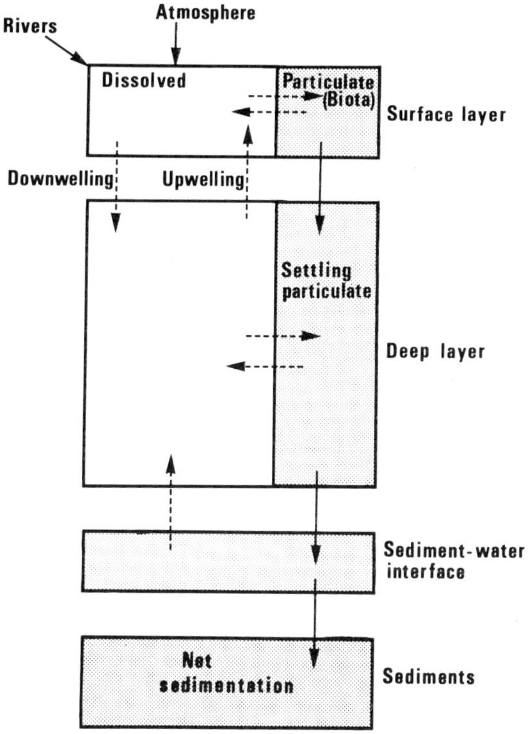

Figure 14.16. Four-box model of the ocean.

3. *The Sediment–Water Interface.* The composition of particulate matter recovered or sampled in sediment traps above the ocean floor differs from that of the sediments. In particular, the organic matter concentrations in the sediment traps are higher. This shows that at the sediment–water interface important degradation processes are taking place which possibly result in the release of trace metals. In addition, the diffusion of trace metals in the pore waters of ocean sediments (enhanced by bioturbation and consolidation) provides the overlying surface waters with some trace metals. The occurrence of nepheloid layers, the region of increased suspended matter near the seabed, on the other hand, provides large surface areas for adsorption and may promote the removal of trace metals.

4. *The Ocean Sediment.* The ocean sediments are the ultimate sink for the trace metals in the hydrological cycle. In the sediments, solid-phase transformations take place which affect the mode of occurrence of the trace metals in them.

This four-box model of the ocean is a simplification of processes occurring in the vertical profiles. The processes in the polar areas are different due to

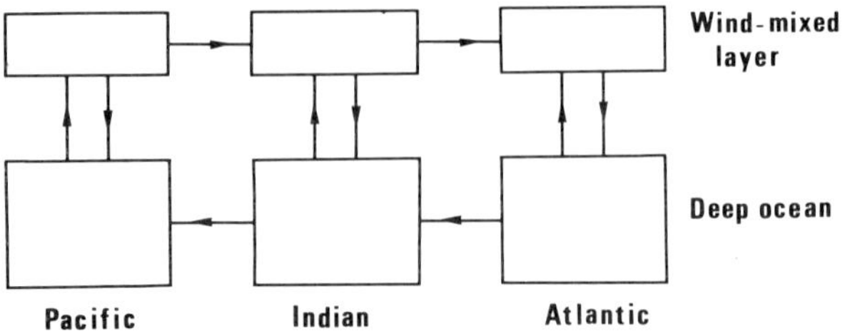

Figure 14.17. Schematic diagram of the lateral movement of watermasses in the ocean.

absence of a well-defined surface layer and less biological activity than midlatitude areas. Also, the vertical movement of water is not taken into account (Fig. 14.17). The Pacific receives relatively large amounts of water from the Atlantic. As a consequence, the deep waters in the Pacific are older and hence have been subject to a large passing through of particulate matter and as a consequence have higher nutrient levels; also, the carbonate compensation depth differs between the Atlantic and the Pacific.

With regard to the impact of the atmosphere on trace metal concentrations in the oceans, it is necessary to include a fifth box (i.e., the atmosphere–seawater interface). At the atmosphere–seawater interface, flux from the atmosphere to the seawater and vice versa are taking place. The oceans play a special role in the hydrological cycle since they are able to influence the composition of the atmospheric aerosol while the atmosphere influences the composition of the surface waters. Therefore, it is necessary to discuss first the processes at this interface before reviewing the available evidence for the impact the atmosphere has on metal cycling in the oceans.

5.2. Metals in the Oceanic Aerosol

5.2.1. Formation of the Oceanic Aerosol

The oceanic aerosol originates from continental sources and from sea spray. It has been estimated that the sea produces between 1,000 and 10,000 t/yr of atmospheric seasalt particles (Eriksson, 1959, 1960; Blanchard, 1963; Petrenchuk 1980). The influence of continents on the oceanic aerosol has been well demonstrated for the North Atlantic Ocean which is influenced by Saharan dust storms (Schuetz et al., 1981), and also for the Pacific Ocean where certain areas are influenced periodically by dust storms originating from the mainland of China (Duce et al., 1980).

The formation of sea spray is caused by processes following the entrainment of air bubbles in the water by wave action, their subsequent rise to the surface and bursting. Other processes, like the direct shearing of wave crests by the wind, are of minor importance (Wu, 1981a). In coastal waters, bubble generation by decaying organic matter on the sea floor may also be significant (Mulhearn, 1981).

The minimum size of the bubbles is about 14–18 μm (Wu, 1981), and the maximum size may be up to several millimeters. The size spectrum of the bubbles is very similar in shape to that of the water droplets in the sea spray, except that the absolute size of the latter is about 10–20% that of the bubble (Wu, 1981a). The processes associated with the bubbles bursting at the water–air interface are complex. Intensive studies have been carried out by Woodcock and Blanchard (Woodcock et al., 1953; Blanchard and Woodcock, 1957; Blanchard, 1963; Blanchard and Woodcock, 1980; Cipriano and Blanchard 1981) and by MacIntyre (MacIntyre, 1970, 1972, 1974). When the bubble is at the surface of the water, water drains from the film cap which is exposed to the water surface and the bubble subsequently breaks producing very small film drops. With the breaking of the bubble, a high-speed jet of water is produced which moves upward from the bottom of the bubble cavity. This jet rises rapidly, becomes unstable, and produces between two and five drops (jet drops) whose size is about 10–20% the size of the original bubble (Wu, 1981). The processes involved in the breaking of the bubble at the water surface and the formation of an aerosol are schematically shown in Figure 14.18.

Cipriano and Blanchard (1981) have shown that most of the smaller sea salt particles are film drops. These are derived from large bubbles (1 mm and larger), which may produce as many as 10 or more film drops per bubble. Film drop production reaches a maximum for bubbles with roughly 2-mm diameter; bubbles smaller than 1 mm or larger than 5 mm are relatively

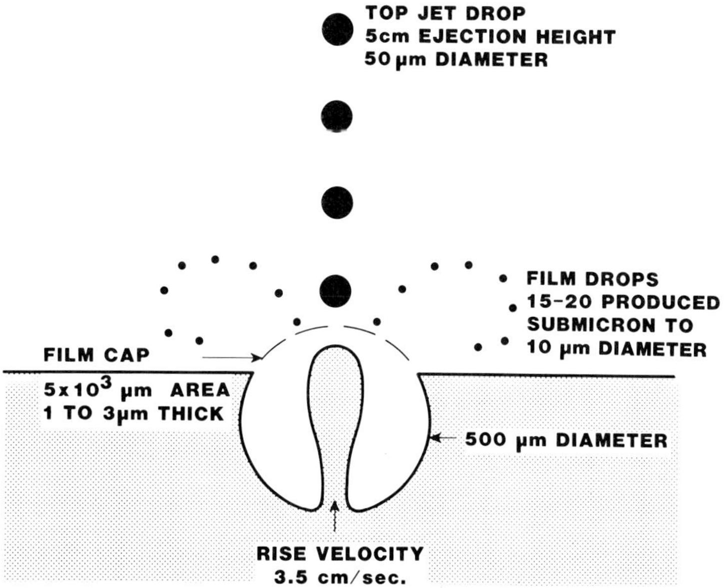

Figure 14.18. Processes involved in the breaking of an air bubble at the water surface and the formation of an aerosol (Winchester and Duce, 1977).

inefficient. The production of jet drops falls off rapidly as the bubble diameter exceeds 1 mm. The total flux of salt to the air in the model experiments of Cipriano and Blanchard (1981) was dominated by the jet drops, suggesting that most of the oceanic salt mass flux derived from it. It should be noted, however, that although the total mass transported into the atmosphere is dominated by jet drops, the film drops will have the largest residence time. Important with respect to the composition of the oceanic aerosol is the fact that the film cap skims off a portion of the surface layer of the water, and in the formation of the jet drops the surface layer of the water is also skimmed off (interior of the bubble cavity). MacIntyre (1970) estimated that the top jet drop is composed of material from the interior of the bubble surface to a depth of 0.05% of the bubble diameter. Therefore, the composition of the sea surface microlayer determines to a large extent the composition of the marine component of the oceanic aerosol.

5.2.2. Sea Surface Microlayer

The sea surface microlayer may be operationally defined to extend from 3 Å (the diameter of a water molecule) to about 3 mm, which is near the extreme limit of nonturbulent kinetics with no wind and also the penetration depth of jets from small bubbles (MacIntyre 1974). Various ingenious methods have been designed to sample the surface microlayer, by removing the layer varying in depth from 1 mm to 300 Å (Lion and Leckie, 1981a). This variation in sampling depth makes it difficult to compare reported concentrations and enrichments of metals with the concentrations in the bulk seawater. However, a recent survey showed that despite the large differences in sampling methods, in most cases the concentrations of trace metals and organic matter in the surface microlayer are enriched compared with the bulk seawater (Lion and Leckie, 1981b).

The absolute amount of trace metals in the surface layer may be quite high, and in marsh areas high enough to necessitate taking it into account for flux calculations (Pellenbarg and Church, 1979). The surface microlayer in oceans receives material from the water column by rising bubbles (flotation) and from the atmosphere. The flotation process is similar to the one used for the recovery of ore minerals in the mining industry. Organic matter and particulate matter become attached at the water–air interface of the bubble and are transported in this way to the surface.

In laboratory studies, Wallace and Duce (1975) showed that over 50% of the particulate Al, Mn, Fe, V, Cu, Zn, Ni, Pb, Cr, and Cd present in Narrangansett Bay water could be scavenged by bubbles and transported to the surface microlayer. These results were similar to those obtained for the Sargasso Sea (Wallace and Duce, 1978). The upward bubble flux of trace metals was in the same order of magnitude as the flux of atmospheric trace metals to the water surface.

In the experiments, large particles were formed which could not be broken up by vigorous shaking. This shows that bubbles in the sea may concentrate particulate trace metals and increase the size of particulate matter and hence the flux. The downward flux from the sea surface layer was in remarkable agreement with the independently estimated flux of particulate trace metals to deep waters associated with sinking organic matter (Wallace et al., 1977). Bacon and Elzerman (1980) compared $^{210}Po/^{210}Pb$ ratios in bulk seawater, the surface layer, and atmospheric fallout. The atmospheric ratio was much lower than the ratio in bulk seawater, whereas the ratio in the surface layer showed only small deviations from that in bulk seawater. This comparison suggests that the ocean is an important source of trace elements in the ocean surface microlayer.

Hunter (1980) made a detailed study of various processes affecting metal concentrations in the surface microlayer by considering atmospheric deposition, Brownian diffusion, gravitational settling, bubble flotation, and mixing. A schematic presentation of these processes is shown in Figure 14.19.

Both iron and manganese were strongly depleted in the surface microlayer of the North Sea on all occasions studied by Hunter (1980). The phases bearing these two trace metals and probably also part of the other particulate trace metals (PTM) consist of river-derived terrigenous material present on large particles. These particles are transported to the microlayer by mixing processes but settle relatively rapidly. The fact that other trace metals are enriched (Ni, Cu, Zn, Pb, and Cd) shows that for these trace metals, the gravitational settling process is minor compared with the other processes controlling these element concentrations in the surface microlayer. By comparing residence times for both flotation and atmospheric deposition

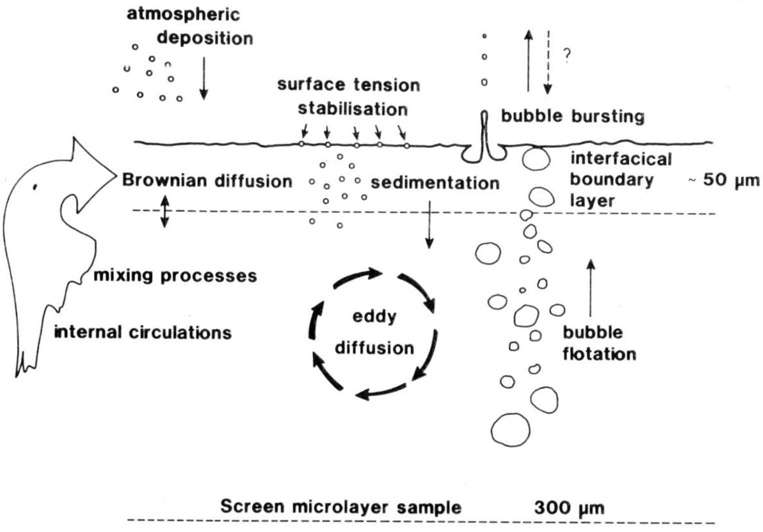

Figure 14.19. Schematic presentation of the processes affecting trace metals in the surface microlayer (Hunter, 1980).

necessary to cause a twofold enrichment (times varying from 1 to 28 min for the enriched elements by flotation and 24 to 1500 min for the atmospheric deposition), it was shown that bubble flotation was the major process for the enrichment of particulate trace metals in the surface layer. Foggy conditions which accelerated the scavenging of suspended particles in the atmosphere enhanced the atmospheric flux. After the fog had cleared, the metal concentrations in the surface microlayer declined (Hunter, 1980).

Metals in aerosols are partly soluble (Figs. 14.7 and 14.26). The solubility of metals in aerosols in seawater has been studied by Crecelius (1980) and Hodge et al. (1978). The deposition of aerosol particles, therefore, contributes both to the particulate and dissolved metals in ocean surface waters.

Model calculations on the speciation of trace metals in the sea surface microlayer have been carried out by Lion and Leckie (1981). In the bulk seawater, both Cd and Hg are strongly complexed with chloride. Copper and Pb compete more effectively for dissolved complexing agents. If transport to the surface layer takes place by dissolved organics, then the transport of Cu and Pb is favored over Cd and Hg. These results support the fact that in surface layers, the enrichment by Cu and Pb has been more consistently observed than enrichment of Cd and Hg (Lion and Leckie, 1981).

5.2.3. Composition of the Oceanic Aerosol

The composition of the marine components of the aerosol depends on the composition of the film and jet drops. They both skim off part of the surface layer, but additionally the bubbles scavenge the water column during their rise to the surface. Piotrowicz et al. (1979) and Weisel (1981) observed a strong dependence between the EF(sea) (the enrichment factor in the aerosol with respect to the composition of the bulk seawater) and the depth at which bubbles were generated in an experimental setup in the field for Zn, Cd, and Pb. Copper gave a positive intercept in the graph depicting the relation of EF (sea) value to the bubble generating depth, which was attributed to microlayer enrichment of this trace metal (Piotrowicz et al., 1979). Very high EF(sea) values for copper were found in areas with visible slicks. It was concluded that the enrichment of copper on sea salt aerosols may be related to biological and/or organic processes and that Cu and Zn may be scavenged by rising bubbles (Piotrowicz et al., 1979). The data from Weisel (1981) on lead also suggest that rising bubbles play an important role in its enrichment in the aerosol. Therefore, it appears that the marine component of the aerosol depends not only on the composition of the surface layer, but also on the bubble generation depth.

If the EF(sea) value for trace elements is known, the contribution of the seasalt spray to the marine aerosol can be easily calculated from data on the amount of sea spray produced annually. Since data are not available on the total number of bubbles generated and the depth of submergence, Weisel (1981) made calculations for minimum EF(sea) values of metals by using the data obtained from experiments conducted just below the sea surfaces and for maximum EF(sea) values from experiments conducted at a depth of two meters. This depth was chosen because studies by Blanchard and Hoffmann (1978) suggested that at 2 m the bubble surface appears to be saturated and ceases to scavenge additional material. The results of calculations by Weisel (1981) are shown in Table 14.10.

The results show that more than 10% of the total global atmospheric concentrations of cadmium may come from the oceans; for vanadium there also may be a significant marine input. It should be noted, however, that anthropogenic sources are localized and that their influence may be less for pristine marine environments than global data suggest.

Interesting results have been obtained on the relative strength of the various sources contributing to the aerosol at Eniwetok in the North Pacific. The composition of the atmospheric aerosol at Eniwetok shows strong seasonal changes (Fig. 14.20). The concentration of salt with a MMD of 3–4 μm was nearly constant, whereas the dust concentration showed strong variations. The MMD of the dust particles ranged from 0.7 to 1 μm, which is consistent with a very long atmospheric transport path (Merrill and Duce, 1982). The alterations could be explained by seasonal changes in large-scale wind patterns over the North Pacific and the seasonal character of dust storms in desert regions of China (Duce et al., 1980). The dust storm activity is apparently greatest in the spring in China due to the combined effects of low rainfall, the increased occurrence of high surface winds, and wind erosion of soil freshly plowed for planting (Merrill and Duce, 1982).

Some trace metals (e.g., Al, Sc, Mn, Fe, Eu, Ni, Co, V, Hf, Cr, Th, Cu, and Rb) were also found to be associated with the dust storms. Their EF (crust) values varied between 1 and 3; zinc, Cs, Sb, Ag, Pb, Cd, and Se had EF(crust) values higher than 4, and apparently had some source other than continental weathering. The MMD data, the EF(crust) data, and the mean atmospheric concentrations at Eniwetok are shown in Table 14.11.

The elements associated with crustal material showed seasonal cycles that reflected the input of dust from mainland China. However, the enriched elements did not show these strong seasonal variations. These elements

Table 14.10. Source Strength for Trace Metals in the Marine Aerosol ($\times 10^9$ g/yr)[a]

	Ocean		
Element	Mean	Min	Max
Cadmium	0.4	0.1	7
Lead	5	0.2	20
Vanadium	9	5	20
Zinc	10	0.5	100

[a] From Weisel (1981).

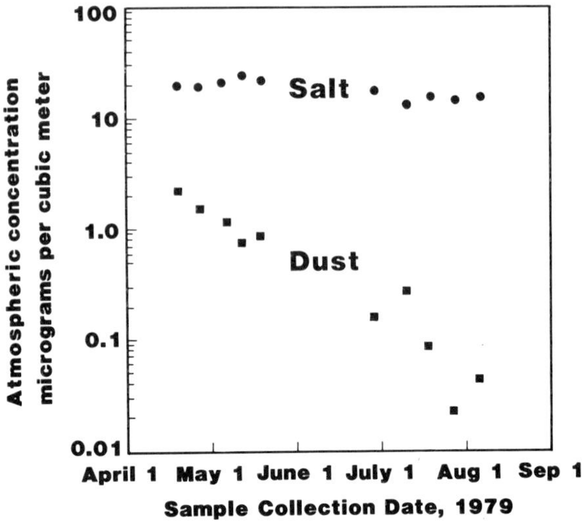

Figure 14.20. Atmospheric concentrations of seasalt and dust at Eniwetok (Merrill and Duce, 1982).

Table 14.11. Selected Data for Trace Elements in the Atmosphere at Eniwetok

			Mean Atmospheric Concentration	
Metal	MMD	EF(crust)	March–June (ng/m^3)	Rest of the Year (ng/m^3)
Al	0.8	1	75	3
Fe	0.72	1	50	2
Mn	0.88	0.9	1	0.04
			(pg/m^3)	*(pg/m^3)*
Sc	0.70	0.8	20	1
Cr		1.8	200	30
Co	0.75	1.0	25	1
Cs	0.79	4.8	15	0.5
Rb	1.1	3.0	200	40
Th	0.84	2.0	20	1
V	0.76			
V	0.76	1.6	120	20
Zn	0.04	4.6	250	80
Cd		57	10	2
Cu	0.96	2.3	50	10
Pb	0.25	45	150	100

[a] From Merrill and Duce (1982).

apparently have continental sources which are not as seasonal in nature, but are more evenly distributed throughout the year (Merrill and Duce, 1982). The MMD of the enriched elements is <0.5 μm, which is consistent with a possible pollution source. If they were associated with sea spray, the MMD would be higher.

The sources for lead were unraveled by Settle and Patterson (1982) by using lead isotope ratios. The $^{206}Pb/^{207}Pb$ ratios (1.184) in the surface waters differed from that (1.226) in the dry deposition and showed that most of the dry deposition originated from lead-rich particles which are transported from land. Experiments with a bubble-generating device confirmed this conclusion: both methods indicated that only 1–15% of lead in air at Eniwetok originates from seawater lead in sea spray. The deposition velocity of the lead indicated that it must be associated with large diameter (7 μm) particles. However, the MMD diameter of lead in the atmosphere was only 0.25 μm (Table 14.11). The small lead-rich submicron particles are transferred to large sea salt particles during cloud-forming condensation processes (Settle et al., 1981). These processes make aerosols injected into the atmosphere by bursting bubbles as efficient as plant foliage on land in sequestering small anthropogenic lead-rich particles and make the dry deposition flux of lead to the oceans important (Settle and Patterson, 1982).

The $^{206}Pb/^{207}Pb$ ratio also showed seasonal changes. During the dry season, when the silicate dust concentrations were highest, the ratio was similar to the ratio in lead aerosols emitted from Asia and Japan. During periods of low dust concentrations, the ratio of $^{206}Pb/^{207}Pb$ was more like the ratio in lead aerosols emitted from the United States. In a comparison of experimental data with the atmospheric concentrations at remote regions like Eniwetok, Weisel (1981) concluded that up to 10% of the lead is from a marine source, for cadmium and vanadium this contribution is in excess of 30%, and for zinc 10 to 20%. The marine component at Eniwetok is found mostly on large particles.

Recent information on sea to land transfer of elements in sea spray aerosols has been gained by studies on radionuclides (Fraizier et al., 1977; Martin et al., 1981; Eakins et al., 1982). Since 1957, an excess of 16,000 Ci of plutonium alpha activity and 650,000 Ci of ^{137}Cs activity has been discharged in the Irish Sea. The deposition of Pu in relation to fall out in Cumbria has increased at the majority of grassland sampling areas situated near the coast, but shows a general decrease with distance inland from the coast. Enrichments from plutonium in sea spray ranged from 70 to 800 Ci relative to filtered seawater. On the other hand, ^{137}Cs was only slightly enriched in the sea spray (Eakins et al., 1982). It was estimated that 2 Ci of $^{230+240}Pu$ may be transferred from the sea to land by sea spray over the past 14 yr.

Weisel (1981) estimated that for the oceans as a whole, the oceanic contribution is 10% for Cd, 3% for Zn and 1% for Pb present in the oceanic aerosol.

5.3. Vertical and Horizontal Distribution of Trace Metals in the Oceans

Since the mid-1970s, important advances have been made in solving the problems associated with sampling, storage, and analysis of trace metals at the low levels found in the oceans. The pioneering work of Patterson at the California Institute of Technology (Goldberg, 1981) is particularly noteworthy. The amount of reliable data on the metal distribution in the oceans is still limited, and the present discussion merely gives a state-of-the-art overview of the available knowledge in a field in which rapid advances are being made.

Metal distribution in the Pacific Ocean, which has been studied by independent groups, is the best known at present.

Profiles of Cd, Zn, Cu, Ni together with those for the nutrients phosphate, nitrate, and silicate are presented in Figure 14.21.

All four metals studied are depleted in the surface ocean layers, as are the nutrients nitrate, silicate, and phosphate.

A comparison of the cadmium profile with those for the nutrients shows a strong resemblance. In fact, a high correlation is observed between phosphate and cadmium (Boyle et al., 1976; Bruland et al., 1978; Bruland, 1980). The linear correlation can be explained by a removal of cadmium from the surface waters by organisms. The organic tissue carrier is regenerated in the thermocline, releasing cadmium and the labile nutrients.

The profile of zinc shows a regeneration at greater depths, as is also the case with silicate. A good linear correlation is observed between zinc and silicate, showing that the release of zinc is associated with the regeneration of hard skeletal material.

The profile of nickel has also been reported by Sclater et al. (1976). In the upper 800 m, it appears to be similar to phosphate (Bruland, 1980), while at greater depths it exhibits a maximum similar to silicate. Nickel appears to undergo both a shallow and deep water regeneration cycle.

The profile for copper (and silver, Martin et al., 1983) differs from those for the other three metals; it shows a surface minimum, but the Cu concentration increases with depth. Similar observations have been made by Boyle et al. (1977). The curves are interpreted as the result of vertical advection of bottom waters high in copper and diffusive mixing with overlying intermediate waters. Results on copper in pore water show that pore waters in oceanic sediments are able to provide the overlying bottom waters with copper (Klinkhammer, 1980; Klinkhammer et al., 1982). The curvature of the copper profile implies scavenging in deep and intermediate waters. Similar scavenging may also take place for Cd, Zn, and Ni. However, their distributions are strongly influenced by the internal biogeochemical cycles, which makes it difficult to observe these removal processes (Bruland, 1980).

The transfer of airborne material to the deep ocean is relatively fast as is shown by the occurrence of fallout radionuclides in sediments and in sediment traps (Lal, 1980; Livingston and Anderson, 1983) two to three decades after they were delivered to the sea surface. The horizontal distribution of trace

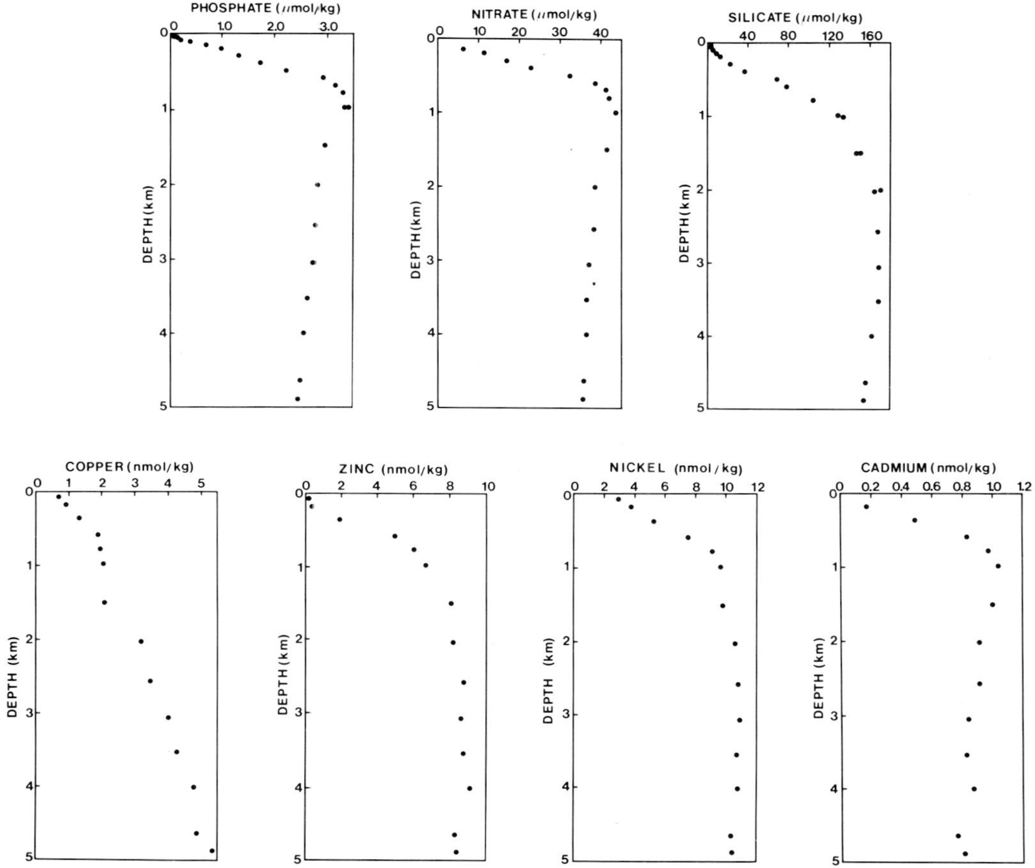

Figure 14.21. Metal concentrations along profiles in the central gyre of the Pacific Ocean (Bruland, 1980).

metals in the surface waters of the Pacific (Bruland, 1980; Boyle et al., 1981) and in the Atlantic (Boyle, 1981) shows large differences in concentrations. The lowest concentrations of Cd, Ni, and Cu were observed in nonupwelling open ocean areas; metal concentrations are high in cool nutrient-rich waters (Boyle et al., 1981). The distinctly higher copper concentrations in the coastal waters north of the Gulf of Panama and in shelf waters north of the Gulf Stream may be caused by copper remobilized from mildly reducing shelf sediments (Boyle et al., 1981). As an example of horizontal variations in the Pacific, data for a transect from Hawaii to Monterey, Calif., are presented in Figure 14.24.

Although for Cu, Ni, Zn, and Cd a large removal is found in the surface layers, a large part is regenerated in the deep water or at the sediment–water interface (notably Cu) and returned to the water masses and through upwelling enters the surface layer again. Comparatively little loss occurs to the sediments during each stirring time. This means that the metals can participate in the vertical redistribution processes under steady-state conditions for a long time, requiring only small external inputs to the water column, just sufficient to balance losses to the sediments (Schaule and Patterson, 1981).

Crecelius (1982) calculated the input from the atmosphere and by upwelling into a 100-m thick surface layer for an open ocean area remote from river inputs and land masses. Some of his results are presented in Table 14.12.

Although these calculations give only order-of-magnitude estimates on the fluxes, they clearly show the important contribution from the atmosphere for Pb and Zn, and the importance of upwelling for Ni, Cr, Cu, and Co. It should be noted, however, that the ocean at present is not in a steady state. The

Table 14.12. Mean Concentrations in the Surface (0–100 m) and Subsurface Layers, the Input and Output from the Surface Layer and the % Contribution of the Atmosphere to the Total Input to the Surface Layer[a]

	Mean Metal Concentrations		Flux (mg/m^2 yr)			
Metal	Surface	Sub-surface	Up Welling	Down Welling	Atmosphere	Input Atmosphere to Total Input (%)
Mn	165	55	0.28	0.85	0.22	44
Pb	30	10	0.05	0.15	1.3	96
Zn	7	70	0.35	0.035	1.3	79
Cu	80	130	0.65	0.40	0.28	30
Ni	140	180	0.90	0.70	0.076	8
Cr	200	150	0.75	1.0	0.10	12
Cd	1	10	0.050	0.0050	0.050	50
Co	10	10	0.050	0.050	0.018	26

[a] From Crecelius (1982).

concentrations in the deep layers still partly reflect the preindustrial period. Especially for lead (and possible for cadmium), the ocean cannot be considered to be in a steady state. Undoubtedly, this type of calculation will be refined in the future when more data become available.

Schaule and Patterson (1981) compared the natural flux of lead to sediments in the central North Pacific with present-day fluxes. The natural flux varied between 1.2 and 7.5 ng/cm^2/yr with a mean value of about 3 ng/cm^2/yr, whereas the present-day flux to the surface waters was estimated to be 68 ng/cm^2/yr, which is a 10-fold increase. Due to larger industrial emissions in North America than in Asia the fluxes in the Atlantic are even higher (170–330 ng/cm^2/yr) (Schaule and Patterson, 1983).

The high atmospheric input of atmospheric lead in the surface waters of the oceans is reflected in the depth profiles for both the Atlantic and the Pacific (Fig. 14.22).

A notable feature of Figure 14.22 is the difference in Pb concentrations between the Atlantic and the Pacific, the values for the Atlantic being significantly higher. This is not the case of all trace metals. In the Pacific, the concentrations of Cd, Cu, and Zn (Bruland and Franks, 1983) exhibit two- to fourfold decreases. Examples of these differences are shown in Figure 14.23.

This observation can be explained by the continuous supply of some trace metals from degrading organic particles to deep waters along the trajectory of the watermasses (e.g., the deep waters of the Pacific are derived partly from the Atlantic) in a way similar to that for the nutrients (Broecker, 1974). Since the residence time of the trace metals (except Pb) is much longer than the stirring time of the ocean, this leads to a continuous buildup of metal concentrations. The residence time of lead, which is 100–150 yr for the

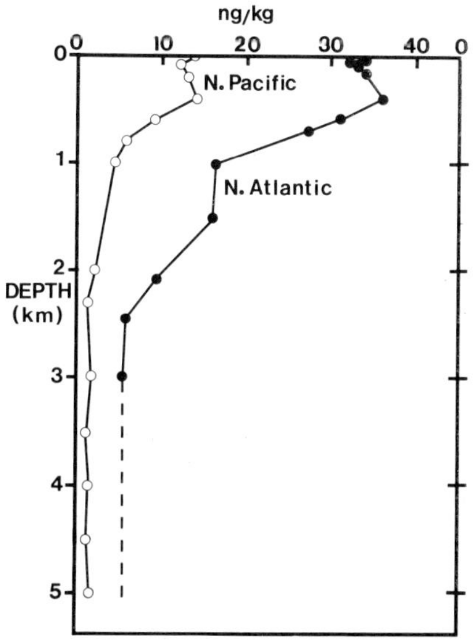

Figure 14.22. Depth profile for lead in the Atlantic and the Pacific (Schaule and Patterson, 1983).

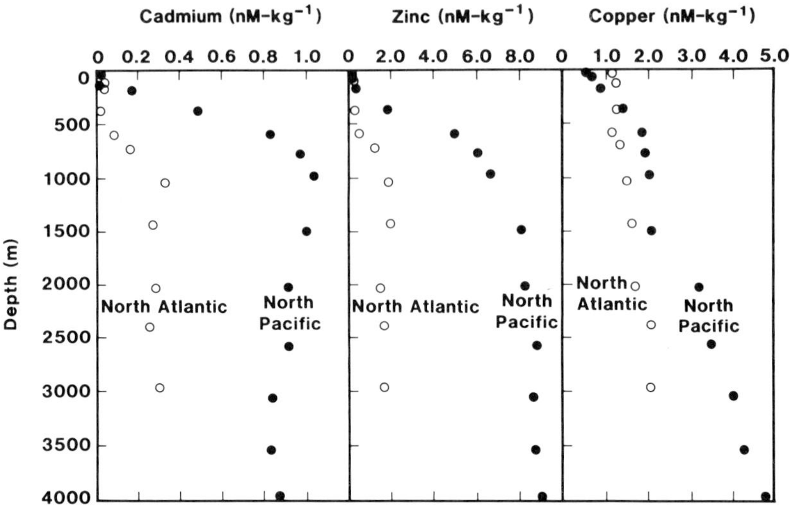

Figure 14.23. Cadmium, zinc, and copper profiles in the Atlantic and Pacific (Bruland and Franks, 1983).

Northwestern Atlantic and 80 yr for the Northeast Pacific (Schaule and Patterson, 1982), is much lower than the transit time of the deep water masses, and no buildup occurs.

As a consequence, the concentrations in these waters are increasing in response to large increases of lead input (see Fig. 14.4) during the past decades. The longer residence time in the Atlantic causes a slower response than in the Pacific, and concentrations have certainly not reached steady state (Schaule and Patterson, 1982). In the surface waters, the residence time of lead is about 2 yr and steady-state conditions prevail, so that the present larger industrial lead input to the North Atlantic compared to that in the North Pacific is recorded directly as higher concentrations in its surface waters.

A few other elements also exhibit pronounced maxima in the surface layers. This is especially true for cobalt and manganese, and in particular in profiles taken close to the continent and reflecting riverine inputs (Martin and Knauer, 1980; Schaule and Patterson, 1981; Knauer and Martin, 1982). The influence of riverine inputs on the distribution of manganese is also shown by increased concentrations in surface water close to the continents in transects from the open ocean to the continent (Schaule and Patterson, 1981) and by the correlation between salinity and cobalt concentrations (Knauer et al., 1982). Lead concentrations, on the other hand, decrease (Fig. 14.24).

This removal of lead on the continental shelves is probably associated with the increased biological activity and elevated loads of organic and inorganic particles. Turekian (1977) showed that very little lead escapes the estuaries and coastal areas. Li et al. (1981) studied seasonal cycles of ^{210}Pb concentra-

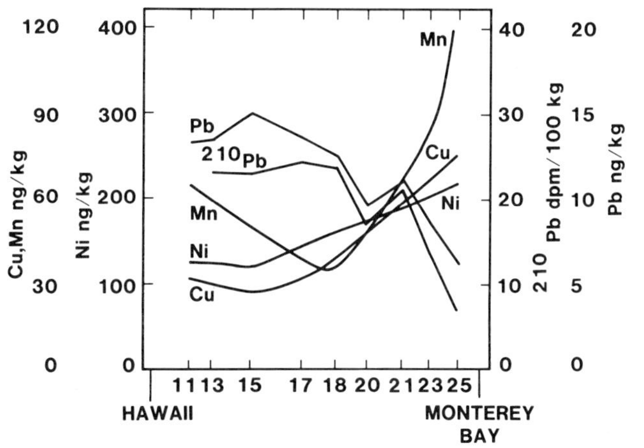

Figure 14.24. Metal concentrations in surface waters in a transect from Hawaii to Monterey, Calif. (Schaule and Patterson, 1981).

tions in waters of the New York Bight. They found that the pathway of phytoplankton–zooplankton–fecal pellets is not important for the removal of ^{210}Pb in the shelf, but in the slope areas. Also, during winter periods, they observed a regeneration of ^{210}Pb from the sediments and a transport back to the surface waters. The most likely overall removal of ^{210}Pb from the shelf surface water is adsorption onto suspended particles and subsequent setting to the bottom.

5.4. The Coastal Zone

The coastal zone is subject to an input of trace metals from riverine sources; discharges of solid and liquid waste (e.g., dredged material, sewage sludge); atmospheric inputs; and release from bottom sediments.

In this section the relative importance of atmospheric trace metals as a pollutant input in coastal areas will be discussed with emphasis on the North Sea. This sea is a relatively well-studied area, making it possible to discuss in detail the problems associated with assessing atmospheric inputs.

Data on the atmospheric input from numerous studies have recently been compiled, and new estimates are available on trace metal inputs both by wet and dry deposition (van Aalst et al., 1983). Some of the results are summarized in Table 14.13.

Despite the large ranges in estimates due to a large variability in the available data, a comparison with more remote areas shows that atmospheric trace metal inputs in the coastal zone are higher than in open ocean areas. A number of authors have compared the riverine transported metals directly

Table 14.13. Flux (ng/cm^2 yr) of Trace Metals from the Atmosphere to the Sea Surface

	North Sea[a]	Western Mediterranean[b]	Tropical North Atlantic[c]	Tropical North Pacific[d]
Cr	14–280	49	14	6
Ni	72–720		20	
Cu	280–2000	96	25	8
Zn	1440–11600	1080	130	22
As	44–144	54		
Cd	22–86	13	5	B1
Pb	720–2600	1050	310	12

[a] From van Aalst et al. (1983).

[b] From Buat-Menard (1983).

[c] From Buat-Menard and Chesselet (1979).

[d] From Duce (1982).

with atmospheric inputs for coastal areas. However, the actual amount of material transported by rivers to the open sea depends on the processes in estuaries (Salomons and Forstner, 1984). For the North Sea, a direct comparison of riverine inputs (the river Rhine/Meuse accounts for 50% of the fluvial transport) shows that the atmospheric flux is equal to the amount of trace metals transported by rivers. Taking into account the amounts of trace metals held back in estuaries (in the case of the river Rhine only 30–50% of its trace metal load enters the North Sea (Salomons and Eysink, 1981)), the atmospheric input is actually even higher. However, fluvial inputs and direct discharges into coastal areas are not the only fluxes which have to be taken into account. The data on copper in the ocean and in continental shelf show that bottom deposits may act as a source for copper in the surface waters. Strong evidence for a flux of trace metals from bottom deposits in the North Sea has been presented by Kremlin (1983). He measured metal concentrations on a transect between the open Atlantic Ocean and the German Bight (Fig. 14.25). A pronounced increase in trace metal levels was found after station 25 (boundary between Scottish coastal waters and oceanic waters). Such an increase could not be attributed to atmospheric inputs (the boundary is too sharp) or to riverine inputs (no major rivers present). Kremlin (1983) concluded that remobilization of trace metals from partly reduced sediments and subsequent mixing into the surface waters caused the increase in trace metal levels. Although the atmosphere is recognized as an important source for contaminants in coastal waters, its relative importance is difficult to assess because retention processes in estuaries are far from being understood and the role of the bottom deposits as a source of trace metals has only recently been recognized.

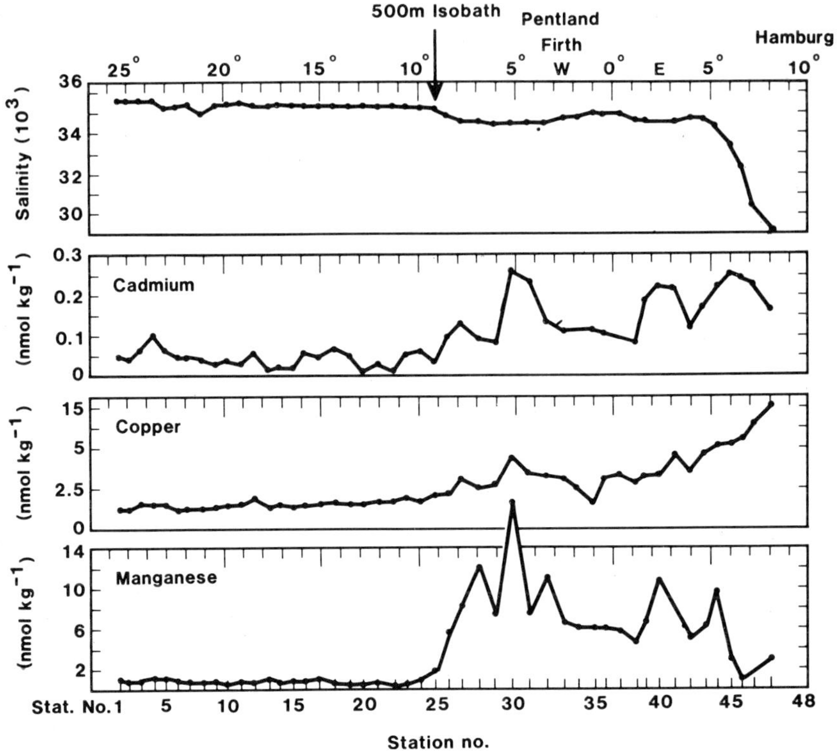

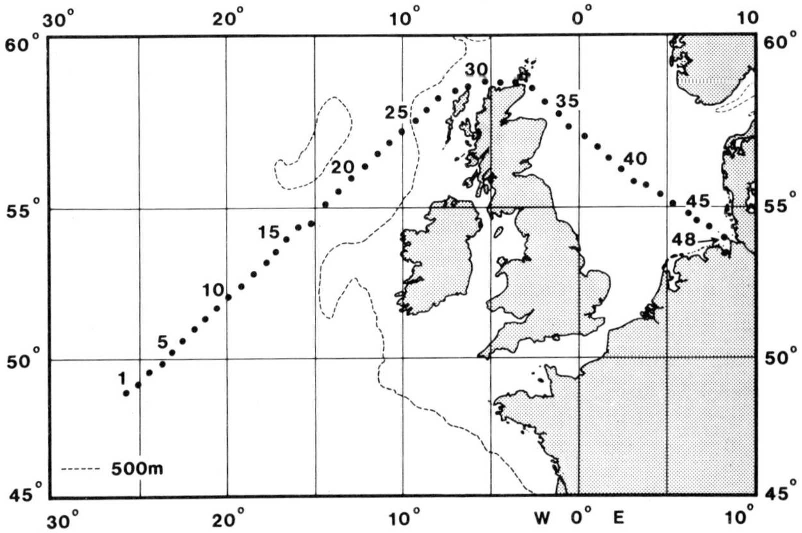

Figure 14.25. Metal concentrations in surface waters in a transect from the Atlantic Ocean to the German coast (Kremlin, 1983).

6. TERRESTRIAL ECOSYSTEMS

In forest ecosystems a number of pathways for the atmospheric trace metals can be distinguished, such as (1) the dry deposition of trace metals directly on the forest soil and on the leaves of the trees and other vegetation and (2) the wet deposition of trace metals which will also remove the dry deposition from leaves.

Trace metals deposited by both dry and wet deposition will interact with the vegetation. Some metals appear to be taken up through the leaves, whereas others are leached from the leaves (foliar leaching).

Within the forest ecosystems, trace metals are incorporated in the aboveground cycling fraction (leaves, needles, etc.), in the noncycling fractions of the vegetation (wood, bark), in the organic-rich surface layers, and in the soil. Part of the metals accumulated in the forest ecosystem will be leached away from the soil and enter streams and rivers. However, if the rate of input is not matched by the rate of output, an accumulation of metals will occur.

Areas which have been studied in detail include the Walker Branch Watershed (WBW) (Lindberg et al., 1979; Lindberg and Turner, 1983) in the United States and the Solling Forest in the FRG (Mayer, 1981). In addition, detailed records of metal concentrations in atmospheric deposition are available for the Federal Republic of Germany (Schaldot and Nuernberg, 1982; Nurnberg et al., 1983).

The Walker Branch Watershed (surface area, 97.5 ha) is located in Eastern Tennessee and is situated within 20 km of two coal-fired power plants (annual coal consumption about 7×10^6t) and within 350 km of 22 other coal-fired power plants. The Solling Forest is located close to the major industrial region of Europe (the Ruhr area). Using a mass balance approach, it was shown (Andren and Lindberg, 1977) that no more than 5% of the total composition of the aerosol in the Walker Branch Watershed could be accounted for by the local coal-fired power plants. However, they accounted for up to 20% of the concentrations of individual elements. The enrichment of lead in the aerosol could be accounted for by automotive emissions, and zinc and possibly mercury by the three nearby power plants. However, these local sources still could not account for most of the As, Cd, Cu, Hg, Se, V, and Zn in the aerosol. Recent atmospheric trace metal concentrations at WBW, together with the results of a speciation scheme are given in Table 14.14.

The fractionation scheme consists of an extraction with distilled water and an acid extraction at pH 1.2. The results show the very high relative solubility of trace metals in the atmospheric aerosol. The relative solubility increased with decreasing particle size (Fig. 14.26) (Lindberg and Harriss, 1981), a feature also shown by coal combustion emissions (Lindberg and Harriss, 1980).

The dry deposited particles on the leaves contained significant fractions larger than 5 μm in diameter and also numbers of particle aggregates. The relatively low solubility of the deposited particles (for lead) is in accordance

Table 14.14. Characteristics of Atmospheric Particles above the Forest Canopy at Water Branch Watershed[a]

Metal	MMD (μm)	Atmospheric concentration[b] Total acid Leacheable (ng/m^3)	Water Soluble Alone (ng/m^3)	Relative Solubility[c] (%)
Cd	1.5	0.17	0.14	82
Mn	3.4	9.4	8.6	83
Pb	0.5	112	84	76
Zn	0.6	9.9	8900	89

[a] From Lindberg and Harriss (1981).

[b] The acid leaching was carried out at a pH of 1.2 and includes both soluble and acid leachable components.

[c] Relative solubility is defined as (water-soluble concentration) × 100/(acid-leachable concentration).

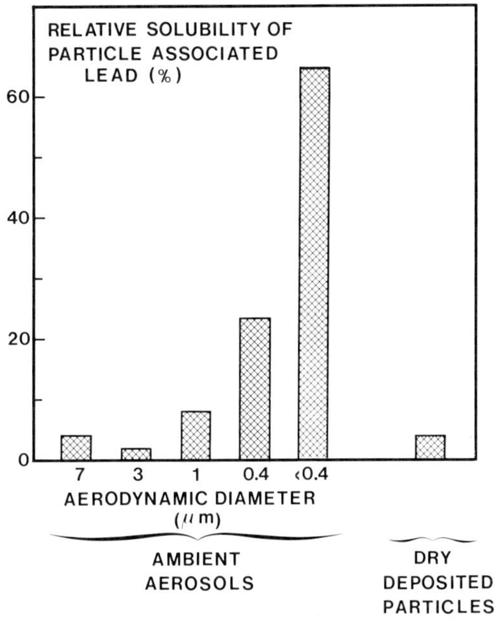

Figure 14.26. The relative solubility of lead in aerosol over the Walker Branch Watershed as a function of particle size and for the dry deposited particles (Lindberg and Harris, 1981).

with the relatively coarse particle sizes. Similar high solubilities of metals in aerosols have been found by Gatz et al. (1982) who studied the solubility in rainwater. The distribution of the percent soluble metals showed the following order: Zn=Cd>Cu>Pb; differences for all metals between sample types follow the order: wet>bulk=dry. The dry deposition rate, the dry deposition velocity, the trace metal concentrations in the precipitation above and below the canopy (throughfall), and the total atmospheric deposition for the Walker Branch Watershed are shown in Table 14.15.

The dry deposition rate follows the general relationship in which the higher the MMD, the higher the deposition rate. Also, the mean deposition velocities (as determined from deposition on inert flat surfaces) reflect the influence of MMD. It should be noted, however, that the deposition velocities increase by a factor of about 2–3 if the full canopy effect (larger surface area) of the forest during summer is taken into account (Lindberg and Harriss, 1981). Mayer and Ulrich (1982) observed higher deposition rates in a spruce forest compared with a beech forest. This deposition excess of the spruce forest is mainly found during the winter period, when beech has lost its leaves and therefore has a reduced surface area compared with the evergreen spruce canopy. The dry deposition particles or their water soluble metals are washed from the leaves during rain periods and become part of the flux to the forest floor. Important with respect to potential negative effects for the vegetation are dew periods and small rainfall events (Wisniewski, 1982). In such cases, the dry deposition particles are not washed off from the leaves, but instead dissolve partly in the water film on the leaf. The dry deposition particles contain high concentrations of soluble metals and as a consequence elevated metal concentrations occur in the water film and increase after subsequent evaporation (Table 14.16).

The metal concentrations in the water film are up to several hundred times higher than typical rain concentrations in this area (Lindberg, et al., 1982). The physiological effects of surface-deposited materials on vegetation either in

Table 14.15. Summary of Mean Data for the Precipitation at the Walker Branch Watershed[a]

	Dry Deposition		Precipitation		
Metal	Rate ($1/\mu g^2 d$)	Velocity (cm/s)	Incident ($\mu g/L$)	Throughfall	Atmospheric Deposition (mg/m^2)
Cd	0.60	0.37	0.44	1.41	0.5
Mn	28	6.8	3.46	135	35
Pb	6.8	0.06	6.85	11.8	15
Zn	1.5	0.46	6.13	15.4	9.3
pH			4.13	4.26	

[a] From Lindberg and Harriss (1981).

Table 14.16. Concentrations of Metals on Leaf Surfaces after 1.3 mm of Rain[a]

Metal	Concentration ($\mu g/L$)
Cd	13
Zn	120
Pb	230
Mn	1300

[a] Deposited on a 50-cm^2 leaf, about 6 mL of water (Lindberg et al., 1982).

particulate or dissolved form require considerable study, as they are poorly understood (Lindberg and McLaughlin, 1982). Possible effects may damage microorganisms on the leaf surfaces, epiphytes, or the leaves themselves (Mayer and Ulrich, 1982).

The flux of metals to the forest floor differs from the sum of wet and dry precipitation, showing the effects of interactions between the precipitation and leaf surfaces (Lindberg and Harriss, 1981; McColl and Bush, 1978; Heinrichs and Mayer, 1980; McColl, 1981). The enrichment in manganese in particularly high (see Table 14.15); for zinc and cadmium smaller enrichments are observed. Also, the pH increases in the throughfall precipitation.

The uptake of H^+ by the leaf surfaces and the concomitant loss of several elements suggests a combination of ion-exchange and leaching processes. This foliar leaching is important for manganese, zinc, and cadmium in the Walker Branch Watershed (Fig. 14.27).

If the flux of cadmium by the annual leaf fall is considered representative of the total cadmium in the leaves, then almost 20 times as much Cd present in the leaves at any time is leached from it to the forest floor during the summer period. This indicates the high replenishment of Cd (and also of Mn and Zn) in the leaves.

The behavior of lead is different from the other metals studied. For lead, the annual flux to the canopy exceeds the flux to the forest floor by a factor of 1.3. This balance study, as well as leaf washing experiments, suggests that the canopy assimilates some of the deposited lead (Lindberg and Harriss, 1981). In the Solling Forest (Heinrichs and Mayer, 1980), an assimilation of lead was also observed.

Part of the total flux to the forest floor will accumulate in the soil. However, part will be removed by leaching processes.

A compartment model and mass balances were constructed for two experimental forest systems in Western Germany (Mayer 1981). Part of the results of the mass balance study are presented in Table 14.17.

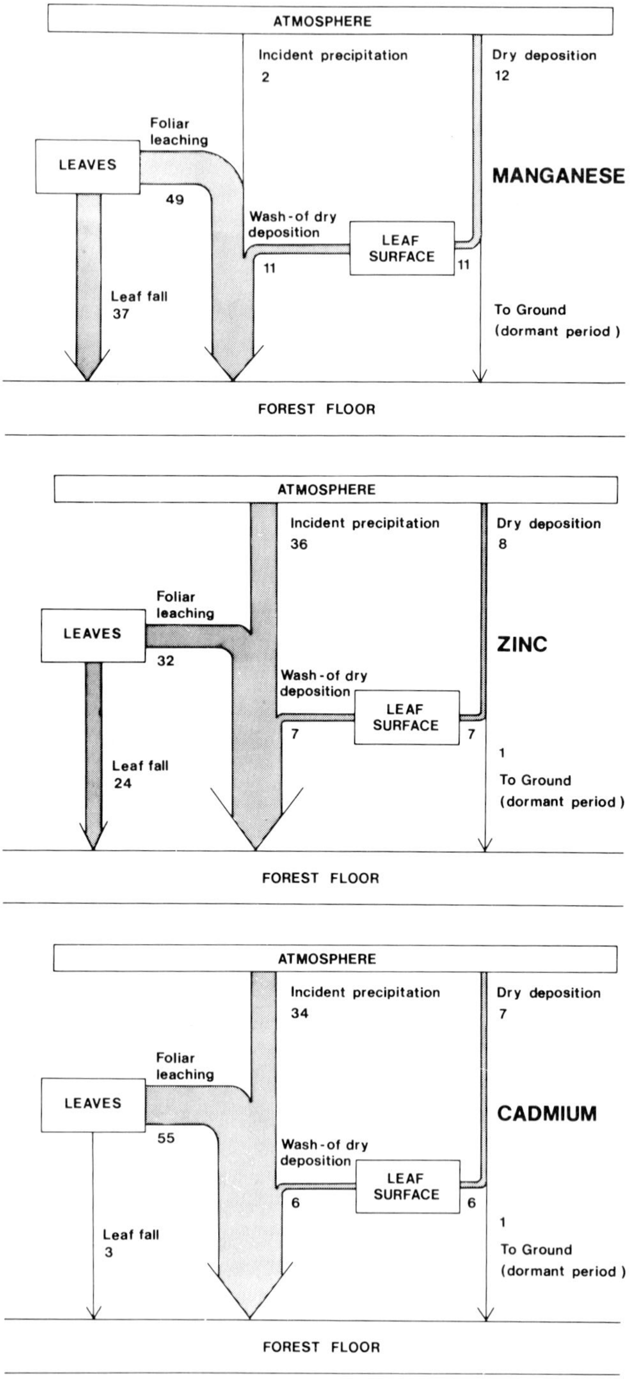

Figure 14.27. Processes affecting the annual flux of cadmium, zinc, and manganese to the forest floor in the Walker Branch Watershed. Based on data from Lindberg and Harris (1981). Values represent the fraction of the total annual flux to the floor that a given process contributes.

Table 14.17. Metal Balance for Two Experimental Forest Systems in Western Europe[a]

	Mn		Ni		Cu		Zn		Cd		Pb	
	A	B	A	B	A	B	A	B	A	B	A	B
Input	1750	5150	123	140	470	659	1632	1732	16	20.1	437	733
Through fall	3840	5190	33	39	162	227	2169	2121	12.6	20.1	302	467
Accumulation in leaves and roots	2900	1300	88	78	310	300	110	246	1.4	3.2	49	76
Soil accumulation	−6790	−8230	14	−4	54	249	397	−878	−2	−9.3	413	720
Seepage	5900	11100	21	66	106	110	1125	2364	16.5	26.2	24	13
Balance	−4150	−5950	102	74	364	549	507	−632	−0.5	−6.1	413	720

[a] All data in g/ha yr. A, Beech forest. B, Spruce forest. From Mayer (1981).

The input given in Table 14.17 is the total deposition from the atmosphere on the ecosystem. The throughfall is the input of trace metals to the forest floor. The accumulation in the soil is the balance of inputs through the throughfall, litter fall, and losses through uptake by roots and to the groundwater. The total balance is the atmospheric input minus the losses to the groundwater. It differs from the accumulation in the soil, since this term also incorporates the storage in growing plants. The balance (input > output) is positive for Cr, Fe, Ni, Cu, and Pb, negative for Mn and Co, and fairly even for Zn and Cd. The considerable seepage losses of Mn, Co, Zn, and Cd are caused by anthropogenic factors (e.g., the acid precipitation). Mayer (1981) showed that within forest ecosystems, the microorganisms and vegetation with its root system are endangered by the heavy metal accumulation at the soil surface, the accumulation having reached levels where deleterious effects are expected. Mobilization of the accumulated metals may follow any kind of ecosystem manipulation like clearcutting, fertilization, deposition of wasted products, and spreading of deicing salts into roadside forests.

The significance of long-range atmospheric transport on the supply of trace metals to terrestrial ecosystems may be illustrated by some recent data from Norway.

In a national study of trace element deposition using moss analysis (Rambaek and Steinnes, 1980), it was found that areas in the southernmost part of the country receive about 10 times as much lead, arsenic, and antimony as most areas north of the 62° N. A similar but somewhat less pronounced trend was evident for zinc, cadmium, silver, vanadium, and selenium. In a simultaneous study (Hanssen et al., 1981), these elements were shown to be strongly intercorrelated in aerosols collected at Birkenes, Southern Norway. A sector analysis confirmed that the highest air concentrations of these elements in Southern Norway occur with trajectories from the south and southwest, clearly indicating a major contribution from areas in Central and Western Europe.

In another national survey, the geographic distribution of lead in natural surface soils, shown in Figure 14.28, was found to be very similar to that of the airborne deposition of lead (Allen and Steinnes, 1979). Other volatile trace elements such as cadmium, arsenic, and antimony were also found to be strongly enriched in surface soils in the southernmost part of Norway as compared to more northerly areas of the country, whereas less volatile elements such as copper and nickel did not exhibit this kind of trend. Most probably, this enrichment of the volatile-group elements is due to deposition of metals from long-distance atmospheric transport and subsequent strong fixation in the organic fraction of the soils. This conclusion is supported by similar findings from a study of trace element profiles in ombrotrophic bogs from different parts of Norway (Hvatum et al., 1983).

Not only the soils are affected by the regional differences in trace element deposition, but also organisms living in the areas concerned. The content of lead and cadmium in forest trees and heather is about 3–5 times higher in

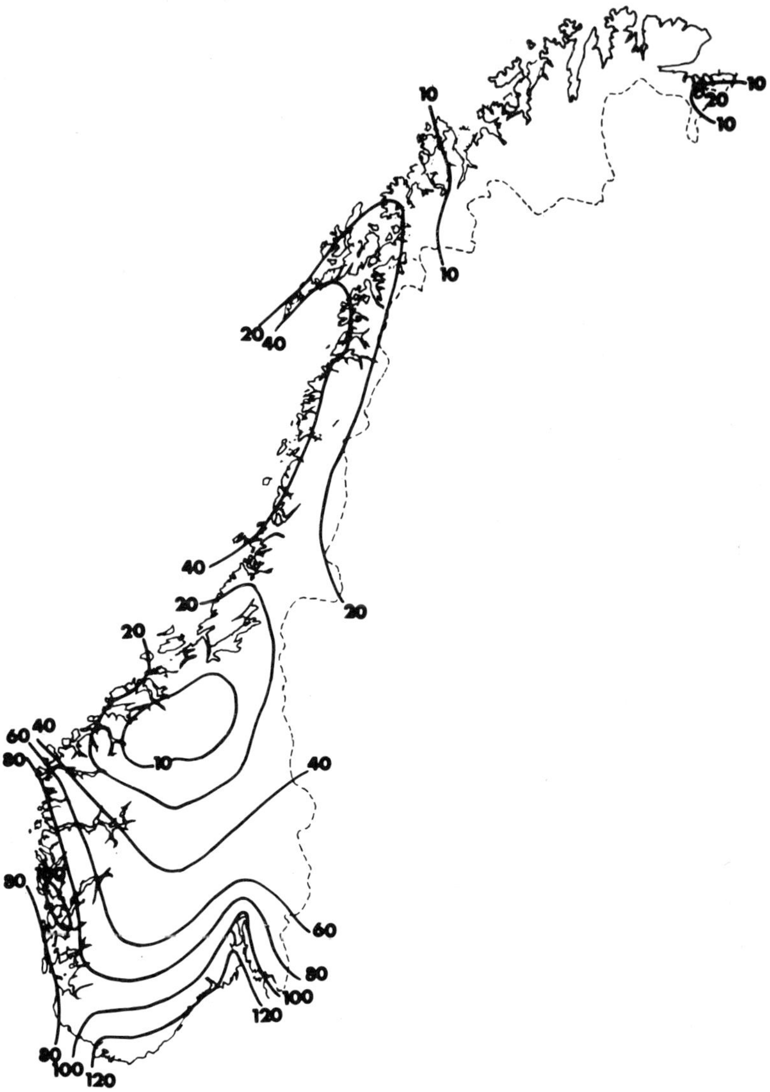

Figure 14.28. Regional distribution of lead in surface soils in Norway, indicating strong contribution of atmospheric deposition from distant sources. (Allan and Steinnes, 1979).

southernmost Norway than in ecologically equivalent localities in Middle Norway, whereas no significant differences are evident for the essential trace elements copper and zinc (Solberg and Steinnes, 1983). Lambs from different parts of Norway, slaughtered after having spent their first summer season on natural pasture land, showed liver concentrations of lead and cadmium in good accordance with the patterns of airborne deposition, whereas other trace metals investigated did not show clear regional differences (Froslie et al.,

1983). Atmospheric supply from natural and anthropogenic processes may considerably affect the exchangeable metal fraction and in some cases even the total content of the element concerned in the surface layer of soils. This has been clearly shown for elements associated with the marine environment such as Na, Mg, Cl, Br, and I (Lag, 1968; Lag and Steinnes, 1976). As mentioned before, the supply of Pb, Cd, As, and Sb from long-distance atmospheric transport strongly affected their regional distribution in Norwegian soils (Allen and Steinnes, 1979) as well as their vertical distribution in the soil profile (Solberg and Steinnes, 1983). For selenium, the regional distribution in Norway indicates atmospheric supply from a natural source, possibly marine, in addition to the anthropogenic input (Lag and Steinnes, 1974; Steinnes, 1982; Hvatum et al., 1983).

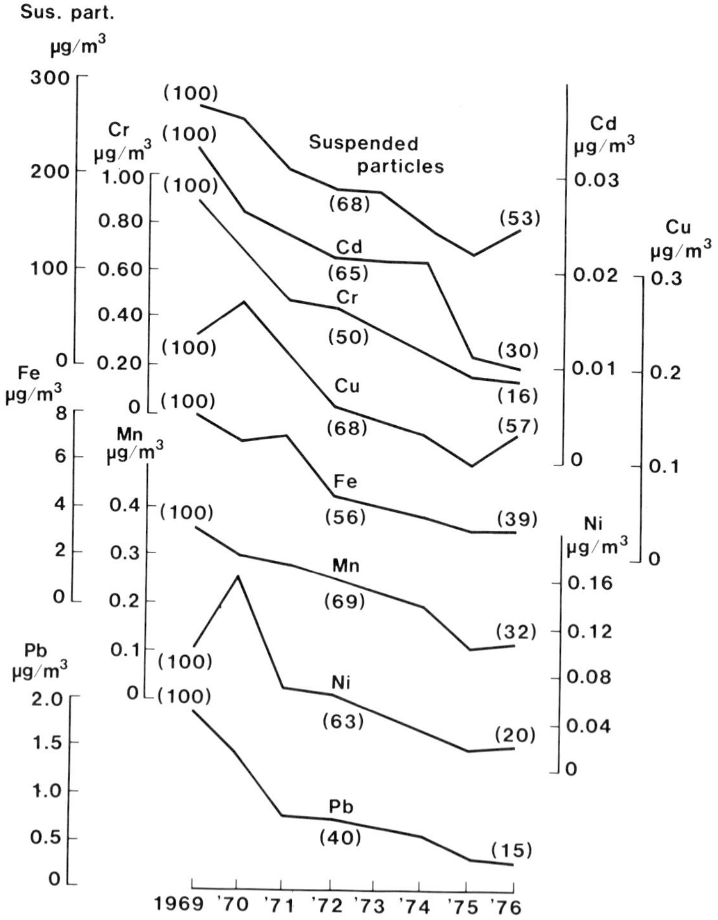

Figure 14.29. Changes in the annual concentrations of heavy metals in the atmosphere of Tokyo. Figures in parentheses show the percentages of values in 1969 (Komai, 1981).

Table 14.18. Estimation of Soil Pollution Caused by Heavy Metals in Falling Dust in Tokyo[a]

	Concentration in Air ($\mu g/m^3$)	Falling Amount (mg/m^2 yr)	Annual Increment in Soil, 0–20 cm (ppm)
Lead	0.922	109.2	0.55
Iron	4.239	338.4	16.8
Manganese	0.183	98.4	0.49
Cadmium	0.018	9.8	0.05

[a]From Komai (1981).

Other terrestrial areas in which the input of atmospheric trace metals is dominant are urban soils. Detailed studies have been carried out in Japan (Komai, 1981). As an example of the metal concentrations in urban atmosphere, the data for Tokyo over the period 1969–1976 are given in Figure 14.29. The high concentrations in urban air cause a flux of trace metals to urban soils which are sometimes used for agriculture. In Table 14.18, the metal concentrations in air, the flux to the soil and the resulting increase in soil concentrations for the layer from 0 to 20 cm are given. This shows that apart from fertilizers, sewage sludge, and other metal additions to agricultural soils, the input from the atmosphere must be a matter of some concern.

Cawse (1982) made calculations for the United Kingdom and showed that for some elements, a 30-year period of continuous atmospheric metal input may pose problems with regard to recommended maximum additions of trace metals to agricultural land.

REFERENCES

Aalst, R.M. van, Ardenne, R.A.M. van, Kreuk J.F., and de, Lems, Th. (1983). Pollution of the North Sea from the atmosphere. Netherlands Organization for Applied Scientific Research. Report No CL/82/152.

Allen, R.O., and Steinnes, E. (1979). "Contribution from long-range atmospheric transport to the heavy metal pollution of surface soil." In *Heavy Metals in the Environment.* CEP Consultants, London, pp. 271–274.

Andren, A.W., and Lindberg, S.E. (1977). "Atmospheric inputs and origin of selected elements in Walker Branch Watershed. Oak Ridge, Tennessee." *Water Air Soil Pollut.* **8,** 199–215.

Bacon, M.P., and Elzerman A.W. (1980). "Enrichment of Pb-210 and Po-210 in the sea surface micro layer." *Nature* **284,** 332–334.

Bertine, K.K., and Goldberg, E.D. (1971). "Fossil fuel combustion and the major sedimentary cycle." *Science* **173,** 233–235.

Blanchard, D.C. (1963). "Electrification of the atmosphere by particles from the sea." *Prog. Oceanogr.* **1,** 71–202.

Blanchard, D.C., and Hoffmann, E.J. (1978). "Control of jet drop dynamics by organic material in sea water." *J. Geophys. Res.* **83,** 6187–6191.

Blanchard, D.C., and Woodcock, A.H. (1957). "Bubble formation and modification in the sea and its meteorological significance." *Tellus* **9,** 145–158.

Blanchard, D.C., and Woodcock, A.H. (1980). "The production, concentration and vertical distribution of sea salt aerosols." *Ann. N. Y. Acad. Sci.* **338,** 330–347.

Borg, H. (1983). "Trace metals in Swedish natural waters." *Hydrobiologia* **101,** 27–34.

Boyle, E.A., Huested, S.S., and Jones, S.P. (1981). "On the distribution of copper, nickel and cadmium in the surface waters of the North Atlantic and North Pacific Ocean." *J. Geophys. Res.* **89,** 8048–8066.

Boyle, E.A., Sclater, F., and Edmond, J.M. (1976). "On the marine geochemistry of cadmium." *Nature* **263,** 42–44.

Boyle E.A., Sclater F.R., and J.M. Edmond (1977). "The distribution of dissolved copper in the Pacific." *Earth Planet. Sci. Lett.* **37,** 38–54.

Broecker, W.S. (1974). *Chemical Oceanography.* Harcourt Brace Jovanovich Inc, New York, 214 pp.

Bruland, K.W. (1980). "Oceanographic distributions of cadmium, zinc, nickel and copper in the north Pacific." *Earth. Planet. Sci. Lett.* **47,** 176–198.

Bruland, K.W., and Franks, R.P. (1983). "Mn, Ni, Ni, Zn and Cd in the Western North Atlantic." In C.S. Wong, E. Boyle, K.W. Bruland, J.D. Burton, and E.D. Goldberg, Eds., *Trace Metals in Sea Water.* NATO Conf. Series. IV Mar. Sci. Plenum, New York, London, pp. 395–414.

Bruland, K.W., Knauer, G.A., and Martin, J.H. (1978). "Zinc in north-east Pacific water." *Nature* **271,** 741–743.

Buat-Menard, P. (1983). "Particle geochemistry in the atmosphere and oceans." In P.S. Slinn and W.G.N. Slinn, Eds., *Air-Sea Exchange of Gases and Particles.* Reidel Publishing Company, Dordrecht, Holland.

Buat-Menard, P., and Chesselet, R. (1979). "Variable influence of the atmospheric flux on the trace metal chemistry of oceanic suspended matter." *Earth Planet Sci. Lett.* **42,** 399–411.

Cameron, E.M., and Ballantyne, S.R. (1975). "Experimental hydrogeochemical surveys of the High Lake and Hackett River areas, Northwest Territories." *Geol. Surv. Can.* Report 75-29; 19 pp.

Campbell, J.A., and Yeats, P.A. (1982). "The distribution of manganese, iron, nickel copper and cadmium in the waters of Baffin Bay and the Canadian Arctic Archipelago." *Oceanol. Acta* **5,** 161–168.

Cawse, P.A. (1981). "Trace metals in the atmosphere of the UK." ESNA meeting, Aberdeen, AERE Harwell. 15 pp.

Cawse, P.A. (1982). "Inorganic particulate matter in the atmosphere." In H.J.M. Bowen, Ed., *Environmental Chemistry,* Vol. 2. Royal Society of Chemistry, London, pp. 1–68.

Cipriano, R.J., and Blanchard, D.C. (1981). "Bubble and aerosol spectra produced by a laboratory "breaking wave"." *J. Geophys. Res.* **86,** 8085–8092.

Crecelius, E.A. (1980). "The solubility of coal fly ash and marine aerosols in seawater." *Mar. Chem.* **8,** 245–250.

Crecelius, E.A. (1983). "The significance of atmospheric input on metal concentrations in oceanic surface water and suspended matter." Paper presented at the NATO Adv. Res. Inst. Symp. Trace. Metals in Sea Water, Erice, Italy, 1981.

Dams, R., and de Jonge, J. (1976). "Chemical composition of Swiss aerosols from the Jungfraujoch." *Atmos. Environ.* **10,** 1079–1084.

Davis A.O., Galloway J.N., and Nordstrom D.K. (1982). "Lake acidification: its effect on lead in the sediments of two Adirondack lakes." *Limnol. Oceanogr.* **27,** 163–167.

Dickson, W. (1980). "Properties of acidified waters." In D. Drablos and A. Tollan, Eds., *Ecological Impact of Acid Precipitation.* Proceedings of an International Conference. Sandefjord, Norway, March 11–14, 1980.

Duce, R.A. (1982). "Sea salt and trace element transport across the sea-air interface." Abstract, Symposium Ocean/Atmosphere Material Exchange, Joint Oceanographic Assembly, Halifax, Nova Scotia Canada. As cited by Buat-Menard (1983).

Duce, R.A., Hoffmann, G.L., and Zoller, W.H. (1975). "Atmospheric trace metals at remote northern and southern hemisphere sites—pollution or natural?" *Science* **187,** 59–61.

Duce, R.A., Unni, C.K., Ray, B.J., Prospero, J.M., and Merrill, J.T. (1980). "Long range atmospheric transport of soil dust from Asia to the tropical North Pacific: temporal variability." *Science* **209,** 1522–1524.

Eakins, J.D., Lally, A.E., Burton, P.J., Kilworth, D.R., and Pratley, F.A. (1982).. "Studies of environmental radioactivity in Cumbria." AERE-R10127. HMSO, London.

Eisenreich, S.J. (1980). "Atmospheric input of trace metals to Lake Michigan." *Water Air Soil Pollution* **13,** 287–301.

Eriksson, E. (1959). "The yearly circulation of chlorine and sulfate in nature: Meteorological, geochemical and pedological implications. Part I." *Tellus* **11,** 375–403.

Eriksson, E. (1960). "The yearly circulation of chlorine and sulfate in nature: Meteorological, geochemical and pedological implications. Part II." *Tellus* **12,** 63–109.

Fraizier, A., Masson M., and J.C. Guary. (1977): "Recherches preliminaries sur le role des aerosols dans le transfer de certains radioelements du milieu marin en milieu terrestre." *J. Rech. Atmos.* **11,** 49–60.

Froslie, S., Norheim, G., Rambaek, J.P., and Steinnes, E. (1983). "Connection between atmospheric deposition and trace metal contents in lamb's liver." In prep.

Galloway, J.N., Eisenreich, S.J., and Scott, B.C. (1980). "Toxic substances in atmospheric deposition: A review and assessment." National Atmospheric Deposition Program Report NC-141. U.S. Environmental Protection Agency Report EPA-560/5-80-001, Washington, D.C. 146 pp.

Galloway, J.E., and Likens, G.E. (1979). "Atmospheric enhancement of metal deposition in Adirondack Lake sediments." *Limnol. Oceanogr.* **24,** 427–433.

Galloway, J.N., Thornton, J.D., Norton, S.A., Volchok, H.L., and McLean, R.A.N, (1982). "Trace metals in atmospheric deposition: a review and assessment." *Atmos. Environ.* **16,** 1677–1700.

Gatz, D.F. (1975). "Pollutant aerosol deposition into southern Lake Michigan." *Water Air. Soil Pollut.* **5,** 239–251.

Gatz, D.F., Warner, B.K., and Chu, L.C. (1982). "Solubility of metal ions in rain water". *Am. Chem. Soc. Div. Environ. Chem. Acid Precipitation Symposium.* Las Vegas, Nevada, April 1982.

Gatz, D.F., Warner K.B., and Lih-Ching Chu (1983). "Heavy metals: Solubility in atmospheric deposition." *Proc. Int. Conf. Heavy metals in the Environment,* Vol. I. CEP Consultants, Edinburgh, England, pp. 183–187.

Glass, N.R., Glass, G.E., and Rennie, P.J. (1980). "Effects of acid precipitation in North America." *Environ. Int.* **4,** 443–452.

Goldberg, E.D. (1981). "Editors and revolutions". *Mar. Pollut. Bull.* **12,** 225.

Graedel, T.E., and Eschler, C.J. (1981). "Chemistry within aqueous aerosols and raindrops." *Rev. Geophys Space Phys.* **19,** 505–539.

Haines, T.A. (1981). "Acidic precipitation and its consequences for aquatic ecosystems: A review." *Trans Am. Fish. Soc.* **110,** 669–707.

Hamilton-Taylor, J. (1979). "Enrichments of Zn, Pb and Dc in recent sediments of Windermere, England." *Environ. Sci. Technol.* **13,** 693–697.

Hanssen, J.E., Rambaek, J.P., Semb, A., and Steinnes, E. (1981). "Atmospheric deposition of some heavy metals in Norway". In *Heavy Metals in the Environment.* CEP Consultants, Amsterdam, pp. 322–325.

Heggie, D.T. (1982). "Copper in surface waters of the Bering Sea." *Geochim. Cosmochim. Acta* **46,** 1301–1306.

Harrison, R.M., Laxen, D.P.H., and Wilson, J.S. (1981). "Chemical associations of lead, cadmium, copper, and zinc in street dusts and roadside soils." *Env. Sci. Technol.* **15,** 1378–1383.

Heinrichs, H., and Mayer, R. (1977). "Distribution and cycling of major and trace elements in two Central European forest ecosystems." *J. Environ. Qual.* **6,** 402–407.

Heinrichs, H., and Mayer, R. (1980). "The role of forest vegetation in the biogeochemical cycle of heavy metals." *J. Environ. Qual.* **9,** 111–118.

Hester, K., and Boyle, E. (1982). "Water chemistry control of cadmium content in Recent bentic foraminifera." *Nature* **298,** 260–262.

Hidy, G.M. (1973). "Removal processes of gaseous and particulate pollutants." In S.I. Rasool, Ed., *Chemistry of the Lower Atmosphere.* Plenum Press, New York.

Hodge, V., Johnson, S.R., and Goldberg, E.D. (1978). "Influence of atmospherically transported aerosols on surface ocean water composition." *Geochem. J.* **12,** 7–20.

Hultberg, H. (1977). "Thermally stratified acid water in the late winter—a key factor inducing self-accelerating processes which increase acidification." *Water Air Soil Pollut.* **7,** 279–294.

Hunter, K.A. (1980). "Process affecting particulate trace metals in the sea surface microlayer." *Mar. Chem.* **9,** 49–70.

Hvatum, O.O., Bolviken, B., and Steinnes, E. (1983). "Heavy metals in Norwegian ombrotrophic bogs." *Ecol. Bull.* **35,** 351–356.

Johannessen, M., and Henrikson, A. (1978). "Chemistry of snow meltwater: changes in concentration during melting." *Water Resour. Res.* **14,** 615–619.

Kemp, A.L.W., Thomas, R.L., Dell, C.I., and Jaquet, J.M. (1976). "Cultural impact on the geochemistry of sediments in Lake Erie." *J. Fish. Res. Board Can.* **33,** 440–462.

Kemp, A.L.W., and Thomas, R.L. (1976). "Impact of man's activities on the chemical composition in the sediments of Lakes Ontario, Erie and Huron." *Water Air Soil Pollut.* **5,** 469–490.

Klinkhammer, G.P. (1980). "Early diagenesis in sediments from the Eastern Equatorial Pacific. II. Pore water metal results." *Earth Planet. Sci. Lett.* **49,** 520–527.

Klinkhammer, G.P., Heggie D.T., and D.W. Graham (1982). "Metal diagenesis in oxic marine sediments." *Earth Planet. Sci. Lett.* **61,** 211–219.

Knauer, G.A., Martin, J.H., and Gordon, R.M. (1982). "Cobalt in north-east Pacific waters." *Nature* **297,** 49–51.

Komai, Y. (1981). "Heavy metal pollution in urban soils." In K. Kitagishi and I. Yamane, Eds., *Heavy Metal Pollution of Soils in Japan.* Japan Scientific Press, Tokyo.

Kramer, J. (1976). "Geochemical and lithological factors in acid precipitation." *U.S. Forestry Service Technical Report,* NE-23, pp. 611–618.

Lag, J. (1968). "Relationships between the chemical composition of the precipitation and the content of exchangeable ions in the humus layer of natural soils." *Acta Agric. Scand.* **18,** 148–152.

Lag, J., and Steinnes, E. (1974). "Soil selenium in relation to precipitation." *Ambio* **3,** 237–238.

Lag, J., and Steinnes, E. (1976). "Regional distribution of halogens in Norwegian forest soils." *Geoderma* **16,** 317–325.

Lal, D. (1980). "Comments on some aspects of particulate transport in the oceans." *Earth. Planet. Sci. Lett.* **49,** 520–527.

Lantzy, R.J., and MacKenzie, F.T. (1979). "Atmospheric trace metals: global cycles and assessment of man's impact." *Geochim. Cosmochim. Acta* **43,** 511–525.

Lerman, A. (1979). *Geochemical Processes. Water and Sediment Environments.* Wiley, New York, 481 pp.

Li, Y-H., Santschi, P.H., Kaufman, A., Benninger, L.K., and Feely, H.W. (1981). "Natural radionuclides in waters of the New York Bight." *Earth Planet. Sci. Lett.* **55,** 217–228.

Lindberg, S.E. (1981). "The relationship between manganese and sulfate ions in rain." *Atmos. Environ.* **15,** 1749–1981.

Lindberg, S.E., and Harriss, R.C. (1980). "Emissions from coal combustion: use of aerosol solubility in hazard assessment." In J.J. Singh and A. Deepak Eds., *Environmental and Climatic Impact of Coal Utilization.* Academic Press, New York.

Lindberg, S.E., and Harriss, R.C. (1981). "The role of atmospheric deposition of an eastern US deciduous forest." *Water Air Soil Pollut.* **16,** 13–31.

Lindberg, S.E., and Harris, R.C. (1983). "Water and acid soluble trace metals in atmospheric particles." *J. Geophys. Res.* 88, 5091–5100.

Lindberg, S.E., and Turner, R.R. (1983). "Trace metals in rain at forested sites in the Eastern United States." *Proceedings of the International Conference of Heavy Metals in the Environment.* Heidelberg, September 1983.

Lindberg, S.E., Harriss, R.C., Turner, R.R., Shriner, D.S., and Huff, D.D. (1979). "Mechanisms and rates of atmospheric deposition of selected trace elements and sulfate to a deciduous forest watershed." ORNL/TM-6674, Oak Ridge National Laboratory, Oak Ridge, Tennessee. 514 pp.

Lindberg, S.E., Harriss, R.C., and Turner, R.R. (1982). "Reports: Atmospheric deposition of metals to forest vegetation." *Science.* **215,** 1609–1611.

Lion, L.W., and Leckie, J.O. (1981). "The biochemistry of the air–sea interface." *Ann. Rev. Earth. Planet. Sci.* **9,** 449–486.

Lion, L.W., and Leckie, J.O. (1981). "Chemical speciation of trace metals at the air–sea interface: The application of an equilibrium model." *Env. Geol.* **3,** 293–314.

Livingstone H.D., and Anderson R.F. (1983). "Large particle transport of plutonium and other fallout radionuclides to the deep ocean." *Science* **303,** 228–231.

Lum, K.R. (1982). "The potential availability of P, Al, Cd, Co, Cr, Cu, Fe, Mn, Ni, Pb and Zn in urban particulate matter." *Environ. Technol. Lett.* **3,** 57–62.

MacIntyre, F. (1970). "Geochemical fractionation during mass transfer from sea to air by breaking bubbles." *Tellus* **22,** 451–462.

MacIntyre, F. (1972). "Flow pattern of bursting bubbles." *J. Geophys. Res.* **77,** 5211–5228.

MacIntyre, F, (1974). "Chemical fractionation and sea surface micro layer processes in the sea." In E. Goldberg, Ed., *Marine Chemistry,* Vol. 5. Wiley, New York, pp. 245–299.

Maenhaut, W., and Zoller, W.H. (1977). "Determination of the chemical composition of the South Pole aerosol by instrumental neutron activation analysis." *J. Radioanalyt. Chem.* **37,** 637–650.

Martin, J.H., and Knauer, G.A. (1980). "Manganese cycling in northeast Pacific waters." *Earth. Planet. Sci. Lett.* **51,** 266–274.

Martin, J.H., Knauer, G.A., and Gordon, R.M. (1983). "Silver distributions and fluxes in northeast Pacific Waters." *Nature* **305,** 306–309.

Martin, J.M., Thomas A.J., and Jeandel, C. (1981). "Transport atmospherique des radionucleides artificiels de la mer vers le continent." *Oceanologica Acta* **4,** 263–266.

Mayer, R. (1981). "Naturlicher und antrapogene komponente des schwermetallhaushalts von waldokosystemen." *Gott. Bodenkdl. Ber.* **70,** 1–152.

Mayer, R., Ulrich, B. (1982). "Calculation of deposition rates from the flux balance and ecological effects of atmospheric deposition upon forest ecosystems." In H.W. Georgii and J. Pankrath, Eds., *Deposition of Atmospheric Pollutants.* Reidel Publ. Co., pp. 195–200, Dordrecht, the Netherlands.

McInnes, G. (1979). "Multi-element survey: analysis of the first two years' results." Report LR 305 (AP). Warren Springs Laboratory, Dept. of Industry, Stevenage.

McColl, J.G., and Bush, D.S. (1978). "Precipitation and throughfall chemistry in San Francisco Bay area." *J. Environ. Qual.* **7,** 352–357.

McColl, J.G. (1981). "Trace elements in the hydrologic cycle of a forest ecosystem." *Plant Soil.* **62,** 337–349.

Merrill, J.T., and Duce, R.A. (1982). "The meteorology and atmospheric chemistry of Enewetak atoll." *The Natural History of Enewetak atoll,* E. Reese et al., Eds., Ch. 3.

Moore, R.M. (1981). "Oceanographic distributions of zinc, cadmium, copper and aluminum in waters of the central Arctic." *Geochim. Cosmochim. Acta* **45,** 2475–2482.

Mulhearn, P.J. (1981). "Distribution of microbubbles in coastal waters." *J. Geophys. Res.* **86,** 6429–6434.

Murozumi, M., Chow, T.J., and Patterson, C.C. (1969). "Chemical concentrations of pollutant lead aerosols, terrestrial dusts and sea salts in Greenland and Antarctic snow strate." *Geochim. Cosmochim. Acta* **33,** 1247–1294.

Ng, A., and Patterson, C. (1981). "Natural concentrations of lead in ancient Arctic and Antactic ice." *Geochim. Cosmochim. Acta* **45,** 2109–2121.

Norton, S.A., Hess, C.T., and Davis, R.B. (1981). "Rates of accumulation of heavy metals in pre- and post-European sediments in New England lakes. In Eisenreich S.J., Ed., *Atmospheric Pollutants in Natural Waters.* Ann Arbor Sci., Ann Arbor, Mich., pp. 409–421.

Nriagu, J.O. (1979). "Global inventory of natural and anthrapogenic emissions of trace metals to the atmosphere." *Nature* **279,** 409–411.

Nriagu, J.O., Kemp, A.L.W., Wong, H.K.T., and Harper, N. (1979). "Sedimentary record of heavy metal pollution in Lake Erie." *Geochim. Cosmochim. Acta* **43,** 247–258.

Nriagu, J.O., Wong, H.K.T., and Coker, R.D. (1982). "Deposition and chemistry of pollutant metals in lakes around the smelters at Sudbury, Ontario." *Environ. Sci. Technol.* **16,** 551–559.

Nuernberg, H.W., Valenta, P., and Nguyen, V.D. (1983). "The wet deposition of heavy metals from the atmosphere in the Federal Republic of Germany." *Proceedings of the International Conference of Heavy metals in the Environment,* Vol. I. CEP Consultants, Edinburgh, England, pp. 115–123.

Ochs, H.T., and Gatz, D.F. (1980). "Water solubility of atmospheric aerosols." *Atmos. Environ.* **14,** 615–616.

Pattenden, N.J. (1974). "Atmospheric concentrations and deposition rates of some trace elements measured in the Swansea/Neath/Port Talbot area." AERE Harwell Report R 7729. HMSO, London.

Patterson, C.C. (1971). "Native copper, silver and gold accessible to early metallurgists." *Am. Antiq.,* 286–321.

Patterson, C.C. (1972). "Silver stocks and losses in ancient and medieval times." *Econ. Hist. Rev. 2nd Ser.* **25,** 205–235.

Pellenbarg, R.E., and Church, T.M. (1979). "The estuarine surface microlayer and trace metal cycling in a salt marsh." *Science* **203,** 1010–1012.

Petrenchuk, O.P. (1980). "On the budget of sea salts and sulfur in the atmosphere." *J. Geophys Res.* **85,** 7439–7444.

Piotrowicz, S.R., Duce, R.A., Fasching, J.L., and Weisel, C.P. (1979). "Bursting bubbles and their effect on the sea-to-air transport of Fe, Cu and Zn." *Mar. Chem.* **7,** 307–324.

Rahn, K.A. (1976). "The chemical composition of the atmospheric aerosol." Kingston (Rhode Island) Graduate School of Oceanography. University of Rhode Island. Technical Report, July 1, 265 pp.

Rahn, K.A. (1981). "Atmospheric, riverine and oceanic sources of seven trace constituents to the arctic ocean." *Atmos. Environ.* **15,** 1507–1516.

Rambaek, J.P., and Steinnes, E. (1980). "Atmospheric deposition of heavy metals studied by analysis of moss samples using neutron activation analysis and atomic absorption spectrometry." Nuclear Methods Environmental Energy Research, USDOE CONF-800433, pp. 175–180.

Reuther, R., Wright, R.F., and Forstner, U. (1981). "Distribution and chemical forms of heavy metals in sediment cores from two norwegian lakes affected by acid precipitation." *International Conference of Heavy metals in the Environment,* Amsterdam, pp. 318–321.

Rippey, B., Murphy, R.J., and Kyle, S.W. (1982). "Anthropogenically derived changes in the sedimentary flux of Mg, Cr, Ni, Cu, Zn, Hg, Pb and P in Lough Neagh, Northern Ireland." *Environ. Sci. Technol.* **16,** 23–30.

Rose, A.W., Hawkes, H.E., and Webb, J.S. (1979). *Geochemistry in Mineral Exploration.* Academic Press, New York, 657 pp.

Salomons, W., and W.D. Eysink (1981). "Pathways of mud and particulate trace metals from rivers to the Southern North Sea." In *Holocene Marine Sedimentation in the North Sea Basin,* S.D. Nio., R.T.E. Schuettenhelm, and Tj. C.E. van Weering, Eds. Blackwell Science Publishers, Oxford-London.

Salomons, W., and Mook, W.G. (1980). "Biogeochemical processes affecting metal concentrations in lake sediments (Ijsselmeer, the Netherlands)." *Sci. Total Environ.* **16,** 217–229.

Salomons, W., and Forstner, U. (1984). "Metals in the Hydrocycle." Springer Verlag, Berlin, Heidelberg, New York.

Schaule, B.K., and Patterson, C.C. (1981). "Lead concentrations in the northeast Pacific: Evidence for global anthropogenic perturbations." *Earth Planet. Sci. Lett.* **54,** 97–116.

Schaule, B.K., and Patterson, C.C. (1983). "Perturbations of the natural depth profile in the Sargasso Sea by industrial lead." *Proceedings of NATO Advances Research Institute of Trace Metals in Seawater. Erice Italy, 1981.* Plenum Press, N.Y., pp. 407–504.

Schladot, J.D., and Nuernberg H.W. (1982). "Atmosphaerische belastung durch toxische metalle in der Bundesrepublik Deutschland, Emission und deposition." Berichte der Kernforschungsanlage Juelich, Nr. 1776.

Schütz, L., Jaenicke, R., and Pietrek, H. (1981). "Saharan dust transport over the North Atlantic Ocean." *Geol. Soc. Am. Spec. Paper* **186,** 87–100.

Sclater, FF., Boyle E., and Edmond, J.M. (1976). "On the marine geochemistry of nickel." *Earth Planet. Sci. Lett.* **31,** 119–128.

Settle, D.M., Duce, R.A., and Patterson, C.C. (1981). "Importance of transfer in the atmosphere of metal-rich sub-micrometer particles to large particles in determining magnitudes of dry deposition fluxes of metals." Presented at the IAMAP Third Scientific Assembly, Symposium on the role of Oceans in Atmospheric Chemistry, Hamburg, FRG, August 1981.

Settle, D.M., and Patterson, C.C. (1980). "Lead in Albacore: Guide to lead pollution in Americans." *Science* **207,** 1167–1176.

Settle, D.M., and Patterson, C.P. (1982). "Magnitudes and sources of precipitation and dry deposition fluxes of industrial and natural leads to the North Pacific at Enewetak." *J. Geophys. Res.* **87,** 8857–8869.

Sievering, H., Dave, M., Dolske, D., and McCoy, P. (1980). "Trace element concentrations over midlake Michigan as a function of meteorology and source region." *Atmos. Environ.* **14,** 39–53.

Sievering, H., Dave, M., Dolske, D., McCoy, P. (1981). "Transport and dry deposition of trace metals over southern lake Michigan." In S.J. Eisenreich, Ed., *Atmospheric Pollutants in Natural Waters.* Ann Arbor Sci., Ann Arbor, Mich., pp. 285–325.

Sigg, L., Sturm, M., Stumm, W., Mart, L., and Nurnberg, H.W. (1982). "Schwermetalle im bodensee—Mechanismen der Konzentrations-regulierung." *Naturwissenschaften* **69,** 546–548.

Solberg, W., and Steinnes, E. (1983). "Heavy metal contamination of terrestrial ecosystems from long-distance atmospheric transport." Proc. Acid Rain and Forest Res. Conf. Quebec, Canada.

Turekian, K.K. (1977). "The fate of metals in the oceans." *Geochim. Cosmochim. Acta* **41,** 1139–1144.

Wallace, G.T., and Duce, R.A. (1975). "Concentration of particulate trace metals and particulate organic carbon in marine waters by a bubble flotation mechanism." *Mar. Chem.* **3,** 157–181.

Wallace G.T., and Duce, R.A. (1978). "Open ocean transport of particulate trace metals by bubbles." *Deep Sea Res.* **25,** 827–835.

Wallace, G.T., Hoffmann, G.L., and Duce, R.A. (1977). "The influence of organic matter and atmospheric deposition on the particulate trace metal concentrations of northwest Atlantic surface seawater." *Mar. Chem.* **5,** 143–170.

Weisel, C.P. (1981). "The atmospheric flux of elements from the ocean." Ph. D. dissertation. University of Rhode Island, Kingston, R.I.

Winchester, J.W., and Duce, R.A. (1977). "The air–water interface: Particulate matter exchange across the air–water interface." In I.H. Suftet, Ed., *Fate of Pollutants in the Air and Water Environment.* Wiley-Interscience, New York.

Winchester, J.W., and Nifong, G.D. (1971). "Water pollution in Lake Michigan by trace elements from pollution aerosol fallout." *Water Air Soil Pollut.* **1,** 50–64.

Wisniewski, J. (1982). "The potential acidity associated with dews, frosts, and fogs." *Water Air Soil Poll.* **17,** 361–377.

Woodcock, A.H. et al., (1953). "Giant condensation nuclei from bursting bubbles." *Nature* **172,** 1144.

Wu, J. (1981a). "Evidence of sea spray produced by bursting bubbles." *Science* **212,** 324–326.

Wu, J. (1981b). "Bubble populations and spectra in the near surface ocean: Summary and review of field measurement." *J. Geophys. Res.* **86,** 457–463.

15

ATMOSPHERIC TOXIC METALS AND METALLOIDS IN THE SNOW AND ICE LAYERS DEPOSITED IN GREENLAND AND ANTARCTICA FROM PREHISTORIC TIMES TO PRESENT

Claude F. Boutron

Laboratoire de Glaciologie et Geophysique de l'Environnement
Centre National de la Recherche Scientifique
GRENOBLE-Cedex, France

---

1. INTRODUCTION

Great efforts have been devoted during the last fifteen years to assessing the occurrence of toxic metals (Wood, 1974; EPA, 1976; Galloway et al., 1982) such as Pb, Hg, Sb, Cd, Ag, Se, As, Cu, and Zn in the successive well-preserved snow and ice layers deposited in the central areas of the large Greenland and Antarctic ice caps (Figs. 15.1 and 15.2), in order to try to obtain historical records of atmospheric concentrations of these elements in the remote polar areas of both hemispheres from prehistoric times to present. Of particular interest is the establishment of reference levels for the former natural state of the atmosphere by analyzing ice deposited several thousand years ago, long before the start of significant human impact on the atmosphere. There is also a need to assess present-day human influence on the atmosphere in the remote polar areas of each hemisphere by analyzing recent snow in Greenland (Northern Hemisphere) and in Antarctica (Southern Hemisphere) and comparing the measured concentrations with those in very old ice.

Due to the extreme purity of polar snow and ice (concentrations of toxic metals and metalloids to be measured range from 10^{-10} g/g to 10^{-14} g/g or less), it is however extremely difficult to collect the samples in the field and to analyze them for these elements in the laboratory without introducing any contamination. Many published data have become questionable (see, for instance, Patterson, 1980). There are, moreover, difficulties in interpreting the

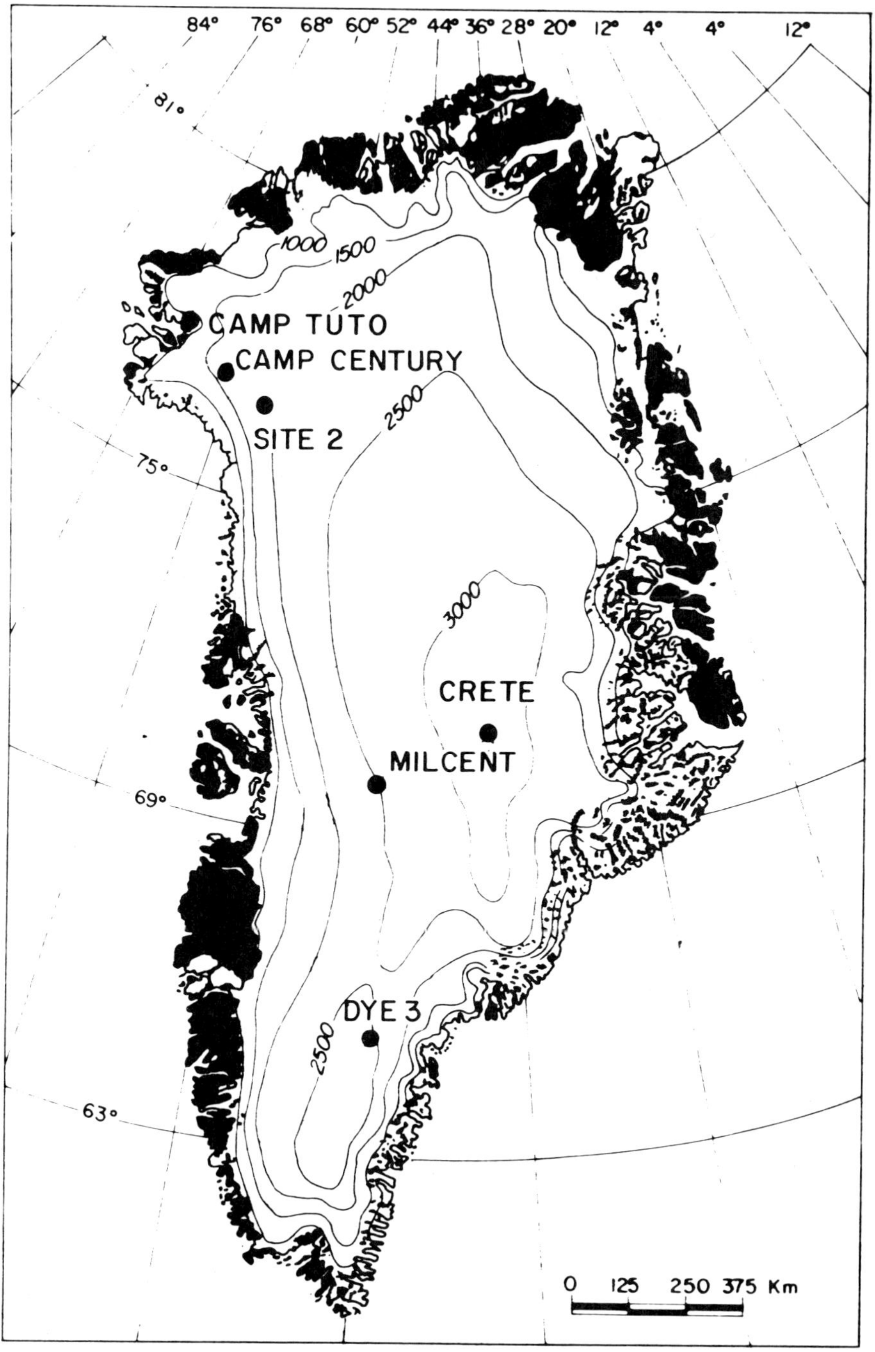

Figure 15.1 Map of Greenland showing the sampling sites referred to in the text.

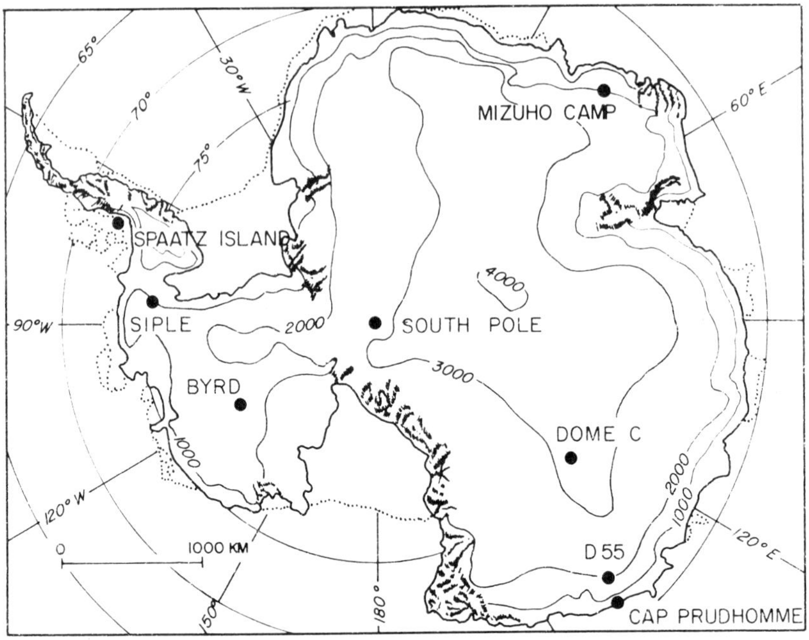

Figure 15.2 Map of Antarctica showing the sampling sites referred to in the text.

few reliable data, as we do not know to what extent chronological changes of these elements in Antarctic or Greenland snow and ice faithfully reflect chronological changes in the local atmosphere.

This chapter presents a succint discussion of the various ultraclean field sampling techniques, laboratory decontamination procedures, and analytical methods which have been developed by various research groups for the analysis of toxic metals and metalloids in Greenland and Antarctic snow and ice. This is followed by a critical review of the data that have been obtained for Greenland and Antarctica. Finally, a discussion of our present knowledge on the atmosphere–snow interactions which are responsible for these elements in the snow and ice layers is presented.

2. FIELD SAMPLE COLLECTION

2.1. Snow Layers from the Surface down to the Firn–Ice Transition

The successive snow layers deposited from the surface down to the firn–ice transition (about 60–100 m deep in the central plateau areas) are especially interesting for toxic metals and metalloids investigations. They typically integrate about the last ten centuries in the central areas of Antarctica (mean annual snow accumulation rates are generally a few grams H_2O per square centimeter per year in these areas) and about one century in the central areas of Greenland. Great care must be taken to eliminate or at least minimize outside contamination of the samples. Snow is highly porous so that any large contamination on the outside of the samples can easily diffuse into the interior to a depth of a number of centimeters, which makes any further laboratory decontamination difficult.

Cores obtained within permeable firn by classic mechanical, electromechanical, or thermal drilling must definitively be rejected as their outer surfaces are always highly contaminated because of the gross uncleanliness of the auger, extension rods, gloves, and clothing of the workers (Murozumi et al., 1969; Boutron, 1979a). The cleaning of such highly contaminated permeable, small size firn cores is hopeless for trace metals studies.

Much cleaner samples have been obtained as follows. Investigators first dig a pit (Herron et al., 1977a; Boutron and Lorius, 1979; Boutron, 1980; 1982; Peel and Wolff, 1982), a trench (Murozumi et al., 1969; Weiss et al., 1975) or an inclined shaft (Murozumi et al., 1969) as cleanly as possible (for instance using electric chain saws powered by generators deployed up to several hundred meters downwind). The operators then put on clean-room clothes, particle masks, and acid-cleaned plastic gloves, and cut back the upwind wall to a distance of about 50 cm with acid-cleaned plastic shovels or saws (Boutron, 1979a; Peel and Wolff, 1982). Samples are then collected from various levels, for instance, by forcing acid-cleaned high-purity plastic containers into the snow (Boutron, 1979a; Peel and Wolff, 1982) or by cutting out large blocks with high mass/surface ratios (Murozumi et al., 1969). The depth thus reached, however, is only 10–20 m at best. This technique is moreover very time consuming (at least two weeks to dig and then sample a 10-m-deep pit).

Recently, very clean samples have also been obtained using a specially designed acid-cleaned mechanical auger entirely made of plastics such as polycarbonate (Boutron and Patterson, 1983). Such an auger is hand-operated by investigators with clean-room garb and acid-cleaned gloves, far away from any contamination sources (camp, vehicles). The cores obtained are transferred without any contact with the gloves of the investigator into acid-cleaned plastic tubes for storage and transport to the laboratory (Boutron and Patterson, 1983). This technique has the advantage of being much faster (drilling to a depth of 15 m can be achieved within one day by three experienced investigators). Maximum depth which can be reached is, however, again limited to 20 m at best with the auger used by Boutron and Patterson (1983), mainly because of the brittleness of the plastic extension rods.

In any case, the cleanliness of the samples obtained is critically dependent on many factors including the quality of the cleaning procedures used for the preparation of the sampling equipment and the experience and dedication of the individuals who are in the field. As seen above, a serious restriction on the sampling procedures described here lies in the limited depth (10–20 m) which can be reached. So far as we are aware, no satisfactory sampling technique has been developed thus far to collect deeper snow samples from 20 down to 100 m or so within permeable firn.

2.2. Ice Layers from the Firn–Ice Transition to the Bedrock

Sampling the old ice layers below the firn–ice transition is essential for establishing reference concentrations of toxic metals and metalloids from the former natural state of the atmosphere before man started to impact significantly on it. This ice, whose age can reach several hundred thousand years, is usually sampled as small diameter (10–12 cm) cores by thermal or electromechanical drilling in a dry hole (intermediate drilling up to 500 to 1000 m) or in a hole filled with a stabilizing fluid of the same density as ice such as diesel fuel mixed with trichlorethylene (for greater depths). Such cores have the advantage of being easily dated and of not being disturbed by percolation. But their outside is unfortunately always heavily contaminated for trace metals and metalloids, especially when the hole is filled with a fluid. Additional contamination is often added during handling, transportation, and storage. For example, Pb contamination has been shown by Ng and Patterson (1981) to be about 10^{-6} g Pb/g in electromechanically drilled Greenland core sections taken from fluid-filled drill holes and about 3×10^{-8} g Pb/g in thermally drilled Antarctic core sections taken from nonfluid-filled holes. These are 10^6- and 10^4-fold higher, respectively, than Pb concentration in the original ice (about 1×10^{-12} g Pb/g). The small diameter of the cores makes any further laboratory decontamination (see Sec. 3) extremely difficult. Provided that sophisticated techniques are used, such a decontamination is not as hopeless as was the case for firn cores. Impermeable ice acts as a much more efficient barrier than firn for the transfer of contamination from the outside to the center of the ice cores, despite the possible existence of minute cracks caused by pressure release and by thermal or mechanical shocks (which later anneal during storage) through which contamination can be forced.

An alternative way to collect old ice is to excavate large blocks of ice in selected ablation zones of coastal areas (see Murozumi et al. (1969) for Greenland and Boutron and Patterson (1983) for Antarctica). The deeply buried ice which emerges at the surface in such areas after traveling hundreds of kilometers originates from the polar plateau (Lorius, 1967; Raynaud et al., 1979), and its age can be up to several hundred thousand years. This ice has the disadvantage of being difficult to date and of being possibly disturbed by percolation during the summer months. But it is logistically very easy to sample, either directly from the surface (after removing the thin top layer disturbed by percolation) or from tunnels excavated for other purposes (Murozumi et al. 1969). Rather clean procedures can easily be used to extract the blocks, resulting in limited outside contamination. As an illustration, Pb contamination on the outside of the coastal Antarctic block analyzed by Boutron and Patterson (1983) was about 1×10^{9} g Pb/g. If large blocks with high mass to surface ratios are excavated, this limited contamination cannot penetrate to the center of the block because several decimeters of ice act as an efficient barrier, so further decontamination will be easy and much more efficient than for small-size cores.

3. LABORATORY DECONTAMINATION OF THE SAMPLES

3.1. Necessity of a Decontamination

From Section 2, it can be anticipated that most samples to be analyzed for trace metals and metalloids in the laboratory are more or less contaminated on their outside, the degree of contamination varying with the cleanliness of the field sampling procedure. Huge contamination of deep ice cores drilled in fluid-filled holes is common with very limited but often significant contamination for snow samples collected from the walls of pits using acid-cleaned plastic containers. It is therefore essential to clean the samples in the laboratory by removing the contaminated outside veneer layers. If ultraclean procedures are used to prevent transfer (by entrainment and handling) of contamination existing at high levels in the outer layers to cleaner inner layers and if the size of the sample is large enough, the decontamination procedures make it possible to get the uncontaminated inner part of the sample whose analysis will give the original metal concentrations in the Antarctic or Greenland snow or ice. The shape of the curves representing the variations of the concentrations measured in the successive veneers from the outside toward the center is probably the only accurate way to clearly determine whether the concentration measured in the final inner part of the sample is still affected by outside contamination or not and then to assess the validity of the data obtained. Examples of such curves are given by Ng and Patterson (1981) and Boutron and Patterson (1983).

3.2. Mechanical Subsampling

Patterson and co-workers have successfully developed a cleaning procedure which involved mechanically chiseling several veneers of ice or snow in progression from the outside to the inside of each sample and analyzing each separate veneer and the remaining inner part of the sample itself (Ng and Patterson, 1981; Boutron and Patterson, 1983). This procedure has been used for the decontamination of (1) highly contaminated deep core sections, 8 to 12 cm in diameter, from Greenland and Antarctica (Ng and Patterson, 1981), (2) a moderately contaminated large size (37 × 37 × 37 cm) blue ice block from an Antarctic coastal site (Boutron and Patterson, 1983), and (3) slightly contaminated snow cores, 7.5 cm in diameter, drilled down to 10 m with an acid-cleaned all-plastic auger in Antarctica (Boutron and Patterson, 1983).

The procedure used in all these studies is basically the same. As an illustration, the procedure used for the Antarctic snow cores can be briefly described as follows (a detailed description is given in Boutron and Patterson, 1983). The sample is taken from its sampling container and placed on an acid-cleaned polyethylene sheet inside a cooled, double-walled, acid-cleaned conventional polyethylene tray flushed with cooled, high-purity nitrogen. The assembly was placed inside an ultraclean laboratory flushed with filtered air (Patterson and Settle, 1976). The operators wear over their clean-room gowns and caps a large acid-cleaned polyethylene bag placed over the body, open side down, with slits cut for the arms and the eyes and shoulder-length acid-cleaned polyethylene gloves. The first veneer layer (about 5–7 mm thick) is removed using an acid-cleaned stainless steel or quartz chisel by making successive shallow chord shavings from one end of the core to the other along the side. The ends of the core are similarly shaved. The chips so obtained fall directly into a specially designed, acid-cleaned quartz tray, and are then transferred into a quartz beaker for analysis. The remaining sample is then transferred to a fresh clean part of the working area, and the next veneer layer is removed by the operators after putting on fresh acid-cleaned polyethylene gloves and cleaning the chisel and the quartz tray with ultrapure acids. The procedure is then continued until an inner core 3 to 4 cm in diameter is obtained. New acid-cleaned polyethylene sheets, gloves, and chisel are used for each veneer.

Figure 15.3 shows the variations of Pb and K in the successive veneer layers obtained by this technique for an ice core section 1490 yr old thermally drilled in a nonfluid filled hole at Byrd Station, Antarctica (Ng and Patterson, 1981). For Pb, an overall 10^4-fold decrease in concentrations is observed (Figure 15.3), changing from a huge exterior value of 2×10^{-8} g Pb/g to an interior value of 1.4×10^{-12} g Pb/g. Continuous decrease of Pb with no plateau indicates that exterior Pb contamination has penetrated to the centers of the core, because Pb concentrations should have leveled off at fixed values if the interior sections were not contaminated. Only an upper limit can then be given for the original concentration of Pb in ice. The situation for K is radically different (see Fig. 15.3). After a 100-fold decrease in going from the outside to about 5 cm from the center, concentrations clearly level off, so that a reliable estimate of the original concentration of K in ice can be made.

Figure 15.4 shows the variations of Pb in two snow cores drilled using an acid-cleaned, all-plastic hand-operated auger at stake D 55 in East Antarctica (Boutron and Patterson, 1983). Pb concentrations of 63×10^{-12} g Pb/g and 32×10^{-12} g Pb/g, respectively, are measured in the exterior layers, showing that very slight but significant Pb contamination occurred on the outside of these snow cores despite the use of exceptionally clean field sampling procedures. As shown in Figure 15.4, Pb concentrations decrease towards the inside of the snow cores, but the shape of the curves obtained is not the same for the two cores. For the 8.56–8.68-m section, the decrease is progressive, without any kind of plateau, which suggests that outside contamination has

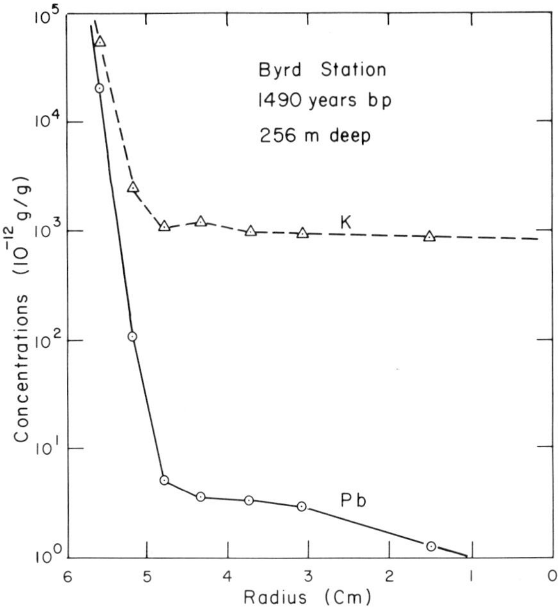

Figure 15.3 Pb and K concentrations as a function of radius in a 1490 year old ice core section obtained by thermal drilling in a nonfluid filled hole at Byrd Station, Antarctica (from Ng and Patterson, 1981).

probably intruded in significant amounts to the center of this section and that the rather high concentration (7.2×10^{-12} g Pb/g) obtained for the corresponding inner core is probably unreliable. For the 9.02–9.40-m section, on the other hand, there is a sharp decrease of Pb concentrations from the first (outside) layer to the second one, followed by rather constant concentrations, although no definitive plateau is observed. For such a core section, it seems likely that only very limited contamination, if any, has intruded to the center of the core, and that the low concentration (2.5×10^{-12} g Pb/g) measured in the corresponding inner core is close to the original one.

Several other authors have tried to decontaminate ice or snow samples by mechanically removing the outside of their samples (Boutron et al., 1972; Appelquist et al., 1978; Landy and Peel, 1981; Herron, 1982; Peel and Wolff, 1982), but their procedures were much less sophisticated. No curves showing the variations of concentrations from the outside to the inside were obtained by these authors, with the exception of Peel and co-workers who have recently obtained very satisfying curves with plateaus for Pb and Al for near-surface snow cores drilled at an Antarctic Peninsula site (British Antarctic Survey, 1983).

It can be seen that most data on mechanical decontamination techniques refer to a single metal, Pb. This is mainly because of the pioneering work on Pb by Patterson's group at California Institute of Technology. Similar

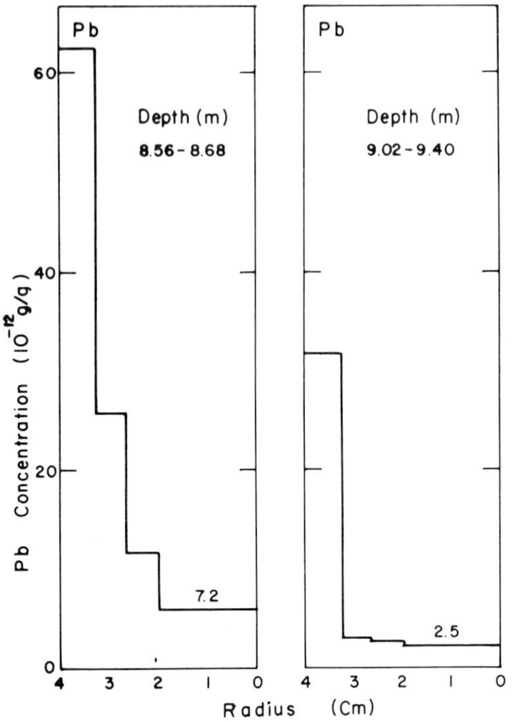

Figure 15.4 Pb concentrations as a function of radius in two snow core sections collected at Stake D 55, Antarctica, using an ultraclean all-plastic hand-operated auger (from Boutron and Patterson, 1983).

difficult work still remains to be done in the future for all other toxic metals and metalloids. Such essential work would certainly lead to a major improvement, which is greatly needed, in the reliability of the data on the concentrations of these other elements in polar snow and ice.

3.3. Decontamination by Rinsing the Outside of the Sample with a Fluid

Such decontamination techniques have been principally used by Langway's group for the study of Pb, Cd, Zn, Cu, and other elements in thermally or electromechanically drilled deep ice cores (Ragone and Finelli, 1972; Ragone et al., 1972; Herron et al., 1973; Langway et al., 1973; Cragin et al., 1975; Herron et al., 1977a; Herron, 1980). They were also used by Petit et al. (1981) for their Antarctic Zn study on the thermally drilled Dome C deep core.

The procedures used by Langway's group vary depending on whether or not the drilling hole was filled with a fluid. For cores from nonfluid filled holes, the procedure (Herron et al., 1973; Langway et al., 1974) involves removing the outer surface of the ice by running ultrapure water over the core while it is held by precleaned stainless steel tongs in a class 100 clean laboratory (Herron, 1980) until about 20% of the core is removed. The remaining inner core only is kept for analysis. For cores from fluid filled holes, the procedure (Herron et al., 1973) involves washing the outer surface with acetone, then following the steps given above. The procedure used by Petit et al. (1981) was similar to the first one described above, except that rinsing with ultrapure water was performed until about 60% of the original sample was melted off.

Unfortunately, none of these authors tried to check the efficiency of their cleaning procedure by studying the variations of measured concentrations of the investigated toxic metals as a function of the percentage of ice melted off, as was done for Na, Mg, K, and Ca in the early work by Ragone and Finelli (1972) and Ragone et al. (1972). It is therefore impossible to evaluate the efficiency of these rinsing decontamination techniques or to assess the quality of the data obtained from cores decontaminated in this manner.

For Pb, the recent work by Ng and Patterson (1981) shows beyond doubt that the deep ice cores decontaminated using rinsing procedures of Langway's group were still highly contaminated; these authors reported Pb concentration values artificially high by at least one order of magnitude.

4. LABORATORY ANALYTICAL PROCEDURES

Due to the extremely low concentrations to be measured, the key problem associated with the analytical procedures is contamination. The simple use of sophisticated heavy metal detection techniques will not, by itself, provide correct analyses at the extremely low concentration levels which are involved (Murphy, 1976; Patterson, 1982). The reliability of the analyses will depend primarily on the control of contamination during the various steps of the analytical procedure (Mitchell, 1973; Murphy, 1976; Patterson and Settle, 1976; Patterson, 1982; Tschopel and Tolg, 1982; Moody, 1982). Many investigators have unfortunately still not fully realized this crucial point.

4.1. Analytical Techniques

Various analytical techniques have been used in analysing metals and metalloids in polar snow and ice. Instrumental Neutron Activation Analysis (INAA) with or without a preconcentration or extraction step before or after the irradiation has been used by Warburton and Young (1968) for Ag; Weiss and Bertine (1973) for Cu, As, Cd, Sb, and Hg; Weiss et al. (1971a) for Se; Appelquist et al. (1978) for Hg; and Buat-Menard and Boutron (1982) for Zn, As, Se, and Sb. Other investigators have relied on Flameless Atomic Absorption Spectrometry (FAAS) without preconcentration (Herron et al.,

1977a; Herron, 1980), or after a preconcentration by nonboiling evaporation (see Boutron and Lorius, 1979; Boutron and Martin, 1979; Boutron, 1979b; Boutron, 1980 and Boutron, 1982 for Pb, Cd, Cu, Zn, and Ag) or after a preconcentration by adsorption onto tungsten wires (Wolff et al., 1981, Peel and Wolff, 1982 for Cd, Pb, and Zn). Some authors have also determined Cd, Pb, and Zn by Differential Pulse Anodic Stripping Voltametry (DPASV) (Landy, 1980, and Landy and Peel, 1981). Finally, Patterson and co-workers have extensively used Isotope Dilution Mass Spectrometry (IDMS) after a chloroform–dithizone extraction for their Pb studies (Murozumi et al., 1969; Ng and Patterson, 1981; Boutron and Patterson, 1983).

All these methods are basically valuable, but IDMS appears to be by far the most trustworthy technique at the very low concentration levels involved. It is indeed an absolute method in the sense that a single measurement suffices, because there are no standards or working curves that are required for comparison purposes.

4.2. Contamination by the Ambient Air, Containers, and Reagents

The use of a high-performance clean laboratory, pressurized with filtered air and designed specifically for ultra low concentration measurements of metals is of paramount importance. Such laboratories have been described in detail by several authors (Patterson and Settle, 1976; Zief and Mitchell, 1976; Shaeffer and Davidson, 1979; Moody, 1982). It must be emphasized that a distinction should be made between a conventional "dust-free" clean laboratory and a clean laboratory specially designed to minimize airborne and laboratory-induced contamination for one or several given metals. This is clearly illustrated by Patterson's laboratory (Patterson and Settle, 1976), which is not a conventional dust-free clean laboratory, but was designed as a low-Pb-blank laboratory by optimizing both air quality and construction materials for Pb only. Many studies on metals and metalloids in polar snow and ice were unfortunately conducted inside conventional dust-free clean laboratories which were not designed to minimize such sources of contamination.

The problems contributed by walls of containers rival those caused by airborne and laboratory-induced contamination. Metals can indeed be leached from the walls of containers by the snow or ice samples after melting and the walls can adsorb metals from melted samples. Leaching problems are by far the most crucial, despite the fact that adsorption problems can be severe for some metals such as Hg and Ag. Systematic studies conducted for Pb have shown that three materials can be used to process melted samples and reagents: FEP Teflon, ultrapure quartz, and conventional polyethylene (Patterson and Settle, 1976; Patterson, 1982). Similar studies conducted for several other metals (Cd, Ag, Se, Zn, and Cu especially) have shown that various teflons and conventional polyethylene are generally the less contaminating materials (Moody and Lindstrom, 1977).

Extensive cleaning of containers is essential. As an illustration, the following cleaning procedures are used by Patterson and co-workers for quartz beakers. They are first rinsed with $CHCl_3$ to remove grease, then immersed successively in 10% reagent grade HF for 10 min, in concentrated reagent grade aqua regia for 24 h, in 25% reagent grade HNO_3 at 55°C for 3 days, in 1% high purity HNO_3 at 55°C for 3 days, and finally in 0.1% ultrapure National Bureau of Standards HNO_3 (Moody and Beary, 1982) diluted in purest water at 55°C for 3 days and until use (Patterson, 1982). After each use, the beakers are recycled successively through the last three steps.

Cleaned laboratory ware must be subjected to a preconditioning treatment immediately before use. This means, for instance, that if the ware is to contain a dilute acid solution, it must be preconditioned with the same type of dilute acid, which is then discarded immediately before use.

4.3. Blank Corrections

It is essential to estimate the level of contamination introduced during the various steps of the analytical procedure, including the laboratory decontamination of the samples. A major fraction of the analyst's efforts should be devoted to the determination of blanks (see Patterson, 1982). As an illustration, Table 15.1 shows typical blank values obtained by Boutron and Patterson (1983) for the IDMS analysis of Pb in a 316-g Antarctic snow veneer. It can be seen that the overall Pb contamination introduced by the chiseling procedure plus the chemical treatment (chloroform–dithizone extraction) plus the transfer to the filament of the mass spectrometer is estimated to be 143×10^{-12} g Pb; this represents about 20% of the Pb content of the sample.

Most authors have unfortunately failed to make such detailed determinations of procedural blanks, so that the concentration values they publish represent only upper limit values.

5. AVAILABLE DATA FOR GREENLAND AND ANTARCTIC SNOW AND ICE

The various metals and metalloids will be discussed successively, according to decreasing atomic weights. The reader is asked to refer to Figs. 15.1 and 15.2 to identify the various sampling locations quoted in the text.

Table 15.1. Estimation of Pb Contamination[a]

Step	Pb Contamination Introduced (10^{-12} g)
1. Decontamination of the sample (mechanical chiselling)	52
2. Chloroform–dithizone extraction	
2.1. Reagents: 0.466 mL HNO_3 (5×10^{-12} g Pb/mL)	2.3
27 drops NH_4OH (15×10^{-12} g Pb/mL)	13.5
1 ml dithizone (11×10^{-12} g Pb/mL)	11.0
16 ml $CHCl_3$ (0.3×10^{-12} g Pb/mL)	4.8
17 ml ultrapure water (0.22×10^{-12} g Pb/mL)	3.7
2.2 Beakers and extraction funnels	31
3. Loading of the sample on the filament of the mass spectrometer	20
Total	142.9
Amount of Pb in the sample	636

[a] Introduced during the successive steps of the IDMS analysis of the 316-g inner part of a typical snow sample collected at stake D 55 in East Antarctica (Boutron and Patterson, 1983).

5.1. Lead-207

Despite the fact that this metal is probably the most difficult to measure accurately because of severe contamination problems, it is the only toxic metal for which definitively reliable—but still very incomplete—data are now available for both Greenland and Antarctica. This is mainly due to the outstanding contribution of Patterson's research group which has resulted in very high quality Pb data thanks to the use of IDMS coupled with stringent contamination-free procedures.

For Greenland, it now seems well established from several good quality data sets that the concentrations in present-day surface snows are in the 150 to 400×10^{-12} g Pb/g range (Table 15.2) without any large geographical variations. This corresponds to crustal enrichment factors EF_{crust} (Rahn, 1976) of about two orders of magnitude with respect to the mean crustal composition of Taylor (1964). On the other hand, there has been strong disagreement between Patterson's group and Langway's group regarding the occurence of Pb in old Greenland ice and on the variations of Pb concentrations from preindustrial times to present. These controversies, however, have been recently resolved by Ng and Patterson (1981). They

Table 15.2. Published Data on the Occurrence of Pb in Greenland Snow and Ice

Location	Description of Samples	Pb Measured Concentrations (10^{-12} g/g)	Reference
I *Recent Snow*			
80 km ESE Camp Century	8 samples 1963–1965	140–420 (mean 230)	Murozumi et al. (1969)
	9 samples 1753–1946	Increase from 11 to 160	
Dye 3	10 samples 1955–1971	300–1200 (mean 850)	Cragin et al. (1973)
Milcent	6 samples 1971–1973	80–220 (mean 144)	Herron et al. (1977a)
7 sites along a 400-km East–West axis via Milcent and Crete	20 samples 1973–1974	130–340 (mean 245)	Boutron (1979b)
Dye 3	16 samples 1978–1979	42–150 (mean 120)	Davidson et al. (1981)
II *Old Ice*			
Camp Tuto	1 blue ice block 2700 B.P.	$\leqslant 1$	Murozumi et al. (1969)
Dye 3	31 samples 1232–1915	$\leqslant$ 40 to 300 (mean 70) no increase trend	Cragin et al. (1973)
Milcent	21 ice core sections 1170–1886	$\leqslant$ 20 to 90 (mean 45) no increase trend	Herron et al. (1977a)
Milcent	75 ice core sections 1866–1883	19 to 2200	Herron (1980)
Camp Century	3 ice core sections 2700 B.P., 4500 B.P., and 5500 B.P.	$\leqslant 4.1$, $\leqslant 3.9$, and $\leqslant 1.6$	Ng and Patterson (1981)

obtained high quality data from three sections of a deep ice core drilled at Camp Century (Table 15.2) and showed beyond doubt that all Pb data previously obtained by Langway's group from the analysis of deep ice cores (Cragin et al., 1973; Herron et al., 1977a; Herron, 1980) were highly erroneous, mainly because of the inefficiency of the rinsing decontamination procedure (see Sec. 3.3) they used. The concentrations in several thousand years old Greenland ice are definitively shown to be close to 1×10^{-12} g Pb/g, which corresponds to EF_{crust} values near unity, as previously suggested by Murozumi et al. (1969) from the analysis of a single block of old ice collected in the coastal ablation area near Camp Tuto (Table 15.2). The Pb concentrations are moreover shown to have continuously increased about 200-fold from prehistoric times to present (Fig. 15.5), that is, a much larger increase factor than the one which was obtained by Cragin et al. (1973) and Herron et al. (1977a). This tremendous 200-fold increase is clearly related to increasing

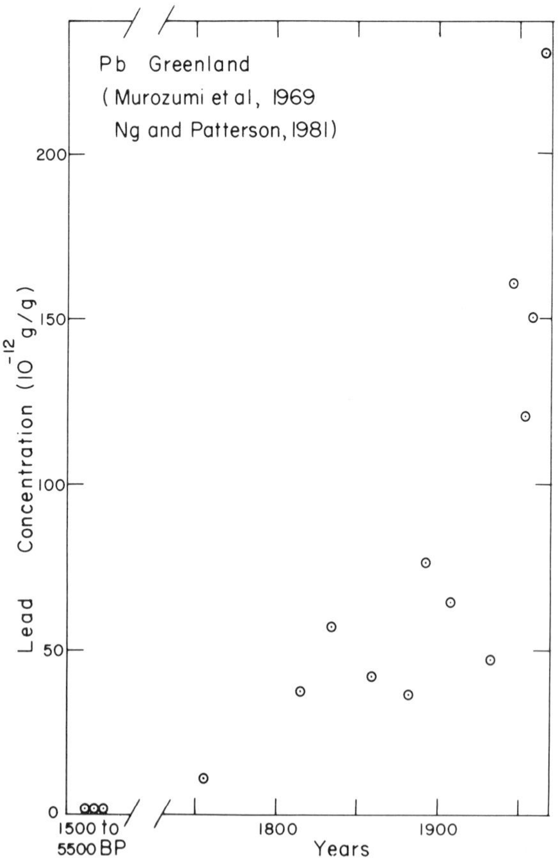

Figure 15.5 Northwest Greenland: measured variations in Pb concentrations as a function of time from prehistoric times to present (from Murozumi et al., 1969; Ng and Patterson, 1981).

human-induced Pb emissions to the atmosphere in the northern hemisphere (Nriagu, 1978, 1979). The high EF_{crust} values obtained for Pb in present-day Greenland surface snow are then caused by anthropogenic effects.

For Antarctic snows, the recent analysis by Boutron and Patterson (1983) of ultraclean samples covering the last two centuries which were collected at stake D 55 in Adelie Land has shown that most previously published data were in high positive error. The possible exceptions may be the lowest values published by Murozumi et al. (1969) for the Byrd Station area and by Boutron (1982) for the geographic South Pole and the recent data obtained by Landy and Peel (1981) and Peel and Wolff (1982) for the Spaatz Island area in the Antarctic Peninsula. It appears that the concentrations in present-day central Antarctic surface snows are in the 1 to 5.10^{-12} g Pb/g range, corresponding to EF_{crust} values smaller than ten. Apparently there has been no significant increase in Pb concentrations during the last few centuries (Fig. 15.6), despite significant fluctuations which have been tentatively ascribed to time variations of the volcanic influence. For old Antarctic ice, the only available data were published recently by Ng and Patterson (1981) (analysis of two ice core sections 1490 and 2010 years old obtained at Byrd Station) and by Boutron and Patterson (1983) (analysis of one block of ice more than 12,000 years old collected in the coastal ablation area at Cap Prudhomme). These data show that the concentrations in very old Antarctic ice range from 1 to 2×10^{-12} g Pb/g, depending on whether the ice has been deposited during Holocene or Wisconsin times. The corresponding EF_{crust} values are of

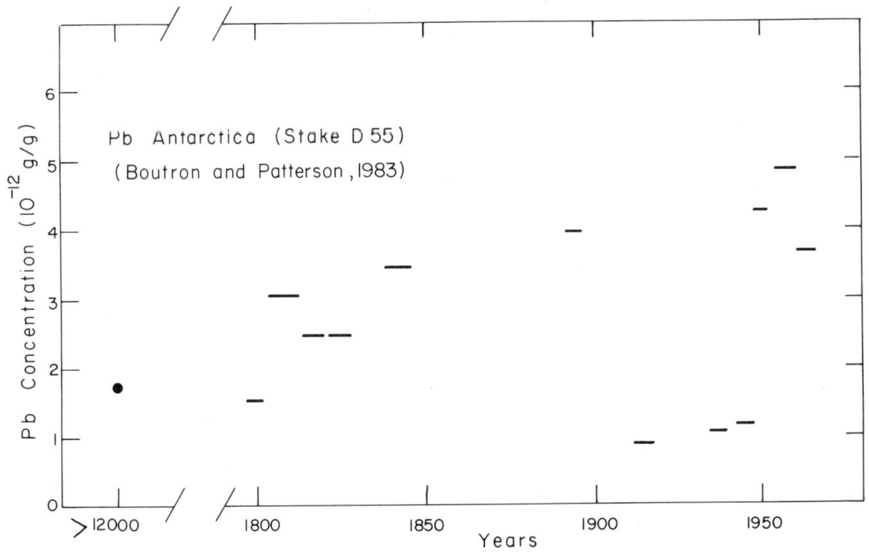

Figure 15.6 Stake D 55 and Cap Prudhomme, Antarctica: measured variations in Pb concentrations as a function of time from prehistoric times to present (from Boutron and Patterson, 1983).

the order of 3 to 5. It appears then that there has been no significant increase of Pb concentrations in Antarctic snow and ice from prehistoric times to present, contrary to what is observed in Greenland. One may conclude that the remote polar areas of the Southern Hemisphere are still little affected by anthropogenic Pb emissions to the atmosphere. This dissimilarity between Greenland and Antarctica is consistent with the fact that more than 90% of anthropogenic Pb emissions to the atmosphere occur in the Northern Hemisphere (Nriagu, 1978, 1979) and that the Equator is a barrier for tropospheric aerosols.

It is interesting to observe that very old Antarctic ice has been found by Boutron and Patterson (1983) to contain about 1×10^{-12} g Pb/g natural excess Pb above silicate dust Pb. This natural excess Pb is suggested by Boutron and Patterson (1983) to originate in part from volcanoes (about 0.1×10^{-12} g Pb/g) and from emissions by the surface Pb enriched microlayer of the oceans (about 0.25×10^{-12} g Pb/g). But there is probably some extra natural excess Pb not provided by volcanoes or sea spray. These other natural contributions could come from gaseous emissions by methylation, direct volatilization from rocks, natural Pb emissions from plant leaves, or other unknown sources. However, none of these possible natural excess Pb sources has been conclusively demonstrated.

5.2. Mercury-201

This metal is extremely difficult to measure at the very low concentration levels which are observed in polar snow and ice. All Hg concentration data presently available for Greenland and Antarctic snow and ice are thought to be somewhat questionable.

As shown in Table 15.3, published data on Hg in present-day Greenland surface snows are very scattered, ranging from 7 to 900×10^{-12} g/g. The most reliable ones are probably those published by Appelquist et al. (1978) for the snow layers deposited at Crete from 1957 to 1971 (Table 15.3) which vary from 7 to 12×10^{-12} g Hg/g corresponding to mean EF_{crust} values of 1300. But even these data are probably questionable mainly because Appelquist and co-workers did not evaluate the efficiency of the mechanical subsampling procedure they used to decontaminate the core sections they analyzed (see Sec. 3.2). Regarding old Greenland ice, there are no data for very old ice; the only available data are for ice 1 to 3 centuries old, but these data are very scattered, ranging from 2 to 820×10^{-12} g Hg/g (see Table 15.3). The most reliable ones are probably again the 2 to 19×10^{-12} g Hg/g corresponding to mean EF_{crust} values of about 800, published by Appelquist et al. (1978) for Crete and Dye 3. Even these last data are questionable for the same reasons as discussed above. These data by Appelquist et al. (1978) do not indicate any clear time trend for Hg during the last three centuries (Fig. 15.7).

Table 15.3. Published Data on the Occurrence of Hg in Greenland Snow and Ice

Location	Description of Samples	Hg Measured Concentrations (10^{-12} g/g)	Reference
I *Recent Snow*			
80 km ESE Camp Century	8 samples 1946–1965	53–230 (mean 60)	Weiss et al. (1971b)
Camp Century, Site 2	2 samples 1930–1940	50–100 (mean 75)	Carr and Wilkniss (1973)
Dye 3	20 samples 1966–1971	29–73 (mean 48)	Weiss et al. (1975)
Milcent	6 samples 1971–1972	290–881 (mean 494)	Herron et al. (1977b)
Crete	6 samples 1957–1971	7–12 (mean 11)	Appelquist et al. (1978)
II *Old Ice*			
Camp Century, Site 2	2 samples 1850–1870	13–169 (mean 91)	Carr and Wilkniss (1973)
Milcent	5 samples 1631–1864	263–823 (mean 513)	Herron et al. (1977b)
Crete, Dye 3	15 samples 1727–1786	2–19 (mean 7)	Appelquist et al. (1978)

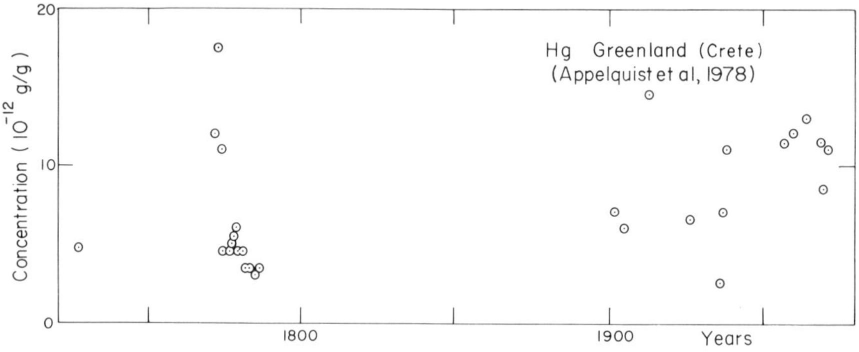

Figure 15.7 Dye 3 and Crete, Greenland: measured variation of Hg concentrations as a function of time from 1700s to present (from Appelquist et al., 1978).

Regarding the Antarctica, the only Hg data presently available are those published by Murozumi et al., 1978. These authors have analyzed various sections of a core drilled from the surface down to 145 m at Mizuho Camp. No age is given for the samples, but a simple calculation using present-day annual accumulation rates at this site suggest that the core integrates approximately the last 800 years. Most concentration values are in the 0.1 to 3×10^{-12} g Hg/g range (EF_{crust} from about 100 to 3000), without any clear time trend. In the upper part of the core, higher concentrations up to 50×10^{-12} g Hg/g are, however, observed for several of the snow core sections. The reliability of all these Hg data by Murozumi et al. (1978) is difficult to assess, since these authors give no indication in their paper on the decontamination technique, if any, they used to clean their core samples. The feeling is that the highest values they obtain entail some contamination of the corresponding core sections, and that their low concentration values in the 0.1 to 3×10^{-12} g Hg/g range are possibly only upper limits of the original concentrations.

5.3 Antimony-122

Almost nothing is known about the occurrence of this interesting toxic metal in Greenland and Antarctic snow and ice. The only available data for Greenland are believed to be those published by Weiss et al. (1975) for 20 surface snow samples covering the 1966–1971 time period. The samples were collected at Dye 3 and the values reported ranged from 8 to 91×10^{-12} g Sb/g, which give a mean EF_{crust} value of 645. For Dome C in Antarctica, Buat-Menard and Boutron (1982) found ≤ 3 to 5×10^{-12} g Sb/g (corresponding to EF_{crust} values of ≤ 800 to 3000) in 9 snow samples whose age ranged from 1881 to 1977. Boutron et al. (1984) gave an upper limit of $\leq 21 \times 10^{-12}$ g Sb/g (EF_{crust} ≤ 2700) for ice more than 12,000 years old collected at Cap Prudhomme. The reliability of all these Sb data is probably questionable.

5.4 Cadmium-112

There are rather numerous data for this metal both for Greenland and for Antarctica, but it is difficult to clearly assess their reliability.

Concentrations of Cd in Greenland surface snows are listed in Table 15.4. The very high values, in the 200 to 600 $\times 10^{-12}$ g Cd/g range, obtained by Weiss et al. (1975) at Dye 3 appear now to be erroneous. The other data suggest that the concentrations in present-day Greenland surface snows are in the 1 to 30 $\times 10^{-12}$ g Cd/g range, which corresponds to mean EF_{crust} values of about two orders of magnitude. The available data show no large geographical variations, but their reliability will need to be checked in the future. Regarding old Greenland ice, there are presently no data with the exception of some values for ice up to 800 years old published by Herron et al. (1977a) (see Table 15.4). These data indicate concentrations in the 5–14 $\times 10^{-12}$ g Cd/g range, that is, about what is found in present-day snows, without any increasing trend from 1170s to present times (see Fig. 15.8). Herron et al. (1977a) suggested that anthropogenic inputs of Cd are still insignificant in Greenland. The reliability of these values by Herron and co-workers is questionable. The highly contaminated deep ice core sections they analyzed for Cd were indeed decontaminated by the rinsing technique described in Section 3.3 which has been subsequently shown by Ng and Patterson (1981) to be insufficient for Pb study. It will be necessary in future to check whether this was also true for Cd or not by drawing curves showing the measured Cd concentrations as a function of the percentage of ice mass removed by the rinsing procedure.

Continuous profiles showing time variations of Cd concentrations in Antarctica snow during the last century have been published by Boutron and Lorius (1979) and Boutron (1980) for Dome C, and by Boutron (1982) for the geographic South Pole (see Fig. 15.9). Mean concentrations in surface snows at these two sites are found to be about 5 $\times 10^{-12}$ g Cd/g, corresponding to mean EF_{crust} values of about three orders of magnitude. No increasing trend is observed during the last century, despite rather large fluctuations of concentrations (see Fig. 15.9). These fluctuations are not clearly related to known volcanic events or other discontinuous phenomena and could be due to contamination problems. Studies by Landy and Peel (1981) and Peel and Wolff (1982) on surface snow deposited in 1971–1979 in the Spaatz Island area in the Antarctic Peninsula show Cd concentrations in the 1 to 10 $\times 10^{-12}$ g Cd/g range, with short-term variations which were ascribed to meteorological processes. For old Antarctic ice, the only available information has been published by Boutron et al. (1984) who analyzed a single block of ice more than 12,000 years old collected at Cap Prudhomme. They found 2.6 $\times 10^{-12}$ g Cd/g (EF_{crust} of 300), a value not significantly different from the ones observed in present-day Antarctic surface snows. They suggested that there is still little evidence of man-made Cd in the remote areas of the Southern Hemisphere. All of these Antarctic Cd values will need to be confirmed in the future.

Table 15.4. Published Data on the Occurrence of Cd in Greenland Snow and Ice

Location	Description of Samples	Cd Concentrations (10^{-12} g/g)	Reference
I *Recent Snow*			
80 km ESE Camp Century	5 samples 1964–1965	≤ 1–7 (mean 3.2)	Weiss et al. (1975)
	6 samples 1815–1960	4–42, no increase trend	
Dye 3	20 samples 1966–1971	140–2870 (mean 639)	Weiss et al. (1975)
Milcent	6 samples 1971–1973	3–10 (mean 7.6)	Herron et al. (1977a)
7 sites along a 400-km East–West axis via Milcent and Crete	20 samples 1973–1974	0.7–33 (mean 8.7)	Boutron (1979b)
Dye 3	16 samples 1978–1979	7.8–< 19 (mean 13.0)	Davidson et al. (1981)
II *Old Ice*			
Dye 3	12 samples 1802–1915	≤ 2–144 (mean 34) no increase trend	Weiss et al. (1975)
Milcent	21 ice core sections 1170–1886	5–14 (mean 8) no increase trend	Herron et al. (1977a)
Milcent	104 ice core sections 1866–1883	5–2400	Herron (1980)

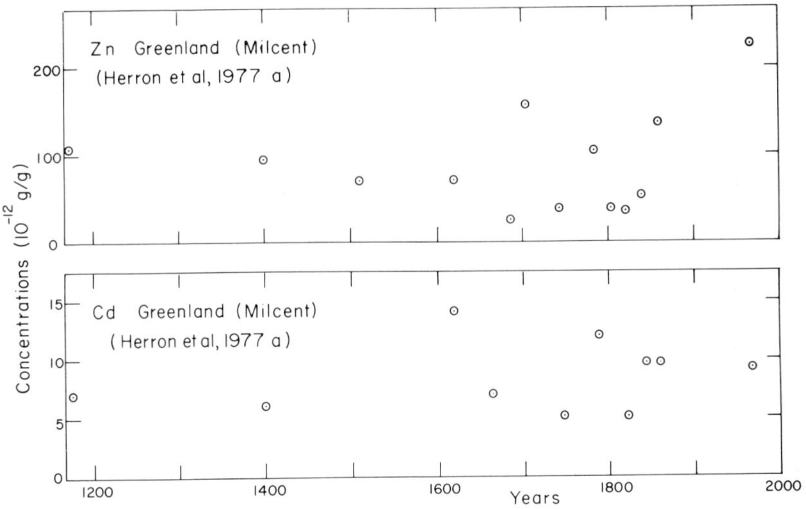

Figure 15.8 Milcent, Greenland: measured variations of Cd and Zn concentrations as a function of time from 1170s to present (from Herron et al., 1977a).

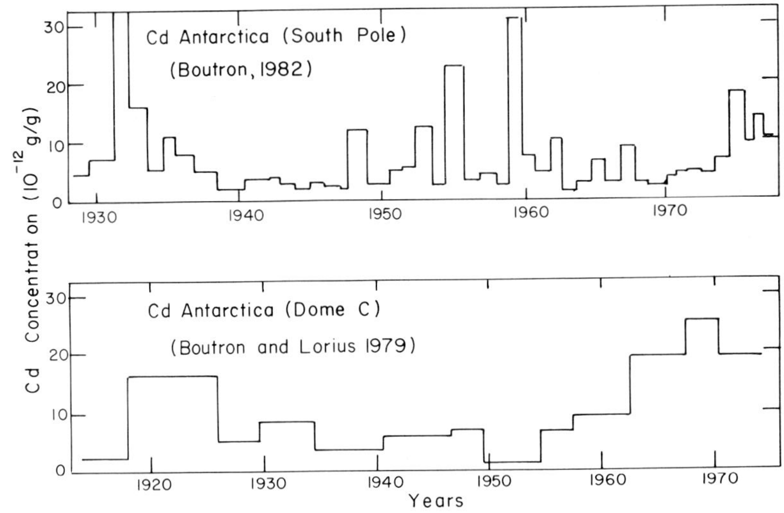

Figure 15.9 Geographic South Pole and Dome C, Antarctica: measured continuous variations of Cd concentrations as a function of time during the last century (from Boutron and Lorius, 1979; Boutron, 1982).

5.5. Silver-108

For Greenland, the only published values are for surface snows. Boutron (1979b) obtained mean concentrations of 8.4×10^{-12} g Ag/g (range 6.8–10.5 $\times 10^{-12}$ g Ag/g) corresponding to EF_{crust} values of 1000. The 20 samples collected in 1973–1974 at 7 sites along a 400-km East–West axis showed no significant geographical variations. Davidson et al. (1981) found about 2.6×10^{-12} g Ag/g in a Dye 3 1978 surface snow sample.

For Antarctica, there are continuous variation profiles for Ag for the last 100 years at Dome C (Boutron and Lorius, 1979; Boutron, 1980) and for the last 50 years at the geographic South Pole (Boutron, 1982), as shown in Figure 15.10. Mean concentrations observed are similar at the two sites, being about 5×10^{-12} g Ag/g (EF_{crust} of about 7000). No significant increase of concentrations is observed, despite fluctuations of concentrations with time which are not clearly understood. These fluctuations are interpreted by Boutron as indicating that there is no significant anthropogenic effect on Ag in the remote polar areas of the Southern Hemisphere. Other Ag values were published by Warburton et al. (1973) for numerous snow samples covering the years 1947–1970 plus one sample about one century old collected at Byrd Station, Siple Station, and the geographic South Pole. They obtained very low values in the 0.3 to 1.6×10^{-12} g Ag/g range (mean EF_{crust} 800), without any clear temporal trend.

The reliability of all these Ag values is difficult to assess and needs to be confirmed.

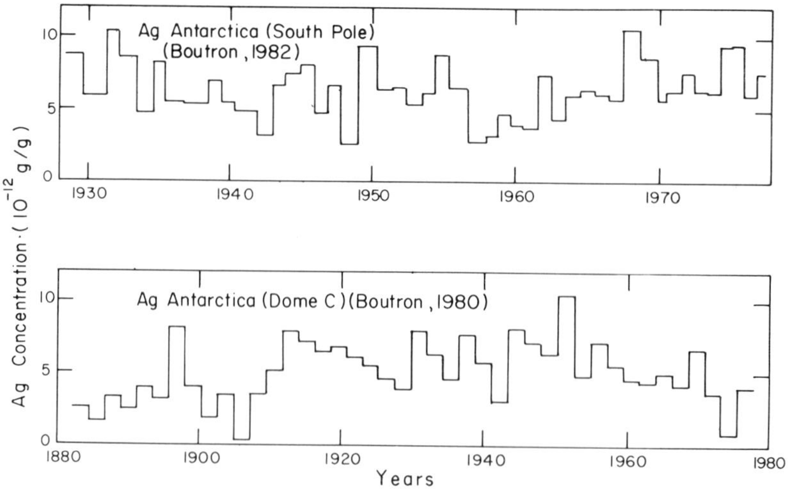

Figure 15.10 Geographic South Pole and Dome C, Antarctica: measured continuous variations of Ag concentrations as a function of time during the last century (from Boutron, 1980, 1982).

5.6 Selenium-79

There are only very few data for this interesting toxic metalloid. Weiss et al. (1971a) have measured Se in various Greenland snow samples ranging from 1815 to 1965 collected 80 km east southeast of Camp Century, and a single sample of ice 2700 years old collected at Camp Tuto. They found a mean concentration of 12×10^{-12} g Se/g (EF_{crust} of 2000) without any significant temporal trend.

For Antarctica, several Se values have been published by Buat-Menard and Boutron (1982) (10 samples in the 1881–1977 time period collected at Dome C), Weiss et al. (1971a) (a single 1724 sample collected at Byrd Station), and Boutron et al. (1984) (a single block of ice more than 12,000 years old). The concentrations observed range from ≤ 8 to 30×10^{-12} g Se/g (EF_{crust} from 3,000 to 20,000), and no temporal trend is observed. The reliability of these Se data are questionable.

5.7 Arsenic-75

There are very few data for this metalloid, and their validity is again difficult to assess. For Greenland, the only available data are for recent snow. Weiss et al. (1975) got a mean value of 10×10^{-12} g As/g (range of 2 to 38×10^{-12}) for the 1966–1971 snow layers at Dye 3. Davidson et al. (1981) obtained 19×10^{-12} g As/g for a single 1978 snow sample at the same location. The corresponding EF_{crust} values are about 50. For Antarctica, Buat-Menard and Boutron (1982) and Boutron et al. (1984) obtained only upper limits (≤ 8 and $\leq 4 \times 10^{-12}$ g As/g) for surface snow collected at Dome C and for ice more than 12,000 years old collected at Cap Prudhomme.

5.8. Zinc-65

Concentrations published for Greenland surface snows are summarized in Table 15.5. These values are very scattered, but there is now some evidence that the highest values reflect unsolved contamination problems. The concentrations in recent Greenland snows are probably in the 100 to 400×10^{-12} g Zn/g range, which correspond to EF_{crust} values of 10 to 50. Apparently, there are no large geographical variations. There are no published data for Greenland ice several thousand years old, but there are some values published by Herron et al. (1977a) for ice deposited from the 1170s to 1886 (see Table 15.5 and Fig 15.8). These indicate concentrations in the 25 to 150×10^{-12} g Zn/g range (mean of 77×10^{-12} g Zn/g; mean EF_{crust} of 14) and no significant long-term temporal trend. Herron et al. (1977a) suggested that these concentrations were substantially smaller than in present-day snows, and that there has been a three- to fourfold increase of Zn concentrations in Greenland snow and

Table 15.5. Published Data on the Occurrence of Zn in Greenland Snow and Ice

Location	Description of Samples	Zn Measured Concentrations (10^{-12} g/g)	Reference
I *Recent Snow*			
Dye 3	20 samples 1966–1971	140–2870 (mean 1047)	Weiss et al. (1975)
Milcent	5 samples 1971–1973	174–270 (mean 224)	Herron et al. (1977a,b)
7 sites along a 400-km East–West axis via Milcent and Crete	20 samples 1973–1974	97–1045 (mean 433)	Boutron (1979b)
Dye 3	16 samples 1978–1979	140–<580 (mean 290)	Davidson et al. (1981)
II *Old Ice*			
Dye 3	3 samples 1807–1859	70–290 (mean 180)	Weiss et al. (1975)
Milcent	12 ice core sections 1170–1886	25–156 (mean 77) no increase trend	Herron et al. (1977a)
Milcent	104 ice core sections 1866–1883	50–11500	Herron (1980)

ice during the last century which can be attributed to increasing anthropogenic emissions of Zn to the atmosphere in the Northern Hemisphere (Nriagu, 1979). These Zn values by Herron and co-workers will, however, need to be checked in the future, as the deep ice core sections they analyzed were decontaminated by the rinsing procedure (see Sec. 3.3) whose efficiency has not been evaluated.

For Antarctica snows, Boutron and Lorius (1979) and Boutron (1980, 1982) have published continuous variation profiles of Zn for time periods up to one century at Dome C and at the geographic South Pole. Landy and Peel (1981) and Peel and Wolff, (1982) have studied Zn concentrations in recent snow in the Spaatz Island area. They obtained mean concentrations of about 50×10^{-12} g Zn/g (EF_{crust} of 50), with short-term variations of concentrations but without any increasing trend during the last century. Regarding old Antarctic ice, Petit et al. (1981) analyzed numerous sections of the Dome C 905-m deep ice core which spans some 32,000 yr. During the 2,700 to 28,000 B.P. time period, they found scattered concentrations in the 5 to 95×10^{-12} g Zn/g range (Fig. 15.11) but without any clear time trend. This corresponds to EF_{crust} values of about one order of magnitude, except for the last stage of Wisconsin (approximately 12,000 B.P. to 23,000 B.P.) during which EF_{crust} values are close to unity probably mainly because of the high Al values observed during this stage (Petit et al., 1981). Besides this work on the Dome C deep core, Boutron et al. (1984) obtained about 60×10^{-12} g Zn/g (EF_{crust} of 22) for a block of ice more than 12,000 years old collected at Cap Prudhomme. The reliability of all these Antarctic data will need to be checked in the future, especially those by Petit et al. (1981) which used contaminated deep cores decontaminated by a rinsing procedure (see Sec. 3.3) whose efficiency is questionable.

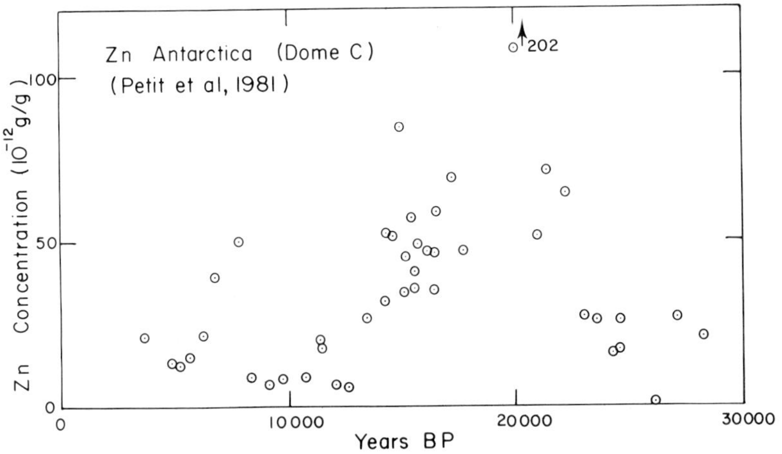

Figure 15.11 Dome C, Antarctica: measured variations of Zn concentrations as a function of time from 3000 to 28,000 B.P. (from Petit et al., 1981; Briat, personal communication).

5.9. Copper-64

Concentrations of this metal in Greenland surface snows are probably in the 30 to 100 × 10^{-12} g Cu/g range (EF_{crust} about 10) and show no large geographical variations (Boutron, 1979b; Davidson et al., 1981). These values are much lower than what was suggested by Weiss et al. (1975) from the analysis of strongly contaminated samples. There are, actually, no reliable values for old Greenland ice, since the very high values published by Weiss et al. (1978) (840 × 10^{-12} g Cu/g for a single block of ice 2700 years old collected at Camp Tuto) and by Herron (1980) (18 to 14100 × 10^{-12} g Cu/g from the analysis of numerous sections about one century old of a Milcent ice core) appear to be highly unreliable.

For Antarctic snows, there are continuous variation profiles for the last century at Dome C (Boutron and Lorius, 1979; Boutron, 1980) and for the last 50 years at the geographic South Pole (Boutron, 1982), shown in Figure 15.12. The Cu concentrations are found to be in the 15 to 60 × 10^{-12} g Cu/g range (EF_{crust} of 25 to 100) without any discernable temporal trend despite significant short-term variations. For old Antarctic ice, the only Cu value is for ice more than 12,000 years old collected at Cap Prudhomme with a concentration of 15 × 10^{-12} g Cu/g (EF_{crust} of 8), determined by Boutron et al. (1984).

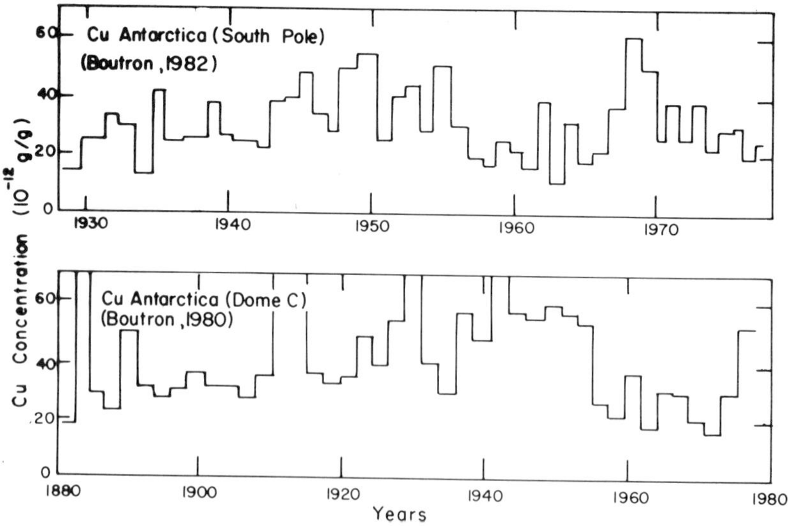

Figure 15.12 Geographic South Pole and Dome C, Antarctica: measured continuous variations of Cu as a function of time during the last century (from Boutron, 1980, 1982).

The reliability of these Cu data is not definitively proven and will need to be confirmed in the future.

5.10. Conclusions

From this brief review, it appears that our understanding of the occurrence of toxic metals and metalloids in Greenland and Antarctic snow and ice layers is still very limited. There are indeed some data, at least for Pb, Cd, Ag, Cu, and Zn, but many of them are highly questionable, mainly because the authors have not clearly demonstrated that they have completely resolved contamination problems during field sampling, laboratory analysis, or both. This is especially the case for old ice, as most data have been obtained through the analysis of highly contaminated deep cores decontaminated using techniques whose effectiveness was not demonstrated.

In Greenland, Pb concentrations are definitively shown to have strongly increased from prehistoric times to present, but surprisingly, no clear increase has been observed for the other metals and metalloids with the possible exception of Zn. This lack of increase is not consistent with global inventories of natural and anthropogenic emissions of toxic metals and metalloids to the atmosphere (Nriagu, 1979; Lantzy and MacKenzie, 1979) which indicate that for the heavy elements, the current anthropogenic fluxes to the atmosphere largely exceed natural rates in the Northern Hemisphere and have strongly increased since the last century. Future work may show that this lack of observed increase in Greenland snow and ice is due to gross unreliability of the data, especially for old ice. Regarding Antarctica, no significant increase is observed for the investigated metals and metalloids, which is in good agreement with the fact that the bulk of man-made emissions occur in the Northern Hemisphere and that the equator is an efficient barrier for tropospheric aerosol transport. This does not imply, however, that the Antarctic data are reliable.

Most available data suggest that toxic metals and metalloids were already enriched with respect to the mean crustal composition (and also with respect to bulk seawater composition) by up to several orders of magnitude in very old Greenland or Antarctic ice deposited long before man started to impact significantly on metal burdens in the atmosphere. These enrichments would then have been natural and are suggested by the various authors to be the result of emissions of metals and metalloids by volcanoes, surface microlayer of the oceans, or plant leaves; of methylation or direct volatilization from rocks; or of some other unknown natural sources. It is believed however that at least part of these enrichment data are erroneous, and more reliable studies will probably show that enrichments in very old polar ice are much lower than those presently published. This is already the case with Pb for which the most reliable data (Murozumi et al., 1969; Ng and Patterson, 1981; Boutron and Patterson, 1983) indicate very low enrichments for ancient Greenland and Antarctic ice, while previous data suggested high enrichments which were claimed to be natural.

6. METAL BURDENS IN PAST AND PRESENT-DAY ATMOSPHERE

Data reviewed in previous section pertain to temporal variations from prehistoric times to present in the concentrations of toxic metals and metalloids in Greenland and Antarctic snow and ice. A major problem now involves interpretation of these snow and ice data in terms of past variations in the concentrations of these elements in the atmosphere. To what extent do chronological changes observed in polar snow and ice reflect parallel chronological changes in atmospheric metal burdens?

6.1. Theoretical Considerations

Junge (1977) has reviewed the processes which are responsible for the trace element content of snow in clean air conditions such as the ones which are observed in the central areas of the Greenland and Antarctic ice caps, that is, areas where the mixing ratio within the troposphere is rather uniform. If we exclude gaseous phases, which are probably unimportant when dealing with toxic metals and metalloids, there are rainout processes (uptake of aerosols as condensation nuclei and attachment of unactivated aerosol particles to cloud droplets), washout processes (scavenging of aerosol particles below the cloud, and evaporation), and processes at the snow or ice surface (dry deposition, evaporation, and sublimation). From a detailed scrutiny, Junge concluded that in polar areas, the uptake of condensation nuclei is by far the most dominant process which controls the concentrations observed in deposited snow. The concentration of a given element in the air, c, should then be proportional to the corresponding concentration in the deposited snow, k, according to the relation: $c = \phi k = \phi\, \epsilon_n \eta / L$ where ϵ_n is the mass fraction of aerosols used for condensation, η is a factor linked with evaporation below the cloud, and L the liquid water content of the cloud. By assuming $0.1 \leq \epsilon_n \leq 1$, $1 \leq L \leq 3$ g/m^3 and η close to 1, the coefficient ϕ was anticipated by Junge (1977) to vary over the range $1.0 < \phi < 6.0$ g/m^3.

Experimental ϕ values for low airborne concentrations and for average samples and conditions reviewed by Junge (1977) agree fairly well with these theoretical ϕ values for a variety of elements or compounds. Most of the experimental data quoted by Junge (1977) refer to rainfall in temperate areas rather than to snowfall in polar areas. Experimental verification of Junge's theoretical predictions is needed for Greenland and Antarctica.

Experimental determination of ϕ for toxic metals and metalloids in Greenland and Antarctica is quite difficult. We have already seen the problems in the measurement of these elements in snow or ice. The same problems apply to air, since concentrations to be measured are extremely low, in the range of 10^{-10} to 10^{-13} g/m^3 air at STP. As was the case for snow, contamination problems are very difficult to solve. High volume samplers are to be used together with filters with very low blanks, and extreme care must be taken to avoid any local contamination from generators, scientific stations, vehicles, and aircrafts. It would moreover be necessary to get air measurements in the clouds, as c refers to concentrations within the cloud, but the few available data refer to concentrations at ground level.

6.2. Experimental Estimate of Present-Day ϕ Values for Greenland

There are numerous data on the concentrations of toxic metals and metalloids in Greenland air. However, most of them refer to coastal ice-free areas (Flyger and Heidam, 1978; Heidam, 1981, 1984). To our knowledge, the only values presently available for the Greenland ice sheet itself are those recently published by Davidson et al. (1981) for Dye 3 in southern Greenland. Davidson and co-workers have performed simultaneous combined air/snow sampling during the summer of 1979. Measurements were made for Pb, Ag, and Zn in the air samples, and Pb, Cd, Ag, As, Cu, and Zn in the snow samples. As shown in Table 15.6, the ϕ values obtained are close to those predicted by Junge (1977). These ϕ values are, however, very preliminary. The sampling periods were very short, during summer only, and local contamination by activities at Dye 3 could have influenced the data despite the fact that samplings were conducted several kilometers from the station and that careful control of wind direction was performed. The analytical reliability of both c and k values will need to be confirmed in the future.

6.3. Experimental Estimate of Present-Day ϕ Values for Antarctica

There are several data sets on the concentrations of toxic metals and metalloids in Antarctic air. Some of them refer to bulk air (sampling of large volumes of air on filters, then analysis of these filters), and include the studies by Zoller et al. (1974), Maenhaut et al. (1979); Cunningham (1979), Cunningham and Zoller (1981), and Peel and Wolff (1982). Others reported the analysis of individual aerosol particles (Parungo et al., 1979, 1981; Shaw, 1983).

There is presently a single study which involved simultaneous combined air/snow sampling programs—the one recently conducted for Pb, Zn, and Cd by Peel and Wolff (1982) at Spaatz Island and at a plateau site located about 100 km of Spaatz Island. However, these two sites are in the Antarctic

Table 15.6. Measured Concentrations in Air and Snow at Dye 3, Greenland[a]

Element	Concentration in Air, c^b (10^{-12} g/m^3 STP)	Concentration in Snow, k^c (10^{-12} g/g)	$\phi = c/k$ (g/m^3 STP)
Pb	$\leqslant$ 120 to 210 (150)	42 to 150 (120)	1.2
Cd		7.8 to $\leqslant$ 19 (13)	
Ag	$\leqslant$ 1.1 to 3.8 (2.3)	1.2 to 3.6 (2.6)	0.9
As		9.4 to $\leqslant$ 37 (19)	
Cu		28 to 65 (50)	
Zn	$\leqslant$ 850 to 2100 (1300)	140 to $\leqslant$ 580 (290)	4.5

[a] During discontinuous periods of summer 1979 (from Davidson et al., 1981).

[b] The values shown in parentheses are tentative mean values calculated from four c values for each element.

[c] The values shown in parentheses are tentative mean values calculated from the sixteen k values for each element.

peninsula, close to the sea coast, and not representative of central Antarctic plateau areas. Data were moreover obtained only for very short time periods during the 1980 austral summer. The c values were measured about 1 m above the snow surface, but Peel and Wolff (1982) argued that during the sampling the cloud bases were close to ground level, so that below-cloud processes were probably limited. As shown in Table 15.7, the ϕ values obtained are in the 0.34–4.27 g/m^3 range, that is, in satisfactory agreement with the theoretical values suggested by Junge (1977).

At the geographic South Pole, Zoller and co-workers have extensively monitored numerous metals and metalloids including Pb, Cd, Sb, Se, As, Zn, and Cu in near-surface air during four austral summer seasons and two overwinter periods ranging from 1970 to 1978 (Zoller et al., 1974; Maenhaut et al., 1979; Cunningham, 1979; Cunningham and Zoller, 1981). During austral summers, sampling was conducted at a remote site about 5 km from the station, but winter sampling was performed at South Pole Station itself. Great care was taken to avoid any local contamination by using wind-directional controllers. Local contamination by the station and by aircraft used to supply it can however be a problem as suggested by Boutron (1982) and Warren and Wiscombe (1980). No simultaneous snow sampling program was conducted by Zoller and co-workers, but their aerosols data can be used with the snow data published by Boutron (1982) for the geographic South Pole (Table 15.8). The ϕ values so obtained are again in good agreement with the theoretical values predicted by Junge (1977). These ϕ values must however be considered as very tentative, since the time periods during which air values were obtained do not correspond exactly to the time periods integrated by the snow samples. The c and k values shown in Table 15.8 are the best ones presently available. Some of them are, however, probably questionable. This is especially the case for Pb. The k value for this metal given in Table 15.8 is probably rather high, and the recent data by Boutron and Patterson (1983)

Table 15.7. Experimental Values of $\phi = c/k$ at Two Coastal Sites in the Antarctic Peninsula[a]

		$\phi = c/k$ (g/m^3 STP)[b]		
Location	Sampling dates	Pb	Cd	Zn
Spaatz Island (72° 58′S, 74° 41′W, altitude 408 m)	10–28 January 1980	0.34	2.51	4.27
Plateau site (74° 00′S, 70° 45′W, altitude 1130 m)	11–16 February 1980	1.89	2.22	0.71

[a] During two periods of the 1979–1980 austral summer (from Peel and Wolff, 1982).

[b] Air concentrations c in 10^{-12} g/m^3 STP; snow concentrations k in 10^{-12} g/g.

Table 15.8. Geographic South Pole, Antarctica: Comparison of Air Data Published by Zoller and Co-workers with Snow Data Published by Boutron and Co-workers.

Element	Concentrations in Air c^a (10^{-12} g/m³ STP)		Concentrations in Snow k^b (10^{-12} g/g)	$\phi = c/k^c$ (g/m³ STP)
	Winter Average	Summer Average		
Pb		27–73[d]	28	0.96[e] 2.6[f]
Hg				
Sb	2.1	0.45	$\leqslant 3$[g]	$\geqslant 0.43$
Cd	$\leqslant 200$	$\simeq 49$	8.5	~ 5.8[h]
Ag	$\simeq 1.0$	$\leqslant 3.0$	7 0.4[i]	2.5[j]
Se	6.9	6.3	16[g]	0.4
As	17	8.4	$\leqslant 8$[g]	$\geqslant 1.6$
Zn	77	35	37	1.5
Cu	79	59	28	2.5

[a] Data from several time periods during 1970–1978 (Cunningham and Zoller, 1981).

[b] Mean of nine snow samples continuously integrating the years 1970–1977 (Boutron, 1982).

[c] Value c calculated by averaging winter average and summer average.

[d] The first value refers to samples collected on Whatman filters, the second one to samples collected on Nuclepore filters (Maenhaut et al., 1979).

[e] From Whatman c values.

[f] From Nuclepore c values.

[g] No value for the South Pole; the value given here is for the 1976–1977 snow layer at Dome C (Buat-Menard and Boutron, 1982).

[h] From the summer c value.

[i] 1970 snow layer (Warburton et al., 1973).

[j] From the winter c value and k value (from Warburton et al., 1973).[i]

suggest that k lies in the 1 to 5×10^{-12} g Pb/g range. The c value for this metal, which is derived from earlier work by Maenhaut et al. (1979), is probably also too high, since it is higher than the Pb concentration recently obtained at American Samoa in the South Pacific Easterlies (Settle and Patterson, personal communication). The geographic South Pole, much more remote than American Samoa and protected by the 50° S polar convergence, should have Pb concentrations in air which are much lower than those at American Samoa.

6.4. Conclusions

The few crude experimental ϕ data presently available for Greenland and Antarctica suggest that, for average samples and conditions, concentrations in deposited snow, the k values, are proportional to concentrations in the air, c, as predicted by Junge's theory. If we assume that such a relationship has not changed markedly with time, this suggests that historical records of toxic metals and metalloids obtained from the analysis of Greenland and Antarctic snow and ice probably give access to past variations of these elements in the air in these polar areas. This relationship will obviously need to be substantiated in the future, by performing high quality measurements for longer time periods at various locations in Greenland and Antarctica, with simultaneous air/snow sampling at ground level and air measurements from the ground to the cloud bases.

We must, however, point out that these experimental data refer to present-day conditions. We cannot assume that the air–snow relationship has remained constant over time. The present-day relationship might be different from preindustrial natural ones, since present-day conditions are characterized by inputs from anthropogenic sources whose influence has possibly modified the size distribution of aerosols (at least in the northern hemisphere), which could have resulted in altered ϕ values. This relationship might also have been different under different climatic conditions such as glacial periods. It has been suggested that the size distribution of aerosols was different during ice ages (Petit et al., 1981).

For lead, Ng and Patterson (1981) have argued from various measurements at several remote sites that Pb/silicate dust ratios are systematically 10-fold smaller in air than in snow or rain collected at the same location and time. This present-day situation is attributed to submicron industrially produced, Pb-enriched aerosols and is probably anomalous to prehistoric natural times during which there was probably little or no difference between Pb/silicate ratios in the dust in air and in precipitation. If this is also true in Greenland and Antarctica, the long-term changes of Pb concentrations observed in polar snow and ice then represent only upper limits for the corresponding changes in the atmosphere. This could be true also for other toxic metals and metalloids.

Another critical point is that the air–snow relationship refers to average conditions. It would be improper to infer from this relationship that short-term variations observed in snow and ice reflect faithfully identical short-term time variations in the air. It will be necessary to establish in the future whether or not such short-term fluctuations of concentrations in the air are faithfully recorded in Greenland and Antarctic snow and ice.

REFERENCES

Appelquist, H., Jensen, K.O., Sevel, T., and Hammer, C. (1978). "Mercury in the Greenland ice sheet." *Nature* **273,** 657–659.

Boutron, C. (1979a). "Reduction of contamination problems in sampling of Antarctic snows for trace element analysis." *Anal. Chim. Acta* **106,** 127–130.

Boutron, C. (1979b). "Trace element content of Greenland snows along an East-West transect." *Geochim. Cosmochim. Acta* **43,** 1253–1258.

Boutron, C. (1980). "Respective influence of global pollution and volcanic eruptions on the past variations of the trace metals content of Antarctic snows since 1880's." *J. Geophys. Res.* **85,** 7426–7432.

Boutron, C. (1982). "Atmospheric trace metals in the snow layers deposited at the South Pole from 1928 to 1977." *Atmos. Environ.* **16,** 2451–2459.

Boutron, C., Echevin, M., and Lorius, C. (1972). "Chemistry of polar snows. Estimation of rates of deposition in Antarctica." *Geochim. Cosmochim. Acta* **36,** 1029–1041.

Boutron, C., Leclerc, M., and Risler, N. (1984). "Atmospheric trace elements in Antarctic prehistoric ice collected at a coastal ablation area." *Atmospheric Environment* **18,** 1947–1953.

Boutron, C., and Lorius, C. (1979). "Trace metals in Antarctic snows since 1914." *Nature* **277,** 551–554.

Boutron, C., and Martin, S. (1979). "Preconcentration of dilute solutions at the 10^{-12} g/g level by non boiling evaporation with variable variance calibration curves." *Analyt. Chem.* **51,** 140–145.

Boutron, C., and Patterson, C.C. (1983). "The occurrence of lead in Antarctic recent snow, firn deposited over the last two centuries and prehistoric ice." *Geochim. Cosmochim. Acta* **47,** 1355–1368.

British Antarctic Survey (1983). *Annual Report 1981–1982.* Natural Environment Research Council, Lavenham Press, Lavenham, UK, pp. 45–46.

Buat Menard, P., and Boutron, C. (1982). "Some evidence of long-distance transport, local sources and fractionation between air and snow for metallic aerosols deposited in Antarctic snows since the 1880's" (abstract). *Ann. Glaciol.* **3,** 348.

Carr, R.A., and Wilkniss, P.E. (1973). "Mercury in the Greenland ice sheet: further data." *Science* **174,** 692–694.

Cragin, J.H., Herron, M.M., and Langway, C.C. (1975). "The chemistry of 700 years of precipitation at Dye 3, Greenland." *U.S. Army Cold Regions Res. Eng. Lab. Res. Rep.* **341.**

Cunningham, W.C. (1979). "The composition, sources and sink of South Polar aerosols." PhD Thesis. University of Maryland, College Park, Md.

Cunningham, W.C. and Zoller, W.H. (1981). "The chemical composition of remote area aerosols." *J. Aerosol Sci.* **12,** 367–384.

Davidson, C.I., Chu, L., Grimm, T.C., Nasta, M.A., and Qamoos, M.P. (1981). "Wet and dry deposition of trace elements onto the Greenland ice sheet." *Atmos. Environ.* **15,** 1429–1437.

EPA (1976). "National interim primary drinking water regulations." Rep. 570/9-76-003. U.S. Environmental Protection Agency, Washington, D.C.

Flyger, H., and Heidam, N.Z. (1978). "Ground level measurement of the summer troposphere aerosols in northern Greenland." *J. Aerosol Sci.* **9,** 157–168.

Galloway, J.N., Thornton, J.D., Norton, S.A., Volchock, H.L., and McLean, R.A. (1982). "Trace metals in atmospheric deposition: A review and assessment." *Atmos. Environ.* **16,** 1677–1700.

Heidam, N.Z. (1981). "On the origin of the Arctic aerosol: A statistical approach." *Atmos. Environ.* **15,** 1421–1427.

Heidam, N.Z. (1984). "The components of the arctic aerosol." *Atmos. Environ.* **18,** 329–343.

Herron, M.M. (1980). "The impact of volcanism on the chemical composition of Greenland ice sheet precipitation." PhD Thesis. State University of New York at Buffalo, N.Y.

Herron, M.M. (1982). "Glaciochemical dating techniques." *Am. Chem. Soc. Symp. Ser.* **176,** 303–318.

Herron, M.M., Cragin, J.H., and Langway, C.C. (1973). "Improved procedures for the removal of surface contaminants from ice cores for chemical analysis." *U.S. Army Cold Regions Res. Eng. Lab. Tech. Note,* 13 pp.

Herron, M.M., Langway, C.C., Weiss, H.V., and Cragin, J.H. (1977a). "Atmospheric trace metals and sulfate in the Greenland ice sheet." *Geochim. Cosmochim. Acta* **41,** 915–920.

Herron, M.M., Langway, C.C., Weiss, H.V., Hurley, P., Kerr, R., and Cragin, J.H. (1977b) "Vanadium and other elements in Greenland ice cores." In *Isotopes and Impurities in Snow and Ice* (Proceedings of the IUGG Symposium, Grenoble, August–September 1975). IAHS Publ. 118, pp. 98–102.

Junge, C.E. (1977). "Processes responsible for the trace content in precipitation." In *Isotopes and Impurities in Snow and Ice* (Proceedings of the IUGG Symposium, Grenoble, August–September 1975). IASH Publ. 118, pp. 63–77.

Landy, M.P. (1980). "An evaluation of differential pulse anodic stripping voltammetry at a rotating glassy carbon electrode for the determination of cadmium, copper, lead and zinc in Antarctic snow samples." *Anal. Chim. Acta* **121,** 39–49.

Landy, M.P., and Peel, D.A. (1981). "Short term fluctuations in heavy metal concentrations in Antarctic snow." *Nature* **291,** 144–146.

Langway, C.C., Herron, M.M., and Cragin, J.H. (1974). "Chemical profile of the Ross ice shelf at Little America V, Antarctica." *J. Glaciology* **13,** 431–435.

Lantzy, R.J., and Mackenzie, F.T. (1979). "Atmospheric trace metals: Global cycles and assessment of man's impact." *Geochim. Cosmochim. Acta* **43,** 511–525.

Lorius, C. (1967). "A physical and chemical study of the coastal ice sampled from a core drilling in Antarctica." In *Commission of Snow and Ice* (Proceedings of the IUGG Symposium, Bern, September–October 1967). IASH Publ. 79, pp. 141–148.

Maenhaut, W., Zoller, W.H., Duce, R.A., and Hoffman, G.L. (1979). "Concentration and size distribution of particulate trace elements in the south polar atmosphere." *J. Geophys. Res.* **84,** 2421–2431.

Mitchell, J.W. (1973). "Ultrapurity in trace analysis." *Analyt. Chem.* **45,** 492A–500A.

Moody, J.R. (1982). "NBS clean laboratories for trace element analysis." *Analyt. Chem.* **54,** 1358 A–1376 A.

Moody, J.R., and Lindstrom, R.M. (1977). "Selection and cleaning of plastic containers for storage of trace element samples." *Analyt. Chem.* **49,** 2264–2267.

Moody, J.R., and Beary, E.S. (1982). "Purified reagents for trace metal analysis." *Talanta* **29,** 1003–1010.

Murozumi, M., Chow, T.J., and Patterson, C.C. (1969). "Chemical concentrations of pollutant lead aerosols, terrestrial dusts and sea salts in Greenland and Antarctic snow strata." *Geochim. Cosmochim. Acta* **33,** 1247–1294.

Murozumi, M., Nakamura, S., and Yoshida, Y. (1978). "Chemical constituents in the surface snow in Mizuho plateau." *Memoirs of National Institute of Polar Research Tokyo,* Special Issue **7,** 255–263.

Murphy, T.J. (1976). "The role of the analytical blank in accurate trace analysis." In P. La Fleur, Ed., *Accuracy in Trace Analysis.* National Bureau of Standards Special Publication 422, pp. 509–539. Washington, D.C.

Ng A., and Patterson, C.C. (1981). "Natural concentrations of lead in ancient Arctic and Antarctic ice." *Geochim. Cosmochim. Acta* **45,** 2109–2121.

Nriagu, J.O. (1978). "Lead in the atmosphere." In J.O. Nriagu, Ed., *The Biogeochemistry of Lead in the Environment,* Vol. 1A. Elsevier, Amsterdam, pp. 137–184.

Nriagu, J.O. (1979). "Global inventory of natural and anthropogenic emissions of trace metals to the atmosphere." *Nature* **279,** 409–411.

Parungo, F., Ackerman, E., Caldwell, W., and Weickmann, H.K. (1979). "Individual particle analysis of Antarctic aerosols." *Tellus* **31,** 521–529.

Parungo, F., Bodhaine, B., and Bortniak, J. (1981). "Seasonal variation in Antarctic aerosol." *J. Aerosol. Sci.* **12,** 491–504.

Patterson, C.C. (1980). "An alternative perspective—Lead pollution in the human environment: Origin, extent and significance." In *Lead in the Human Environment.* National Academy of Sciences, Washington, D.C.

Patterson, C.C. (1982). "Analysis of lead at levels of 10^{-13} to 10^{-9} g/g." *J. Assoc. Off. Analyt. Chem.* **65.**

Patterson, C.C., and Settle, D.M. (1976). "The reduction of orders of magnitude errors in lead analyses of biological materials and natural waters by evaluating and controlling the extent and sources of industrial lead contamination introduced during sample collection and analysis." In P. La Fleur, Ed., *Accuracy in Trace Analysis.* National Bureau of Standards Special Publication 422, pp. 321–351. Washington, D.C.

Peel, D.A., and Wolff, E.W. (1982). "Recent variations in heavy metal concentrations in firn and air from the Antarctic peninsula." *Ann. Glaciol.* **3,** 255–259.

Petit, J.R., Briat, M., and Royer, A. (1981). "Ice age aerosol content from East Antarctic ice core samples and past wind strength." *Nature* **293,** 391–394.

Ragone, S.E., and Finelli, R.V. (1972). "Procedure for removing surface contaminants from deep ice cores." *U.S. Army Cold Regions Res. Eng. Lab. Special Rep.* **167.**

Ragone, S.E., Finelli, R.V., Leung, S., and Wolff, C. (1972). "Cationic analysis of the Camp Century, Greenland, ice core." *U.S. Army Cold Regions Res. Eng. Lab. Special Rep.* **179.**

Rahn, K.A. (1976). "The chemical composition of the atmospheric aerosol." Technical Report. University of Rhode Island, Kingston, R.I. 273 pp.

Raynaud, D., Lorius, C., Budd, W.F., and Young, N.W. (1979). "Ice flow along an IAGP flow line and interpretation of data from an ice core in Terre Adelie, Antarctica." *J. Glaciol.* **24,** 103–115.

Shaeffer, M.D., and Davidson, C.I. (1979). "An energy saving clean laboratory." *Heating, Piping and Air Conditioning* **51,** 61–62.

Shaw, G.E. (1983). "X-ray spectrometry of polar aerosols." *Atmos. Environ.* **17,** 329–339.

Taylor, S.R. (1964). "Abundance of chemical elements in the continental crust: A new table." *Geochim. Cosmochim. Acta* **28,** 1273–1285.

Tschopel, P., and Tolg, G. (1982). "Comments on the accuracy of analytical results in ng and pg trace analysis of the elements." *J. Trace Microprobe Techn.* **1,** 1–77.

Warburton, J.A., and Young, L.G. (1973). "Neutron activation procedures for silver analysis in precipitation." *J. Appl. Meteorol.* **7,** 433–443.

Warburton, J.A., Linkletter, G.O., and Young, L.G. (1973). "Silver concentrations in Antarctic snow and firn." *Antarctic J.* **8,** 342–343.

Warren, S.G., and Wiscombe, W.J. (1980). "A model for the spectral albedo of snow. II: snow containing atmospheric aerosols." *J. Atmos. Sci.* **37,** 2734–2745.

Weiss, H.V., and Bertine, K.K. (1973). "Simultaneous determination of Manganese, Copper, Arsenic, Cadmium, Antimony and Mercury in glacial ice by radioactivation." *Anal. Chim. Acta* **65,** 253–259.

Weiss, H.V., Bertine, K., Koide, M., and Goldberg, E.D. (1975). "The chemical composition of a Greenland glacier." *Geochim. Cosmochim. Acta* **39,** 1–10.

Weiss, H.V., Herron, M.M., and Langway, C.C. (1978). "Natural enrichment of elements in snow." *Nature* **274,** 352–353.

Weiss, H.V., Koide, M., and Goldberg, E.D. (1971a). "Selenium and sulfur in a Greenland ice sheet: Relation to fossil fuel combustion." *Science* **172,** 261–263.

Weiss, H.V., Koide, M., and Goldberg, E.D. (1971b). "Mercury in a Greenland ice sheet: evidence of recent input by man." *Science* **174,** 692–694.

Wolff, E.W., Landy, M.P., and Peel, D.A. (1981). "Preconcentration of Cadmium, Copper, Lead and Zinc in water at the 10^{-12} g/g level by adsorption onto Tungsten wire followed by flameless atomic absorption spectrometry." *Analyt. Chem.* **53,** 1566–1570.

Wood, J.M. (1974). "Biological cycles of toxic elements in the environment." *Science* **183,** 1049–1052.

Zief, M., and Mitchell, J.W. (1976). *Contamination Control in Trace Elements Analysis.* Wiley, New York.

Zoller, W.H., Gladney, E.S., and Duce, R.A. (1974). "Atmospheric concentrations and sources of trace metals at the South Pole." *Science* **183,** 198–200.

16

MONITORING THE ATMOSPHERIC DEPOSITION OF METALS BY USE OF BOG VEGETATION AND PEAT PROFILES

Walter A. Glooschenko

Aquatic Ecology Division
National Water Research Institute
Burlington, Ontario

1. INTRODUCTION

Concern with the long-range atmospheric transport of contaminants, especially metals, has led to the necessity of monitoring programs. In remote northern areas of North America and Europe, precipitation sampling networks have not been in operation until fairly recently, in particular because of the costs of installation and maintenance in isolated areas. Also, until recent concern with the acid rain problem, such isolated areas have been neglected in terms of monitoring atmospheric deposition. This is not to say that remote areas do not have local sources of atmospheric input of metals. Mining and smelting activities and coal-fired electrical power plants which can serve as sources of metal input have been located in northern areas.

In areas without precipitation sampling networks, other means of environmental monitoring have been used. These involve the use of plants which accumulate metals and the use of lake sediments. In terms of plants, emphasis has been placed upon lower plants, especially mosses and lichens. However, higher plants also have been used, including grasses, ferns, shrubs, and trees. Dead plant remains, that is, peat and litter, also have been used. This chapter will examine in detail the use of vegetation in monitoring the atmospheric deposition of metals within peatlands.

The question remains of which vegetation type and ecosystem would be most satisfactory for monitoring metal inputs. Lower plants such as mosses and lichens have been used in numerous studies published in the literature, but they can grow in various habitats. For example, mosses and lichens are found on exposed soils and rocks, growing on other plants, especially tree trunks (epiphytes), or on the forest floor. In the former case, exposed mosses and lichens have the advantage of not being subjected to interception loss of metals by forest canopies. However, plants growing on rocks and soils (epilithic species) could obtain metals from the substrate, complicating the interpretation of atmospheric deposition of such metals. Thus, the choice of vegetation must be made to maximize the atmospheric input of metals and minimize the role of the substrate the plant grows on (or in the case of higher vegetation, rooted in).

The ecosystem that fulfills this requirement is the bog. The definition of a bog is an "ombrotrophic peatland, wet, extremely nutrient-poor, acid with a vegetation in which *Sphagnum* species play a very important role, tree cover less than 25 percent" (Stanek, 1977). The key word in this definition is ombrotrophic, that is, nourished by rain (DuRitz, 1949). In terms of peatlands, two types exist. The first of these are minerotrophic, that is, peatlands which receive nutrients and other chemical inputs from both soil water and precipitation. These include swamps, fens, and some marshes that have limited peat (Zoltai et al., 1973). As peat accumulates in minerotrophic peatlands, compaction leads to a decrease in permeability to water movement (Rycroft et al., 1975; Verry and Boelter, 1978). In fact, the nature of plant distribution in peatlands is basically controlled by hydrology (Ingram, 1967; Moore and Bellamy, 1974; Ivanov, 1981). When enough peat accumulates to essentially cut off the flow of mineral-rich waters from the substrate below, the peatland switches from being minerotrophic to ombrotrophic. The upper, ombrotrophic portion of a peat profile can be delineated by stable isotope techniques (Dever et al., 1982) or by radioisotopes such as tritium (Gorham and Hoffsteter, 1971). The development of peatlands is beyond the scope of this chapter; the reader is referred to other sources such as the book by Moore and Bellamy (1974) and the papers by Clymo (1978) and Tolonen (1979).

Thus the bog, being ombrotrophic, is an ideal ecosystem to measure the atmospheric input of metals. In this chapter, two major aspects will be discussed: (1) the use of living vegetation growing on the bog surface and (2) the use of metal profiles in the peat.

2. THE USE OF BOG VEGETATION IN ENVIRONMENTAL MONITORING

2.1. General Aspects of Bog Vegetation

Bogs are vegetated by different types of plants ranging from lower plants through shrubs and trees. In this section, several general aspects of the vegetation of bogs will be discussed. In general, the bog is characterized by a fairly continuous cover of *Sphagnum* moss consisting of several different species, each occupying a somewhat different habitat in the bog. For example, some *Sphagnum* species are characteristic of slightly raised hummocks on the bog surface while others occur in hollows or pools (Moore and Bellamy, 1974). This will be discussed in more detail later. Other lower plants may occur along with *Sphagnum* including lichens and other species of moss. Higher plants may then root into the peat. This vegetation consists mainly of sedges (*Carex* spp.), forbs, and low shrubs of the family Ericaceae (heaths). The latter include low shrubs such as *Ledum groenlandicum* (Labrador tea) and *Chamaedaphne calyculata* (leather leaf). Trees may also occur, predominantly *Picea mariana* (Black Spruce). Various combinations of these plants may occur, and bogs range from an open *Sphagnum* bog through a low shrub-rich treed bog. However, an ombrotrophic treed bog must not be confused with a swamp, which is a minerotrophic wet forested peatland (Stanek, 1977).

2.2. Lower Bog Plants—Mosses and Lichens

Most studies of metal deposition in bog vegetation have used *Sphagnum* moss as opposed to lichens which have been used in many atmospheric deposition related studies (Nieboer and Richardson, 1981; Puckett and Burton, 1981). In part, *Sphagnum* mosses are more common than lichens (mainly *Cladonia* spp.) in some bogs, especially in more southerly, temperate areas.

The ability of *Sphagnum* to trap metals is related to several properties of the moss and nature of the moss layer including biomass (stem) density, growth rate, structure of *Sphagnum* leaves, cation exchange processes, and strength of metal absorption (Pakarinen, 1978a). These factors, which will be discussed in more detail, lead to variability within a given bog. Regional differences between bogs then would be attributable to air quality, that is, proximity to local sources of metal emissions including industry, mining and smelting activities, power plants, and, in the case of some metals (mainly lead), highways (Pakarinen and Tolonen, 1976b). Such regional aspects will be discussed in more detail later in the paper.

An important factor to consider with *Sphagnum* is interspecies variability. The most definitive work was done by Aulio (1980) in Finland. He sampled 13 *Sphagnum* species from an ombrotrophic bog and analyzed them for the elements Ca, Mg, K, Na, Fe, Mn, Zn, and Cu. The major control appeared to be species habitat, that is, drier hummocks versus wetter hollows. The highest element in concentration was K with Ca $>$ Mg $>$ Na $>$ Fe $>$ Mn $>$ Zn $>$ Cu. He also studied intraspecific versus interspecific variation in four species of *Sphagnum* in ombrotrophic versus minerotrophic habitats (Aulio, 1982). Here, as expected, minerotrophic species had a higher concentration of elements than those from ombrotrophic environments. Thus, this study and others show that species differences are critical and one cannot compare results from different species without previous studies of interspecific variability.

Another problem is the intraspecific variability of metals at one bog site. Aulio (1982) considered the variability of metals in four species of *Sphagnum* moss collected from two sites. He found the following coefficients of variation (CV) in *Sphagnum fuscum* collected from an ombrotrophic site: Cu, 31.3%; Zn, 34.0%; Fe, 9.9%; Mn, 18.8%; Ca, 16.9%; Mg, 14.9%; K, 15.1%; and Na, 22.6%. Thus, the trace metal data exhibited more variability than the major elements. Pakarinen (1981b) measured CVs in Finnish bogs and got values for Mn of 22%; Fe, 11%; Pb and Cu, 8%; and Zn, 7%. Glooschenko et al. (1984) analyzed 10 replicate samples of *Sphagnum fuscum* from a bog in northwest Quebec and determined the following CVs: Cu, 36.1%; Zn, 8.0%; Fe, 16.1%; Pb, 38.0%; and Cr, 61.1%. Major elements were not analyzed. The highest CV found was for Cr, probably due to this metal being near the lower detection limit. In general, most researchers collect single integrated samples from throughout a bog as opposed to replicated samples.

In terms of the physiological ecology of the moss, growth rate and biomass density are an important influence upon metal uptake by mosses. Pakarinen (1977a) reported the higher the production of new plant material, the less concentration of metal within the mosses. He showed that Pb, Zn, and Mn were higher in *Sphagnum* species growing on hummocks than those species in hollows which have a faster growth rate (Pakarinen and Mäkinen, 1976). This was not true for major elements Mg, K, and ash (Pakarinen, 1978b). In another study, Pakarinen (1978c) reported higher Pb, Mn, and Fe in hummock species while Zn and Cu showed no statistically significant difference. In terms of uptake (expressed as mg/m^2 yr), hummock species exhibited a significantly lower uptake of Zn and Cu and higher Mn with no difference in uptake rate of Fe and Pb.

Besides growth rate, another important attribute influencing uptake of metals is cation-exchange capacity. *Sphagnum* possess a very high cation-exchange capacity due mainly to polyuronic acids in cell walls (Clymo, 1963; Schwarzmaier and Brehm, 1975; Kilham, 1982). Species differences occur in *Sphagnum* in terms of cation-exchange capacity (Puustjärvi, 1955). Also, this would cause differential preference for various cations (Bell, 1959). Leaf structure may also affect ion uptake (Pakarinen, 1978a). The strength of adsorption of elements can also be important in terms of resistance of leaching of metal from *Sphagnum* (Pakarinen, 1978a). This will be discussed later in terms of peat profiles of various elements. Little is known about the physiological effect of such metals upon the *Sphagnum* mosses. However, some evidence is present for the selective exclusion of potentially toxic metals by bog vegetation, especially Mn and Al (Small, 1972).

Another important question concerns the active versus passive uptake of different elements. Some elements such as N, P, and K exhibit active uptake by *Sphagnum* while others, such as Pb, are felt to be passive in terms of uptake (Pakarinen, 1978a; Pakarinen and Tolonen, 1977a). A possible method of differentiating surface adsorption versus active cellular uptake would be the use of scanning electron microscopy coupled with an elemental analyzer such as EDAX. Using lichens, Garty et al. (1979) was able to localize the site of uptake of heavy metals and identify the nature of surface metallic particulates; however, this study was not made in peatlands. In terms of ecological significance, the question of active versus passive uptake is probably unimportant. The metal still would end up in the peat upon decomposition of the *Sphagnum,* or if the living moss was consumed by invertebrates, it would make no difference where the metal was found in the moss.

Lichens, namely *Cladonia* spp., have also been used in monitoring metal deposition in bogs (Pakarinen, 1981a,b; Glooschenko et al., 1981). Such species appear to retain less major elements than mosses (Pakarinen, 1981a). The same phenomenon was also observed with metals (Pakarinen, 1981b) where *Sphagnum*/*Cladonia* ratios for the following elements were found in terms of decreasing order: Ca (6.3) > Mn (5.8) > Mg (3.3) > Cu (2.4) > Fe

(1.9) > Pb (1.8) and > Zn (1.2). Glooschenko et al. (1981) found similar results for ratios of metals in *Sphagnum* compared to *Cladonia* in northern Ontario. They reported the following: Cu, 1.4; Pb, 1.1; Zn, 1.2; Cr, 1.7; and Cd, 1.3. However, as will be discussed later, bog lichens are still suitable for regional monitoring of metal deposition.

2.3. Higher Bog Plants

As discussed earlier, low shrubs and trees can also be significant components of bog vegetation. Pakarinen and Mäkinen (1976), studying Finnish bogs, compared metals in two species of pine (*Pinus*) needles, three species of *Sphagnum,* and four species of lichens. In terms of Pb, metal contents decreased from lichens to *Sphagnum* to pine, while for Zn, lichens and mosses were similar (except one lichen species which was higher) and pine needles lower. Manganese was higher in pine than the other two groups of plants.

Twigs and stems of the shrub *Ledum palustre* collected from southern Finnish bogs were analyzed for Mn, Fe, Zn, Pb, and Cu by Pakarinen and Mäkinen (1978). Concentrations of metals were dependent upon age of leaves and stems for most elements. In terms of elements, concentrations decreased in the order Mn > Cu > Zn > Pb > Fe. Compared to peat derived from *Sphagnum fuscum,* Mn was most concentrated with a ratio of 5.75 followed by Cu (1.36). Other elements were more concentrated in the moss-derived peat: Fe, 0.17; Zn, 0.94; and Pb, 0.31. In the northern Ontario study of Glooschenko et al. (1981) the shrubs *Chamaedaphne calyculata* and *Ledum groenlandicum* were analyzed for Cd, Cr, Cu, Ni, Pb, and Zn. Compared with *Sphagnum fuscum,* levels of Cd, Cr, Cu, Pb, and Zn were lower in the leaves of the two higher plants, with Pb being relatively the lowest. In general, no significant difference was found between the two species. They also analyzed the recent year's needles of *Picea mariana* (black spruce). The needles were generally higher in Cd, Ni, and Zn and lower in Pb, while Cu results were variable.

2.4. Choice of Species

Thus, these studies have shown differences among lower versus higher plants and interspecies variability with *Sphagnum.* The choice of plant, therefore, is somewhat dependent upon the goal of the study. In general, the higher plants should be avoided due to variability induced by age of leaf, leaf position, and possibility of the higher plant's roots penetrating surface ombrotrophic peats into deeper minerotrophic peats or even underlying mineral substrates. One problem, however, is that in some areas *Sphagnum* may be absent due to killing off by SO_2 pollution from smelters. This was noted in the Sudbury, Ontario, Canada area by Glooschenko et al. (1981), in the Rouyn-Noranda,

Quebec area by Glooschenko et al. (1984) and in the southern Pennine hills of England (Ferguson et al., 1978; Ferguson and Lee, 1983). Here, one needs to sample other types of vegetation in bogs such as low shrubs.

However, if these problems do not exist, the use of *Sphagnum,* especially *Sphagnum fuscum,* is recommended. This species occurs both in Europe and North America, allowing for intercomparison of results. Other species, especially of higher plants, are more geographically restricted.

3. REGIONAL STUDIES OF ATMOSPHERIC DEPOSITION OF METALS USING BOG ECOSYSTEMS

The use of bog vegetation, especially *Sphagnum,* has been used for the regional mapping of atmospheric metal deposition. In this approach, several precautions must be taken. For example, vegetation sampling sites should be (1) treeless, (2) ombrotrophic, (3) occupied by hummocks with as little shrub cover as possible, and (4) at least 200 m from major highways to minimize metals, especially Pb, from automobile exhausts. Also, metals such as Cd, Cr, Ni, and Pb should be neither actively taken up nor easily leached out. In order to average out seasonal or year-to-year differences, several year's growth of *Sphagnum* should be collected (Pakarinen, 1978a).

3.1. European Regional Studies

The first study to utilize bog mosses was that of Ruhling and Tyler (1971) in Scandinavia. Samples of the moss *Hypnum cupressiforme* were collected from stones and stumps, (i.e., nonombrotrophic bog habitats) from a transect across southern Sweden. Also, *Hylocomium splendens,* a forest floor moss species, was collected from southern and northern Scandinavia, and *Sphagnum magellanicum* from ombrotrophic bogs in southern Sweden. In general, the metal concentrations in *Hylocomium splendens* were much higher in southwestern Sweden than in northern Norway, especially Pb and Cd which were more than 10 times higher. In southern Sweden, concentrations of Cd, Co, Cr, Cu, Ni, Pb, and Zn in *Hypnum cupressiforme* were higher in the southwestern portion of the transect compared to the northeast. No regional differences in Ca, K, and Mg were noted. Concentrations of metals in *Sphagnum magellanicum* coincided well with the forest floor mosses except for Cu. The authors also calculated deposition rate of metals in *Sphagnum magellanicum* from southern Sweden (in g/ha yr) as follows: Zn, 600; Pb, 450; Cu, 30; Ni, 20; Cr, 15; Cd, 7; and Co, 3–4. They also attributed higher concentrations in southern Scandinavia to greater human activity there compared to the north.

Pakarinen and Tolonen (1976a,b) carried out a regional survey of metals in *Sphagnum* mosses, mainly *Sphagnum fuscum,* in Finland. They analyzed samples for Fe, Mn, Zn, Cu, Pb, Cr, Ni, Mo, Cd, and Hg. *Sphagnum* from southern Finland was higher in Fe, Zn, Ni, Hg, and especially Pb (3.4 times higher), while northern sites were higher in Mn. Copper and Cr showed no distinct trend. The authors also estimated the deposition rate from *Sphagnum* data using the following relationship:

$$R = (C \times P) - L + U$$

where R = deposition rate (mg/m^{2}yr)
C = moss metal concentration (mg/kg)
P = average biomass production of moss (kg/m^2 yr)
L = leaching of element from moss
U = uptake of element from older organic matter

Two years of production data was used to minimize L and U, and the sum of $L + U$ was assumed to be zero. Comparing data using *Sphagnum fuscum* versus direct chemical monitoring data, they found moss data to underestimate Cr, Cu, and Fe by factors of 0.5, 0.6, and 0.7, respectively, while Ni, Zn, Pb, and Mn were overestimated by factors of 1.6, 1.9, 2.1, and 6.3. They felt the large discrepancy for Mn was due to biological uptake by mosses and leaching. In general, these researchers stated such moss results were reasonably good except for Mn and the moss method offered a cheaper alternative to direct chemical measurements.

Another study on metals in Finnish *Sphagnum fuscum* samples was done by Pakarinen (1978c). He wanted to determine whether metal gradients occurred in southern Finland between coastal and inland sites. He found Fe, Zn, and Pb to be significantly higher in coastal sites while Mn was lower and Cu showed no difference. This study showed Cu to be definitely lower at northern sites in contrast to Pakarinen and Tolonen (1976a,b). Pakarinen (1981a) also studied regional metal differences in bogs using lichens (three species of *Cladonia*). Regional differences in metal content were found; southern species were significantly higher in N, Fe, Zn, Pb, and Cu. Higher Ca and Mg were found in inland northern areas while no trends in K, P, and Mn were seen. The author stated that *Cladonia* was suitable for regional surveys. The regional distribution of mercury in Finland was also studied by Pakarinen and Hasanen (1983) using bog mosses and lichens. Mercury was somewhat higher in southern Finland *Sphagnum fuscum* with a mean of 40 μg/kg compared to 32 μg/kg in central and northern Finland; *Cladonia* had means of 34 and 20 μg/kg in the two areas.

Pakarinen (1981b) has summarized the metal data for northwestern Europe for *Sphagnum* using his data from Finland, Norway (Hvatum et al. 1983), and Germany (Wandtner, 1981). This data is summarized in Table 16.1.

A regional study on the distribution of the elements Pb, Zn, Cu, Cd, Cr, Ni, Co, Fe, Mn, As, Sb, and Se in Norwegian bogs was made by Hvatum et al. (1983). Peat samples were collected from depths of 3.5, 10, 20, and 50 cm.

Table 16.1. Metal Concentrations in *Sphagnum* Samples[a]

Location	Latitude	Species	Year Sampled	Concentration (mg/kg)				
				Fe	Mn	Zn	Pb	Cu
Finland, minimum	60–70° N	*S. fuscum*	1975–1980	95	53	16.9	3.8	2.6
median	60–70° N	*S. fuscum*	1975–1980	323	261	32.4	14.6	5.4
maximum	60–70° N	*S. fuscum*	1975–1980	1590	653	60.0	64.6	12.2
USSR, Zelenogorsk	60° 12′N	*S. fuscum*	1979	867	186	49.7	23.8	9.4
Norway, Botn	69° 45′N	*S. fuscum*	1978	166	37	18.3	5.3	6.5
Haram	62° 40′N	*S. rubellum*	1978	338	104	18.5	11.9	6.4
Haram	62° 40′N	*S. fuscum*	1975–1980	183	47	16.0	8.7	5.0
Ho	60° N	*S. fuscum*	1978	179	130	19.3	25.3	5.8
Rogaland	58° 45′N	*S. fuscum*	1975–1980	247	185	33.0	31.0	7.3
FRG, Ahlenmoor	53° 45′N	*S. magellanicum*	1979	283	105	53.2	31.9	14.1
Ahlenmoor	53° 45′N	*S. rubellum*	1975–1980	358	73	39.4	29.3	8.9
Lengener Moor	53° 25′N	*S. magellanicum*	1979	462	80	53.4	37.2	10.3
Scotland, Fort William	56° 46′N	*S rubellum*	1979	140	20	18.0	13.9	3.3
Wales, Aberystwyth	52° 29′N	*S rubellum*	1975–1980	225	13	22.7	15.4	3.6
Rhayader	52° 14′N	*S. rubellum*	1975–1980	891	59	68.7	38.8	4.4
Builth Wells	52° 11′N	*S. rubellum*	1975–1980	704	391	28.2	14.6	3.7

[a] Table modified from Pakarinen, 1981b.

Table 16.2. The Annual Rates of Metal Accumulation in *Sphagnum* Mosses (Species of Bog Hummocks)[a]

		Concentration (mg/m^2 yr)					
Geographic Area	Species	Fe	Mn	Zn	Pb	Cu	Reference
N. Finland (mean of A–C)	*S. fuscum*	62	75.8	5.6	2.2	0.9	Pakarinen, 1981b
Kittilä, Sokostovuoma (A)	*S. fuscum*	52	48.2	3.8	2.2	0.7	
Kittilä, Parvavuoma (B)	*S. fuscum*	78	95.4	6.4	2.1	1.1	
Sodankylä, Jänkävuopaja (C)	*S. fuscum*	55	83.8	6.8	2.4	1.0	
S. Finland (6 bogs)	*S. fuscum*	113	33.6	7.2	4.6	1.2	Pakarinen, 1978c
S. Sweden (4 bogs)	*S. magellanicum*	391	24	35.7	31.9	2.0	Ruhling and Tyler, 1971
Denmark (1 bog)	*S. magellanicum*	735	95	42	36	3.6	Aaby and Jacobsen, 1979
FRG, Niedersachsen (4 bogs)	*S. magellanicum*	nd	35.4	17.0	7.1	1.5	Wandtner, 1981
FRG, Harz (2 bogs)	*S. magellanicum*	nd	26.6	29.6	20.7	2.7	Wandtner, 1981
England (1 bog)	*S. magellanicum*	490	nd	42	63	8	Clymo, 1978

[a] The current results from N Finland (samples from 1975–76) are compared with the published data from S. Finland, S. Sweden, Denmark, FRG (Niedersachsen and Harz) and England. (Pakarinen, 1981b).

Surface (3.5-cm) peats showed an enrichment in southern Norway peats of Pb, As, Sb, Cd, Zn, and Se. The authors attributed this to greater atmospheric deposition in the south than in the north. Also, peat samples collected in 1962 to 1964 had similar elemental contents than the 1979 samples. Elements exhibited a sharp decrease in peat profiles.

Summarizing this data, most European data appears to fall into the Finnish range of concentrations; maximum values, especially for Zn, Fe, Pb, and Cu, were found in polluted areas subjected to traffic or located near smelters. Fairly high metals were reported in the FRG (Wandtner, 1981), USSR, and Wales. Pakarinen also calculated deposition rates as summarized in Table 16.2. Note that highest deposition rates occurred mostly in southern Sweden, Denmark, FRG (Harz bog), and Wales. However, different techniques were used for deposition rate measurements by Aaby and Jacobsen (1979) and Clymo (1978), so results may not be too comparable. Pakarinen also repeated analyses in 1979–1980 on *Sphagnum fuscum* sites which were analyzed in 1975–1976. Several metals declined: Fe by 48%, Zn by 39%, Pb by 48%, and Cu by 3%, while Mn increased by 75%. He felt these trends were significant except for Mn which exhibited seasonal variability.

3.2. North American Regional Studies

In contrast to the regional studies previously described for northwestern Europe, North American research on metals in *Sphagnum* has been more site-specific. For example, Gorham and Tilton (1972, 1978) studied major and minor elements in *Sphagnum fuscum* collected in Minnesota and Wisconsin, and northwest Saskatchewan, Canada. Pakarinen and Tolonen (1976a) collected samples of *Sphagnum* from a bog near Great Slave Lake and the Edmonton, Alberta area. Glooschenko and Capobianco (1978) compared metals in *Sphagnum* from sites at Porter Lake, Northwest Territories, Canada and near Moosonee in the James Bay region of Ontario. Furr et al. (1979) analyzed *Sphagnum* samples for 46 elements collected from the Adirondack Mountains of northeastern New York.

Metal data is also available from studies of specific metal emissions such as smelters at Sudbury, Ontario (Glooschenko et al., 1981; Arafat and Glooschenko, 1982), Rouyn-Noranda, Quebec (Glooschenko et al., 1984), uranium mines at Elliot Lake, Ontario (Beckett et al., 1982), and iron mining activities at Sudbury and Atikokan, Ontario (Glooschenko and DeBenedetti, 1983). These latter studies on point sources will be discussed in more detail later.

Percy (1983) has done a regional study of metals in *Sphagnum magellanicum* collected in the Maritime Provinces of Canada (New Brunswick, Nova Scotia, Prince Edward Island, and Newfoundland). He analyzed the moss samples for Cd, Co, Cr, Cu, Fe, Hg, Mn, Ni, Pb, S, and Zn. In terms of metal concentration, the following order was found: Fe $>$ Mn $>$ Zn

Table 16.3. Summary of Metal Data for North America *Sphagnum*

			Element Concentration (mg/kg)											
Location	Reference	Species	As	Cd	Cr	Co	Cu	Fe	Pb	Mn	Mg	Hg	Zn	Ni
Canada														
Great Slave Lake, N.W.T.	Pakarinen and Tolonen, 1976a	*S. fuscum*		0.2	0.5		3.2	82	0.7	51		0.053	79	
Edmonton, Alberta	Pakarinen and Tolonen, 1976a	*S. fuscum*		0.17	2.4		2.9	546	7.3	77		0.085	180	0.8
Porter Lake, N.W.T.	Glooschenko and Capobianco, 1978	*S. fuscum*		0.1	3.0		13	203	6	94		0.062	24	
Kinoje Lake, Hudson Bay Lowland, Ontario	Glooschenko and Capobianco, 1978	*S. fuscum*		0.9	3.6		14	241	23	327		0.058	35	
Sudbury, Ontario	Glooschenko et al., 1981; As, Arafat and Glooschenko, 1982; Fe, Glooschenko and DeBenedetti, 1983	*S. fuscum*	0.03–23	1.2–29 $\bar{x}=1.8$	8–35 $\bar{x}=16$		4–124 $\bar{x}=16$	302–2478 $\bar{x}=1052$	15–78 $\bar{x}=27$				21–87 $\bar{x}=84$	
Atikokan, Ontario	Glooschenko and DeBenedetti, 1983	*S. fuscum*					5	1443–7352 $\bar{x}=3276$		416			23	
Rouyn-Noranda, Quebec	Glooschenko et al., 1984	*S. fuscum*			3–16 $\bar{x}=7$		10–138 $\bar{x}=49$	679–2784 $\bar{x}=49$	20–351 $\bar{x}=140$				125–148 $\bar{x}=75$	10
Atlantic Provinces	Percy, 1983	*S. magellanicum*		0.17	1.2		2.5	353	9.4	245		0.31	24.1	1.5
Manitoba (Central), Ontario, (NE)	Pakarinen and Gorham, 1983	*S. fuscum*					2.5 6.5	580 294	4.0 6.7	177 265			12.2, 27.6	
United States														
Minnesota/Wisconsin	Gorham and Tilton, 1972	*S. fuscum*				10	6.8	620	23	300	15		33	
Adirondack Mountains, New York	Furr et al., 1979	*Sphagnum, not identified to species*	0.1	0.3	227	6.2	27	4182	78	86	12	0.11	87	1.6
Northern Minnesota	Pakarinen and Gorham, 1983	*S. fuscum*		$n=3$			35–50	521–1325	5.9–12.6	369–591			16.8–22.6	

> Pb > Cu > Ni > Cr > Co > Hg > Cd. Highest concentrations of Pb and Hg occurred in proximity to anthropogenic sources. For example, Pb was highest near the cities of St. John, New Brunswick, Halifax, Nova Scotia, and downwind of a smelter at Belledune, New Brunswick. Mercury was high in *Sphagnum* near two chloralkali plants, but several other high Hg values were found where no sources were obvious except possibly geological formations high in Hg. Other metals had a less consistent pattern in *Sphagnum magellanicum*. In general, concentrations of metals were lower in the Maritimes than in northern Ontario (Glooschenko and Capobianco, 1978) and southern Sweden (Ruhling and Tyler, 1971) in the same moss species.

Results from these studies have been summarized in Table 16.3. This allows a regional interpretation of results as Pakarinen (1981b) did for northwest Europe. Unfortunately, a lot of the data are from mining and smelting regions, which would tend to bias numbers. In these studies (Glooschenko et al., 1981; Glooschenko et al., 1984; Glooschenko and DeBenedetti, 1983; Arafat and Glooschenko, 1982) ranges are given and the low values would represent background data remote from the direct influence of the local source as will be discussed later.

3.3. Site-Specific Studies

Many of the examples previously discussed have involved regional studies and long-range atmospheric transport and deposition of metals. However, the analysis of *Sphagnum* mosses in bogs may also be used to determine relatively localized effects of point sources such as mining and smelting activities, and power plants in relatively isolated areas. This, however, can only be utilized in areas where bogs are found. Climatically, this means north temperate to subarctic climatic zones.

A study of metal deposition in relation to the Sudbury, Ontario, Canada nickel smelter was made by Glooschenko et al. (1981). They measured Cd, Cr, Cu, Ni, Pb, and Zn in *Sphagnum* mosses, a lichen (*Cladonia*), leaves from two low shrubs (*Chamaedaphne calyculata* and *Ledum groenlandicum*), and needles of *Picea mariana* (black spruce). A total of 14 sites were sampled in a northern direction from a bog 18 km from the smelter site to a remote bog 940 km away located near Winisk, Ontario on Hudson Bay. Several elements increased as Sudbury was approached (Fig. 16.1). These included Ni, Cu, Pb, and Cr. However, Cd and Zn exhibited no distinct trends. Ni levels in mosses, lichens, and shrubs were generally below detection limits beyond 85 km from Sudbury, but needles of *Picea mariana* were high enough in Ni to develop a better picture of the fallout pattern of this metal. The Cd content of needles also showed a general increase to the south. In the case of Pb, the authors felt that increased automobile activity in the south was another cause of elevated Pb besides smelter activities. The peak at 250 km (Fig. 16.1) is probably due to emissions from the Rouyn-Noranda smelter.

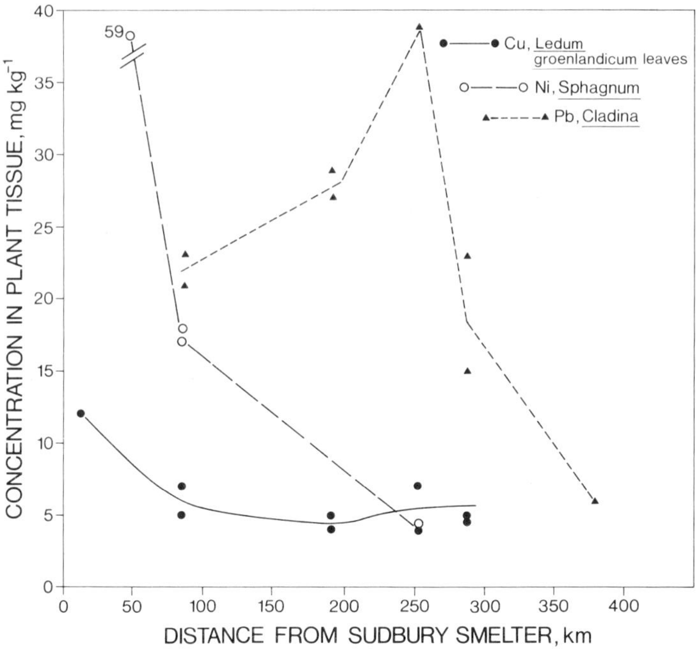

Figure 16.1. Distribution of Cu, Ni, and Pb in plants collected from bogs in Sudbury, Ontario area.

Arafat and Glooschenko (1982) analyzed the *Sphagnum, Cladonia* lichen, and shrub samples previously collected for As. In *Chamaedaphne calyculata* leaves which grew at all sites, As was highly elevated at the site 18 km from the smelter (2.3 mg/kg) with other sites all below 0.1 mg/kg. A secondary increase in As was found in the area near Timmins, Ontario, and Rouyn-Noranda, Quebec. These are sites of mining activity.

In another study related to mining activities, Beckett et al. (1982) studied the accumulation of uranium and lead in lichens and mosses in the vicinity of the Elliot Lake and Agnew Lake, Ontario uranium mining and milling operations (see Boileau et al., 1982, for sampling and analytical procedures). At the two sites, they deliniated a "macro-pollution zone", that is, a distance at which U and Pb levels were above background values. These were 22 km at Elliot Lake and 9 km at Agnew Lake. They also reported localized elevated concentrations of U and Pb near mine exhaust vents. Fe and Ti were also high in the area (Nieboer et al., 1982).

The impact of iron mining activities upon Fe content of *Sphagnum* was studied by Glooschenko and DeBenedetti (1983). Two sites were chosen: Atikokan, an iron mining locality in northwestern Ontario, and Sudbury, Ontario. At Atikokan, a level of Fe of 7352 mg/kg was found in *Sphagnum fuscum* sampled 8 km from the mine sites. However, beyond 40 to 50 km from the town, background levels of 1644 mg/kg were reached. This still was higher than Fe levels from remote areas and was attributed to other iron mining activities located to the south in Minnesota. The influence of Sudbury upon Fe concentrations extended to 200–250 km from the site. The level of Fe in a bog 18 km from the site was 2478 mg/kg, somewhat lower than that at an equivalent distance from Atikokan. However, regional background levels were much lower than Atikokan, only 464 mg/kg. The authors also estimated an Fe deposition rate in the Sudbury area between 42 and 991 mg/m^2 yr, which is of the same order of magnitude as determined by direct precipitation sampling.

Another major source of atmospheric emissions is the copper smelter at Rouyn-Noranda, Quebec. Glooschenko et al. (1984) sampled *Sphagnum fuscum* from 10 bogs within a 70-km radius of the smelter. The samples were analyzed for Cu, Ni, Pb, Zn, Cr, and Fe. The elements Cu, Pb, and Zn were high in concentration near the smelter and decreased exponentially out to approximately 50 km (Fig. 16.2). No trend was seen with Fe, Cr, and Ni except Fe and Cr were high at two sites 49 and 69 km northwest of Rouyn-Noranda. Also, As and Se were quite concentrated in mosses near the smelter.

Thus, the use of *Sphagnum* moss to monitor atmospheric deposition in relation to point emission sources appears to be a good approach in delineating areas of local influence. However, one needs a suitable series of bogs to do this. Often in remote areas, road access to such bogs is nonexistent and helicopter use may be necessary.

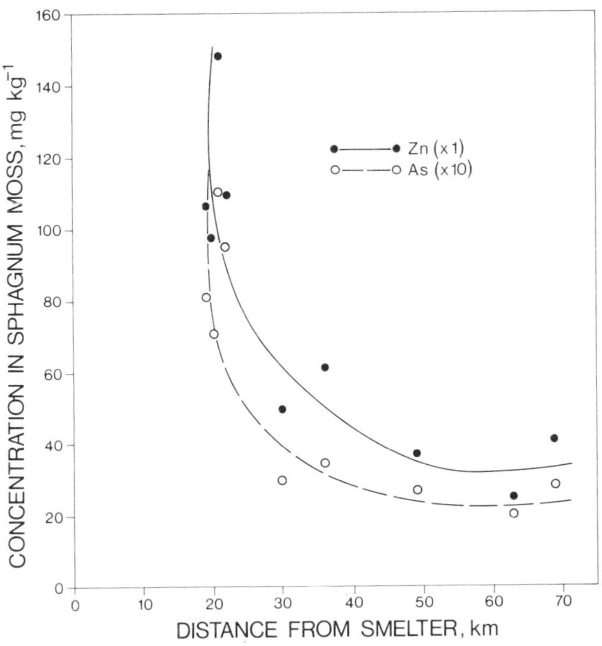

Figure 16.2. Concentration of Zn and As in *Sphagnum fuscum* in relation to distance from Rouyn-Noranda, Quebec smelter.

3.4. Moss Bag Methods

Sphagnum moss has also been utilized for monitoring atmospheric metal pollution in another method—the moss bag approach. Here, moss is suspended in fine nylon nets at various points around sources in order to obtain a synoptic monitor of metal fallout. Little and Martin (1974) used this approach on a monthly basis at 47 sampling sites around a Zn–Pb smelter in England. They were able to construct computer maps of Zn, Pb, and Cd in relation to meteorological factors and topography. Gill et al. (1975) used a moss bag approach to measure airborne metals over a 4500-km^2 area in southwestern England. He felt this approach provided relative rates of deposition and highlighted high emission areas which needed further study. In another study, Ratcliffe (1975) studied the atmospheric deposition of Pb adjacent to a battery factory in England. Goodman et al. (1975) discussed the moss bag approach in further detail, comparing this approach with total deposition and dry deposition gauges and air concentration measurements. They found a highly significant correlation between such moss bags and direct precipitation sampling.

4. THE USE OF PEAT PROFILES IN ATMOSPHERIC DEPOSITION STUDIES

4.1. Behavior of Metals in Peat

Living plants, both lower and higher, will eventually age and die. This may either be as part of the plant (i.e., leaves or roots) or the entire plant in the case of lower plants. The waterlogged, anoxic nature of the bog would then allow only partial decomposition of the plant's organic matter and peat is thus formed. This is a fairly slow process with accumulation rates of peat averaging between 2 and 8 cm per century (Moore and Bellamy, 1974). Other recent studies have gone into more detail on processes and rates of peat accumulation (Clymo, 1978; Tolonen, 1979).

During decomposition of plant material, changes are apparent in the organic composition of the plant material. For example, some constituents decrease in concentration during peat formation (including cellulose, hemicellulose, and lignin) while others increase (such as humic acid and bitumins) (Assarsson, 1961; Walmsley, 1977). Metals associated with the living plant then could be expected to change in concentration and mobility in the peat profile. This could be related to such factors as changes in ion-exchange processes in the peat and release of organic acids. These biogeochemical processes in bogs have been described in some detail in the papers of Hemond (1980) and Kilham (1982). In general, Hemond found that bogs accumulate elements, although the only metal he studied was Pb which had a net accumulation rate of 460 g/m^2 yr of which 46 mg/m^2 was stored in the peat.

Of the metal that remains in the peat (i.e., is not subjected to loss through ground water discharge from the bog), several factors can influence its concentration. First would be the nature of the peatland ecosystem, that is, ombrotrophic versus minerotrophic origins (Mornsjo, 1968; Stanek et al., 1977; Waughman, 1980). In general, in minerotrophic peats from fens, swamps, and marshes, major ions are higher. In ombrotrophic bogs, metals are found to increase (Waughman, 1980). Whether this is due to low pH, increased metals in atmospheric deposition, or a combination of both has not been determined.

Another important factor is the nature of metal adsorption and desorption in ion-exchange processes. Bunzl et al. (1976) studied ion exchange in peat and found a selective order of metals: $Pb^{+2} > Cu^{+2} > Cd^{+2} > Zn^{+2} > Ca^{+2}$. Such processes are rapid with half times of about 5–15 s. Thus, one would expect different metals to vary in behavior. Humic acids are also important, and they behave as natural organic cation exchanger (Szalay and Szilágyi, 1968). Some metals in peat exhibit very high geochemical enrichment factors (ratio for cation in peat to aqueous phase at low concentration, i.e., Fe^{3+} (2.65×10^4), Zn^{2+} (8.6×10^3), Cu^{2+} (2.38×10^3)) while others are lower: Ni^{2+} (4.51×10^2). Other processes could change metal behavior in peat such as redox potential. This important parameter influencing metal form and mobility can also vary in bogs (Urquhart and Gore, 1973).

The most comprehensive study of the distribution and movement of elements in ombrotrophic peat bogs was done by Damman (1978). He studied the distribution of the following elements in cores from these Swedish bogs: N, Na, K, Ca, Mg, P, Al, Fe, Mn, Pb, and Zn. Several observations were made in terms of metal distribution. In hummocks above the high water table, Fe, Zn, and Mn increased with depth. In the zone of water-table fluctuation, Fe accumulated and Mn decreased rapidly. In permanently wet peat, Mn, Pb, Fe, and Zn were very low in concentration. He determined the amounts of metals decreased in the order Fe > Pb, Zn, Mn. In terms of individual metals, Fe was immobile above the water table and accumulates, while in anaerobic peat, Fe becomes mobile as the Fe^{2+} ion accumulating in the water-table fluctuation zone. Manganese behaved somewhat similarly being reduced from the Mn^{4+} to Mn^{2+} state, that is, from insoluble MnO_2 to a mobile cation. It increased with depth above the high water level and is removed both from permanently anaerobic peat and the zone of fluctuating water level. Pb showed high surface concentrations due to atmospheric inputs. It accumulates in the zone of water-table fluctuations and is removed from anaerobic peat. Zn distribution resembled Fe and Pb in behavior except no surface enrichment as seen with Pb was noted. Thus, Damann showed metals to be mobile in peat profiles with water-table location and fluctuation being a major influence on metal behavior.

4.2. Case Studies—Europe

With these previous studies in mind, the use of metal profiles in peat profiles will be discussed. Earliest studies emphasized major elements, Fe and Mn. Chapman (1964) studied Fe in several raised bogs in England. He found fairly consistent values with sampling depth until minerotrophic-derived peat was reached. Here, Fe increased to much higher values. Thus Chapman's research indicated the Fe content of peat reflected its ombrotrophic versus minerotrophic origins. Mörnsjo (1968) measured Fe and Mn in peat profiles in two southern Swedish peatlands. Again, minerotrophic peats contained higher metal levels. His results indicated a slight surface enrichment with decreased Fe and Mn until lower depths where minerotrophic peat was present in the profile.

Several studies involving metals were conducted in Finland. Tanskanen (1976) studied metals in 103 peat profiles collected in Lapland and attempted to relate results in relation to peat pH, degree of decomposition, and ash content. He found pH dependency of several metals. For example, V, Co, Ni, and Cr were positively correlated with pH while Cu, Zn, and Pb were higher at lower pH values. In terms of degree of humification, Zn and Pb were highest in more decomposed peats. In relatively undecomposed peats Pb was found to be higher in surface layers. Sillanpää (1976) studied 13 elements in two peat profiles including Al, Co, Cr, Cu, Fe, Mn, Mo, Ni, Pb, Sn, Sr, V, and Zn. The general depth distribution was similar with surface enrichment, a decrease to mid-profile and strong increase at the peat–mineral substrate transition zone to a maximum in the underlying mineral soil. He felt the metals originated in the underlying mineral soil. Plants then growing on the peat deplete lower peat layers of metals causing an enrichment in upper layers where surface evaporation assisted in concentrating them. He discounted aerial inputs due to his sites being away from industrial or populated areas. Yliruokanen (1976) studied metals in 130 Finnish bogs. Surficial enrichment of Zn and Pb was found while Cu and Ni increased with depth. He also found high V and As levels in some bogs. He cautioned about the use of peat as an energy source due to problems with the metal content of flue gases and ash during disposal. Sapek (1976) also noted surficial enrichment of Pb and Cd in Polish peats and attributed it to atmospheric pollution. Thus, by the mid-1970s researchers had noted a surficial metal enrichment in bogs and the possibility of atmospheric deposition as the cause was advanced.

Further research was continued on metals in Finnish peats. Pakarinen and Tolonen (1977a) studied the distribution of Pb in *Sphagnum fuscum* and peat profiles (Fig. 16.3*b*). They found higher Pb in dead moss of several years age compared to the living moss. This was interpreted as proof that living moss did not actively take up Pb from the underlying peat substrate. A secondary peak in Pb at 20–50-cm depth occurred, and the authors suggested this was due to leaching of Pb and subsequent accumulation at the aerobic/anaerobic interface. They also felt that the cation-exchange capacity of the peat could

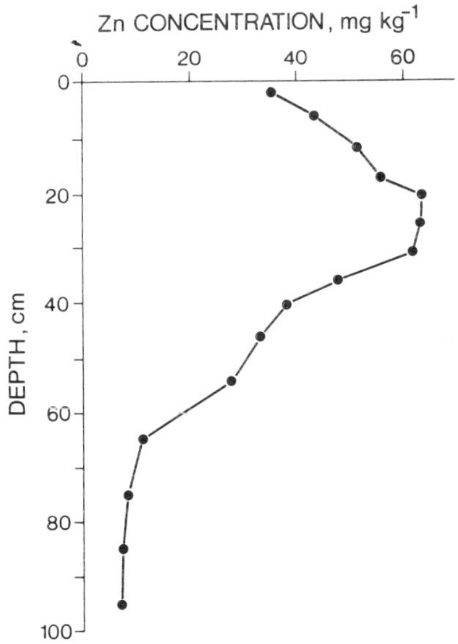

Figure 16.3(*a*) Vertical distribution of Pb in a dated peat profile from England. Figure redrawn from Livett et al., 1979. (***b***) Vertical distribution of Pb from a Finnish bog. Figure redrawn from Pakarinen and Tolonen, 1977a.

affect Pb retention. At lower depths, peat becomes more decomposed with an increased cation-exchange capacity. Their findings tend to negate the idea of Sillanpää (1976) that surface vegetation in peatlands takes up elements from deep in the peat.

Pakarinen and Tolonen (1977b) studied Pb and Zn distribution with depth. Again, Pb showed a secondary enrichment with depth. With Zn, a slight surface enrichment was seen in some profiles, but a major peak was found at 15–35 cm (Fig. 16.4). Pakarinen et al. (1981) investigated the vertical distribution of Fe, Mn, Cu, Zn, and Pb in southern Finnish bogs. The elements Mn and Cu were highest at the surface with decreasing concentrations at depth, that is, no secondary peaks at depth were seen in contrast to Zn and Fe. Then two elements were enriched at the aerobic–anaerobic interface at 20–60 cm. Lead again showed surface enrichment but did not exhibit a subsurface enrichment in all cases. In terms of mean accumulation rates, the following in units of mg/m^2/yr were found based upon the 0- to 10-cm layer of living *Sphagnum fuscum:* Fe, 125; Mn, 37; Zn, 10; Pb, 5.5; and Cu, 1.1.

In another study, Pakarinen et al. (1983) studied the long-term accumulation rate of Mn, K, Zn, Cu, Pb, Fe, Ca, and Mg in deeper anaerobic peat layers and compared it with aerobic surface peats. They dated both surface and deep peat layers and calculated an enrichment ratio: mg metal/m^2 yr in surface peat compared to deep peat. The ratios decreased in the following order: Mn, K $>$ Zn, Cu, Pb $>$ Fe, Ca $>$ Mg. They concluded that Mn and K were depleted by recycling and leaching while Fe, Ca, and Mg were stored in peat. They also cautioned that differential mobility of elements must be considered when interpreting peat profiles.

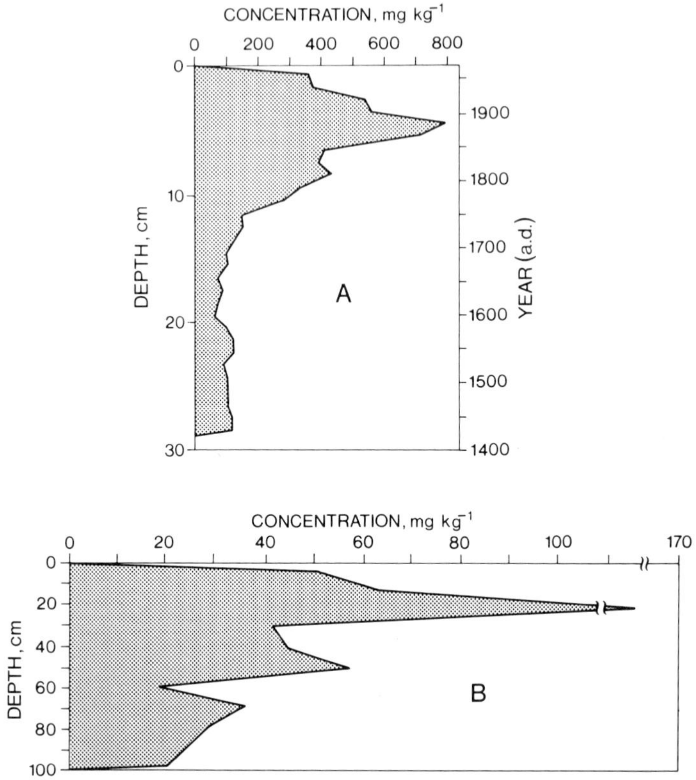

Figure 16.4. Vertical distribution of Zn in a Finnish bog. Figure redrawn from Pakarinen and Tolonen, 1977b.

In other studies in Europe, Aaby and Jacobsen (1979) studied a bog in Denmark to determine metal deposition rates. They found the following rates in units of mg/m^{2}yr: Fe, 735; Mn, 95, Zn, 42; Pb, 36; and Cu, 3.6. In England, Livett et al. (1979) analyzed Pb, Zn, and Cu in surface peats, vegetation, and peat profiles in blanket bogs. In general, they found deposition rates of Pb and Cu in surface peats to be comparable to rates determined by direct measurements of atmospheric deposition using gauges. However, Zn was lower, which was probably due to leaching from surface layers. Also, Zn was more mobile in plants and susceptible to leaching. Patterns of Pb corresponded to the intensity of Pb mining and smelting, and industrialization (Fig. 16.3*a*). The authors suggest use of peat has the advantage of smoothing out yearly and seasonal fluctuations giving a mean estimate over a 10–20 yr period. They also felt peat to be superior over other deposition estimates such as those made in snow and ice, fluvial sediments, and lake sediments. Clymo (1978) also determined accumulation rates in a blanket bog in Moor House, UK. He found the following rates in mg/m^{2}yr: Fe, 1960; Cu, 5.8; Pb, 235; and

Zn, 80. His rates are two to three times those found by Livett et al. (1979). He used a dating method based upon the ^{137}Cs peak compared to the use of pollen analysis and geochronology by Livett et al. (1979). It is possible that the ^{137}Cs method could have errors due to movement in the peat profile while pollen dating could underestimate age for less compacted surface peats. Mercury deposition was studied by Madsen (1981) in two Danish bogs. He found rates of 31 and 53 $\mu g/m^2yr$ compared to 54 and 65 $\mu g/m^2yr$ by direct precipitation analysis. Also, a general increase in Hg deposition was found over the past 100–200 yr.

A somewhat different approach to atmospheric deposition in peat was taken by Oldfield et al. (1978). They measured the saturated isothermal remanent magnetization (SIRM) at depth in three English bogs. They found a sharp increase in SIRM at a time corresponding to the late 18th to early to middle 19th century and attributed this to industrial/urban sources of magnetic particles. This increase was two to three orders of magnitude above background. Oldfield et al. (1979a) found small-scale variations in bogs of SIRM. Hummocks had a SIRM at least an order of magnitude higher than hollows. Oldfield et al. (1981) measured SIRM in Finnish peat profiles. Increases began around 1860 at the beginning of the industrial revolution and peaked after World War II. Most of this later SIRM was due to magnetite spherules due to fossil-fuel combustion. In general, the major source of these spherules is major industrial complexes although local sources (i.e., steelworks), could act as sources.

One of the big problems in the use of peat profiles is an accurate chronology. In the past, ^{14}C dating has been used but has problems of unreliability earlier than 300–500 yr B.P. (Pakarinen and Gorham, 1983). Pollen analysis, as used by Livett et al. (1979), requires historical data relating to land use and vegetation changes. Lead-210 (Hemond, 1980) has been used to date recent peats up to approximately 140 years ago. This technique was also used by Aaby and Jacobsen (1979) and Madsen (1981). Oldfield et al. (1979b) compared the radionuclide-based geochronology methods (^{210}Pb, ^{137}Cs, and ^{239}Pu) in two bogs. They determined the ^{137}Cs and ^{239}Pu methods to have serious limitations in peat dating. The ^{210}Pb method showed some evidence of sublateral mobility which could lead to an underestimation of age. But in general, the ^{210}Pb dates agreed fairly well with dates derived from pollen analyses. Much more research is necessary on recent geochronological methods and their limitations in peat profiles.

4.3. Case Studies—North America

Little information is available on the metal distribution in peat profiles in North America. Glooschenko and Capobianco (1982) collected peat samples at depths of 0 to 20 and 20 to 40 cm from 11 bogs in the James Bay region of Canada. These depths represented a minimum age of 0 to 300 and 300 to 600 yr, respectively. The peats were analyzed for Zn, Pb, Cu, Cr, Cd, and Hg. No differences were found between the two depths for Cu, Cr, and Hg. Zn was somewhat higher in surface peats, but the difference was not statistically significant. However, Pb was over twice as high in surface peats, a statistically significant difference. This was attributed to fossil-fuel combustion and subsequent atmospheric deposition. The authors showed such peat metal concentrations were similar to those found in U.S. coals and cautioned about potential pollution problems. Pakarinen and Gorham (1983) collected peat profiles from three sites in northern Minnesota, one in central Manitoba, and one in northeastern Ontario. Metals analyzed included Fe, Mn, Zn, Cu, and Pb. Surface enrichment of Mn was found in the living *Sphagnum* layer while Pb and Fe were richer in underlying peats. Cu showed only a small change with depth. In terms of deep to surface ratios, the following metal order was found: Fe $>$ Zn $>$ Pb $>$ Cu $>$ Mn. As previously discussed, Fe and Pb were enriched at the zone of water-table fluctuation. These authors also mentioned regional differences. For example, Pb and Cu were lowest at the remote Manitoba site, while Fe was lowest in northeastern Ontario. However, this site was enriched in Cu, probably from mining activity at Sudbury or Timmins. Fe was highest in Minnesota near iron mining areas, an observation made by Glooschenko and DeBenedetti (1983).

5. CONCLUSIONS

Two approaches have been used in environmental monitoring of atmospheric deposition of metals using peatland: (1) living vegetation and (2) distribution of metals in peat profiles. In conclusion, the relative merits and limitations to both approaches will be discussed.

In terms of vegetation, "first you need a bog." Emphasis was placed upon the use of ombrotrophic ecosystems, which are limited to northern temperate to subarctic climates. Thus the approach mentioned would be limited to certain geographic areas. The use of lower plants (mosses and lichens) alleviates problems with roots penetrating into minerotrophic substrates. Species growing on peat hummocks, such as *Sphagnum fuscum,* are better accumulators of metals, and choice of either a single species or few species reduce problems of interspecies variation in metal accumulation. This does not preclude a better understanding of intraspecies variability within a given bog site. In areas of high SO_2 emissions such as near smelters, one must resort to higher vegetation if mosses are killed. This leads to another question: What are the effects of metals upon mosses and lichens? How would this affect metal concentration near a point source? Another problem is separating natural atmospheric particulates such as dustfall and forest fire-derived ash from atmospheric pollution.

If rates of deposition are required, data on moss productivity is needed. Such data is time-consuming and tedious to collect and is quite rare in North

America as compared to northern Europe. Also, more information is required on the leaching of metals out of bog vegetation and possible movement upward from peat below the living vegetation layer.

Also, more baseline metal data is needed, especially from remote areas in North America such as Alaska and the Northwest Territories of Canada. Data from bogs in the Southern Hemisphere, such as those located in Tierra del Fuego, Tasmania, and New Zealand, would assist in a further understanding of global metal deposition. Another problem here is standardization of sampling and analytical methodology so results from different investigators are truly comparable.

As for peat profiles, limitations appear to be much greater. Metal mobility can lead to accumulations of metals at depths corresponding to the zone of water-table fluctuations as opposed to the surface of the peat. Converting this profile into an atmospheric deposition rate would seem extremely difficult. Even if a well-defined surface metal peak is present, dating of peat at depth appears to be a problem with limitations using geochronological methods such as the ^{210}Pb technique to get dates over the past 100 to 200 yr. Traditional approaches using palynology lead to average peat accumulation rates over long periods of time and may not give the resolution necessary over the past 200 yr or so of the industrial revolution. This is especially true since peat accumulation rates are about 0.1–1 mm/yr and surficial peats being less decomposed tend to have lower bulk densities. This leads to problems in compaction during coring and sectioning.

Thus, the use of vegetation growing in ombrotrophic bogs appears to have less problems than peat profile approaches. If such concerns exist as choice of standard sampling depths (top 2–3 cm excluding surface peat) and use of same species and analytical methodology, then an "environmental specimen bank" as proposed by Pakarinen (1982) could be implemented. This, of course, would require the conservation of bogs for such purposes, especially in areas subjected to intensive land-use pressures such as drainage and conversion to agriculture or power generation.

ACKNOWLEDGMENTS

I wish to thank Lynne Holloway for her proofreading and helpful suggestions on the manuscript.

REFERENCES

Aaby, B., and Jacobsen, J. (1979). "Changes in biotic conditions and metal deposition in the last millenium as reflected in ombrotrophic peat in Draved Mose, Denmark." *Danm. Geol. Unders.*, Årbog 1978, Copenhagen, pp. 5–43.

Arafat, N.M., and Glooschenko, W.A. (1982). "The use of bog vegetation as an indicator of atmospheric deposition of arsenic in northern Ontario." *Environ. Pollut. Ser. B* **4**, 85–90.

Assarsson, G. (1961). "Sodra Sveriges Torvillgowgar." *Kemska Analyser Sveriges Geologiska Undersokning. Stockholm Ser. C* **578.**

Aulio, K. (1980). "Nutrient accumulation in *Sphagnum* mosses. I. A multivariate summarization of the mineral element composition of 13 species from an ombrotrophic raised bog." *Ann. Bot. Fennici* **17,** 307–314.

Aulio, K. (1982). "Nutrient accumulation in *Sphagnum* mosses. II. Intra- and interspecific variation of four species from ombrotrophic and minertrophic habitats." *Ann. Bot. Fennici* **19,** 93–101.

Beckett, P.J., Boileau, L.J.R., Padovan, D., and Richardson, D.H.S. (1982). "Lichens and mosses as monitors of industrial activity associated with uranium mining in northern Ontario, Canada—Part 2: distance dependent uranium and lead accumulation patterns." *Environ. Pollut. Ser. B* **4,** 91–107.

Bell, P.R. (1959). "The ability of *Sphagnum* to absorb cations preferentially from dilute solutions resembling natural waters." *J. Ecol.* **47,** 351–355.

Boileau, L.J.R., Beckett, P.J., Lavoie, P., and Richardson, D.H.S. (1982). "Lichens and mosses as monitors of industrial activity associated with uranium mining in northern Ontario, Canada—Part 1: field procedures, chemical analysis and interspecies comparisons." *Environ. Pollut. Ser. B* **4,** 69–84.

Bunzl, K., Schmidt, W., and Scinsoni, B. (1976). "Kinetics of ion exchange in soil organic matter. IV. Adsorption and desorption of Pb^{2+}, Cu^{2+}, Cd^{2+}, Zn^{2+} and Ca^{2+} by peat." *J. Soil Sci.* **27,** 32–41.

Chapman, S.B. (1964). "The ecology of Coom Rigg Moss, Northumberland. II. The chemistry of peat profiles and the development of the bog system." *J. Ecol.* **52,** 315–321.

Clymo, R.S. (1963). "Ion exchange in *Sphagnum* and its relation to bog ecology." *Ann. Bot., N.S.* **27,** 309–324.

Clymo, R.S. (1978). "A model of peat bog growth." In O.W. Heal and D.F. Perkins, Eds., *The Ecology of Some British Moors and Montane Grasslands.* Ecological Studies, Vol. 27. Springer-Verlag, Berlin, pp. 187–223.

Damman, A.W.H. (1978). "Distribution and movement of elements in ombrotrophic peat bogs." *Oikos* **30,** 480–495.

Dever, L., Lathier, M., and Hillaire-Marcel, C. (1982). "Caractéristiques isotopiques (^{18}O, $^{13}CO_2$, ^{3}H) des écoulements dans une tourbière sur pergélisol au Nouveau-Quebec." *Can. J. Earth Sci.* **19,** 1255–1263.

DuRitz, E.G. (1949). "Huvudenheter och huvugränser i svensk myr vegetation." *Svensk Botanisk Tidskrift* **43,** 274–309.

Ferguson, P., and Lee, J.A. (1983). "Past and present sulphur pollution in the Southern Pennines." *Atmos. Environ.* **17,** 1131–1137.

Ferguson, P., Lee, J.A., and Bell, J.N.B. (1978). "Effects of sulphur pollutants on the growth of *Sphagnum* species." *Environ. Pollut.* **16,** 151–162.

Furr, A.K., Schofield, C.L., Grandolfo, M.C., Hofstader, R.A., Guntenmann, W.H., St. John, L.E., Jr., and Lisk, D.J. (1979). "Element content of mosses as possible indicators of air pollution." *Arch. Environ. Contam. Toxicol.* **8,** 335–343.

Garty, J., Galun, M., and Kessel, M. (1979). "Localization of heavy metals and other elements accumulated in the lichen thallus." *New Phytologist* **82,** 159–177.

Gill, R., Martin, M.H., Nickless, G., and Shaw, T.L. (1975). "Regional monitoring of heavy metal pollution." *Chemosphere* **2,** 113–118.

Glooschenko, W.A., and Capobianco, J.A. (1978). "Metal content of *Sphagnum* mosses from two northern Canadian bog ecosystems." *Water Air Soil Pollut.* **10,** 215–220.

Glooschenko, W.A., and Capobianco, J.A. (1982). "Trace element content of northern Ontario peat." *Environ. Sci. Technol.* **16,** 187–188.

Glooschenko, W.A., and DeBenedetti, A. (1983). "Atmospheric deposition of iron from mining activities in northern Ontario." *Sci. Total Environ.* **32,** 73–79.

Glooschenko, W. A., Holloway, D. L., and Arafat, N. (1985). "The use of mires in monitoring the atmospheric deposition of heavy metals." *Aquatic Botany,* paper submitted.

Glooschenko, W.A., Sims, R.A., Gregory, M., and Mayer, T. (1981). "Use of bog vegetation as a monitor of atmospheric input of metals." In S.J. Eisenreich, Ed., *Atmospheric Pollutants in Natural Waters.* Ann Arbor Science, Mich. pp. 389–399.

Goodman, G.T., Smith, S., Inskip, M.J., and Parry, G.D.R. (1975). "Trace metals as pollutants: monitoring aerial burdens." In *Symposium Proceedings of the International Conference on Heavy Metals in the Environment,* Vol. II, pt. 2. Toronto, Ont., 623–642.

Gorham, E., and Hofstetter, R.H. (1971). "Penetration of bog peats and lake sediments by tritium from atmospheric fallout." *J. Ecol.* **52,** 898–902.

Gorham, E., and Tilton, D.L. (1972). "Major and minor elements in *Sphagnum fuscum* from Minnesota, Wisconsin and Northwest Saskatchewan." *Bull. Ecol. Soc. Am.* **53,** 33.

Gorham, E., and Tilton, D.L. (1978). "The mineral content of *Sphagnum fuscum* as affected by human settlement." *Can. J. Bot.* **56,** 2755–2759.

Hemond, H.F. (1980). "Biogeochemistry of Thoreau's Bog, Concord, Massachusetts." *Ecol. Monogr.* **50,** 507–526.

Hvatum, Ø., Bølviken, B., and Steinnes, E. (1983). "Heavy metals in Norwegian ombrotrophic bogs." *Environmental Biogeochemistry. Ecol. Bull. (Stockholm)* **35,** 351–356.

Ingram, H.A.P. (1967). "Problems of hydrology and plant distribution in mires." *J. Ecol.* **55,** 711–724.

Ivanov, K.E. (1981). *Water Movement in Mirelands.* Academic Press, London, 226 pp.

Kilham, P. (1982). "The biogeochemistry of bog ecosystems and the chemical ecology of *Sphagnum.*" *Mich. Bot.* **21,** 159–168.

Little, P., and Martin, M.H. (1974). "Biological monitoring of heavy metal pollution." *Environ. Pollut.* **6,** 1–19.

Livett, E.A., Lee, J.A., and Tallis, J.H. (1979). "Lead, zinc, and copper analyses of British blanket peats." *J. Ecol.* **67,** 865–891.

Madsen, P.P. (1981). "Peat bog records of atmospheric mercury deposition." *Nature,* **293,** 127–130.

Moore, P.D., and Bellamy, P.J. (1974). *Peatlands.* Springer-Verlag, N.Y., 221 pp.

Mörnsjo, T. (1968). "Stratigraphical and chemical studies on two peatlands in Scania, South Sweden." *Bot. Not.* **121,** 343–360.

Nieboer, E., and Richardson, D.H.S. (1981). "Lichens as monitors of atmospheric deposition." In S.J. Eisenreich, Ed., *Atmospheric Pollutants in Natural Waters.* Ann Arbor Science, Mich., pp. 339–388.

Nieboer, E., Richardson, D.H.S., Boileau, L.J.R., Beckett, P.J., Lavoie, P., and Padovan, D. (1982). "Lichens and mosses as monitors of industrial activities associated with uranium mining in northern Ontario, Canada—Part 3: Accumulations of iron and titanium and their mutual dependence." *Environ. Pollut. Ser. B* **4,** 181–192.

Oldfield, F., Thompson, R., and Barber, K.E. (1978). "Changing atmospheric fallout of magnetic particles recorded in recent ombrotrophic peat sections." *Science* **199,** 679–680.

Oldfield, F., Brown, A., and Thompson, R. (1979a). "The effect of microtopography and vegetation on the catchment of airborne particles measured by remanent magnetism." *Quat. Res.* **12,** 326–332.

Oldfield, F., Appleby, P.G., Cambray, R.S., Eakins, J.D., Barber, K.E., Battarbee, R.W., Pearson, G.R., and Williams, J.M. (1979b). "^{210}Pb, ^{137}Cs and ^{239}Pu profiles in ombrotrophic peat." *Oikos* **33,** 40–45.

Oldfield, F., Tolonen, K., and Thompson, R. (1981). "History of particulate atmospheric pollution from magnetic measurements in dated Finnish peat profiles." *Ambio* **10,** 185–188.

Pakarinen, P. (1978a). "Element contents of *Sphagna:* variation and its sources." *Bryophytum Bibliotheca* **13,** 751–762.

Pakarinen, P. (1978b). "Production and nutrient ecology of three *Sphagnum* species in Southern Finnish raised bogs." *Ann. Bot. Fennici* **15,** 15–26.

Pakarinen, P. (1978c). "Distribution of heavy metals in the *Sphagnum* layer of bog hummocks and hollows." *Ann. Bot. Fennici* **15,** 287–292.

Pakarinen, P. (1981a). "Nutrient and trace metal content and retention in reindeer lichen carpets of Finnish ombrotrophic bogs." *Ann. Bot. Fennici* **18,** 265–274.

Pakarinen, P. (1981b). "Metal content of ombrotrophic *Sphagnum* mosses in NW Europe." *Ann. Bot. Fennici* **18,** 281–292.

Pakarinen, P. (1982). "On the trace element and nutrient ecology of the ground layer species of ombrotrophic bogs." *Publ. Dept. Bot., Univ. Helsinki* **10,** 1–32.

Pakarinen, P., and Gorham, E. (1983). "Mineral element composition of *Sphagnum fuscum* peats collected from Minnesota, Manitoba and Ontario. *Proceedings of the International Peat Symposium*, Bemidji, Minnesota, Oct. 11–13, 1983.

Pakarinen, P., and Häsänen, E. (1983). "Mercury concentrations of bog mosses and lichens." *Suo* **34,** 17–20.

Pakarinen, P., and Mäkinen, A. (1976). "Comparison of Pb, Zn and Mn contents of mosses, lichens and pine needles in raised bogs." *Suo* **27,** 77–83.

Pakarinen, P., and Mäkinen, A. (1978). "Trace metal distribution in *Ledum palustre*." *Suo* **29,** 93–98.

Pakarinen, P., and Rinne, R.J.K. (1979). "Growth rates and heavy metal concentrations of five moss species in paludified spruce forests." *Lindbergia* **5,** 77–83.

Pakarinen, P., and Tolonen, K. (1976a). "Regional survey of heavy metals in peat mosses (*Sphagnum*)." *Ambio* **5,** 38–40.

Pakarinen, P., and Tolonen, K. (1976b). "Studies on the heavy metal content of ombrotrophic *Sphagnum* species." In *Proceedings of the 5th International Peat Congress,* Poznan, Poland, pp. 264–275.

Pakarinen, P., and Tolonen, K. (1977a). "Distribution of lead in *Sphagnum fuscum* profiles in Finland." *Oikos* **28,** 69–73.

Pakarinen, P., and Tolonen, K. (1977b). "Vertical distributions of N, P, K, Zn and Pb in *Sphagnum* peat." *Suo* **28,** 95–102.

Pakarinen, P., Tolonen, K., and Soveri, J. (1981). "Distribution of trace metals and sulfur in the surface peat of Finnish raised bogs." In *Proceedings of the 6th International Peat Congress,* Duluth, Minnesota, pp. 645–648.

Percy, K.E. (1983). "Heavy metal and sulphur concentrations in *Sphagnum magellanicum* Brid. in the Maritime Provinces, Canada." *Water, Air, Soil Pollut.* **19,** 341–349.

Puckett, K.J., and Burton, M.A.S. (1981). "The effect of trace elements on lower plants." In N.W. Lepp, Ed., *Effect of Heavy Metal Pollution on Plants,* Vol. 2 of *Metals in the Environment.* Applied Sci. Publ., London, pp. 213–238.

Puustjärvi, V. (1955). "On the colloidal nature of peat-forming mosses." *Arch. Soc. Vanamo,* **9**(suppl.), 257–272.

Ratcliffe, J.M. (1975). "An evaluation of the use of biological indicators in an atmospheric lead survey." *Atmos. Environ,* **9,** 623–629.

Rühling, Å., and Tyler, G. (1971). "Regional differences in the deposition of heavy metals over Scandinavia." *J. Appl. Ecol* **8,** 497–507.

Rycroft, D.W., Williams, D.J.A. and Ingram, H.A.P. (1975). "The transmission of water through peat I. review." *J. Ecol.* **63,** 535–556.

Sapek, A. (1976). "Contamination of peat soils with lead and cadmium." In *Proceedings of the 5th International Peat Congress,* Vol. 2. Poznán, Poland, pp. 284–294.

Schwarzmaier, V., and Brehm, K. (1975). "Detailed characterization of the cation exchanges in *Sphagnum magellanicum* Brid." *Z. Pflanzenphysiol., Bd.* **75,** 250–255.

Sillanpää, M. (1976). "Distribution of trace element in peat profiles." In *Proceedings of the 4th International Peat Congress,* Vol. 5. Finland, pp. 185–191.

Small, E. (1972). "Ecological significance of four critical elements in plants of raised *Sphagnum* peat bogs." *Ecology* **53,** 498–503.

Stanek, W. (1977). "A list of terms and definitions." In N.W. Radforth and C.O. Brawner, Eds., *Muskeg and the Northern Environment in Canada.* University of Toronto Press, Toronto, Canada, pp. 367–387.

Stanek, W., Jeglum, J.K., and Orloci, L. (1977). "Comparisons of peatland types using macronutrient contents of peat." *Vegetatio* **33,** 163–171.

Szalay, A., and Szilágy, M. (1968). "Accumulation of microelements in peat humic acids and coal." In P.A. Schenck and J. Havenaar, Eds., *Advances in Organic Geochemistry 1968.* Pergamon, Oxford, pp. 567–578.

Tanskanen, H. (1976). "Factors affecting the metal contents in peat profiles." *J. Geochem. Explor.* **5,** 412–414.

Tolonen, K. (1979). "Peat as a renewable resource: long-term accumulation rates in North European mires." In *Proceedings of the International Symposiums on Classification of Peat and Peatlands.* International Peat Society, Helsinki, pp. 282–296.

Urquhart, C., and Gore, A.J.P. (1973). "The redox characteristics of four peat profiles." *Soil Biol. Biochem.* **5,** 659–672.

Verry, E.S., and Boelter, D.H. (1978). "Peatland hydrology." In P.E. Greeson, J.R. Clark, and J.E. Clark, Eds., *Wetlands Functions and Values: The State of Our Understanding.* American Water Resources Assoc., Minneapolis, Minn. pp. 389–402.

Walmsley, M.E. (1977). "Physical and chemical properties of peat." In N.W. Radforth and C.O. Brawner, Eds., *Muskeg and the Northern Environment in Canada.* University of Toronto Press, Toronto, Canada, pp. 82–129.

Wandtner, R. (1981). "Indikatoreigenschaften der vegetation von Hochmooren der Bundes Republik Deutschland für schwermetallimmissionen." *Dissertationes Botanicae* **59,** 1–90, Vaduz.

Waughman, G.J. (1980). "Chemical aspects of the ecology of some South German peatlands." *J. Ecol.* **1980,** 1025–1046.

Yliruokanen, I. (1976). "Heavy metal distributions and their significance in Finnish peat bogs." In *Proceedings of the 5th International Peat Congress.* Poznán, Poland, pp. 276–283.

Zoltai, S.C., Pollett, F.C., Jeglum, J.K., and Adams, G.D. (1973). "Developing a wetland classification for Canada." In B. Bernier and C.H. Winget, Eds., *Forest Soils and Forest Land Management.* Le Presses de l'Université Laval, Quebec City, pp. 497–511.

17

MERCURY VAPOR IN THE ATMOSPHERE: THREE CASE STUDIES ON EMISSION, DEPOSITION, AND PLANT UPTAKE

S. E. Lindberg

Environmental Sciences Division
Oak Ridge National Laboratory
Oak Ridge, Tennessee

1. INTRODUCTION

Mercury is unique among the heavy metals typically studied in the atmosphere. Because of its high vapor pressure, mercury is emitted to the atmosphere primarily in vapor form, as opposed to aerosol form, and it remains there largely in this physical state. Its tendency for volatilization and its toxicity (in the elemental form, Hg^0, and in organometallic compounds) have, of course, been long recognized (Giese, 1940; Bidstrup, 1914). As a result of some highly publicized epidemiological incidents of Hg poisoning (Nelson, 1971), several large-scale studies of the environmental chemistry, cycling, and effects of mercury were completed in North America during the 1970s (Young, 1971; Lindberg et al., 1975; Kudo et al., 1976; Windom, 1977; NRC, 1978; Hildebrand et al., 1980). Because the most obvious discharges of Hg were directly into waterways, most of these studies dealt primarily with the aquatic environment. Similarly, in response to many of the epidemiological incidents cited above, early regulations were aimed at limiting direct aquatic discharges.

Before 1970 the amount of Hg discharged to the environment from anthropogenic sources was estimated to be comparable with quantities derived from continental weathering processes (Klein and Goldberg, 1970). As a result of new regulations, most large industrial sources have been identified and the aquatic discharges of Hg reduced to negligible amounts relative to natural sources. Current atmospheric emissions of mercury are estimated to exceed releases to waterways by more than an order of magnitude, yet there are still no legislated ambient air quality standards for mercury (Harriss and Hohenemser, 1978). It is estimated that ~35% of the global atmospheric emission of mercury from human activities results from coal combustion, with the ratio of anthropogenic to natural emissions to the atmosphere ranging over several orders of magnitude on regional scales (Harriss, 1979). Mercury enters the atmosphere primarily in gaseous forms, but ultimately, various transport and scavenging processes return airborne Hg to the aquatic and terrestrial environment, where it undergoes the same physicochemical reactions as Hg discharged directly into these environments.

Given the existing regulations on aquatic discharges and the increased use of coal for power generation, it seems that a situation of local water pollution has been traded for one of potential regional air pollution. The National Research Council (1978) concluded in an assessment document that the effects of chronic exposure to airborne mercury have been insufficiently evaluated and may pose a long-term threat. The number of recent studies on emission, chemistry, and deposition of airborne mercury indicate that the potential threat has been recognized and is being investigated. Two recent workshops recommended further research on the role of mercury vapor in air and the development of standardized collection and analysis procedures for airborne and deposited Hg (Elder, 1979; Galloway et al., 1980).

This chapter is not intended to be a comprehensive review of the biogeochemical cycle of atmospheric Hg, which has been recently published

(Nriagu 1979). Rather, the approach in organizing this discussion has been to highlight three separate but related research projects dealing with the sources and fate of atmospheric mercury vapor. These projects, completed in 1980, involved research on three of the most important industrial sources of Hg vapor to the atmosphere, the chloralkali, mercury mining, and electric power generation industries. In summarizing these case studies, these findings are updated as appropriate in light of recently published information.

2. METHODOLOGIES

The techniques for collection and detection of mercury vapor in air are highly varied, including absorption into a liquid (e.g., acidic permanganate or iodine monochloride solutions), adsorption onto or reaction with a solid sorbent (e.g., activated charcoal), and amalgamation onto a noble metal surface (e.g., Au, Ag). Use of these techniques generally involves time-integrated sampling, allowing accumulation of sufficient mercury vapor for subsequent extraction and analysis by conventional methods. However, real-time sensors have been developed and successfully applied to the determination of elevated concentrations of mercury vapor in air (Hadelshi et al., 1975; Dowd et al., 1976; Jepsen and Langan, 1971). Many of the above methods are not specific to individual species of Hg vapor (which may include the elemental metal vapor, gaseous inorganic compounds, and various organomercurials), although selective adsorption methods have been applied on a limited scale to determine Hg vapor speciation (e.g., Braman and Johnson, 1974; Soldanov et al., 1975). It is beyond the scope of this paper to discuss the details of the above techniques. The reader is referred to the recent review by Schroeder (1982) for further information.

The specific methods used in the three studies discussed below are described in the literature. Briefly, they involved collection of Hg vapor using activated charcoal traps (Moffitt and Kupel, 1971) through which air was pumped for variable time periods at a rate of 1 L/min, collection of particulate Hg using standard aerosol filtration techniques, and analysis of all samples by flameless atomic absorption using the methods of Feldman (1974). Further details on sampling efficiency, extraction techniques, analytical precision and accuracy, and the design of the chamber for measurement of surface emission rates of Hg are available in a report (Hildebrand et al., 1980) and in the literature (Lindberg, 1980).

3. CASE STUDIES

3.1. Mercury Emission from Chloralkali Solid Wastes

The role of the chloralkali industry in global and regional Hg cycling has been widely discussed (e.g., Caban and Chapman, 1972; Van Horn, 1975; Jernelöv and Wallin, 1973; Högström et al., 1979), with the general conclusion that their contribution may be of limited importance globally but of particular concern to local air quality. Although limited studies have been published on the emission of Hg vapor from production processes in active plants (Wallin, 1976; Caban and Chapman, 1972; Van Horn, 1975), none have considered the problem of atmospheric emissions from waste products ponded or otherwise stored near both active and inactive plants. The case study reported here deals with loss of Hg vapor from solid wastes stored near an inactive plant in the eastern United States (Lindberg and Turner, 1977).

The study area is at the site of a large chemical manufacturing facility in Saltville, Va., which ceased operation in 1972 because of an inability to meet newly imposed state and federal water quality standards. Sodium bicarbonate and related by-products were produced by the ammonia soda process since 1894. The plant began production of chlorine and caustic soda using the Hg-cell process in 1952. Between that time and 1972, waste products from both processes were disposed of in large sludge basins, a procedure relatively common in this industry (Caban and Chapman, 1972). The disposal site, with a surface area of ~44 ha, is situated adjacent to a river in an area enclosed on one side by a natural ridge and on the river side by dikes built of $CaCO_3$ waste and boiler ash generated in the $NaHCO_3$ process.

The volatility of metallic Hg (Hg^0) and many of its compounds suggests that its presence in large-scale waste deposits such as these may have a potential influence on local air quality. Waste material collected from the surface of the waste ponds was used in laboratory experiments to determine the emission of Hg^0 from a known surface area at controlled temperatures. Approximately 4 kg of waste material (with a total Hg concentration of 155 μg/g) was slurried into a 19-L glass chamber and allowed to dry to a moisture level similar to that observed for the surface of the actual waste ponds (approximately 75% water by weight). Mercury-free air was pumped through the chamber over the waste surface and the filtered output air sampled for Hg^0. The emission of elemental Hg vapor from the waste material was clearly related to surface temperature as expected (Fig. 17.1). The nature of the relationship is similar to that between temperature and the saturation concentration of Hg^0 in air (Wallace et al., 1971), suggesting that the mercury occurs in the waste material to a large extent in the elemental state. We are not aware of any published data on emission rates from similar material for comparison. Natural degassing rates of Hg on regional or global scales have been estimated based on input/output calculations as ranging from 0.02 to 0.03 $\mu g/m^2$ h (Weiss et al., 1971; Kothny, 1973), and by rock volatility experiments as ranging from 0.04 to 0.08 $\mu g/m^2$ h (Desaedeleer and Goldberg, 1978), while measured rates for mineralized soils are as high as 1.7 $\mu g/m^2$ h (McCarthy et al., 1969). These values are one to three orders of magnitude lower than those we have measured at 25°C for the waste material (~40 $\mu g/m^2$ h).

If the waste deposits are considered as an area source, a simple Gaussian diffusion model (Gifford, 1968) can be used to estimate 24-h mean concentra-

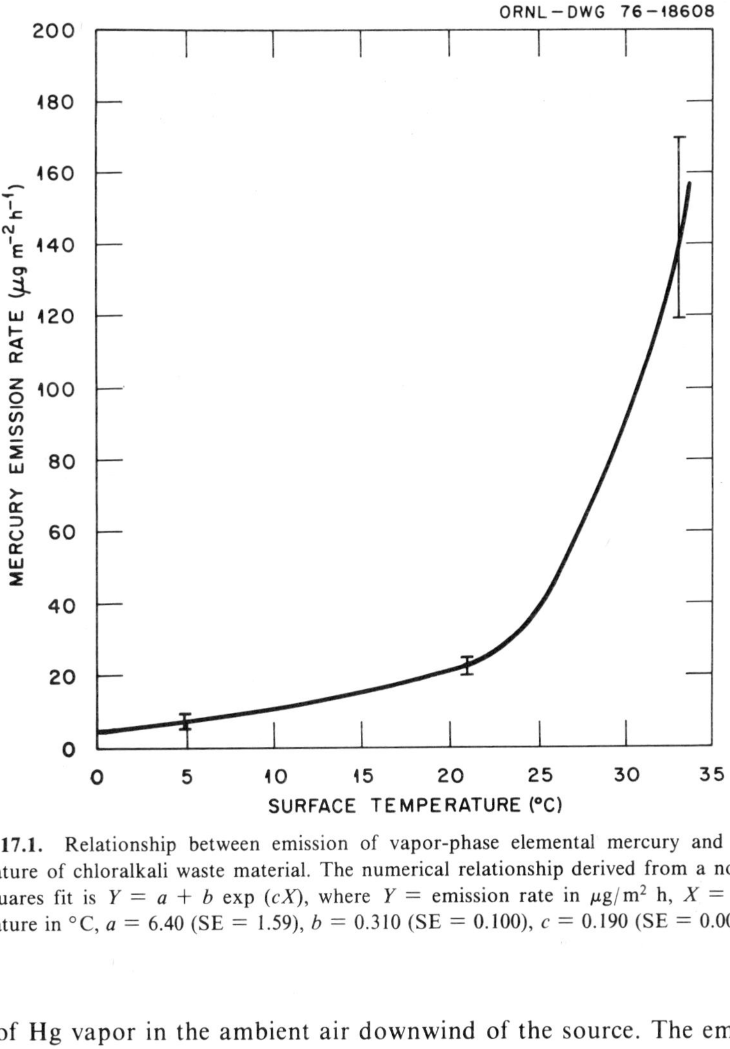

Figure 17.1. Relationship between emission of vapor-phase elemental mercury and surface temperature of chloralkali waste material. The numerical relationship derived from a nonlinear least-squares fit is $Y = a + b \exp(cX)$, where Y = emission rate in $\mu g/m^2$ h, X = surface temperature in °C, $a = 6.40$ (SE = 1.59), $b = 0.310$ (SE = 0.100), $c = 0.190$ (SE = 0.00982).

tions of Hg vapor in the ambient air downwind of the source. The emission rates measured under controlled conditions (Fig. 17.1) were used in the model; measured mean wind speed directly over the pond area was 1 m/s; and the background air Hg concentration was taken as 10 ng/m^3 based on measured values. For comparative purposes, air sampling stations were established near the ponds on two occasions. These were equipped to measure Hg^0 in air, Hg on suspended particles, wind speed and direction, air temperature, and bulk deposition. Table 17.1 presents a comparison of measured and predicted air concentrations of mercury vapor. Given the simplicity of the modeling approach, the close agreement between the measured and predicted 24-h mean air concentrations under similar conditions suggests that the entire surface area of the waste pond was emitting Hg vapor at a rate comparable to that measured in controlled experiments.

Table 17.1. Daily Mean Concentrations of Mercury Vapor (Hg°) in Air in Vicinity of Chloralkali Waste Ponds under Various Meteorological Conditions

Distance from Center of Waste Ponds (km)	Percent of Time Downwind[a]	Mean Air Temperature	Concentration of Mercury Vapor (Hg°) in Air: Measured (ng/m^3)	Concentration of Mercury Vapor (Hg°) in Air: Predicted (ng/m^3)
0.4	78	6.3°C	90	65
0.5	99	6.3	64 ± 6[b]	72
1.5	67	6.3	18	26
1.9	21	6.3	12	15
0.5	96	29.4	991 ± 2[b]	778

[a] Expressed as a percent of the total sampling period (24 h), during which the sample location was downwind of the waste pond.
[b] $\overline{X} \pm \sigma$, $n = 3$.

The levels of Hg^0 in air measured near the pond during periods of low air temperature (Table 17.1) were slightly elevated above those reported for large U.S. cities, 5–50 ng/m^3 (Johnson and Brahman, 1974; Van Horn, 1975), and for areas near active chloralkali plants, 10–50 ng/m^3 (Högström et al., 1979). The fact that these concentrations decreased to rural background levels, 3–9 ng/m^3 (Kothny, 1973; Ferrara et al., 1982, Brosset, 1982) within ~2 km of the waste disposal area, is attributable to the ability of a well-mixed atmosphere to rapidly dilute gaseous pollutants. However, these levels are still an order of magnitude higher than those reported for remote regions over the Atlantic and Pacific oceans, 1 to 3 ng/m^3 (Slemr et al., 1981; Fitzgerald et al., 1983). During periods of higher air temperature, the concentration of Hg measured near the pond area approached the U.S. Environmental Protection Agency guideline level for total Hg in ambient air (1000 ng/m^3) (Federal Register, 1973). This compares with levels of 30–1600 ng/m^3 for air over natural Hg deposits and 10–40,000 ng/m^3 for natural degassing and geothermal emissions (Kothny, 1973; Siegel and Siegel, 1975). The concentration of particulate Hg during this sampling period was surprisingly low (0.25–0.36 ng/m^3), considering the area in question, and reflected the stability of the waste pond surface to dust resuspension. In fact, these values are comparable with those reported for urban areas in the Mediterranean, 0.2–0.3 ng/m^3 (Ferrara et al., 1982), but exceed levels measured at remote oceanic sites by two orders of magnitude, 0.4–2.0 pg/m^3 (Fitzgerald et al., 1983). The indication that essentially 100% of the Hg in air is in vapor form agrees with reported results for ambient air in remote and urban areas (Johnson and Braman, 1974; Fitzgerald et al., 1983) and industrial power plant emissions (see Sec. 3.3).

Given the elevated levels of Hg^0 measured near the pond during the high temperature period (Table 17.1), the diffusion model (Gifford, 1968) can also be used to determine approximate isopleths of Hg^0 concentration in residential areas near the pond during a typical hot weather period. The results of this calculation are presented in Figure 17.2 which illustrates mean hourly Hg^0 concentrations in air at ground level. The time modeled was a daytime period in August when the pond surface temperature was taken to be 32 to 34°C from 0900 to 1900 hours; the mean wind speed was measured as 0.27 m/s (~0.6 mph); the prevailing wind was from the southwest (see windrose in Fig. 17.2); and only the largest of the two waste ponds (Pond 1) was assumed to be a source of Hg^0 to simplify the model. Since the model is not generally applied to short-distance dispersion of an air pollutant from an area source, the contributing pond must be approximated as a point source (at the pond center). Although the model does not consider terrain effects, the influence of local terrain on the wind direction and hence atmospheric dispersion of mercury emitted from the ponds is obvious. Of most interest is the small areal extent of the >1000-ng/m^3 concentration range and the rapid decrease in Hg concentration to estimated regional background levels within ~1.5 km of the center of the waste ponds, even under meteorological

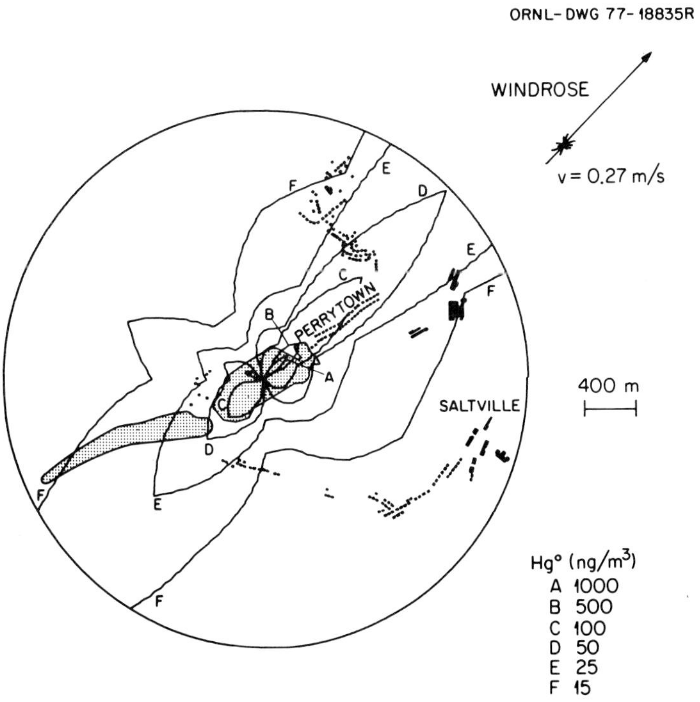

Figure 17.2. Isopleths of Hg^0 concentrations expected at ground level given one waste pond as the primary Hg source; other assumptions as described in the text.

conditions chosen to maximize the effect of the source area on local air quality.

The deposition of Hg was also measured during the period of air monitoring. Duplicate bulk deposition collectors were situated 3 m from the pond edge to sample wet plus dry deposition for a 68-h period. During this time, one rain event was sampled that lasted approximately 1 h and resulted in 0.25 cm of precipitation. The concentrations of Hg in the replicate bulk samples were 1.1 and 1.4 $\mu g/L$. These levels are comparable to values reported for urban areas, 0.05–4.0 $\mu g/L$ (Galloway et al., 1980), but are approximately an order of magnitude higher than concentrations of Hg in rain (wetfall-only) collected in eastern Tennessee (0.08–0.54 $\mu g/L$), a rural area influenced by emissions from regional coal-fired power plants (Andren and Lindberg, 1977); ~25 times higher than Hg in bulk precipitation in a remote area of New Hampshire, 0.06 $\mu g/L$ (Schlesinger et al., 1974); and ~100 times higher than Hg in rain collected on the Atlantic coast in Connecticut, 2–20 ng/L (Fitzgerald, 1976; Fogg and Fitzgerald, 1979). The deposition rates of Hg during this period, as calculated from the above two concentrations, were 7.4 and 9.0 $\mu g/m^2$ day, which are from 15 to 20 times higher than the estimated deposition rate of Hg around five active Swedish chloralkali plants, ~0.5 $\mu g/m^2$ day (Jernelöv and Wallin, 1973). However, in their calculations, Jernelöv and Wallin purposefully excluded the area within a radius of 500 m of the plants because of the large variability associated with the data, stating only that the deposition rates in this area were "very high." More recent data from active plants confirm this variability near the source, with mean bulk deposition rates ranging from ~1 to 40 $\mu g/m^2$ day at distances of 50 to 500 m, while deposition rates at distances of 500 m to 50 km were more uniform, ranging from ~0.5 to 1 $\mu g/m^2$ day (Högström et al., 1979). Values summarized as representative of rural areas not influenced by local emissions range from 0.2 to 3 $\mu g/m^2$ day (Schlesinger et al., 1974; Galloway et al., 1980), while estimates of the total deposition to remote sea surfaces are on the order of 0.02 to 0.06 $\mu g/m^2$ day (Fitzgerald et al., 1983).

Using simple regression analyses we have obtained functions which estimate surface emission of Hg^0, based on air temperature. Applying these relationships to hourly temperature data for the waste pond area for a typical year, it is possible to calculate the estimated annual loss of mercury from the waste disposal site by the atmospheric pathway. The sum of the predicted hourly fluxes over the course of a year is 36 ± 4 kg Hg/yr (sum ± estimated standard error of the sum of predicted fluxes). Interestingly, one estimate of annual aquatic loss of Hg by leaching and runoff from the waste ponds into the adjacent river was similar (39 ± 2 kg Hg/yr; Turner and Lindberg, 1978).

The estimated total annual flux of Hg from this inactive plant, 75 ± 5 kg/yr, is approximately 6% of the U.S. Environmental Protection Agency standard for emission of Hg to air and water from active chlorine plants situated in riverine environments. Early reported losses of Hg to the atmosphere and surface waters by Swedish chloralkali plants were in the range of 150–1400 kg/yr (Wallace et al., 1971; Wallin, 1976), while more recent estimates are on the order of 300 kg/yr (Högström et al., 1979). Losses from a typical, active plant in the United States were ~1100 kg/yr in 1973 (830 kg to the atmosphere), with estimates of losses using 1983 technology at ~510 kg/yr (490 kg to the atmosphere) (Van Horn, 1975). For comparison, Hg vapor emissions from power plants range from 60 to 120 kg/yr for geothermal units (Robertson et al., 1977), and from 500 to 2700 kg/yr for coal-fired units (see Sec. 3.3).

At present, the most significant source of mercury emissions to the environment in the chloralkali industry involves disposal of large volumes of solid waste materials. As the industry reduces direct losses to air and water, losses to landfill areas increase. Of the total Hg consumption by the industry, $\sim 6 \times 10^5$ kg/yr during the period 1965–1975 for the United States alone (Harriss and Hohenemser, 1978), an estimated 60 to 90% is deposited in landfill sites and holding ponds. As the chloralkali industry modernizes by the increasing conversion of Hg-cell plants to the more efficient diaphragm process, which does not use Hg, it is apparent that these waste deposits will become the major source of Hg emissions from this industry. Given the volatility of Hg in these wastes, the quantity of Hg deposited in waste ponds (2.3×10^5 kg during 1973 in the United States alone; Van Horn, 1975) and the residence time of Hg in the wastes studied here (conservatively estimated at 100 yr by Hildebrand et al., 1980), it seems that the role of solid wastes in the local, and perhaps regional, Hg cycle in the atmosphere will increase in importance.

3.2. Emission and Plant Uptake of Mercury from Soils near an Active Mercury Mine

Natural emissions of Hg to the atmosphere from ore-rich areas can be enhanced by the presence of mining and refining operations. These sources have been estimated to contribute ~12% of the global anthropogenic emissions of Hg to air (Van Horn, 1975). Cinnabar deposits are usually refined near the mine site by ore-roasting, followed by condensation of the released Hg vapor. Resultant fugitive emissions, combined with ventilated air from mine shafts, can supply a significant quantity of Hg^0 to the atmosphere. Local deposition of emissions can further increase the Hg burden of vegetation growing on mercuriferous soils. Reports in the literature confirm that plants growing near active mining areas are exposed to elevated soil and airborne Hg (Hitchcock and Zimmerman, 1957; Ross and Stewart, 1962; Gilmour and Miller, 1973). However, there have not been any reported measurements of natural Hg emission rates from such soils or of direct plant uptake of the emitted vapor.

This case study (Lindberg et al., 1979) involved an examination of agricultural soils from an area near a mercury mine and smelter, with the following objectives: (1) comparing the distribution of Hg in soils located near the mine to that in soils unaffected by the mining emissions, (2) quantifying the uptake and distribution of Hg by alfalfa as affected by soil and air Hg concentration, and (3) determining the effect of soil surface temperature and plant cover on volatile Hg emission rates from soil. The soils used in these experiments were collected in Almadén, Spain, site of the largest and oldest mercury mining/refining operation in the world. Moderate winds flow predominantly east and west within a shallow valley surrounding the site. The most significant release of Hg from the mine/smelter operations is Hg^0 vapor discharged into the atmosphere from the ore-roasting operation via a 30-m stack and from the mine-ventilation system. Another source is windblown dust, including HgS from mine tailings. Estimates of the total annual atmospheric emissions of Hg from this site are unavailable. Surface soils were collected from agricultural fields in and surrounding Almadén (Fig. 17.3), air dried, sieved through a 2-mm screen, sealed in plastic bags, and returned to the laboratory. For purposes of Hg uptake and emission studies, two soils were chosen from the above set: one representing a Hg-enriched soil collected 1 km west of the mine site, termed Almadén soil; the other representing a control soil collected 20 km east of the mine. Emission rates as a function of temperature and plant cover were determined using the flow-through chamber method cited earlier, while plant uptake studies were done using alfalfa grown directly in these soils and in an inert substrate (an expanded silicate material termed perlite). All plants were exposed to airborne Hg vapor in a controlled environment chamber.

In the environment these soils are exposed to three major sources of Hg: chemical weathering of residual cinnabar, wet and dry deposition of Hg^0 released during the mining/roasting operations, and deposition of resuspended mine tailings. The effect of the predominant wind patterns in dispersing airborne Hg is evident in the order-of-magnitude higher surface soil concentrations 15 km east and west of the mine (sites G and H, total Hg concentration 3.1 to 5.3 μg/g) compared to 15 km north and south (sites I and J, concentration 0.32–0.48 μg/g; Fig. 17.3).

The pattern of dispersion is similar within 2 km of the mine where local turbulence also plays an important role. In addition, it is possible that the Hg levels in soils near the mine area are more strongly influenced by residual minerals. Within this area, proximity to the mine has an obvious influence on the surface soil Hg levels as is evident in the three adjacent stations on a transect running southeast to northwest near the mine (i.e., stations A (0.5 km from the mine), B (1.0 km), and D (2.0 km): 263, 97, and 68 μg/g Hg, respectively). All of the soil concentrations measured during this study were considerably elevated above the average crustal abundance of 0.07 μg/g Hg (Vinogradov, 1959) and above uncontaminated topsoils collected in Sweden (0.07 μg/g), Africa (0.02 μg/g), Eastern Europe (0.03–3 μg/g), and the United States (0.02–0.04 μg/g; U.S. Geological Survey, 1970). However, they are comparable to values reported for soils collected near other mercury sources (2.5–10 μg/g; U.S. Geological Survey, 1970; Andersson, 1979).

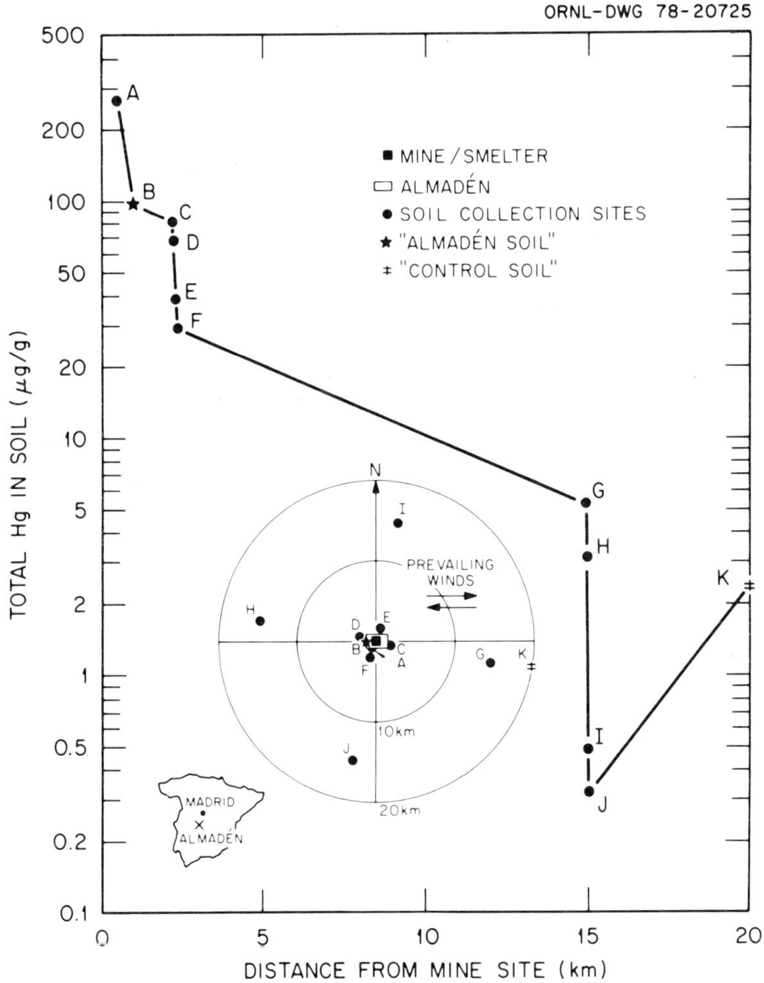

Figure 17.3. Study area and concentrations of total Hg in surface soils exposed to atmospheric emissions from the Almadén cinnabar mine.

Also illustrated in Figure 17.3 are the locations of the Almadén (Hg = 97 ± 7 μg/g soil) and control (2.3 ± 0.4 μg/g) soils chosen for detailed chemical analysis and used in the plant uptake studies and emission rate experiments. The two soils are similar with respect to various characteristics known to influence soil Hg concentrations (e.g., pH, organic content; Andersson, 1979). The distribution of Hg in these two soils was characterized by density gradient fractionation (Francis et al., 1972). A major portion of the fractionated Hg in each soil (40% Almadén, 80% control) was associated with the most dense inorganic soil fractions and likely represents cinnabar; the Hg concentration in this fraction of the Almadén soil (490 μg/g) exceeded that in this fraction of the control soil by a factor of 25, reflecting the importance of weathered mineral deposits near the mine. However, the least dense soil fraction (which includes organoclay complexes and surface-soil-deposited Hg^0) comprised 50% of the Almadén soil Hg but only 10% of the control. The mercury concentration in this fraction of the Almadén soil exceeded that in the control by two orders of magnitude, reflecting the additional influence of atmospheric deposition on the Hg content of the Almadén soil.

The results of the Hg emission experiments are summarized in Figure 17.4 and illustrate the variation in emission rates with soil surface temperature for both soils with and without alfalfa cover. For all cases the emission rates are clearly related to surface temperature, as expected, given the dependence of Hg vapor pressure on temperature. This relationship has also been noted for soils experimentally amended with $Hg(NO_3)_2$ (Landa, 1978). In general, the emission rate of Hg from Almadén soils was greater than that from control soils at both elevated temperatures. These trends may be related to the greater proportion of atmospherically deposited Hg^0 in the Almadén soils compared to controls. At 25°C the emission rate of Hg from bare Almadén soils (0.32–0.34 $\mu g/m^2$ h) was significantly greater ($P < 0.05$) than that for planted soils (0.07–0.09 $\mu g/m^2$ h). This trend was evident at 35°C as well, although not significant. For control soils, the effect of plant cover in reducing the

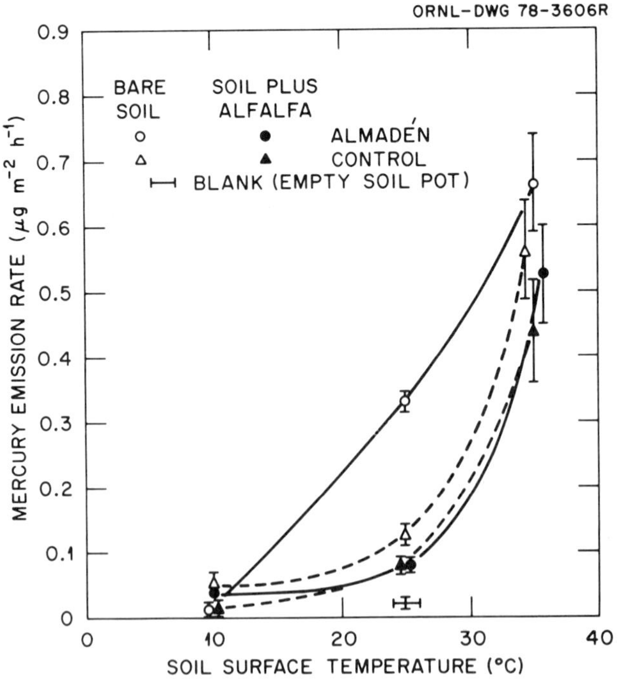

Figure 17.4. Relationship between emission of Hg vapor and soil surface temperature for Almadén and control soils with and without plant cover.

emission of Hg was most apparent at 35°C. The effect of plant cover may simply be to reduce mixing of Hg-rich air at the soil surface with overlying air being sampled above the alfalfa canopy.

Comparative data on Hg emission rates from natural soils are scarce, although some researchers have reported observing release of Hg from soils experimentally enriched with organic and inorganic Hg (Williston, 1968; Gilmour and Miller, 1973; Johnson and Braman, 1974; Landa, 1978). Natural degassing rates described earlier range from 0.02 and 0.03 $\mu g/m^2$ h, while rates estimated for mercuriferous soils range from 0.2 to 1.7 $\mu g/m^2$ h. Although not stated in the original references, it is assumed that the reported emission rates were estimated for the temperature range 20–30°C. Thus, at 25°C the measured Hg emission rates for bare Almadén (0.32–0.34 $\mu g/m^2$ h) and control (0.12–0.14 $\mu g/m^2$ h) soils exceed natural rates by factors of 4 to 10 but are within the range reported for mineralized soils.

The experimental technique used here also provided data on the time-averaged gaseous Hg concentration over soil in both the emission rate chamber and the plant growth chamber. For the control soils, the 24-h average air concentration of Hg vapor ranged from 4 to 30 to 220 ng/m^3 at 10, 25, and 35°C, respectively, while for the Almadén soils the concentrations ranged from 20 to 130 to 250 ng/m^3. In the growth chamber, which contained several pots of both soils, the air concentration ranged from 10 to 20 ng/m^3 in the 5 to 10°C range, 50 to 100 ng/m^3 at 20 to 25°C, and 220 to 520 ng/m^3 at 30 to 35°C. The time-weighted average air concentration of Hg vapor in the growth chamber during the 16-week plant uptake studies was approximately 80 ng/m^3. As discussed earlier, the measured concentration in air over nonmercuriferous areas ranges from 3 to 9 ng/m^3, while surface air in mineralized areas ranges from 100 to 20,000 ng/m^3 (U.S. Geological Survey, 1970).

Our measurements of Hg accumulation in alfalfa grown on these soils suggest two mechanisms of Hg uptake by plants under these conditions: one controlling aboveground, foliar levels of Hg and being independent of soil Hg levels, and another influencing root uptake which is related to soil concentrations. The Hg concentrations in alfalfa foliage from plants on both soil types were similar (~1–2 $\mu g/g$), both exceeding world average levels in grasses by an order of magnitude (Wallace et al., 1971). On the other hand, root concentrations in Almadén plants exceeded those in control plants by a factor of 20 and represented 80% of the total Hg accumulated by plants in Almadén soils. For control soils, 80% of the absorbed Hg occurred in aboveground foliage.

The phenomenon influencing Hg content in alfalfa foliage is apprently direct absorption of Hg vapor by leaves. As discussed above, the plants in the growth chamber were exposed to elevated levels of Hg vapor throughout the uptake experiments. To test the ability of aboveground portions of alfalfa to absorb gaseous Hg, a replicate set of alfalfa plants was grown from seed on an inert substrate (perlite) in the growth chamber and then exposed to Hg vapor emitted from Almadén soils. These plants were planted and harvested simultaneously with plants grown in Almadén and control soils. Although the perlite contained considerably less Hg than the experimental soils (<0.0005 μg/g), the alfalfa grown on perlite exhibited a Hg concentration of 2.0 ± 0.1 μg/g in the aboveground plant, and 0.35 ± 0.01 μg/g in the roots. This aboveground concentration of Hg is comparable to the levels measured in the plants grown on both soil types (pooled mean $= 1.4 \pm 0.6$ μg/g, $n = 18$). The root concentration, on the other hand, is a factor of 20 lower than the Almadén roots ($\overline{X} = 8.0 \pm 2.4$, $n = 9$) but not significantly different from the control roots ($\overline{X} = 0.38 \pm 0.14$, $n = 9$).

The comparability of Hg concentrations in all aboveground plants maintained in the growth chamber supports the idea that the primary leaf uptake mechanism for Hg is direct absorption of Hg vapor. A small fraction of the absorbed Hg may then be translocated to the roots, explaining the relatively high concentration of Hg in perlite roots compared to the perlite itself (this could also be explained by direct deposition of Hg vapor to the perlite followed by root absorption). The results of previous plant uptake studies support the hypothesis of direct foliar absorption of Hg vapor. In experiments with amended soils, little translocation of Hg from roots to aboveground foliage has been reported for trees and grasses (Ross and Stewart, 1962; Gilmour and Miller, 1973). The ability of plants to absorb Hg vapor from the atmosphere is well known (Hitchcock and Zimmerman, 1957) and has been used in surveys of Hg air pollution with trees and mosses (Huckabee, 1973; Wallin, 1976; Lodenius and Laaksorvirta, 1979). Detailed chamber exposure studies with wheat and labelled Hg vapor suggest uptake to be stomatally controlled but not influenced by air temperature; surface absorption was considered negligible (Browne and Fang, 1978). In addition, the rate of uptake increased with increasing air concentration, total uptake increased with duration of exposure, and the absorbed Hg was confined solely to leaves. Vascular plants growing in Hg-rich areas have also been reported to release elemental Hg vapor directly to the atmosphere in a stomatally controlled process (Kozuchowski and Johnson, 1978).

Although the activated charcoal collection technique used in our studies results in collection of both elemental and chemically bound Hg vapor, our data and that of others suggest the speciation of the soil-emitted Hg vapor to be elemental Hg (Hg^0), while its source is thought to be chemically or biologically reduced mercury compounds in the soil. Frear and Dills (1967) demonstrated that, in Hg-treated soils, increasing amounts of organic matter, higher temperature, increased pH, and decreased moisture content all accelerate the reduction of inorganic Hg to Hg^0. Experiments by Gilmour and Miller (1973) indicated that volatilization of Hg was the dominant factor in the loss of Hg from turfgrass and soil amended with Hg_2Cl_2 and $HgCl_2$, with minor contributions from plant uptake and soil leaching. Several volatile Hg species have been measured in ambient air as well as over undisturbed soils and soils amended with $HgCl_2$ (Johnson and Braman, 1974). Amended soils resulted in immediate release of Hg^0 and subsequent release of methyl Hg(II) compounds; emissions from natural soils were measured for 8 days, the predominant species detected being Hg^0 (ranging from 17 to 100% of the total vapor), methyl Hg(II) type (0 to 66% of total), and inorganic Hg(II) type compounds (generally <30% of total).

An unresolved question is whether mercurials are directly transformed into Hg^0 or first decomposed to inorganic Hg followed by chemical or biological methylation or reduction to Hg^0. Hitchcock and Zimmerman (1957) mixed six organic and seven inorganic Hg compounds separately with soil plus soil organic matter with the result that all species tested reacted with soil organic matter to release Hg^0. The case for direct, abiotic methylation of Hg in soils has been presented by Rogers (1977) and Rogers and MacFarlane (1979) who amended soils with $Hg(NO_3)_2$. Within 1 week the soils contained measurable quantities of methyl Hg which was removed by volatilization. However, when these soils were sampled for atmospheric emission of Hg over a 2-day period, Hg^0 was found to be the predominant vapor species (~70%) followed by methyl Hg (~20%), and small amounts of Hg^{2+} and dimethyl Hg. Landa (1978) recently reported Hg volatilization rates from amended soils, concluding Hg^0 to be the predominant species lost from the soils. The role of soil microorganisms in the volatilization process was suggested by a marked reduction in Hg loss rates from two of five autoclaved soils. However, the degree to which microorganisms influenced loss rates was not easily discernible due to the inability to suppress microbial recolonization in autoclaved soils. Thus, in the majority of work reported to date, the major species of Hg vapor emitted from soils appears to be Hg^0. The mechanism of volatilization, however, is not as clear. Reported data favor reduction of various Hg compounds or soil–Hg complexes to Hg^0 in the soil zone, followed by loss of the vapor to the atmosphere, where it apparently remains in this same form (Fitzgerald et al., 1983; Sec. 3.3).

3.3. Mercury Behavior in a Coal-Fired Power Plant Plume

Combustion of fossil fuels for power generation is estimated to account for nearly 30% of the total release of Hg to the atmosphere by human activities in the United States (Harriss and Hohenemser, 1978). The release of Hg vapor during coal combustion is well recognized, with estimates of the fraction of the feed-coal mercury discharged as a vapor ranging from 90 to 97% (Billings and Matson, 1972; Klein et al., 1975; Anderson and Smith, 1977). Although mercury vapor emission rates and concentrations in stack gases and particles have been measured and reported, to our knowledge only limited data have been published on the concentration of Hg vapor in power plant plumes. These reports have not considered particle-associated mercury, however

(Jepsen and Langan, 1971). The study summarized here involved an investigation of mercury partitioning between particulate and vapor forms in the plume of a modern coal-fired power plant (Lindberg, 1980).

Samples were collected in the plume of the Tennessee Valley Authority Cumberland Power Plant, located 80 km northwest of Nashville, Tenn. The plant includes two horizontally opposed, pulverized coal-burning boilers with a total electric power generation capacity of 2600 MW. Each unit is equipped with a 305-m stack and an electrostatic precipitator with a design efficiency of 99%. During our research only one of the two units was operating.

A Sikorsky S-58 helicopter was used to collect samples during flights across the path of plume flow during early morning hours while the plume was visually well defined. Samples were drawn through two 5-cm Teflon tubes which extended $\cong$1 m in front of the helicopter body (out of the influence of the downwash of the main rotor). Isokinetic sampling was achieved by adjusting the velocity of sampling to match the velocity of the main stream, by using probes designed to offer minimum disturbance, and by directing the probes into the stream of flow (Meagher et al., 1978). Samples were collected at altitudes ranging from 250 to 450 m and distances of 0.25, 7, and 22 km downwind of the source stack during horizontal passes perpendicular to the wind direction in an attempt to sample similarly aged aerosols at each downwind distance. For the experiments reported here, a single set of gas and aerosol samples was collected at each location. In-plume sampling times ranged from 5 to 35 min and required 20 to 50 passes. Gas samples for elemental Hg vapor (Hg^0) were collected using activated charcoal absorption traps as described earlier, while aerosol samples were collected using standard filter methods (0.1 μm Teflon membranes).

The relationships between plume age (as indicated by distance from source), particulate Hg concentration (Hg_P), Hg vapor to particulate Hg concentration ratio (Hg^0/Hg_P), and total suspended particle concentration (TSP) in the plume are illustrated in Figure 17.5. As expected, Hg in the plume sample collected 0.25 km from the stack was dominated by the vapor phase. Approximately 92% of the total mercury in the plume was present as Hg^0 at this point at a measured concentration of 1700 ng/m^3. During plume travel from 0.25 to 7 to 22 km, the measured concentrations of Hg^0 decreased to 1000 ng/m^3 at 7 km and 200 ng/m^3 at 22 km (background concentrations were about 10–20 ng/m^3). Calculation of plume dispersion factors from ancillary data on SO_2 and condensation nuclei concentrations (CN) in the plume suggested some discrepancies may exist in the data for the point closest to the stack. Dispersion factors between 7 and 22 km calculated from SO_2, CN, and Hg^0 data are in good agreement. However, between 0.25 and 7 km the factor for Hg^0 was seven times higher than that calculated for CN. Unfortunately, because of instrumentation problems a value could not be calculated for SO_2. The problems at the 0.25-km sampling location result from a combination of the response time of the instrumentation, the narrow cross section of the plume at this distance, and the resulting short sampling

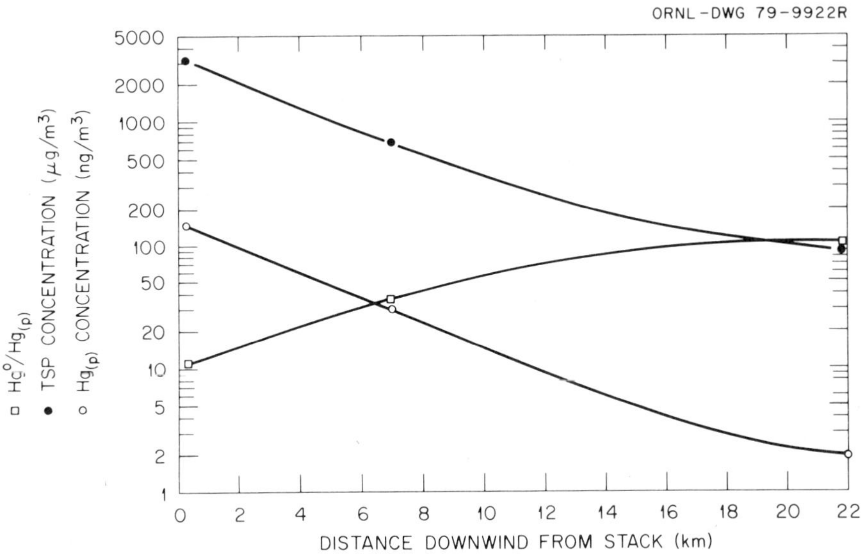

Figure 17.5. Concentration of total suspended particles (TSP, ●) and particulate mercury (Hg_p, ○), and Hg vapor to particulate Hg concentration ratio (Hg^0/Hg_p, □) in the plume of a coal-fired power plant.

time in the plume during each pass (Meagher et al., 1978). Because of this, the absolute concentrations of Hg vapor in the plume should be regarded cautiously, particularly at the point closest to the smokestack. It will be more meaningful in the remainder of this discussion to consider the ratios of the measured concentrations.

If significant adsorption or condensation of Hg vapor were occurring as the plume cooled and mixed with ambient air, as suggested elsewhere (Williston, 1968; Staff, 1971; Billings and Matson, 1972; Lockeretz, 1974), the vapor/particle ratio should decrease, assuming little settling loss for particles in the size range encountered in the plume. However, this ratio increased with distance from 11 at 0.25 km to 33 at 7 km and 100 at 22 km. Similarly, the fraction of the total Hg present as a vapor increased to 97% at 7 km and 99% at 22 km. This increase in the Hg^0/Hg_P ratio indicates not only the absence of any measurable gas-to-particle conversion but also the loss of some fraction of the initial particulate Hg population.

Because of the high vapor pressure of Hg at ambient temperatures (~20 mg/m³) one would not expect condensation to influence the gas-phase concentrations. Rather, the process must involve adsorption/desorption phenomena. The loss of particulate Hg could be the result of desorption of Hg vapor from the aerosol or physical removal of some fraction of the particles during plume travel. As illustrated in Figure 17.5, particulate mercury was apparently influenced by the removal process to a similar extent as the total plume aerosol. Calculated plume dispersion factors from the particle concentration data in Table 17.2 are 1, 0.21, and 0.027 for the total suspended particulate load at 0.25, 7, and 22 km, respectively. The dispersion factors calculated from the particulate Hg concentrations are similar at 7 km (0.20), but somewhat lower at 22 km (0.013). Considering plume travel just from the 7- to 22-km distance, the CN data yield a dispersion factor (0.12) in good agreement with that of TSP (0.13), but higher than that of Hg_P (0.07).

The close agreement in the dispersion factors for CN and TSP at this distance suggests that the decrease in TSP concentration is primarily determined by dispersion, while the decrease in Hg_P concentration is somewhat enhanced by another process. Loss of aerosols by gravitational settling over these distances would not be expected for particles in the size range encountered in the plume (~0.14–1.5 μm diameter; Lindberg and Harriss, 1979; Meagher et al., 1978). Assuming settling loss of Hg_P to be negligible over the course of these measurements, the additional decrease in the concentration of Hg_P in the plume between 7 and 22 km, indicated by the lower dispersion factor, the increased Hg^0/Hg_P ratio at 22 km, and the decreased total Hg concentration in the solid at 22 km (Table 17.2), must be explained by desorption or displacement of Hg from the particle surfaces. Because of the dominance of Hg^0 in the plume, displacement of a relatively small quantity of Hg from the particulate fraction could account for a large increase in the Hg^0/Hg_P ratio. This displacement may be the result of reactions between the suspended particulate matter and gaseous oxides of S and N in the plume. Reactions between SO_2, for example, and surfaces of metal-containing particles are well known (Fennelly, 1975), although the effects of these interactions on adsorbed species have not been documented.

Table 17.2. Mercury Content of Coal, Precipitator Ash (ESP Ash), and Air Samples at a Coal-Fired Power Plant

Sample	Hg Vapor Concentration in Air (ng/m^3)	Particulate Hg Concentration in Air (ng/m^3)	Total Suspended Particle Concentration in Air (μg/m^3)	Total Hg Concentration in the Solid (μg/g)	MCF[a]
Coal				0.28[b]	1
ESP ash				0.0037[c]	0.01
Plume, 0.25 km	1700[d]	150	3460	43	150
Plume, 7 km	1000	30	740	40	140
Plume, 22 km	200	2	95	20	70
Background	12	0.1	17	6	—

[a] MCF = mass concentration factor = $\dfrac{\mu\text{g Hg/g particle}}{\mu\text{g Hg/g coal}}$.

[b] $\sigma = \pm 0.04$, $n = 5$ analyses.

[c] $\sigma = \pm 0.0009$, $n = 4$ samples.

[d] $n = 1$.

The relationships between Hg concentration of the feed coal, electrostatic precipitator (ESP) ash, and the suspended particles collected at each plume location are also presented in Table 17.2. These concentrations have been normalized to the feed-coal Hg levels to calculate a mass concentration factor (MCF) equal to Hg content per gram of particulate matter divided by Hg content per gram of feed coal. The depletion of Hg in ESP ash relative to coal (MCF=0.01) indicates that Hg passes through the precipitator primarily as a vapor or is associated with the submicrometer particles which are not efficiently collected in the ESP. A similar depletion in ESP ash Hg content relative to coal was reported by Gladney et al. (1976). The MCF is considerably higher for the particulate sample collected at 0.25 km downwind in the plume, but is somewhat lower for subsequent downwind samples. This is a further indication that no significant gas-to-particle conversion has occurred in the plume between 0.25 and 22 km. The lower MCF at the 22-km location may also reflect mixing of the plume with background aerosols of lower Hg content in addition to some displacement of particle-bound Hg, as discussed above. These data suggest that the major Hg adsorption reactions have been essentially completed by the time the plume has traveled 0.25 km from the stack. Evidence that such reactions are occurring within the power plant comes from the work of Gladney et al. (1976) whose data indicated a 17-fold Hg enrichment relative to coal for in-stack particles (concentrations normalized to aluminum content).

The fraction of the total incoming feed-coal Hg which is discharged to the atmosphere from the power plant as well as the source strength can be estimated, assuming the 0.25-km downwind plume data to represent the maximum particle/vapor concentration ratio and knowing the coal feed rate (4.8×10^5 kg/h), the precipitator efficiency (99%), the composite coal Hg content, and the ESP ash Hg concentration. Less than 1% of the initial feed-coal Hg is retained in the plant (in collected ash) while ~7% is released in particulate form and 92% in vapor form in the stack emissions. The estimated source strength for the single unit sampled was ~3.5 kg Hg/d as a vapor and ~0.3 kg Hg/d in particulate form. Because the major fraction of the feed-coal Hg is emitted from the stacks, total Hg emission rates should be proportional to power plant coal utilization rates as reflected by megawatt capacity. For example, the total Hg emission rate for this 1300-MW plant (3.8 kg/day) is higher than that estimated for a 200-MW power plant (~1.5 kg/day) by Anderson and Smith (1977), similar to that estimated for a 1000-MW plant (4.3 kg/day) by Lockeretz (1974), but less than that estimated for a 2100-MW plant (7.5 kg/day) by Billings and Matson (1972).

IMPLICATIONS

The indication that essentially all of the Hg in the various types of samples we have analyzed is present as a vapor is in agreement with reported measurements of Hg partitioning in ambient air (Federal Register, 1973; Johnson and Braman, 1974; Fitzgerald and Gill, 1979; Slemr et al., 1981; Fitzgerald et al., 1983). The implications of this vapor-phase dominance are many. As discussed earlier, crop plants have been shown to absorb and retain Hg through leaf uptake, while the absorption of particulate forms is considered less likely (Hosker and Lindberg, 1982). In addition, inhaled metallic Hg vapor is able to diffuse much more extensively into blood cells and various tissues than inorganic particle-associated Hg (Magos, 1968).

Other important implications involve atmospheric transport, residence time, and deposition. The occurrence of airborne Hg primarily as the vapor-phase species is conducive to long-range transport from the source and a long atmospheric residence time. Several studies of the dispersion and deposition of Hg near point sources have confirmed this hypothesis (Jernelov and Wallin, 1973; Lockeretz, 1974; Anderson and Smith, 1977, Crockett and Kinnison, 1979; and Högström et al., 1979). Slemr et al. (1981) recently reappraised earlier estimates of the mean global tropospheric residence time of atmospheric Hg, based on numerous measurements of Hg in the atmosphere over the Atlantic Ocean. Using box model calculations and the relationship between average residence time and the variability of gas concentrations in the troposphere proposed by Junge (1974), these authors calculate residence times in the range of 0.7 to 2 yr, considerably larger than values of 0.03 to 0.1 yr proposed earlier (Weiss et al., 1971; Kothny, 1973; Wollast et al., 1976; Andren and Nriagu, 1979). More recent estimates based on measurements from both Atlantic and Pacific oceans (0.4–0.8 yr) confirm these longer times (Fitzgerald et al., 1983).

Minimal data exist on the rates and mechanisms of Hg removal from the atmosphere, particularly dry deposition of the vapor species. Theoretical estimates (not measurements) of the contribution of dry deposition to the overall wet-plus-dry transport of Hg to the earth's surface are as follows:

1. 4 to 40% within 2 km of a power plant and 40 to 90% at 20 km for this same source (Lockeretz, 1974);
2. ~99% within 0.2 km of a chloralkali plant and 94% at 5 km from the plant (Högström et al., 1979);
3. <20% to a forest canopy in the eastern United States (Andren and Lindberg, 1977);
4. ~50% to the sea surface in the north Pacific (Fitzgerald et al., 1983); and
5. from <1% to "possibly significant" on the global scale (Lantzy and Mackenzie, 1979; and National Research Council, 1978, respectively).

Clearly, empirical data on dry removal rates of atmospheric Hg are needed.

Once Hg is dispersed from the source, precipitation scavenging may favor greater removal rates for the vapor than for the particulate forms because ambient aerosol Hg is concentrated in the 0.6- to 1.1-μm size range in ambient air (Lindberg, unpublished data), a size range for which precipitation scavenging efficiencies are at a minimum (Beard, 1977). Precipitation removal of particles is largely a physical process. Following incorporation in a raindrop, a particle containing Hg, depending on its speciation, may or may not release the metal to solution. Thus, the initial composition of the raindrop has little influence on the scavenging efficiency for particulate Hg, although it may obviously influence the ultimate dissolved Hg concentration in the droplet. However, scavenging of the vapor is highly dependent on its solubility in the raindrop, and any characteristic of the initial drop which increases the solubility of vapor-phase Hg will enhance the removal rate.

Reported solubilities of Hg^0 in pure water range from 30 to 60 $\mu g/L$ for a saturated atmosphere (Onat, 1974; Sanemasa, 1975). The solubility of vapor-phase Hg in water increases significantly as the concentrations of 0_2 or H^+ increase. Since published concentrations of Hg in precipitation, 0.001–1.0 $\mu g/L$ (Nriagu, 1979); National Research Council, 1978), do not approach these levels, one might assume that the dissolution of Hg vapor in the raindrop is not a limiting step in its removal from the atmosphere. This would agree with reports by McCarthy et al. (1969) that mercury is completely washed out of air by rain, independent of the initial concentration in the air, and by McLean (1976) that mercury concentrates in falling precipitation. However, because the above water solubilities apply to an atmosphere saturated with Hg^0, the equilibrium solubilities for ambient air concentrations of Hg^0 are much lower ($\sim 3 \times 10^{-5}$ $\mu g/L$ for a background air concentration of 10 ng/m^3, Fogg and Fitzgerald, 1979). Even at the highest ambient air concentrations measured during the case studies reported here ($\sim$1000 ng/m^3), the equilibrium concentration of Hg^0 in pure water is several orders of magnitude below reported levels in precipitation. This supports other field observations that precipitation does not completely remove Hg vapor from the atmosphere (Johnson and Braman, 1974; Fogg and Fitzgerald, 1979). Possible explanations for the discrepancy between measured concentrations of Hg in rain and values calculated for Hg^0 solubility from Henry's Law include (1) Hg^0 solubility is greatly enhanced by the presence of 0_2 and H^+ in rain, (2) Hg^0 is rapidly transformed to considerably more soluble vapor species in an atmosphere saturated with water vapor (i.e., during a precipitation event), and (3) most of the Hg in rain originates from particle removal from the air. The last hypothesis seems least likely because of the low particle/gas Hg concentrations in air and their low removal efficiency by rain. Testing of the other hypotheses awaits further analytical and field data.

The global wet deposition of Hg from the atmosphere has been variously estimated as ranging from 10^6 to 10^8 kg/yr (Weiss et al., 1971; Garrels et al., 1975; Fitzgerald, 1976; Slemr et al., 1981), suggesting considerable uncertainty in wet removal rates. Lack of measurements of dry deposition rates of both vapor and particle phases of Hg hinders the assessment of the importance of this process as a removal mechanism. However, precipitation scavenging of the vapor (which should theoretically increase in efficiency as precipitation acidity increases) appears to be the major removal process on a global scale. These points should be considered in future research in light of reports of relationships between acid precipitation and elevated Hg levels in fish from remote locations (Brouzes et al., 1977; Brosset and Svedung, 1977).

ACKNOWLEDGMENTS

Research assistance with the case studies summarized in this chapter was provided by R. R. Turner, J. W. Huckabee, J. Meagher, J. R. Lund, D. R. Jackson, S. A. Janzen, and M. Levin. Assistance with sample collection was provided by the Tennessee Valley Authority and the Consejo de las Minas de Almadén. Financial support was provided in part by the National Science Foundation, the Electric Power Research Institute, the Environmental Protection Agency, and the U.S. Department of Energy to the Oak Ridge National Laboratory, operated by Union Carbide Corporation under contract W-7405-eng-26 with the U.S. Department of Energy. Publication No. 2223, Environmental Sciences Division, ORNL. By acceptance of this article, the publisher or recipient acknowledges the U.S. Government's right to retain a nonexclusive, royalty-free license in and to any copyright covering the article.

REFERENCES

Anderson, W. L., and Smith, K. E. (1977). "Dynamics of mercury at coal-fired power plant and adjacent cooling lake." *Environ. Sci. Technol.* **11,** 75–80.

Andersson, A. (1979). "Mercury in soils." In *The Biogeochemistry of Mercury in the Environment.* Elsevier/North Holland Biomedical Press, New York, pp. 79–112.

Andren, A. W., and Lindberg, S. E. (1977). "Atmospheric input and origin of selected elements in Walker Branch Watershed, Oak Ridge, Tennessee." *Water Air Soil Pollut.* **8,** 199–215.

Andren, A. W., and Nriagu, J. O. (1979). "The global cycle of mercury." In *The Biogeochemistry of Mercury in the Environment.* Elsevier/North Holland Biomedical Press, Amsterdam.

Beard, K. V. (1977). "Rain scavenging of particles by electrostatic-inertial impaction and Brownian diffusion." In R. G. Semonim and R. W. Beadle, Eds., *Precipitation Scavenging—1974.* ERDA Symposium Series 41, CONF-741003. Technical Information Center, Oak Ridge, Tenn.

Bidstrup, L. P. (1914). *Toxicity of Mercury and Its Compounds.* Elsevier, Amsterdam.

Billings, C. E., and Matson, W. R. (1972). "Mercury emission from coal combustion." *Science* **176,** 1232–1233.

Braman, R. S., and Johnson, D. L. (1974). "Selective absorption tubes and emission technique for determination of ambient forms of Hg in air." *Environ. Sci. Technol.* **12,** 996–1003.

Brosset, C., and Svedung, I. (1977). *Preliminary Study of the Possibility of a Relationship Between High Acidity in Lakes and High Hg Content of Fish Populations.* Report B378, Swedish Water and Air Pollution Res. Lab., Gothenburg, Sweden.

Brosset, C. (1982). "Total airborne Hg and its possible origin." *Water Air Soil Pollut.* **17,** 37–50.

Brouzes, R. J. P., McLean, R. A. N., and Tomlinson, G. H. (1977). "Mercury—the link between pH of natural waters and the mercury content of fish." Paper presented at the meeting of the U.S. National Academy of Sciences, National Research Council Panel on Mercury, Washington, D.C.

Browne, C. L., and Fang, S. C. (1978). "Uptake of mercury vapor by wheat: An assimilation model." *Plant Physiol.* **61,** 430–433.

Caban, R., and Chapman, T. (1972). "Losses of Hg from chlorine plants: A review of a pollution problem." *Am. Inst. Chem. Eng. J.* **18,** 892–903.

Crockett, A. B., and Kinnison, R. R. (1979). "Mercury residues in soil around a large coal-fired power plant." *Environ. Sci. Technol.* **13,** 465–476.

Desadeleer, G., and Goldberg, E. D. (1978). "Rock volatility—some initial experiments." *Geochem. J.* **12,** 75–79.

Dowd, G., Carte, G., and Monkman, J. L. (1976). "A Hg vapor detector based upon the automatic adjustment of large intensity to the mercury concentration." *J. Air Pollut. Control Assoc.* **26,** 678–679.

Elder, F. C. (1979). *Atmospheric mercury deposition workshop report.* National Water Research Institute, Canadian Center for Inland Waters, Burlington, Ontario.

Federal Register (1973). *Fed. Regist.* **38,** 8820–8845. U.S. Government Printing Office, Washington, D.C.

Feldman, C. (1974). "Perchloric acid procedure for wet-ashing organics for the determination of mercury (and other metals)." *Anal. Chem.* **46,** 1606–1609.

Fennelly, P. F. (1975). "Primary and secondary particulates as pollutants: A literature review." *J. Air Pollut. Control Assoc.* **25,** 697–704.

Ferrara, R., Petrosino, A., Maserti, E., Seritti, A., and Bavghigiani, C. (1982). "The biogeochemical cycle of mercury in the Mediterranean Part II." *Environ. Technol. Lett.* **3,** 499–456.

Fitzgerald, W. F. (1976). "Mercury studies of seawater and rain: Geochemical flux and implications." In *Marine Pollutant Transfer.* Lexington Books, D.C. Health Co., Lexington, Mass.

Fitzgerald, W. F., and Gill, G. A. (1979). "Subnanogram determination of mercury by two-stage-gold amalgamation and gas phase detection applied to atmospheric analysis." *Anal. Chem.* **51,** 1714–1720.

Fitzgerald, W. F., Gill, G. A., and Hewitt, A. D. (1983). "Air–sea exchange of mercury." In *Trace Metals in Sea Water.* Plenum Publishing Corp, New York.

Fogg, T. R., and Fitzgerald, W. F. (1979). "Mercury in southern New England coastal rains." *J. Geophys. Res.* **84,** 6987–6989.

Francis, C. W., Bonner, W. P., and Tamura, T. (1972). "An evaluation of zonal centrifugation as a research tool in soil science. I. Methodology." *Soil Sci. Soc. Amer. Proc.* **36,** 366–372.

Frear, D. E. H., and Dills, L. E. (1967). "Mechanism of the insecticidal action of mercury and mercury salts." *J. Econ. Entomol.* **60,** 970–974.

Galloway, J. N., Eisenreich, S. J., and Scott, B. C. (1980). *Toxic Substances in Atmospheric Deposition: A Review and Assessment.* Environmental Protection Agency Report EPA 560/5-80-001, Washington, D.C.

Garrels, R. M., McKenzie, F. T., and Hunt, C. (1975). *Chemical Cycles and the Global Environment. Assessing Human Influences.* Report published by W. Kaufman, Inc., Los Altos, Calif.

Gladney, E. S., Small, J. A., Gordon, G. E., and Zoller, W. H. (1976). "Composition and size distribution of in-stack particulate material at a coal-fired power plant." *Atmos. Environ.* **10,** 1071–1077.

Giese, A. C. (1940). "Mercury poisioning." *Science* **91,** 476.

Gifford, F. A. (1968). *Meteorology and Atomic Energy.* United States Atomic Energy Agency Report 66116, Washington, D.C.

Gilmour, J. T., and Miller, M. S. (1973). "Fate of a mercuric-mercurous chloride fungicide added to turfgrass." *J. Environ. Qual.* **2,** 145–148.

Hadelshi, T., Church, D. A., McLaughlin, R. D., Zak, B. D., Nakamura, M., and Chang, B. (1975). "Mercury monitor for ambient air." *Science* **187,** 348–349.

Harriss, R. C., and Hohenemser, C. (1978). "Mercury: Measuring and managing the risk." *Environment* **20,** 25–36.

Harriss, R. C. (1979). "Coal combustion as a global source of volatile heavy metals. In J. J. Singh and A. Deepak, Eds., *Environmental and Climatic Impact of Coal Utilization.* Academic Press, New York.

Hildebrand, S. G., Lindberg, S. E., Turner, R. R., Huckabee, J. W., Lund, J. R., and Andren, A. W. (1980). *Biogeochemistry of Mercury in a River-Reservoir System: Impact of an Inactive Chloralkali Plant in the Holston River—Cherokee Lake.* ORNL/TM-6141. Oak Ridge National Laboratory, Oak Ridge, Tenn.

Hitchcock, A. E., and Zimmerman, P. W. (1957). "Toxic effects of vapors for mercury and compounds of mercury on plants." *Ann. N.Y. Acad. Sci.* **65,** 474–497.

Hogstrom, U., Enger, L., and Svedung, I. (1979). "A study of atmospheric Hg dispersion." *Atmos. Environ.* **13,** 465–476.

Hosker, R. P., and Lindberg, S. E. (1982). "Review: Atmospheric deposition and plant assimilation of gases and particles." *Atmos. Environ.* **16,** 889–910.

Huckabee, J. W. (1973). "Mosses: Sensitive indicators of airborne Hg pollution." *Atmos. Environ.* **7,** 749–754.

Jepsen, A. F., and Langan, L. (1971). *Monitoring Mercury Vapor near Pollution Sites.* EPA Report 10620 GLY 05/71. Water Pollution Control Research Series, Office of Research and Monitoring, Environmental Protection Agency, Washington, D.C., 66 pp.

Jernelöv, A., and Wallin, T. (1973). "Air-borne mercury fallout on snow and around five Swedish chloralkali plants." *Atmos. Environ.* **7,** 209–214.

Johnson, D. L., and Braman, R. S. (1974). "Distribution of atmospheric mercury species near ground." *Environ. Sci. Technol.* **8,** 1003–1009.

Junge, C. E. (1974). "Residence time and variability of tropospheric trace gases." *Tellus* **26,** 477–488.

Klein, D. H., and Goldberg, E. D. (1970). "Mercury in the marine environment." *Environ. Sci. Technol.* **4,** 765–768.

Klein, D. H., Andren, A. W., and Bolton, N. E. (1975). "Trace element discharge from coal combustion for power production." *Water Air Soil Pollut.* **5,** 71–77.

Kothny, E. L. (1973). "The three-phase equilibrium of Hg in nature." In *Trace Elements in the Environment.* Advances in Chemistry Series 123, American Chemical Society, Reinhold, New York, pp. 48–79.

Kozuchowski, J., and Johnson, D. L. (1978). "Gaseous emissions of Hg from an aquatic vascular plant." *Nature* **274,** 467–469.

Kudo, A., Miller, D. R., Mortimer, D. C., and DeFreitas, A. S. W. (1976). "Distribution and transport of pollutants in flowing water ecosystems." Ottawa River Project Report 3, National Resources Council, Ottawa, Canada.

Landa, E. R. (1978). "Microbial aspects of the volatile loss of applied mercury (II) from soils." *J. Environ. Qual.* **7,** 84–86.

Lantzy, R. J., and Mackenzie, F. T. (1979). "Atmospheric trace metals: Global cycles and assessment of man's impact." *Geochim. Cosmochim. Acta* **43,** 511–525.

Lindberg, S. E., Andren, A. W., and Harriss, R. C. (1975). "Geochemistry of mercury in the estuarine environment." In *Estuarine Research,* Vol. I of *Chemistry Biology and the Estuarine System,* Academic Press, New York.

Lindberg, S. E., and Turner, R. R. (1977). "Mercury emission from chlorine-production solid waste deposits." *Nature (London)* **268,** 133–136.

Lindberg, S. E., and Harriss, R. C. (1979). "Emissions from coal combustion: Use of aerosol solubility in environmental assessment." In J. J. Singh and A. Deepak, Eds., *Environmental and Climatic Impact of Coal Utilization.* Academic Press, New York.

Lindberg, S. E., Jackson, D. R., Huckabee, J. W., Jansen, S. A., Levin, M. J., and Lund, J. R. (1979). "Atmospheric emission and plant uptake of mercury from agricultural soils near the Almadén mercury mine." *J. Environ. Qual.* **8,** 572–578.

Lindberg, S. E. (1980). "Mercury partitioning in a power plant plume and its influence on atmospheric removal mechanisms." *Atmos. Environ.* **14,** 227–231; **15,** 631–634.

Lockeretz, W. (1974). "Deposition of airborne mercury near point sources." *Water Air soil Pollut.* **3,** 179–193.

Lodenius, M., and Laaksorvirta, K. (1979). "Mercury content of *Hypogymnia physodes* and pine needles affected by a chloralkali works at Kuusankoski, Finland." *Ann. Bot. Fennici* **16,** 7–10.

Magos (1968) cited in Wallace, R. A., Fulkerson, W., Shultz, W. D., and Lyon, W. S. (1971). *Mercury in the Environment: The Human Element.* ORNL/NSF/EP-1. Oak Ridge National Laboratory, Oak Ridge, Tenn.

McCarthy, J. R., Vaughan, W. W., Learned, R. E., and Mueschke, J. L. (1969). *Mercury in Soil, Gas, and Air—A Potential Tool in Mineral Exploration.* United States Geological Survey Circular 609, Washington, D.C.

McClean, R. A. N. (1976). *The Determination of Hg in the Environment in the Quevillon Area of North-Western Quebec.* Mimeo. Rep. Res. Center, Domtar, Ltd., Senneville, Quebec.

Meagher, J., Stockburger, L., Bailey, E. M., and Huff, O. (1978). "The oxidation of SO_2 to sulfate aerosols in the plume of a coal fired power plant." *Atmos. Environ.* **12,** 2197–2203.

Moffitt, A., and Kupel, R. (1971). "Rapid method employing impregnated charcoal and atomic absorption for determination of Hg." *Amer. Ind. Hyg. Assoc. J.* **32,** 614–620.

National Research Council. (1978). *An Assessment of Mercury in the Environment.* Printing and Publishing Office, National Academy of Sciences, Washington, D.C., 185 pp.

Nelson, N. (1971). "Hazards of mercury: Report of the study group on mercury hazards." *Environ. Res.* **4,** 1–69.

Nriagu, J. O. (1979). *The Biogeochemistry of Mercury in the Environment.* Elsevier/North Holland Biomedical Press, New York.

Onat, E. (1974). "Solubility studies of metallic mercury in pure water at various temperatures." *J. Inorg. Nucl. Chem.* **36,** 2029–2032.

Robertson, D. E., Crecelius, E. A., Fruchter, J. S., and Ludwick, J. D. (1977). "Mercury emissions from geothermal power plants." *Science* **196,** 1094–1097.

Rogers, R. D. (1977). *Abiological Methylation of Mercury in Soil.* EPA-600/3-77-007. U.S. Environmental Protection Agency, Washington, D.C. 11 pp.

Rogers, R. D., and MacFarlane, J. C. (1979). "Factors influencing volatilization of Hg from soil." *J. Environ. Qual.* **8,** 255–260.

Ross, R. G., and Stewart, D. K. R. (1962). "Movement and accumulation of mercury in apple trees and soil." *Can. J. Plant Sci.* **42,** 280–285.

Sanemasa, I. (1975). "The solubility of elemental mercury vapor in water." *Bull. Chem. Soc. Japan* **48,** 1795–1798.

Schlesinger, W. H., Reiners, W. A., and Knopman, D. S. (1974). "Heavy metal concentrations and deposition in bulk precipitation in montane ecosystems of New Hampshire, U.S.A." *Environ. Pollut.* **6,** 39–47.

Schroeder, W. H. (1982). "Sampling and analysis of Hg and its compounds in the atmosphere." *Environ. Sci. Technol.* **16,** 394A–400A.

Siegel, S. M., and Siegel, B. Z. (1975). "Geothermal hazards: Mercury emission." *Environ. Sci. Technol.* **9,** 473–474.

Slemr, F., Seiler, W., and Schuster, G. (1981). "Latitudinal distribution of Hg over the Atlantic Ocean." *J. Geophys. Res.* **86,** 1159–1166.

Soldanov, B. A., Bien, P., and Kwan, P. (1975). "Air-borne organo-mercury and elemental mercury emissions with emphasis on central sewage facilities." *Atmos. Environ.* **9,** 941–944.

Staff, Environmental Magazine. (1971). "Mercury in the air." *Environment* **13,** 24–35.

Turner, R. R., and Lindberg, S. E. (1978). "Behavior and transport of mercury in a river-reservoir system downstream of an inactive chloralkali plant." *Environ. Sci. Technol.* **12,** 918–923.

U.S. Geological Survey. (1970). *Mercury in the Environment.* Department of the Interior, Geological Survey Professional Paper 713, 67 pp.

Van Horn, W. (1975). *Materials Balance and Technology Assessment of Mercury and its Compounds on National and Regional Bases.* EPA 560/3-75-007. U.S. Environmental Protection Agency, Washington, D.C., 293 pp.

Vinogradov, A. (1959). *Geochemistry of rare and dispersed chemical elements in soils.* Chapman and Hall Publishing, London.

Wallace, R. A., Fulkerson, W., Schults, W. D., and Lyon, W. S. (1971). *Mercury in the Environment.* ORNL/NSF-EPI. Oak Ridge National Laboratory, Oak Ridge, Tenn., 61 pp.

Wallin, T. (1976). "Deposition of airborne mercury from six Swedish chloralkali plants surveyed by moss analysis. *Environ. Pollut.* **10,** 101–114.

Weiss, H. V., Koide, M., and Goldberg, E. D. (1971). "Mercury in the Greenland ice sheet: Evidence of recent input by man." *Science* **174,** 692–694.

Williston, S. H. (1968). "Mercury in the atmosphere." *J. Geophys. Res.* **73,** 7051–7055.

Windom, H. L. (1977). *Trace Element Geochemistry of the South Atlantic Bight.* Progress Report, June 1976–May 1977. Skidaway Institute of Oceanography, Savannah, Ga.

Wollast, R., Billen, G., and McKenzie, F. T. (1976). "Behavior of Hg in natural systems and its global cycle." In *Ecological Toxicity Research,* Plenum, New York.

Young, D. R. (1971). *Mercury in the Environment: A Summary of Information Pertinent to the Distribution of Hg in the Southern California Bight.* Southern California Coastal Water Research Report, Los Angeles, Calif.

18

BIOGEOCHEMICAL CYCLING OF ORGANIC LEAD COMPOUNDS

W.R.A. De Jonghe and F.C. Adams

Department of Chemistry
University of Antwerp
Wilrijk, Belgium

1. INTRODUCTION

Trace metals and metalloids occur in nature as chemical compounds (species), and as such they interact with biologically important molecules and exert their beneficial or detrimental effects on the basis of distinct chemical and/or physical properties. The toxicity and availability to organisms may be totally different for different chemical species of the same element. A well-known example is the compound-specific activity of arsenite, arsenate, and the various organoarsenic derivatives (Luh, 1973). The green copper arsenite pigment (Scheele's green and Paris or emerald green), introduced industrially in about 1780, was widely used in paints and wallpapers. Throughout the nineteenth century many people were affected, even killed, when molds on damp wallpaper (e.g., S *brevicaulis*) metabolized the arsenic compound to volatile arsenic trimethyl (Challenger, 1945; Hunter, 1978; Sanger, 1893). This source of arsenic has recently been invoked to explain the abnormally high arsenic concentration in some samples of Napoleon's hair (Forshufvud and Smith, 1964; Forshufvud et al., 1961; Jones and Ledingham, 1982; Leslie and Smith, 1978; Lewin et al., 1982; Smith et al., 1962). Molecular characterization and quantification of the individual species in which the element of interest is present in the sample are thus prerequisites for the evaluation of its health hazard.

With the identification of methylmercury chloride as the chemical species primarily responsible for the Minamata disease (Irukayama et al., 1961), an increasing interest developed for the selective determination of organometals and organometalloids. The presence of these substances in environmental media may be caused by a widespread commercial use (Frey and Shapiro, 1971; Lexmond et al., 1976; Woolson, 1975; Zuckerman, 1976) or by natural alkylation processes (Ridley et al., 1977; Saxena and Howard, 1977) (Table 18.1); usually they are much more toxic than the respective inorganic metallic or metalloidal compounds (Carter and Fernando, 1979; Röderer, 1982a; Thayer, 1974).

Since their introduction as antiknock agents in the combustion of gasoline, the tetraalkyllead compounds undoubtedly play a very important role in our society. Few organic chemicals are produced in greater quantity (Frey and Shapiro, 1971), and it has been claimed that automotive traffic accounts for 98% of the total amount of lead disseminated into the environment (Fishbein, 1974). Due to the use of leaded gasoline, in the early 1970s about 18 t Pb entered the atmosphere of the metropolitan area of Los Angeles daily (Huntzicker et al., 1975). The impact of this pollution source on the ecosystem

Table 18.1. Organometals and Organometalloids in the Environment

Compounds in Commercial Use[a]	Elements Known to Be Naturally Alkylated	Elements Possibly Alkylated
Tetraalkyllead (gasoline additive)	Hg	Pb
Organomercurials (fungicide, herbicide disinfectant)	As	Pd
Trialkyltin salts (biocide)	Se	Pt
Dialkyltin salts (PVC stabilizer)		Au
Methylarsenic acids (herbicide)		Sn
Methylcyclopentadienyl (gasoline additive)		Tl
Manganesetricarbonyl		Te
		Cd
		Sb

[a] Methylation and decomposition products of all these may give rise to several other organometallic species.

is impressive, as illustrated by the historical record of the lead content of Greenland ice (Murozumi et al., 1969) and Southern California Basin sediments (Bertine, 1977; Bruland et al., 1974) (Figs. 18.1 and 18.2).

Although the anthropogenic lead burden is usually entirely attributed to the inorganic compounds, the hazard from organic lead is considered to be nonnegligible (Grandjean and Nielsen, 1979; Harrison and Perry, 1977). Data referring to its concentration in ambient matrices are scarce, and no clear consensus emerges from the surveys performed so far. In general, there is a lack of suitable analytical techniques (Chau et al., 1980; Harrison and Perry, 1977; Van Loon, 1981). The present survey of the literature was set up against this background to ascertain in a critical way the precise nature and extent of the burden caused by organic lead known as of this date.

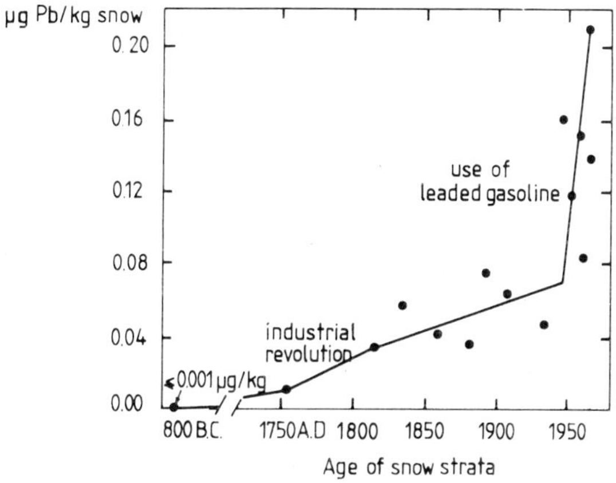

Figure 18.1. Increase of lead content in snow at Camp Century, Greenland, since 800 B.C. (Murozomi et al., 1969).

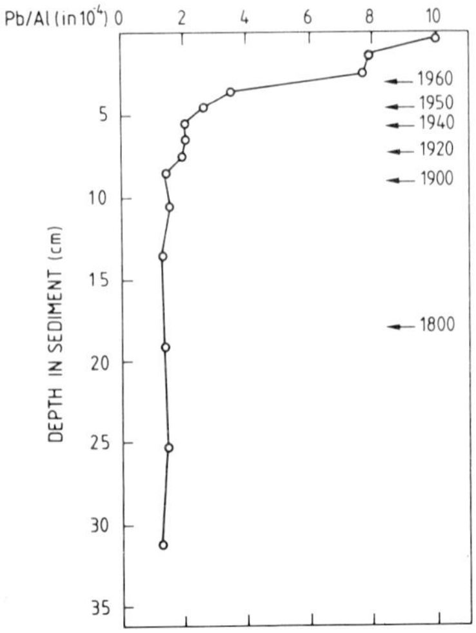

Figure 18.2. Pb/Al ratio in sediment from San Pedro Basin, Southern California (Bertine, 1977; Bruland et al., 1974).

2. TETRAALKYLLEAD AS ANTIKNOCK ADDITIVE

In high-compression internal combustion engines, the burning front may be responsible for the detonation of the rest of the gas mixture compressed in the cylinders; this gives rise to a decrease in thermal efficiency and high mechanical strains on various parts of the motor (the so-called knocking). A successful remedy, applied since 1923, consists in the addition of liquid tetraethyllead (TEL) to the gasoline. The antiknock action is shown by many other compounds, but TEL is by far the most effective (Frey and Shapiro, 1971) (Table 18.2). The only commercial competition is provided by tetramethyllead (TML) and the mixed methylethyllead compounds, trimethylethyllead (TMEL), dimethyldiethyllead (DMDEL), and methyltriethyllead (MTEL). Although the intrinsic antiknock effect of these species is less, their increased volatility is beneficial for fuels containing large amounts of aromatic compounds. Under certain conditions of engine operation such as rapid acceleration, the lighter hydrocarbon fractions would pass into the combustion chamber deficient in lead content if TEL alone was used (Barry, 1975). Since 1960, a mixture of the five tetraalkyllead (TAL) compounds is usually added.

The nature and mechanism of the antiknock effect is still not well understood, but it appears that the TAL compounds must decompose to be active (Frey and Shapiro, 1971). Many different lead compounds are eventually produced, among which lead monoxide is the major one. However, because of the inclusion of 1,2-dichloroethane and 1,2-dibromoethane scavengers, most of the exhausted lead is in the form of halide salts. Once in the atmosphere, these are altered chemically into carbonates, oxides, oxycarbonates, sulfates, and oxysulfates (Ter Haar and Bayard, 1971). Approximately 70% of the lead used in the gasoline is emitted through the tail pipe, the remainder being retained within the exhaust system and lubricating oil

Table 18.2. Relative Effectiveness of Various Compounds as Antiknock Additive[a,b]

Tetraethyllead	118
Tetraphenyllead	73
Iron pentacarbonyl	50
Nickel carbonyl	35
Diethyltelluride	27
Triethylbismuth	24
Diethylselenide	7
Stannic chloride	4.1
Tetraethyltin	4
Triphenylarsine	1.6
Xylidine	1.6
Diphenylamine	1.5
N-Methylaniline	1.4
Dimethylcadmium	1.2
Ethanol	0.1

[a] Aniline = 1 (on mole basis) (Frey and Shapiro, 1971).

[b] In recent years methylcyclopentadienylmanganesetricarbonyl has also found some application, especially as synergist for TEL.

(Barry, 1975). Of the lead emitted, the bulk is present in particulates with an aerodynamic diameter below 1 μm (Müller, 1969).

In 1974 the production of TAL amounted to 375,000 t as Pb in Western Europe and 250,000 t in the United States (Grandjean and Nielsen, 1979). Owing to the increased number of cars equipped with catalytic converters, however, in succeeding years the use of leaded gasoline in the United States has dropped to about 30% of the total consumption (Anderson, 1978; Dartnell, 1980). In Western Europe the TAL production will drop in relation to the lowering of the maximum permissible lead content in gasoline. At present, in most countries the upper limit is in the range 0.15–0.40 g Pb/L (Rohbock and Schmitt, 1981); unleaded gasoline is still allowed to contain up to 13 mg Pb/L (Koizumi et al., 1979). In Eastern Europe the lead content in gasoline may exceed 0.5 g Pb/L (Rohbock and Schmitt, 1981), but in the large cities of the USSR the use of leaded gasoline is prohibited (Grandjean and Nielsen, 1979).

3. HEALTH EFFECTS OF ORGANIC LEAD

The metabolism and toxicology of organic lead compounds are fundamentally different from those of inorganic lead salts. Inorganic lead is present in the atmosphere as fine particulates, and uptake via the inhalation pathway is regulated by the mechanisms of particle retention in the lungs, as well as by the chemical form of the lead. Significant concentrations of organic lead are found in foodstuffs and drinking water and, in general, ingestion represents the major uptake pathway. It is estimated that the average daily amount of lead absorbed into the blood of adults is 29 μg, of which 6.4 μg is derived from the air, 1.5 μg from drinking water, and 21 μg from food intake (Settle and Patterson, 1980). Organic lead, on the other hand, appears to be principally present as TAL in the gas phase of the atmosphere, hence its uptake pathway is dominated by inhalation. Nevertheless it should be noted that marine organisms also may contain appreciable concentrations of organic lead (Birnie and Hodges, 1981; Chau et al., 1979, 1980; Cruz et al., 1980; Mor and Beccaria, 1977; Sirota and Uthe, 1977).

3.1. Uptake and Metabolism

Like other lipophylic compounds, TAL is fairly readily absorbed through the skin. Rabbits absorb lethal quantities of TEL in the course of an hour when pure TEL is applied to their naked skin (Kehoe and Thamann, 1931); the absorption of TML seems to be somewhat slower (Davis et al., 1963). It is reasonable to assume that TAL compounds are relatively rapidly absorbed also through the alveoli and the gastrointestinal tract. Using human volunteers, the retention of inhaled vapor in the lungs has been found to amount to 37% for TEL and 51% for TML (Heard et al., 1979). The initial uptake and distribution of TAL in the body is governed by a reversible gas/liquid phase transfer between the air in the lungs and the blood, which implies that it can be lost again from the blood by exhalation. About 40% of the TML initially retained and about 20% of TEL may be exhaled during the 48 h following exposure. In contrast, the excretion of lead from inhaled TAL in urine and feces is limited (Heard et al., 1979).

Lead, following TAL uptake, is found principally in the liver, kidneys, and brain. In the liver TAL is metabolized to trialkyllead (TriAL), the ultimate toxic agent in case of TAL intoxication (Cremer, 1959, Stevens et al., 1960). For oral intake TriAL may be formed under the influence of the gastric hydrochloric acid before absorption into the blood (Grandjean and Nielsen, 1979). *In vitro* findings suggest that triethyllead (TriEL) cations inhibit oxidative phosphorylation and glucose oxidation of brain slices, but not of kidney or liver slices, and that they inhibit human serum acetylcholinesterase (Aldridge et al., 1962, 1977; Galzigna et al., 1969). These effects are not exhibited by inorganic lead or TEL (Aldridge et al., 1962; Cremer, 1965). From these investigations it might be anticipated that TriEL ions—and presumably also trimethyllead (TriML) ions—have a rather selective effect on the nervous system.

3.2. Toxicology

Convincing evidence indicates that TAL, in spite of its high toxic potential to living organisms, is not at all toxic itself. It must be converted to the lower analogue TriAL to become toxic. A striking example of this peculiar property is shown in Figure 18.3. When TEL-containing cultures of the phytoflagellate *Poterioochromonas malhamensis* are kept in darkness, even at high concentrations of the agent there is a complete lack of toxic effect (Röderer, 1980). Upon illumination, however, the corresponding parallel cultures are profoundly inhibited by the photolytically generated TriEL. For aquatic species higher up the tropic scale the situation is more complicated. Whereas the hydrophylic TriAL ions have difficulty in penetrating the gill membranes of the animal, the hydrophobic TAL compounds are readily absorbed and metabolized (Wood, 1977). Hence, in this case the TriAL compounds are apparently less toxic than the TAL compounds (Maddock and Taylor, 1977) (Fig. 18.4). In mammals, the internal administration of TEL and TriEL gives rise to the same acute symptoms and nearly identical LD_{50} values (the lethal dose or LD_{50} is defined as the quantity of substance, administered as a single dose required to kill half of the exposed population within a period of 14 days). Yet, the toxicity of TML is smaller than that of TriML (Table 18.3). This is due to a faster dealkylation of TEL than of TML (Waldron and Stofen, 1974); for the same reason TEL is more toxic than TML. The further degration products, dialkyllead (DiAL) compounds, are considerably less toxic, their LD_{50} being approximately the same as that of inorganic lead.

The lethal dose for man is not known but it can be estimated on the basis of the LD_{50} for experimental animals. In fact, the LD_{50} should not be stated in relation to body weight, but to the metabolic rate which is approximately

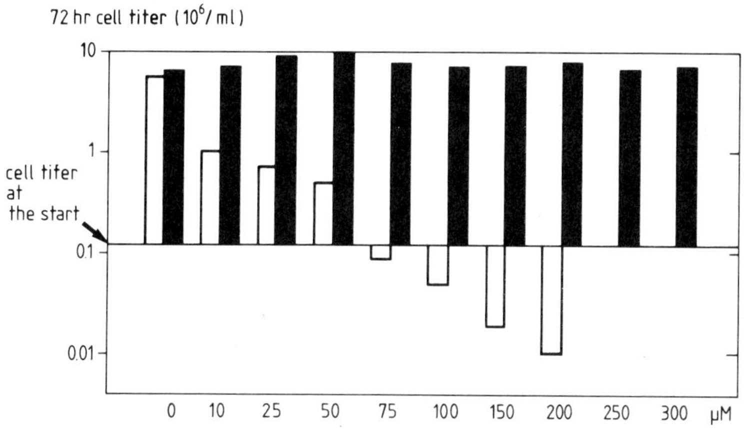

Figure 18.3. Inhibition of cellular growth of *Poterioochromonas malhamensis* following treatment with TEL under light (open bars) and dark (closed bars) conditions (Röderer, 1980).

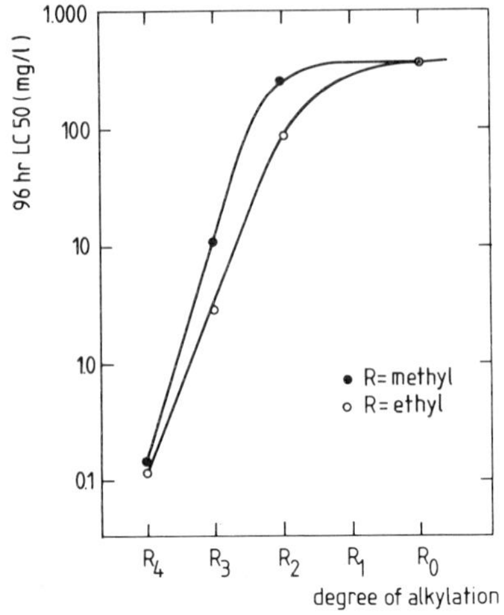

Figure 18.4. Toxicity of lead compounds to marine animals (averaged for shrimp, mussels, and plaice) (Maddock and Taylor, 1977).

proportional to body surface, that is, corresponding to the body weight to the power 0.75 (Grandjean and Nielsen, 1979). From experiments on rats it may be calculated that the lethal dose for an adult person is about 0.25 g TEL and more than 1 g TML. From observations with inorganic lead, however, it might be anticipated that prolonged exposure to subacute doses is also hazardous. In this context it is important to note that, so far, no effective treatment for TAL intoxication has been found. No effect is obtained by BAL (British-Anti-Lewisite), EDTA, or penicillamine, which are used in the

Table 18.3. LD_{50} for Different Organic Lead Compounds in the Rat[a]

Compound	Administration[b]	LD_{50} (mg Pb/kg)
$(C_2H_5)_4Pb$	i.g.	15
	i.v.	15
	i.p.	15
$(C_2H_5)_3PbCl$	i.g.	20
	i.v.	11
$(C_2H_5)_2Pb(OOCCH_3)_2$	i.g.	130
$(C_2H_5)_2PbCl_2$	i.g.	120
$(CH_3)_4Pb$	i.g.	80
	i.v.	88
	i.p.	90
$(CH_3)_3PbCl$	i.g.	≤ 36
$(CH_3)_2Pb(OOCCH_3)_2$	i.g.	120
Pb^{2+}	i.v.	150

[a] From Grandjean and Nielsen (1979).

[b] i.g. = intragastric, i.v. intravenous, i.p. = intraperitoneal.

management of inorganic lead poisoning (Boyd et al., 1957; Cremer, 1959; Haley, 1969; Röderer, 1982b). For symptomatic (mostly neurological) treatment, the use of barbiturates has been recommended (Grandjean and Nielsen, 1979). One case of toxic psychosis was treated with electroshock therapy (Boyd et al., 1957).

4. SOURCES OF ORGANIC LEAD IN THE ENVIRONMENT

4.1. Emission into the Atmosphere

Most organic lead enters the atmosphere during manufacture, transfer of leaded gasoline, and use in vehicles. Precise information on the relative importance of different emission sources is scanty, because until a few years ago the only available methods of analysis required sampling large volumes of air, and precision and specificity of methods were insufficient for environmental research.

It has been estimated that in 1973 ~140 t Pb entered the atmosphere of the United Kingdom as TAL during the manufacture and subsequent transfer of leaded gasoline (Central Unit of Environmental Pollution, 1974). In the United States in 1970, about 11% (1900 t) of the total annual lead emissions from industries resulted from the processing of gasoline additives at the six manufacturing plants currently in operation, but the fraction due to inorganic lead was not reported (Nielsen, 1983).

The relatively high vapor pressures (Shapiro and Frey, 1968) (Table 18.4), especially that of TML, imply that handling of leaded gasoline at filling stations inevitably presents a number of TAL emissions sources, such as direct evaporation, displaced fuel-tank vapors, entrained fuel droplets in the displaced vapors, and gasoline spillage. Huntzicker et al. (1975) have suggested that the loss of TAL from filling stations is about 1.3%. In the 1960s, studies already showed that these losses were more serious for TML than for TEL. Handling of gasoline containing both TML and TEL entailed three times more TML in the surrounding air than TEL (Kehoe et al., 1963).

Table 18.4. Physicochemical Properties of the Different TAL Compounds[a]

TAL Compound	Boiling Point (°C)	Vapor Pressure at 20°C (mm Hg)	Water Solubility (mg Pb/L)
TML	110; 6(10 mm Hg)	26	15
TMEL	130; 28(11 mm Hg)	7.3	
DMDEL	51(13 mm Hg)	2.2	
MTEL	70(16 mm Hg)	0.75	
TEL	202; 82(13 mm Hg)	0.26	<0.1

[a] From Grove (1977) and Shapiro and Frey (1968).

However, the significance of filling stations as point emission sources has not yet been fully elucidated.

A minor part of the TAL used in vehicles is not converted to inorganic lead and is emitted to the ambient air as a result of evaporation from the fuel tank and carburetor, and the presence of uncombusted TAL in the blowup gas from the crankcase and in the exhaust. About 20% of the hydrocarbons emitted by older cars result from evaporation from the fuel tank and carburetor, 25% from the so-called blow-by, and 55% from unburned gasoline in the exhaust (Nielsen, 1983). The same may be assumed to apply to TAL. Since about 1970 the pollution from blow-by gases and carburetors has been eliminated in many of the newly built cars owing to the use of positive crankcase ventilation systems and modifications of the carburetor design.

It is most difficult to evaluate exactly the amount of atmospheric TAL resulting from automotive traffic because the emission depends on several factors, such as composition and content of the antiknock additives in the gasoline used, the ambient air temperature, the condition of the engines, and the driving pattern. Cold starts, low speed, and irregular driving patterns result in a less complete combustion of TAL and, therefore, a higher emission rate of the compounds during city driving than during highway driving. In view of its higher thermal stability (Ryason, 1963), incomplete combustion is most likely to occur for TML. It has been estimated that the total amount of unconverted TAL escaping a car is 1.1% of the input lead in Western Europe (Grandjean and Nielsen, 1979).

In addition to the emission of TAL, vehicles may also provide for an emission of other organic lead compounds. The decomposition of TAL starts by its dissociation into a trialkylplumbyl and an alkyl radical (Gilroy et al., 1972) : $R_4Pb \rightarrow R_3Pb\cdot + R\cdot$. One possible pathway for the trialkylplumbyl radical is its transformation to a TriAL salt. A single study indicates that diethyl- and triethyllead salts may be present in modest quantities in the exhaust (<23% as compared with the quantity of unchanged TEL) (Rifkin and Walcott, 1956). The analytical procedure used was, however, of questionable accuracy. Heating of a gasoline containing TEL and 1,2-dibromoethane caused the formation of triethyllead bromide (Widmaier, 1953).

4.2. Pollution of the Hydrosphere

Dry and wet deposition processes of the atmosphere obviously may constitute a source of organic lead in the hydrosphere. As TAL compounds are in the gaseous form and nearly insoluble in water (Table 18.4), pollution is only likely to occur for the degradation products of TAL, that is, TriAL and DiAL salts. Accidental spills, possibly occurring during transport of antiknock fluids or leaded gasoline by ship (Harrison, 1977), are probably a more important source. Due to the high density, such TAL spills deposit as a comparatively stable pool (Grove, 1977; Cleaver, 1977); solubilization is low but may be increased by turbulences and emulsification (Cleaver, 1977).

Wastewater from refineries may contain organic lead compounds and thus pollute the receiving water. Also, sludge from leaded gasoline tanks is sometimes buried, in which case it may contaminate subsoil water (Grandjean and Nielsen, 1979). It is expected that organic lead, if present in the aquatic environment, is predominantly in the form of the highly water-soluble TriAL compounds.

4.3. Natural Alkylation Processes

There are indications that natural methylation of inorganic lead could provide an additional source of organic lead, especially of TML, in the environment. Evidence was claimed by Harrison and Laxen (1978b) in view of their findings of elevated organic/inorganic lead concentration ratios in the atmosphere at rural locations in northwestern England when sampling coincided with periods during which the air masses had passed over "clean" sea and coastal areas. The high concentrations of organic lead found in marine fauna from unpolluted waters (Birnie and Hodges, 1981, Chau et al., 1979, 1980; Cruz et al., 1980; Mor and Beccaria, 1977; Sirota and Uthe, 1977) similarly may suggest the existence of environmental or *in vivo* methylation of inorganic lead. On the other hand, measurements of organic lead at rural continental sites in China (Jiang, 1982) were below the detection limit of 0.1 ng/m^3, this precluding terrestrial sources of any practical importance. It is uncertain whether a natural methylation process for inorganic lead exists, but if so, it is doubtful whether it will be responsible for emission rates which could explain the abnormal enrichment in the marine or the continental environments (Rah, 1976).

At present, the mechanism of the possible alkylation process is a controversial subject about which there have been a number of studies. It has been established that incubation of trimethyllead acetate with marine sediments readily generates TML (Craig, 1980; Jarvie et al., 1975; Silverberg et al., 1977; Thompson and Crerar, 1980). The lead is, however, already in its tetravalent state prior to its conversion, and a chemical disproportionation suffices to produce TAL (Jarvie et al., 1975; Craig, 1980; Reisinger et al., 1981). What is less certain is whether divalent lead can also be methylated in nature. Wong et al. (1975) claim to have methylated lead nitrate and lead chloride, but not lead hydroxide, cyanide, oxide, or bromide, when incubated with certain sediments; similar observations were made by several other workers (Dumas et al., 1977; Huber et al., 1978; Schmidt and Huber, 1976; Thompson and Crerar, 1980). On the other hand, there have also been reports of failure to repeat these experiments (Craig, 1980; Jarvie et al., 1975; Ridley et al., 1977). In one case the lead(II) salts were methylated to the TriML stage, but the formation of TML was not detected (Jarvie and Whitmore, 1981). Ahmaad et al. (1980) observed the synthesis of TAL from lead(II) salts in water upon addition of methyliodide, but according to Snyder and Bentz

(1982), this was catalyzed by the aluminum foil-wrapped stoppers, thus reducing the lead(II) salts to finely divided metallic lead. The relatively rapid methylation of lead powder has also been noted by other workers (Jarvie and Whitmore, 1981).

5. FATE OF TETRAALKYLLEAD IN THE ENVIRONMENT

Convincing evidence indicates that, eventually, in every environmental matrix TAL compounds are converted into inorganic lead through TriAL and DiAL salts; monoalkyllead salts are too unstable to permit their isolation, but are probably also formed as an intermediate in the final conversion step to inorganic lead (De Vos and Wolters, 1980; Shapiro and Frey, 1968). These species may again give rise to TAL, however, via a disproportionation reaction (R = alkyl, X = anion) (Haupt et al., 1972; Huber et al., 1978; Shapiro and Frey, 1968) :

$$4R_3PbX \rightleftharpoons 2R_2PbX_2 + 2R_4Pb \qquad (1)$$

$$2R_2PbX_2 \rightarrow R_3PbX + RX + PbX_2 \qquad (2)$$

$$3R_3PbX \rightarrow 2R_4Pb + RX + PbX_2 \qquad (3)$$

Reaction (2) is much faster than reaction (1), and therefore TriAL compounds are probably the only alkyllead salts occurring in measurable quantities in nature. The stoichiometry of their decomposition can be described practically by reaction (*3*).

5.1. Transformations in the Atmosphere

The principal reaction pathways of TAL breakdown under atmospheric conditions are homogeneous, and typical of those applying to hydrocarbons in general. Breakdown is due primarily to attack by reactive species generated by photochemical activity. Although the TAL compounds have only a small absorbance in the tropospheric solar uv region, there is in addition a pathway for direct photolytic decomposition (Harrison and Laxen, 1978a; Riccoboni, 1941). Surface reactions on atmospheric particulates appear to be of secondary importance. Yet, there are some indications that a reversible physical adsorption is possible (Edwards and Rosenvold, 1974; Edwards et al., 1975; Rohbock et al., 1980). In Table 18.5 the maximum decay rates for the different reaction pathways of TML and TEL in bright sunshine in a moderately polluted atmosphere are summarized. During the night the decay rates are less than 1 %/h and mainly the results of ozone attack (Harrison and Laxen, 1978a).

Table 18.5. Estimated Upper Limit Rates of TML and TEL Decay in the Day in a Moderately Polluted Irradiated Atmosphere[a]

	Concentration of Reactive Species[b]		Decay Rate (%/h)			
			TML		TEL	
Decay Path	Summer	Winter	Summer	Winter	Summer	Winter
OH attack	$(1-3)\times10^{-7}$ mm	$(1-2)\times10^{-8}$ ppm	8–21	1–1.5	51–88	7–13
Photolysis	$z \sim 40°$	$z \sim 75°$	8	2	26	7
O_3 attack	100–200 ppb	40 ppb	1–2	0.5	9–17	4
$O(^3P)$ attack	10^{-8} ppm	10^{-9} ppm	<0.1	<<0.1	0.1	<0.1
Particulates	200 $\mu g/m^3$	200 $\mu g/m^3$			0.03	<0.03
Total			16–29	3–4	67–93	17–23

[a] From Harrison and Laxen (1978a).

[b] z = solar zenith angle.

TML is clearly more stable than TEL. Taking into account the diurnal cycle of the concentration of photochemical generated species and the possible cloudcover during the day, Nielsen (1983) estimated the total decomposition rates of TML and TEL at conditions typical for a 24-h period at summertime to be about 8 and 30 %/h, respectively. Hence the half-life of TML is of the order of 9 h, whereas that of TEL is only 2 h. In winter, the atmospheric residence time is appreciably longer and can range over several days. This implies that TAL compounds may be transported from the sources of emission to rather remote areas downwind.

To our knowledge, no investigations have been performed on the atmospheric fate of TriAL and DiAL compounds. As the sublimation point of trimethyllead chloride and triethyllead chloride is 190 and 166°C (Müller, 1975; Shapiro and Frey, 1968), respectively, the dominant proportion of TriAL will probably remain gaseous. Several organic compounds with boiling points up to 400°C (e.g., anthracene and fluoranthene) are also present in urban atmospheres, mainly in the gaseous phase (Alsberg and Stenberg, 1979; Cautreels and Van Cauwenberghe, 1978; Pedersen et al., 1980). Nevertheless, a certain fraction can also be expected to condense onto particulate matter. DiAL compounds on the other hand are likely to occur only in the particulate matter fraction. Due to their high solubility in water, washout is probably a major sink for airborne TriAL and DiAL. In addition, the latter species may be significantly removed by dry deposition.

5.2. Transformations in the Hydrosphere

TAL compounds may disappear from aqueous systems as a result of both photolytical breakdown and volatilization losses. The decomposition is

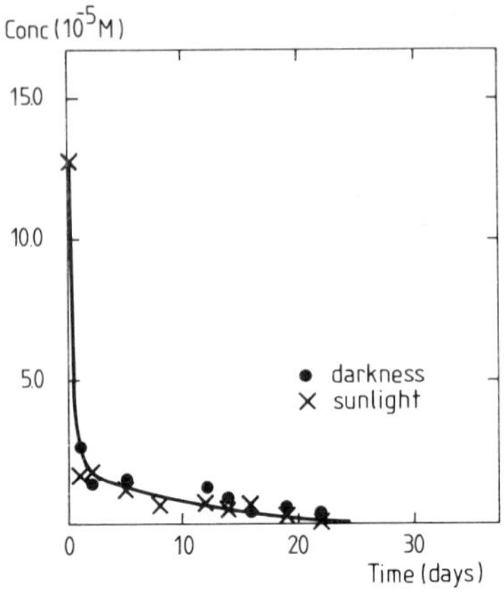

Figure 18.5. Removal of TML from aqueous suspension (Jarvie et al., 1981).

promoted by suspended particulate material and various cations (Jarvie et al., 1981; Robinson and Rhodes, 1980). Because of the similar removal rate of TML in darkness and light (Fig. 18.5), evaporation must be the major sink for the most volatile TAL species. For TEL, on the other hand, light-induced degradation is predominant (Jarvie et al., 1981) (Fig. 18.6). According to Grove (1977), removal of TAL from seawater takes place at rates which give half-lives measurable in days.

The main products of degradation are in the form of highly water-soluble TriAL carbonates, hydroxides, halides, and so forth, which are much more persistent than the TAL compounds themselves (Grove, 1977). As for the TAL compounds, breakdown is promoted by sunlight; trimethyllead species are more stable than triethyllead species (De Jonghe, 1983; Jarvie et al., 1981) (Fig. 18.7). The reaction with cations and silica is minimal, but adsorption onto sediment appears to be important (De Jonghe, 1983; Jarvie et al., 1981). It has been found that the disproportionation of TriAL into TAL

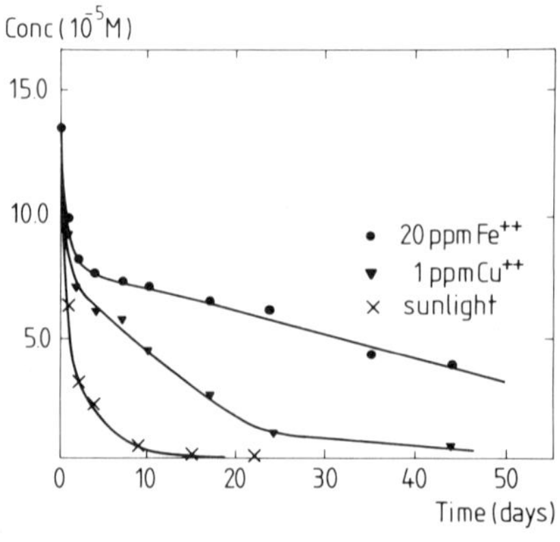

Figure 18.6. Removal of TEL from aqueous suspension (Jarvie et al., 1981). In darkness, TEL was recovered unchanged over 77 days.

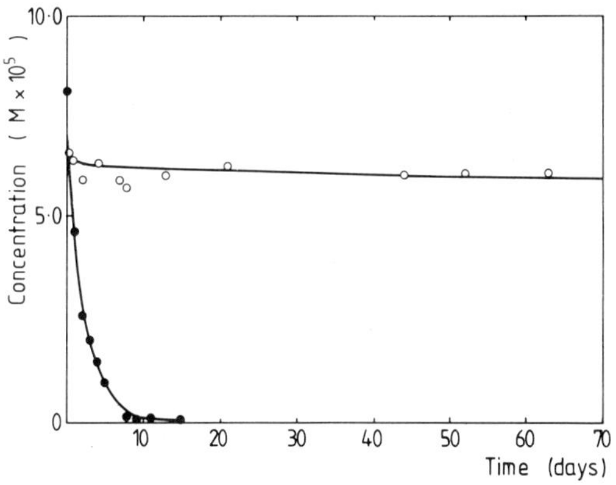

Figure 18.7. Removal of TriML(○) and TriEL(●) chloride from aqueous solution (De Jonghe, 1983; Jarvie et al., 1981).

is catalyzed by various anions, among which the sulfide ion is particularly effective (Huber et al. 1978; Jarvie et al., 1981; Schmidt and Huber, 1976). The concentration of DiAL, generated through degradation of TriAL, is always small.

5.3. Biogeochemical Cycle

Based on the above considerations, the presumed fate of TAL and of its derivatives TriAL, DiAL, and inorganic lead is outlined in Figure 18.8. Organic lead can be introduced into the environment by several routes: via anthropogenic emissions in the air, surface water and wastewater, spills of leaded gasoline or antiknock fluid, and possibly natural alkylation of inorganic lead. The environmental behavior and the toxic potential is largely determined by the physicochemical characteristics of the compounds in the various states of alkylation, such as water solubility, volatility, density, stability, and half-life, and by the prevailing geological factors, such as solar irradiation, air and water movements, and presence of decomposing agents. Most probably the lead compounds can be taken up by aquatic organisms and can become accumulated through the food chain (Wong et al., 1981; Wood, 1977). A small part of the bioaccumulated lead is withdrawn from the aquatic system and can be ingested by humans. The remaining lead pool gradually deposits with excreta, dead organisms, or as particulate-bound lead.

Although there is *in vitro* evidence for different stages of the biogeochemical cycle, it still remains obscure whether all these processes are significant in natural environments. Crucial points in this respect are the importance of natural alkylation and bioaccumulation processes. Nevertheless, the conclusion must be drawn that all organisms in TAL-polluted biotopes could come into contact not only with TAL, but also with the highly toxic

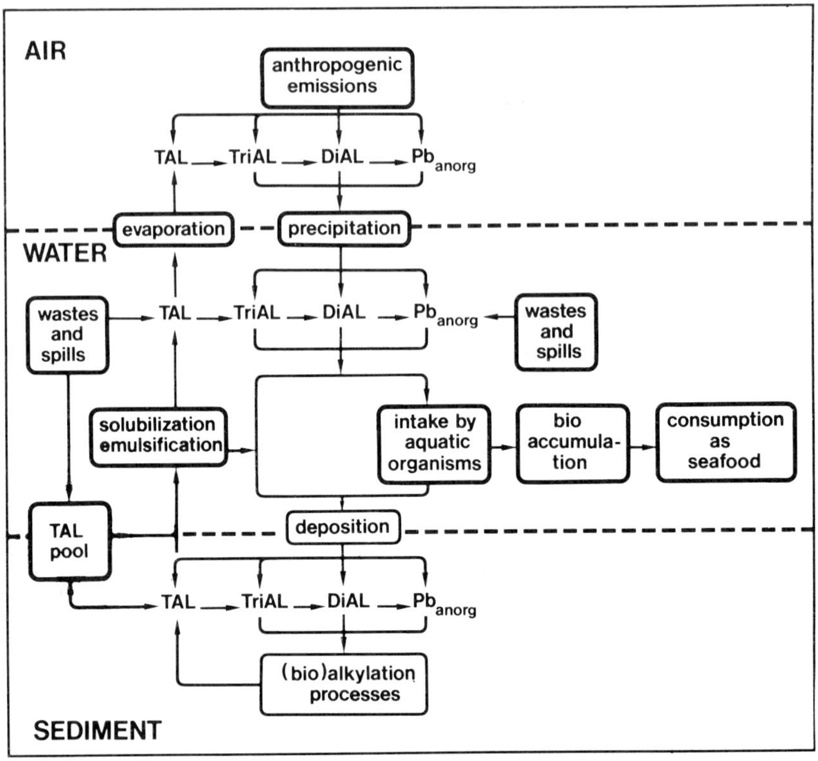

Figure 18.8. Simplified biogeochemical cycle for TAL compounds in the environment. The actual pathways are likely to be more complex in view of disproportionation reactions possibly occurring at different stages of the cycle and the equilibria that may exist between organic lead in the particulate matter and in the gas phase.

TriAL and the other derivatives. This has to be taken into account when evaluating the potential toxic hazards for humans from a TAL-polluted biotope.

6. MEASUREMENTS OF ORGANIC LEAD IN THE ATMOSPHERE

In most of the procedures commonly used for the determination of traces of organic lead in the atmosphere, separation of organic from inorganic lead is based on filtration. The particulate matter, containing most if not all of the inorganic lead, is assumed to remain quantitatively on the filter while the organic lead vapors pass on for concentration and subsequent analysis of the nonfilterable lead.

A distinction should be made between three different analytical methodologies:

1. methods that provide information on the "total" organolead content of the non-filterable fraction of the air particulate system,
2. procedures that are based on the determination of the "volatile" alkyllead compounds, and
3. speciation methods for individual TAL compounds.

The distinction between 1 and 2 is not always clearcut, however. Often it is not clearly stated in the literature data to which category the results of a particular method belong.

A number of methods suffer from lack of specificity for organic lead and present concentrations which are probably too high, because of inclusion of inorganic lead that escapes trapping by the prefilter. Discrepancies could also be expected between the results obtained with short-term sampling procedures and those obtained with extended sampling techniques, in view of the marked diurnal fluctuations in concentration (Harrison et al., 1979). A direct comparison of the results obtained with different analytical procedures is therefore not straightforward. Finally, it should be emphasized that nearly all methodologies that have so far been developed for measuring airborne alkyllead, pay attention exclusively to the nonfilterable species, though a fraction of the organic lead present in the atmosphere may be associated with the particulate matter. This fraction probably contains mostly the degradation products TriAL and DiAL and few TAL itself. The term "total" organic lead should hence be used with caution when referring to results based only on the nonfilterable fraction.

6.1. Analytical Methods

One of the approaches for collecting total organic lead from the air consists in the conversion of the volatile TAL compounds into nonvolatile lead salts through reaction with iodine. This is achieved by passing the air through an aqueous iodine solution, an iodine monochloride solution or through a tube filled with iodine crystals. Alkyllead can then be detected after extraction with dithizone or another reagent, using spectrophotometry of atomic absorption spectrometry. Particulate lead may cause an interference unless it is masked with a suitable complexing agent such as EDTA (Hancock and Slater, 1975).

Another approach is based on the adsorption of the compounds on activated carbon or on iodine impregnated carbon. The TAL compounds can be assayed directly, for example, with X-ray fluorescence, or they can be first eluted from the adsorbent, for example, with ethanol.

Volatile organolead can also be collected from the air on cooled gas chromatography (GC) column packing material. The compounds can be thermally desorbed and determined by atomic absorption spectrometry. The major problem in such sampling procedures is due to adsorption of water vapour. Nevertheless detection limits of 0.04 $\mu g/m^3$ have been obtained for 1-h sampling periods (Harrison et al., 1974).

Similar sampling procedures have been combined with GC separation for species-specific determinations. Specificity for organolead compounds is obtained by detection methods based on mass spectrometry (GC/MS) (Laveskog, 1970), atomic absorption (GC/AAS) (Chau et al., 1976), especially with electrothermal atomization (De Jonghe et al., 1980) or microwave plasma detection (GC/MWPD) (Reamer et al., 1978).

A full recent account of the analytical methodology is given elsewhere (De Jonghe and Adams, 1982).

6.2. Concentration Levels

Tables 18.6 and 18.7 summarize a few recent analytical results from surveys performed with the different methods described. The concentration found for organic lead is usually expressed as a fraction of the total lead concentration, as this figure is less dependent than the absolute level on the total lead use at a particular location and the sampling site topography and meteorology (Harrison and Perry, 1977). However, the total composition of atmospheric lead is not yet definitely known; reliable data are only available for filter-collected particulate lead and gaseous lead alkyls. The variable proportions of gaseous inorganic lead (Robinson, 1978; Robinson and Rhodes, 1979) and nonfilterable particulate lead in Aitken particles (Georgii et al., 1978), as well as the organic lead present in the particulate matter, have not been sufficiently investigated. As total lead is thus more or less a fictitious quantity, it is better to use the ratio of nonfilterable organic to filtered particulate lead (Rohbock et al., 1981).

A direct comparison of the alkyllead data so far available is still rather problematic. Mean concentrations in ambient air range from 2 ng/m^3 at a rural site to up to 2000 ng/m^3 in a parking garage; individual measurements at specific sites cover a wide range. Owing to the diurnal fluctuations, there is a far greater variability in the concentrations measured over shorter averaging periods.

Until now, no specific maximum permissible concentrations have been issued for the organic lead content of environmental air. For inorganic lead the U.S. Environmental Protection Agency (1978) has proposed a monthly mean concentration of 1.5 $\mu g/m^3$. The EC-directive is 2 $\mu g/m^3$ as the maximum yearly average (Directive C 151-22, 1975). In view of the dissimilar toxic properties of inorganic and organic lead, it might be expected that regulations for the latter form, when enacted, are much more stringent.

Important questions remain unanswered regarding the contribution of the bio-methylation processes to the atmospheric concentrations observed. Also, the TAL degradation products (TriAL, DiAL) have not yet been directly determined in atmospheric samples, although the discrepancies between the total organolead results and those obtained with species-specific TAL methods may support the idea of substantial quantities of TriAL and DiAL compounds in the air (Jiang et al., 1981).

Table 18.6. Recently Reported Concentrations of Total and Volatile Organic Lead in Ambient Air

Location and Date	Type of Site	No. of Samples	Averaging Time	Organic Lead Conc (ng/m^3) Range	Organic Lead Conc (ng/m^3) Mean	Organic/Particulate Lead (%) Range	Organic/Particulate Lead (%) Mean	References
Lancaster, 1978	Rural	33	1–2 days	0.5–230	20	1.5–49	11	Laxen, 1978
	Residential			1.7–20	9	0.8–16	4.7	
London, 1980	Urban, 5 m height	7	1 day	24–190	94	3.3–11	6.6	Birch et al., 1980
	Urban, 14 m height			16–130	65	3.6–15	7.2	
Glasgow, 1981	Urban, 0.3 m height	1	1 day		96		11	Gibson and Farmer, 1981
	Urban, 7.5 m height	28		7.2–7.7	32	2.5–25	6.7	
	Residential	10		1.5–54	16	0.9–6.7	3.0	
	Rural	8		1.6–6.5	3.9	1.5–64	14	
	Urban, overnight	5	14 h	2.7–30	11	0.8–4.2	2.5	
	Urban, daytime		10 h	21–144	53	2.2–6.4	4.0	
	Urban, 3 m height	8	7 h	37–110	73	2.5–6.3	3.7	
	Urban, 12 m height	5		31–53	43	2.5–6.0	3.5	
	Urban, 5–30 m height	6		2.9–52	24	1.7–8.1	4.9	
	Petrol station	3		47–274	195	15–47	25	
Antwerp, 1980	Residential	5	6 h	8–20	13	2.6–13	7.9	De Jonghe and Adams, 1980
	Urban, 1 m height	3		77–262	166	14–26	20	
	Urban, 20 m height	2		76–112	94	19–24	22	
	Tunnel	2		99–112	106	1.7–6.1	3.9	
	Petrol station	2		192–213	203	31–32	3.2	
Frankfurt, 1980	Urban, 3.5 m height	40	2–3 h	5–170	45		7.1	Rohbock et al., 1980
	Urban, 20 m height	20		2–130	20		9.0	
	Residential	30		1–90	24		7.8	
	Rural	4		1–7	3		3.5	
	Highway	10		3–15	8		0.7	
	Parking garage	4		450–1000	678		28	

Table 18.7. Recently Reported Concentration of TAL in Ambient Air

Location and Date	Type of Site	No. of Samples	Averaging Time	Species Determined	Organic Lead Conc. (ng/m^3)		Organic/Particulate Lead (%)		References
					Range	Mean	Range	Mean	
Stockholm, 1981	Urban	12	10 h	TML	11–77	39			Nielsen et al., 1981
Copenhagen, 1981	Urban	2	2 h	TML, TEL	185–195	195			Nielsen et al., 1981
	Residential	6	2–26 h	TML, TEL	5.3–60	34			
	Rural	6	16–26 h	TML	0.5–2.5	1.5			
Toronto, 1979	Urban		18 h	TMEL, DMDEL, MTEL, TEL		14		2.2	Radziuk et al., 1979
Antwerp, 1981	Urban	9	1 h	TML, . . . , TEL	49–109	83	4.6–12	8	De Jonghe et al., 1981
	Residential	7		TML, TMEL, DMDEL	3.2–14	7	0.6–3.4	2	
	Highway	10		TML, . . . , TEL	14–44	24	0.8–3.0	1.7	
	Tunnel	15			12–162	39	0.2–1.4	0.5	
	Car-repair shop	6			100–290	205	5.9–14	10	
	Petrol station	21			17–410	149	2.7–35	12	
	Rural	6		TML, TMEL	0.3–3.9	2	0.1–0.7	0.5	
Baltimore, 1978	Highway	6	2 h	TML, . . . , TEL	26–75	53	1.4–5.2	3.2	Reamer et al., 1978
	Tunnel	4			57–130	92	0.4–0.8	0.6	

7. MEASUREMENTS OF ORGANIC LEAD IN THE HYDROSPHERE AND BIOSYSTEMS

7.1. Analytical Methods

Measurement of organic lead in ambient water samples has received much less attention than determinations in air. On the occasion of the Cavtat accident, in which a shipload of ~300 t of antiknock fluid sunk in the Adriatic sea, the lack of adequate analytical methodology was emphasized (Brondi et al., 1981, Harrison, 1977; Tiravanti and Boari, 1979; Tiravanti et al., 1980). Of the procedures which in principle could be applied, only those with environmental significance will be cited here. A more detailed review has been given elsewhere (Crompton, 1974).

7.1.1. Determination of Tetraalkyllead

Tetraalkyllead compounds can be enriched from aqueous samples by means of solvent extraction. Using petroleum ether as the organic phase and gas chromatography with electron capture detections (GC/ECD) for the quantification, Potter and co-workers (1976, 1977) could detect the analytes in water down to concentrations of 2 μg/L. Similar detection limits were reported by Noden (1977) for the GC/MS determination of TAL following separation with hexane; when GC/MS was applied with the head-space technique, the sensitivity was improved by an order of magnitude. Quantitative extraction of TAL from fish tissue was achieved with an aqueous EDTA/benzene solution (Sirota and Uthe, 1977). After digestion of the organic extract, the residue was analyzed for Pb by graphite furnace AAS.

Chau et al. (1980) determined TAL in fish, sediment, water, and vegetation by a method originally developed by Cruz et al. (1980). In this procedure the TAL compounds are thermally desorbed from the sample and preconcentrated in a cooled GC adsorption tube, after which they are volatilized together into a GC/AAS apparatus or, alternatively, directly into the AAS graphite furnace. For water analysis this approach gives a limit of detection of about 0.5 μg/L, the same as obtained by Chau et al. (1979) using GC/AAS in combination with solvent extraction; for sediment, vegetation, and fish samples this figure amounted to 0.1 ppb. In the method developed by Beccaria et al. (1978) and Mor and Beccaria (1977), the TAL compounds were extracted by vacuum distillation at room temperature from sediment or fish tissue, together with the adsorbed water and subsequently condensed in a liquid nitrogen cooled trap. After extraction of the compounds from the condensed liquid with benzene, the measurement was completed either by GC with flame ionization detection (FID) or by direct GFAAS. Limits of detection were about 0.2 ppm.

7.1.2. Determination of Ionic Alkyllead Species

The ionic alkyllead species are usually determined following the removal of any TAL present in the sample by solvent extraction. A number of workers adopted the method of Pilloni and Plazzogna (1966), based on the spectrophotometric determination of DiAL using 4-(2-pyridylazo)-resorcinol (PAR) (Noden, 1977; Potter, 1976; Potter et al., 1977; Schmidt, 1977; Schmidt and Huber, 1978). This reagent does not react with TriAL, and therefore the analysis can be made specific for DiAL after masking inorganic lead. Complexation can also be achieved with dithizone. Although in this case the extraction is less selective (Brondi et al., 1981; Crompton, 1974; Henderson and Synder, 1961; Irving and Cox, 1961; Noden, 1977), the latter approach has the advantage that the complexes formed—unlike those with PAR—are soluble in organic solvents. Hence a favorable organic/aqueous phase ratio allows limits of detection which are, in general, 10 times lower (2 $\mu g/L$) than is possible with PAR.

TriEL can be transferred as the undissociated chloride to an organic solvent, after saturation of the aqueous phase with NaCl (Bolanowska, 1967). However, according to investigations performed by Noden (1977), with toluene only 20% of TriML is extracted. The recovery for the latter species becomes quantitative when potassium iodide is also added, but under these conditions DiAL interferes severely. After phase separation, the TriAL can be back-extracted into a dilute acid solution and measured as DiAL following reaction with iodine monochloride. The NaCl-saturation technique was found inadequate for the analysis of sediment (Potter, 1976; Potter et al., 1977).

A simple extraction procedure for the sensitive determination of traces of trialkyllead compounds in water is described by Chakraborti et al. (1984). After enrichment of the sample by a fast vacuum distillation and saturation of the residual volume with NaCl, the analytes are extracted in chloroform. By incorporating specific purification steps, interference from other forms of inorganic and organic lead is completely eliminated. The final chloroform extract is treated with a sulfuric acid solution to transfer the trialkyllead back into an aqueous phase. The analysis is then completed by graphite furnace atomic absorption spectrometry. A detection limit of 20 ng in a 1-L water sample is achieved.

Hodges and Noden (1979) determined the DiAL and TriAL species in seawater by differential pulse anodic stripping voltammetry (DPASV). Without pretreatment of the sample, these authors could achieve a detection limit as low as 0.2 $\mu g/L$. The method was further developed by Birnie and Hodges (1981) to allow its application to the analysis of marine fauna. Colombini et al. (1981) used the DPASV technique in combination with a series of specific solvent extractions, and in this way succeeded in determining TAL, TriAL, DiAL, and inorganic lead simultaneously down to concentrations of ~0.3 $\mu g/L$. The latter procedure suffers from a lack of quantitative recovery at several stages of the extractions and involves many steps of calculation by difference.

The use of gas chromatographic techniques has also been reported. Robinson et al. (1979) described a method in which both TEL and TriEL were determined directly in seawater using GC/AAS. With a detection limit of 1 mg/L the sensitivity was modest, however. A similar method was adopted by Chau and Wong (1981) for the simultaneous determination of TriML and TriEL, although in this case the water sample was first extracted (in the presence of NaCl) with iso-amyl alcohol. The sensitivity was reported to be much less than that obtainable for TAL compounds. As a result of thermal rearrangement reactions, gas chromatography with packed columns is, in general, not recommended (De Jonghe and Adams, 1983; Estes et al., 1980). In the method described by Estes et al. (1981) the organic extract was vacuum reduced and then analyzed with GC/MPWD; the use of fused silica columns was claimed necessary. The extraction vacuum reduction procedure was determined to be approximately 50% efficient for TriEL and was probably even less efficient for TriML. The detection limit in tap water was about 20 ppb (TriEL).

A real breakthrough in species-specific water analysis is provided by the recently published methods in which the alkyllead compounds are butylated by Grignard reagent to the tetraalkyl form, R_nPbBu_{4-n} (R = Me,Et), and inorganic lead to Bu_4Pb. The determination is completed with GC/AAS or GC/MWPD (Chau et al., 1983; Estes et al., 1982). A detection limit of 0.1 μg/L can be obtained. Other promising species-specific methods include the use of thin-layer chromatography or paper chromatography, but none of them currently possess the required sensitivity (Barbieri et al., 1958; Crompton 1974; Giustiniani et al., 1964; Potter, 1976; Potter et al., 1977).

7.2. Concentration Levels

Only a few surveys revealed the presence of organic lead in the aquatic environment. Van Cleuvenbergen and Adams (1985) could repeatedly detect significant amounts of TriAL in rainwater at concentration levels of 100 μg/L. Positive results for organic lead were also obtained in road drainage water, whereas the concentration of these compounds in other surface water samples was nearly always below the limit of detection of the respective methods (Chau et al., 1979, 1980; Cruz et al., 1980; De Jonghe, 1983; Potter, 1976; Potter et al., 1977). The extremely low concentration of TriAL in most water samples appears to be due to adsorption on particulates and solar radiation-induced breakdown (De Jonghe, 1983; Jarvie et al., 1981).

In marine sediments and fauna, on the contrary, organic lead was detected on several occasions. The reported concentrations for marine fauna are summarized in Table 18.8. In general, the concentration of TAL is low, typically representing less than 10% of the total lead. The values of volatile and solvent extractable lead, which should include TAL and possibly a fraction of TriAL, are generally much higher. It should be stressed, however, that for these analyses there is always a risk of contamination by inorganic lead. The type of TAL species present in the samples appears to be specific for each location (Chau et al., 1980).

Table 18.8. Reported Concentrations of Organic Lead in Marine Samples

Location and Date	Type of Samples	No. of Samples	Lead Species Determined	Organic Lead Conc. (μg/g)		Organic/Total Pb (%)		References
				Range	Mean	Range	Mean	
Halifax (Canada), 1977	Fish	7	Benzene extractable	0.01–4.79	0.75	9.5–90	40	Sirota and Uthe, 1977
Mediterranean sea, 1977	Mussel	4	Volatile + benzene Extractable	9.7–48	22	4.7–20	13	Mor and Beccaria, 1977
Ontario lakes, 1980	Fish	4	TAL	0.003–0.007	0.004	1.8–6.7	5.1	Cruz et al., 1980
			Hexane extractable	0.021–0.023	0.022	8.4–38	23	
			Volatile	0.003–0.016	0.009	2.2–29	10	
Ontario lakes, 1980	Fish	17	TAL	0.001–0.016	0.004	0.3–14	4.3	Chau et al., 1980
		13	Hexane extractable	0.002–0.072	0.017	0.8–80	20	
		15	Volatile	0.001–0.009	0.004	1.3–14	4.0	
N.W. coast of England, 1981	Fish	2	TriAL	0.03–0.05	0.04	2.7–2.8	2.8	Birnie and Hodges, 1981

The occurrence of organic lead compounds in fish pose immediate questions as to the validity of the allowable lead level in food without considering its chemical forms, especially since some of the analyses revealed concentrations in excess of the 2 μg/g tolerance limit permitted by the American Medical Association (Boudène, 1978). Organic lead concentrations in water apparently are much less than the U.S. EPA recommendation of 50 μg/L as the maximum concentration of lead in seawater (U.S. Environmental Protection Agency, 1972).

8. CONCLUSIONS

As appears from the previous discussions, the determination of individual alkyllead species is generally a difficult task. The analytical techniques required must be capable of separation, selective detection, identification, and quantitation of trace element compounds at and below the ppb level in a complex matrix. In general, a chromatographic separation, especially by GC, combined with a specific spectrometric detection of lead (e.g., by AAS or MWPD) is optimum for the TAL compounds; similar methods can be applied for the ionic alkylleads after butylation to the tetraalkyl form.

With regard to the importance of organic lead in environmental air, it can be concluded from the surveys performed that the use of leaded gasoline gives rise to a serious impact only in the immediate vicinity of anthropogenic sources. Both direct evaporation and exhaust gases appear to be significant emission sources. It seems that gasoline filling stations contribute only to a minor extent, as the atmospheric alkyllead concentrations observed close to these do not differ markedly from the levels encountered in a city street. At downtown locations, the average alkyllead concentration in the air amounts to 50–200 ng Pb/m^3, corresponding to an organic/inorganic lead concentration ratio of 5–15%.

Although there are indications that TriAL and/or DiAL are also present in measurable quantities in the air, gaseous TAL compounds are probably of paramount importance; the TAL content in the particulate matter is extremely small. From limited studies it also became clear that natural alkylation processes, if existing, require very specific topographic and climatological circumstances (e.g., the estuarine and coastal mud flats at the northwest-coast of England) as in the absence of automotive traffic the organic lead content of the air is extremely low (Harrison and Laxen, 1978b; Jiang, 1982).

Despite the presence of significant concentrations of TriAL in rainwater, the organic lead content of surface water appears to be negligible. The analyte could only be detected in the drainage water from highways, and it can reasonably well be assumed that subsequent dilution and decomposition reactions prevent measurable quantities from occurring elsewhere. Accumulation of organic lead in waterways as a result of environmental alkylation was not observed, but in fish and other marine fauna this possibility can not be excluded.

Most important is whether alkyllead compounds are present at significant concentration levels in humans especially in the brain, the most critical organ for lead poisoning. The only literature data available on organic lead (as TriAL) are the results of 22 autopsy samples from Denmark (Nielsen et al., 1978). These results indicate a considerably higher TriAL concentration in residents of lower floor city buildings in Copenhagen, than in residents of upper floor buildings, suburb dwellers, and villagers. No extractable (organic) lead could be detected using the same procedure in an archeological sample of mummified brain (Grandjean et al., 1975; Nielsen et al., 1978). The significance of this latter result obviously depends on the long-term stability of the organolead compounds in human tissue.

There is criticism in the literature on the conclusions of Nielsen et al. (1978). On the basis of the extent of measured TAL variation with vertical height above street level and the relative magnitude of urban and rural TAL levels, it was claimed that urban upper and lower floor dwellers would experience similar environmental exposure to the compounds, with rural dwellers forming a distinctly separate low-exposure group (Birch et al., 1980; Gibson and Farmer, 1980).

Indirectly, the influence of organic lead on living systems is much larger than would be inferred from the measured concentration ratio of organic or inorganic lead. Owing to the instability of the tetraalkyllead compounds, the largest part of the lead emitted from cars is indeed in the particulate form, whereas in the environment they are also quickly transformed into inorganic lead. Earlier measurements state that up to 98% of the environmental lead pollution is due to the use of leaded gasoline (Fishbein, 1974). Information on the measurement of organolead compounds is scarce, but interesting conclusions are to be expected from the isotope lead experiment (Lubinska, 1982). This study involving thousands of people in the area of Turin (Italy) and backed by the European Commission, petrol companies, and the International Lead and Zinc Research Organisation, will finish in December 1983. By submitting lead with a special isotope composition, mined from Broken Hill in New South Wales (Australia), for the normal lead additive in gasoline distributed in the area, the lead derived from the gasoline can be identified in blood samples. About 25% of lead in the blood of the city residents appears to result from the Broken Hill isotopes. Among country dwellers the proportion is halved. On the other hand, attention should also be given to the conclusion of the report by the Central Unit of Environmental Pollution (1980). It states that airborne lead, including that derived from gasoline, is usually a minor contributor to the body burden and that food and water are more important. Finally, there is a long-standing scientific controversy about the fact that even low concentrations of lead in air could have harmful effects on the nervous system of critical groups such as children and pregnant women (Budiansky, 1981).

In order to issue a maximum permissible concentration specific for organic lead (either in the air or the gasoline), authorities have to take into account the economical aspects of the problem. As a direct result of the dwindling reserves of crude oil and the oil embargo of 1973 and the consequent increasing cost, it has been recognized that efforts should be made to reduce automotive gasoline consumption. The combination of the energy penalty in both the vehicle and refinery with the use of 100% unleaded gasoline at 92 RON (research octane number) is estimated in the United States at 660,000 barrels per day or 3.5% of the total U.S. oil requirements based on the 1978 consumption (Dartnell, 1980). To date, no other substance has been found which, if lead alkyls are removed, offers a more economical route to high octane numbers than refinery processing. Moreover, the substituting compounds are not necessarily less toxic.

Clearly, much more research is necessary on the environmental behavior of lead compounds, in particular with regard to organic lead. The possibility of natural alkylation should be investigated further, whereas bioaccumulation in fish and other biological systems should be considered in more detail. Selective methods for the determination of the tetraalkyllead degradation products in the air have to be developed further. Also, the identity of other possible alkyllead species, such as the additional lead containing peak occasionally observed in GC/AAS chromatograms (Cruz et al., 1980; De Jonghe, 1983), deserves closer attention. In fact, the environmental cycle proposed in Figure 18.8 has been investigated only to a very limited extent. Despite the trend towards reduction of the lead content in gasoline and the decreasing consumption, monitoring of organic lead in the environment should continue.

REFERENCES

Ahmad, I., Chau, Y.K., Wong, P.T.S., Carty, A.J., and Taylor, L. (1980). "Chemical alkylation of lead(II) salts to tetraalkyllead(IV) in aqueous solution." *Nature* **287,** 716–717.

Aldridge, W.N., Cremer, J.E., and Threlfall, C.J. (1962). "Trialkylleads and oxidative phosphorylation: A study of the action of trialkylleads upon rat liver mitochondria and rat brain cortex slices." *Biochem. Pharmacol.* **11,** 835–846.

Aldridge, W.N., Street, B.W., and Skilleter, D.N. (1977). "Halide-dependent and halide-independent effects of triorganotin and triorganolead compounds on mitochondrial functions." *Biochem. J.* **168,** 353–364.

Alsberg, T., and Stenberg, U. (1979). "Capillary GC-MS analysis of PAH emissions from combustion of peat and wood in a hot water boiler." *Chemosphere* **8,** 487–496.

Anderson, E.V. (1978). "Phasing lead out of gasoline: Hard knocks for lead alkyls producers." *Chem. Eng. News* 12–16 (Feb. 6).

Barbieri, R., Belluco, U., and Tagliavani, G. (1958). "Separazione cromatografica su carta di composti metallorganici di piombo e di stagno." *Ann. Chim.* **48,** 940–949.

Barbieri, R., Faraglia, G., and Giustiniani, M. (1964). "Complessi di composti organometallici." *Ric. Sci.* **3,** 109–117.

Barry, P.S.I. (1975). "The current lead pollution problem." *Postgrad. Med. J.* **51,** 783–787.

Beccaria, A.M., Mor, E.D., and Poggi, G. (1978). "A method for the analysis of traces of inorganic and organic lead compounds in marine sediments." *Ann. Chim.* **68,** 607–617.

Bertine, K.K. (1977). "Lead and the historical sedimentary record." Proc. Int. Exp. Disc. Meeting, *Lead-Occurrence, Fate and Pollution in the Marine Environment,* Rovinj, Yugoslavia, pp. 319–324.

Birch, J. Harrison, R.M., and Laxen, D.P.H. (1980). "A specific method for 24–48 hours analysis of tetraalkyllead in air." *Sci. Tot. Environ.* **14,** 31–42.

Birnie, S.E., and Hodges, D.J. (1981). "Determination of ionic alkyllead species in marine fauna." *Environ. Technol. Lett.* **2,** 433–442.

Bolanowska, W. (1967). "Metoda oznaczania trojetylku olowiu w krwi i moczu." *Chem. Analityczna* **12,** 121–129.

Boudène, C. (1978). "Food contamination by metals." Proc. CEC Research Seminar *Trace Metals: Exposure and Health Effects,* Guilford, England, pp. 163–183.

Boyd, P.A., Walker, G., and Henderson, I.N. (1957). "The treatment of tetraethyllead poisoning." *Lancet* **1,** 181–185.

Brondi, M., Dall'Aglio, M., Ghiara, E., Mignuzzi, C., and Tiravanti, G. (1981). "Environmental studies on lead alkyl release in sea water by the Cavtat wreck." *Sci. Tot. Environ.* **19,** 21–31.

Bruland, K.W., Bertine, K., Koide, M., and Goldberg, E.D. (1974). "History of metal pollution in Southern California coastal zone." *Environ. Sci. Technol.* **8,** 425–432.

Budiansky, S. (1981). "Lead: the debate goes on, but not over science." *Environ. Sci. Technol.* **15,** 243–246.

Carter, D.E., and Fernando, Q. (1979). "Chemical toxicology—part II. Metal toxicity." *J. Chem. Ed.* **56,** 490–495.

Cautreels, W., and Van Cauwenberghe, K. (1978). "Experiments on the distribution of organic pollutants between particulate matter and the corresponding gas phase." *Atmos. Environ.* **12,** 1133–1141.

Central Unit of Environmental Pollution (1974). *Lead in the Environment and Its Significance to Man.* Department of the Environment, HMSO, London.

Central Unit of Environmental Pollution, (1980). *Lead and Health: The Report of a DHSS Working Party on Lead in the Environment.* Department of the Environment, HMSO, London.

Chakraborti, D., De Jonghe, W.R.A., Van Mol, W.E., Van Cleuvenbergen, R.J.A., and Adams, F.C. (1984) "Determination of Ionic Alkyllead Compounds in Water by Gas Chromatography/Atomic Absorption Spectrometry." *Anal. Chem.* **56,** 2692–2697.

Challenger, F. (1945). "Biological methylation." *Chem. Rev.* **36,** 315–361.

Chau, Y.K., and Wong, P.T.S. (1983) "Direct speciation analysis of molecular and ionic organometals." In *Trace Elements Speciation in Surface Waters and Its Ecological Implications,* G.G. Leppard, Ed., Plenum, pp. 87–103.

Chau, Y.K., Wong, P.T.S., Bengert, G.A., and Kramar, O. (1979). "Determination of tetraalkyllead compounds in water, sediment and fish samples." *Anal. Chem.* **51,** 186–188.

Chau, Y.K., Wong, P.T.S., and Kramar, O. (1983). "The determination of dialkyl-, trialkyl-, tetraalkyllead and lead(II) species in water by chelation/extraction and gas chromatography-atomic absorption spectrometry." *Anal. Chim. Acta* **146,** 211–217.

Chau, Y.K., Wong, P.T.S., Kramar, O., Bengert, G.A., Cruz, R.B., Kinrade, J.O., Lye, J., and Van Loon, J.C. (1980). "Occurrence of tetraalkyllead compounds in the aquatic environment." *Bull. Environ. Contam. Toxicol.* **24,** 265–269.

Chau, Y.K., Wong, P.T.S., and Saitoh, H. (1976). "Determination of tetraalkyllead compounds in the atmosphere." *J. Chromatogr. Sci.* **14,** 162–164.

Cleaver, J.W. (1977). "Dispersion of lead alkyls from pools located on the sea-bed." Proc. Int. Disc. Meeting *Lead-Occurrence, Fate and Pollution in the Marine Environment,* Rovinj, Yugoslavia, pp. 325–343.

Colombini, M.P., Corbini, G., Fuoco, R., and Papoff, P. (1981). "Speciation of tetra-, tri-, dialkyllead compounds and inorganic lead at nanomolar levels (sup-ppb) in water samples by differential pulsed electrochemical techniques." *Ann. Chim.,* 609–629.

Craig, P.J. (1980). "Methylation of trimethyllead species in the environment, an abiotic process?" *Environ. Technol. Lett.* **1,** 17–20.

Cremer, J.E. (1959). "Biochemical studies on the toxicity of tetraethyllead and other organolead compounds." *Br. J. Ind. Med.* **16,** 191–199.

Cremer, J.E. (1965). "Toxicologie et biochimie des composés de plomb alkylés." *Rev. Hyg. Profess.* **17,** 15–20.

Crompton, T.R. (1974). *Chemical Analysis of Organometallic Compounds,* Vol. 3. Academic Press, London, pp. 99–189.

Cruz, R.B., Lorouso, C., George, S., Thomassen, Y., Kinrade, J.D., Butler, L.R.P., Lye, J., and Van Loon, J.C. (1980). "Determination of total, organic extractable, volatile and tetraalkyllead in fish, vegetation, sediment and water samples." *Spectrochim. Acta* **35B,** 775–783.

Dartnell, P.L. (1980). "Lead in petrol 1—energy conservation." *Chem. Brit.* **16,** 308–310.

Davis, R.K., Horton, A.W., Larson, E.E., and Stemmer, K.L. (1963). "Inhalation of tetramethyllead and tetraethyllead." *Arch. Environ. Health* **6,** 473–482.

De Jonghe, W.R.A. (1983). "Selective determination and concentration levels of organic lead in the environment." Ph.D. thesis. University of Antwerp, Belgium.

De Jonghe, W.R.A., and Adams, F.C. (1980). "Organic and inorganic lead concentrations in environmental air in Antwerp, Belgium." *Atmos. Environ.* **14,** 1177–1180.

De Jonghe, W.R.A., and Adams, F.C. (1982). "Measurements of organic lead in air—a review." *Talanta* **29,** 2057–2067.

De Jonghe, W., and Adams, F. (1983), "Gas chromatography with flame ionization detection for the speciation of trialkyllead halides." *Fres. Zeit, Anal. Chem.* **314,** 552–554.

De Jonghe, W.R.A., Chakraborti, D., and Adams, F.C. (1980). "Sampling of tetraalkyllead compounds in air for determination by gas chromatography/atomic absorption spectrometry." *Anal. Chem.* **52,** 1974–1977.

De Jonghe, W.R.A., Chakraborti, D., and Adams, F.C. (1981). "Identification and determination of individual tetraalkyllead species in air." *Environ. Sci. Technol.* **15,** 1217–1222

De Vos, D., and Wolters, J. (1980). "Synthesis, reactions and physicochemical properties of mono-organolead(IV) compounds." *Rev. Silicon Germanium Tin Lead Compds.* **4,** 209–243.

Dumas, J.P., Pazdernik, L., Belloncik, S., Bouchard, D., and Vaillancourt, G. (1977). "Methylation du plomb en milieu aquatic." *Water Pollut. Res.—Can.* **12,** 91–100.

Edwards, H.W., and Rosenvold, R.J. (1974). "Uptake of tetraethyllead vapour by atmospheric dust compounds." Proc. 2nd Ann. NSF-RANN Trace Contam. Conf. *Trace Contaminants in the Environment,* Asimolar, Pacific Grove, California, pp. 59–63.

Edwards, H.W., Rosenvold R.J., and Wheat, H.G. (1975). "Sorption of organic lead vapor on atmospheric dust particles." In D.D. Hemphill, Ed., *Trace Substances in Environmental Health.* University of Missouri Press., Columbia, pp. 197–205.

Estes, S.A., Poirier, C.A., Uden, P.C., and Barnes, R.M. (1980). "Gas chromatography with plasma emission spectroscopic detection of Friedel-Crafts catalyzed alkyl group redistribution products among Si, Ge, Sn and Pb atoms." *J. Chromatogr.* **196,** 265–277.

Estes, S.A., Uden, P.C., and Barnes, R.M. (1981). "High-resolution gas chromatography of trialkyllead chlorides with an inert solvent venting interface for microwave excited helium plasma detection." *Anal. Chem.* **53,** 1336–1340.

Estes, S.A., Uden, P.C., and Barnes, R.M. (1982). "Determination of *n*-butylated trialkyllead compounds by gas chromatography with microwave plasma emission detection." *Anal. Chem.* **54,** 2402–2405.

European Economic Community (1975). *Off. J. EEC* (July 7). Directive C 151–22.

Fishbein, L. (1974). "Mutagens and potential mutagens in the biosphere. II. Metals-mercury, lead, cadmium and tin." *Sci. Tot. Environ.* **2,** 341–371.

Forshufvud, S., and Smith, H. (1964). "Napoleon's illness 1816–1821 in the light of activation analysis of hair from various dates." *Arch. Tox.* **20,** 210–219.

Forshufvud, S., Smith, H., and Wassen, A. (1961) "Arsenic content of Napoleon's hair probably taken immediately after his death." *Nature* **192,** 103–105.

Frey, F.W., and Shapiro, H. (1971). "Commercial organolead compounds." *Topics Curr. Chem.* **16,** 243–297.

Galzigna, L., Corsi, G.C., Saia, B., and Rizzoli, A.A. (1969). "Inhibitory effect of triethyllead on serum cholinesterase *in vitro.*" *Clin. Chim. Acta* **26,** 391–393.

Georgii, H.-W., Müller, J., and Rohbock, E. (1978) "Messung von gasformigen bleiverbindungen in der atmosphere und ihre beziehung zum partikelförmig gebundenen blei." *GAF-Berichte* **6,** 359–364.

Gibson, M.J., and Farmer, J.G. (1981). "Tetraalkyllead in the urban atmosphere of Glasgow." *Environ. Technol. Lett.* **2,** 521–530.

Gilroy, K.M., Price, S.J., and Webster, N.J. (1972). "Determination of $D[(CH_3)_3Pb\text{-}Ch_3]$ by the toluene carrier method." *Canad. J Chem.* **50,** 2639–2641.

Giustiniani, M., Faraglia, G., and Barbieri, R. (1964). "Complexes of organometallic compounds. VII. Paper electrophoresis of $(C_2H_5)_2Pb^{2+}$ and $(C_2H_5)_3Pb^+$ in chloride solutions." *J. Chromatogr.* **15,** 207–210.

Grandjean, P., Fjerdingstad, E., and Nielsen, O.V. (1975). "Lead concentration in mummified Nubian brains." Proc. Int. Conf. *Heavy Metals in the Environment,* Toronto, pp. 171–180.

Grandjean, P., and Nielsen, T. (1979). "Organolead compounds: Environmental health aspects." *Residue Rev.* **72,** 97–148.

Grove, J.R. (1977). "Investigations into the formation and behaviour of aqueous solutions of lead alkyls." Proc. Int. Exp. Disc. Meeting *Lead-Occurrence, Fate and Pollution in the Marine Environment,* Rovinj, Yugoslavia, pp. 45–52.

Haley, T.J. (1969). "A review of the toxicology of lead." *Air Quality Monogr.* No. 69-7. Am. Petroleum Institute, New York, p. 22.

Hancock, S., and Slater, A. (1975). "A specific method for the determination of trace concentrations of tetramethyl- and tetraethyllead vapours in air." *Analyst* **100,** 422–429.

Harrison, G.F. (1977). "The Cavtat incident." Proc. Int. Exp. Disc. Meeting *Lead-Occurrence, Fate and Pollution in the Marine Environment.* Rovinj, Yugoslavia, pp. 305–317.

Harrison, R.M., and Laxen, D.P.H. (1978a). "Sink processes for tetraalkyllead compounds in the atmosphere." *Environ. Sci. Technol.* **12,** 1384–1391.

Harrison, R.M., and Laxen, D.P.H. (1978b). "Natural source of tetraalkyllead in air." *Nature* **275,** 738–740.

Harrison, R.M., and Perry, R. (1977). "The analysis of tetraalkyllead compounds and their significance as urban air pollutants." *Atmos. Environ.* **11,** 847–852.

Harrison, R.M., Laxen, D.P.H., and Birch, J. (1979). "Tetraalkyllead in air: Sources, sinks and concentrations." Proc. Int. Conf. *Heavy Metals in the Environment.* London, pp. 257–261.

Harrison, R.M., Perry, R., and Slater, D.H. (1974). "An adsorption technique for the determination of organic lead in street air." *Atmos. Environ.* **8,** 1187–1194.

Haupt, H.J., Huber, F., and Gmehling, J. (1972). "Zersetzungsreaktionen von dimethylbleidichlorid in lösungen mit und ohne fremdsalzzusatz." *Z. Anorg. Allg. Chem.* **390,** 31–40.

Heard, M.J., Wells, A.C., Newton, D., and Chamberlain, A.C. (1979). "Human uptake and metabolism of tetraethyl- and tetramethyllead vapour labelled with ^{203}Pb." Proc. Int. Conf. *Heavy Metals in the Enmvironment.* London, pp. 103–108.

Henderson, S.R., and Snyder, L.J. (1961). "Rapid spectrophotometric determination of triethyllead, diethyllead and inorganic lead ions, and application to the determination of tetraorganolead compounds." *Anal. Chem.* **33,** 1172–1175.

Hodges, D.J., and Noden, F.G. (1979). "The determination of alkyllead species in natural waters by polarographic techniques." Proc. Int. Conf. *Heavy Metals in the Environment.* London, pp. 408–411.

Huber, F., Schmidt, U., and Kirchman, H. (1978). "Aqueous chemistry of organolead and organothallium compounds in the presence of microorganisms." *ACS Symp. Ser.* **82,** 65–81.

Hunter, D. (1978). *Diseases of Occupation.* Hodder & Staughton, London, pp. 348–363.

Huntzicker, J.J., Friedlander, S.K., and Davidson, C.I. (1975). "Material balance for automobile emitted lead in Los Angeles basis." *Environ. Sci. Technol.* **9,** 448–457.

Irukayama, K., Kondo, T., Kai, F., and Fujiki, M. (1961). "Studies on the origin of the causative agent of Minamata disease. I. Organic mercury compounds in fish and shellfish from Minamata Bay." *Kumamoto Med. J.* **14,** 157–169.

Irving, H.J., and Cox, J.J. (1961). "Studies with dithizone. Part VIII. Reactions with organometallic compounds." *J. Chem. Soc.,* 1470–1479.

Jarvie, A.W.P., and Whitmore, A.P. (1981). "Methylation of elemental lead and lead(II) salts in aqueous solution." *Environ. Technol. Lett.* **2,** 197–204.

Jarvie, A.W.P., Markall, R.N., and Potter, H.R. (1975). "Chemical alkylation of lead." *Nature* **255,** 217–218.

Jarvie, A.W.P., Markall, R.N., and Potter, H.R. (1981). "Decomposition of organolead compounds in aqueous systems." *Environ. Res.* **25,** 241–249.

Jiang, S. (1982).. "Determination of alkyllead and alkylselenide compounds by gas chromatography-graphite furnace atomic absorption spectrometry." PhD thesis. University of Antwerp, Belgium.

Jiang, S.G., Chakraborti, D., De Jonghe, W., and Adams, F. (1981). "Atomic-absorption spectrometric determination of volatile organolead compounds in the atmosphere." *Fres'. Z. Anal. Chem.* **305,** 177–180.

Jones, D.E.H., and Ledingham, K.W.D. (1982). "Arsenic in Napoleon's wallpaper." *Nature* **299,** 626–627.

Kehoe, R.A., and Thamann, F. (1931). "The behaviour of lead in the animal organism, II. Tetraethyllead." *Am. J. Hyg.* **13,** 478–498.

Kehoe, R.A., Cholak, J., Spence, J.A., and Hancock, W. (1963). "Potential hazard of exposure to lead, Part I. Handling and use of gasoline containing tetramethyllead." *Arch. Environ. Health* **6,** 239–255.

Koizumi, H., Mc Laughlin, R.D., and Hadeishi, T. (1979). "High gas temperature furnace for species determination of organometallic compounds with a high pressure liquid chromatograph and a Zeeman atomic absorption spectrometer." *Anal. Chem.* **51,** 387–392.

Lantzy, R.J., ,and Mackenzie, F.T. (1979). "Atmospheric trace metals: Global cycles and assessment of man's impact." *Geochim. Cosmochim. Acta* **43,** 511–523.

Laveskog, A. (1970). "A method for the determination of tetramethyllead (TML) and tetraethyllead (TEL) in air." *Proceedings of the 2nd International Clean Air Congress,* Washington, D.C., pp. 549–557.

Laxen, D.P.H. (1978). "Sink processes for tetraalkyllead compounds in the atmosphere." Ph.D. thesis. University of Lancaster, England.

Leslie, A.C.D., and Smith, H. (1978). "Napoleon Bonaparte's exposure to arsenic during 1816." *Arch. Tox.* **41,** 163–167.

Lewin, P.K. Hancock, R.G.V., and Voynovich, P. (1982). "Napoleon Bonaparte—no evidence of chronic arsenic poisoning." *Nature* **299,** 627–628.

Lexmond, T.M., de Haan, F.A.M., and Frissel, M.J. (1976). "On the methylation of inorganic mercury and the decomposition of organomercury compounds—a review." *Neth. J. Agric. Sci.* **24,** 79–97.

Lubinska, A. (1982). "Turin experiment reveals lead threat." *New Scientist* **96,** 281–281.

Luh, M., Barker, R.A., and Henley, D.E. (1973). "Arsenic analysis and toxicity—a review." *Sci. Tot. Environ.* **2,** 1–12.

Maddock, B.G., and Taylor, D. (1977). "The acute toxicity and bioaccumulation of some lead alkyl compounds in marine animals." Proc. Int. Exp. Disc. Meeting *Lead-Occurrence, Fate and Pollution in the Marine Environment,* Rovinj, Yugoslavia, pp. 233–261.

Mor, E.D., and Beccaria, A.M., (1977). "A dehydratation method to avoid loss of trace elements in biological samples." Proc. Int. Exp. Disc. Meeting *Lead-Occurrence, Fate and Pollution in the Marine Environment,* Rovinj, Yugoslavia, pp. 53–59.

Müller, E. (ed.) (1975). *Methoden der Organischen Chemie (Houben-Weyl),* Vol. 13, Part 7, G. Thieme Verlag, Stuttgart, p. 87.

Müller, J. (1968). "Messung der grössenverteilung von schwebstaubgebundenen schwermetallen." *Staub-Reinhaltung Luft* **2,** 69–70.

Murozumi, M., Chow, T.J., and Patterson, C. (1969). "Chemical concentrations of pollutant lead aerosols, terrestrial dusts and sea salts in Greenland and Antarctic snow strata." *Geochim. Cosmochim. Acta* **33,** 1247–1294.

Nielsen, T. (1983). "Atmospheric occurrence of organolead compounds." In P. Grandjean, Ed., *Biological Effects of Organolead Compounds,* CRC Press, Boca Raton, Fla., pp. 43–62.

Nielsen, T., Egsgaard, H., Larsen, E., and Schroll, G. (1981). "Determination of tetramethyllead and tetraethyllead in the atmosphere by a two-step enrichment method and gas-chromatographic-mass spectrometric isotope dilution analysis." *Anal. Chim. Acta* **124,** 1–13.

Nielsen, T., Jensen, K.A., and Grandjean, P. (1978). "Organic lead in normal human brains." *Nature* **274,** 602–603.

Noden, F.G. (1977). "The determination of tetraalkyllead compounds and their degradation products in natural water." Proc. Int. Exp. Disc. Meeting *Lead-Occurrence, Fate and Pollution in the Marine Environment,* Rovinj, Yugoslavia, pp. 83–91.

Pedersen, P.S., Ingwersen, J., Nielsen, T., and Larsen, E. (1980). "Effects of fuel, lubricant and engine operating parameters on the emission of polycyclic hydrocarbons." *Environ. Sci. Technol.* **14,** 71–79.

Pilloni, G., and Plazzogna, G. (1966). "Spectrophotometric determination of diethyllead and diethyltin ions with 4-(2-pyridylazo)-resorcinol." *Anal. Chim. Acta* **35,** 325–329.

Potter, H.R. (1976). "The nature and concentration of organometallics in natural waters." PhD thesis. University of Aston, Birmingham, England.

Potter, H.R., Jarvie, A.W.P., and Markall, R.N. (1977). "Detection and determination of alkyllead compounds in natural waters." *Wat. Pollut. Control.* **76,** 123–128.

Radziuk, B., Thomassen, Y., Van Loon, J.C., and Chau, Y.K. (1979). "Determination of alkyllead compounds in air by gas chromatography and atomic absorption spectrometry." *Anal. Chim. Acta* **105,** 255–262.

Rahn, K.A. (1976). "The chemical composition of the atmospheric aerosol." Technical Report of the Graduate School of Oceanography, University of Rhode Island, Kingston, R.I.

Reamer, D.C., Zoller, W.H., and O'Haver, T.C. (1978). "Gas chromatograph–microwave plasma detector for the determination of tetraalkyllead species in the atmosphere." *Anal. Chem* **50,** 1449–1453.

Reisinger, K., Stoeppler, M., and Nürnberg, H.W. (1981). "Experimental evidence for the absence of biological methylation of lead in the environment." *Nature* **291,** 228–230.

Riccoboni, L. (1941). "Spettri di assorbimento di alcuni composti metallorganici dello stagno e del piombo." *Gazz. Chim. Ital.* **71,** 696–713.

Ridley, W.P., Dizikes, L.J., and Wood, J.M. (1977). "Biomethylation of toxic elements in the environment." *Science,* **197,** 329–332.

Rifkin, E.B., and Walcott, C. (1956). "Decomposition of TEL in an engine." *Ind. Eng. Chem.* **48,** 1532–1541.

Robinson, J.W. (1978). "The analysis of tetraalkyllead compounds and their significance as urban air pollutants—further comments." *Atmos. Environ.* **12,** 1247–1248.

Robinson, J.W., and Rhodes, L.J. (1979). "Sources of inorganic molecular lead in the ambient atmosphere." *Spectrosc. Lett.* **12,** 781–807.

Robinson, J.W., and Rhodes, I.A.L. (1980). "Solubility of TEL in sea water. The effect of suspended particles." *J. Environ. Sci. Health* **A15,** 201–209.

Robinson, J.W., Kiesel, E.L., and Rhodes, I.A.L. (1979). "Studies of interactions between tetraethyllead and sea water using GC-AA." *J. Environ. Sci. Health* **A14,** 65–85.

Röderer, G. (1980). "On the toxic effects of tetraethyllead and its derivatives on the chrysophyte *Poterioochromonas Malhamensis.* I. Tetraethyllead." *Environ. Res.* **23,** 371–384.

Röderer, G. (1982a). "Biological effects of inorganic and organic compounds of mercury, lead, tin and arsenic." Proc. 16th Ann. Conf. *Trace Substances in Environmental Health,* Missouri (Columbia).

Röderer, G. (1982b). "On the toxic effects of tetraethyllead and its derivatives on the chrysophyte *Poterioochromonas Malhamensis.* IV. Influence of lead-antidotes and related agents." *Chem. Biol. Interact.*

Rohbock, E., and Schmitt, G. (1981). "Vergleich des aromaten- und bleigehaltes Europäischer benzine." *Environ. Technol. Lett.* **2,** 263–270.

Rohbock, E., Georgii, H.-W., and Müller, J. (1980). "Measurements of gaseous lead alkyls in polluted atmospheres." *Atmos. Environ.* **14,** 89–98.

Rohbock, E., Georgii, H.-W., and Müller, J. (1981). "Measurements of gaseous lead alkyls in polluted atmospheres—Discussion." *Atmos. Environ.* **15,** 423–424.

Ryason, P.R. (1963). "Thermal stabilities of the methylethyl lead alkyls." *Combustion Flame* **7,** 235–243.

Sanger, C.R. (1983). "On chronic arsenical poisoning from wall papers and fabrics." *Proc. Am. Acad. Arts Sci.* **29,** 148–177.

Saxena, J., and Howard, P.H. (1977). "Environmental transformation of alkylated and inorganic forms of certain metals." *Adv. Appl. Microbiol.* **21,** 185–226.

Schmidt, U. (1977). "Biomethylierung von Pb^{2+} und alkylbleiverbindungen und untersuchungen zur wachtstumshemmung von bakterien durch alkylbleiverbindungen." PhD thesis. University of Dortmund, FRG.

Schmidt, U., and Huber, F. (1976). "Methylation of organolead and lead(II) compounds to $(CH_3)_4Pb$ by microorganisms." *Nature* **259,** 157–158.

Schmidt, U., and Huber, F. (1978). "Spectralphotometrische bestimmung von blei(II)-sowie dialkylblei- und trialkylbleiverbindungen in geringen konzentrationen." *Anal. Chim. Acta* **98,** 147–149.

Settle, D.M., and Patterson, C.C. (1980). "Lead in albacore: guide to lead pollution in Americans." *Science* **207,** 1167–1176.

Shapiro, H., and Frey, F.W. (1968). *The Organic Compounds of Lead.* Wiley, New York.

Silverberg, B.A., Wong, P.T.S., and Chau, Y.K. (1977). "Effect of tetramethyllead on freshwater green algae." *Arch. Environ. Contam. Toxicol.* **5,** 305–313.

Sirota, G.R., and Uthe, J.F. (1977). "Determination of tetraalkyllead compounds in biological materials." *Anal. Chem.* **49,** 823–825.

Smith, H., Forshuvfund, S., and Wassen, A. (1962). "Distribution of arsenic in Napoleon's hair." *Nature* **194,** 725–726.

Snyder, L.J., and Bentz, J.M. (1982). "Alkylation of lead(II) salts to tetraalkyllead in aqeous solution." *Nature* **296,** 228–229.

Stevens, C.D., Feldhake, C.J., and Kehoe, R.A. (1960). "Isolation of triethyllead ion from liver after inhalation of tetraethyllead." *J. Pharmacol. Exp. Ther.,* **128,** 90–94.

Ter Haar, G.L., and Bayard, M.A. (1971). "Composition of airborne lead particles." *Nature* **232,** 553–554.

Thayer, J.S. (1974). "Organometallic compounds and living organisms." *J. Organometall. Chem.* **76,** 265–295.

Thompson, J.A.J., and Crerar, J.A. (1980). "Methylation of lead in marine sediments." *Marine Pollut. Bull.* **11,** 251–253.

Tiravanti, G., and Boari, G. (1979). "Potential pollution of a marine environment by lead alkyls: the Cavtat incident." *Environ. Sci. Technol.* **13,** 849–854.

Tiravanti, G., Rozzi, A., Dall'Aglio, M., Delaney, W., and Dadone, A. (1980). "The Cavtat accident: evaluation of alkyl lead pollution by simulation and analytical studies." *Prog. Wat. Technol.* **12,** 49–65.

U.S. Environmental Protection Agency, (1972). *Water Quality Criteria.* National Academy of Sciences, Washington, D.C., p. 249.

U.S. Environmental Protection Agency, (1978). "National ambient quality standard for lead." *Fed Regis.* **43,** 46246–46277.

Van Cleuvenbergen, R.J., and Adams, F.C. (1985). "Occurrence of tri- and dialkyllead species in environmental water." *Environmental Science and Technology*, submitted.

Van Loon, J.C. (1981). "Review of methods for elemental speciation using atomic spectrometry detectors for chromatography." *Canad. J. Spectrosc.* **26A,** 22–32.

Waldron, H.A., and Stofen, D. (1974). *Sub-clinical Lead Poisoning.* Academic Press, London, p. 61.

Widmaier, O. (1953). "Reaktionszerfall von bleibenzin." *Brennstoff Chem.* **34,** 83–87.

Wong, P.T.S., Chau, Y.K. Kramer, O., and Bengert, G.A. (1981). "Accumulation and depuration of tetramethyllead by rainbow trout." *Wat. Res.* **15,** 621–625.

Wong, P.T.S., Chau, Y.K., and Luxon, P.L. (1975). "Methylation of lead in the environment." *Nature* **253,** 263–264.

Wood, J.M. (1977). "Lead in the marine environment: some biochemical considerations." Proc. Int. Exp. Disc. Meeting *Lead-Occurrence, Fate and Pollution in the Marine Environment,* Rovinj, Yugoslavia, 299–303.

Woolson, E.A. (Ed.) (1975). *Arsenical Pesticides,* ACS Symp. Series, No. 7, American Chemical Society, Washington D.C.

Zuckerman, J.J. (Ed.) (1976). "Organotin compounds: New Chemistry and Applications." *Adv. Chem. Ser.* No. 157. American Chemical Society, Washington D.C.

19

AIRBORNE LEAD IN THE ENVIRONMENT IN FRANCE

Jean Servant

Laboratoire d'Aerologie
Universite Paul Sabatier
Toulouse, France

1. INTRODUCTION

Like other trace metals, lead is introduced into the atmosphere by human activities. Its sources are widespread, since it is used as a gasoline additive and it is found in the atmosphere in particulates of submicron diameter. So Pb particles from pollution sources are found at all continental sites and are dispersed by the atmospheric circulation over thousands of kilometers, including the remote oceanic and the polar regions. Since lead is a poison for humans, it is necessary to understand the distribution of Pb pollution in ecosystems and its importance in comparison with the natural level of Pb. The total quantity of Pb emitted in the atmosphere (around 1973) was 4 or 4.5 $\times 10^{11}$ g/yr (Nriagu, 1979; Servant, 1982). Both estimates assume that 75% of the lead used in gasoline is dispersed into the atmosphere. Direct measurements of fallout near highways give a 90% dispersion (Chamberlain et al., 1978), and measurements over a 10-yr period show only 60% (Little and Wiffen, 1978). In a very detailed study, Nriagu (1979) estimated that the other industrial emissions amount to about 1.71 $\times 10^{11}$ g/yr and emissions from natural sources are roughly 0.18 $\times 10^{11}$ g/yr. It is noteworthy that major contributors to airborne lead include the production of steel and base metals, a fact which had not formerly been fully appreciated. It is well known that the highest Pb concentrations in the air are observed near emission centers such as towns, highways, and some industrial centers. The lead concentrations in the air near the ground decrease with distance from these local emissions, and the lead is deposited progressively by dry deposition on the vegetation and the surfaces near the ground, and by wet deposition with rain. According to the data of the UK monitoring network (Cawse, 1974), the residence time of lead in the atmosphere is about 7.5 days for average rainfall intensity of 2 mm/day (Servant, 1982). Therefore, lead could be carried from one country to another in western Europe, for example, via the atmospheric circulation. The present study in France was carried out to (1) measure the deposition of lead over the different regions, (2) assess the importance of the dry fallout by measuring the deposition speed on the vegetation, (3) estimate the fraction of atmospheric lead carried by three rivers in southwest France, and (4) find out the natural level of lead in human blood with the view of establishing the level of contamination of the population. The complexity of the transfer of lead from one compartment to another in the biosphere as well as the research of the origins of lead in the environment and in the diet call for geochemical studies, as pointed out by Patterson (1965).

2. SITES, MATERIALS, AND METHODS

2.1. Sites

Rainwater was collected at 11 stations, and tree leaves at 5 sites. These sites are displayed in Figure 19.1. Tree leaves were also gathered in a forest in the Congo about 200 km from Brazzaville. The pluviometers were located in a residential area of Toulouse, and 2–3 km from the center of various other villages.

The rivers sampled run in the alluvial couloirs in a hilly region. The Hers is a very small river (60-km long), and its source is located on the plain, and not in the mountains, contrary to the two others. The lowland region is devoted to agriculture, and forests are abundant in the mountains. The flow of the rivers is about 200 m^3/s for the Garonne, between 50 and 100 m^3/s for the Ariege,

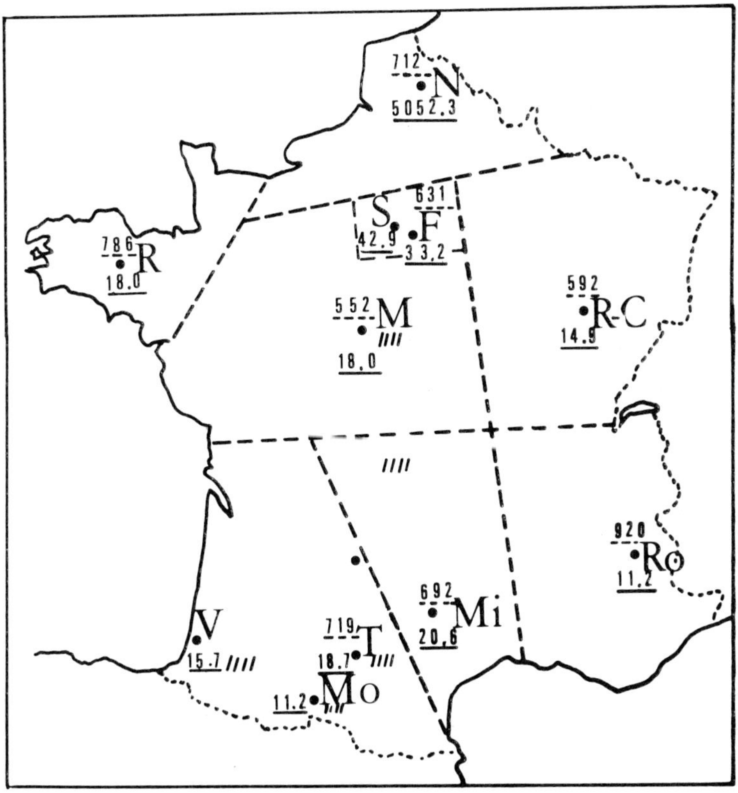

Figure 19.1. Lead concentration in the rainwater in France (1972–1973). Sites for sampling of rainwater and tree leaves: N, (62) Noyelles-Godault, pop. 5,200; S, (91) Saclay, 570; F, (91) Fontenay le Viconte, 326; R, (22) Rostrenen, 2,950; M, (41) Maslives, 250; R-C, (88) Rouvres la Chetive, 423; V, (40) Vieux Boucau, 1,500; Mo, (09) Moulis, 1,013; T, (31) Toulouse, 400,000; Mi, (12) Millau, 23,442; and o, (05) Rosans, 370. /// Tree leaves. France is divided into 8 regions; —, yearly mean of the lead concentration in rain water $\mu g/L$; ...., pluviometry (mm).

and about 3 m^3/s for the Hers. The range of variation in the flow of each river is maximum in May and June (de Martonne and Demangeon, 1947). The widths of the drainage basins are 60 km for the Ariege and the Garonne and 6 km for the Hers.

2.2. Materials

In Toulouse, the leaves were collected along the Canal du Midi at two different places, 4-km apart, to avoid local contamination. The nearest busy road was always at a distance of more than 1 km. All tree leaves were collected from the outer portion of the lower crown, approximately 2-m above ground. Plastic gloves were used while gathering plant specimens *in situ*.

The river waters were sampled by pail from a bridge over the Ariege and the Hers and the bank on the Garonne. The sampling points were 11-km upstream from the agglomeration of Toulouse.

2.3. Methods

Lead analysis was made by colorimetry at $\lambda = 510\ \mu m$ (Charlot, 1966; Gilbert, 1964) and then by conventional atomic absorption. The rainwater and river water (1 L) were preconcentrated by slow evaporation for 17 h. The blanks showed 0.25 μg Pb on 50 evaporations of deionized water. The highest laboratory contamination was less than 0.8 μg Pb; this value was observed twice. The blank was reduced to $\leq 0.14\ \mu g$ Pb by solvent extraction with no preconcentration, the final volume being 2 mL dilute acid. The ^{210}Pb in river water was recovered from 10Lby iron hydroxide precipitation (0.5 g Fe) after the addition of 0.010 g of old lead (free of ^{210}Pb). The iron was separated by three ether extractions. The lead was purified by absorption (2 *M* HCl) on an Amberlite resin (25 mL) and desorption by deionized water (50 mL). The lead was precipitated as the bichromate and then deposited in a stainless steel cup ($S = 6.6\ cm^2$). The background of the low-level flow counter was equal to 0.56 ± 0.03 cpm. Sample activities were higher than this value; total counting stopped at 5000 counts, and the statistical error was 2%. Microquantitative determinations of Pb and ^{210}Pb purification were performed in two different rooms.

The lead on surfaces of tree leaves was desorbed by soaking 50 g (wet) of sample for 5 h in 2 L HCl solution.

Surface soil samples were taken at depths of 0–20 cm. Soluble lead was determined by digesting 30 g dry weight with 0.1 *M* HCl for 24 h. Total lead and lead-210 were dissolved by Hf and $HClO_4$ attack. To study the lead and lead-210 content as a function of granulometry, the soil particles were separated by classical sedimentation methods. Ultrasound was used to break down aggregates: 5 g of soil + 5 mL NH_4OH + 95 mL H_2O, treated for 20 min with an ultrasonic probe (100 W) dipped into the solution. Analyses of lead in food were performed on 100 g (wet) samples, after roasting at 500°C and digesting the residues with acid.

Blood samples were taken at the University's Rangueil Hospital from nonsmoking male and female personnel of Paul Sabatier University. Within the hour following the sampling, 100 mL was roasted at 550°C after adding K_2SO_4 in order to avoid sputtering. Dissolution and analysis for Pb and ^{210}Pb were performed, as previously described (Servant, 1975; Servant and Delapart, 1978). We first compared our analyses of calcinated samples with the USPHS method for the determination of lead in airborne and biological samples (National Academy of Sciences, 1972). Since results are comparable, the more convenient chemical treatment of 100 mL of blood was adopted. Using an identical procedure, three samples of 100 mL of physiological fluid were checked for lead contamination during sampling and subsequent handling; the results were found to be negative.

3. DEPOSITION OF LEAD OVER FRANCE

3.1. Wet Deposition

Average annual concentrations in rainwater for 1972 and 1973 are shown in Figure 19.1. Regular patterns in the variations of different meteorological parameters are not observed, and large differences from site to site occur (see Table 19.1).

The highest mean concentration was observed at Noyelles-Godault in northern France near an industrial center, where the mean value was equal to 5052 μg/L. In the Paris region with dense traffic, the mean concentration was 43 μg/L; in the different rural zones it was 16 μg/L; and in the mountainous regions, only 9 μg/L. In the latter regions the lowest value, 2 μg/L, was observed.

The annual deposition of lead by rainwater is calculated for eight regions shown in Figure 19.1. For a given region, the deposition is given by the

Table 19.1. Levels of Lead in the Rainwater in France (1972–1973)

	Concentration		
Site	Mean	Min	Max
Industrial	5092	75	8100
Paris (region)	43	6	67
Rural zones	16	4	38
Mountainous regions	9	2	27

Table 19.2. Wet Deposition of Lead over France

Year	Rainfall (mm)	Wet Deposition (t)
1972	805	5560
1973	636	5860

multiplication of the monthly concentration of lead in rainwater and the rainfall rate. The results are presented in Table 19.2. For 1972 and 1973, the wet deposition values are 5.6 and 5.9 t, respectively. These values were from rural areas with low pollution. Near the great urban centers, dry deposition may be higher.

3.2. Dry Deposition

The measurement of natural dry deposition in a quantitative manner is a difficult task. Usually, dry deposition is measured by particle collection on a horizontal surface of filter paper or on the walls of a pluviometer. This type of collection does not correspond exactly to that on any natural surfaces such as soil, grass, or tree leaves. Therefore in this study, an effort has been made to measure the deposition of lead on tree leaves directly. One difficulty of this method is to distinguish between the lead originating from the soil and the lead deposited from the atmosphere. Lead-210 (22-yr half-life) is a radionuclide produced by the decay of ^{222}Rn (3.8-day half-life) emanating from the ground. It is associated with submicron particles so that it resides in aerodynamically similar types of particles as those of atmospheric lead. For a mass unit of lead

$$y_X X_a = (1 - y)X_s = X_l \tag{1}$$

where

X_a = ^{210}Pb/Pb values in air (a),
X_s = ^{210}Pb/Pb values in soil (s),
X_l = ^{210}Pb/Pb values in leaves (l), and
y = fraction of lead of atmospheric origin.

To show the usefulness of this procedure, a comparison has been made for lead deposition on tree leaves in French rural forests and in a Congolese forest (Tables 19.3 and 19.4).

We can assume that the X_s values in these different sites have the same value but that the X_a value is about 10 times higher in the Congo (13.0 dpm/μg Pb compared to 1.3 dpm/μg Pb), since the lead levels in the air are 10 times lower. The total lead obtained by washing the leaves is twofold higher in

Table 19.3. Atmospheric Lead Deposition on Tree Leaves in French Rural Forests (September 1974)[a]

Forest	$^{210}Pb/Pb$ (dpm/μg Pb)	Total Pb ($\mu g/400\ cm^2$)	Atmospheric Pb ($\mu g/400\ cm^2$)
Mt. Marsan	0.60	14.2	3.7
Boulogne	0.94	7.0	6.4
Chambord	0.89	12.0	10.7
Brive	0.71	13.5	9.6
Mean	0.78	11.7	7.6

[a]Calculations are made with $X_s = 0.35$ dpm/μg Pb and $X_a = 1.3$ dpm/μg Pb.

Table 19.4. Atmospheric Lead Deposition on Tree Leaves in a Tropical Forest (Rural Area of the Congo, May 1974)[a]

Tree	$^{210}Pb/Pb$ (dpm/μg Pb)	Total Pb ($\mu g/400\ cm^2$)	Atmospheric Pb ($\mu g/400\ cm^2$)
Milletia versi	4.6	1.7	0.60
Byrsocarpus sp.	0.81	6.1	0.37
Nanclea catifolia	1.6	7.4	0.89
Milletia versi	2.3	3.7	0.67
Mean	2.3	4.7	0.64

[a]Calculations are made with $X_s = 0.035$ dpm/μg Pb and $X_a = 13.0$ dpm/μg Pb.

France than in the Congo, but when correction is made with the $^{210}Pb/Pb$ index the aerial deposition of lead in France is 10 times higher than in the Congo—7.6 $\mu g/400\ cm^2$ compared to 0.64 $\mu g/400\ cm^2$. The fraction of deposited atmospheric lead is 65% in France and 14% in Congo, and the amounts of lead originating from the soils in both countries are about the same.

3.2.1. Pluviometers

The monthly averages of dry and wet depositions were measured from April to December in a residential section of Toulouse (Fig. 19.2). The dry depositions were collected by a pluviometer in normal position and by a pluviometer placed upside down. The mean monthly dry deposition in the first pluviometer is about one half that of the second pluviometer, 14.8 μg Pb compared to 28.8 $\mu g\ Pb/400\ cm^2$ per month. It can be concluded that half of the lead is dissolved by rainwater. This fact seems reasonable in view of the dissolution of atmospheric lead particles in deionized water. A solubility of 60 $\pm$ 20% was obtained after 24 h of soaking particulate lead collected on 6 filters (162 μg Pb on a Whatman 41 filter). On the other hand, some of the rains, especially light rains, may evaporate on the inner surfaces of the collector and give rise to dry deposition. Observation shows that this

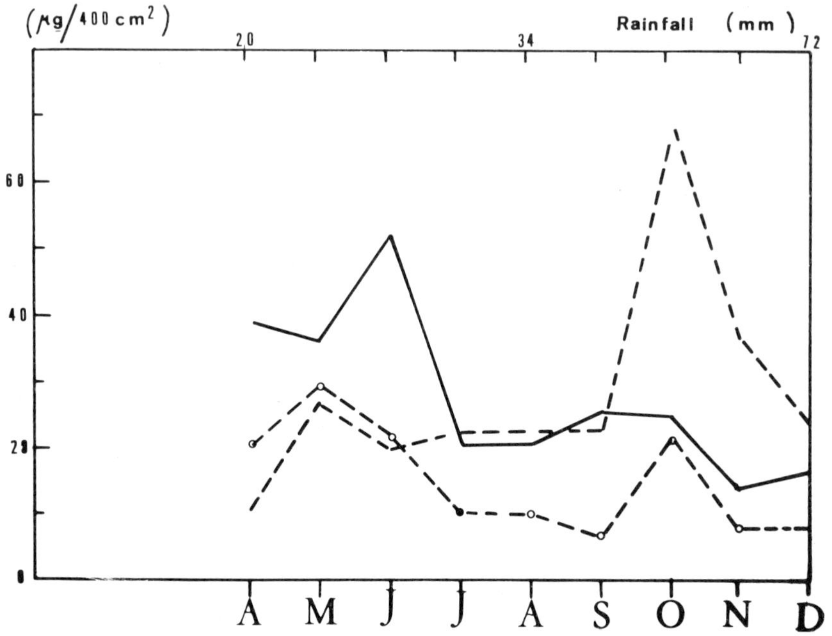

Figure 19.2. Residential site (Toulouse) : lead deposition of atmospheric lead particles on pluviometer; ...., wet deposition; ——, dry deposition; —O—, dry deposition washed by rainwater (1973).

phenomenon is quite rare at this site. Dry deposition was high from April to June, and wet deposition was high in October and December; however, their monthly means were about the same: 28.8 and 25.7 µg Pb/400 cm^2 per month, respectively. The velocity of dry deposition, usually called v_g, is equal to 0.13 cm/s. This value is higher than those given by Chamberlain (1967). From experiments with grass in a wind tunnel, Chamberlain found $v_g = 3 \times 10^{-2}$ cm/s for droplets of tricresyl phosphate ($\Phi = 1$ µm) and Aitken nuclei ($\Phi = 0.08$ µm). There is no value reported for particles with a diameter between 0.1 and 1 µm typical of lead particles. The wind speed measured at a height of 15 cm above the surface of the grass (5 m/s in the wind tunnel experiment) is similar to the wind speed observed in nature. Our value of 0.13 cm/s is comparable to those given by different authors, for example, 0.22 to 0.69 cm/s in the UK for five sites, and 0.29 cm/s in Los Angeles. All the studies have shown that the mass median equivalent diameter for lead aerosols at remote continental locations as well as in urban areas are submicron, the range being between 0.12 and 0.70 µm (Nriagu, 1978). It has also been shown that 23% of the mass of lead is on particulates with diameter $\geq$2.3 µm (Cawse, 1978); 78 to 92% of the deposition is due to these particules. Assuming a density of 1 to 5 g/cm^{-3}, they would have $0.15 \leq v_d \leq 0.45$ cm/s. In addition to this, a very accurate study with isokinetic sampling shows that lead deposition in an urban area of Pasadena (California) is controlled by sedimentation of particulates with diameter $\geq 10\ \mu m$, which represents 14% of the total mass (Davidson et al., 1975). Our value, which is two times lower than that of Davidson et al., can be explained by the upsidedown position of our pluviometer which did not catch the sedimenting component. Our value is therefore probably more representative of rural zones. On the other hand, the speed of deposition of ^{210}Pb atoms presumably fixed on small particles of the atmosphere is about the same as that of lead particles, because the ratio $^{210}Pb/Pb = 0.092$ dpm/μg in air. Greater deposition of lead in large particles with diameter $>1\ \mu m$ would give a lower value of the $^{210}Pb/Pb$ ratio in the deposits. On the other hand, the possibility of a cloud of submicron particles above the rural zones is probable according to a study in the Los Angeles basin (Huntzicker et al., 1975).

3.2.2. Deposition on Natural Surfaces

The lead desorbed by a dilute HCl solution (3% v/v) from tree leaves (Fig. 19.3) shows that an apparent accumulation of lead on the leaves from June to October of 14 μg Pb to 70.4 μg Pb/400 cm^2 per month. In November a decrease appears on the four trees with no apparent explanation. In December

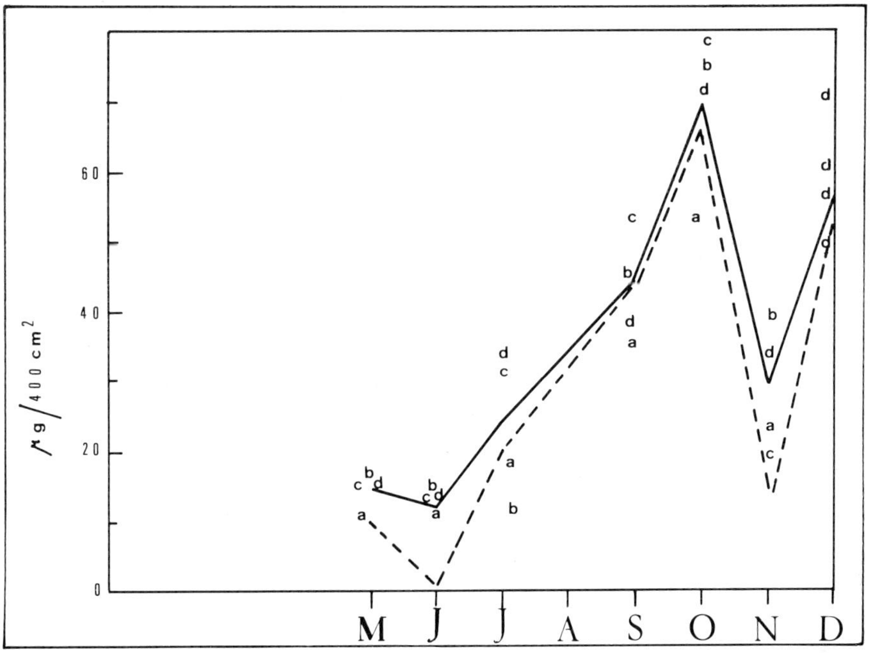

Figure 19.3. Residential site (Toulouse) : lead deposition of atmospheric lead particles to natural surfaces; ——, lead content desorbed by a dilute HCl solution (3%) for 5 h; ...., atmospheric lead input calculated by $^{210}Pb/Pb$ values (Table 19.1 and Eq. (1). Species : a, plane; b, polar; c, elm; d, oak (1973).

the oak leaves are completely dry, and again one notices an accumulation of lead which is about twofold higher than in November. The accumulation during winter is shown for the year 1972, by two samples of pine needles, to be 186 μg Pb and 200 μg Pb/400 cm^2, per month. In a given month, lead deposition on different trees varies by a factor of two and even three in July. Generally, the elm is the best receptor and the plane tree the worst. The difference could be due to aerodynamic conditions surrounding the large surface area of plane tree leaves.

True atmospheric deposition is obtained by correcting the total lead in order to measure the part originating from the soils; this is done by Eq. (1). At this site $X_a = 0.13$ dpm/μg Pb; the latter value was obtained from the lead in young 15-day-old leaves. The correction, month by month, from the X_l values measured (Fig. 19.4) is serious only for June where the lead is mostly of soil origin. For the annual mean, 82% of the lead is of atmospheric origin.

It is not known if the washing of tree leaves by rainwater can be compared with the washing of surfaces like those of a pluviometer. The exposure of the leaves sampled was rather advantageous for receiving raindrops. But the frequently sticky surface of the leaves could, to some extent, prevent the dissolution of atmospheric lead. Therefore, two alternative removal rates can be considered: a dissolution of one half of the lead or no dissolution at all. From April to October, the mean dry deposition on leaves in October compared with that of the pluviometer is 88% assuming the first removal rate

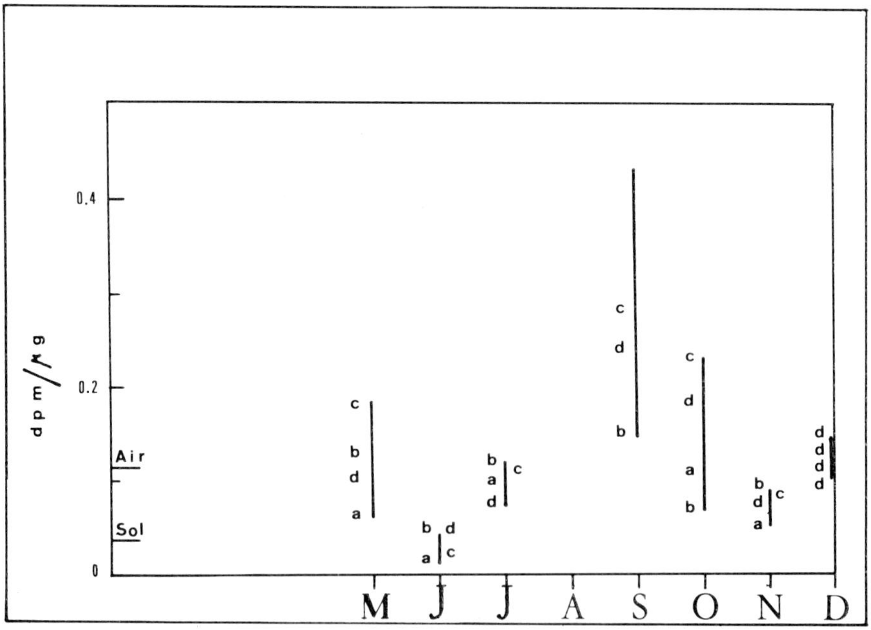

Figure 4. Residential site (Toulouse) : ^{210}Pb/Pb values of lead in tree leaves. Species : a, plane; b, poplar; c, elm; d, oak (1973).

and 44% according to the second rate. For oaks, from April to December, this ratio is 50% for the first alternative rate and 25% assuming the second alternative rate. The maximum efficiency of a natural surface is only 44% of that of a pluviometer because the leaves have two sides, and 22% seems to be the minimum rate.

In conclusion, the rate of deposition on the leaves of the four trees that were analyzed gives $v_d = 0.086 \pm 0.024$ cm/s. These are isolated trees, but in forests it has been shown that lead deposition is nearly the same. Theoretical considerations could explain this result. The deposition velocity depends on the horizontal speed of the wind. In nature this speed varies by a factor of five for different vegetations, such as a forest or a maize field. Let us consider two classes of particles of diameter: $\Phi < 1$ μm and $\Phi \geq 1$ μm. The laws of deposition (Belot and Gauthier, 1974) are given by Eqs. (2) and (3) and illustrated by Figure 19.5.

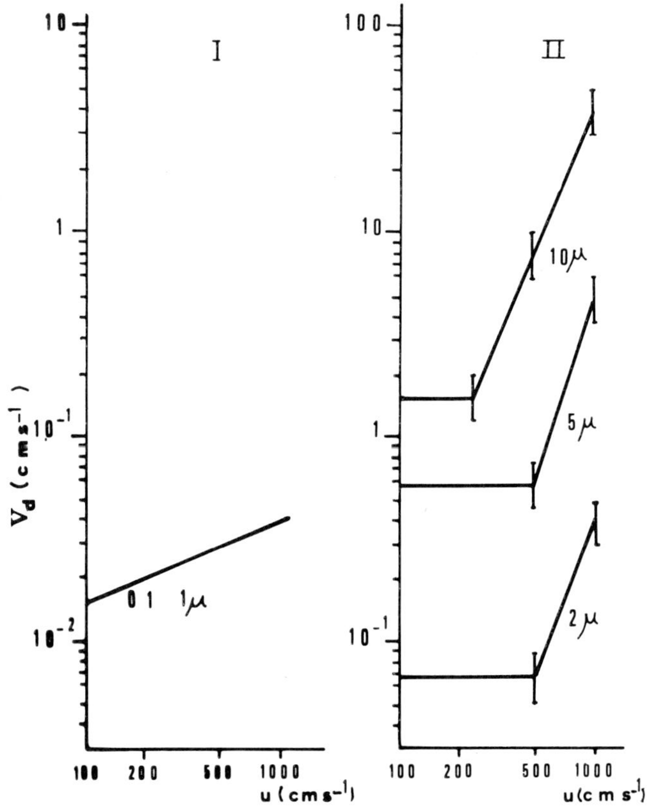

Figure 5. Deposition speed (v_d) of the particles with wind speed. I, Submicron particles (theoretical values); II, Micronic particles collected by the leaves of an oak (the horizontal corresponds to the sedimentation speed v_s). The particles are generated from a uranine solution in methanol and methylene blue (Belot and Gauthier, 1974).

$$v_d = k\,u^{0.5}, \qquad \phi < 1 \quad \mu m \tag{2}$$

$$v_d = a\,u^n \qquad \text{with } 1<n<3; \qquad \phi \geq 1 \quad \mu m \tag{3}$$

where

v_d is deposition speed; u is horizontal speed of the wind and a, k are constants.

The difference between the two classes is clear: For $\Phi < 1$ μm, a variation of u by a factor of 10, from say 100 to 1000 cm/s, gives a threefold variation of v_d from 0.014 to 0.042 cm/s. For $\Phi \geq 1$ μm, the variation of v_d is 600 times (0.05–30 cm/s). It's noteworthy that for $\Phi \geq 1$ μm, the sedimentation speed v_s is not negligible compared to v_d. In fact, it controls the deposition under a moderate wind speed of 5 m/s for $\Phi < 5$ μm. In conclusion, the effects of the variation of the wind speed between 100 and 500 cm/s on v_d are the following: (1) for the submicron particles, 0.1 μm $< \Phi <$1 μm, an increase by a factor of 2, and (2) for the micron particles, 1 μm $< \Phi <$10 μm, a change by a factor of 100.

It follows that in the rural zones where the lead particles are probably in the submicron range, their deposition speed to the natural surfaces would be independent of the variations of the wind speed, at least in France where u is about equal to 5 m/s or less. This explains the comparable rate of lead deposition on an isolated tree and in a forest with an area of 50 km^2. A calculation based on an exponential wind profile in which the value at the base is one-fifth or one-third that of the summit (Bayton et al., 1965; Allen, 1968; Meroney, 1968), and assuming the variation of the foliar density with height shows that the ratio of the lead deposition in a forest tree to an isolated tree should be 0.70 ± 0.05.

3.2.3. *Total Deposition over France*

The total deposition is the sum of wet and dry depositions. From our measurements in the different stations, the wet deposition is $D_h = 5.7 \times 10^9$ g Pb /yr. In the rural zones with a lead concentration in the air of 0.04 μg/m^3, a v_g of 0.060 ± 0.014 cm/s on tree leaves corresponds to a dry deposition of 0.076 ± 0.019 μg Pb/cm^2 yr. A quarter of France is covered with forests and the rest with meadows assumed to have foliar surfaces of 21 and 9 m^2, respectively. Taking a deposition time of 7 months over the forests and 12 months for the meadows per year, the total dry deposition during the 1972 and 1973 period is equal to

$$D_s = (0.060 \pm 0.014)\left(\frac{1}{4} \times 21 \times \frac{7}{12} + \frac{3}{4} \times 9 \times \frac{12}{12}\right) = (4.1 \pm 1.0) \times 10^9 \text{g Pb/yr}$$

Therefore, dry deposition accounts for about 40% of the total fallout in the rural zones.

For polluted towns, highways, and industrial zones, the dry deposition has also been estimated (Table 19.5). The deposition is less than 3×10^8 g/yr, that is, only a few percent of the total deposition in rural zones and may therefore be regarded as negligible.

Around industrial centers such as in Noyelles-Godault, the local pollution could be important, but there are no data to assess their importance on a nationwide scale.

3.2.4. Discussion

The consumption of leaded gasoline in France in 1973 was 1.5×10^{10} L/yr. From the permitted lead content of 0.64 g/L, the emission of lead could be 9.6×10^9 g/yr maximum. If 75% of this lead is dispersed into the atmosphere, the amount emitted would be 7.2×10^9 g/yr.. In fact, the lead content of gasoline was found to be notably higher than 0.64 g/L—up to 1.3 g/L (Anonymous, 1973). A 20% higher emission therefore seems reasonable, perhaps 8.6×10^9 g/yr. On the other hand, base metal production per square kilometer is relatively low in France, and therefore, so is the lead emission associated with it. For example, in Belgium, with a steel production of 47.7×10^7 g/km^2 (Demuynck, 1975), this source was responsible for about 40% of the lead emitted annually. In France, the annual steel production was only 4.4×10^7 g/km^2, so base metal production could be responsible for 4% of the emitted lead which is negligible in the budget. The scavenging of the 0.04 μg/m^3 atmospheric lead by rainfall to give a lead content of 15 μg/L in rainwater corresponds to a washout factor of 375. This value is about twofold higher than that of the two clean stations, Leiston and Plynlimon, of the UK network (Cawse,, 1974). The value 0.04 μg/m^3 may be compared to 0.023 μg/m^3 reported in Lerwick, Shetland Islands (Cambray et al., 1975) and to 0.05 μg/m^3 in the rural areas of the United States (McMullen, 1970). The range of lead levels found in rainwater in the different countries is from 0.96 μg/L at Livermore, Calif. (Volchock, 1979; Galloway et al., 1982) to 97 μg/L at Swansea, England (Pattenden, 1974); these extreme values represent either very clean conditions or notable pollution. Values such as 7.5 and 15.5 μg/L recorded at Plynlimon and Wraymires, UK, seem more representative of rural zones (Cawse, 1974).

Table 19.5. Dry Deposition of Lead in Polluted Zones[a]

Site	Surface (m^2)	Concentration in Air (μg/m^3)	Dry Deposition (t)
Towns	2.5×10^8	1.2	25–50
Highways	1.6×10^8	20.0	230

[a]The concentrations in the air are from Servant (1975b) for the towns, and from Chow (1972) for the highways. $V_d = 0.24$ cm/s.

The dry deposition velocities for lead particulates were estimated to be between 1.26 cm/s for forested zones and 0.54 cm/s for meadows. This range is two- to fourfold higher than the reasonable average of 0.3 cm/s (Nriagu, 1978). However, it seems probable that the real deposition velocities on a ground surface covered with vegetation must be greater. The increase is not directly proportional to the natural surface areas because one leaf can hide others. It must also be said that these higher values are consistent with the intensity of the emissions, considering that the emission rate of 8.6×10^9 g Pb/yr and the estimated fallout of 9.6×10^9 g Pb/yr are in good agreement. It is also possible that lead is transported over the countries in Western Europe, and monitoring this long-range transport would require many more stations.

4. IMPORTANCE OF THE ATMOSPHERIC CONTRIBUTION OF LEAD AND LEAD-210 TO SOME RIVER WATERS OF THE SOUTHWESTERN PART OF FRANCE

Extensive pollution of plants and soils has been noted in the vicinity of highways and towns in many countries. The lead has increased by a factor of 100 to 700 in leaves (Smith, 1973; Lerche and Breckle, 1974; Impes et al., 1972) and by a factor of 30 in soils (NAS, 1972; Chow, 1970). Atmospheric lead is thought to be retained in rural areas since it accumulates in soils close to highways. The ^{210}Pb profiles (Benninger et al., 1975; Fisenne, 1968; Moore and Poet, 1976) have shown that this is a general phenomenon. River waters nevertheless can be affected because a fraction of lead deposited on the highways can be washed out and carried away in the rivers. Lead is usually trapped in the organic-rich topsoil, and very little flows off into stream channels. Under such conditions, the lead content of river water may be very low since soils are not saturated with the lead. A reconnaissance survey (Durum et al., 1971) in the United States showed some regional differences in the lead content of river waters. The highest values were measured in the Northeast (median value 6 μg/L; range 1–890 μg/L) and in the Southeast (median value 4 μg/L; range 1–44 μg/L). Hem and Durum (1973) noticed that these regions are also those which receive the greatest fallouts of atmospheric lead (Lazrus et al., 1970).

4.1. Results

The Pb and ^{210}Pb contents were measured in the surface waters of three rivers during the period December 1973 to July 1974. The results are given in Table 19.6.

The variation of individual values of lead, lead-210, and ^{210}Pb/Pb ratios span the range from 1 to 10. The highest values—up to 52.5 μg/L in the Garonne—were found during floods. The median values of lead (range

Table 19.6. Lead, Lead-210, and Lead-210/Lead Concentrations in River Waters of Southwestern France

River	Concentration Pb (μg/L)		Concentration ^{210}Pb (dpm/L)	
	Average	Range	Average	Range
Hers	11.6	3.5–33.5	0.44	0.13–0.74
Ariège	14.4	4.5–43.0	0.40	0.13–1.29
Garonne	13.2	6.0–52.5	0.32	0.12–0.94

11.6–14.4 μg/L), of lead-210 (range 0.32–0.44 dpm/L) are comparable for all the rivers.

Further analyses were made during high and low flow periods in order to determine lead solubility. The results are presented in Table 19.7. The data show the existence of two rates for the lead values: high flow rates with high concentrations of about 10 μg/L and low flow rates with low concentrations of about 0.2 μg/L. When the pH = 7.2, the soluble fraction of lead amounts to 18%, approximating the value given by Hem (1976) for the cation-exchange capacity of suspended material of between 10^{-4} and 10^{-5} mole/L. With high flow rates the lead content in particles is high (183 ppm) and comparable to the value given for the Garonne (231 ppm). With low flow rates the lead content in particles is lower (<50 ppm).

The lead content of suspended sediments is significantly higher than the average for local soils which averaged 26.8 ± 4.2 ppm for 10 analyses. Yet the lead content of such soils varies with the size of particles and the depth (Table 19.3). It is higher in the very finely grained fraction ($\Phi < 0.5$ μm) and higher also in the surface soil (0–10 cm) than at depth (60–180 cm). These high values are unrelated to atmospheric Pb since the atmospheric pollution is known to have no influence at such a depth. Furthermore, the ^{210}Pb/Pb ratio in these samples is close to that observed in soils and much smaller than that of

Table 19.7. Soluble and Insoluble Lead in River Waters of Southwestern France

River	Analysis 1, Pb (μg 1^{-1}) Average		Analysis 2, Pb (μg 1^{-1}) Average	
	Sol.	Insol. (ϕ<0.1 μ)	sol.	Insol. (ϕ<0.1 μ)
Hers	<0.14	<0.20	2.7	10.0
Ariège	0.27	<0.20	1.4	8.0
Garonne	<0.14	<0.20	1.3	6.5

Table 19.8. Lead Content and Lead-210/Lead Ratio of Soils as a Function of Particle Size and of the Depth of the Sampling

	Depth, 0–10 cm		Depth, 60–180 cm	
Particle size	Pb (ppm)	^{210}Pb/Pb (dpm/μg)	Pb (ppm)	^{210}Pb/Pb (dpm/μg)
0.5 μ	431	0.042	323	0.051
0.5<φ<2 μ	57	0.044	45	0.055
2<φ<10 μ	21	0.039	15	0.047

atmospheric samples (Table 19.8). The values are 1.3 dpm/μ g in air and 0.038 dpm/μ g in soils. Therefore, the very finely grained fraction which represents 5% of the total weight of soil samples contains most of the Pb.

The rather high lead content in the three rivers during the high flow period may be derived from natural sources. The presence of 50 mg/L of small particles in suspension with a lead content of 183 ppm gives a lead content of 9.1 μg/L in the water. In any case, the values are higher than the average values for English and American rivers of 0.92 μg/L and 6 μg/L, respectively (Abdullah and Royle, 1972; Hem, 1972). This suggests that the atmospheric pollution is not to be excluded *a priori*. The pollution component will be estimated in two different ways.

4.2. Discussion of Atmospheric Pollution

Atmospheric pollution may be due to lead contained in rainwater and to lead deposited on roadways which has been swept along by rainwash.

4.2.1. Rain

The average ^{210}Pb content in rainwater in the area is 12.5 dpm/L (Marenco et Fontan, 1972); in river water, it amounts to 0.40 dpm/L, that is, 3.2% of the former, Thus, almost all ^{210}Pb and Pb from rainfall is retained by soils. If we consider the lead content to average 16 μg/Liter of rainwater, the lead concentration in rivers from this source will amount to 0,5 μg Pb/L. As a matter of fact, part of the ^{210}Pb found in rivers might originate from soils and would lower the atmospheric contribution of ^{210}Pb and Pb.

4.2.2. Roads

No values are available with respect to lead deposit on roadways. However, if the lead content in the air is taken as 20 μg/m^3 and the deposit speed as 0.29 cm/s (Davidson et al., 1975), this deposit is 4800 μg/m^2 day. Such a calculation is correct for major traffic roads (traffic = 20,000 to 30,000 vehicles per day). If the rainfall is 2 mm/day, the width of the valley is 6 km (Garonne and Ariege), and the width of roadways is 20 m, then the pollution of rainwash water with lead is from 7.8 (Hers) to 0.78 $\mu g/L$ (Garonne and Ariege). The first figure is quite high.

In the Hers valley one would expect, during the low flow season, the pollution figure to reach 23.4 $\mu g/L$, in other words, a value three times higher than the annual average. We assume a constant annual flow of traffic in this region and we note that one third of the average annual rainfall occurs during this period. We detect no more than 0.2 $\mu g/L$ Pb. Thus, it may be that during the summer at least, the lead deposited on roadways and swept along by rainwash settles in soils before it reaches rivers. As a matter of fact, water is collected via small ditches that do not reach the larger rivers directly, thus allowing time for soil filtration processes to be significant.

In the Ariege and Garonne valleys, rainwash pollution is minimal during the high flow season. During the low flow season, Pb from the same source of pollution may well reach 2.4 $\mu g/L$ but with no effect because of fixation by the soil for the same reasons given previously.

4.3. Conclusion

During the year, the variation of the lead content in some river waters analyzed ranged from 14.4 to less than 0.2 $\mu g/L$. Precautions in the analytical procedure show the highest values to be accurate. The lead fallout from the atmospheric is higher than lead transport in rivers. The circulation of ^{210}Pb suggests that at least 87% of this atmospheric lead is retained by soils. During the high flow period, this pollution probably accounts for a maximum lead content of 0.4 $\mu g/L$ in river waters.

Dry deposition of lead on roads, which can then be carried on to rivers, can result in a significant increase in lead content of the water. This increase is not observed during the low flow season, indicating that the lead is retained by the soils. During the high flow season the effect of pollution from atmospheric sources is insignificant in the Ariege and Garonne Rivers. It seems likely that most of the Pb in these rivers originates from natural sources. It is easily accounted for by the presence of particles with high lead content, up to 200 ppm during high flow rate periods. During low flow periods, the lead content of river waters is reduced, since both the suspended load of waters and their lead content are lower, the lead content of the particles then being less than 50 ppm.

5. BLOOD LEAD AND LEAD-210 ORIGINS IN RESIDENTS OF TOULOUSE

5.1. Introduction

The amount of skeletal lead retained by humans increases with the age of the individual, and the average skeletal lead in U.S. adults varies between 100 and 400 mg (Kehoe, 1961). Geochemical distribution patterns for heavy metals have led Patterson (1965) to conclude that a maximum skeletal lead burden acquired from soils should only be 2 mg. This figure indicates that the average individual in U.S. is subjected to heavy lead insult from pollution sources.

Recent studies on New York residents (Bogen et al., 1975; Bogen et al., 1976) have revealed an average adult skeletal lead content of 100 mg with 57% estimated to be of atmospheric origin, and the balance from the food chain. This figure is close to that reported by Kehoe (1961) and that adopted by Holtzman (1962) for the calculation of the biological half-life of lead in body and skeleton. A great difference, nevertheless, exists in Pb ingestion: 460 μg/day for Holtzman (1962) and 84 μg/day for Bogen et al. (1975). To derive the pollution component, one can consider the natural body content of ^{210}Pb. The amount of ^{210}Pb in 18 cadavers was measured by Hursh (1960); it varied from 77 to 664 dpm with a mean of 231 dpm. The variation from one person to another was sevenfold and even after considering the weight of wet bones, it was still fivefold. In his compilation, Holtzman (1978a) reports mean values of 1100 and 1540 dpm. The body content of ^{210}Pb depends upon the ingestion of ^{226}Ra during the lifetime. In the United States, the daily radionuclide intake from diets shows a variation factor of ten (Holtzman, 1978a), which is comparable to the variation in ^{226}Ra contents of soils–between 0.21 and 1.8 dpm/g (Lowder and Solon, 1956). To judge whether this variation is from natural sources or from contamination, it is more convenient to compare the ^{210}Pb/Pb concentration ratios in the blood and also in foods and atmospheric particulates. In the study on New York residents (Bogen et al., 1975), it was shown that these ratios were comparable to those of the skeleton, strongly suggesting that the subjects were not exposed to other lead contamination. The only point the authors failed to discuss was the possibility of human food-chain contamination. We now report our findings on this matter by comparing the Pb and ^{210}Pb concentrations in the blood of subjects in the residential areas of Toulouse (pop. 400,000) and in a contamination-free environment.

5.2. Theoretical Model

A simple theoretical model can help to estimate the relative magnitude of Pb and ^{210}Pb concentrations in humans (Servant and Delapart, 1981). In a contamination-free environment, especially under clean atmospheric conditions, and using the lowest lead concentrations known for food, the ICRP (ICRP60) model provides the results in Table 19.9 (Servant and Delapart, 1981). The ingestion of ^{210}Pb from air particulates and foods are close to those calculated from the mean concentrations reported in 1976 for the United States, that is, 0.62–1.15 dpm/μ g for air and 3.0–3.6 dpm/day for food (Magno et al., 1960). Lead content in water is the minimum value measured in river water in France during low flow rate (Table 19.7). The dietary intakes of ^{210}Pb vary a little in the United States; the higher levels of these nuclides are

Table 19.9. Ingestion, Assimilation, and Blood Lead and Lead-210 Levels in Humans in an Uncontaminated Environment

	Pb (μg/day)	^{210}Pb (dpm/day)	^{210}Pb/Pb (dpm/μg)
Ingestion			
Air[a]	0.04	0.72	18
Food[b]	80	3	0.03
Water[c]	0.14	0.004	0.03
Assimilation (blood)			
Air	0.02	0.21	
Food	6.4	0.24	
Water	0.011	0.0003	
Blood,[d]			
5 L	0.19[e]	13.2	0.07

[a] Air: concentration Pb, 0.002 μg/m^3; ^{210}Pb = 0.036 dpm/m^3 inhalation 20 m^3/day − assimilation = 29%.

[b] Food: concentration Pb, 0.05 μg/g; ^{210}Pb = 0.0015 dpm/μg absorption 2000 g/day − assimilation = 8%.

[c] Water: concentration Pb, 0.14 μg/L; ^{210}Pb, 0.004 dpm/L.

[d] Blood: total Pb and ^{210}Pb content is calculated for a daily intake and residence time of 30 days (Rabinowitz et al., 1973).

[e] In mg/day.

due to particular sources, such as the consumption of reindeer and caribou meat by some residents of the Arctic, and the consumption of fish in Japan (Holtzman, 1978b).

The mean blood lead concentration is 4 μg/dL and the specific activity (^{210}Pb/Pb) is slightly greater than that found in food, due to atmospheric ^{210}Pb inhalation. The atmosphere contributes less than 2% of lead, but nearly half of the ^{210}Pb. The value of the ^{210}Pb/Pb concentration ratio is higher than that of New York city because there is no dilution by atmospheric lead.

5.3. Results

The analyses performed during 1977 on soil and human blood samples yield the results shown in Table 19.10. The blood lead concentrations reach four times the maximum concentrations estimated with the ICRP model for contamination-free environments, the ^{210}Pb/Pb concentration ratios being comparable to those given by this model. The ^{210}Pb/Pb concentration ratios

Table 19.10. Human Blood Lead Concentration and $^{210}Pb/Pb$ Concentration Ratios for Soluble Soil Lead and Blood Lead in Southwestern France

Samples[a]	Pb (μg/dL)	$^{210}Pb/Pb$ (dpm/μg)
Soil		0.055 ± 0.012
Blood	19.7 ± 5.8	0.073 ± 0.037

[a] Average of 40 soil samples and 10 blood tests.

for human blood are 33% higher than the ratios for the lead that plants pick up from the soil.

5.4. Discussion

The difference in the values of $^{210}Pb/Pb$ concentration ratios in the blood and the soils is so small that a discussion with data obtained in a limited area is imperative. The 10-fold variation in the ^{226}Ra content is soils, and therefore in ^{210}Pb content, reported in the literature therefore excludes any generalization.

The relatively high human blood lead concentration may arise from a contamination of the food chain, a high lead concentration in soils, or from atmospheric pollution via dry deposits on plants.

Food contamination with Pb free of ^{210}Pb can be excluded because this would lead to $^{210}Pb/Pb$ concentration ratios lower than those found in soils, which is not the case. The soil lead available to plants is equal to 4.5 μg/g. This is near the highest value in the 0.05-5 μg/g range reported in the literature (National Academy of Sciences, 1972).

In residential areas of Toulouse, inhalation of air with Pb concentrations of 0.3 μg/m^3 corresponds to a daily assimilation of 1.8 μg, and a blood lead concentration of 1.2 μg/dL. This assumes that steady-state conditions are reached after constant inhalation during 200 days (Rabinowitz et al., 1973). Subtracting this contribution leaves 17.5 μg/dL to be accounted for. This corresponds to a ^{210}Pb assimilation of 2 dpm/day, 10 times that absorbed in the respiratory tract. The ^{210}Pb can come from food sources. The increase in $^{210}Pb/Pb$ ratios in blood compared to soils can be explained by dry deposits of airborne lead with high ^{210}Pb levels on plants in rural zones. In our case, 86% of the lead comes from the soil ($^{210}Pb/Pb = 0.055$ dpm/μg) and 14% from the atmosphere ($^{210}Pb/Pb = 1.3$ dpm/μg).

In order to illustrate the importance of this atmospheric lead contribution to the total lead in plants, different crops were sampled on a farm. The results are given in Table 19.11. The $^{210}Pb/Pb$ concentration ratios in crops vary by a factor of 4; the mean is 0.35 dpm/μg and the mean of the soils is 0.083 dpm/μg. Thus, the increase in isotopic ratios is noticeable and corresponds to

Table 19.11. Lead Contents and $^{210}Pb/Pb$ Concentration Ratios in Food and Soils from a Farm Southwest France[a]

Samples	Pb, Wet Weight (ppm)	$^{210}Pb/Pb$ (dpm/μg)
Potatoes	0.026	0.35
Carrots	0.076	0.59
Turnips	0.082	0.23
Wheat	0.055	0.31
Barley	0.064	0.27
Maize	0.084	0.16
Green vegetables	0.096	0.40
Fruits	0.093	0.49
Mean	$0.072 \pm {}^{0.021}_{0.046}$	$0.35 \pm {}^{0.24}_{0.19}$
Soils mean	2.3 ± 0.4	0.083 ± 0.021

[a] Weight of food samples, 100 g; weight of soil samples, 30 g. The mean lead concentration is that of soluble lead calculated from 10 analyses of surface soils.

a 25% contribution from atmospheric lead. The mean lead concentration is 0.172 ppm (wet sample) and corresponds to 0.054 ppm concentration in the absence of atmospheric pollution. In this case, the concentration is very similar to that given for food reaching the New York market, that is, 0.040 ppm (Bogen et al., 1976).

Actually, the atmosphere is a much more intense potential source of lead. In rural zones, for example, with airborne lead particles depositing at a rate of 0.086 cm/s (Servant, 1975), an air concentration of 0.030 $\mu g/m^3$, and a productivity of 0.12 g/cm^2, a 6-month exposure period brings the contamination in plants to 4.6 ppm, which is 80 times the 0.054 ppm value we previously mentioned. If this is not seen in the food, it must be that most of the airborne lead is retained in plant foliage, only a small fraction of which reaching the fruit and vegetables. This result agrees with the studies of Ter Harr (1972) and Hemphill and Rule (1975).

Subtracting 14% of the atmospheric lead contamination, one finds blood lead concentrations of about 14.9 μg/dL. This is still three times greater than that calculated with the ICRP (ICRP 60) model. We think that this discrepancy is related to natural dietary intake, namely, a daily lead absorption of 304 μg and an assimilation of 24.3 μg.

Recent analyses of 103 individuals in a populated Nepal Himalayan valley show a mean blood lead level of 3.4 μg Pb/dL (Piomelli et al., 1980). The low concentration of Pb in air of 0.86 ng/m^3 (STP) is typical of a remote region. This blood lead concentration confirms those given in the model (Table 19.9). On the other hand, Hecker et al. (1974) have reported a lower concentration in blood (0.83 μg Pb/dL) in uncultured Yanomana Indians. This fact is in line with Patterson's theory of a concentration in blood of only 0.1 μg Pb/dL in an uncontaminated environment (Settle and Patterson, 1980). In such a case, the lead concentration in human diet could be 0.002 ppm or 20 times less than the value used here; the level in water of 0.020 ppm, is about 7 times less. For food, the low value is based on the biopurification of calcium with respect to lead and barium in the different successive consumer stages. For example, the concentration ratio Ba/Ca equals 3000×10^{-6} in crustal rocks and 2×10^{-6} to 7×10^{-6} in the skeletons of Americans, Britons, and ancient Peruvians. Barium offers the advantage that its pollution in the environment is an order of magnitude less than that of lead; its worldwide production and atmospheric emission are, respectively, 0.42×10^{12} and 0.041×10^{12} g/yr (Servant, 1982). The biopurification for lead is thought to be an order of magnitude greater than that for barium. Therefore the main problem is the lead content of human food. Our value given in Table 19.11 of 0.054 ppm after correction for input of atmospheric lead is relatively high, but the concentration of soluble lead in soil is also high. Since we can only measure quantities of lead greater than 0.14 μg, it was imperative that the weight of the sample analyzed be at least 100 g. In most samples the lead content averaged 7.2 μg or more than 50-fold the blank. This implies that the lead measured did not come from contamination in the laboratory. However, if there was a matrix effect, the measurement could be in error. At the same time the lead-210 content measured is such that if one assumes that the lead content is only 0.14 μg/100 g in food, then the concentration ratio ^{210}Pb/Pb equals 18 dpm/μg. This value is typical of the air in remote areas; in the atmosphere around the laboratory it is only 0.13 dpm/μg. So it is difficult to find a source with such a high ratio of ^{210}Pb/Pb in a food sample of a rural zone. Since there is agreement between the different researchers about the assimilation factor of lead in food, the correct answer to the problem of natural lead level in humans lies in the measurement of the lead content of food harvested in rural zones, which will help to define and to measure the efficiency of the biopurification of calcium with respect to lead. It is safe to say that the concentration of lead-210 in blood is not a good index of lead contamination because one half of this radionuclide intake, at any location, arises through breathing the air.

5.5. Conclusion

Excluding any atmospheric pollution from industrial or domestic sources, minimum blood lead concentrations in humans should be near 3.8 μg/dL, according to the ICRP (ICRP60) model calculation. By contrast we found in 10 men and women subjects living in residential areas of Toulouse noticeably higher values of 19.7 $\pm$ 5.8 μg/dL. We estimate that atmospheric pollution accounts for 20% of the total blood lead: 6% comes from absorption in the respiratory tract and 14% by gastrointestinal absorption via the food chain.

Subtracting this contribution from pollution, 15 μg/dL remains and can be attributed to normal lead exposure from the environment in the absence of pollution. This is confirmed by comparing $^{210}Pb/Pb$ ratios for soil and human blood. This high natural lead intake is related to the high concentrations (4.5 μg/g) of soluble lead found in the soils of southwestern France. Therefore, a blood lead concentration of 15 μg/dL can be taken as the limiting value for humans in a contamination-free environment where the soluble lead content of the soils, outside mining areas, is one the highest reported in the literature.

REFERENCES

Abdullah, M.I., and Royle, L.G. (1972). "The occurence of lead in natural waters—International Symposium "Environmental Health Aspects of Lead." Amsterdam, Oct. 2–6, Public Commission of the European Communities, pp. 113–124.

Allen, L.H., Jr. (1968). "Turbulence and wind spectra within a Japanese larch plantation." *J. Appl. Meteorol.* **7,** 73–78.

Anonymus (1973). "Test de 20 marques d'essence." *Que choisir?* **80,** 10–21.

Baynton, H.W., Biggs, W.G., Hamilton, H.L. Jr., Sherr, P.E., and Worth, J.J.B. (1965) "Wind structure in and above a tropical forest." *J. Appl. Meteorol.,* 670–675.

Belot, Y., and Gautheir, D. (1974) "Transport of micronic particles from atmosphere to foliar surfaces." International Seminar on Heat and Mass Transfer. Dubrovnick, Yugoslavia, August.

Benninger, L.K., Lewis, D.M., and Turekian, K.K. (1975). "Marine chemistry in the coastal environment". *Am. Chem. Soc. Symp. Serv.* **18,** 201–210.

Bogen, D.C., Welford, G.A., and Morse, R.S. (1975). *General Population Exposure of Stable Lead and ^{210}Pb to Residents of New York City.* Harley, J.H., Ed. U.S. AEC Rep. HASL-299, 17pp.

Bogen, D.C., Welford, G.A., and Morse, R.S. (1976). "General population exposure of stable lead and ^{210}Pb to residents of New York City." *Health Phys.,* 359–361.

Cambray, R.S., Jeffries, D.F., and Topping, G. (1975). "An estimate of the input of atmospheric trace elements into the North Sea and the Clyde Sea (1972–73)." A.E.R.E. Rep. R-7669, H.M.S.O., London, 30 pp.

Chamberlain, A.C. (1967). "Deposition of particules to natural surfaces." In *Symposia of the Society for General Microbiology. Nb XVII Airborne Microbes,* pp 138–164.

Chamberlain, A.C., Heard, M.J., Little, P., Newton, D., Wells, A.C., and Wiffen, R.D. (1978). "Investigations into lead from motor vehicules." A.E.R.E. Rep. R-9198. H.M.S.O., London, 151 pp.

Charlot, G. (1966). *Les Methodes de Chimie Analytique,* 5th Ed. Masson et Cie, Paris, pp. 872–874.

Cawse, P.A. (1974) "A survey of atmospheric trace elements in the U.K. (1972–73)." A.E.R.E. Rep. R-7669. H.S.M.O. London, 84 pp.

Chow, T.J. (1970). "Lead accumulation in roadside and grass." *Nature (London)* **225,** 295–296.

Davidson, C.I., Hering, S.V., and Friedlander, S.K. (1975). "The deposition of Pb-containing particles from the Los Angeles atmosphere." International Conference *Environmental Sensing and Assessment.* Las Vegas, Nev., Sept. 18.

Demuynck, M. (1975). "Compilation of an inventory for particulate emissions in Belgium." *Water Air Soil Pollut.* **5,** 3–10.

Durum, W.H., Hem, J.D., and Heidel, S.G. (1971). "Reconnaissance of selected minor elements in surface waters of US." U.S. Geological Survey Circular 643.

Fisenne, I.M. (1968). "Distribution of ^{210}Pb and ^{226}Ra in soil." Rep. UCRL 18140. U.S. Atomic Energy Commission, Washington D.C., pp. 145–158.

Galloway, J.N., Thornton, J.D., Norton, S.A., Volchok, H.L., and McLean, R.A.N. (1982). "Trace metals in atmospheric deposition: A review and assessment." *Atmos. Environ.* **16,** 1677–1700.

Gilbert, T.W. (1964). *Treatise on Analytical Chemistry,* 6th Ed., Vol. II, I.M. Kolthoff and P.J. Elving, Eds., Interscience, New York, pp. 69–175.

Hecker, L., Allen, H.E., Dinman, D.D., Neel, J.V. (1974). "Heavy metals levels in acculturated and unacculturated populations." *Arch. Environ. Health* **29,** 181–185.

Hem, J.D. (1976). "Geochemical controls on lead concentrations in stream water and sediments." *Geoch. Cosmo. Acta* **40,** 599–609.

Hem, J.D., and Durum, W.H. (1973). "Solubility and occurence of lead in surface water." *J. Am. Water Work Assoc.* **63,** 562–568.

Hemphill, D.D., and Rule, H.J. (1975). "Foliar uptake and translocation of ^{210}Pb and ^{109}Cd by by plants." *International Conference of Heavy Metals in the Environment,* Hutchinson, T.C., Ed. Vol. II, pp. 77–86.

Holtzman, R.B. (1962). "Desirability of expressing concentrations of mineral-seeking constituents of bone as a function of ash weight." *Health Phys.* **8,** 315–319.

Holtzman, R.B. (1978a). *Application of Radiolead to Metabolic Studies in the Biogeochemistry of Lead in the Environment.* J.O. Nriagu, Ed. Vol. 1B: Elsevier/North Holland, New York pp. 37–96.

Holtzman, R.B. (1978b). "Normal dietary levels of ^{226}Ra, ^{228}Ra, ^{210}Pb and ^{210}Po for man." *The Natural Environment,* Vol. III, Houston, Tx, April 23–28.

Huntzicker, J.J., Friedlander, S.K., and Davidson, C.I. (1975). "Material balance for automobile-emitted lead in the Los Angeles basin." *Env. Sci. Technol.* **9,** 448–457.

Hursh, J.B. (1960). "Natural lead-210 content of man." *Science* **132,** 1666–1667.

Impes, R., N'Vunzu, Z., and Nangiot, P. (1972). "Détermination du plomb sur les végétaux croissant en bordure des autoroutes." International Symposium *Environmental Health Aspects of Lead.* Oct. 2–6, Ed. Commission of the European Communities, Amsterdam pp. 135–143.

Kehoe, R.A. (1961). "The metabolism of lead in health and disease." The Harben Lectures, 1960. *J. Roy. Instit. Public Health Hyg.* **24,** 1–81, 101–120, 129–143, 177–203.

Lazrus, A.L., Lorange, E., and Lodge, Jr P.J. (1970). "Lead and other metal ions in United States precipitation." *Env. Sci. Technol.* **9,** 448–457.

Lerche, H., and Breckle, S.W. (1974). "Blei im Ökosystem Autobahnrand." *Die Naturwissen.* **5,** 218.

Little, P., and Wiffen, R.D. (1978). "Emission and deposition of lead from motor exhaust. II. Airborne concentration, particle size and deposition of lead near motorways." *Atmos. Environ.* **12,** 1331–1343.

Lowder, W.M., and Solon, L.R. (1956). "Background radiation environment: A literature search." U.S. AEC Rep. NYO-4712.

McMullen, T.B., Faoro, R.B., and Morgan, G.B. (1970). "Profile of pollutant fractions in nonurban suspensed particulate matter." *J. Air Pol. Control Assoc.* **20,** 369–372.

Magno, P.J., Groulx, P.R., and Apidianakis, J.C. (1960). "Lead-210 in air and total diets in the United States during 1966." *Health Phys.* **18,** 383–388.

Marenco, A., and Fontan, J. (1972). "Etude par simulation sur modèles numériques du temps de séjour des aérosols dans la troposphère." *Tellus* **24,** 429–441.

De Martonne, E., and Demangeon, A. (1947). *Géographie Universelle*. A. Colin, Ed., Vol. VI. Paris, pp. 376–381.

Meroney, R.N. (1968). "Characteristics of wind and turbulence in and above model forests." *J. Appl. Meteorol.* **7,** 780–788.

Moore, H.E., and Poet, S.E. (1976). "^{210}Pb fluxes determined from ^{210}Pb and ^{226}Ra soil profiles." *J. Geophys. Res.* **81,** 1056–1058.

National Academy of Sciences (1972). "Lead-airborne lead in perspective." Division of Medical Sciences, Washington D.C., pp. 34.

Nriagu, J.O. (1978). "Lead in the atmosphere." In J.O. Nriagu, Ed. *The Biogeochemistry in the Environment.* Elsevier/North Holland Biomedical Press, New York.

Nriagu, J.O. (1979). "Global inventory of natural and anthropogenic emissions of trace metals to the atmosphere." *Nature* **279,** 409–411.

Pattenden, N.J. (1974). "Atmosphere concentrations and deposition rates of some trace elements measured in the Swansea/Neath/Port Talbot area." A.E.R.E. Rep. R-7729. H.M.S.O., London.

Patterson, C.C. (1965). "Contaminated and natural lead environments of man." *Arch. Environ. Health* **11,** 344–363.

Piomelli, S., Corash, L., Corash, M.B., Seaman, C., Mushak, P., Glover, B., and Padgett, R. (1980). "Blood lead concentrations in a remote himalayan population." *Science* **210,** 1135–1136.

Rabinowitz, M.B., Wetherill, G.W., and Kopple, I.D. (1973). "Lead metabolism in the normal human: Stable isotope studies." *Science* **182,** 725–727.

Servant, J. (1975a). "Deposition of atmospheric lead particles to natural surfaces in field experiments." *Atmosphere-Surface* Exchange of Particulate and Gaseous Pollutants. Symposium, Richland, Wash., Sept., pp. 87–95.

Servant, J. (1975b). "The deposition of lead over France (1972–1973) consideration about the budget importance of the dry deposition." *International Conference of Heavy Metals in the Environment,* Toronto, Ontario, Canada, Oct. 27–31, pp. 975–986.

Servant, J. (1982). "Atmospheric trace elements from natural and industrial sources." Marc Rep. 27, Chelsea College, London, 37 pp.

Servant, J., and Delapart, M. (1978). "Lead and lead-210 in some river waters of southwestern part of France-Importance of the atmospheric contribution." *Environ. Sci. Technol.* **13,** 105–107.

Servant, J., and Delapart, M. (1981). "Blood lead and lead-210 origins in residents of Toulouse." *Health Phys.* **41,** 483–487.

Smith, W.H. (1973). "Metal contamination of urban woody plants." *Environ. Sci. Technol.* **7,** 631–636.

Ter Harr, G. (1972). "The sources and path-ways of lead in the Environment." *International Symposium on Environmental Health Aspect of Lead.* Oct. 2–6, Commission of the European Communities, Amsterdam, pp. 59–76.

INDEX